PHYSICS
Including Human Applications

PHYSICS
Including Human Applications

HAROLD Q FULLER
University of Missouri—Rolla

RICHARD M. FULLER
Gustavus Adolphus College

ROBERT G. FULLER
University of Nebraska—Lincoln

HARPER & ROW, PUBLISHERS
New York / Hagerstown / San Francisco / London

Sponsoring Editor: Malvina Wasserman
Project Editor: Cynthia Hausdorff
Designer: Emily Harste
Production Supervisor: Stefania J. Taflinska
Compositor: Ruttle, Shaw & Wetherill, Inc.
Printer and Binder: Halliday Lithograph Corporation
Art Studio: J & R Technical Services, Inc.
Cover Illustration: Bill Greer

PHYSICS
Including Human
Applications

Library of Congress Cataloging in Publication Data

Fuller, Harold Q   Date-
    Physics, including human applications.

    1.   Physics.   I.   Fuller, Richard M.,
joint author.   II.   Fuller, Robert G., joint
author.   III.   Title.
QC21.2.F85   530   77-26255
ISBN 0-06-042214-9

Dedicated to
our students who provided insights
and motivation for this book.

Contents

Kinematics   41

Forces and Newton's Laws   71

† Denotes Enrichment section found at the end of most chapters.

Fluid Flow   181

Goals
Prerequisites

Transport Phenomena   207

Goals
Prerequisites

Temperatures and Heat   225

Goals
Prerequisites

Thermal Transport   249

Goals
Prerequisites

Thermodynamics   265

Goals
Prerequisites

Elastic Properties of Materials   293

Goals
Prerequisites

Molecular Model of Matter   305

Goals
Prerequisites

Simple Harmonic Motion 325

Goals
Prerequisites

Traveling Waves 347

Goals
Prerequisites

Basic Electrical Measurements 495

Goals
Prerequisites

Magnetism 525

Goals
Prerequisites

Electromagnetic Induction   551

Goals
Prerequisites

Alternating Currents   571

Goals
Prerequisites

Bioelectronics and Instrumentation   593

Goals
Prerequisites

Quantum and Relativistic Physics   607

Goals
Prerequisites

Atomic Physics   629

Goals
Prerequisites

Molecular and Solid-State Physics   651

Goals
Prerequisites

X Rays   669

Goals
Prerequisites

Nuclear Physics   687

Goals
Prerequisites

Applied Nucleonics   709

Goals
Prerequisites

List of Tables

Preface

In our many years of teaching we have found that students are always interested in the applications of physics to human beings. In addition, physical principles have been used to bring important advances to our understanding of life processes. Today at the interface of physics and the biological sciences there are many interesting and important problems to be solved. We hope that a textbook that attempts to show the relationships between physics and its human applications will encourage students to tackle the tough problems that lie in that interface and which demand expertise in both physics and bioscience for their solution.

For the past several years we have been involved in teaching physics to life science students, and we have developed an interest in a number of topics not treated in many general physics courses. We have attempted to integrate these topics into this book. We have also received suggestions from physiologists, former students who have gone to medical schools, articles in the *American Journal of Physics*, papers presented at American Association of Physics Teachers meetings, and from the physics topics listed in *The New MCAT Student Manual*. We have written a comprehensive text. We have covered all the principles of physics usually found in a general physics text, and we have related them to as many human applications as we could.

In the past few years a number of physics teachers have been using individualized, self-paced instruction. This strategy of instruction often makes use of learning objectives, repeatable testing, and student-student interactions. In many cases the instructor in such courses must spend many hours writing additional materials to help the students study the chosen textbook. All three of us have had experience teaching physics in a self-paced, repeatable-testing situation. We believe that the learning objectives and interactive testing that helps students learn physics in these courses can be put into a conventional physics text. These pedagogical tools will help the students in a traditional lecture course, while making the book easier to use in a self-paced course. We have tried to develop a format for this text that will offer these aids for learning to the students using this book.

Each chapter has the same format, and starts with a double page. On the left-hand page are the *chapter goals* and the *chapter prerequisites*. You may use the prerequisites to develop a chapter sequence for your course. In the instructor's manual that accompanies this textbook we have given a number of ways that this book can be adapted to various kinds of physics courses now being offered to bioscience, medical, and allied health students.

A *chapter summary*, which consists of a self-check with short answers is presented at the end of the narrative portion of each chapter. This interactive *summary* enables the student to determine whether or not he or she has mastered the chapter goals set out at the beginning of the chapter.

Algorithmic problems follow the chapter summary. The set of algorithmic problems is grouped together with a list of the important equations in the chapter. These one-step problems are designed to help the student translate words into equations and to solve equations by the direct substitution of numbers into equations. The answers to all the algorithmic problems are given at the end of that section of the chapter.

Exercises that can be answered using the content from a single section of the chapter follow the algorithmic problems. These exercises allow the student to put into practice the concepts of the text. The answers to most of these exercises are given at the end of the exercise.

Problems, which are more comprehensive than the exercises and may require the use of two or more concepts or equations, follow the exercises. The answers to the problems are given at the end of each problem.

An *enrichment section* is included in most chapters at the end of the narrative portion of the chapter. In many cases the enrichment section is a treatment of certain topics within the chapter using calculus. The problems marked with a dagger (†) require the enrichment material for their solution.

This book was intended to be used in a variety of courses in general physics with students who have been exposed to algebra and trigonometry. Since student exposure to such mathematics is known to produce a wide variety of effects upon the students, we have included a self-study *mathematical background for physics* supplement at the back of the book.

Recently the American Association of Physics Teachers has supported a number of workshops on physics teachings and the development of reasoning. These workshop materials, based upon the findings of Jean Piaget, raise the following question: In what way can physics be used to develop the reasoning of students? We believe that proceeding from concrete specific examples to general principles is the best way to make physics accessible to students who typically use concrete reasoning patterns. These students should also find the algorithmic problems to their liking. But we believe that a textbook needs to be supported by manipulative experiences if the physics course is really going to help students achieve cognitive growth. So we have provided a laboratory manual, using learning cycles, to accompany this text.

In summary, we have tried to develop a complete package of textbook, student study guide, laboratory manual, and instructor's guide — instructional materials that make physics and its human applications interesting and accessible to every college student. We hope that our textbook and its format will enable students in either traditional lecture courses or individualized, self-paced courses to study and learn effectively. The laboratory manual offers experiments in the exploration-invention-application learning cycle to encourage the development of reasoning as well as the mastery of the content of physics.

The metric system of units is used throughout the text, and in most cases, the International System of Units is used. Both calorie and joule are used as units of heat energy. We have not used the units of Pascal and Siemens. At the introductory textbook level we believe it is sometimes more appropriate to use derived units instead of the recent SI names, e.g., N/m^2 for pressure instead of Pascals. Many of these newer names are slowly becoming universally used in the scientific literature.

We are very grateful to our wives for their patience and support during the period of development of these materials and to our friends and colleagues who have used the material in preparatory forms and, in particular, William H. Bessey, Butler University, Indianapolis, Ind. Valuable suggestions have been made by the following reviewers: Dean Dragsdorf, Kansas State University of Agriculture and Applied Science, Manhattan; W. H. Kelly, Michigan State University, East Lansing; Terrill Mayes, University of North Carolina, Charlotte; Teymoor Gedayloo, California State Polytechnic College, San Luis Obispo; James P. Lincoln, San Antonio College, San Antonio, Tex.; Victor Cook, University of Washington, Seattle; Stanley A. Williams, Iowa State University, Ames; Larry L. Abels, University of Illinois, Chicago Circle; Quinton Bowles, University of Missouri, Kansas City. The staff of Harper & Row have been very cooperative and have made many worthwhile contributions to the final products, in particular, the Project Editor, Cynthia Hausdorff, and the Copy Editor, Caroline Eastman. We appreciated the assistance of Malvina Wasserman in the later phases of the project development.

Harold Q Fuller
Richard M. Fuller
Robert G. Fuller

Dear Physics Student,

We hope you will find the study of this book interesting and rewarding. We believe that physics is an essential subject for everyone to study. This belief is not based solely upon the enjoyment we have gotten from physics. The mental processes used in physics are used in many other areas of human life. These processes include using concrete experiences as a basis for understanding the present and predicting the future; inventing definitions, developing concepts, or constructing models to impose order on these experiences; and testing these models, or mental constructs, by the accumulation of additional experiences. We see these processes as universal human activities. A wide variety of human activities have been justified by reliance upon such abstract models as justice, national security, democracy, communism, capitalism, and Christianity. These models are considerably more complex than those you will study in this physics book. However, the same mental skills are required in all of these cases. We hope your attention to these physics materials will include your awareness of a development of your own reasoning skills. We believe that such cognitive development is a prerequisite for effective living in our modern world.

In addition to our belief in the universality of reasoning skills used in physics, we discover physics everywhere, in the kitchen as well as in the laboratory. For us, the physics problems encountered in the repair of a faulty electric frying pan are as challenging as those met in solving a complex physics research problem. We are sympathetic to your desire to have physics materials that are relevant. After all, since physics is everywhere, why should physics books not contain more everyday examples and applications? We have tried to illustrate the physics principles with examples drawn from the applications of physics to human systems. We have developed these materials to meet the need for a textbook flexible enough to be used in the physics courses required for students preparing for careers in bioscience, medicine, or allied health.

A definite learning sequence which begins with concrete experiences and progresses to abstract ideas is appropriate for a general physics course. Each new topic is introduced with an exploration of concrete experiences of nature common to most people, then the necessary definitions, concepts, or models are invented to help you impose order on your common experiences of nature, and finally applications of these concepts, or models, are provided in the examples, the exercises, and the problems.

This book has been designed to encourage you to be actively involved with its contents. Each chapter begins with a list of the chapter's goals so you will know where the chapter is headed. Each chapter includes a list of prerequisites so you know if you are adequately prepared for it. The early chapters that require some specific mathematics skills offer a short mathematics self-check for you to use to assess your skills. If you need an additional mathematics background, you will find it in the back of the book in the section entitled Mathematics Background for

Physics. Throughout the book many questions are asked to encourage you to reflect upon what you are reading and studying. At the end of each chapter is a chapter summary in the form of a number of short answer questions keyed to the chapter goals. You will find it helpful to review the contents of the chapter by working through these chapter summary questions. The answers to these questions and a reference to the related section of the chapter are provided to help you achieve the chapter goals you may have missed.

A reasoning skill you need to develop is the ability to translate a written problem, or question, into an experimental situation, or perhaps into a quantitative relationship. At the end of each chapter you will find a series of algorithmic problems. These problems are grouped together with all the quantitative relationships you need to solve them. This activity will enable you to become familiar with the translation of a word question into a quantitative question, the selection of the proper algebraic relationship to answer the question, and the mathematics manipulation required to obtain the correct numerical answer.

At the end of each chapter a series of exercises have been organized according to the section of the textbook to which they are related. Each exercise can be answered using the material of only one section of the book. The exercises will help you consolidate your understanding of each individual section of the book.

Finally, each chapter concludes with a number of problems that may require you to use more than one concept to answer them. These problems will help you bring together the concepts from many different parts of the book. As you work on these problems, we hope you will be putting it all together so that it makes sense for you.

Learning is a do-it-yourself activity. No one else can learn physics for you. You are the ultimate test of this book. What happens to you as you interact with these materials and think anew about your experiences of nature? Do you develop your reasoning skills as you study these materials? Do you feel good about yourself and about physics? Do you have a feeling that you understand nature better?

Best wishes for success in physics!
The Fullers

PHYSICS
Including Human Applications

GOALS

When you have mastered the contents of this chapter, you will be able to achieve the following goals:

Definitions
Define the following terms, which can be used to describe a relationship between you and your environment:

system	energy
state	force
variable	field
interaction	intensity
transducer	

Stimuli
List the major external stimuli that are detectable by humans.

Human Responses
Describe an elementary threshold measurement experiment, and interpret the data obtained.

Models
Use a mental construct, or model, to explain a common human experience.

PREREQUISITES

You may find it necessary to review your knowledge of the powers of ten notation before you study this chapter. To assess your use of powers of ten notation, you may want to use the following powers of ten self-check.

Powers of Ten Self-Check

Solve each of the following problems, and give the answer in powers of ten notation.

1. $0.252 \times 0.000000700/0.0360 =$ _____.
2. $6.380 \times 10^3 \times 5.00 \times 10^4/2.50 \times 10^{-5} =$ _____.
3. $3.20 \times 10^7 + 6.83 \times 10^6 - 9.90 \times 10^5 =$ _____.

Powers of Ten Self-Check Answers

If you had difficulty in correctly solving these problems, please study the Section A.2 on the powers of ten notation, page 731.

1. 4.90×10^{-6}
2. 1.28×10^{13}
3. 3.78×10^7

If you gave your answers with more accuracy than given here, please study Section A.3 on the use of significant figures, page 734.

1

Human Senses

1.1 Introduction

You are the beginning and the end of science. In scientific experiments and medical diagnoses the final analysis of experience is done by a human being. Your senses, your central nervous system, and your brain are the essential tools of science.

How precise are your detecting skills? Hold your breath, and listen. What can you hear? Shut your eyes and concentrate on listening. What can you hear? Now put your ear on your study table. What can you hear? More? Less? Same? Different? Sense? Nonsense? Music? Noise? Voices?

How precise is your sense of location, your ability to know where you are? Close your eyes, and walk to the nearest light switch. How easily were you able to find it? With your eyes closed, stand erect, and lean slightly forward and then slightly backward. How easily can you tell when you are standing up vertically?

How precise are your detection systems? Have you ever called a dog with a whistle that you could not hear? Are your eyes equally good at seeing all colors? How well trained is your sense of touch? Is your sense of smell better than your sense of taste? How do you explain the operation of your various detectors, or senses?

1.2 The Human System

You can think of yourself as a human system interacting with your environment. A *system* is defined as a whole entity. The heating system of your home consists of the furnace, the duct work, the thermostat control, and the chimney. The effectiveness of the heating system is determined by the organization of the system and its operation as a whole entity. The public schools of your community form another system. The effectiveness of this educational system depends on its internal organization and on its interaction with other systems in the world.

An *interaction* is defined as an action or influence exerted between systems. The heating system of your home interacts with the air temperature in the house, the fuel supply system, and the actions of the people in the house. The public school system interacts with the community through its program, the tax system, the election system, and the teacher supply system.

You as a physical, living being are a system. You are endowed with a set of sensors for interacting with your environment. These sensors detect the stimuli of light, sound, taste, smell, temperature and tactile contact.

A *variable* is a quantity that is subject to change. The opening and closing of windows in a house represents a variable that affects the heating system of the house. The tax base of the community is a variable that affects the public school system.

We are particularly interested in the interactions involving human sense stimuli. These stimuli involve such physical variables as temperature, pressure, light intensity, sound intensity, color, and sound frequency.

A physical system exists in a particular form, or condition, that is defined as the physical *state* of the system. For example, ice is the solid state of water. The physical variables that determine the state of water are temperature and pressure. The state of a home heating system is determined by the condition of its various parts and by the quality of its maintenance as well as by the temperature of the house. "Sitting in a chair" may be a way of describing your state now.

The physical state of a system can be altered by changing one or more of the physical variables that determine the state of a system. For example, a change in temperature can alter the state of water, and a change in the tax base can alter the state of the educational system of a community.

Questions

1. What are some of the states you might ascribe to the human system?
2. What are the variables that determine your current state? Are all of these physical variables?

1.3 The Human Detector

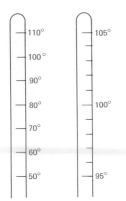

FIGURE 1.1

Two glass mercury thermometers with the same length of scale but with different calibrations. The thermometer on the right has greater sensitivity because a temperature change of one degree produces a greater change in the length of the mercury column in this thermometer.

Consider two thermometers that have the same length of scale but different calibrations covering different temperature ranges (see Figure 1.1). In the temperature range 95° to 105° a temperature change of 1° produces a greater change in the length of the liquid column in thermometer B than in thermometer A. The *sensitivity* of a detector indicates its ability to respond to changes in a variable stimulus. The ratio of response to stimulus is larger for a more sensitive detector than for a less sensitive detector.

Suppose that you have a rubber band, and you find that it will support eight paper clips before you are able to detect an increase in its length. This load is then called the threshold for stretching the rubber band. The *threshold* is defined as the minimum value of the stimulus or physical variable that produces a detectable response by a sensor or detector.

You may be able to distinguish between two lights on the basis of their color or, if they are the same color, on the basis of their brightness. Your ability to distinguish between two different stimuli is called *discrimination*.

Any detector or sensor can be described in terms of threshold, sensitivity, and discrimination. These characterizations also apply to human detectors.

The process of transforming a change in one physical variable to a change in another is called *transduction*. For example, the mercury thermometer transforms a change in temperature to a change in the height of a column of mercury in a narrow capillary tube. A sensor or detector that performs in this manner is called a *transducer*. Solar batteries used on orbiting satellites transform radiant energy from the sun into the electrical energy used to run instruments and transmitters on the satellite. Your sense transducers respond to the external stimuli and produce an electrical signal that is transmitted to the brain, the main control center and data handling system of the human body.

The interaction between you and your physical environment involves either *direct contact* with another system or *interaction at-a-distance*, when there is no actual contact between you and the stimulus source. In either of these interactions there is an *energy* transfer between you and your environment. *Energy* can be defined as a property of a system that causes changes in its own state or the state of its surroundings. The heat energy supplied to the popcorn by the popcorn popper changes the state of popcorn. The mechanical energy you supply to the lawn mower changes the state of the lawn. Two forms of energy that you are continually exchanging with your environment are heat energy and chemical energy. The food you eat contains chemical energy that is used to maintain your body processes. This chemical energy is transformed partially to heat, which enables you to maintain your body temperature, and partially to energy of motion associated with body movements. Energy of motion is called *kinetic energy*.

Questions

3. What are other forms of energy involved in your interaction with your environment?
4. Which of these forms of energy are transferred by direct contact, and which involve interaction-at-a-distance?

1.4 Direct-Contact Interactions

The sense of touch involves direct contact between you and your environment. The energy transferred by touch is mechanical energy produced by the force of contact. *Force* is a measure of the strength of an interaction. You exert a force on an object when you push or pull it. The push you use to slide away from the dinner table is a force. The pull you exert on a window shade is another force. Whenever *work* is done there is either a change in location, displacement, or a change in the state of motion of a system, or both. The energy transferred in a displacement interaction is defined as *work*. The work done in lifting an object is determined by the product of its weight (force) times the height (displacement) through which the object is lifted. Hence, the energy trans-

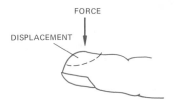

FIGURE 1.2
The sense of touch is a response of your body to a force that displaces your skin. The amount of work a constant force does is equal to the product of the size of the force times the displacement of your skin.

ferred to one of your touch receptors is determined by the force of the touch contact and by the displacement of the receptor (see Figure 1.2). A heavy book displaces the skin of your hand more than a light book does.

From your own experience you may be able to rank the *sensitivity* of touch detectors distributed over your body. The most sensitive touch receptors are those that are able to detect the smallest contact force or the smallest mechanical energy input. Your *threshold* of touch can be expressed in terms of the minimum force or the minimum energy required for you to have the sensation of touch contact. For example, you would detect a fly much more easily on your forehead than on your neck. From similar experiences, you may be able to rank different points on your body in order of their thresholds for touch.

With touch you may distinguish between two touch contacts due to different contact forces. The ability to discriminate between two distinct contacts varies at different parts of the body according to the density of touch receptors in the area of contact. The tip of your tongue is capable of discriminating between two contacts a few microns (micron $= 10^{-6}$ m) apart. On your back a space of several centimeters between contacts is required before you can discriminate between two contacts. You can perform a simple discrimination experiment by using a caliper as your variable spaced stimuli source as shown in Figure 1.3.

Distances can be estimated by using sensory data from touch and from muscles controlling the parts of the body used, from fingers or arms, for example. This is a crude method, but with practice it can become fairly dependable. Machinists and carpenters frequently have perfected these measuring skills.

Other direct contact interactions are involved in the senses of taste and temperature. Your taste and thermal detectors *discriminate* between different kinds of input stimuli as well as between different levels, or intensities, of input stimuli.

There are four basic taste sensations: sweet, sour, bitter, and salty. Experiments suggest that any complex taste can be duplicated by various combinations of these four basic tastes. The exact basis for the generation of the taste sense message is not yet known, but there is evidence that it is based on combined electrical and chemical processes. The tongue is the main taste detector, and with practice you can calibrate your tongue to some extent. Professional wine tasters, through diligent practice, develop their ability to distinguish between various kinds of wine. A more simple technique can be used for improving your tasting skills. Try tasting solutions with different, known concentrations of dissolved material. (See question 6 below.) Changes in body chemistry and temperature are known to affect this calibration.

Warmth and cold are sensed by two different types of receptors. The distribution of these temperature sensitive receptors varies widely over the body. For example, the forearm has approximately 10 times as many cold receptors as warm receptors per square millimeter.

FIGURE 1.3
Two-point discrimination test for touch. Calipers may be used to measure the arrangement of touch receptors in your body. They can be used to find the minimum distance required to feel two touches instead of one when two pointed objects are placed gently against your skin. Figures (a) through (e) show a typical range of values for various parts of the body.

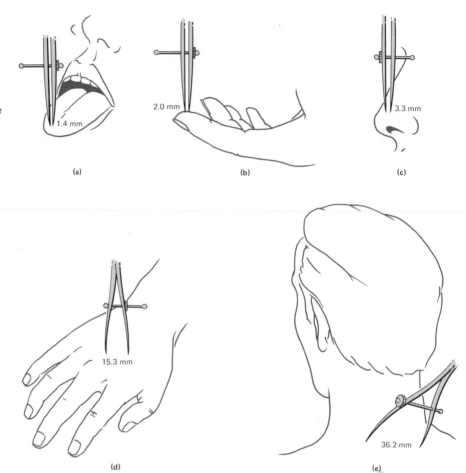

Questions

5. Which parts of your body are least sensitive to touch?
6. Try an experiment with four different concentrations of salt water, and see if you can rank them according to their salt content by taste. How would you calibrate your salt taste detection system?
7. Thermal detection usually involves contact interaction. There are temperature transducers in the skin, some being quite sensitive to hot and cold. What two areas of the skin have the most sensitive skin thermal detectors?

SIMPLE EXPERIMENT

Check out your temperature sensing ability using your fingers. Place your finger in ice water and then in room temperature water. Measure the room temperature (ice water temperature is 0°C). Now place your "calibrated" finger in unknown temperature samples (water from a refrigerator and warm water from a faucet), and estimate the temperature of the unknown samples. Compare your estimates with actual temperatures of the unknowns.

a. Perform a simple threshold experiment on the temperature sensing ability of your right index finger. Make a hot and cold water mixture that feels to be just the same as the temperature of your finger. Measure and record the actual temperature of the water. Now add small amounts of hot water until you can just tell that the water is hot. Record the temperature. The difference between these two temperatures is the temperature threshold for hot water for your right index finger.

b. Determine the temperature sensitivity of your right index finger by making a water mixture that feels warm. Then add hot water to the mixture until it feels twice as warm, then three times as warm, etc. Record the temperature each time. Does your index finger seem to give you consistent responses? What is the relationship between the temperature your index finger feels and the water temperature according to the thermometer readings?

c. Test your ability to discriminate between nearly equal temperatures by making a number of water baths of different, but almost the same, temperatures. Can you, by using your index finger, arrange them in order from coldest to hottest?

To become more aware of some aspects of your set of sensors, estimate the volume and weight of an object, such as this book. What sensors do you use for this task? Compare your estimated values with measured values using a ruler and a weight measuring device.

1.5 Interaction-at-a-Distance

If a system called "supper on the stove" is burned in the kitchen, then your odor sensing system in the living room will be influenced. If a construction crew is tearing up the street outside your window, then your hearing system will be influenced, even though the street may be some distance from you. The sun shines its light upon you and influences your total body system, even though the sun is a great distance away from you. These are all specific examples of systems that interact-at-a-distance. *Interaction-at-a-distance* is used to describe any action influencing one system not in direct contact with a second system.

Since we seem to feel most at home with direct contact interactions we can treat interaction-at-a-distance as a special type of direct-contact interaction. We can pretend that all of the distance between two interacting systems is in a *field* that connects the two systems. Then anything that is done to the first system is said to change its field. The second system is influenced by the changes in the first system through the changes in its field. A *field* can be defined whenever it is possible to assign a value of a physical variable to every point in space.

One of the most common interactions-at-a-distance is that of a magnetic compass with the earth's magnetism. The compass needle aligns itself with the *magnetic field* at the location of the compass. This alignment is produced by the magnetic interaction between the earth and the compass needle.

Sight involves a noncontact interaction between your human system and your environment. Sight involves the detection of light transmitted to you from a light source some distance from your eyes.

The weight of an object is determined by the gravitational interaction between the object and the earth. The *gravitational field* is defined as the gravitational force per unit mass at a point. Thus you can determine the gravitational field for all points. If you know the field, you can calculate the weight of any mass at any point (mass × gravitational field = weight) due to the gravitational interaction between the earth and the object at the given point.

The *electric field* is defined as the electric force per unit charge at each point in space. Energy is transferred by the transmission, or *propagation,* of fields. This transmission is called *radiation,* and if electric and magnetic fields are involved, as in the case of light, it is called *electromagnetic radiation.*

Electromagnetic radiation is characterized by the *wavelength* and the *intensity* of the electromagnetic waves. *Wavelength* is defined as the length of one electromagnetic wave in meters. *Intensity* is defined as the energy crossing a unit area per second. The wavelength of electromagnetic radiation is related to the color of the visible light which you detect. Blue light has a shorter wavelength than red light. A chart of the entire electromagnetic spectrum is shown in Figure 1.4. The intensity of solar electromagnetic radiation at the surface of the earth as a function of wavelength is shown in Figure 1.5. The sensitivity of a typical human eye to electromagnetic radiation is shown in Figure 1.6. The range of wavelengths that you detect with your primary light detectors, your eyes, is limited to a small part of the electromagnetic spectrum. In particular you normally see only wavelengths between 3.9×10^{-7} m (blue) and 7.0×10^{-7} m (red) in length. Your eyes are not uniformly sensitive to all wavelengths or colors within this "visible" spectrum. They are most sensitive to yellow-green light approximately 5.5×10^{-7} m in wave-

FIGURE 1.4

The regions of the electromagnetic spectrum as a function of wavelength in meters or frequency in hertz where 1 hertz is equal to one cycle per second. The visible region from 4.0×10^{-7} m to 7.0×10^{-7} m in wavelength is a very small portion of the total spectrum.

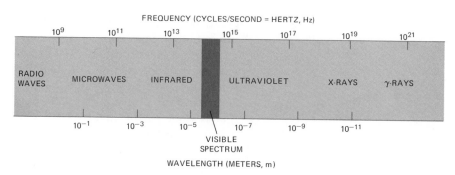

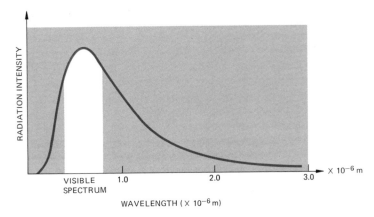

FIGURE 1.5
The intensity of solar radiation at the earth's surface as a function of wavelength in microns (10^{-6} meters). Note the peak of this intensity curve corresponds to the visible region.

length. Your eyes are extremely sensitive to light within the visible region. You can see in very low levels of light intensity, or brightness. You are also able to discriminate between the different wavelengths, or colors, within the range of sensitivity of your eyes.

In addition to color and intensity discrimination, your visual system can be used for distance and depth discrimination. The estimation of distance and depth is accomplished by the use of cues. Apparent size is one external cue in distance estimation for objects of known size. Binocular (two-eyed) vision provides an internal cue for these estimations. In binocular vision the images of the two eyes are slightly different, and you learn through experience to interpret these data in terms of distances.

There are other electromagnetic wave receptors in the human body besides the eyes. The sun and the earth's atmosphere provide us with the spectrum of electromagnetic energy at the surface of the earth as shown in Figure 1.5. This energy comes to us in a band that includes not only the visible spectrum but the infrared (longer wavelengths) and ultraviolet (shorter wavelengths) as well. The human body responds to the wavelengths outside the visible spectrum primarily through skin receptors. These *receptors* transform the ultraviolet energy into chemical changes associated with skin pigment changes, such as suntan and sunburn. The temperature sensors in the skin are excited by the infrared radiation. Other portions of the electromagnetic spectrum are also common in our environment, for example radio waves and microwaves. At this time, the effects of radio and microwaves on humans are not well known.

Sound is energy transmitted by pressure or density variations propagated through matter. *Pressure* is defined as the force per unit area. Sound is generated by mechanical vibrations set up in matter. The speaker of a hi-fi system vibrates at the frequency of the original music (if it is truly high fidelity). This vibrating speaker sets up vibrations of air molecules which is the sound propagated to your ears. Sound is characterized by its frequency and intensity.

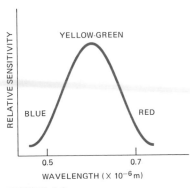

FIGURE 1.6
The relative sensitivity of typical human vision as a function of wavelength in microns (10^{-6} m). The typical human eye is more sensitive to yellow-green light than to red or blue light.

FIGURE 1.7
The typical threshold for human hearing in decibels (dB) as a function of frequency. Middle C on a piano has a frequency of 256 Hz. A sound of the frequency of 2000 Hz, about three octaves above middle C, is most easily heard by most humans.

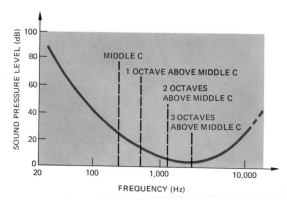

You may be able to detect sound with your touch detectors since they are pressure change detectors, but your ear is a much more sensitive sound detection system. This detector system operates over a limited frequency range (that is, the number of pressure cycles per second). Your ear responds to frequencies from about 30 Hz to about 20,000 Hz in incoming sound waves. In your ears most sensitive frequency range (they are not equally sensitive to all frequencies), you can detect displacements of air molecules in the sound wave as small as 4×10^{-12} m, which is the order of 1/100 of the size of an atom. (A threshold of hearing curve is illustrated in Figure 1.7.)

Your sense of smell represents an interaction-at-a-distance from the source of the odor, but it also involves direct contact with molecules that carry the odor to your nose. The sense of smell is more sensitive than the sense of taste, and your ability to identify odors can be greatly improved with practice. Olfactory sensitivity is estimated to be 10,000 times the sensitivity of taste, but the threshold of odor perception is extremely difficult to measure.

Questions

8. List the interactions-at-a-distance you have experienced today.
9. Classify these experiences as electromagnetic, gravitational, sound, or odor.
10. Which interactions-at-a-distance are most common?

SIMPLE EXPERIMENT

Compare your tactile distance sensing with your visual distance sensing:

a. While blindfolded, estimate the lengths of different small blocks using the fingers of one hand. Record your estimates for each unknown.
b. Repeat the procedure with larger blocks using two hands.
c. Remove the blindfold, and use your vision to estimate the lengths of the unknown solids.

d. Measure the actual lengths of the unknown and determine the differences in measurement for your two distance sensing systems.

1.6 Other Detectors and Interactions

This covers your normal senses, but are there other physical variables that you can detect? There are some interesting possibilities to consider. What about gravitational changes? You have a balance system that enables you to detect your orientation with respect to gravitational vertical. What about magnitude changes in gravitational field? One of the results of space lab research may be additional insight concerning gravitational sensitive systems of humans. What about electrostatic and magnetic fields? Some researchers argue that it is an ability of the human to detect electrostatic and magnetic fields that allows them to synchronize body time with the physical environment and to display behavior that suggests "biological clocks."[1] In any case, biomagnetism has become an active area of research, and there is potential for exciting new discoveries in this area of the human interaction with the physical environment.[2]

In addition there is experimental evidence that supports each of the following effects produced by electromagnetic forces:[3]

1. stimulation of bone growth
2. stimulation of tissue regeneration
3. influence of basic level of nerve activity and function

This evidence suggests that the human system responds to electric and magnetic fields, even though we are not conscious of these phenomena.

1.7 Threshold for the Sense of Touch

We have discussed the different human detectors for physical variables. Let us discuss the design of an experiment to measure the threshold for the sense of touch. In designing an experiment there are some useful steps to follow:

1. Define clearly the variable to be manipulated (*independent*, or *manipulated, variable*) and the variable which will change in response to the manipulations (*dependent*, or *responding, variable*).
2. Design the experiment to maximize your control of all other variables, and thus make your results as reproducible as possible.

[1] Frank A. Brown, Jr., J. Woodland Hastings, and John D. Palmer, *The Biological Clock: Two Views* (New York: Academic Press, 1970).
[2] Madeleine F. Barnothy, ed., *Biological Effects of Magnetic Fields*, 2 vols. (New York: Plenum Press, 1964).
[3] R. O. Becker, "Electromagnetic Forces and Life Processes," *Technology Review*, 75 (December), 32 (1972).

3. If the response of the system under study is totally unknown, allow for measurement of as many variables of the state of the system as possible.

We now apply these steps to our problem of touch threshold measurement. The variable we will change will be the mechanical force of contact. A stack of books on a table exerts a contact force equal to the weight of the books on the table. The contact force can be varied by removing books from the stack. This will also vary the mechanical energy involved in the interaction. To make the experiment as clear as possible, we want to control such other variables as temperature, area of contact, and location of contact.

The response we expect in this experiment is the subject's acknowledgement of the touch contact. It might also be useful to monitor the physiological response of the subject. Such measurements might include electric brain wave pattern (electroencephalogram, EEG), heart rate, and electric heart activity patterns (electrocardiogram, EKG or ECG), eye pupil size, and measurement of electrical resistance of the skin due to sweat gland activity (galvanic skin response, GSR), and the measurement of the electrical pattern produced by muscle action (electromyogram, EMG).

The experiment can be carried out by using various size nylon fibers, and noting the smallest diameter producing the detection of touch. A simple way to control the contact force is to use different diameter fibers as the contact agent. Each fiber can be calibrated by pressing it against one pan of a laboratory scale balance and measuring the force required for noticeable bending. The pressure exerted by all the fibers will be the same since the force required to bend a fiber is proportional to the cross sectional area of the fiber, i.e., pressure = bending force/unit area. Since the bending force is proportional to the area, the pressure is a constant. Hence, only the force of the stimulus will change.

Questions

11. What variables, other than temperature, area of contact, and location of contact, might be important to control in this experiment?
12. What are the potential sources of error in this experiment?
13. Design an experiment for measuring the discrimination distance between distinct contact points.

1.8 Model Building

A primary part of science is the construction of definitions, concepts, theories, and models to help give us a feeling of understanding in some areas of our experiences. Because of the importance of model building in science, and in physics in particular, we will spend considerable time in this course building models, that is, mental constructs, to help explain the common characteristics of a variety of experiences. For example, who would have thought that an apple falling on Newton's head was in any way related to the motion of the moon in an orbit around the earth? Yet, after Newton constructed a model of gravity, these two experiences were both understood as examples of the gravitational interaction of matter.

The basic ingredients used in model building are simple intuitive ideas. Intuitive ideas that allow us to visualize a variety of experiences in a simple way can be put together by our brain to build a model. Model building in physics makes use of a wide variety of intuitive ideas, many of which may not, in fact, actually apply to the experiences we are seeking to explain. For example, consider a molecular model of matter that assumes that all of matter is made of very small, perfectly elastic, constantly moving spheres called molecules. If we assume that these molecules interact with each other in some way, then we can use this model to explain the existence of the three states of matter, of, for example, H_2O which we experience as ice, water, and steam. It turns out that the elastic sphere model for a molecule is hopelessly inadequate to explain how molecules are constructed, how they interact with each other, or how they seem to be always in motion. We often tend to build models, like this one, that focus on the major aspects of our experiences. The finer details of experience often wait many years for an explanation in terms of a more refined model.

Model building is the primary means of explaining experiences in physics, but remember that the true test of nature is not our model, but our experiences. Model building has some positive features that we seek to exploit in physics. A model allows us to unify a variety of experiences and to impose order on some of our experiences. It enables us to make predictions so that we can extend our imposed order in time, to predict what results future experiments will give, and in experience, to decide what new experiments should be performed. Maybe best of all, model building allows us to feel at home in our universe because our models provide us with a feeling of understanding.

The use of models has some dangers. Some people may substitute the model for experience. They may choose to rely upon their mental construct rather than upon an experiment. The experimental data may not be used to test their model. The corrections necessary in a model may never be made. The history of science includes the names of many persons whose experimental data exceeded the models popular in their time. Galileo, Tycho Brahe, and Ernest Rutherford each gathered data that proved to be the downfall of some previously held model. Today many common examples of the use of models to replace experience are found in areas other than physics. The Marxist belief in the "labor theory of value" for the structure of a modern society is one example. During the gasoline crisis of 1974 there were numerous calls for a return to the "free market value" of oil. To anyone the least bit familiar with the history of oil production by large corporations such a model as "free market value" seems ludicrous when applied to oil. Perhaps, this substitution of model for experience is best illustrated by Lucy, the character in the Peanuts comic strip, when she reported, "I love mankind, it's people I can't stand."

Another danger to be recognized in our use of models is our ability to hide some basic presuppositions about our experiences in our model. If they are not carefully examined, our presuppositions might limit our ability to formulate useful models. For example, science is based upon the presupposition that human experiences of nature will be such that we can find order in them. Two results of this presupposition of science may interest you: Cultures that presuppose nature to be capricious have not developed any "science" in the commonly understood sense of that word. Secondly, to argue from the order of science to the existence of an Ultimate Order Producer in the universe is a meaningless, circular argument. Science starts with a presupposition for finding order, and scientists proceed to impose order on all manner of human experiences. Scientists are always seeking to expand the collection of ordered experiences by claiming previously unexplained events.

Model building seems to be a universal human activity. All of us build a variety of mental images of our world. We use these images to understand ourselves, our roles in society, as well as to explain natural events. The models we use in physics almost always have a quantitative feature. We nearly always use a physical model to help formulate a quantitative relationship using the language of mathematics. The use of models in physics to derive formulas, to perform numerical calculations, and to reduce the results of an experiment to numbers is in strong contrast to the more descriptive models used in medicine and biology. It is the quantitative aspects of physical models that is now being copied by social sciences in their thirst for numerical data and statistical analyses.

In subsequent chapters we will be introducing you to model building in physics by having you build some models of your own and by having you apply the models built by other people. But please remember that all of our models are tentative. Each model is only able to help explain a small portion of our total experiences. We must always be open to additional experiences that will require us to modify our models.

Questions

14. Consider the following experiences. What are some aspects of the models that were used to explain their experiences. What actions were taken?
    a. The United States government interacting with the American Indians during most of the nineteenth (and twentieth) century.
    b. The Nazi leaders interacting with the German-Jewish people.
    c. The American government interacting with the American-born Japanese during World War II.

15. What aspects of human experience do you presently consider to be outside of the ordering of science? Can you remember any experiences that have recently become ordered?

SUMMARY

Use these questions to evaluate how well you have achieved the goals of this chapter. The answers to these questions are given at the end of this summary with the section number where you can find the related content material.

Definitions

1. Use one of the following terms in each of the blanks below:

system   energy
state   force
variable   field
interaction   intensity
transducer

a. The retina of the eye, which converts light into nerve impulses, is an example of a _____.
b. A pendulum clock can be adjusted by changing the length of its pendulum, so we call the length of the pendulum a _____ of the clock as an isolated _____.
c. The weight of an object applies a _____ to your hand as you lift it. The greater the distance you lift the object, the larger is the amount of _____ you use.
d. The influence of the earth on the moon and of the moon on the earth is an example of a gravitational _____ which we attempt to understand by assigning a value for the gravitational _____ at every location in space.

Stimuli and Human Responses

2. Place a check mark by each of the following external stimuli that you can easily detect. Relate the concepts of threshold, sensitivity, and discrimination to each stimulus you check:

high temperature   microwaves
noise   x-rays

visible light   odor
touch   salty taste
gravitational field   bitter taste
electrical field   sour taste
magnetic field   sweet taste
ultraviolet light   low temperature
infrared light

Threshold Experiments

3. Describe how you could measure the threshold of salty taste and of bitter taste. How would you compare these two thresholds?

Models

4. Suppose you have an uneducated kitchen helper drying your dinner dishes. He notices that the dry piece of corn husks jump off of the counter top onto the recently dried plastic drinking glasses when the glasses are placed near the corn husks. Develop a model to explain this phenomenon to him.

Answers

Definitions

a. transducer (Section 1.3)
b. variable (Section 1.2), system (Section 1.2)
c. force (Section 1.4) energy (Section 1.3)
d. interaction (Section 1.2), field (Section 1.5)

Stimuli and Human Responses

Qualitative answers, see Sections 1.3, 1.4, 1.5, 1.6 and 1.7.

Threshold Experiments

See Section 1.7 for an example.

Models

See Section 1.8 on model building.

ALGORITHMIC* PROBLEMS

Equations

Consult Figures 1.4 through 1.7 for data.

Problems

1. What do you expect to happen to the displacement when the contact force is reduced? See Figure 1.2.
2. What experimental conditions are necessary for discrimination studies using the calipers shown in Figure 1.3?
3. a. The radiation of the higher frequency next to the visible spectrum will produce sunburn. What is this radiation called?
   b. The radiation of the lower frequency next to the visible spectrum is used in thermal therapy. What is this radiation called?
4. Is there a threshold for solar radiation reaching the surface of the earth? If so, estimate the threshold wavelength.
5. What color of visible light is most easily seen?
6. What frequency of sound would you expect to hear most easily?

Answers

1. Displacement decreases to zero at threshold force
2. Blindfold the subjects; vary one touch and two touches; ask subject to identify number of touches

3. a. ultraviolet
   b. infrared
4. yes, 0.2 microns
5. yellow-green
6. 2000 Hz

* See the Preface.

EXERCISES

These exercises are designed to help you apply the ideas of a section to physical situations. When appropriate, the numerical answer is given at the end of the exercise.

Section 1.2

1. a. What are the systems interacting in an automobile stalled along a highway?
   b. What is the state of this automobile? What physical parameters might be used to define the state of the automobile? How could the state of the automobile be changed? [a. The fuel injection system, internal combustion chambers, and exhaust system are examples of systems that must interact properly for the car to run; b. The car is at rest: Its state can be described by its location, color, temperature, etc. Its state can be changed by moving it, painting it, heating it, etc.]

2. For each of the systems listed below give an appropriate interaction and indicate the variable(s) involved in the interaction.
   a. thermometer-environment
   b. phonograph-record
   c. piano-pianist
   d. teacher-class
   e. minister-congregation
   f. thunderstorm-radio
   g. living plant-sun
   h. ocean-moon
   [a. thermal, contact; b. mechanical, contact; c. touch,

contact; d. verbal, at-a-distance; e. caring, both contact and at-a-distance; f. electrical, at-a-distance; g. light, at-a-distance; h. gravity, at-a-distance.]

Section 1.3

3. For each transducer listed below, describe the variable it detects and its output signal.
   a. mercury thermometer
   b. mercury manometer
   c. litmus paper
   d. exposure meter
   e. phonograph cartridge
   f. microphone
   g. plumb bob
   h. photographic film
   [a. temperature, change in length of column; b. pressure, change in length of column; c. acidity of a solution, color of paper; d. light intensity, pointer movement; e. mechanical vibrations, electrical signal; f. sound, electrical signal; g. gravity, direction of hang; h. light intensity and color, image formation on emulsion.]

4. For each transducer in problem 3, tell whether the interaction is a contact interaction or an interaction-at-a-distance interaction. [a. contact; b. contact; c. contact; d. at-a-distance; e. contact; f. at-a-distance; g. at-a-distance; h. at-a-distance.]

Section 1.4

5. What would be the necessary procedure for measuring the sensitivity of touch? Contrast this procedure with a threshold measurement experiment. [A sensitivity measurement will seek to discover the minimum difference between two touches that is required to make the touches detectably different. Once again a fiber experiment could be done. This time with fibers all of the same area, but made from different materials so as to apply a range of forces well above the threshold value on the skin. Then various combinations of two fibers could be tested. Do the two feel the same? Which one feels heavier? Which one lighter? The subject should be blindfolded for the experiment, and several touches, where both fibers are identical, should be made to test validity of the subject's responses. How does this procedure differ from the threshold measurements?]

6. Draw a graph of the touch displacement versus contact force.

Section 1.5

7. Estimate what fraction of the whole electromagnetic spectrum between 10^8 Hz and 10^{20} Hz lies in the visible region. What explanation can you offer for the selectivity of human vision? See Figures 1.4 and 1.5.

$$[\text{Fraction visible} = \frac{7.5 \times 10^{14} - 4 \times 10^{14}}{(10^{20} - 10^8)\text{Hz}} \text{ Hz}$$

$$= \frac{3.5 \times 10^{14}}{10^{20}} = 3.5 \times 10^{-6}$$

or 0.00035% of the total spectrum. The intensity of radiation from the sun that reaches the earth peaks in the region of human vision.]

8. Compare the sound level required for human hearing at 100 Hz and 2000 Hz (Figure 1.7) [Sounds at 100 Hz, about one octave below middle C, must be about 50 dB louder than sounds at 2000 Hz, about three octaves above middle C, to be heard.]

Section 1.8

9. Choose a familiar model from each of the following broad categories of human experience—natural science, social science, and religion. Answer the following questions for each model:
   a. What human experiences does the model explain for you?
   b. What various experiences does it unify?
   c. What are the major aspects of experience upon which it focuses? Do you know some finer details that are not explained by this model?
   d. What does it allow you to predict without experiencing? [*Natural science*: the planetary model of our solar system. a. the seasons of the year, the apparent motion of the planets; b. the apparent motion of the sun, the apparent motion of the planets, the apparent motion of the stars; c. the experience of seeing the apparent irregular motion of the planets in the night sky while the sun, moon, and stars seem to move in simple orbits; finer details: bending of light as it passes the sun; d. predicts future locations of heavenly objects, e.g., sunrise and sunset; *social science*: the labor theory of value (Marxism). a. how the prices of various market items are determined. b. a wide range of economic experiences "Money begets money," "The rich get richer, and the poor get children"; c. the inequalities in a capitalistic economy; finer details: the "law" of supply and demand contradicts the labor theory of value; d. predicts the

eventual slavery and rebellion of the capitalist workers. *religion:* the Exodus, or Easter. a. the experience of a new future, new freedom, new responsibilities, of new possibilities for life that can grow out of seemingly hopeless experiences of slavery, illness, or death; b. a variety of persons who have found a new victory after a defeat, a new freedom after slavery, a new life after death; c. the new possibilities are actualized for each person; finer details: how these new possibilities are actualized for each person; d. no matter how dark the hour, a new dawn is possible.]

10. Consider some of the more obvious characteristics of day and night, the apparent motion of the sun and moon, and the seasonal changes in the amount of daylight. How are these experiences explained by the Ptolemaic geocentric model of the solar system? Which experiences of yours cause you to believe the Copernician, heliocentric model rather than the Ptolemaic model? [According to the Ptolemaic model, the earth is at rest and the sun and moon in orbit around it. The stars and planets make various irregular and complex orbits around the earth. You probably cannot name any experience that makes the Copernician model "true" for you. The differences between the two models, in terms of human *experiences,* are rather small. After all these two models were developed to explain the same data, which they both do effectively.]

11. It has been suggested that the behavior of a soft drink dispensing machine may be explained by a model that assumes a small being lives inside the machine. Examine the behavior of the soft drink machine in terms of this model. How can you prove or disprove the existence of this being without opening the machine? By opening the machine? [The being lives on electricity, water, and money. It converts electricity, money and water into soft drinks and containers. If you are sufficiently clever in your model of "the being," it may not be possible to disprove its existence by any means, including opening the machine.]

GOALS

When you have mastered the content of this chapter, you will be able to achieve the following goals:

Definitions

Define each of the following terms, and use each term in an operational definition:

equilibrium	restoring force	conservation laws
inertia	oscillatory motion	feedback
gradient	linear system	natural frequency
current	superposition	resonance

Inertia

Give an example of a physical system that has mechanical, thermal, and electrical inertia.

Energy Transfer

Explain how you would maximize the transfer of energy at the interface between two systems.

Superposition

Solve problems making use of the superposition principle — given the proper physical variables of the systems.

PREREQUISITES

Before beginning this chapter you should have achieved the goals of Chapter 1, Human Senses. If you have not recently been working with cartesian coordinate graphs and dimensional relations, you may wish to review the material on graphs and dimensional analysis in the mathematical background supplement in the appendix (page 729). To help you assess your readiness for this chapter you may use the following self-check.

Graphing and Dimensional Analysis Self-Check

1. The following table is taken from a drivers manual and shows data for stopping an automobile on dry pavement.

Velocity (m/sec)	Thinking Distance (m)	Total Stopping Distance (m)
8.8	6.7	14
13	10	27
18	13	43
23	17	61
27	20	86

   a. Draw a graph of thinking distance (y-axis) versus velocity (x-axis), and find the slope of the curve at the point on the curve where $x = 15$ m/sec.

   b. Draw a graph of the total stopping distance (y axis) versus the velocity (x axis), and find the slope of the curve at the point where $x = 20$ m/sec.

2. We can define length, mass, and time as fundamental dimensions in a system of measurement. What are the SI (System International) units for

   a. Length

   b. Mass

   c. Time

   The SI units are related to each other by multiples of ten, and the units are represented by the fundamental unit with the proper prefix. What are the relationships between the fundamental unit and the following common prefixes?

   d. The prefix *centi-* means _____, so one tesla = _____ centiteslas. [10^{-2}; 10^2]

   e. The prefix *milli-* means _____, so one liter = _____ milliliters. [10^{-3}; 10^3]

   f. The prefix *kilo-* means _____, so one watt = _____ kilowatts. [10^3; 10^{-3}]

Graphing and Dimensional Analysis Self-Check Answers

If you had difficulty in correctly solving these problems, please study Section A.4, Cartesian Graphs, and A.5, Dimensional Analysis, on pages 734–741.

1. a. 0.60 sec; b. 4.8 sec
2. a. meter; b. kilogram; c. second; d. 10^{-2}, 10^2; e. 10^{-3}, 10^3; f. 10^3, 10^{-3}

2
Unifying Approaches

2.1 Introduction

As you reach for the mug of coffee on the table in front of you, has it ever suddenly evaded your grasp? It is not likely that you have had that happen. In fact, your experiences with coffee mugs at rest upon the table have taught you quite a bit about what to expect from natural events. The mug seems perfectly content to remain at rest as long as you do not try to push it around. In fact, a gentle push on the mug is met with the cup's determination to remain at its present location. If you reach across the top of the mug to get a doughnut and accidently tip the mug slightly, you know that it will bang back down on the table, the surface of the coffee will slosh around for a short time, probably spilling out some, and then it will settle back down to a restful state.

While you would be quite surprised if some invisible, mysterious power quickly removed the mug from your grasp, you think nothing at all of leaving your hot coffee mug on the table only to find that some time later the mug and coffee have cooled noticeably. Has a mysterious power been quickly removing something from your mug to make it feel cool?

Suppose you reach out to pour some more coffee from a coffee pot into your mug. How does your brain manage to get your hand to go to the proper location to perform that task? After you have poured more coffee into your mug, what has happened to the coffee in the pot?

What is the point of all this rhetoric? Since birth you have been continuously interacting with the physical environment in which you have lived. You already have considerable knowledge about how nature works. It is the purpose of this chapter to introduce you to some mental constructs that you can use to unify your approach to studying nature. These mental constructs, or models, are introduced first in the use of words that can be explained by operations performed on a specific system. These words are to be defined in terms of specific experiences, not in terms of other abstract words. We call such definitions *operational definitions*. In giving an operational definition for a word you should choose a specific system, a hot coffee mug sitting on a table, for example, and explain the characteristics of the system that are described by the word you are defining.

There are several unifying approaches that can be made in the study of physical and living systems. These approaches make use of the qualitative relationships that exist between the various properties of a system. Your own physical experiences and observations must be used to incorporate these approaches into your understanding of your environment.

2.2 Equilibrium

Your physics textbook at rest upon your desk is said to be in equilibrium. The book is in *equilibrium* because the sum of all forces acting on

FIGURE 2.1
Two examples of a body in equilibrium: (a) a book resting on a table; (b) a weight hanging at rest on a spring.

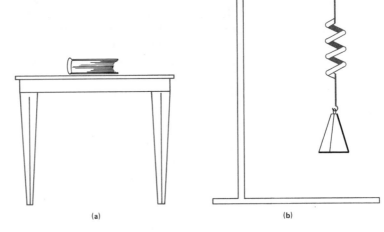

(a)   (b)

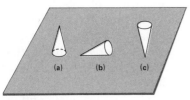

FIGURE 2.2
Three equilibrium positions of a right circular cone: (a) stable equilibrium, (b) neutral equilibrium, (c) unstable equilibrium.

the book is zero, and there are no forces tending to make the book rotate (see Figure 2.1a). Another example of equilibrium is an object hanging on a spring (see Figure 2.1b). In daily life you have seen many systems in equilibrium. In general a system is in a state of equilibrium when all the influences acting on the system are cancelled by others, resulting in a balanced, unchanging system.

Consider a solid right circular cone like the one in Figure 2.2. The cone may rest on a table on its base, on its side, or on its apex in a vertical position. In each case the cone is in equilibrium. However, we find these positions are different types of equilibrium. We distinguish between these different types of equilibrium by the answer to the following question: If the system is changed slightly, what is the tendency of the system to return to its original state? With slight displacement from base position as in Figure 2.2a the cone tends to return to its original position. From side position (Figure 2.2b) there is no tendency to return to original position. From apex position (Figure 2.2c) there is a tendency to go further from its original position. When resting on its base, the cone is said to be in *stable equilibrium;* on its side, to be in *neutral equilibrium;* and on its apex, to be in an *unstable equilibrium* condition.

Physical systems in stable equilibrium tend to return to their original state if they are slightly changed from their equilibrium state. This indicates that there is a force tending to restore the changed system to equilibrium.

2.3 Inertia

Lift one rock in each hand, rocks that are alike except in size. If rock A is larger than rock B, a larger force is required to lift rock A than to lift rock B. Now shake the two rocks back and forth rapidly. What difference do you notice in the feel of the rocks? Do you notice that rock A is more difficult to start moving in one direction and that after it is moving it is more difficult to stop rock A and cause it to move in the opposite direction? You probably find it is much easier to shake rock B back and forth than it was rock A. The property of these rocks that measures their resistance to changes in their state of motion is called *mechanical inertia*. The rock with the larger mechanical inertia has the larger mass. We can think of the mass of an object as a measure of its mechanical inertia. The inertia of a system tends to maintain it in an equilibrium condition. A given system may have several different inertial variables, each appropriate to a specific physical property of the system. For example, a rock has mechanical inertia (called mass) which resists a change in its state of motion. A rock has *thermal inertia;* the larger its thermal inertia (called *heat capacity*), the more difficult it is to change its temperature. A rock has *electrical inertia*. Again, the larger its electrical inertia (*reactance*), the more difficult it is to get electrical charge to move through it. In general, *inertia* is the property of a system that is a measure of the system's resistance to change. The inertial property of a system determines how fast the system responds to external forces. That is, the inertial property for a given physical variable determines the response time of the system for changes in that variable.

Early in their mastery of the physics of pool, players discover another characteristic of inertial properties. Consider the example of the head-on collision between two pool balls. You will observe that ball A approaches ball B, which is at rest, with a certain speed and that after a collision, ball A comes to rest and ball B moves away with essentially the same speed as ball A originally had. This is an example of the behavior of a contact interaction at the interface between the two balls (systems). In this case, the mechanical inertia of both systems, that is, the mass of both pool balls, is the same and nearly all the original energy of motion of ball A is transferred to ball B. On the other hand, consider the collision between a moving bowling ball and a stationary tennis ball. In this case the mechanical inertia (mass) of the moving bowling ball is much, much greater than the mass of the resting tennis ball. The motion of the bowling ball is hardly changed at all by a collision with a tennis ball. The amount of energy of motion transferred from the bowling ball to the tennis ball is small. The general rule that applies to such interactions between systems is that the maximum energy transfer occurs between the two systems when their inertial properties for the transferred variable are equal. This is a useful rule with many applications. The transfer of electrical energy is greatest when the source has the same reactance (similar to inertia in a mechanical system) as the external circuit. The maximum energy is transferred to a stereo speaker when its reactance is equal to that of the input amplifier. You can now see that it is necessary to study the inertial properties of a system if you wish to understand its physical behavior.

FIGURE 2.3
Sketch of a body in equilibrium and
displaced from equilibrium.

EQUILIBRIUM
(a)

DISPLACED
(b)

Questions

1. Name some systems in equilibrium.
2. Name some systems which tend to return to equilibrium if slightly displaced. What is the "restoring force" in each system?
3. Consider a social system such as your physics class. List possible inertial properties for the class and some possible consequences of these forms of inertia.

2.4 Oscillatory Motion and Restoring Force

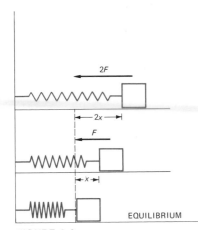

FIGURE 2.4
Diagram showing the relationship
between displacement and restoring
force on a linear system.

Consider a child at rest in a swing. If she is displaced (Figure 2.3) from the equilibrium position (the position where she is nearest to the ground) and released, she will tend to return to equilibrium. Why? There is a force acting on her to return her to the equilibrium position. This force is called a *restoring force*. In this case the restoring force is a gravitational force. The pull of gravity tends to bring her back to her equilibrium position.

In many systems the restoring force is directly proportional to the displacement x of the system from equilibrium position. Then there is a restoring force F acting to return the system to equilibrium. (See Figure 2.4.)

$$F = -kx \tag{2.1}$$

where k is the *force constant* of the system. If the system is moved to a distance twice as far from equilibrium, $2x$, then the restoring force will be twice the original restoring force, or $2F$.

This system is then said to be *linear*. An example of a linear system with which you are acquainted is a spring balance. This means that the force required to stretch the spring is directly proportional to the stretch. This is shown in the graph, Figure 2.5.

Think of a weight hanging on a coil spring. The weight is moved slightly from equilibrium position and released. The system responds to

FIGURE 2.5
A linear system consisting of a spring balance and weight. The relationship between load and displacement (stretch) is shown in the curve.

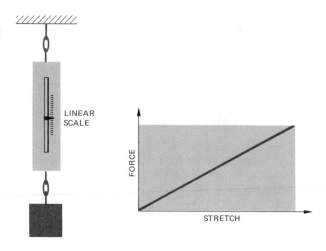

the linear restoring force by oscillating about the equilibrium position and, in all real systems, gradually comes back to rest at equilibrium.

When an automobile is traveling at a high speed on a smooth highway and hits a sharp bump in the road the auto begins to oscillate up and down. However, the shock absorbers subtract out a portion of the impact of the bump, and on each oscillation the car vibrates less widely. The auto finally returns to smooth motion until it strikes another bump. Many physical systems exhibit this same kind of behavior. Nonlinear forces cause this damped oscillatory motion. Friction is one example of nonlinear force.

Each of the examples, the girl on the swing, the weight on a spring, and the undamped car, illustrates the *oscillatory motion* that is characteristic of systems that are linear, those in which the restoring force is proportional to the displacement of the system from equilibrium. The frequency of the to and fro motion of such linear systems is completely determined by the *restoring force constant* (force per unit displacement) and the inertial property of the system for the displacement involved. This frequency, the number of complete oscillations per second, is called the *natural frequency* of the system. Almost every physical system behaves as a linear system when it is subjected to small displacements from equilibrium. Hence, most systems show oscillatory motion for small displacements from equilibrium.

2.5 *Current and Gradient*

What happens to rainwater when it hits the ground? What happens to the rainwater that collects in your backyard, or that collects at the upper end of your street, or that falls on the mountains? How do you explain what you see happening to rainwater as it flows along the ground?

What happens to the handle of a silver spoon when it is placed in a bowl of hot gravy? How do you explain the flow of heat energy through a solid?

What happens to a storage battery when you connect it to a light bulb? How do you explain the flow of charge in an electrical circuit?

These are only a few of the examples of the motion of something from one part of the universe to another. This type of motion occurs in many aspects of nature and exhibits similar properties in its many different appearances in nature.

If all parts of a system are the same, nothing happens, and the status quo is maintained. The flow of something always happens in response to a difference in some property between two different parts of the system. In the above examples, the water flows as a result of a difference in elevation between two locations; the heat flows as result of a difference in temperature between two places; the electrical charge flows as a result of a difference in electrical potential between two different points in a circuit.

The rate of flow, or *current*, is related to the changes of some property from one part of the system to another. In general,

$$\text{current} = \frac{\Delta \text{ (quantity)}}{\Delta \text{ (time)}} \qquad \text{where } \Delta \text{ mean } \textit{change in} \qquad (2.2)$$

For example, the greater the temperature difference across a given length of silver spoon handle the faster the heat energy flows through the handle. Because of the importance of the change of the properties of systems with location, let us define a new term called a *gradient*.

Gradients have both size and direction. Such quantities are called vectors, and they will be discussed in Chapter 3. The size of the gradient of a physical quantity is determined by how rapidly the quantity changes with position.

$$\text{magnitude of gradient} = \frac{\Delta \text{ (physical quantity)}}{\Delta \text{ (position)}} \qquad (2.3)$$

The direction of the gradient is in the direction of greatest positive change. For example, consider a metal bar with one end in ice water 0°C) and the other end in boiling water (100°C). The graph of the temperature along the bar as a function of distance from the ice is shown in Figure 2.6b. The slope of the temperature vs. distance curve is defined as the magnitude of the gradient. The direction of the gradient is positive from the cold end to the hot end. This direction is opposite to the direction of the current produced by the gradient. For this example, the current is flow of heat energy from the 100°C end to the ice bath end. (Note that cold is the absence of heat, and consequently there is no cold current.) In many cases the magnitude of a current, that is, the rate of flow, is directly proportional to a gradient. (We will discuss specific examples of such transport phenomena in later chapters.) The proportionality constant K between the current and gradient is a basic physical property of the material in which the current occurs.

FIGURE 2.6
Diagram of a conducting rod between
two temperatures of 0° and 100°C
and a plot of temperature vs. position
along the rod. Note the method for
getting the temperature gradient.

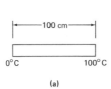

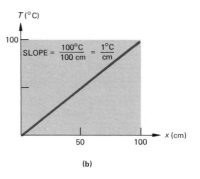

(a)

(b)

$$\text{current} = -\text{K} \times \text{gradient} \qquad (2.4)$$

The minus sign indicates that the direction of the current is opposite the direction of the gradient. For our example this becomes

$$\frac{\Delta Q}{\Delta t} = -K \times \frac{\Delta T}{\Delta x}$$

where ΔQ is the change in heat energy, ΔT, the change in temperature, K, the thermal conductivity, Δt, the change in time, and Δx, the change in distance.

2.6 Conservation Laws

An important property of physical systems is that certain properties are neither created or destroyed within the system. This property is the basis of the *conservation laws* of physics. These conservation laws are expressed by the simple mathematical statement:

$$\frac{\text{change in physical property}}{\text{change in time}} = 0 \qquad (2.5)$$

The concept of conservation is one of the early mental developments of human beings. Several tasks are used with young children to test their conservation reasoning. In one of them two horizontal rows of checkers, one row of six black and one row of six red checkers, are made. After the child is convinced that there are the same number of each, the red checkers are stacked in a vertical pile. The child is asked if there are the same number of red and black checkers. In other words, is the number of checkers of one color independent of their arrangement in space? This is conservation of number,

$$\frac{\text{change in number}}{\text{change in arrangement}} = \frac{\Delta N}{\Delta a} = 0 \qquad (2.6)$$

In another test six matches are laid end to end as in Figure 2.7a, and six other matches are laid in some other configuration as shown in Figure 2.7b. A child is asked if the distance along the matches from A to B

FIGURE 2.7
Conservation of length. The length of six matches laid end to end (a) in a straight line, and (b) not in a straight line.

(a)

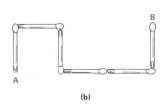

(b)

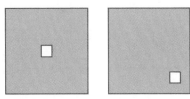

FIGURE 2.8
Diagram of the area within large square and outside small square.

FIGURE 2.9
Diagram of the area within large square and outside a number of small squares (6 in this case) for two distributions. Figures 2.8 and 2.9 are used to demonstrate conservation of area.

(a)   (b)

FIGURE 2.10
Conservation of volume and mass. Two masses of clay each having the same volume and mass but, in (b), of different shapes.

is the same. This is conservation of length,

$$\frac{\text{change in length}}{\text{change in arrangement}} = \frac{\Delta L}{\Delta a} = 0 \qquad (2.7)$$

Suppose you give a child two green sheets of paper of equal size, representing two pastures of grass. You also give the child a number of blocks of equal size, representing barns. The child places one square on each sheet (Figure 2.8). When the child is convinced that each sheet has equal areas of pasture for the two configuration shown in Figure 2.8, a number of blocks are used, and the child is given two configurations such as shown in Figure 2.9. The child is then asked, "In which pasture will a horse have more grass to eat? Which one has more green grass or are they the same?" This is an example of conservation of area,

$$\frac{\text{change in area}}{\text{change in arrangement}} = \frac{\Delta A}{\Delta a} = 0 \qquad (2.8)$$

Consider another example. A child is shown two balls of modeling clay of equal mass (Figure 2.10a). The questioner then shapes one of the balls into a pancake form (Figure 2.10b) and asks, "Which one contains the more clay?" This is an example of conservation of mass,

$$\frac{\text{change in mass}}{\text{change in shape}} = \frac{\Delta M}{\Delta S} = 0 \qquad (2.9)$$

The laws of conservation in physics refer to the constancy of a system variable in time. A variable that does not change in time is said to be conserved. For example, in classical physics we find that the mass, the energy, the momentum, and the electrical charge of a system are conserved. In relativity physics, the conservation laws of mass and energy become one combined law.

2.7 Feedback

You can buy a camera that will automatically adjust the exposure for the intensity of light and the type of film you are using. The film requires a given amount of light energy for the proper exposure. The amount of energy reaching the film depends upon the size of the aperture, time of

FIGURE 2.11
A feedback system in the human body.

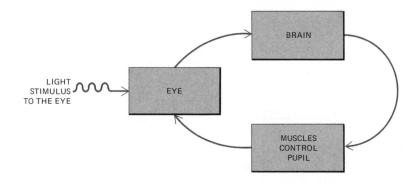

exposure, and the intensity of light. In using the camera, you set the aperture. Then the intensity of light produces an input to the electrical circuit that controls the exposure time. In another type of camera the exposure is set, and the aperture automatically adjusts for the speed of the film. This second type is parallel to the system of the human eye. In the eye the light incident upon the eye is the stimulus to the muscles that control the size of the pupil of the eye (Figure 2.11). You can observe the change in the size of the pupil of your eyes by looking in a mirror in a darkened room and then turning on the lights.

Feedback is said to exist in a given system if part of the output from the system is returned as an input into the same system. This feedback may be either positive or negative, that is, additive or subtractive. In general, if the feedback is negative, the system will return to a stable situation. If the feedback is positive, the system will oscillate or go out of control.

Figure 2.12 illustrates positive and negative feedback loops for two cases in which one automobile follows another and the lead vehicle slows down or stops. In Figure 2.12a the lead automobile is sighted, and the message is delivered to brain of the following car. The second driver's brain commands that brakes be applied. Brakes are applied, and the second automobile is slowed or stopped so that a safe distance between the cars is maintained. Thus a steady system is approached. In

FIGURE 2.12
Two feedback systems with different types of responses: (a) negative feedback leading to stability; (b) positive feedback leading to a system out of control.

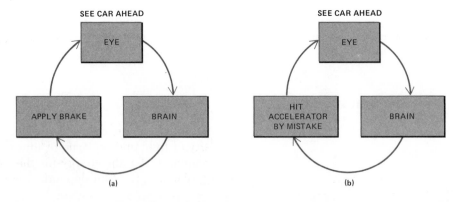

Figure 2.12b the proper execution is not made. The visual stimulus results in more gasoline going into the engine so that the car speeds up, and the distance between the cars decreases. A collision will occur if this feedback continues. That is, the system goes out of control, and a collision occurs.

The thermal regulation systems for a house and for the human body are examples of feedback used to maintain equilibrium. In a house, a thermal sensing element in the thermostat provides a feedback signal. When the temperature falls below a set point, the feedback signal turns on the furnace to supply more heat. The heat raises the temperature of the house above the set point, and the feedback signal is shut off. In a similar way the body's thermal sensors provide data to a portion of the brain (the hypothalamus) that regulates the body metabolism to maintain the skin temperature of the body in the equilibrium range of about 33 to 38°C in a room of 27°C.

Questions

4. Consider yourself as a system in the physics classroom environment. What are the "input" and "feedback" for you? For your instructor?

2.8 Superposition Principle

Picture a rope with two people holding opposite ends of the rope. Each person independently snaps his end of the rope and sends a pulse down the rope (Figure 2.13). What is the resultant displacement of the rope as the result of the activities of the pair? Let y_1 be the displacement of a given point in the rope at a given time as given by Alan and y_2 be the displacement of the same point at the same time as given by Barbie. The resultant displacement then is the sum of y_1 and y_2, that is,

$$y = y_1 + y_2 \tag{2.10}$$

This procedure assumes that the pulses do not interact with one another, that each propagates as though the other were not present. This summation process is called the *superposition principle*. We can use the superposition principle in an analysis of a situation only when the sys-

FIGURE 2.13
A simple system demonstrating the principle of superposition.

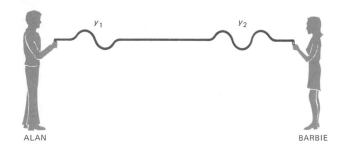

A pulse of same magnitude is started from each end of a spring and travels toward the other end. This is an example of superposition of two pulses. Note that the separate displacements add (see the sixth frame) when they are together. (Picture from *PSSC Physics,* D.C. Heath and Company, Lexington, Mass., 1965.)

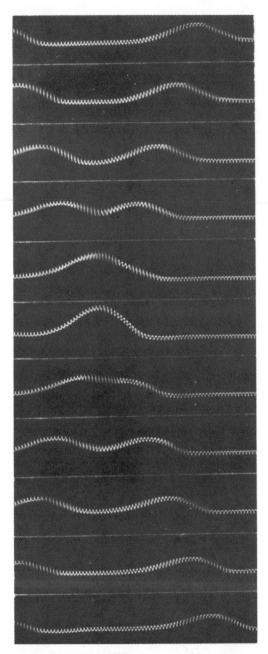

tems involved are independent and interact linearly. For cases in which these criteria are satisfied, the analysis of complex systems is possible by means of the superposition principle. The complex system can then be broken into simpler parts, and the sum of the properties of the parts is taken to be the behavior of the complex system. For example, the interaction of the earth and the solar system is equal to the sum of the

FIGURE 2.14
The principle of superposition as applied to the driving of a stake for a circus tent by three men. Parts (a), (b), and (c) show their individual contributions. Part (d) shows the total of all contributions.

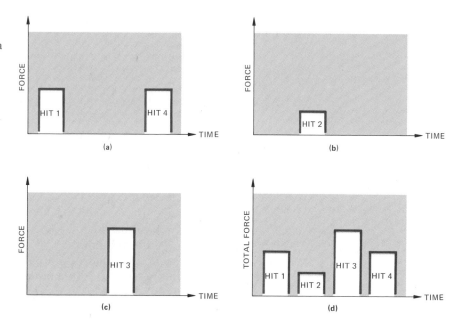

interactions of the earth with each individual constituent of the solar system. The pressure in a gas filled container is equal to the sum of the partial pressures produced by each of the gases in the container.

Perhaps you have seen the stakes for a circus tent being driven. Let us assume that a stake is driven by three men, each hitting the stake in succession. The force contributed by each man is a function of time. See Figure 2.14. The total force exerted is the sum of the separate contributions (see Figure 2.14d).

There are many nonlinear physical systems. Electronic amplifiers, for example, can be operated in a nonlinear way. Such an amplifier distorts the input signal by amplifying different frequencies by different amounts. Most of the human senses seem to respond nonlinearly.

Questions

5. Consider a married couple as a system. Does this system obey the superposition principle? Is this system equivalent to the sum of individual interactions? Explain your answer.
6. List some physical, biological and social systems, and determine whether they are linear systems with respect to the significant variables for the given system.

2.9 Applications of Unifying Approaches

Consider a cork floating in a motionless pool of water. It is in equilibrium because it remains at rest. We drop another cork onto the first, producing a small displacement of the first cork. This cork will oscillate about its equilibrium position. The bouyant force of the water is a restoring force. The amplitude of the oscillation, the amount of maximum displacement from the equilibrium position, will depend upon the inertia of the cork (its mass) and on the inertia of the dropped cork. If the corks have the same mass, we will obtain a maximum oscillation amplitude. The oscillation of the cork sets up a water wave, which transports energy through the water.

If a repeating external force is applied to the floating cork, we will observe another phenomena. As we change the repetition rate of the applied force, we will find a particular repetition rate at which the amplitude of the cork reaches a maximum creating turbulent water waves. This condition of maximum oscillation under the influence of a repeating force is called *resonance*. Resonance occurs when the energy transferred from the applied force to the oscillating system is maximum. Resonance is observed in many physical situations, and in each case the externally applied resonance frequency matches the *natural frequency* of the oscillating system. The natural frequency of any system is determined by its inertial properties and by its restoring force constant. The restoring force constant is the ratio of the restoring force to displacement.

$$\text{restoring force constant} = \frac{\text{restoring force}}{\text{displacement}} \qquad (2.11)$$

For a linear system the restoring force is a constant given by the slope of the force versus the displacement shown in Figure 2.5. You have undoubtedly experienced a resonance phenomena in your automobile. At a particular speed on a bumpy road your automobile can experience large vibrations. This occurs when the repeating forces applied to the automobile by the road have the same frequency as the natural vibration frequency of the suspension system of your automobile.

2.10 Homeostasis

The broad applicability of the various principles discussed in this chapter becomes clear when we consider the biological concept of homeostasis. Higher animals are able to survive in a wide variety of external environments because they carry their own environment with them in the form of fluids that bathe their cells. The constancy of this internal environment in spite of variation in the external environment is characteristic of all higher forms of life. The term *homeostasis* refers to the stability of the internal environments of organisms. The concept of homeostasis, or steady-state control, is now recognized as one of the fundamental principles of biology. Multicellular animals maintain homeostasis by means of feedback.

Consider the homeostasis and feedback involved in human hunger. What determines your hunger equilibrium point? How do you determine your hunger inertia? It is now known that the feeling of hunger is stimulated by the concentration of glucose in the blood, which varies from high levels just after eating to lower levels several hours later. What is an important gradient in the feedback cycle for hunger control? The stimulation of the hunger nerve center in the brain causes the feeling of hunger. Eating raises the blood sugar level and reduces the stimulation to the hunger center. At the same time it provides stimulation to the satiety nerve center and actively counteracts the feeling of hunger. Hence, we alternate between hunger and its absence, and we eat enough to maintain the proper glucose concentrations in the blood for the healthy functioning of the body cells.

SUMMARY

Use these questions to evaluate how well you have achieved the goals of this chapter. The answers to these questions are given at the end of this summary with a reference to the section where you can find the related content material.

Definitions

1. The property of a system that measures resistance to change is called
   a. gradient
   b. current
   c. equilibrium
   d. inertia
   e. superposition
2. When a system is displaced from _____ and the _____ is proportional to the displacement, the system is called a _____. Such systems will display oscillatory motion with a _____ determined by its restoring force constant and inertial property.
3. Fluid flow through a tube is proportional to the pressure difference between one end of the tube to the other. The current in this example is _____. The gradient magnitude would be _____, and the direction of the gradient is _____.
4. If a measurable quantity is thought to obey a conservation law, how could you verify that the quantity is conserved?
5. Give an example of feedback as it applies to human vision.
6. On a TV commercial, Ella Fitzgerald is shown shattering a glass with her voice. This is an example of resonance. What are the conditions necessary for this occurrence?

Inertia

7. Resistance to _____ change would be the thermal inertia of a metal rod.
8. Resistance to _____ change would be the mechanical inertia of a metal rod.
9. Resistance to _____ would be the electrical inertia of a metal rod.

Energy Transfer

10. If you plotted the energy transfer of two interacting systems against the inertial property ratio of the two systems, what kind of qualitative curve would you expect to get?

Superposition

11. If two interactions produce the following responses when acting alone on a linear system, sketch the response they produce when both interactions act on the same system at the same time.

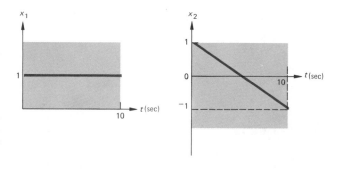

Answers

1. d (Section 2.3)
2. equilibrium, restoring force, linear system, natural frequency (Sections 2.2 and 2.4)
3. fluid flow rate, pressure difference/tube length, from low pressure to high pressure (Section 2.5)
4. Measure the quantity at different times, and find the same value within experimental error (Section 2.6)
5. The iris closes rapidly in bright light to protect the retina and opens (much more slowly) to improve dark vision. Feedback loops control this response. (Section 2.9)
6. Voice frequency must equal the natural frequency of vibration of glass (Section 2.7)
7. temperature (Section 2.3)
8. state of motion (Section 2.3)
9. electrical charge flow (Section 2.3)

10.

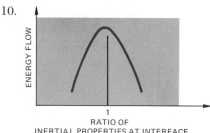

11.

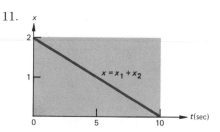

ALGORITHMIC PROBLEMS

Listed below are the important equations from this chapter. The problems following the equations will help you learn to translate words into equations and to solve single concept problems.

Equations

$$\text{restoring force} = -kx \tag{2.1}$$

$$\text{current} = \frac{\Delta \,(\text{quantity})}{\Delta \,(\text{time})} \tag{2.2}$$

$$\text{gradient} = \frac{\Delta \,(\text{physical quantity})}{\Delta \,(\text{position})} \tag{2.3}$$

$$\text{current} = -K \times \text{gradient} \tag{2.4}$$

$$y = y_1 + y_2 \quad (\text{superposition for linear system interactions}) \tag{2.10}$$

Problems

1. A spring has a force constant of 10.0 newton/meter (N/m). What is the force necessary to stretch the spring 0.250 m?
2. A force of 200 N is required to draw a bow 0.50 m. What is the value of k for the bow? What are the units of k?
3. A teakettle holds water at 100°C, the surface of the stove is at 200°C, and the thickness of the bottom of the teakettle is 0.150 cm. What is the temperature gradient in the bottom of the teakettle?

4. If the temperature of the air in a room is 20.0°C and the air outside the window is −4.00°C, what is the temperature gradient in the glass of 3.00 mm thickness?

5. The difference in potential between two metal plates 2.00 mm apart is 300 volts. What is the potential gradient in a glass dielectric that fills the space between the plates.

6. The pressure is measured at two points in a horizontal water system. It is found that pressure reading at one position is 80.0 N/cm² and at the second position it is 60.0 N/cm². These positions are 10.0 m apart. What is the average pressure gradient between the two points?

7. Two people push on opposite sides of a box with forces of 90.0 N to the right and 60.0 N to the left. Find the resultant force due to these two people. That is, give both the magnitude and the direction of the force.

8. In Figure 2.13, the displacement of a point P in the rope by Alan is 1.00 cm and the displacement by Barbie is 2.00 cm. What is the resultant displacement?

Answers

1. 2.50 N
2. 400 N/m
3. 667° C/cm
4. 8.00°C/mm
5. 150 V/mm

6. a decrease of 2.00 N/cm² for each meter of distance
7. 30.0 N to the right
8. 3.00 cm

EXERCISES

These exercises are designed to help you apply the ideas of this chapter to physical situations. Where appropriate the quantitative answer is given at the end of the exercise.

Section 2.1

1 Reread the introduction to this chapter, and discuss the questions asked in the introduction in terms of the words and concepts that have been defined in this chapter.

Section 2.2

2. Describe a system, other than the one given in this text, that exhibits stable, unstable, and neutral equilibrium.

Section 2.3

3. Use the mechanical inertial properties of persons to

explain the characteristics of large and small football players, include starting, stopping, and collision actions.

Section 2.4

4. In carrying out an experiment in the laboratory a student obtains the following data when she loads a pan suspended on a spring balance.

Load	Index Reading
0.00	0.00
10.0 grams	1.20 cm
20.0 grams	2.30 cm
30.0 grams	3.60 cm
40.0 grams	4.90 cm
50.0 grams	6.00 cm

Plot the curve with *load* as ordinate (vertical) and

FIGURE 2.15
Exercise 5.

index reading as abscissa. What is the value of the slope of the curve? What is the physical significance of the slope? Describe the motion of the pan if it is displaced slightly. [slope = 8.33 g/cm; slope = spring constant.]

Section 2.5

5. Consider three triangular supports with equal altitudes as shown in Figure 2.15, and with bases in ratio of 3:2:1. Compare the slopes of the three. In which case is the gradient the largest? If equal drops of water, that is, equal inertia are placed on each slope at the highest point, in which case will the drop reach the bottom first? Why? [The *a-a* triangle has steepest slope; *a-3a* triangle has the smallest slope. Gradient is largest for *a-a*. Drop will reach bottom first on triangle *a-a*.]

Section 2.6

6. Use the conservation of mass to calculate the amount of water produced when 4 g of hydrogen combine with 32 g of oxygen. [36 g H_2O]

Section 2.7

7. Construct a feedback loop for light reaching the film in a camera. Include shutter speed and aperture size.

(Refer to any set of encyclopedias for further information.)

8. Construct a feedback control loop for regulating the size of the iris in the human eye. [Refer to any introductory biophysics or physiology book.]

9. Construct a feedback loop for both positive and negative feedback for change in the socioeconomic population spectrum of your home community. Indicate the nature of the feedback in each case.

Section 2.8

10. A man buys a new house at a cost of Q_0 dollars. He decides that in order to maintain the quality of his house, he must spend money at the yearly rate of three percent of the original value. The total money spent on upkeep is then $Q = 0.03\ Q_0 T$ where T is years. The aging process will decrease the value of his house. He estimates that his house decreases in value with age according to

$$Q_a = Q_0 \times \sqrt{\frac{(40 - T)}{40}}$$

for the first 40 years where T is in years. Use the superposition principle to draw a graph of the value of his house as a function of T for 40 years. Estimate the age of his house when it has its maximum value. Assume there is neither inflation or recession during the 40 years! [value = $Q_a + Q$]

11. Two independent disturbances are simultaneously impressed upon body A. These disturbances were given by: $y_1 = 8$ and $y_2 = 4(1 - t)$. What is the form of the y-axis displacement of body A as a function of t? Plot the curve.

PROBLEMS

The following problems involve more than one physical concept.

12. Describe a simple experiment to illustrate:
   a. the property of inertia of a body
   b. a system in equilibrium
   c. the action of a restoring force
   d. the principal of superposition
13. a. Heat flows from a stove burner through the bottom of a pan. The gradient producing the heat current

will be _____. [(burner temp minus inside temp)/thickness of pan]
   b. When the switch of a flashlight is turned on, the gradient across the lamp is _____ and the current is given in _____. [electric potential/m; electric charge/sec]
   c. Diffusion is a process involving mass flow (current = mass transferred/sec). What kind of gradient might produce such a current? [mass density/cm]
   d. What kind of gradient produces the water flow

from the root system of a tree up into its limbs and leaves? [pressure/distance]

14. There are a wide range of homeostasis systems within the human body. Use the operational definitions you have learned in this chapter to describe some of the physical parameters of these systems. Consider the following examples:
    a. gases in the blood
    b. water content of the blood
    c. excretion
    d. thirst
    e. temperature control of the body
    f. circulation
    g. respiration

15. Construct a model using feedback to explain threshold phenomena for a biological system.

GOALS

When you have mastered the content of this chapter, you will be able to achieve the following goals:

Definitions
Use the following terms to describe the physical state of a system:
displacement
velocity
acceleration
uniformly accelerated motion
projectile motion

uniform circular
  motion
radial acceleration
tangential
  acceleration

Equations of Motion
Write the equations of motion for objects with constant velocity and for objects with constant acceleration.

Motion Problems
Solve problems involving freely falling and other uniformly accelerated bodies, projectile motion, and uniform circular motion.

Acceleration Effects
List the effects of acceleration on the human body.

PREREQUISITES

Before beginning this chapter you should have achieved the goals of Chapter 1, Human Senses, and Chapter 2, Unifying Approaches. You must also be able to use the properties of right triangles to solve problems.

Mathematics Self-Check

If you can solve the following problem easily and correctly, you are prepared for this chapter: A surveyor wishes to determine the distance between two points A and B, but he cannot make a direct measurement because a river intervenes. He steps off a line AC at a 90° angle to AB and 264 meters long. With his transit, at point C he measures the angle between line AB and the line formed by C and B. Angle BCA is measured to be 62°. What is the distance from A to B? [497 m]

If you had difficulty getting this answer, you will find additional information in Section A.6, Right Triangles, of the appendix.

3

Kinematics

3.1 Introduction

For the greater part of your life, you have been engaged in the process of getting from here to there. First you learned to crawl, then to walk, and later to run. These are examples of motion and change of position. In these motions you were concerned with distances, directions, rates of motion, and time, or duration, of motion. This same concern with motion is true for change of position by a mechanical device such as a bicycle, an automobile, or an airplane.

How would you describe your present state of motion? How would you describe the motion of an Olympic sprinter? What would be your description of a professional figure skater's motion when she does a spin on ice skates? Have you seen pictures of an astronaut moving about in the "zero gravity" environment of space? What concepts do you need to describe the astronaut's motion? You will be introduced to the concepts of motion in various forms in this chapter. This study of motion (without concern for its causes) is known as *kinematics*.

3.2 Characteristics of Distance and Displacement

In order to develop the relationships and characteristics of motion, it is necessary for us to define some terms. If a body is moved from one place to another, it is said to be displaced. This *displacement* is specified by both *magnitude* and *direction*. If you move your coffee cup along the table top 10.0 cm to the east and then 10.0 cm to the north, you will have displaced your cup 14.1 cm to the northeast. The coffee cup will have traveled a distance of 20.0 cm and will be a distance of 14.1 cm from its starting point. You will notice that *distance* has only magnitude. Such a physical quantity is called a *scalar* quantity. A scalar quantity is completely specified by a number and its proper dimensional unit. Can you think of other scalar quantities with which you are familiar?

A quantity such as displacement, that is only completely determined when you have given **both** its magnitude and its direction is called a *vector*. A **vector** quantity can be represented graphically by an arrow in which the shaft of the arrow represents the line of action and the arrow head is the direction of action along the line. Vector A will be shown in **boldface** type **A**. The *length* of the line gives the *magnitude* of the vector and is represented in the usual type style *A*. Thus a displacement of 8 km northeast is represented by a vector making an angle of 45° with the easterly direction and 8 units long (see Figure 3.1).

Suppose you ride a bicycle from your home to the physics building, a straight line distance of 2 km. The total distance you traveled was 2 km. After you ride back home, you will have traveled a distance of 4 km, but your net displacement is zero (Figure 3.2). The addition of distances (scalars) follows the usual rule of addition. The addition of displacements (vectors) must take into account the directions of the displace-

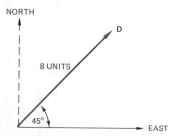

FIGURE 3.1
Vector representation of a displacement of 8 km northeast of origin.

FIGURE 3.2
Graphical representation indicating
how one can travel 4 km and have
zero displacement.

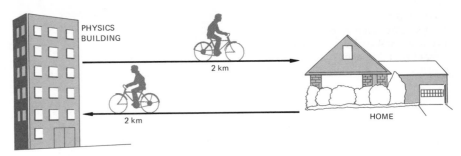

FIGURE 3.3
Three simple vector additions: (a)
parallel in same direction, (b) anti-
parallel, and (c) at 60° for the case of
equal magnitudes.

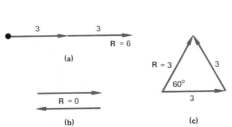

ments involved. In this example the first displacement (from your home to the physics building) and the second displacement (from the physics building to your home) are in opposite directions. The addition of these two vector displacements gives a zero net displacement.

Vectors do not obey the simple algebraic properties of scalars. For example, when you add the two scalars, 2 plus 2, you obtain 4. If you add two vectors, both of magnitude three, you may obtain any number from 0 to 6 for the magnitude of the sum of these two vectors (see Figure 3.3). Add the two displacements 3 km east and 3 km east. What is the net displacement result? If you said 6 km east, you got it. Now add the two displacements, 3 km east and 3 km west. What is the net displacement? If you said 0 km, you are right. How can you add two displacements, each of which has a magnitude of 3 km, and obtain a final displacement whose magnitude is 3 km?

3.3 Graphical Method for Adding Vectors

Suppose you dropped a contact lens from your eye and it rolled across the tile floor. It rolled 10.0 m across the floor in a direction 37° north of east. There it struck the wall and bounced 8.0 m in a direction 30° west of north. What is the displacement of the contact lens?

To use the graphical method for adding vectors, we represent the first displacement by an arrow pointing 37° north of east and scaled to represent 10.0 m in length, as we have drawn vector **A** in Figure 3.4a. We represent the rebound displacement by the vector **B**. To add the vectors **A** and **B**, we draw the vector **B** extending from the tip of vector **A** as shown in Figure 3.4c.

FIGURE 3.4

Vector representation of two displacements, **A** and **B**. (c) The resultant **R** of displacements **A** and **B**.

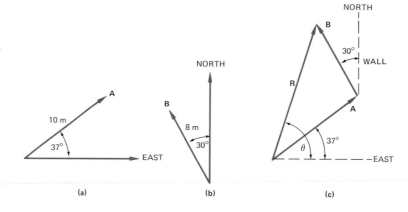

(a)   (b)   (c)

The sum of vector **A** and vector **B** is called the *resultant* and is shown by the vector **R**. The magnitude of the resultant displacement **R** can be measured with a ruler, and the angle between **R** and east can be measured with a protractor. For this example **R**, the final location of the dropped lens, is found to be given by a vector 13.7 m at an angle 73° north of east.

Now suppose we wish to add vector **C** to vectors **A** and **B** given above. Vector **C** has a magnitude of 6.0 m and points to the west. We draw vectors **A** and **B** as above and then from the tip of **B** draw vector **C** as shown in Figure 3.5. We can find the resultant **R**, or the vector sum, of **A** + **B** and **C** by drawing a vector from initial point of **A** to the tip of **C**. We can obtain the magnitude of **R** by scaling, that is, by measuring the length and the direction of **R** by measuring the angle from the east-west reference axis with a protractor. Then we have all the data needed to define the vector **R**, direction and magnitude. For this example the vector **R** is given by a displacement of 13.1 m in a direction 81° north of west. This procedure can be used for addition of any number of vectors.

You can find the difference between two vectors by using the same procedure. To find the value of the vector (**A** − **B**), you add the vector − **B** to the vector **A**. The vector − **B** has the same magnitude as the vector **B** but the opposite direction. The vector (**A** − **B**) is shown in Figure 3.6. The method for the finding of the magnitude of the vector **R** and its direc-

FIGURE 3.5

Graphical representation of the addition of three vectors (displacements).

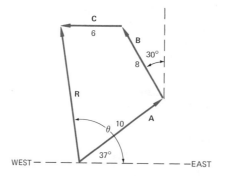

FIGURE 3.6
The vector difference between
two given vectors.

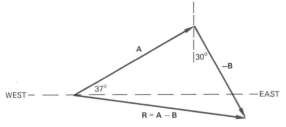

tion is as given above. For this case the vector **R** is given by a vector about 7° south of east with a magnitude of 12.0 m.

In Section 3.2 we asked how you might add two displacements, each of 3 km magnitude and obtain a final displacement of 3 km. From this graphical method of adding vectors, you see that the resultant **R** and the two vectors **A** and **B** form a triangle. If the three vectors **A**, **B**, **R** each have a length 3, then the vectors, **A**, **B** and **R** must form an equilateral triangle. Hence vector **R** makes an angle of 60° with vector **A** (see Figure 3.3c).

3.4 An Algebraic Method for Adding Vectors

Another way of designating the displacement of the contact lens dropped in the previous section is to specify the number of the floor tile on which it is lying from a designated corner of the room. This method of locating a point in a plane with two *coordinates* is the visual technique incorporated in the cartesian model of space.

By analogy, the cartesian model enables us to state the position vector of the final displacement of the contact lens (from the designated corner) as the sum of a north component vector and an east component vector. In our example the lengths of these component vectors would have units of tile length. In general, the vector in a plane can be specified by its horizontal x and vertical y components in any chosen coordinate system. For instance, vector **A** given in Figure 3.5 has an east-west component of $A \cos 37°$ and a north-south component of $A \sin 37°$. Let us make the substitution of the x-axis for the east-west direction and the y-axis for the north-south direction and proceed to find the resultant of **A**, **B**, and **C** in Figure 3.5. First we find the x component and the y component of each vector **A**, **B**, and **C**. The x component of the resultant **R** is equal to the algebraic sum of the x components of **A**, **B**, **C**, and the y component of **R** is equal to the algebraic sum of the y components of **A**, **B**, and **C**. The method is outlined in Table 3.1:

TABLE 3.1

Vector	x Component	y Component
A	$A \cos 37° = 10.0\,(0.800) = 8.00$	$A \sin 37° = (10.0)(0.600) = 6.00$
B	$B \cos 120° = (8.00)(-0.50) = -4.00$	$B \sin 120° = (8.00)(0.866) = 6.93$
C	$C \cos 180° = (6.00)(-1.00) = -6.00$	$C \sin 180° = (6.00)(0.00) = 0.00$
R	$R \cos \theta = (8.00 - 4.00 - 6.00)$ $= -2.00$	$R \sin \theta = (6.00 + 6.93 + 0.00)$ $= 12.93$

The magnitude of **R** is found by using the pythagorean theorem,

$$R = \sqrt{(2.00)^2 + (12.9)^2}$$

$$R = 13.1 \text{ m}$$

We can find the direction of **R** by using the definition of the tangent of an angle,

$$\tan \theta = \frac{R \sin \theta}{R \cos \theta} = \frac{12.9}{-2.00} = -6.45$$

$$\theta = 98.8°$$

In this case, the resultant vector **R**, which is the sum of **A**, **B**, and **C** is given by a vector of length 13.1 m in a direction 98.8° counterclockwise from the x-axis.

3.5 Characteristics of Motion

In the above examples please notice that the displacement and the distance traveled may be given by different numerical values. In understanding problems of motion it is very important to have clearly in mind what information you have been given: Is it a distance or a displacement? What you are seeking: Is the answer to be a scalar or a vector?

In many cases we are interested not only in whether a body has moved but also in how fast the body moved. If we measure how much time is required to move an object a given distance or through a given displacement, we can calculate the rate of change of distance with time or the time rate of change of displacement. *Speed* is defined as the time rate of change of the distance traveled. Since both distance traveled and time are scalars, speed is a scalar quantity with the dimensions of length divided by time, or, for our purposes, with units of meters divided by seconds. *Velocity* is defined as the time rate of the change in displacement. The average velocity shown by the symbol $\bar{v}$ is found as follows:

$$\bar{v} = \frac{s_1 - s_0}{t_1 - t_0} = \frac{\Delta s}{\Delta t} \tag{3.1}$$

where s_0 is the displacement of the body at time t_0 and s_1 is the displacement of the body at a later time t_1. We have used Δs to represent the change in displacement that occurred in the time of Δt.

Since velocity is the ratio of the change in displacement (a vector) to the change in time (a scalar), velocity is a vector quantity. What are the dimensions of velocity, and what units does it have? You may notice that velocity is the ratio of a quantity measured in meters (displacement) to a quantity measured in seconds. *Instantaneous velocity* is velocity at any given instant in time and is discussed in the Section 3.11. Can you change the velocity of a moving object without changing its speed? If you change only the direction of the velocity of an object and not its magnitude, then the speed, the magnitude of velocity, does not change. If you are walking along the sidewalk with a velocity of 3 km/hr east, and turn a corner to go 3 km/hr north, what is your speed? Your speed remains the same 3 km/hr, but your velocity has changed from 3 km/hr east to 3 km/hr north.

The simplest motion that we can have is that of constant velocity. That means neither the direction nor the magnitude of the time rate of change of the displacement is changing. This is motion in a straight line at a constant rate. One example is walking at a rate of 5 km/hr east. The displacement that occurs when a body is moving with constant velocity is computed from the equation

$$\mathbf{s} = \mathbf{v}t \tag{3.2}$$

Another simple kind of motion is to travel at a constant speed. The direction of displacement may change, but the time rate at which the distance traveled changes is constant. An example of such motion is traveling along the highway at 89 km/hr (55 mph). In this kind of motion the distance traveled is given by the product of the speed and the time of travel.

You recognize the difficulty in always traveling with either constant velocity or constant speed: It does not permit you to stop moving if you are moving already or to start moving if you are presently at rest. Clearly, then we need to consider other kinds of motion in which the velocity changes.

3.6 Linear Motion

To begin let us simplify our discussion of motion with changing velocity by restricting it to motion of objects along a line. This includes a number of common experiences such as a runner on a track, an automobile on the highway, or a toy car rolling down an inclined table. In these cases, the object is moving either forward or backward, either away from or toward the starting point. Hence, velocity can have only two directions which we can designate as positive and negative. In these common situations, you notice that the difference between velocity and speed appears to be of minor importance; only the sign may be different. So you can understand why in ordinary conversation the distinction between velocity and speed is not carefully preserved. However, the sign in front of the magnitude of the velocity is highly significant. It tells you whether an object is going forward or backward, up or down, right or left, depending upon the direction that you have chosen as positive.

When a body starts moving from rest, its velocity changes. If we choose forward as the positive direction, as you back your car out of the garage you decrease the velocity of your auto, that is, you start from rest ($\mathbf{v} = 0$), and give it a negative (backward) speed. As you start your car

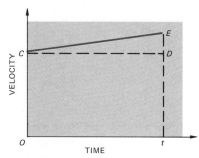

FIGURE 3.7
A velocity-time graph for constant acceleration.

forward down the street, you increase the velocity of your car. The change of the velocity of an object in a unit of time is called the *acceleration*. The average acceleration is given by the equation

$$\bar{a} = \frac{v_1 - v_0}{t_1 - t_0} = \frac{\Delta v}{\Delta t} \tag{3.3}$$

where v_0 is the velocity at t_0 and v_1 is the velocity at a later time t_1, and where Δv represents the change in velocity during the time interval Δt. Since velocity is a vector, the acceleration is also a vector quantity. What are the dimensions of acceleration, and what units does it have? You may notice that acceleration is given by the ratio of a quantity measured in meters divided by seconds to a quantity measured in seconds. Instantaneous acceleration is the acceleration at any given instant in time and is treated in the Section 3 11 of this chapter.

Questions

Figure 3.7 shows a plot of velocity as a function of time for constant acceleration. Study the curve and answer the following questions:

1. What does the intercept C on the velocity axis represent?
2. What does the slope of CE represent?
3. What does DE represent?
4. What does tE represent?
5. Note that the area of $OtDEC$ is made up of a rectangle $OtDC$ and triangle CDE. What is the area of $OtDC$, and what does it represent?
6. What is the area of CDE, and what does it represent?
7. What is the total area $OtDEC$, and what does it represent?

Answers

1. original velocity
2. acceleration
3. change in velocity
4. velocity at time t
5. displacement for constant velocity v_0 and time t

6. displacement in time t resulting from the change in velocity
7. total displacement in time t

3.7 Uniformly Accelerated Motion

Let us develop the relationships for motion in which the time rate of change of the velocity, the acceleration, is constant. This is, of course, an idealization since in no real system is it possible to keep the rate of change of the velocity a constant for all times. But the motion of many systems approximates this idealization. For example, you pull away from the curb in your automobile, and after one second you are traveling at a forward speed of 10 km/hr (6 mph). Then after two seconds you are traveling at $+20$ km/hr (12 mph), after three seconds at $+30$ km/hr (19 mph), and so on. The rate at which you are changing your velocity is a constant. That is, the acceleration is constant and has the value of (30 km/hr − 0 km/hr)/(3 sec), or +10 km/hr per second. This is known as uniformly accelerated motion. From our definition of acceleration we get

$$a = \frac{\Delta v}{\Delta t} = \frac{v_f - v_0}{t} \tag{3.3}$$

or solving for v_f gives

$$v_f = v_0 + at \tag{3.4}$$

in which v_f is the velocity at any time t, if the original velocity is v_0 and the acceleration **a** is constant. If the acceleration is constant, the average velocity **v** is given by one-half of the sum of the final and initial values of velocity,

$$\bar{v} = \frac{v_f + v_0}{2} \tag{3.5}$$

For the specific example above the average velocity is to be (30 km/hr + 0 km/hr)/2 = +15 km/hr. The change in displacement during time t is given by the product of the average velocity times the time,

$$\Delta s = s - s_0 = \bar{v}t = \left(\frac{v_f + v_0}{2}\right) t \tag{3.6}$$

For the automobile pulling away from the curb above, the displacement, or forward distance traveled, is 15 km/hr × 3 sec. Converting kilometers to meters and hours to seconds, we obtain (15,000 m/hr)(1 hr/3600 sec) (3 sec) = +12.5 meters. Substituting the value of v_f from Equation 3.4 in Equation 3.6, we obtain an expression for the displacement in terms of the initial displacement, initial velocity, acceleration, and time:

$$s - s_0 = v_0 t + \tfrac{1}{2} at^2 \tag{3.7}$$

Using the example of the automobile, we choose the original displacement at the curb as the position where $s_0 = 0$. The starting velocity v_0 is zero since the auto starts from rest. Since we found the acceleration to be given by 10 km/hr/sec, after 3 sec the displacement is given by

$$s - 0 = 0(3) + \tfrac{1}{2}(10 \text{ km/hr-sec})(3 \text{ sec})^2$$

$$= +45 \text{ km-sec/hr} = 45,000 \text{ m/3600}$$

$$= +12.5 \text{ m}$$

Another equation for uniformly accelerated motion (u.a.m.) is obtained by eliminating time from Equations 3.4 and 3.7 to obtain an expression for the velocity as a function of acceleration and distance.

$$2a(s - s_0) = v_f^2 - v_0^2 \tag{3.8}$$

This product of two vectors is known as a scalar product since it yields a scalar quantity. It will be discussed in Section 5.2. In this section, since all the vectors are along the same line, this product can be treated as the usual algebraic multiplication.

If you consider the initial position as the origin, or zero displacement, then the three basic equations of u.a.m. become

$$v_f = v_0 + at$$

$$s = v_0 t + \tfrac{1}{2} at^2$$

$$2a \cdot s = v_f^2 - v_0^2 \tag{3.9}$$

Starting from rest is a special case in which the initial velocity is zero, $v_0 = 0$. If the initial displacement is also zero, the equations can be reduced to the following shortened forms:

$$v_f = at$$

$$s = \tfrac{1}{2} at^2$$

$$2a \cdot s = v_f^2 \tag{3.10}$$

when the initial velocity is zero and the initial displacement is zero.

Consider a low-friction toy car rolling down a slightly inclined table. Shown below in Tables 3.2 and 3.3 are the experimental data. Compute the missing items in the table. Can you determine what type of motion is represented by this physical situation?

Experimental data for a 53.6 g toy car rolling down an incline are given in Table 3.2. The experiment was repeated with a 50 g mass added to the toy car, and the data in Table 3.3 were obtained.

What can you say about the influence of mass on the motion of the car down the incline? Since mass is a measure of the inertia, the tendency to resist changes in motion of an object, how do you explain the fact that although the mass of a moving object is almost doubled, the data are changed very little?

TABLE 3.2
Table of Data and Calculations

Time	Distance Down the Incline	Average Velocity	Average Acceleration
1 sec	12.4 cm	_____	_____
2 sec	49.2 cm	_____	_____
3 sec	111.8 cm	_____	_____
4 sec	198.8 cm	_____	_____
5 sec	310.0 cm	_____	_____

TABLE 3.3
Table of Data and Calculations

Time	Distance Down the Incline	Average Velocity	Average Acceleration
1 sec	12.0 cm	_____	_____
2 sec	48.3 cm	_____	_____
3 sec	109.8 cm	_____	_____
4 sec	198.2 cm	_____	_____
5 sec	310.0 cm	_____	_____

EXAMPLE

An automobile starts from rest and acquires a forward velocity of 36 km/hr in 5 sec. What is its acceleration, and what is the change in position during this time? Take the forward direction as positive.

$$\mathbf{v}_0 = 0$$

since the car starts from rest.

$$\mathbf{v}_f = +36 \text{ km/hr} = 10 \text{ m/sec}$$

$$t = 5 \text{ sec}$$

How do you find the acceleration? (Hints: 1 km = 1000 m; 1 hr = 3600 sec.) From Equation 3.3

$$\mathbf{a} = \frac{\mathbf{v}_f - \mathbf{v}_0}{t} = \frac{10 - 0}{5} = +2 \text{ m/sec}^2 \tag{3.3}$$

From Equation 3.10

$$\mathbf{s} = \tfrac{1}{2}\mathbf{a}t^2 = \tfrac{1}{2}(2)(5)^2 = +25 \text{ m} \tag{3.10}$$

Does this example describe a realistic situation?

Plot a curve similar to the one in Figure 3.7 for the sample problem worked above. Then plot the displacement as a function of time. What type of curve did you get? Which equation describes the curve?

There are many examples of almost uniformly accelerated motion, but perhaps the most familiar example is a body falling freely through the air when the air resistance is neglected. For such an idealized falling body the acceleration **a** is constant and is directed vertically downward, that is, follows the direction of a plumb bob line. This constant downward acceleration is called the *acceleration due to gravity* and is designated by **g.** All of the equations developed above for uniformly accelerated motion apply for an ideal falling body. One normally replaces **a** by **g,** which has a numerical value of about 9.80 m/sec² downward near the surface of the earth.

You can get an approximate value for the magnitude of **g** by the following simple experiment. Toss a ball straight up, estimate the time the ball is in flight and the height to which the ball is thrown above your hand. You may estimate the time by counting "thousand-and-one, thousand-and-two, . . ." Each count is approximately one second. You should be able to estimate the height in meters that the ball rises. You then can calculate the value of **g** by dividing twice the height by the square of one-half the time of flight. From Equation 3.10, we know that

$$\mathbf{h} = \tfrac{1}{2}\mathbf{g}\left(\frac{t}{2}\right)^2$$

Solving this for **g,**

$$\mathbf{g} = \frac{2\mathbf{h}}{\left(\frac{t}{2}\right)^2} = \frac{8\mathbf{h}}{t^2} \text{ m/sec}^2 \tag{3.11}$$

What value did you get? _____ m/sec². In any case you will find your value to be nearer 10 m/sec² than to either 1 or 100 m/sec². Thus your determined value is the proper order of magnitude.

EXAMPLES

1. A person hangs from a diving board so that his feet are 5 m above the water level in the pool. He lets go of the board. Assuming idealized falling motion, how much·time passes before his feet strike the water, and what is their velocity at that time?

   Given: If we take downward as the positive direction, then s $=+5$ meters down from the board. At the beginning ($t = 0$), the person is at rest. This implies $v_0 = 0$. Then the person begins to fall with an acceleration of $g = +9.80$ m/sec² (positive downward).
   Find:
   a. time of fall $= t$
   b. velocity $= v_f$
      Relationships: $s = v_0 t + \frac{1}{2} g t^2$

   $$v_f = v_0 + gt$$

   Substituting numerical values,

   $$5m = 0t + \frac{1}{2}(9.80 \text{ m/sec}^2)t^2$$

   $$t^2 = \frac{10}{9.80} \text{ sec}^2/m \; t = 1.01 \text{ sec}$$

   $$v_f = 0\left(\frac{m}{sec}\right) \times 1.01 \text{ sec} + 9.80 \frac{m}{sec^2} \times 1.01 \text{ sec}$$

   $$v_f = 9.80 \text{ m/sec downward}$$

2. A ball is thrown vertically upward with a velocity of 30 m/sec. Assuming idealized motion, how high will it rise, and when will it reach its peak of flight? How long will it be before it returns to the starting point, and what will the velocity be at that time?

   Given: $v_0 = +30$ m/sec (positive upward). At its peak the ball stops; this implies that the speed at the peak is zero, $v_1 = 0$ when $t_1 =$ time to reach the peak. Then the ball starts downward with $g = -9.80$ m/sec².
   Find:
   a. time of rise $= t_1 =$ time in flight/2
   b. height $= s$
   c. final speed $= v_f$
      Relationships: $v_f = v_0 + gt$

   $$s = v_0 t + \frac{1}{2} g t^2$$

   $$2g \cdot s = v_f^2 - v_0^2$$

   Using the first of these and inserting values for the final speed v_1, the original speed v_0 and g, we get

   $$0 = 30.0 \text{ m/s} - 9.80 \text{ m/sec}^2 t_1$$

and solve for t_1. Note that if the original velocity is positive upward, $\mathbf{g}$ is then negative (downward).

$$\text{time to reach the peak} = t_1 = \frac{30.0 \text{ m}}{9.80 \text{ m/sec}^2} = 3.06 \text{ sec}$$

Substitute this value for the time into the equation for the displacement:

$$\mathbf{s} = \mathbf{v}_0 t + \tfrac{1}{2}\mathbf{g}t^2$$

$$\mathbf{s} = 30 \text{ m/sec} \times 3.06 \text{ sec} - \tfrac{1}{2}(9.8 \text{ m/sec}^2)(3.06 \text{ sec})^2$$

$$\text{height} = 91.8 \text{ m} - 45.9 \text{ m} = +45.9 \text{ m}$$

or using the other equation

$$2\mathbf{a} \cdot \mathbf{s} = \mathbf{v}_f^2 - \mathbf{v}_0^2 \qquad (3.9)$$

$$+2(-9.80 \text{ m/sec}^2)\mathbf{s} = 0^2 - (30 \text{ m/sec})^2$$

$$\mathbf{s} = \frac{900 \text{ m}^2/\text{sec}^2}{19.6 \text{ m/sec}^2} = +45.9 \text{ m}$$

It will take the ball the same time to fall as to rise; so total time of flight = $2(3.06 \text{ sec}) = 6.12 \text{ sec}$. Therefore, the final velocity

$$\mathbf{v}_f = 30 \text{ m/sec} - 9.80 \text{ m/sec}^2 \times 6.12 \text{ sec}$$

$$\mathbf{v}_f = -30.0 \text{ m/sec}$$

Note the negative sign. When the ball returns, it will be going down (negative), with the same speed with which it started up.

3. A record run in the 200-m dash by Jesse Owens in the 1936 Olympics is well approximated by assuming that Owens started from rest and accelerated at the constant rate of 6.7 m/sec^2 for a time of 1.6 sec. He then ran the remainder of the race at a constant speed. Draw a graph of Owens' acceleration as a function of time. Draw a graph of his speed as a function of time. Draw a graph of the distance he has run as a function of time. [See Figure 3.8 for the solution.]

FIGURE 3.8
Plot of velocity-time curve for Jesse Owens' winning 200-m run in 1936 Olympics.

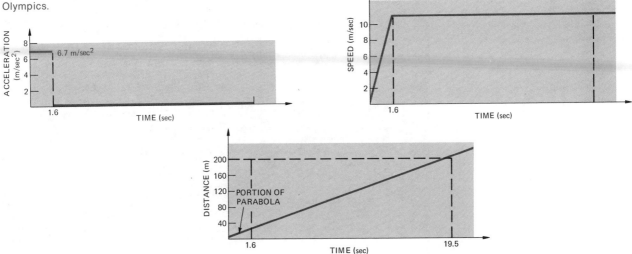

3.8 Projectile Motion

Your friend throws you a tennis ball, which you catch and return to him. The motion of the ball is motion in a vertical plane. It is a common type of motion, but it is more complicated than linear motion. It is motion in which the object has an almost constant velocity in one direction and has almost uniform acceleration in a direction at right angles to the constant velocity (see Figure 3.9). This type of motion is called *projectile motion*. The tennis ball, when we neglect the effects of spin and air

Two golf balls were released simultaneously, one dropped freely and the other one given a horizontal velocity of 2 m/sec. They were photographed at 1/30-sec intervals. The ball that was dropped executes uniformly accelerated motion, and the one with an initial horizontal velocity executes projectile motion. The horizontal white lines in picture are a series of strings placed behind the golf balls at 6-inch intervals. From this you should be able to scale the photograph and determine the vertical velocity for any interval and also the horizontal velocity. (Picture from *PSSC Physics,* D.C. Heath and Company, Lexington, Mass., 1965.)

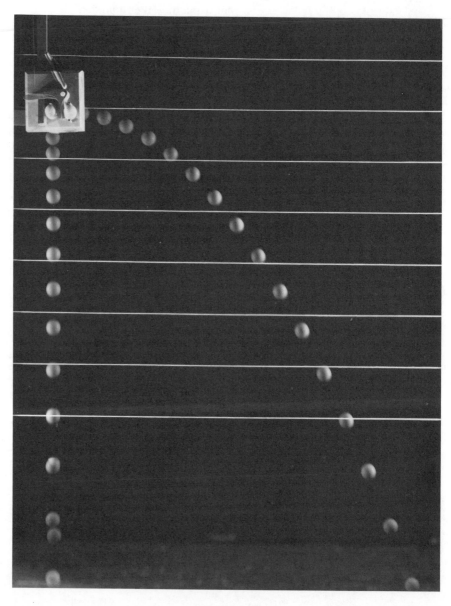

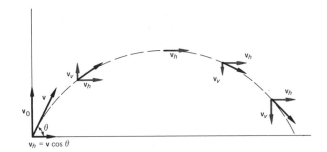

resistance, is moving with a constant velocity in the horizontal direction. (The horizontal components of motion are shown in the figure by the subscript h.) It has the acceleration due to gravity in the vertical direction. (The vertical components of motion are shown by the subscript v.) We shall treat projectile motion as two separate sets of scalar equations using only positive and negative signs to indicate directions, up and down in the vertical direction or forward and backward in the horizontal direction.

Because the motion in the horizontal direction in this idealized case is motion of constant velocity, the horizontal displacement of the tennis ball is given by the product of the horizontal velocity and the time.

$$s_h = v_h t \qquad (3.12)$$

where v_h is the horizontal velocity and t is the time.

This expression neglects both air resistance and spin, and so v_h is a constant since the acceleration in the horizontal direction is zero.

In the vertical direction, the equations of uniformly accelerated motion Equation 3.9 hold true. That is, where v_v = vertical velocity at time t, v_0 is the vertical velocity at $t = 0$, and s_v is the vertical displacement.

$$v_v = v_0 + gt \qquad [a] \qquad (3.13)$$
$$s_v = v_v\, t + \tfrac{1}{2}gt^2 \qquad [b]$$
$$v_v{}^2 = v_0{}^2 + 2gs_v \qquad [c]$$

Note that if we choose the upward direction as positive, then acceleration due to gravity is $g = -9.80$ m/sec².

Suppose a projectile is fired with a velocity of **v** at an angle θ to the ground. The horizontal component is constant during flight but the vertical component is changing because the acceleration is constant in a downward direction: $g = -9.80$ m/sec². The horizontal component of velocity is $v_h = v \cos \theta$, and the original vertical component of velocity is $v \sin \theta$. Hence the vertical component at any time t after firing is

$$v_v = v \sin \theta + gt \qquad (3.14)$$

What is the vertical component of velocity at peak of flight? At the peak of its flight the projectile is moving only in a horizontal direction, so the vertical component of its velocity is zero. Substituting zero for v_v in Equation 3.13a and solving for the time required to reach the peak of flight, we get the following algebraic equations:

$$v_v = 0 = v_0 + gt_{peak} = v \sin \theta + gt_{peak} \tag{3.15}$$

$$t_{peak} = \frac{-v \sin \theta}{g} \tag{3.16}$$

The horizontal displacement relative to position of firing **x** at any time t is given by Equation 3.12 where v_h is given by $v \cos \theta$ and s_h is given by **x**,

$$x = (v \cos \theta)t \tag{3.17}$$

and the vertical displacement s is given by Equation 3.13b where $s_v = y$ and $v_v = v \sin \theta$. With these substitutions, Equation 3.13b becomes

$$y = (v \sin \theta)t + \tfrac{1}{2} gt^2 \tag{3.18}$$

The vertical displacement $y = 0$ at two times, $t = 0$ and $t = -2v \sin \theta/g$. Notice that this later result is two times the time required to reach the peak. You can also use Equation 3.13b to calculate the peak height.

We can show by substituting the value of t_{peak} given by Equation 3.16 that the peak height y_p is

$$y_p = \frac{-v^2 \sin^2 \theta}{g} + \frac{1}{2}\frac{(v \sin \theta)^2}{g} = -\frac{1}{2}\frac{v^2 \sin^2 \theta}{g} \tag{3.19}$$

What is the projectile's range, that is, how far from where it is fired will it strike the ground? If $y = 0$, $x = 0$, at $t = 0$, then the range R is found when y is again zero:

range = horizontal velocity × time of flight

$$R = (v \cos \theta)\left(\frac{-2v \sin \theta}{g}\right) = \frac{-v^2 \sin 2\theta}{g} \tag{3.20}$$

In above equations the value of g is -9.80 m/sec².

At what angle should the projectile be fired to give maximum range for a given firing velocity? The maximum value of the sin 2θ occurs when 2θ is $90°$. Hence R is greatest when θ is $45°$.

You can obtain the equation for the path of the projectile by combining Equations 3.17 and 3.18 and eliminating t. What is the path of a projectile under ideal conditions?

EXAMPLE

Patty Berg, a professional golfer, drives a ball from the tee with a velocity of 37.2 m/sec (120 ft/sec) at an angle of 37° and with no spin. The fairway is straight, and the ball strikes the ground in the same horizontal plane as the tee. What is the horizontal component of the velocity? How long is the ball in flight? How far down the fairway does the ball first strike the ground? What is the angle at which it strikes the fairway? (Neglect air resistance.)

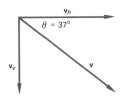

FIGURE 3.10
Graphical representation of the velocity of a projectile at the time it strikes the ground at the end of its flight.

a. The horizontal component of velocity is $v \cos \theta$.

$$v_h = 37.2 \cos 37° = 29.8 \text{ m/sec or } 96 \text{ ft/sec}$$

b. The original vertical component of velocity $= v \sin \theta$.

$$v_v = v \sin \theta = 37.2 \sin 37° = 37.2 \times 0.6 = 22.3 \text{ m/sec or } 72 \text{ ft/sec}$$

At the peak of the flight the vertical component of velocity is 0. In order to find time to reach the peak, we use Equation 3.15 and set $0 = v \sin \theta + gt$.

$$t_1 = \frac{-v \sin \theta}{g} = \frac{-37.2 \times 0.6}{-9.80} = \frac{22.3}{9.80} = 2.28 \text{ sec}$$

Total time of flight is equal to two times the time it takes to reach peak, so $t_t = 2t_1 = 4.56$ sec.

c. range $= (v_0 \cos \theta)t_1 = 37.2 \times 0.80 \times 4.56 = 136$ m.
d. At the point that the ball hits the ground, the horizontal component of velocity is 29.8 m/sec. We can calculate the vertical component because we know that the ball drops for 2.28 sec.

$$v_v = gt = (-9.8)(2.28) = 22.3 \text{ m/sec downward}$$

e. Because we now know both the vertical and horizontal components of velocity when the ball strikes the ground, we can find the angle θ at which it hits.

$$\tan \theta = \frac{v_v}{v_h}$$

$$= \frac{-22.3}{29.8}$$

$$= -\frac{3}{4} \qquad \theta = 37° \text{ below the horizontal.}$$

Is this what you expected? (Look again at Figure 3.9 and at Figure 3.10.)

3.9 *Uniform Circular Motion*

Do you remember taking a ride on a merry-go-round? This is an example of circular motion. First, consider the ideal situation when the merry-go-round is rotating at a constant rate, that is, moving with constant speed. This is known as *uniform circular motion*. Does a body executing this type of motion have an acceleration? Consider the velocity at two points A and B on the circle in Figure 3.11. At each point the velocity is tangent to the circle at that point. We see that v_A and v_B are not the same since the vector v_A and v_B do not point in the same direction. Hence there must be an acceleration even though the magnitudes of velocity v_A and velocity v_B are equal.

Since the magnitude of the velocity in a direction tangent to the circular path is constant, the value of acceleration in the tangential direction must be zero. Hence, if there is an acceleration, and if the component of acceleration along the direction of motion is zero, the acceleration must

FIGURE 3.11
Vectors $\mathbf{v}_A$ and $\mathbf{v}_B$ represent the velocity at two nearby points of a body traveling in a circular path with a constant speed. The direction of acceleration is radial and its in magnitude is v^2/r.

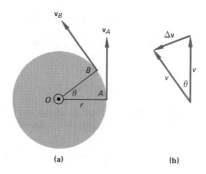

(a)   (b)

be entirely perpendicular to the direction of motion. If the motion is circular, the acceleration must always be directed toward the center of the circle. This acceleration is called the *radial acceleration*.

Let us turn to Figure 3.11 to derive an expression for the magnitude of the radial acceleration. For very small angles the triangles OAB and the velocity triangles in Figure 3.11b are similar triangles. Hence the ratio of the sides is equal,

$$\frac{\Delta v}{\overline{AB}} = \frac{v}{r}$$

in which $\overline{AB}$ is the length of the arc from A to B and v represents the magnitude of $\mathbf{v}_B$ and $\mathbf{v}_A$. But the distance from A to B is given by the velocity times the change in time, for small time changes, $\overline{AB} = v \times \Delta t$. So,

$$\frac{\Delta v}{v \Delta t} = \frac{v}{r} \quad \text{and} \quad \frac{\Delta v}{\Delta t} = \frac{v^2}{r}$$

But $\Delta v/\Delta t$ is equal to the radial acceleration a_r. Hence

$$a_r = \frac{v^2}{r} \tag{3.21}$$

In the merry-go-round example, the circular motion in starting and stopping is, of course, not uniform.

When the circular motion is not uniform, the tangential component of acceleration is not zero. In these cases, we have both a tangential component of acceleration a_t and a radial component of acceleration a_r. The total acceleration is then the vector sum of the a_t and a_r. The tangential acceleration is positive as the merry-go-round starts and negative as it stops.

EXAMPLE

If you are riding on a merry-go-round at a distance of 6 m from the axis of rotation and are making one revolution in 12 sec, what is your radial acceleration?

The tangential velocity of the merry-go-round is the distance traveled, the circumference, divided by the time for one revolution. Thus

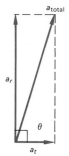

FIGURE 3.12
Graphical representation of the acceleration for a body in circular motion with a variable speed.

$$v = 2\pi \times 6 \text{ m} \times \frac{1}{12} \text{ sec} = \pi \frac{\text{m}}{\text{sec}}$$

Using Equation 3.21, we can find the radial acceleration.

$$a_r = \frac{v^2}{r} = \frac{\pi^2}{6} \approx \frac{10}{6} \text{ m/sec}^2$$

What is the tangential acceleration if the merry-go-round reaches this rate of rotation in 6 sec?

$$v_t = a_t t$$

Thus

$$\pi = a_t t \quad \text{or} \quad a_t = \frac{\pi}{6} \text{ m/sec}^2$$

Total acceleration (see Figure 3.12) is the resultant of a_r and a_t:

$$a_r = \frac{10}{6} \text{ m/sec}^2 \quad \text{and} \quad a_t = \frac{\pi}{6} \text{ m/sec}^2$$

$$\tan \theta = \frac{\dfrac{10}{6}}{\dfrac{\pi}{6}} = \frac{10}{\pi}$$

$$\theta = 72.6°$$

$$a_{total} = \sqrt{\left(\frac{\pi}{6}\right)^2 + \left(\frac{10}{6}\right)^2} = \frac{1}{6}\sqrt{\pi^2 + 100} = +1.75 \text{ m/sec}^2$$

during start-up at the end of 6 sec.

3.10 *Effects of Acceleration*

There are examples of acceleration and deceleration within the human body itself. One example has to do with the passage of food through the body. Another is in the blood circulation system. Can you list the positions of acceleration and deceleration for each of these? Can you think of another human system in which there is acceleration and deceleration?

A number of different accelerations may act upon the human body. These vary in duration, magnitude, rate of onset and decline, and direction. Some acceleration exposures may be so mild that they produce no physiological or psychophysiological effects. On the other extreme they may be so severe that they produce major disturbances such as blackouts. The effects also vary a great deal from individual to individual. Undoubtedly, you have observed this in your childhood play and in your reactions to various types of rides at amusement parks.

The field of acceleration research has produced a number of general principles concerning the effects of acceleration on human physiology

FIGURE 3.13
Sketch of human being accelerated
(a) upward, (b) downward, (c) in
forward direction, and (d) in a backward direction.

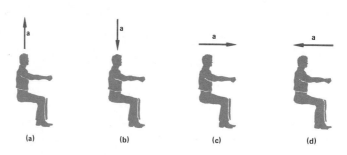

(a)   (b)   (c)   (d)

TABLE 3.4
**Effects of Acceleration on the
Human Body**

Type of Acceleration	Magnitude in Units of $g = 9.8$ m/sec^2	Effect
Positive vertical acceleration (up) (Figure 3.13a)	$1g$	Normal condition at sea level
	$2.5g$	Difficult to raise oneself
	$3g$ to $4g$	Impossible to raise oneself; difficult to raise arms and legs; vision dims after 3-4 sec
	$4.5g$ to $6g$	Diminution of vision; progressive to blackout after 5 sec
Negative vertical acceleration (down) (Figure 3.13b)	$-1g$	Unpleasant facial congestion
	$-2g$ to $-3g$	Severe facial congestion; throbbing headache, and blurring vision
Forward horizontal acceleration (Figure 3.13c)	$2g$ to $3g$	Progressive difficulty in focusing and slight spatial disorientation (2 g tolerable for at least 24 hours)
	$3g$ to $6g$	Progressive tightness in chest; loss of peripheral vision; blurring of vision; difficulty in breathing and speaking
Backward horizontal acceleration (Figure 3.13d)	$2g$ to $3g$	Similar effects to forward acceleration with the modification that chest pressure is reduced and breathing is easier
	$3g$ to $6g$	

and performance. For additional details and information see the *Bio-astronautics Data Book*, Scientific and Technical Information Division, National Aeronautics and Space Administration, from which is derived Table 3.4 showing the effects on humans due to sustained acceleration.

EXAMPLE

Find the stopping acceleration (average) in units of g for a person striking a snow drift at the terminal velocity of 54 m/sec if 1 m of snow brings the person to rest Figure 3.14. From Equation 3.9,

$$v_f^2 - v_i^2 = 2as$$

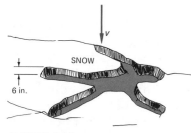

SNOW

6 in.

FIGURE 3.14

Substituting the given values of $v_f = 0$, and $v_i = 54$ m/sec,

$$a = \frac{-(54 \text{ m/sec})^2}{2 \text{ m}}$$

$$= -1458 \text{ m/sec}^2 = 148.8g$$

There is a documented case of a paratrooper free falling without a chute and surviving such a fall without major injuries!

ENRICHMENT
3.11 Instantaneous Velocity and Acceleration

In Equation 3.1 we defined average velocity $\bar{v} = \Delta s/\Delta t$. As the change in time approaches zero $(\Delta t \to 0)$, the instantaneous velocity is the limiting value of $\Delta s/\Delta t$ and is written ds/dt. This is called the derivative of **s** with respect to t,

$$v_{\text{inst}} = \lim_{\Delta t \to 0} \frac{\Delta s}{\Delta t} = \frac{ds}{dt} \qquad (3.22)$$

Thus if $s = f(t)$, then

$$v_{\text{inst}} = \frac{ds}{dt} = \frac{df(t)}{dt} = f'(t) \qquad (3.23)$$

Similarly average acceleration is $\bar{a} = \dfrac{\Delta v}{\Delta t}$, and the instantaneous acceleration is

$$a_{\text{inst}} = \lim_{\Delta t \to 0} \frac{\Delta v}{\Delta t} = \frac{dv}{dt} = \frac{d^2 s}{dt}$$

EXAMPLES

1. Suppose a body moves along the x-axis in accord with the relationship

   $x = 4 - 3t + 2t^2$ meters

   Find the instantaneous velocity, instantaneous acceleration, and when the body is at rest.

   $$v_{\text{inst}} = \frac{dx}{dt} = -3 + 4t \text{ m/sec} = v_0 + at$$

   a. Explain this relationship; that is, what do the -3 and 4 represent?   [the initial velocity and the instantaneous acceleration]
   b. What does the fact that $a_{\text{inst}} = 4$ m/sec^2 indicate about the motion of this body?   [It is uniformly accelerated motion.]
   c. For a body at rest $v_{\text{inst}} = 0$. When does this occur?   [when $-3 + 4t = 0$ or $t = 3/4$ sec]

2. Develop the equation of motion of a particle from the following informa-

tion: An object starts from $y = 2$ m with a velocity of 3 m/sec and a constant acceleration of -0.5 m/sec².

$$a = \frac{dv}{dt} = -0.5 \text{ m/sec}^2$$

Thus,

$$\int dv = \int a \, dt$$

so $v = -0.5t +$ constant. At $t = 0$, $v = +3$ m/sec. So the constant is 3 m/sec. What is v at any time t? $v = -0.5t + 3$ and $v = dy/dt$, so

$$\int dy = \int v \, dt$$

What is the value y at any time t? $y = -(0.5t^2)/2 + 3t +$ constant. $y = 2$ at $t = 0$, so the constant is 2 m. $y = 2 + 3t - 0.25t^2$.

SUMMARY

Use these questions to evaluate how well you have achieved the goals of this chapter. The answers to these questions are given at the end of this summary with the section number where you can find the related content material.

Definitions

1. The slope of a displacement versus time curve is called _____.
2. The speed of an object is constant when it undergoes _____.
3. Uniform circular motion implies that _____ is always zero.
4. If the _____ is constant, then uniformly accelerated motion is observed.
5. The idealized motion of projectiles near the surface of the earth has a constant _____ in horizontal direction and a constant _____ in vertical direction.
6. When a fly wheel on an electric motor starts, it has a positive _____ and its motion is not uniform.
7. An object moving with constant speed around a circle has _____ acceleration.

Equations of Motion

From the equations of uniformly accelerated motion (see Section 3.7) you should be able to answer the following questions.

8. An object whose motion is described as uniformly accelerated always has which of the following properties?
   a. the speed is constant
   b. acceleration is proportional to time
   c. displacement is a quadratic function of the time
   d. the velocity vector does not change its direction
9. The velocity of a uniformly accelerated bicycle
   a. increases linearly with distance
   b. increases linearly with time
   c. increases linearly with acceleration
   d. increases linearly with gravitation
10. For uniformly accelerated motion which of the following quantities must be zero?
   a. the initial acceleration
   b. the initial velocity
   c. the initial displacement
   d. the time rate of change of the acceleration
   e. the time rate of change of the velocity
   f. the time rate of change of the displacement

Motion Problems

From the equations in Section 3.5 you can solve problems about idealized freely falling objects.

11. A parachutist jumped from a helicopter at rest at a height of 78.4 m above the ground. His parachute failed to open. Neglecting air resistance, how long

did it take him to hit the ground, and what was his speed at impact?

a. 2 sec, 4.9 m/sec
b. 4 sec, 19.6 m/sec
c. 6 sec, 58.8 m/sec
d. 4 sec, 39.2 m/sec
e. 2 sec, 9.8 m/sec

From the equations in Section 3.8 you can solve problems of projectile motion.

12. A swimmer leaps horizontally from the edge of the swimming pool with a velocity of 6 m/sec. If he is 2.4 m above the surface of the water when he leaves the edge of the pool, assuming idealized motion, how long will it be before he hits the water? How far will he be from the edge of the pool?

a. 0.9 sec, 5.4 m
b. 0.8 sec, 4.8 m
c. 0.7 sec, 4.2 m
d. 0.6 sec, 3.6 m
e. 0.5 sec, 3.0 m

From the equations in Section 3.9, you can solve problems on uniform circular motion.

13. During premission training, the astronauts are placed in the large NASA centrifuge. They are swung in a horizontal circle at a constant speed of 10 m/sec on the end of a 3.4 m long support rod. What is the horizontal acceleration the astronauts must endure?

a. 4.9 m/sec² or $\frac{1}{2}g$
b. 9.8 m/sec² or $1g$
c. 19.6 m/sec² or $2g$
d. 29.4 m/sec² or $3g$
e. 39.2 m/sec² or $4g$

Acceleration Effects

14. From your reading of Section 3.10 describe the acceleration effects on a space shuttle transporter occupant
   a. when it takes off with a $3g$ acceleration
   b. when it lands with a $2g$ braking acceleration

Answers

1. velocity (Section 3.6)
2. uniform circular motion (Section 3.9)
3. tangential acceleration (Section 3.9)
4. acceleration (Section 3.7)
5. velocity, downward acceleration (Section 3.8)
6. tangential acceleration (Section 3.9)
7. radial (Section 3.9)
8. c (Section 3.7)
9. b, c (Section 3.7)
10. d (Section 3.7)
11. d (Section 3.5)
12. c (Section 3.9)
13. d (Section 3.10)
14. a. impossible to raise oneself, vision dims (Section 3.10)
    b. facial congestion

ALGORITHMIC PROBLEMS

Listed below are the important equations from this chapter. The problems following the equations will help you learn to translate words into equations and to solve single concept problems.

Equations

$$\bar{v} = \frac{\Delta s}{\Delta t} \tag{3.1}$$

$$s = \bar{v}t \tag{3.2}$$

$$\bar{a} = \frac{\Delta v}{\Delta t} \tag{3.3}$$

$$v_f = v_0 + at \tag{3.4}$$

$$\bar{v} = \frac{(v_f + v_0)}{2} \tag{3.5}$$

$$\Delta s = \frac{v_f + v_0}{2} t \tag{3.6}$$

$$s - s_0 = v_0 t + \tfrac{1}{2} a t^2 \tag{3.7}$$

$$v_f^2 - v_0^2 = 2a \cdot s \tag{3.9}$$

$$v_v = v \sin \theta + gt \tag{3.14}$$

$$v_h = v \cos \theta \qquad \text{(definition)}$$

$$x = (v \cos \theta) t \tag{3.17}$$

$$y = v \sin \theta \, t + \tfrac{1}{2} g t^2 \tag{3.18}$$

$$y_p = \frac{-\tfrac{1}{2}(v^2 \sin^2 \theta)}{g} \tag{3.19}$$

$$a_r = \frac{v^2}{r} \tag{3.21}$$

$$v_t = 2\pi r n \qquad \text{(definition)}$$

Problems

1. Find the average speed of a sprinter who runs 100 m in 9.1 sec.
2. If a skier reaches a speed of 20 m/sec in 10 sec after starting from rest, find the acceleration of the skier.
3. A ball is dropped from a window 19.6 m above the ground. Assuming idealized motion, how long does it take the ball to reach the ground?
4. An experimental bumper system is designed to bring a car to rest from an initial speed of 4.0 m/sec. The stopping distance of the bumper is 0.50 m. Find the negative acceleration necessary to make such a stop.
5. A train is originally moving at a speed of 20.0 m/sec when it is accelerated at 2.00 m/sec² for 5.00 sec. Find the distance the train travels during the time of acceleration.
6. A toy train goes around a circular track (radius 1.00 m) at a constant speed of 1.50 m/sec. Find the radial acceleration of the train.
7. The wheel of a moving bicycle is 71.1 cm in diameter and is making 2.00 revolutions per second. Assume that the wheel does not slip on the ground. How fast is the bicycle traveling?
8. A student hits a ping-pong ball at the back edge of the table so that the ball leaves the paddle with a velocity of 2.0 m/sec at 30° above the horizontal. Assume idealized projectile motion. What is the horizontal velocity as the ball leaves the paddle? When is the velocity of the ball entirely horizontal?
9. An object is moving along a straight line such that its displacement is as shown below. What is the average velocity for each second and the entire 3 seconds?

x	t	v_{ave}
0	0	_____
−2	1	_____
0	2	_____
12	3	_____

10. A golf ball is projected at 45° to the horizontal with an initial velocity of 40 m/sec.
    a. Find the horizontal and vertical speed of the ball 5.0 sec after it is projected.
    b. Find the horizontal and vertical position of the ball after the first 5.0 sec of flight.
11. Find the peak height of the golf ball in the flight described in problem 10.
12. Find the initial velocity of a projectile launched at an angle of 30°, if its peak height is 25 m.
13. Compute the tangential velocity and the radial acceleration of an object resting on the edge of a long playing phonograph record ($r = 15$ cm, $n = 33\frac{1}{3}$ rpm).

Answers

1. 11 m/sec
2. 2 m/sec²
3. 2.00 sec
4. 16 m/sec²
5. 125 m
6. 2.25 m/sec²
7. 447 cm/sec
8. 1.7 m/sec, 0.10 sec

9. −2, 2, 12, 4
10. a. 28 m/sec, −21 m/sec
    b. 140 m, 19 m
11. 41 m
12. 44 m/sec
13. 52 cm/sec, 180 cm/sec²

EXERCISES

These exercises are designed to help you apply the ideas of a section to physical situations. When appropriate, the numerical answer is given in parentheses at the end of each exercise.

Section 3.2

1. A body undergoes the following displacements: 6 m in northwest direction, 10 m at an angle of 37° south of west, and 12 m at angle 30° south of east. What is the final position of the body relative to the original position?   [8.0 m, 260°]
2. The weight of a body is a vector quantity, and its direction is vertically downward. If a block of marble weighing 500 newtons (N) is resting on a 20° incline, what are the components of the weight parallel to the incline and perpendicular to the incline?   [171 N, 470 N]

Section 3.6

3. A city bus starts from rest at a bus stop and accelerates at the rate of 4.0 m/sec² for 10 sec. It then runs at this constant rate for 30 seconds and decelerates at 8.0 m/sec² until it stops. Draw a graph of the displacement versus time. What is the displacement of the bus between stops?   [1500 m]

Section 3.7

4. A geology student is trying to determine the depth of a ravine by dropping rocks from a cross-walk. He finds by a stopwatch that 2.50 sec is required for a rock dropped from the bridge to strike the water. Assuming idealized motion, how deep is the ravine?   [30.6 m]
5. The reaction time of an alert automobile driver is 0.700 sec. (The reaction time is the interval between stimulus to stop and application of brakes.) After application of the brakes an automobile can decelerate at 4.9 m/sec². If a car is traveling at 48.4 km/hr (30 mph), what total distance does it travel after the driver sees a stop signal? How far does it travel if the car is traveling with a velocity of 96.8 km/hr (60 mph)? Does this seem realistic to you? What difference would it make if the driver had been drinking and had a slower reaction time of 1.50 sec?   [27.8 m, 92.5 m, 10.8 m farther, 21.5 m farther]

6. There are cases in which the human body has withstood very large accelerations under proper conditions. The following is based on an actual incident. A female, age 21, height 1.7 m (5 ft 7 in.), mass 52.3 kg (weight 115 lbs), jumped from a tenth-story window and fell 28.4 m (93 ft) into a freshly plowed garden where she came to rest with a deceleration distance of 15.3 cm (6 in.). She landed on her right side and back, and her head struck the soft earth (see Figure 3.14). The woman survived, sustaining only a fractured rib and right wrist. Apparently there was no loss of consciousness or concussion. Assume a freely falling body. What were the velocity of impact and the deceleration in g's? [23.6 m/sec, 185.6g]

Section 3.8

7. A ping-pong ball rolls with a speed of 0.60 m/sec toward the edge of a table top which is 0.80 m above the floor. The ball rolls off the table. Assuming idealized motion, how long was it in flight, and how far from the edge of the table did the ball hit the floor? [0.40 sec, 0.24 m]
8. An aviator drops a heavy object from his plane at a height of 490 m while he is moving with a constant horizontal velocity of 30 m/sec. How long does it take for the object to strike the ground? Where is the plane when the object strikes the ground? Where does the object strike the ground relative to the point directly under the plane at the instant the object was dropped? [Neglecting friction, 10 sec, plane vertically above object, 300 m]

9. A baseball leaves the bat of Hank Aaron at a height of 1.22 m (4 ft) above the ground at an angle of 37° with such velocity that it would have a range of 122 m at the height of 1.22 m. However, at a distance of 106.7 m (350 ft) from homeplate there is a 9.15 m (30 ft) high fence. Does Aaron get a home run? [yes, ball is higher than 9.15 m at 106.7 m]
10. A punter kicks a football at an angle of 53° above horizontal. It is observed to be in the air 4 sec. How high did it go, and how long was the kick? Would you want this player for a punter on your football team? [19.6 m, 58.9 m or ~64 yards]

Section 3.9

11. An aviator is said to be doing a 4g circle. What does this mean? What kind of a circle would he be doing if the resultant acceleration at the top of the circle is 0? [$v^2/r = 4g$, $1g = v^2/r$]
12. At what speed must an automobile round a curve with a radius of curvature of 39.2 m to have a radial acceleration equal to g? Now suppose the car continues down the road at the same speed and goes over the top of a hill with the same radius of curvature. What sensation would you experience if you were a passenger in the car? [19.6 m/s]
13. As a result of the earth's rotation, objects at rest on the surface of the earth have a radial acceleration. What is this acceleration at the equator? The radius of the earth is 6.38×10^6 m. What is this acceleration at your latitude? How does this compare with g? [$a_{equator} = 3.37 \times 10^{-2}$ m/sec$^2 = 3.45 \times 10^{-3}g$]

PROBLEMS

Each problem may involve more than one physical concept. A problem that requires material from the enrichment section, Section 3.11, is marked by a dagger (†). The numerical answer is given at the end of each problem.

14. A baseball outfielder can throw a baseball a maximum distance of 78.4 m over the ground before reaching the height from which it was thrown. Assuming idealized motion, with what velocity does he throw it? How long will the ball be in flight? How many bases can a runner safely take during this time if he can run the 100-meter dash in 11 sec? The distance between bases is 27.4 m or 90 ft. [27.7 m/sec, 4.00 sec, 1 base]
15. An open automobile starts from rest with a uniform acceleration at 4.00 m/sec^2. A premed student stands

on the other side of the parking lot, exactly opposite the starting point of the car and 20.0 m from it. As it starts, he throws an apple with a horizontal velocity of 20.0 m/sec to a classmate in the car. In what direction must he throw the apple so his classmate can catch it easily? Assume it passes over the shortest possible distance. [~5.75° to the line between the student and the original position of the car]
16. A small rocket is shot vertically into the air with a speed of 30.0 m/sec. In addition to the acceleration due to gravity there is an average retarding acceleration of 2.20 m/sec^2. How long does it take for the rocket to reach maximum height? What is this height? [2.50 sec, 37.5 m]
17. A helicopter is ascending at a constant rate of 15.0 m/sec. A doctor drops a weighted package of band-

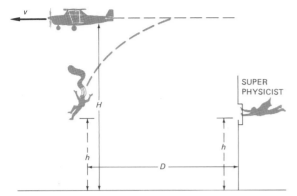

FIGURE 3.15
Problem 19.

ages from the helicopter at a height of 60.0 m to a nurse below. What is the time of flight of the bandages as observed by the nurse on the ground? [5.35 sec]

18. A med-tech student drops a stone from a bridge 19.6 meters above a river. A premed friend throws a stone 0.500 sec later vertically downward so that both stones strike the river at the same time. With what velocity did the premed throw the stone? [5.73 m/sec]

19. Superphysicist (SP) dives out of a window h meters above the ground to save a freely falling sky diver whose chute failed to open. He leaves the window horizontally when his laser eyes see the diver at the same level as the window at distance of D meters away. If the diver fell from a plane at an altitude of H with a velocity v (m/sec) directed horizontally away from SP's building, find the average horizontal velocity SP must have to catch the diver at ground level (see Figure 3.15).

$$\left[V_{ave} = v + \frac{Dg}{\sqrt{2gH} - \sqrt{2g(H-h)}} \right]$$

(What is the physical meaning of $\sqrt{2gH}$ and $\sqrt{2g(H-h)}$?

20. A well-conditioned astronaut can jump 1.5 m on earth. The gravitational acceleration on the moon is one-sixth that of the earth. Find:
a. his initial vertical velocity when jumping on the earth
b. his initial vertical velocity when jumping on the moon
c. how high he jumped on the moon   [$v=5.4$ m/sec, $v = 5.4$ m/sec, 9.0 m]

21. A slingshot launches a projectile with speed v at an angle θ with respect to the ground. Find the velocity of the projectile at the top of its path. Find the acceleration at the top of its path. Since acceleration is perpendicular to velocity at the top, the motion is "instantaneously circular." Find the radius of this circular arc.   [$v \cos \theta$, g down, $v^2 \cos^2 \theta / g$]

22. A river is flowing south 3.0 km/hour. A canoer can paddle 4.0 km/hr relative to water.
a. Where must he head his canoe if he wants to go across the river in a direction perpendicular to the bank?
b. How long will it take him to cross the river if it is one-half kilometer wide?
c. If he wishes to cross the river in minimum time, where should he head the canoe?
d. How long will it take him to cross the river, and where would he land? [a. 41° with bank upstream; b. 0.19 hr; c. perpendicular to bank; d. $\frac{1}{8}$ hr, $\frac{3}{8}$ km downstream]

23. Suppose a ferris wheel with a radius of 9.6 m and a constant tangential speed of 10 m/sec loses a chair at the top of its path. Find the horizontal distance the chair will travel before hitting the ground. Assume center of ferris wheel is 10 m from the ground. [20 m]

24. Given the velocity-time graph in Figure 3.16 for the 400-m run,
a. find the maximum acceleration for this run.
b. find the distance traveled during the positive and negative acceleration periods. [a. 5.00 m/sec²; b. 10.0 m during positive acceleration; 18.0 m during negative acceleration]

25. What is the component of the gravitational accelera-

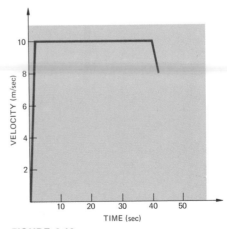

FIGURE 3.16
Problem 24. Velocity versus time record for a 400 m run.

tion parallel to an inclined plane 13 m long that is elevated 5.0 m at one end. Neglecting friction, how long would it take a body starting from rest to slide 9.8 m down the incline?   [3.8 m/sec², 2.3 sec]

† 26. For a particle whose position is given by

$$x = 2.00t^2 - \frac{25.0}{300}t^3$$

as a function of time, t,

a. Find the velocity and acceleration as a function of time.

b. Find the maximum value of x.   [$v = 4.00t - 0.250t^2$, $a = 4.00 - 0.500t$; $x_{max} = 171$]

† 27. $x = 10 + 20t^2 - 30t^4$ is the position of a particle as a function of time, where x is in meters and t in seconds.

a. Find the velocity and acceleration as a function of time.

b. Find the time when acceleration is zero.

c. Find the maximum value of x.   [$v = 40t - 120t^3$, $a = 40 - 360t^2$, $a = 0$ at $t = \frac{1}{3}$ sec, $s_{max} = 13\frac{1}{3}$ m]

† 28. A body moves along the x axis according to the relationship $s = 4 - 6t + 3t^2$ cm.

a. Where is the body at $t = 0$?

b. Where is the body at $t = 2$?

c. What is the original velocity?

d. What is the velocity at time $t = 3$?

e. When is the body at rest?

f. Where is the body when it is at rest?

g. What is the acceleration of the body?

h. Through what distance did the body travel between $t = \frac{1}{2}$ sec and $t = 2$ sec?   [a. 4 cm; b. 4 cm; c. −6 cm/sec; d. 12 cm/sec; e. 1 sec; f. 1 cm; g. 6 cm/sec²; h. 3.75 cm]

GOALS

When you have mastered the concepts of this chapter, you will be able to achieve the following goals.

Definitions
Define each of the following terms, and use each term in an operational definition:

force	coefficient of friction
weight	centripetal force
frictional force	weightlessness

Newton's Laws
State Newton's laws of motion and gravitation.

Resolution of Forces
Given the force acting on a system, draw a force diagram and/or resolve forces into their components and/or solve for an unknown force.

Newton's Second Law Problems
Given any two of the three variables in Newton's second law, solve for the third.

Centripetal Force Problems
Given any three of the four variables in the centripetal force equation, calculate the value of the fourth variable.

PREREQUISITES

Before you begin this chapter you should have achieved the goals of Chapter 3, Kinematics, including uniformly accelerated motion and uniform circular motion.

4

Forces and Newton's Laws

4.1 Introduction

You have experienced many examples of forces. You are familiar with a muscular exertion producing a push or a pull, which are forces. There are many other examples such as: one solid object acting on another through contact; the gravitational force of the earth on an object; the attraction of a magnet for an iron object; the attraction between a plastic rod that has been rubbed with cat's fur and bits of paper; and the force of a stretched spring. One effect of a force is to alter the state of motion of a body.

In this chapter you will study forces. Acceleration, uniformly accelerated motion, projectile motion, and the other kinds of motion require force. If you are to determine these motions, then you must first determine the various forces acting on a body. The study of forces and the associated changes in motion is called *dynamics*. Of particular interest are the conditions leading to equilibrium.

4.2 Newton's First and Third Laws of Motion

In this chapter we will introduce you to the mental constructs devised by Sir Isaac Newton in the seventeenth century. Newton's model seeks to explain how forces acting on an object are related to the motion of an object. This model, sometimes called newtonian mechanics, does not give the correct results in all cases. For example, when objects travel at speeds near the speed of light (3.00×10^8 m/sec) newtonian mechanics is augmented by Einstein's special theory of relativity. In the realm of distances the order of atomic radius (10^{-10} m) quantum theory must be used to give the proper picture of physical phenomena. Nevertheless, in the world of everyday experiences, Newton's mechanics provides us with a good model for understanding forces and motion.

Consider the following experiences you may have had:

1. If you are riding along in an automobile and the brakes are suddenly applied, you experience a tendency to bang into the windshield.
2. If you are riding a galloping horse and he suddenly stops, you may find yourself on the ground in front of him. Why?
3. Your book is resting on a sheet of paper on the table. You go away for several hours and when you return your book is still resting on a sheet of paper on the table. Why?

Newton's first law of motion may provide you with some understanding of such experiences.

Every body persists in its state of rest or of uniform linear motion unless it is acted upon by a unbalanced force.

The first law is sometimes spoken of as the *inertial law*. Recall the concept of inertia introduced in Chapter 2—that is, any object has a

tendency to resist change. In mechanics this property of an object to resist a change in motion is called its *inertia* and is measured by the mass of the object.

Newton's first law helps to unify your experiences with objects at rest not suddenly moving and with moving objects having difficulty in stopping suddenly. Nature seems to prefer to have things change slowly. The measure of the slowness of change in the motion of an object is its mass.

Have you ever faced off to wrestle another person? You grab each others hands and push as hard as you can and for just an instant you are locked in a motionless struggle. You are pushing against your opponent. Your opponent is pushing against you with an equal force. According to Newton's third law that is exactly the same as if you would push against a brick wall (Figure 4.1). Your push against the wall is balanced by the wall's push against you, just as your push was counteracted by that of your wrestling opponent.

Consider the case of your physics book resting on your study desk. The physics book acts on the table with a downward force and the study desk acts on the book with a force in the upward direction. We then have a pair of forces that are equal in magnitude and oppositely directed. One is the book acting on the desk, and the other is the desk acting on the book. One force is called the *action force*, and the other is called the *reaction force*. They form an action-reaction pair of forces. These are experiences unified by Newton's third law of motion:

For every action there is an equal and an opposite reaction.

Suppose you tie a rope around a tree, and you pull on the rope. You exert a force on the rope and the rope exerts a force on you. This is an action-reaction pair. At the other end of the rope we have another action-reaction pair as the rope acts on the tree and the tree acts on the rope. If one has two interacting bodies, A and B, the action-reaction pair can be described by A acting on B and B acting on A.

Consider a tug of war between sides A and B, as shown in Figure 4.2. The rope is acting to the left on group A, $\mathbf{R}_A$, and group A is acting to the right on the rope, $\mathbf{A}_R$. Group A is acting to the left on the earth, and the earth is acting to the right on the group A. These are two examples of action and reaction. What are the action-reaction pairs for group B? Now suppose group B suddenly wraps their end of the rope around a tree, and group A continues to exert the same force as before. Has the tension in

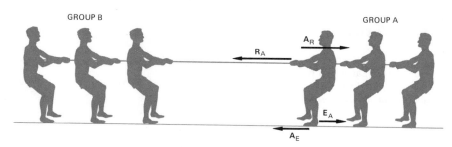

FIGURE 4.2
A tug of war. The force of team members on the rope and of the rope on team members and the force of the team members on the ground and of the ground on them are examples of Newton's third law.

the rope changed? How many pairs of action and reaction forces can you identify around you?

In all of the examples note that each force has magnitude and direction, and thus force is a vector quantity and is so treated.

The action-reaction pair of forces are vectors of equal magnitude but opposite directions. A vector representation of the action-reaction forces for the tug of war in Figure 4.2 is given as: $\mathbf{F}_{AB} = -\mathbf{F}_{BA}$. The rules of vector addition and subtraction can be applied to a force system. Some methods and examples of vector addition were given in Chapter 3.

In accordance with the definition of *equilibrium*, an object at rest experiences no net force.

The vector sum of all forces acting on an object in mechanical equilibrium is zero.

Another way of defining equilibrium is that the vector sum of the force components in any direction is equal to zero. Let us look at a particular case. The leg in Figure 4.3 is held in traction by the cord OA and the weight W. Consider the forces acting on the foot at point O. We will call the tension in the cord OA, T_A; the tension in the leg, T_B, and the tension in the cord OC, T_C. The tension T_C is equal to the weight, say 10.0 newtons (N), where newtons are the SI units of force. The first step in solving any mechanical equilibrium problem is to draw a force diagram like the one in Figure 4.3b. A force diagram is an idealization of the physical problem. All the forces acting on the system are drawn. Each force is represented by a vector. The length of the arrow is proportional

FIGURE 4.3
A pulley traction device. (a) The leg is in equilibrium, (b) the resultant of the forces acting at O is zero.

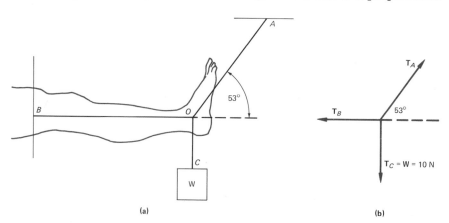

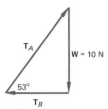

FIGURE 4.4
The graphical, closed polygon method for finding equilibrium forces.

to the magnitude of the force. The direction of the arrow represents the direction in which the force is acting. In many cases, drawing a correct force diagram will essentially solve a force problem for you. The best way to learn how to draw force diagrams is to practice. For each problem and example in this chapter, draw your own force diagram, and then compare it with ours.

We can add these vector forces graphically (see Section 3.3) as shown in Figure 4.4. The fact that the three vectors form a closed polygon means the vector sum of the forces is 0. Can you explain why?

What are the magnitudes of $\mathbf{T}_B$ and $\mathbf{T}_A$? Let us use the component method to answer this question. First find all the components of forces acting in the horizontal direction:

horizontal components $= T_A \cos 53° - T_B = 0$

Then sum the components in the vertical direction:

vertical components $= T_A \sin 53° - W = 0$

Substituting values for the sine and cosine of 53°, $0.600\ T_A = T_B$, and $0.800\ T_A = 10.0$. Solving these two equations gives us the values for the magnitudes of $\mathbf{T}_A$ and $\mathbf{T}_B$, $T_A = 12.5$ N, and $T_B = 7.50$ N. The directions are as shown in Figure 4.3, $\mathbf{T}_B$ acting horizontally to the left and $\mathbf{T}_A$ acting to the right and upward 53° above horizontal.

EXAMPLES

1. Given the tension on the teeth bands as 1.00 N (a reasonable tension as applied by an orthodontist), find the net force, $\mathbf{F}$, that tends to realign the out-of-line tooth as shown in Figure 4.5a. First draw the diagram of the forces acting at the contact between the tooth and the teeth bands. Represent each force by an arrow of about the proper length and pointing in what you assume is the proper direction. See Figure 4.5b. Once you have completed the force diagram you can use the method of vector addition as described in Section 3.3 to form the closed polygon that results when all the vectors add to zero (Figure 4.5c). Notice that this problem is a bit tricky. What is asked for is *not* the force the tooth exerts on the teeth bands, not the force which keeps the system in equilibrium, but rather the force that tends to realign the tooth. However, from Newton's third law we know that the force of the tooth against the bands and of the bands against the tooth are an action-reaction pair, equal in magnitude, but opposite in direction.

   From Figure 4.5c we can see what equations we need to find the numerical value for the magnitude of $\mathbf{F}$. The closed polygon formed by $\mathbf{T}_1$, $\mathbf{T}_2$ and $-\mathbf{F}$ is

FIGURE 4.5
(a) Teeth bands apply force to an out-of-line tooth. (b) The force diagram for the teeth bands acting on the tooth. (c) A graphical method for finding the force on the tooth produced by the tension in the teeth bands.

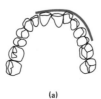

(a)

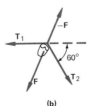

(b)

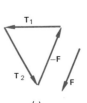

(c)

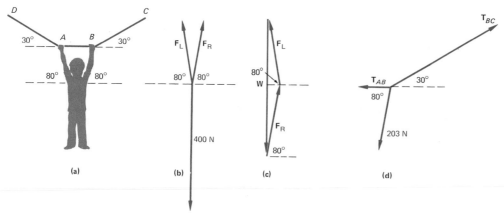

(a)   (b)   (c)   (d)

FIGURE 4.6
(a) Youth hanging from a clothesline.
(b) This force diagram shows the tension in the arms produced by the weight of the youth hanging from the line. (c) The closed polygon for the forces acting on the youth who is in equilibrium. (d) The force diagram for the point where the youth's hand is holding onto the line.

an isoceles triangle with F the length of its base and T the length of the equal sides. Therefore, in terms of the magnitude of $-\mathbf{F}$ and $\mathbf{T}$

$$F = 2T \sin 30°$$

$$F = 2(1.00 \text{ N})(0.500) = 1 \text{ N}$$

So the force that tends to realign the tooth has a magnitude of 1 N and acts inwardly at an angle of 60° from the teeth bands.

2. Consider the case of a 400 N youth hanging on a rope so that his arms make an angle of 80° above the horizontal (Figure 4.6). The rope is horizontal between his hands and makes an angle of 30° above the horizontal beyond his hands.

What are the forces exerted by the youth's arms and by the tension in the rope?

First let us draw a force diagram. We know the weight of the youth is a vector pointing vertically downward with a length proportional to 400 N. The only other forces acting on the youth are the forces exerted by his right arm $\mathbf{F_R}$ and by his left arm $\mathbf{F_L}$. These two forces are of equal size, as is implied by the statement of the problem, and act at an angle of 80° to the horizontal. The force diagram for the situation is shown in Figure 4.6b. Once again use the method of vector addition to draw the closed polygon shown in Figure 4.5c. Now we can use algebraic and trigonometric procedures to solve for the magnitudes of the forces, F_R and F_L. If we consider just the sum of the horizontal forces, we obtain the equation, $F_R \cos 80° = F_L \cos 80°$, since the 400 N weight of the youth has no horizontal component it does not appear in this equation. The result of the horizontal component equation is that the two forces, $\mathbf{F_R}$ and $\mathbf{F_L}$, are of equal magnitude, $F_R = F_L$. From the sum of the vertical components, we obtain

$$400 \text{ N} = F_R \sin 80° + F_L \sin 80° = 2F_R \sin 80°$$

Substituting the value of sin 80°, we obtain

$$203 \text{ N} = F_R = F_L$$

To complete this problem we can draw the force diagram for the forces acting at point B, where the hand of the youth is holding tightly to the rope. Remembering Newton's third law, we know that the force in his arm, which we showed pulling upward to the right in Figure 4.5b, will be pulling downward and to the left on the rope. See Figure 4.5d. From this figure we can use algebraic-trigonometric methods to find the magnitudes of the tensions T_{BC} and T_{AB}. Since $\mathbf{T}_{AB}$ is acting only in the horizontal direction, the vertical upward component of $\mathbf{T}_{BC}$ must just be balanced by the downward pull of the youth's arm:

$$T_{BC} \sin 30° = F_L \sin 80° = 203 \text{ N } \sin 80°$$
$$T_{BC} = 400 \text{ N}$$

Finally, from the equation for the horizontal components of the forces, we can obtain the value for the magnitude of $\mathbf{T}_{AB}$.

$$T_{AB} + F_L \cos 80° = T_{BC} \cos 30°$$

$$T_{AB} + 203 \text{ N } \cos 80° = 400 \text{ N } \cos 30°$$

$$T_{AB} = 346 \text{ N} - 35.3 \text{ N}$$

$$T_{AB} = 311 \text{ N}$$

3. Let us examine two different exercise systems used to develop and strengthen muscles in the leg of a patient. First consider the system in which an 80.0 N weight is placed on the seated patient's foot. The patient then straightens his leg to a horizontal position (see Figure 4.7). What is the component of the force that acts perpendicular to the patient's leg so as to oppose the extension of the leg? We will calculate the magnitude of this component for the situations where the lower leg is vertical, 30° from vertical, 60° from vertical and 90° from vertical (horizontal).

   The force applied to the leg by the 80.0 N weight will always be vertically down. The four different force diagrams are shown in Figure 4.7b through e. In the first position, when the leg is vertical, the applied force is also vertical, and there is no component of the force perpendicular to the extension of the leg. In the second position, the force acts at an angle of 30° to the direction of the leg, so the force opposing the horizontal extension of the leg is 80.0 N sin 30°, or 40.0 N. In the third position, the force acts at an angle of 60° to the leg, so the magnitude of the force perpendicular to the line of the leg is given by 80.0 N sin 60°, or 69.3 N. In the horizontal position all 80.0 N of the weight is acting perpendicular to the leg extension.

   In the second exercise system, an exercise load of 80.0 N is applied to the foot through a pulley system so that the load is maximum when the lower leg

FIGURE 4.7
An exercise system for developing leg muscles. (a) A load hangs on a collar attached around the ankle. Force diagrams can be drawn for four positions of the leg, at angles of (b) 0, (c) 30°, (d) 60°, and (e) 90° with the vertical.

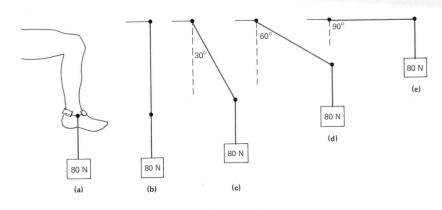

FIGURE 4.8
An exercise system to apply a resistance load to the leg for various positions of the leg.

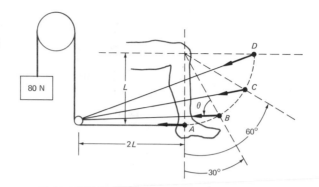

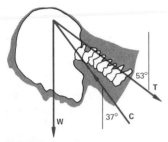

FIGURE 4.9
The forces acting upon a head held in a flexed position, where **W** is weight of the head, **T** is the tension of neck extensor muscles, and **C** is the compression force at the atlanto-occipital joint.

is in a vertical position (see Figure 4.8). Once again let us find the force perpendicular to lower leg opposing the extension of the leg at four positions, lower leg vertical, 30° from vertical, 60° from vertical, and horizontal. Using the geometry given in Figure 4.8, we can calculate the angle between the lower leg and the rope from the exercise apparatus. For the four positions A, B, C, and D the angles Θ between the lower leg and the applied force are 90°, 63°, 40°, and 17°, respectively.

4. As a final example, let us consider the forces required to hold your head in a flexed position (see Figure 4.9). The atlanto-occipital joint exerts a force on the skull at an angle of 37° from the vertical and the neck extensor muscles exert a force at an angle of 53° from the vertical. The vertical components of the muscle tension and weight are balanced by the vertical force exerted by the joint.

$$W + T \cos 53° = C \cos 37°$$

The horizontal component of the muscle tension is balanced by the horizontal force of the joint.

$$T \sin 53° = C \sin 37°$$

where W is magnitude of the downward vertical force of your head, T is magnitude of the tension in the neck extensor muscles and C is magnitude of the compressive force exerted at the atlanto-occipital joint. The solutions to these equations are $T = 2.14W$; so $C = 2.85W$. So if your head weighs 50.0 N (about 11 lb), your neck muscles will have to exert a force of 107 N when you incline your head at a 37° angle.

You may need to practice drawing force diagrams. This is an important problem-solving technique that is based upon the superposition concept. The individual parts of a problem are isolated and solved separately. Then the final result is obtained by adding up the answers obtained for the individual partial solutions.

4.3 Newton's Second Law of Motion

Recall your experiences moving furniture. The massive pieces of furniture required great efforts to get them to move. The smaller pieces could be moved more rapidly as you increased the force you exerted against them. These are not quantitative experiences, but they may make you willing to accept the assertion that the force required to give an object a certain acceleration is proportional to its mass, and that the acceleration you are able to give an object is proportional to the force that you exert upon it. The objects that you move have an unbalanced force acting on them. As a result the objects are not at rest and are not moving at constant velocity but are traveling with a changing velocity. An object has an acceleration in the direction of the unbalanced force as described by Newton's second law:

The time rate of change in velocity of an object is proportional to the net unbalanced force acting upon it.

You can quantify your experiences with moving objects by using a spring balance, a clock, and a low-friction object in the physics laboratory. You will be able to explain your data by assuming that the force required to produce a given acceleration is proportional to the mass:

$$F \propto m \qquad \text{for constant acceleration}$$

For a given mass the acceleration is proportional to the force $a \propto F$ for a given object of constant mass.

These two proportionalities can be combined into one equation of the form $F = kma$, where k is a proportionality constant. It is possible to make k unity, if one chooses two of the variable units and defines the third in terms of these two. That is the customary approach.

In SI we define the *newton* as the magnitude of force that gives a mass of one kilogram an acceleration of one meter per second per second (m/sec²). The vector equation for Newton's second law becomes:

$$\vec{F} = m\vec{a} \qquad \text{thus } \vec{F} \text{ and } \vec{a} \text{ are parallel} \tag{4.1}$$

where F is the net force in newtons, m is the mass in kilograms and a is the acceleration in meters per second per second.

The weight of a body in newtons is then given by multiplying the mass of the body in kilograms by g, the acceleration due to gravity in meters per second per second, $W = m\mathbf{g}$.

The mass of an object is the same everywhere. Because the acceleration due to gravity depends on position of the mass on the earth, the object's weight varies accordingly. For the numerical calculations in this book, we will use a value of 9.80 m/sec² for g near the surface of the earth.

According to Newton's second law the effect of a force is to alter the state of motion of a body. In Chapter 3 we dealt with accelerated motion. Let us now look at some of those problems to see what unbalanced force is acting.

EXAMPLES

1. Chapter 3, exercise 3. Assume a bus has a mass of 2000 kg, and it has an acceleration of 4.00 m/sec². The magnitude of the force needed to accomplish

this acceleration becomes $F_{net} = ma = 2000 \text{ kg} \times 4.00 \text{ m/sec}^2$. Then the net force equals 8000 N in the direction of the acceleration.

2. Chapter 3, exercise 6. A 52.3 kg female drops 28.4 m into a freshly plowed garden and is stopped in 15.3 cm. Find the deceleration force. During the fall the force acting is her weight of *m***g**, and the acceleration is that of gravity, 9.80 m/sec². In order to get the deceleration, we need to find the velocity at impact, that is, the velocity with which the woman strikes the ground. We can compute this velocity using Equation 3.9,

$$2gs = v_f{}^2 - v_0{}^2$$

The initial velocity is zero since the woman started from rest, $s = 28.4$ m, and $g = 9.80$ ft/sec². Then we can compute v_f, the velocity at impact.

$$v_f{}^2 - v_0{}^2 = 2gs$$
$$v_f{}^2 - 0 = 2gs$$
$$= 2 \times 9.80 \times 28.4$$
$$= 557 \text{ m}^2/\text{sec}^2$$
$$v_f = 23.6 \text{ m/sec, vertically downward}$$

During deceleration, the woman travels 15.3 cm, and we must now find the force of deceleration. Once again we can use Equation 3.9 where this time the initial velocity, v_0 is 23.6 m/sec, since the body has this velocity at the instant before impact. The distance traveled is the 15.3×10^{-2} m she dents the garden dirt. The final velocity v_f is zero since the woman is finally at rest.

$$2as = v_f{}^2 - v_0{}^2$$
$$2 \times 15.3 \times 10^{-2} \times a = 0^2 - (23.6)^2 \tag{3.9}$$

$a = -1.82 \times 10^3$ m/sec², an acceleration of nearly 200 g! The minus sign indicates an upward acceleration or a downward deceleration. The net retarding force is found using Newton's second law.

$$F = ma$$
$$F = (52.3 \text{ kg})(-1.82 \times 10^3 \text{ m/sec}^2)$$

An upward impact $F_{net} = -9.52 \times 10^4$ N, a force of more than 10 tons!

The data for this problem are from a real case, and the woman survived.

3. Find the force exerted on the body if a 50.0-kg person jumps stiff-legged from a porch 1.00 m onto a concrete walkway. Assume that the padding on the foot compresses 6.00 mm in stopping the fall (see Figure 4.10).

This problem is similar to the previous example. We must use Equation 3.9 twice. First, given the vertical distance of the jump and the acceleration of gravity, we can find the velocity at impact. Then using our computed value for the velocity at impact and the known distance of foot compression, we can calculate the stopping acceleration. Finally, using our calculated value for stopping acceleration, we use Newton's second law to find the force exerted on the body. If we choose the vertical downward as positive, then first,

$$2gs = 2 \times 9.80 \text{ m/sec}^2 \times 1\text{m} = v_w{}^2$$
$$v_w = 4.43 \text{ m/sec}$$

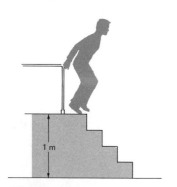

FIGURE 4.10
Jumping stiff-legged from a height of as little as 1 m can produce a decelerating force large enough to damage leg joints.

1 m

Second, upon stopping, we find

$$2 \times 6 \times 10^{-3} \text{ m} \times a = -(4.43 \text{ m/sec})^2$$

where a is the stopping acceleration. Solving for a,

$$a = -1.63 \times 10^3 \text{ m/sec}^2$$

To find the force exerted on the body, we substitute this value for a in the equation for Newton's second law.

$$F = ma = 50 \text{ kg} \times (1.63 \times 10^4 \text{ m/sec}^2) = -8.17 \times 10^4 \text{ N}$$

This force would be transmitted to the leg joints where it would be potentially dangerous. How does flexing the legs at the knees reduce the danger of such a jump?

Let us review the prescription followed in solving these example problems. These problems asked for the determination of a force that was known to provide a certain acceleration for an object of known mass. This unknown force is equal to the mass of the object times the acceleration specified in the problem. The direction of the force will be in the direction of the acceleration. The acceleration called for in the problem can be determined by using the initial and final velocities and the distance traveled during the acceleration. In this case, we use Equation 3.9,

$$2as = v_f^2 - v_0^2$$

to find the acceleration. It is also possible to give the initial and final velocities and the time of such acceleration. In this case we use Equation 3.4

$$\mathbf{v_f} = \mathbf{v_0} + \mathbf{a}t$$

to find the acceleration.

4.4 Force of Friction

In all real motion there is a force that opposes the motion. This is the force of friction. The properties of this frictional force may be complex. The source of the frictional force may be obscure.

Try some experiments in which you slide one surface over another. Some of the variables you may change are the different kinds of surfaces, the conditions of surfaces, the area of contact, and the velocity. Compare starting force and the force for uniform motion, and tabulate your observations.

In doing these experiments you may observe that

1. the starting force is greater than the force required to keep an object moving uniformly after it is started.
2. the frictional force is independent of area of contact.

3. the frictional force is independent of the velocity (for low velocities only).
4. the frictional force is dependent upon the types of surface in contact.
5. the frictional force is dependent upon the condition of the surfaces in contact.
6. the frictional force is dependent upon the forces pressing the surfaces together.

To explain such observations we assume that the friction force can be represented mathematically as a force whose magnitude is a constant number multiplied by the force pushing the two surfaces together.

$$f_s \leq \mu_s N \tag{4.2}$$

where f_s is the *static frictional force*, μ_s is the *coefficient of static friction*, and N is the normal force, the force perpendicular to the surfaces, that press the surfaces together. The direction of the friction force is opposite to the direction of impending motion. The force of static friction can be any value from zero to $\mu_s N$, its maximum value just before motion begins.

The coefficient of friction is greatest between two surfaces at rest with respect to one another (observation 1 above). When the two surfaces begin to move with respect to each other the friction force is reduced. As soon as sliding begins, a *kinetic frictional force* opposes the motion. The coefficient of friction for this force is the *coefficient of kinetic friction* μ_k. The kinetic frictional force is given by the product of μ_k and the normal force, N, pressing the surfaces together.

$$f_k = \mu_k N \tag{4.3}$$

Some approximate values of μ_s and μ_k are given in Table 4.1. Notice that each case $\mu_k < \mu_s$, the kinetic friction is less than the static friction.

In many cases it is desirable to reduce the frictional force to a minimum. In mechanical systems the contact surfaces are polished to reduce the frictional forces. Another way to reduce the effects of friction is to lubricate the surfaces. Can you think of a model to explain how oil works as a lubricant? In the case of the membranes and joints in the human body, the problem is also solved by the use of lubricants. Note in

TABLE 4.1
Coefficient of Static Friction (μ_s) and of Kinetic Friction (μ_k) for Various Surfaces

Surfaces	μ_k	μ_s
Automobile tire on clean concrete	0.8	1.0
Automobile tire on muddy concrete	0.2	0.7
Automobile tire on icy concrete	0.02	0.3
Wood on wood	0.3	0.5
Leather on metal	0.5	0.6
Metal on metal (dry)	0.2	0.6
Metal on metal (lubricated)	0.04	0.1
Metals on wood	0.2	0.6
Joints of human body	.0016 to 0.005	0.02

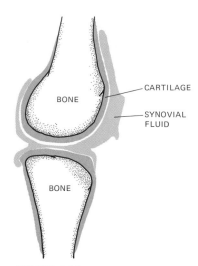

FIGURE 4.11
Cross section of human joint showing the synovial fluid.

Table 4.1 the value of μ in the human body. How can the joints in the human body have such a small coefficient of friction?

The answer to this question is provided by the synovial joint. In a synovial joint the ends of the bones, each covered by cartilage, rest against an enclosed membrane sack (Figure 4.11). Inside the membrane sack is the synovial fluid, which is similar in physical properties to blood plasma. The synovial fluid helps to support the forces exerted by the bones and provides a medium to reduce the frictional forces.

Not all frictional forces are detrimental. Make a list of examples in which frictional forces are absolutely necessary. How could you operate in a frictionless world?

EXAMPLES

Consider a patient with neck traction (T) applied by calipers or tongs attached to the skull (see Figure 4.12). Assume that the head W weighs 50.0 N and that the coefficient of friction μ between the back of the head and the hospital bed is 0.200.

1. Assume that the bed and patient are in a horizontal position and the applied traction force is also horizontal. What is the maximum frictional force that must be overcome before the traction pull will be effective in stretching the neck structures? For a horizontal surface; the normal force N is equal in magnitude to the weight. There is no vertical motion, so the net vertical force is equal to zero (see Figure 4.13a).

$$W = N$$

$f_{max} = \mu N = \mu W$. To stretch the neck structures the force T must be greater than the maximum friction force, $T > f_{max} = (0.200)(50.0 \text{ N}) = 10.0 \text{ N}$.

FIGURE 4.12
Neck traction on a horizontal surface.

FIGURE 4.13
Force diagram (a) for a head in horizontal traction and (b) for a head on a horizontal surface being acting upon by a force at an angle of 37° above the horizontal.

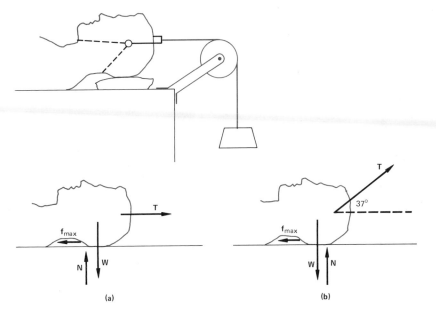

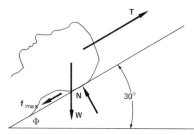

FIGURE 4.14
Neck traction on a surface inclined upwards at 30° with a force parallel to the inclined surface.

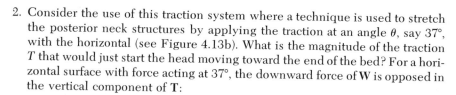

2. Consider the use of this traction system where a technique is used to stretch the posterior neck structures by applying the traction at an angle θ, say 37°, with the horizontal (see Figure 4.13b). What is the magnitude of the traction T that would just start the head moving toward the end of the bed? For a horizontal surface with force acting at 37°, the downward force of **W** is opposed in the vertical component of **T**:

$$N = W - T \sin \theta$$

$f_{max} = \mu N = T \cos \theta$; horizontal component of T must at least equal the maximum value of f to start the motion.

$$f_{max} = T \cos 37° = \mu(W - T \sin 37°)$$

$$0.800T = 0.200(50.0 - 0.6T)$$

$$0.800T + 0.120T = 10.0$$

$$0.92T = 10.0 \text{ N}$$

$$T = 10.87 \text{ N}$$

3. A patient needs neck traction using this system, but his lungs fill with fluid when he is lying horizontally. You solve this by strapping the patient to the bed and inclining the bed at an angle Φ of 30° with the horizontal. What is the maximum force your traction system must overcome if it is to pull the patient parallel to the inclined bed (see Figure 4.14)? In this case the traction will have to pull against both the maximum frictional f_{max} and the component of the weight of the head that is parallel to the inclined plane of the bed. For an incline of 30° with a force parallel to the incline:

$$N = W \cos \Phi$$

$$T = f_{max} + W \sin \Phi$$

From the equation that defines friction

$$f_{max} = \mu N = \mu W \cos \Phi$$

$$T = \mu W \cos \Phi + W \sin \Phi$$

$$T = 0.200(50.0) \cos 30° + 50.0 \sin 30°$$
$$= 10.0\ (0.866) + (50.0)(0.500) = 8.66 + 25.0$$

$$T = 33.7 \text{ N}$$

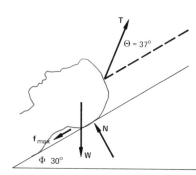

FIGURE 4.15
Neck traction on a surface inclined upward at 30° with a force inclined at 37° to the plane of the inclined surface.

4. Now assume the patient who requires an inclined bed also needs to have the posterior neck structures stretched. What traction T acting at 37° to the bed inclined at 30° will be required to just slide the patient's head (see Figure 4.15). For an incline of 30° with a force 37° above the incline:

$$N + T \sin \theta = W \cos \Phi$$

$$f_{max} = \mu N = \mu(W \cos \Phi - T \sin \theta)$$

$$T \cos \theta = W \sin \Phi + f_{max}$$

$$T \cos \theta = W \sin \Phi + \mu(W \cos \Phi - T \sin \theta)$$

$$0.800T = 50.0(0.500) + 0.200[50.0(0.866) - T(0.600)]$$

$$0.92T = 33.7$$

$T = 36.6 \text{ N}$

5. If the only unbalanced force acting on a moving body is the frictional force, then it opposes the motion and produces a deceleration. A 15,000 N automobile is traveling 39.2 km/hr on a horizontal road. The brakes are applied, and it slides to rest with $\mu_k = 0.500$. What is the force acting? What is the acceleration? How far did the car slide?

$F = f_{max} = \mu_k N$

$F = 0.500 \times 15,000 \text{ N} = 7500 \text{ N}$

$F = ma \text{ or } a = F/m$

$a = F/(W/g) = 7,500/(15,000/g) = \frac{1}{2} \times 9.80 = 4.90 \text{ m/sec}^2$

$v_0 = 39,200 \text{ m}/3,600 \text{ sec} = 10.9 \text{ m/sec}$

$2as = v_f^2 - v_0^2$

$a = -4.9 \text{ m/sec}^2$

$-2(4.9 \text{ m/sec}^2)s = 0 - (10.9 \text{ m/sec})^2$

$s = \dfrac{119(\text{m/sec})^2}{9.80 \text{ m/sec}^2} = 12.1 \text{ m}$

6. a. A water skier (whose mass is 70.0 kg) is pulled through the water at a constant velocity of 20.0 m/sec by a forward force on the skier of 100 N. Find the frictional drag force on the skier.

   Since the skier moves at a constant speed $(a = 0)$, the net force on the skier is zero. Thus the drag force must be 100 N.

   b. A velocity dependent drag force is common for motion through fluids. Assume that the frictional drag force is proportional to the speed of the skier, and find the frictional drag coefficient.

   $f_d = kv$

   Thus

   $k = \dfrac{f_d}{v} = \dfrac{100 \text{ N}}{20.0 \text{ m/sec}} = 5.00 \text{ N–sec/m}$

7. Consider the system illustrated in Figure 4.16. The masses are $M_A = 10.0 \text{ kg}$ and $M_B = 20.0 \text{ kg}$, and the coefficient of sliding friction is 0.200 for the incline. Find the acceleration of each block.

   The first step in solving such problems is to draw a free-body diagram (see

FIGURE 4.16
A body resting on an inclined plane and attached to a freely hanging mass via a frictionless pulley. (b) The free body diagram for mass A, and (c) the free body diagram for mass B.

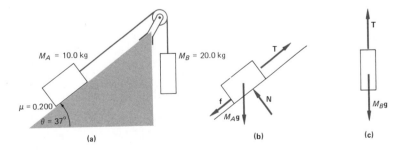

Figure 4.16b) to show clearly all forces acting on each body. The bodies are connected by a rope; if the rope doesn't stretch or break, both bodies will have the same acceleration. Therefore, the net force on each body must equal its mass times the acceleration. Let us solve the problem:

For mass A: There is no motion perpendicular to the plane, thus, the magnitude of the normal force **N** must equal the component of the weight acting vertically:

$$N - M_A g \cos \theta = 0$$

or

$$N = M_A g \cos \theta = 10.0 \text{ kg} \times 9.80 \text{ m/sec}^2 \times 0.800 = 78.4 \text{ N}.$$

For mass A to move up the plane, we must have a net positive force acting parallel to the plane:

$$T - (M_A g \sin \theta + \mu N) = M_A a$$

where a is positive for acceleration up the plane.

For mass B, the tension in the rope acts to oppose its weight, thus the net force equation is

$$M_B g - \mathbf{T} = M_B \mathbf{a}$$

where a is positive for weight B falling down.

We must now solve the equations for the magnitude of a. This can be done by solving the M_B equation for T and substituting this into the equation for mass A. Thus we get:

$$M_B g - M_A g \sin \theta - \mu M_A g \cos \theta = (M_A + M_B)a$$

$$a = \frac{M_B g - M_A g \sin \theta - \mu M_B g \cos \theta}{M_A + M_B}$$

$$a = \frac{20.0 \times 9.80 - 10.0 \times 9.80 \times 0.600 - 0.200 \times 10 \times 9.80 \times 0.800}{30.0}$$

$$a = \frac{196 \text{ N} - 50.8 \text{ N} - 15.7 \text{ N}}{30.0 \text{ kg}} = 4.05 \text{ m/sec}^2$$

Thus, $T = M_B(g - a) = 115 \text{ N}$

4.5 The Law of Attraction Between Two Bodies

In Chapter 3 on kinematics you learned that a freely falling body has an acceleration **g**. In accordance with Newton's second law there must be an unbalanced force acting on the falling body. This force is the weight of the body and results from the attraction between the earth and the falling body. Sir Isaac Newton stated the law of gravitational attraction between two idealized point masses:

any two point masses attract each other by a force that is proportional to the product of their masses and inversely proportional to the square of the distance between the masses. The direction of this force is along the line between the point masses.

This is called the *universal law of gravitation* because it is assumed to apply to point mass interactions throughout the universe.

We can write this law as an equation using a proportionality constant which we will signify by *G:*

$$F = G \frac{m_1 m_2}{r^2} \tag{4.4}$$

where *G* is constant known as the *universal gravitation constant.* It has the same value for any pair of masses, and its value in SI units is $G = 6.6732 \times 10^{-11}$ N–m²/kg².

The universal law of gravitation can be extended to all spherical bodies that have mass distributions that depend only on distance from the center of the body. The superposition principle applied to these bodies shows us that such a body may be treated as if its entire mass were concentrated at its center.

Consider a mass, *m*, freely falling near the surface of the earth. The force of gravity produces the acceleration *g*. Thus we have: (M_e = mass of earth and r_e = earth radius)

$$mg = G \frac{M_e m}{r_e^2} \quad \text{or} \quad g = \frac{GM_e}{r_e^2} \tag{4.5}$$

Gravitation is another example of interaction-at-a-distance since an object does not have to be in contact with the earth to be acted upon by the force of gravity. It is reasonable to construct the concept of a gravitational field that surrounds the earth. At any point a distance *r* from the center of the earth, the gravitational field can be defined to have the magnitude of GM_e/r^2 newtons and be directed toward the center of the earth. If we place an object of mass *m* at that point, we will have its weight, or the gravitational force acting upon it, of GmM_e/r^2 newtons. We can use the change in gravitational field strength with distance as a way of explaining the changing weight of an object as it recedes from the earth. Furthermore, we can ascribe gravitational fields to other objects such as the sun and moon, and we can use the properties of these fields to explain such phenomena as tides and planetary orbits.

4.6 Centripetal Force

Probably you have tied an object to a string and whirled it around. If you have not, do the experiment now. What did you experience?

In Chapter 3 on kinematics you learned that in uniform circular motion, there is an acceleration directed toward the center. Newton's second law states that if there is an acceleration of a body, there is an unbalanced force acting upon the body. The vectors of the acceleration and the unbalanced force are directed toward the center of the circle.

In uniform circular motion the magnitude of the unbalanced force acting on the body toward the center of rotation is given by:

$$F = ma_r = \frac{mv^2}{r} \tag{4.6}$$

since the radial acceleration is given by Equation 3.21. The speed of an object traveling in a circle of radius r is equal to $2\pi rn$, where n is the number of revolutions per sec. This expression can be substituted for v in Equation 4.6 to yield,

$$F = 4\pi^2 n^2 mr \tag{4.7}$$

This force is called the centripetal force. In Equation 4.7 the expression $4\pi^2 n^2$ is the square of $2\pi n$, where 2π is the number of radians in one revolution (2π radians $= 360°$), and n is the number of revolutions per second. The quantity $2\pi n$ is called the *angular velocity*. It is measured in radians per second and is designated by ω,

$$\omega = 2\pi n$$

So Equation 4.7 can be rewritten as

$$F = mr\omega^2 \tag{4.8}$$

In the space age we are aware of the concept of weightlessness as experienced by astronauts in orbiting satellites. For these people and for the entire contents of the spacecraft, the gravitational attraction of the earth is precisely equal to the centripetal force of the body in its orbit,

$$F = \frac{GmM_e}{r^2} = \frac{mv^2}{r} \tag{4.9}$$

where M_e is the mass of the earth, G is the gravitation constant and r is the radius of the orbit. Hence the speed of the spacecraft in orbit is given by

$$v = \sqrt{GM_e/r} \tag{4.10}$$

Notice the difference in experience between the spacecraft observers on the ground and the orbiting astronauts. To us on the earth the astronauts and all the orbiting equipment are falling around the earth acted upon by a centripetal force whose force is given by Equation 4.8. However, in the orbiting spacecraft, all objects are at rest with respect to one another. The astronaut, when he steps away from the dinner table, does not fall toward the earth but continues in an orbit around the earth. To the astronaut the force of gravity on objects is canceled by an imaginary force, called *centrifugal force*, which arises from the orbital motion of spacecraft. Actually, the astronauts are only a few hundred kilometers above the earth. The gravitational force at that altitude is reduced by only about ten percent. Hence, the experience of weightlessness occurs in the frame of reference of the astronauts who can account for the objects that float through the spacecraft by assuming these objects are acted upon by a fictitious outward force equal in magnitude to the gravitational force given by Equation 4.8.

EXAMPLES

1. A merry-go-round is turning at the rate of 10.0 revolutions per minute. A 50.0 kg youth is located 5.00 m from the axis of rotation. What is the centripetal force acting on him? From Equations 4.6 and 4.7 we note that

$$F = m\,\frac{v^2}{r} = 4\pi^2 n^2 m r$$

Since n must be in revolutions per second,

$$n = \frac{10.0 \text{ rpm}}{60 \text{ sec/min}} = \frac{1}{6.00}\text{ rps}$$

Then,

$$F = \frac{4\pi^2}{(6.00)^2} \times 50.0 \times 5.00$$

$$= \frac{1}{9.00} \times \pi^2 \times 250 = 2.74 \times 10^2 \text{ N}$$

2. An ultracentrifuge is used to separate parts of biological samples. It has become an important tool in biochemistry and molecular biology. The rotors in these machines rotate at rates as high as 60,000 rpm. We wish to determine the centripetal force on a microgram sample at a radius of 10.0 cm from the axis of the rotor.

Letting F_c represent the force and a_c the centripetal acceleration.

$$F_c = m a_c \quad \text{and} \quad a_c = \omega^2 r$$

Substituting $r = 0.100$ m and the given rate of rotation,

$$\omega = 2\pi n = (60{,}000 \times 2\pi/60)\text{ rad/sec}$$

FIGURE 4.17
An ultracentrifuge.

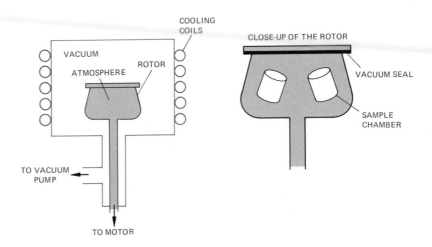

Thus

$$F_c = 10^{-9} \text{ kg} \times (2\pi \times 10^3 \text{ rad/sec})^2 \times 0.100 \text{ m} = 4.00\pi^2 \times 10^{-4} \text{ N} = 3.95 \times 10^{-3} \text{ N}$$

Note that this force is about 4×10^5 times the weight of the sample. It is said that the sample is subjected to a $4 \times 10^5 g$ centripetal force. The tangential velocity of the sample at a 10.0-cm radius is given by

$$v = \omega r = 2\pi \times 10^3 \text{ rad/sec} \times 0.100 \text{ m}$$
$$= 2.00\pi \times 10^2 \text{ m/sec} = 628 \text{ m/sec}$$

This speed is greater than the velocity of sound in air. In order to avoid shock waves due to supersonic speeds, the rotor of the ultracentrifuge spins inside a vacuum chamber. A diagram of an ultracentrifuge is shown in Figure 4.17.

4.7 Inertial and Gravitational Mass Equivalence

A fundamental question concerning the inertia of an object was raised by Newton. He wondered if the property determining the weight of an object was the same as the inertial parameter that determines the acceleration produced by force applied to the object. This can also be stated as follows: Is the gravitational mass the same as the inertial mass of an object? In freely falling object experiments, is the acceleration a constant, or does it depend on the properties of the falling object (such as temperature and chemical composition)? If the inertial mass equals the gravitational mass, we should expect all objects to fall with the same acceleration. Likewise if a nongravitational force equal to the weight of the object is applied, the acceleration of the object should be equal to the acceleration due to gravity. Such experiments and many others have been carried out since the seventeenth century, and all experiments show that within experimental error the gravitational mass and the inertial mass have the same value. Most recent experiments have shown this equivalence to be accurate to within one part in 10^{12}.

ENRICHMENT
4.8 Escape Velocity From the Earth

If the mass is constant, the force is the product of mass and acceleration. In general, an equation of motion is given by

$$F_{net} = ma = m\frac{dv}{dt} = m\frac{d^2r}{dt^2} \tag{4.11}$$

where the acceleration a is given by the time rate of change of the velocity $a = dv/dt$, and where the velocity v is given by the time rate of change of the distance, $v = dr/dt$.

If we know the expression for F for a given situation we can equate it to $m(d^2r/dt^2)$, and solve the differential equation that results. F may have various functions of distance r or time t. For example, F may be given by the law of gravitational attraction of the earth,

$$G\frac{M_e m}{r^2} = m\frac{d^2r}{dt^2} \tag{4.12}$$

To solve this differential equation, the acceleration is expressed as a function of the distance derivative of the velocity by multiplying by dr/dr and rearranging the variables,

$$a = \frac{dv}{dt} = \frac{dv}{dt}\cdot\frac{dr}{dr} = \frac{dr}{dt}\cdot\frac{dv}{dr} = v\frac{dv}{dr} \tag{4.13}$$

For constant mass, Newton's second law may be rewritten as

$$F = mv\frac{dv}{dr} \tag{4.14}$$

This expression can be set equal to the gravitational attraction of the earth,

$$G\frac{M_e m}{r^2} = mv\frac{dv}{dr} \tag{4.15}$$

where M_e is the mass of the earth. We can integrate this equation to obtain an expression for the minimum vertical velocity needed for an object to escape from the earth.

$$GM_e m\int\frac{dr}{r^2} = m\int v\,dv \tag{4.16}$$

$$-\frac{GM_e m}{r} = \frac{1}{2}mv^2 + \text{constant} \tag{4.17}$$

We can evaluate the constant for the escape velocity by requiring the velocity of the object to be zero when the object is an infinite distance from the earth. Then the constant is zero. We find that the vertical velocity the object must have at the surface of the earth is v_{escape}

$$v_{\text{escape}} = \sqrt{2GM_e/r_e} \tag{4.18}$$

where r_e is the radius of the earth.

Questions

1. Assume an asteroid has the same density as the earth, 5.50×10^{-3} kg/m^3. Calculate the size of the asteroid from which you could launch yourself into space by running with the escape velocity on the surface of the asteroid. (Use your estimated maximum running speed on earth for the escape velocity.)

SUMMARY

Use these questions to evaluate how well you have achieved the goals of this chapter. The answers to these questions are given at the end of this summary with the number of the section where you can find the related content material.

Definitions

1. There must be no unbalanced force on a body
   a. at rest
   b. with constant velocity
   c. with no friction
   d. with constant acceleration
   e. all of these
2. The weight of a body may vary with its location because the
   a. mass varies
   b. temperature varies
   c. acceleration due to gravity varies
   d. pressure varies
   e. all of these
3. The coefficient of friction is
   a. inversely proportional to normal force
   b. greater for kinetic than static case
   c. greater than 1
   d. theoretically determined
   e. experimentally determined
4. The force due to friction always
   a. is greater than normal force
   b. opposes motion
   c. is an unbalanced force
   d. is zero for bodies at rest
   e. equals body weight
5. Centripetal force in circular motion accounts for acceleration
   a. toward the center
   b. away from the center
   c. tangent to the path
   d. opposing gravity
   e. equal to zero
6. Weightlessness in orbiting satellites is equivalent to motion in
   a. a centrifuge
   b. free fall
   c. a vacuum
   d. rocket launching
   e. none of these

Newton's Laws

7. According to Newton's law of gravitational interaction the gravitational force between particles is _____ (*attractive/repulsive*) with a magnitude proportional to the product of _____ and the square of the _____.
8. Newton's third law postulates the existence of _____ pairs of forces which are _____ and _____ to one another.
9. Newton's second law helps us understand the motion of objects because it relates the _____ of an object to the _____ acting on the object with the inertial property of the object called _____ as the proportionality constant.
10. The SI unit of force is a _____ and is equivalent to a mass of _____ being accelerated at a constant rate _____.
11. Newton's first law may help explain the persistence of the status quo because it states that an object at rest tends to _____ and moving objects tend to _____ unless acted upon by _____.

Resolution of Forces

12. A force of 13.0 N acting upward at an angle of 23° to the horizontal has an upward vertical component of _____ and a horizontal component of _____.
13. A football sled of downward weight 500 N is pushed horizontally by a force of 400 N. The resultant force is _____ in a direction _____ from horizontal.

Newton's Second Law

14. If a body is acted on by an unbalanced horizontal force of 40.0 N and experiences an acceleration of 2.00 m/sec², its mass is
   a. 80.0 kg
   b. 0.050 kg
   c. 20.0 kg
   d. 2.00 kg
   e. 40.0 kg

15. A body with a weight of 10.0 N is pulled across a horizontal surface with coefficient of kinetic friction $\mu = 0.200$ by a force of 5.00 N making an angle of 37° with the horizontal. The acceleration of the body will be
    a. 1.00g
    b. 0.140g
    c. 0.200g
    d. 2.60g
    e. 0.260g
16. A mass of 2.00 kg moves in a circular path (radius 1.00 m) with an angular velocity of 10.0 rad/sec. The centripetal force on the mass is
    a. 2.00g
    b. 20.0 N
    c. 10.0 N
    d. 200 N
    e. 50.0 N
17. Friction between a block at a distance d from the center of a turntable rotating at n rad/sec keeps the block on the turntable. The coefficient of friction between the block and turntable must be at least
    a. nd/g
    b. g

c. n^2d/g
d. gn^2d
e. g/n^2d

Answers

1. a, b (Section 4.2)
2. c (Section 4.3)
3. e (Section 4.4)
4. b (Section 4.4)
5. a (Section 4.6)
6. b (Section 4.6)
7. attractive, masses of the particles, inverse of the distance between the particles (Section 4.6)
8. action-reaction, equal, opposite (Section 4.2)
9. acceleration, unbalanced forces, mass (Section 4.3)
10. newton, 1 kg, 1 m/sec² (Section 4.3)
11. stay at rest, continue moving in a straight line at constant speed, unbalanced forces (Section 4.2)
12. 5.00 N, 12.0 N (Section 4.2)
13. 640 N, 51.3° down (Section 4.2)
14. c (Section 4.3)
15. e (Section 4.3)
16. d (Section 4.6)
17. c (Sections 4.4 and 4.6)

ALGORITHMIC PROBLEMS

Listed below are the important equations from this chapter. The problems following the equations will help you learn to translate words into equations and to solve single concept problems.

Equations

$$v_f^2 = v_0^2 + 2as \tag{3.9}$$

$$\mathbf{F} = m\mathbf{a} \tag{4.1}$$

$$f_k = \mu_k N \tag{4.3}$$

$$F = \frac{Gm_1m_2}{r^2} \tag{4.4}$$

$$g = \frac{GM_e}{r_e^2} \tag{4.5}$$

$$F = ma_r = \frac{mv^2}{r} = 4\pi^2n^2mr = m\omega^2\mathbf{r} \tag{4.6 and 4.7}$$

$$G = 6.67 \times 10^{-11} (\text{N-m}^2/\text{kg}^2)$$

FIGURE 4.18
Algorithmic problem 5.

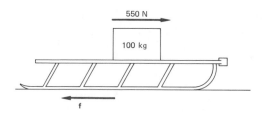

Problems

1. Calculate the braking force needed to give a 2.00 m/sec² negative acceleration to a 1000-kg car.
2. Compute the force needed to move a 1.00-kg block across the floor at a constant speed if the coefficient of friction is 0.300.
3. Find the centripetal force on a satellite of mass m in a circular orbit at a distance D from the center of the earth. The velocity of the satellite is C.
4. The centripetal force in problem 3 is equal to the gravitational attraction between the earth and the satellite. If the mass of the earth is M, show that the value of C must be $(GM/D)^{1/2}$. (Hint: Use the answer to problem 3.)
5. Given that the force of friction on a sled of mass 100 kg is 50.0 N, find the acceleration of the sled when a 550 N force is applied horizontally in an easterly direction (see Figure 4.18).
6. Find the force necessary to accelerate a 1000-kg elevator at 1.00 m/sec² upward.
7. Assume you have a mass of 60.0 kg and are riding in an elevator which is accelerating upward with an acceleration of 2.20 m/sec². What force must your legs withstand?
8. What is the force necessary to hold a 100-g mass in a 0.100-m circle at 1200 rpm? How does this compare with its weight?
9. What is the force of attraction between the earth and a 2.00-kg mass at the surface of the earth?
10. Find the coefficient of friction that gives a force of friction of 5.00 N between a box and the floor if the box weighs 50.0 N.
11. Two boys are holding a box weighing 400 N above the floor. One boy exerts a 225 N force upward. Find the upward force exerted by the other boy.

Answers

1. 2000 N
2. 2.92 N
3. mC^2/D
5. 5.00 m/sec²
6. 10,800 N

7. 720 N
8. 78.9 N, 80.6 times its weight
9. 19.6 N
10. 0.100
11. 175 N

EXERCISES

These exercises are designed to help you apply the ideas of a section to physical situations. Then appropriate numerical answers are given at the end of each exercise.

Section 4.2

1. A picture weighing 10.0 N is hung on a hook at the midpoint of a supporting wire. Each side of the wire

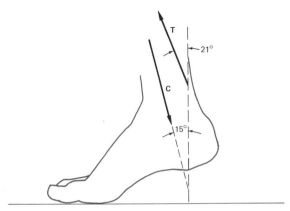

FIGURE 4.19
Force and tension in an isolated foot and ankle. Exercise 4.

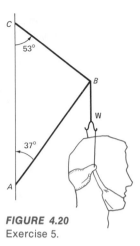

FIGURE 4.20
Exercise 5.

makes an angle of 10° below the horizontal. What is the tension on the wire?   [29 N]

2. A 120–N child in a swing is held by a horizontal force so that the ropes make an angle of 37° with the vertical. What is the tension on each rope and the horizontal force?   [75 N, 90 N]

3. On a field trip, a geologist gets his automobile into a soft spot. He attaches a taut cable to a tree 30.0 m from the car. He then hangs a 50.0-kg rock at the midpoint of the cable, and it sags 30.0 cm. ~~Assume there is no stretch in the cable~~. What is the force acting on the car?   [12,300 N]

4. Consider the foot and ankle as an isolated body, and apply the principle of statics at the moment the ball of one foot is in contact with the ground. Use your weight as W. The tibia **C** makes an angle of 15° with the vertical, and the Achilles tendon **T** makes an angle of 21° with the vertical (see Figure 4.19). Calculate the magnitude of the compressional force in the tibia and of the tension in the Achilles tendon.   [$C = 3.42w$, $T = 2.48w$]

5. A weight **W** is supported by cervical neck traction apparatus consisting of a boom AB and a cable BC as shown in Figure 4.20. Neglecting the weight of the boom, what is the compression of the boom and the tension of the cable?   [$T = 0.600W$, $C = 0.800W$]

Section 4.3

6. A 50.0-kg student is riding in an elevator. What is the compressive force in her legs, if the elevator is moving
   a. upward at a constant rate?
   b. with an upward acceleration of 3.20 m/sec²?

c. with a deceleration of 2.40 m/sec²?   [490 N, 650 N, 370 N]

7. In walking you have to accelerate and decelerate your foot. If you walk at a constant speed of about 6.00 km/hr, the average acceleration of your foot is about 40.0 m/sec². What is the force necessary to give each kilogram of your foot an acceleration of this amount? What would this force be if you were running a 100-meter dash in 10.0 sec?   [40.0 N, 240 N]

8. A girl of mass 45.0 kg (weight ≈ 100 lb) escapes a fire by sliding down an improvised rope made of sheets tied together. The maximum upward force the "rope" can stand is 270 N (~ 60 lb).
   a. Can she slide down the rope at a constant speed? If not, find the least acceleration with which she can slide down the "rope."
   b. If the window is 5 m above the ground, how long will it take her to reach the side of the anxious young man below her window.   [no, 3.80 m/sec², 1.62 sec]

9. Compare the kinetic frictional forces for the same normal force pressing the following surfaces together:
   a. two dry iron surfaces
   b. two wood surfaces
   c. lubricated metal surfaces
   d. two human body joint surfaces   [a:b:c:d = 2:3: 0.4:0.05]

10. A 50.0–N box is sliding down a 30° inclined plane at a constant rate. What is the coefficient of kinetic friction? What force would be required to move this box up the inclined plane at a constant rate?   [0.580, 50 N]

11. A floor polishing block weighing 10.0 N is being pushed at a constant rate over the floor by a 10.0–N

force exerted on the handle which makes an angle of 60° with the horizontal. What is the coefficient of kinetic friction?   [0.28]

12. Find the magnitude of force **F** for each case shown in Figure 4.21. All motion is at a uniform rate. Body A weighs 10.0 N. Body B weighs 20.0 N. The coefficient of kinetic friction is 0.300 for all surfaces.   [a. 9.00 N; b. 12.0 N; c. 15.0 N]

Section 4.5

13. How far from the center of the earth must a rocket be for its weight to be 9/10 its weight on earth?   [$1.05r_e$]

Section 4.6

14. A centrifuge is revolving at the rate of 10,000 rpm. The radius of revolution for a 0.100–g particle being separated is 8.00 cm. What is the centripetal force acting on it?   [8.77 N]

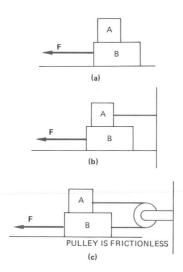

(a)

(b)

PULLEY IS FRICTIONLESS

(c)

FIGURE 4.21
Exercise 12.

PROBLEMS

Each problem may involve more than one physical concept. A problem that requires material from the enrichment section is marked by a dagger †. The numerical answer to the problem is given in brackets at the end of the problem.

15. A train of total mass M is traveling with velocity v when the throttle is suddenly opened and kept open for a distance d. If the accelerating force is F, find the velocity at the end of the distance d.
[$\sqrt{v^2 + 2dF/m}$]

16. In Figure 4.22, block A has a mass of 10.0 kg, and block B has a mass of 4.00 kg. The coefficient of kinetic friction is 0.300. The pulley is massless and

frictionless. Neglecting the mass of the cord, what is the acceleration of the system and the tension in the cord?   [0.700 m/sec², 36.4 N]

17. A weightless, frictionless pulley supports two masses of 90.0 g and 100 g connected by a massless cord (Figure 4.23). The masses are released. What is the acceleration of the system? What is the tension in the cord?   [0.516 m/sec, 0.928 N]

18. An 80.0–kg high jumper just clears the bar at 1.80 m. What vertical velocity did he have at the instant he left the ground? If he acquired this velocity in 0.500 sec, what was his acceleration? What vertical force must his leg have experienced?   [5.94 m/sec, 11.9 m/sec², 1740 N]

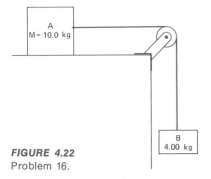

FIGURE 4.22
Problem 16.

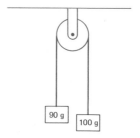

FIGURE 4.23
Atwood's machine. A frictionless and massless pulley supporting two unequal masses. Problem 17.

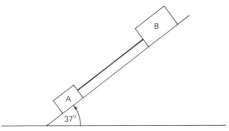

FIGURE 4.24
Problem 20.

19. A 1500–kg automobile climbs a hill at a speed of 12.0 m/sec. If the hill rises 1.00 m in 10.0 m and there is a frictional resisting force of 500 N, how much force must the driving wheels exert? [1970 N]

20. In Figure 4.24, object A weighs 50.0 N, and object B weighs 100 N. The coefficient of friction between object A and the plane is 0.500, and the coefficient of friction between object B and the plane is 0.200. A and B are connected by a rigid rod. (Neglect its mass.) What is the acceleration of the system, and the compression in the rod connecting A and B. [3.53 m/sec², 8.00 N]

21. A student is swinging a bucket partially filled with water in a vertical circle. Her arm is 0.700 m long, and the surface of the water is 0.100 m from the bucket's handle. What is the minimum velocity that the bucket must have to avoid spilling the water? [2.80 m/sec or 3.50 rad/sec]

22. What angle of bank should a road curve of 80.0–m radius have for safety at a speed of 100 km/hr? (Assume no friction between the tires and the road.) [44.5°]

23. A 30.0–kg child is sitting on a horizontal rotating platform at a distance of 2.00 m from the axis of rotation. The coefficient of static friction between the surfaces in contact is 0.200. What is the maximum rate at which the platform can rotate without the child sliding off? [0.990 rad/sec]

24. A track meet is theoretically scheduled on an asteroid which has a radius of 100 km. Assume that the density of the earth and asteroid are the same and that the radius of the earth is 6700 km. How high could a jumper jump on the asteroid if he could jump 2.00 m on the earth? What other assumptions have you made in the solution? [134 m]

25. The mass of the earth is 81.0 times that of the moon. The radius of the earth is 3.75 times that of the moon. On the surface of the earth the weight of an unknown mass as determined by a spring balance and also by

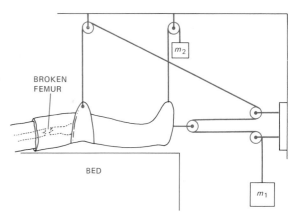

FIGURE 4.25
The Russell traction apparatus. Problem 26.

an equal arm balance with a standard set of weights is 100 N. All of this equipment is taken to the moon. What is the weight of the unknown mass on the moon as measured by both the spring balance and the equal arm balance? [17.4 N, 100N]

26. The Russell traction apparatus (Figure 4.25) is used for a fracture of a femur. The mass of the leg is 4.00 kg, and the tension required is 60.0 N. Find the values of masses m_1 and m_2. [3.06 kg, 0.940 kg]

27. The force exerted on the pilot by his seat is 4.00 mG (m = the pilot's mass) when his plane pulls out of a dive at 480 km/hr. Assume that the plane makes a circular arc at the bottom of the dive as shown in Figure 4.26. Find the radius of the arc and the time required to go through a $\pi/2$ rad arc. [605 m, 7.15 sec]

28. Many of the data available concerning human impact studies have been gathered from rocket and track experiments. If a rocket is brought to rest on 0.250 sec from a speed of 100 m/sec, find the average acceleration and force on an 80.0–kg subject. How far does the rocket travel while braking? [400 m/sec², −3.20 × 10³ N, 12.5 m]

29. Consider the two exercise systems described in

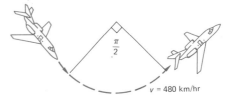

FIGURE 4.26
An airplane in a circular pull-out from a dive. Problem 27.

Example 3 of Section 4.2. Suppose that both systems were simultaneously used on the same patient. Use the superposition principle to find the total force perpendicular to the lower leg so as to oppose the extension of the leg. Draw a graph of the total force as a function of the angle θ that the lower leg makes with vertical. What is the angle of maximum total force?   [$F_p = 80$ N[cos $(\theta - \varphi) + \sin \theta$] where $\tan \phi = (1 - \cos \theta)/(2 + \sin \theta)$ and $\theta =$ leg angle with vertical]

30. Assume Evil Knievel, whose mass is 80 kg, wishes to launch himself into space from a 60° incline riding on a motorcycle of 170 kg mass. What must be the speed of the motorcycle as it leaves the sea level ramp? $M_e = 5.97 \times 10^{24}$ kg; $r_e = 6.37 \times 10^6$ m.   [12.9 km/sec]

GOALS

When you have mastered the contents of this chapter, you will be able to achieve the following goals:

Definitions
Define each of the following terms, and use it in an operational definition:

work   lever systems
energy   theoretical mechanical advantage
potential energy   actual mechanical advantage
kinetic energy   efficiency
power

Conservative and Nonconservative Systems
Establish the difference between a conservative and a nonconservative system.

Conservation of Energy
Explain the principle of conservation of energy.

Energy Problems
Apply the principle of conservation of energy to solve mechanics problems.

Lever Systems
Determine the theoretical mechanical advantage of human body lever systems.

Efficiency
Calculate the efficiency of a machine or human in action.

Power
Determine the power required for a given process or activity.

PREREQUISITES

Before beginning this chapter you should have achieved the goals of Chapter 3, Kinematics, and Chapter 4, Forces and Newton's Laws.

5
Energy

5.1 Introduction

Suppose you release a low-friction car from rest on an inclined track. What do you observe? The car seems to be moving faster, the farther it goes down the track.

Now suppose you place a 50-g mass on the car and perform the experiment over again. What is the influence of mass on the performance of the car? At first, it seems the car should move more slowly since you are increasing its inertia, or maybe it should move more rapidly since you are increasing the force of gravity upon the car. It turns out that the two effects cancel each other, and in the idealized case the speed is not affected by the mass. If you do not believe the preceding statement, try a simple experiment to test it for yourself.

Now let us change our experiment and push the car up the incline. Observe how far up the incline the car rolls. This distance depends upon how strong a push you give the car. That is easily tested by giving different pushes to the car and observing the different heights the car rolls up the incline. What if the mass of the car is increased? If you perform these experiments, you will discover that the distance the car rolls up the incline for a constant push is influenced by the mass of the car. If you double the mass of the car, the distance up the incline is reduced to one-half the initial distance for a constant push. A careful experiment will also reveal that the speed of the car as it starts up the plane also increases with increasing strength of the push. This speed decreases with increasing car mass for a constant push.

We conclude that a push has the ability to give the car a certain speed at the bottom of the incline and to move the car up the incline. This ability to change the state of the car is called *energy*. Energy is a mental construct developed to help explain a wide variety of phenomena in nature. Energy is never directly measured. In our example, the constant push times its distance of operation represented an energy unit. We always calculate the energy of a system from other measured quantities. In the case of the rolling car we measured its mass and the distance up the incline that it rolls and multiply them together to obtain a number proportional to the energy of the car. For other forms of energy, such as heat, we will combine other measured quantities (mass and temperature change in the case of heat) to calculate the change in energy of a system.

The energy construct plays a fundamental role in our understanding of natural processes. In this chapter you will be introduced to energy in a quantitative way and to its use in the analysis of mechanical systems. In the words of an unknown, but no doubt talented, physicist, "If you can learn to feel at home with 'energy,' you will feel at home with the universe."

5.2 Work

You are accustomed to using the word "work" to describe any form of activity that requires the exertion of a muscular or mental effort. The meaning of the term "work" in physics is much more restricted. We shall say that *work* is done when the applied force produces a displacement and when the force has a component parallel to the displacement. If the force is perpendicular to the displacement, no work is done. If a force is exerted to lift an object, work is done, but no work is done in holding an object at a constant height. The centripetal force in circular motion is doing no work. Can you give examples of other forces doing work and not doing work? In accordance with the above definition, we can represent work by the relationship:

work = force component parallel to displacement ×
magnitude of displacement

$$w = (F \cos \theta)(s) = \mathbf{F} \cdot \mathbf{s} \qquad \mathbf{F} = \text{force, } \mathbf{s} = \text{displacement} \qquad (5.1)$$

Equation 5.1 contains a product of vectors that is a scalar quantity. The *scalar* or *dot product* of two vectors is a compact notation devised to represent the product defined as follows: the component of the first vector that is parallel to the second vector is multiplied by the magnitude of the second vector.

EXAMPLE

See Figure 5.1. $\mathbf{F} \cdot \mathbf{s} = (F \cos \theta)(s)$ where $F \cos \theta$ is the component of force $\mathbf{F}$ parallel to $\mathbf{s}$. Note that the scalar product of a vector with itself is equal to the square of the magnitude of the vector, $\mathbf{a} \cdot \mathbf{a} = a^2$. The dot, or scalar, product of two vectors perpendicular to each other is zero.

EXAMPLES

1. Suppose a northeast wind exerts a force of 100 newtons on an airplane flying east. The work done by the wind on the airplane over a 100-km flight would be:

$$\mathbf{F} \cdot \mathbf{s} = 100 \text{ N} \times \cos 45° \times 10^5 \text{ m} = 0.707 \times 10^7 \text{ N-m}$$

$$= 7.07 \times 10^6 \text{ N-m}$$

What are the dimensions of work in terms of the fundamental dimensions of mass M, length L, and time T? In the SI system the unit of work is the *joule* (J) which is equivalent to a force of one newton acting through a displacement

FIGURE 5.1
The work done by a force **F**, at an angle of θ degrees above the horizontal, in moving an object **s** m is given by **F** cos θ · **s**, the product of the magnitude of the component of the net force parallel to the direction of motion times the magnitude of the displacement.

FIGURE 5.2
(a) A boy pulls a 5-kg sled 10 m along the ground at a constant speed. (b) The force diagram. **F**$_f$ is the friction force and is equal in size to μN where N is the magnitude of the normal force. The size of the normal force is given by the weight W minus the upward component of the pull, $P \sin 37°$.

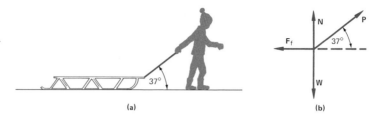

(a)   (b)

of one meter. The *erg* is the unit of work in the cgs system and is equivalent to a force of one dyne acting through a displacement of one centimeter.

2. A boy pulls a 5.00-kg sled 10.0 m along a horizontal surface at a constant speed. What work does he do if the coefficient of kinetic friction is 0.200 and his pull is 37° above the horizontal? The forces acting on the sled are shown in Figure 5.2. The problem can be solved using scalar equations for the horizontal and vertical components of the forces. For constant horizontal speed the forward and backward components must be equal

$$P \cos 37° = F_f$$

where $F_f = \mu N$. For horizontal motion the upward and downward components of the forces must be equal

$$N = W - P \sin 37°$$

So,

$$P \cos 37° = 0.200(W - P \sin 37°)$$

$$0.800P = 0.200(5.00 \times 9.80 - 0.600P)$$

Work done on the sled by the boy is

$$w = (P \cos \theta)(10.0 \text{ m})$$

$$= 10.6 \text{ N} \times 0.800 \times 10.0 \text{ m} = 84.8 \text{ joules (J)}$$

Would the boy do more, less, or the same amount of work if he pulled horizontally on the sled? Would he exert the same force?

5.3 Energy

If a body has the ability to do work, it is said to possess energy. *Energy* is defined as the ability to do work and can be measured by the work done. Accordingly, energy is expressed in the same units as work, that is, joules. There are a number of ways in which a body may possess energy. Can you list some?

In our introduction to this chapter it appeared that the energy of the car was gained from the push. The energy of the car could be studied in terms of its speed or its distance of roll up the incline. We will now take up the forms of energy for a physical system that depend on state of motion and position.

5.4 Kinetic Energy

In the introduction of this chapter we considered an example of a toy car responding to a push. The energy imparted to the car by a push increased the speed of the car.

The energy of motion of a body is called kinetic energy.

Think of some examples of kinetic energy.

If a body has a velocity, it must be decelerated to bring it to rest. There is then a force acting on it. For simplicity let us assume the deceleration is constant and consequently the force is constant. The force will act through a distance s in bringing the body to rest. In this case *work is done on* the body by the braking force, and the kinetic energy of the body decreases. If a body is at rest and it is accelerated by a force to a constant velocity, *work is done on* the body by the accelerating force, and the kinetic energy of the body increases.

Suppose a cyclist starts from rest and accelerates at a constant rate until reaching a speed v. We wish to find the work done by the cyclist to produce the kinetic energy possessed at speed v. Let the total mass of the cyclist and cycle be m. The magnitude of the accelerating force is given by Newton's second law: $F = ma$, but $\bar{a} = v/t$, where t is the time required to reach velocity v. In this same time the cyclist moves a distance s which is equal to the product of the average velocity and the time:

$$s = \frac{v}{2} t$$

where $v/2$ is the average velocity for this case. The work done by the accelerating force of the cyclist is given by the product of the magnitude of the force times the distance through which it acts

$$w = Fs = mas = M \frac{v}{t} \times \frac{v}{2} t = \frac{1}{2} mv^2 \equiv \text{kinetic energy at speed } v \qquad (5.2)$$

The work done by the cyclist to reach speed v is used to define his kinetic energy at speed v.

Consider the braking of the cyclist reducing speed from v to v' in time t. The magnitude of the braking force $\mathbf{F}$ (assumed constant) produces a deceleration given by Newton's second law as

$$F = ma' \quad \text{where} \quad a' = \frac{v' - v}{t}$$

and the braking distance s' is given by

$$s' = \frac{v' + v}{2} t$$

Thus the work done by the braking force is given by Equation 5.1 as:

$$w' = Fs' = m\left(\frac{v' - v}{t}\right)\left(\frac{v' + v}{2}\right)t$$

$$w' = \tfrac{1}{2} mv'^2 - \tfrac{1}{2} mv^2 = \Delta KE \tag{5.3}$$

The work done is equal to the change in the kinetic energy of the cyclist. Note that in this case since kinetic energy decreases, the work done is negative. This sign convention is used throughout physics. Positive work on a system increases the energy of the system. Negative work is work by a system that reduces its energy.

EXAMPLE

What is the kinetic energy of a 50-kg girl running at 4 m/sec?

$$KE = \tfrac{1}{2} mv^2 = \tfrac{1}{2} \times 50 \times 4^2 = 400 \text{ J}$$

This represents the work that must be done to stop the girl.

5.5 Potential Energy

If a body has the ability to do work as a result of its position or configuration, it is said to possess *potential energy*. Suppose we consider the following: A rock of mass m rests at the top of a cliff of height h. In this position it has the ability to work relative to the foot of the cliff. We call this potential energy. The potential energy that it possesses is equivalent to the work which you would have to do to take the rock from the foot of the cliff to the top. The magnitude of the force which you would have to exert to lift the rock at a constant rate is mg. This force must act through a distance of h, and the work is done then

$$w = F \times \cos \theta \times s = mgh \tag{5.4}$$

So we can say that at the top of the cliff the potential energy of the rock is

$$PE = mgh \tag{5.5}$$

This is sometimes called gravitational potential energy. Can you give other examples of gravitational potential energy?

EXAMPLE

If the rock in Figure 5.3 has a mass of 3.00 kg and the height of the cliff is 10.0 m the potential energy is calculated as follows:

$$PE = mgh = (3.00)(9.80)(10.0) = 294 \text{ J}$$

An example of potential energy of a configuration is a stretched spring. You are aware that a stretched door spring has the ability to close the open door and thus possesses energy. The energy that it possesses is given by

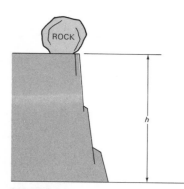

FIGURE 5.3
The amount of work done to place the rock on the cliff is *mgh* so the rock is said to possess potential energy in the amount *mgh* with respect to the bottom of the cliff.

FIGURE 5.4
The spiral spring has a linear restoring force. The work done in stretching this spring against the conservative restoring force results in potential energy being stored in the spring. This work is equal to the average force times the distance through which it acts and is represented by the shaded area.

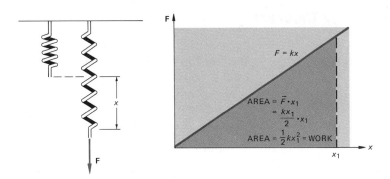

$$\text{PE} = \tfrac{1}{2} kx^2 \tag{5.6}$$

where k is a constant called the spring constant, the force required to stretch a spring unit length, and x is the stretch (see Figure 5.4).

The work that the stretched spring can do is equal to work done to stretch the spring. The work required to stretch a spring is given by the product of the average force and the length of stretch. For a linear spring the force is 0 for zero stretch, and for a stretch of x the force is given by kx where k is the force required to stretch the spring a unit length. The magnitude of the average force F_{ave} is given by $\dfrac{0 + kx}{2}$. Note that this method of finding the average value is valid only in general for a linear variance, that is, when the force varies linearly with the stretch. Thus the work is given by the product of the average force F_{ave} times the stretch x,

$$w = F_{ave}x = \left(\frac{0 + kx}{2}\right)x = \tfrac{1}{2} kx^2$$

EXAMPLE

A certain spring is stretched 10 cm by a 10-N force. Find the energy stored (potential energy) when the spring is compressed 2 cm. First we find the force constant of the spring:

$$k = \frac{10 \text{ N}}{0.1\text{m}} = 100 \text{ N/m}$$

Thus

$$\text{PE} = \tfrac{1}{2} kx^2 = \tfrac{1}{2} \times 10^2 \text{ N/m} \times (2 \times 10^{-2}\text{m})^2$$

$$\text{PE} = 2 \times 10^{-2} \text{ J}$$

In summary, the potential energy of a body is defined as its ability to do work resulting from its position or configuration. Potential energy is a scalar quantity as it has magnitude only relative to an accepted base. In case of gravitational potential energy this base is the height reference, and in case of a stretched spring it is usually the unstretched condition of the spring.

5.6 Work-Energy Theorem

In accordance with Equation 5.3, we may state that the work done by the net force acting on a body is equal to the change in kinetic energy of the body. Even though the above result was derived for a constant force, it holds true whether the resultant force is constant or variable. (This fact will be derived in the enrichment section, Section 5.14, for variable force.) In general then we may state,

$$\text{work of net force} = \text{change in kinetic energy} = \Delta KE \tag{5.7}$$

This is known as the work-energy theorem for a single body. The units of energy are the same as work. Energy is also a scalar quantity. What must be the dimensions of energy?

If a body has several forces acting on it, the net force $\mathbf{F}$ is the vector sum of the group. That is,

$$\mathbf{F} = \mathbf{F}_1 + \mathbf{F}_2 \cdots + \mathbf{F}_n \tag{5.8}$$

The work done by the net force is the algebraic sum of the work done by individual forces. Equation 5.7 may then be rewritten,

$$w_1 + w_2 + \cdots + w_n = \text{change in kinetic energy} = \Delta KE \tag{5.9}$$

5.7 Conservative and Nonconservative Forces

The forces acting on a body will, in general, be one of two types of forces, conservative or nonconservative. A force is said to be *conservative* if no net work is done when a body makes a round trip from a starting point and returns to this point. A force is said to be *nonconservative* if work is done in the round trip. Consider a system that goes from a to b in Figure 5.5 by one route and returns by another.

$w_{ab} + w_{ba} = 0$ for a conservative system

$w_{ab} + w_{ba} \neq 0$ for the nonconservative system

Systems that include frictional forces are nonconservative. All real systems seem to be nonconservative since they always have some frictional forces acting on them, and energy is lost in nonproductive ways. However, many practical systems are well approximated by idealized, frictionless, hence conservative systems.

The paths ab and ba above may indicate any path between points a

FIGURE 5.5

The work done between a and b along path 1 plus the work done in going from b to a along path 2 results in *no net work done* for a conservative system. Thus we have

$W_{ab(1)} + W_{ba(2)} = 0$ or $W_{ab(1)} = -W_{ba(2)}$. Since $-W_{ba(2)} = +W_{ab(2)}$ and $W_{ab(1)} = W_{ab(2)}$, the work is independent of the path.

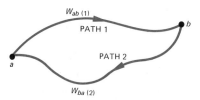

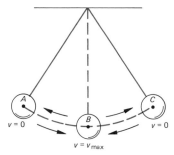

FIGURE 5.6
The simple pendulum illustrates the conservation of energy and the exchange between potential and kinetic energy during its motion. PE + KE = constant total energy.

and *b*. This leads to another equivalent condition for a conservative system: The work done in going from *a* to *b*, or *b* to *a* is independent of the path and depends only upon the end points *a* and *b*. Work done for a nonconservative system depends on the path over which the force acts while doing work on the body.

EXAMPLES

1. A sphere is supported by a light cord and hung by a frictionless support as shown in Figure 5.6. The sphere is originally at rest at A. In this position it possesses potential energy. As it goes from A to B, it loses potential energy and gains kinetic energy. In going from B to C, it loses kinetic energy and gains potential energy. At C in the idealized case it has the same potential energy as it had at A. In the return trip from C to B, the sphere loses potential energy and gains kinetic energy. From B to A the kinetic energy decreases, and the potential energy increases. It returns to A, and the condition is the same as it was originally. The force exerted by the cord does no work as it is always perpendicular to the path of the sphere. This is an example of a conservative system. What statement are you now able to say about the total energy of a conservative system? Note, that whenever negative work is done by a conservative force, this work is equal to the gain in the potential energy of the system.

2. A sled is at the top of a snow covered hill. The sled is released, and it slides down the hill and comes to rest. The sled is pulled back up the hill to the initial point. Then the sled has the same potential energy as it had originally. Is this a conservative system?

    Let us analyze the energy situation. At the top of the hill the sled has potential energy. As it goes down the hill the potential energy decreases, and the kinetic energy increases. Work is also being done against the frictional force. After the sled reaches the lower level, it slows down and loses kinetic energy. Why? Someone then pulls the sled up the hill by a different route than it went down the hill. In going up the hill potential energy is increased. In this round trip *work is done against the frictional forces, and the system is not conservative.*

Questions

1. Suppose we had a valley between two identical hills and that a boy starts on a sled at the top of one hill, slides down through the valley and up the hill on the other side. When the sled comes to rest, he turns it around and makes a return trip. He continues to repeat the cycle. What will be the final state of the sled?
2. What statement can you make about the total energy of a nonconservative system?
3. Which kind of system do you most often encounter, conservative or nonconservative?
4. What statement can you make about the energy state for final equilibrium of a system?

5.8 *Conservation of Energy*

Recall from Chapter 2 that conservation laws result when quantities remain constant (do not change in time). We are now ready to state the principle of conservation of energy. Namely, *in a conservative system the energy remains constant*, that is, *the total energy of the system in state 1 is the same as the total energy of the system in state 2*. If we consider the gravitational energy of the system, its kinetic energy, and take U to represent any other energy of the system, this can be represented as follows:

energy of state 1 = energy of state 2

$$U_1 + mgh_1 + \tfrac{1}{2} mv_1^2 = mgh_2 + \tfrac{1}{2} mv_2^2 + U_2 \qquad (5.10)$$

This is a very important principle of physics and is most useful in the solution of many problems. A corollary to this may be stated as follows: *The work done on a system goes into a change in potential energy, into a change in kinetic energy, or into work against frictional forces.*

For mechanical systems this can be written in equation form:

$$w = \text{work done on system} = \Delta\text{PE} + \Delta\text{KE} + w_f \qquad (5.11)$$

where ΔPE is change in potential energy, ΔKE is change in kinetic energy, and w_f is work done against dissipative (frictional) forces.

EXAMPLES

1. In a conservative system, suppose a 5.00-kg mass starts from *rest* and slides down a frictionless 15.0-m incline. The height of incline is 5.00 m (see Figure 5.7). What is the velocity at the base? From the conservation of energy principle, we can state that the PE at the top of the incline (where KE = 0) = KE at the base (where PE = 0). So

    PE = mgh

    KE = $\tfrac{1}{2} mv^2$

    Thus

    $mgh = \tfrac{1}{2} mv^2$

    $$v = \sqrt{2gh} = \sqrt{2(9.80) \times (5.00)} = 9.90 \text{ m/sec}$$

2. In a nonconservative system, assume that the coefficient of sliding friction is 0.200 (see Figure 5.7). What is the velocity of the block when it reaches the foot of the incline? The kinetic energy at base equals the potential energy at top less the work done against friction.

    work against friction = force × distance

    $$= (\mu N) \times \text{distance}$$

    $$= (\mu mg \cos\theta)(15.0 \text{ m}) \qquad \text{where } \cos\theta \text{ equals } \sqrt{200}/15.0$$

    $$= 0.200\,(5.00)(9.80)\,\frac{\sqrt{200}}{15.0}\,(15.0 \text{ m}) = 139 \text{ J}$$

FIGURE 5.7

A block rests at the top of an inclined plane. In this position the block possesses potential energy. As the block slides down the plane, it gains kinetic energy until it reaches the bottom. If the plane is smooth, the loss in potential energy equals the gain in kinetic energy. If the plane is rough, work is done against friction, and the loss in potential energy is then equal to the work which is done against friction plus the gain in kinetic energy.

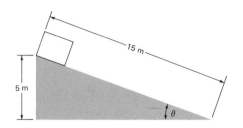

$$PE = mgh$$
$$= (5.00 \text{ kg})(9.80 \text{ m/sec}^2)(5.00 \text{ m})$$
$$= (25.0)(9.80) \text{ J} = 245 \text{ J}$$

We can find the speed at the base from the definition of the KE at the base: $KE = \frac{1}{2} mv^2$.

$$KE_{base} = PE - \text{work against friction}$$
$$= 245 \text{ J} - 139 \text{ J} = 106 \text{ J}$$
$$KE_{base} = \frac{1}{2} mv^2 = 106 \text{ J}$$
$$v^2 = \frac{2 \times 106 \text{ J}}{5 \text{ kg}} = 42.3 \text{ m}^2/\text{sec}^2$$
$$v = 6.51 \text{ m/sec}$$

5.9 Power

Power is a term which has many meanings. In physics power is defined as the time rate of doing work. For example, to fell a tree of a given size may take a beaver all night, a good man with an axe 10 minutes, and a lumberjack with a chain saw 15 seconds. In each case the result is the same, and loosely speaking, one can say the same amount of work is accomplished, but the power rating for each of the three cases is entirely different. We define power as the amount of work done in a unit of time,

$$\text{power} = \frac{\text{work done}}{\text{time to do it}} = \frac{\text{work}}{\text{time}} = \frac{(\text{force})(\text{distance})}{\text{time}} \qquad (5.12)$$

What are the dimensions of power? In the mks system the *watt* is the power when one joule of work is done in one second.

EXAMPLE

What is the power expended by a girl of mass 55.0 kg who takes 6.00 sec to run up stairs that have a vertical height of 5.00 m?

$$\text{work done} = 55.0 \text{ kg} \times 9.80 \text{ m/sec}^2 \times 5.00 \text{ m}$$

$$P = \frac{\text{work}}{\text{time}} = \frac{55.0 \times 9.80 \times 5.00 \text{ J}}{6.00 \text{ sec}} = 449 \text{ watts}$$

As a special case, consider the situation when the force is constant, thus equal to the displacement divided by the time; then we have:

$$P = \frac{\mathbf{F} \cdot \mathbf{s}}{t} = \mathbf{F} \cdot \mathbf{v} = Fv \cos \theta \tag{5.13}$$

where θ is the angle between $\mathbf{F}$ and $\mathbf{v}$.

EXAMPLE

What is the pulling power rating of a tractor that is capable of pulling a 22,250-N force load at 5.00 km/hr?

$$P = Fv = 22,250 \text{ N} \times \frac{5.00 \times 10^3 \text{ m}}{3.6 \times 10^3 \text{ sec}}$$

$$P = 30.9 \times 10^3 \text{ watts} = 30.9 \text{ kilowatts (kW)}$$

5.10 *Simple Machines*

We may think of a machine as a device that alters the magnitude and/or direction of an applied force. The complex machines in industry are generally made up of combinations of certain elements that are called simple machines. These simple machines include the lever, wheel and axle, pulley, inclined plane, and screw.

In any machine there is an energy input, which is always more than the work done by the machine. The loss of energy is caused by friction. The efficiency of a machine is defined as the ratio of the work output to the energy input,

$$\text{efficiency} = \frac{\text{work done by machine}}{\text{energy supplied to machine}} \tag{5.14}$$

Two other important properties of machines are the theoretical mechanical advantage and the actual mechanical advantage. The *theoretical mechanical advantage* (TMA) of a machine depends upon its geometry and is equal to the ratio of distance s through which the applied force F acts to the distance d through which the load W moves,

$$\text{TMA} = s/d \tag{5.15}$$

The *actual mechanical advantage* (AMA) depends upon the working condition of the machine, the amount of friction present, for example, as well as on the geometry of the machine. It is equal to ratio of the load W to the applied force F,

$$\text{AMA} = W/F \tag{5.16}$$

Show that the ratio of AMA to TMA equals the efficiency:

$$\frac{\text{AMA}}{\text{TMA}} = \text{efficiency} = \frac{\text{work done by load}}{\text{work done by applied force}}$$

5.11 *The Human Body as an Engine*

The human body and an engine share many common characteristics. Like most engines the body consumes oxygen in the process of extracting energy from its fuel. Like engines, the human body exhausts carbon compounds. The terms work, efficiency, and energy expenditure apply as well to the human body activities as they do to engines. In spite of analogies with simple engines, it is important to point out that the human body is very complicated. It is a heat engine, chemical engine, electrical engine, and manufacturing system, all intricately coordinated by feedback control systems. The body is even capable of reproducing some of its own parts when they break down.

The human skeleton provides a series of levers that are activated by muscles. These muscles are sources of stored energy (*chemical potential energy*) which is converted to *kinetic energy* of body motion by a series of chemical reactions. We can divide muscle contractions into three classes related to shortening, lengthening, and no change of muscle fibers. Shortening contractions produce positive work done by the muscles, and energy must be supplied to the muscles. *Isometric* muscle work corresponds to the condition of no motion of body limbs, for example, holding a weight at rest or pushing against a wall. According to our physics definition of work, since there is no displacement of the body limbs, there is no external work done by the muscle. A closer look at isometric work reveals that internal energy is expended by the muscles in internal motions balancing the weight. Lengthening of muscles is accomplished by external forces acting on a limb. In this case work is done on the muscles and the rate of energy supplied to the muscles decreases.

5.12 *The Physics of the Body Lever Systems*

The human body is designed for motion. We will consider some basic physics of the lever systems and the muscles that energize these systems. In physics, lever systems are divided into three classes according to their geometry. First-class lever systems are characterized by having the *fulcrum* (axis of rotation for the lever) located between the applied force (muscle force) and the load. Second-class levers are characterized by having the load between the applied force and the fulcrum. Third-class levers have the applied force acting between the fulcrum and the load. These lever classes and an example of each are shown in Figure 5.8. For an idealized lever system the work input should be equal to the work output. It follows that the work input is equal to the product of the applied force F_a times the distance the applied force moves. This distance equals the product of the length of the lever arm, the distance L from force F_a to the fulcrum, times the angle through which the lever rotates θ, in radians. Similarly, the work output is equal to the product of

FIGURE 5.8
The three classes of levers with an
application of each type.

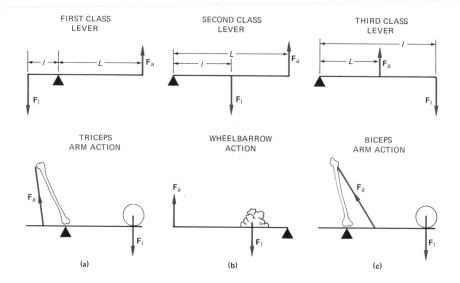

the load force F_l times the distance the load moves,

$$\text{work input} = F_a\,(L\theta) \tag{5.17}$$

$$\text{work output} = F_l\,(l\theta) \tag{5.18}$$

Equating these two expressions and solving for the ratio of the output
force (load) to input force (muscle), we get the equations for the theoretical and actual mechanical advantage of the system:

$$\text{TMA} = L/l \tag{5.19}$$

$$\text{AMA} = F_l/F_a \tag{5.20}$$

These two equations are equal for ideal levers, and TMA = AMA. Third-class levers are the predominant lever systems in the human body. These lever systems are designed for increasing the speed of motion rather than increasing load capabilities. Note from Figure 5.8 that the TMA of third-class levers is always less than or equal to one.

The lifting strength of a muscle is approximately proportional to the cross-sectional area of the muscle. This is a good approximation because the number of muscle fibers per unit area is almost constant for a given muscle, and the total force that muscle can exert is equal to the sum of the individual muscle fiber forces. The work done by a given muscle is equal to total force exerted multiplied by the distance of contraction of the muscle (parallel to the force). A muscle moves a distance proportional to its length. Thus the work done by a muscle is proportional to its area times its length or its volume.

5.13 Energy and the Human Body

Homeostasis is possible only with proper nutrition. The proper diet will supply the necessary ingredients and energy to maintain normal activity for an individual. Physiologists use the kilocalorie (kcal) as their unit of energy (1 kcal = 1000 cal = 4187 J). (The calorie used in nutrition tables is the kcal.)

Body weight control in humans is accomplished by controlling energy intake and energy output. The energy output is determined by the kinds of activity in which a person engages. Each person has a particular metabolic efficiency that determines the ability of the body to extract useful energy from food. If your energy output is less than your energy intake from food, you will convert the excess energy into fat tissue. An average person has an energy output of about 2300 kcal/day. This energy output is coupled with a loss of about 225 g of carbon and 16 g of nitrogen each day. The necessary energy, carbon, and nitrogen can be supplied by an appropriate combination of food materials. A typical energy conversion table for some fuels for humans is given in Table 5.1.

Proteins are needed mostly for their tissue building qualities and only secondarily as energy sources. Carbohydrates and fats supply the major part of the human body energy requirements. Your basic energy requirements are determined by your basal metabolic rate, the amount of physical work you perform, and the temperature of your environment. From

TABLE 5.1
Energy Value of Foodstuffs

Material	Energy Conversion
Carbohydrates	4 kcal/g
Fat	9 kcal/g
Protein	4 kcal/g
Ethanol	7 kcal/g
Gasoline (autos)	11.4 kcal/g

TABLE 5.2
Rate of Energy Consumption for
Human Activities

Activity	Energy Use Rate (kcal/min)	Oxygen Use Rate (liter/min)
Sitting	2.0	0.40
Resting	1.2	0.24
Sleeping	1.1	0.22
Walking	3.8	0.76
Bicycling	6.9	1.38
Swimming	8.0	1.60
Skiing	9.0	1.80
Running	11.4	2.28
Climbing up stairs	12.0	2.4

Table 5.1 you can see that fats are 2.25 times more effective in terms of energy per gram than carbohydrates. Carbohydrates are usually the cheapest source of energy. It should be noted that the human diet must also contain certain minerals and vitamins that are essential in order to maintain homeostasis.

Although efficiencies show individual differences, a 25 percent efficiency for the ratio of work out to metabolic energy increase is a typical average value for human activities. (Actual efficiency studies show that energy expenditure rates depend on the activity, but each activity shows a minimum energy expenditure rate as a function of the speed of the activity for each individual.)

The rate at which additional energy is used for different human activities by an adult of average surface area (1.75 m tall and a mass of 76 kg) is shown in Table 5.2.

Experimental studies show that about 5 kcal of energy is released from fuel for every liter of oxygen consumed by an average adult. It is common practice to measure the oxygen consumption (liters/min) in the study of human energy usage. Individual differences (after adjusting for the body weight) may be as high as 30 percent. It is possible to classify persons according to their metabolic efficiency, that is, their ability to extract energy from food.

The rate of energy use in animals needed to maintain a resting condition is found to be proportional to the three-fourths power of their mass. This empirical relationship is called Kleiber's law. A plot of metabolic rate (kcal/day) against mass (kg) in a log-log plot is shown in Figure 5.9.

EXAMPLES

1. Two people, Bert and Ernie, have metabolic efficiencies of 75.0 percent and 60.0 percent respectively. If both people take in food evaluated at 6000 kcal per day and both perform external work of 3600 kcal/day, what is the result?

   The energy that Bert receives from the food is 0.750×6000 kcal $= 4500$ kcal, and the energy Ernie receives from the same food is 0.600×6000 kcal $= 3600$ kcal. The excess energy that Bert receives will result in an increase in

FIGURE 5.9
A log-log plot of the metabolism against mass for several living species. The empirical relationship is called Kleiber's Law.

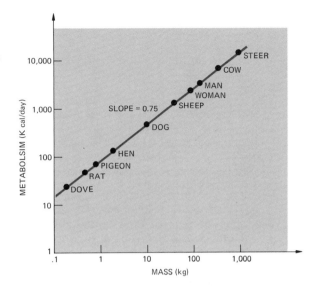

fat for Bert while Ernie balances input energy with output energy. Bert will have to decrease food input to 4800 kcal/day in order to maintain a constant weight.

2. Suppose that 3000 kcal of work in the previous example is performed by scuba divers in underwater work in 5.00 hours time. How much oxygen is needed for this work?

3000 kcal/5.00 hr = 600 kcal/hr

300 kcal/hr is equivalent to 1 liter of oxygen per hour. Thus the diver needs 2 liters of oxygen per hour, or 10 liters for the 5-hour dive.

3. Let us compare the energy required to cycle 10.0 kilometers with the energy required to drive the same distance in a car that gets 20.0 kmpg of gasoline. We assume that the cyclist operates at a level that allows a 15.0 percent conversion of his input energy to mechanical energy output. (This is a typical conversion efficiency. What happens to the rest of the energy that is gained from burning of food?) Assume cyclist rides at 10.0 kmph.

Cyclist: intake energy × 0.15 = energy output = 6.90 kcal/min × 60.0 min = 414 kcal

Thus,

intake energy = 414 kcal ÷ 0.150 = 2760 kcal

Car: At 20.0 kmpg, it takes 0.500 gallon to travel the 10.0 km. One gallon of gasoline corresponds to 2700 g thus we have for the car energy:

car energy = 0.500 × 2700 g × 11.4 kcal/g = 15,390 kcal

We see that the car takes 5.6 times as much energy as the cyclist. (Although this is only the first step in a complete energy analysis of this problem of transportation, it does point out the fact our society has moved to an energy intensive life style in order to save time. This fact deserves careful consideration as we study the energy demands of the world for the future.)

4. A 50.0-kg girl makes a vertical jump using a spring time of 0.250 sec and a

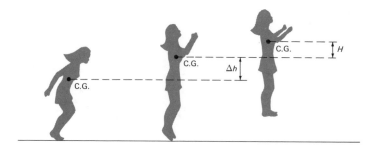

springing length Δh of 0.370 m (see Figure 5.10). Find the jump height H and the additional metabolic energy required for the jump if 25.0 percent of the increased metabolic energy goes into the jump.

During her spring, assume a constant force upward that results in the take-off velocity at the end of the spring. This take-off velocity should be equal to twice the average velocity during the spring:

$$v = 2 \times \frac{\text{spring distance}}{\text{spring time}} = 2 \times \frac{\Delta h}{t} = 2 \times \frac{0.370}{0.250} \text{ m/sec}$$

$$v = 2.96 \text{ m/sec}$$

Using this speed and the conservation of energy, we can find the jump height as follows:

$$\text{take-off kinetic energy} = mgH$$

$$\tfrac{1}{2}mv^2 = mgH$$

Thus,

$$H = v^2/2g = \frac{(2.96)^2}{2 \times 9.8} = 0.447 \text{ m}$$

The energy required for the jump is given by

$$\text{energy} = mgH = \tfrac{1}{2}mv^2$$

$$= 50.0 \times 9.80 \times 0.447 \text{ J} = 219 \text{ J}$$

The metabolic energy (ME) required at 25 percent efficiency is given by

$$0.250 \text{ ME} = 219 \text{ J}$$

Thus ME = 876 J.

When considering the energy cost of human activities, it is important to consider the effect of frictional resistance forces acting to oppose the body motion through air or water. These frictional forces include head winds for walking, running, and cycling and water drag for swimming. All of these fluid frictional forces are proportional to the surface area that is perpendicular to the motion. In addition they depend on the speed of motion relative to the fluid. For low speeds (streamline flow), the frictional force is proportional to the speed. For high speed (turbulent flow), the frictional force is proportional to the square of the speed with respect to the fluid. The actual shape of the body also is a factor in determining the frictional force. Considerable energy savings can be achieved by minimizing these fluid frictional forces. It has been reported that gasoline consumption of semi-trailer trucks could be improved by 30 percent with proper streamlining.

EXAMPLE

Let us find the increased power necessary to overcome air resistance when running a 100-m dash in 10.0 sec.

At this speed the frictional force is proportional to the square of the speed. The proportionality constant for an effective surface area of 0.75 m² is about 0.66 kg/m. Thus the frictional force for this can be written:

$$f = 0.66 \text{ kg/m} \times v^2$$

where v is the speed of the runner, and $v = 100$ m/10 sec $= 10.0$ m/sec.

Recall from Equation 5.12 that the power required to balance this frictional force is

$$P = f \times v = 0.66 \times 10^2 \times 10.0 \text{ J/sec} = 660 \text{ watts}$$

ENRICHMENT
5.14 Work and Energy Using Calculus

The differential element of work is defined by $dw = F \cos \theta \, ds$, where F is the magnitude of the force and θ is the angle between the force and the differential path length ds. If F and θ are constant, then

$$w = F \cos \theta \int_{s_1}^{s_2} ds$$

If the force is a scalar function of the distance, $F = f(s)$ (see Figure 5.11), then

$$w = \int_{s_1}^{s_2} f(s) \, ds$$

This integral is represented by the area under the curve. As an example

FIGURE 5.11

Graphical representation of calculus method for determining work done by a variable force.

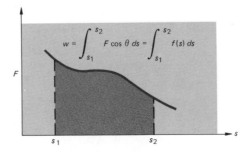

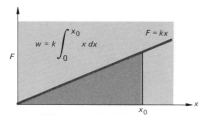

FIGURE 5.12
Graphical representation of work
done by force $F = kx$.

consider the case where $F = kx$ (Figure 5.12). Then

$$w = k \int_0^x x \, dx$$

$$w = \tfrac{1}{2} kx^2$$

Consider the scalar equation

$$F = ma = m\frac{dv}{dt} \cdot \frac{ds}{ds} = m\frac{ds}{dt} \cdot \frac{dv}{ds} = mv\frac{dv}{ds}$$

$$F \, ds = mv \, dv$$

But by the definition of work

$$w = \int F \, ds$$

Then

$$\frac{w}{v} = m \int_{v_1}^{v_2} v \, dv$$

$$w = \tfrac{1}{2}(mv_2{}^2 - mv_1{}^2)$$

That is, work done is equal to the change in kinetic energy. Instantaneous power is given by

$$P = \frac{dw}{dt} = \frac{d(\mathbf{F} \cdot \mathbf{s})}{dt}$$

If $\mathbf{F}$ is constant, the instantaneous power

$$P = \mathbf{F} \cdot \frac{ds}{dt} = \mathbf{F} \cdot \mathbf{v}$$

Using calculus we can derive the potential energy of a body of mass m at a distance r from the center of a planet of mass M and radius R. Newton's law of universal gravitation gives the force between the planet and the mass as:

$$F = G\frac{Mm}{r^2}$$

directed toward the center of planet, where $G = 6.67 \times 10^{-11}$ N–m^2/kg^2. The work done to move the body from distance r to infinity against the gravitational force is defined to be the potential energy at distance r. Since $\mathbf{F} \cdot d\mathbf{r} = +F \, dr$ and $\Delta \mathrm{PE} = -(\mathbf{F} \cdot d\mathbf{r}) = -F \, dr$,

$$\mathrm{PE}(r) = -\int_r^\infty \frac{GMm}{r^2} \, dr = +\frac{GMm}{r}\bigg|_r^\infty = -\frac{GMm}{r}$$

The negative sign is characteristic of potential energy in an attractive force field. A system whose total energy ($\mathrm{KE} + \mathrm{PE}$) is less than zero is called a bound system. Can you think of other bound systems? The potential energy difference between a point on the surface of the planet $(r = R)$ and a point h meters above the planet $(r = R + h)$ is given by:

$$\Delta\text{PE} = \text{PE}\,(R + h) - \text{PE}(R) = -\frac{GMm}{R + h} - \left(-\frac{GMm}{R}\right)$$

$$\Delta\text{PE} = \frac{GMm}{R}\left(1 - \frac{1}{1 + h/R}\right)$$

The weight of a person on the planet is defined as the force of gravity at the surface of the planet,

$$mg = \frac{GMm}{R^2} \quad \text{or} \quad g = \frac{GM}{R^2} \quad \text{or} \quad \frac{GM}{R} = Rg$$

Thus

$$\Delta\text{PE} = mgR\left(1 - \frac{1}{1 + h/R}\right)$$

for $h/R \ll 1$, the binomial expansion of $(1 + h/R)^{-1} \simeq 1 - h/R$. Therefore, $\Delta\text{PE} \simeq mgR(1 - 1 + h/R) \simeq mgh$.

This is the equation used in this chapter, and we see that it is an approximation that is valid for $h/R \ll 1$. Since $R = 6371$ km for earth and $h \leq 10$ km, the height of earth's highest mountains, the condition is satisfied for many potential energy calculations on the earth.

The escape velocity for a bound system can be determined by setting the total energy equal to zero. This corresponds to the minimum kinetic energy to just free one particle from the system. Consider the escape velocity for a particle from the earth. On the surface of the earth the total energy of the particle is

$$E_{\text{total}} = \tfrac{1}{2}mv^2 - \frac{GMm}{R}$$

Setting this equal to zero, we get

$$\tfrac{1}{2}mv_{\text{escape}}{}^2 = \frac{GMm}{R}$$

$$v_{\text{escape}} = \left(\frac{2GM}{R}\right)^{1/2} \simeq 11.2 \text{ km/sec}$$

SUMMARY

Use these questions to evaluate how well you have achieved the goals of this chapter. The answers to these questions are given at the end of this summary with the section number where you can find the related content material.

Definitions

1. Mechanical energy is defined as
   a. heat
   b. ability to do work
   c. temperature
   d. force
   e. none of these
2. When a box is pushed up an incline, it gains
   a. weight
   b. inertia
   c. kinetic energy
   d. potential energy
   e. none of these
3. When a box slides down a frictionless incline, it gains
   a. weight
   b. inertia
   c. kinetic energy
   d. potential energy
   e. none of these
4. Power can be expressed in terms of force F, displacement s, and time t as

   a. $\dfrac{Ft}{s}$

   b. Fst

   c. $\dfrac{F}{st}$

   d. $\dfrac{st}{F}$

   e. $\dfrac{F \cdot s}{t}$

5. The lever system that is most common in the human body is
   a. first class
   b. second class
   c. third class
   d. none of these
6. The TMA of a lever system is equal to
   a. (force out)/(force in)
   b. (force in)/(force out)
   c. (output distance)/(input distance)
   d. (input distance)/(output distance)
   e. none of these
7. Efficiency of a machine is equal to
   a. (work in)/(work out)
   b. (force in)/(force out)
   c. (work out)/(work in)

   d. (force out)/(force in)
   e. none of these

Conservative and Nonconservative System

8. For a conservative force, the work done is independent of _____ .
9. For a nonconservative force, the work done between points A and B is determined by _____ .
10. The force of gravity acting on a box as it is pushed up an incline is a _____ (*conservative/nonconservative*) force.

Conservation of Energy

11. The principle of conservation of energy means that for a conservative system if E is total energy, KE = kinetic energy, and PE = potential energy, then the following is true
    a. $\Delta E/\Delta t = 0$
    b. $\Delta KE + \Delta PE = 0$
    c. $\Delta KE/\Delta t = 0$
    d. $\Delta PE/\Delta t = 0$
    e. none of these
12. For a simple pendulum of energy E at the bottom of its swing, the following is true:
    a. $KE = E$
    b. $KE = PE$
    c. PE is minimum
    d. KE is minimum
    e. KE is maximum
13. If an applied force for a lever system moves through 5 cm while the load moves through 1 cm, the TMA is _____ .
14. If the efficiency of the human body is 80 percent for an activity, an output energy of 3200 kcal requires an input energy of _____ kcal.

Power

15. The power expended by a bee is given as 1.25×10^{-3} watts and the velocity of the bee is 0.1 m/sec. The average force exerted by the bee must be _____ N.

Answers

1. b (Section 5.3)
2. d (Section 5.5)
3. c (Section 5.4)
4. e (Section 5.9)

5. c (Section 5.12)
6. d (Section 5.10)
7. c (Section 5.10)
8. path (Section 5.7)

9. path AB (Section 5.7)
10. conservative
    (Section 5.7)
11. a, b (Section 5.8)
12. a, c, e (Section 5.8)

13. 5 (Section 5.10)
14. 4000 kcal
    (Section 5.13)
15. 12.5×10^{-3} N
    (Section 5.9)

ALGORITHMIC PROBLEMS

Listed below are the important equations from this chapter. The problems following the equations will help you learn to translate words into equations and to solve single concept problems.

Equations

$$w = F(\cos \theta)s = \mathbf{F} \cdot \mathbf{s} \tag{5.1}$$

$$KE = \tfrac{1}{2}mv^2 \tag{5.2}$$

$$PE = mgh \tag{5.5}$$

$$PE = \tfrac{1}{2}kx^2 \tag{5.6}$$

$$U_1 + mgh_1 + \tfrac{1}{2}mv_1{}^2 = U_2 + mgh_2 + \tfrac{1}{2}mv_2{}^2 \tag{5.10}$$

$$P = \frac{\Delta w}{\Delta t} = \mathbf{F} \cdot \mathbf{v} \tag{5.12 and 5.13}$$

$$\text{efficiency} = \frac{\text{work done by machine}}{\text{energy supplied to machine}} \tag{5.14}$$

$$TMA = L/l \tag{5.19}$$
$$AMA = F_l/F_a \tag{5.20}$$

Problems

1. Find the work done in lifting a 20.0-kg box from the floor to a shelf 1.50 m above the floor.
2. If the work in problem 1 is done in 10.0 sec, find the power expended.
3. If the box in the first problem falls to the floor from the shelf, find the kinetic energy of the box as it hits the floor assuming idealized free fall.
4. Find the TMA of each lever system shown in Figure 5.13.

FIGURE 5.13
Algorithmic problem 4.

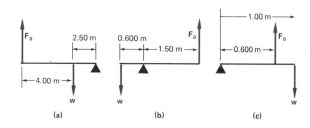

(a)   (b)   (c)

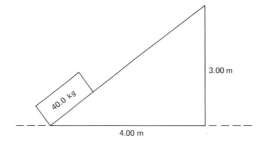

FIGURE 5.14
Algorithmic problem 6.

3.00 m

40.0 kg

4.00 m

5. A 2.00-kg ball is falling freely and passes an open window with a speed of 5.00 m/sec. Find the power expended by the ball as it passes the window.
6. a. Find the work required to push the 40.0-kg block up the imaginary frictionless inclined plane shown in Figure 5.14.
   b. Find the speed of the block at the bottom of the plane if it is released from rest at the top of the plane.

Answers

1. 294 J
2. 29.4 watts
3. 294 J
4. a. 2.60
   b. 2.50

c. 0.600
5. 98.0 watts
6. a. 1180 J
   b. 7.67 m/sec

EXERCISES

These exercises are designed to help you apply the ideas of a section to physical situations. When appropriate the numerical answer is given at the end of each exercise.

Section 5.2

1. How much work is done removing 1200 liters of water from a coal mine 130 m deep? [15.3×10^5 J]
2. Find the work done in drawing a 50.0-kg box up a plane 15.0 m long to a platform 9.00 m high if the coefficient of kinetic friction is 0.250. [5880 J]

Section 5.4

3. What is the kinetic energy of a 0.500-kg ball at the peak of its flight if it is projected at an angle of 37° with a velocity of 40.0 m/sec? (Assume ideal projectile motion.) How does this compare with original kinetic energy? How high was the ball at the peak of flight? [256 J; 16/25; 29.4 m]
4. At a low level of activity a heart ejected 3.00 kg of blood at 30.0 cm/sec. At a higher activity it ejects 6.00 kg at 60.0 cm/sec. What is the KE in each case?

How much work does the heart do at each activity? [0.135 J; 1.08 J]

Section 5.5

5. What is the potential energy of an 800-kg elevator car at the top of the Empire State Building, 380 m above the street level as base line? [2.98×10^6 J]
6. A vertical spring is 10.0 cm long when supporting a mass of 20.0 kg, and it is 12.0 cm long when supporting a 32.0-kg mass. Assume the stretch is linear with the load.
   a. What is the length of spring with zero load?
   b. How much work would be done in stretching the spring from 10.0 cm to 15.0 cm? [a. 6.67 cm; b. 17.1 J]

Section 5.10

7. A man raises a 250-kg stone with a 2.00-m bar. The distance from the stone to the fulcrum is 20.0 cm.
   a. What is the ideal mechanical advantage of the lever?

b. What force will the man have to exert? [a. 9.00; b. 272 N]

8. A lever system of the first class is 1.80 m long.
   a. If the load is 30.0 cm from the fulcrum, find the TMA for the system.
   b. If the actual load lifted is 40.0 N for an applied force of 10.0 N, what is the AMA?
   c. Find the efficiency of the system. [a. 5.00; b. 4.00; c. 80.0 percent]

9. A third-class lever system using a 1.80-m lever has an application arm of 0.300 m. Find its TMA and mechanical disadvantage (MD = (distance of load/input distance). [0.167, 6.00]

Section 5.13

10. The data show that in the Gemini space flight each astronaut expended about 2400 kcal/day. How much oxygen would be needed for an astronaut for the four-day flight? [1920 liters]

11. Assuming 100-g food packages of carbohydrates, fats, and proteins are available for the Gemini flight described in problem 10, find the minimum food allotment that will provide the energy as well as the 225 g of carbon and the 16.0 g of nitrogen needed by each astronaut each day. Find the excess energy that such a diet would supply each day. Use Table 5.4 for carbon and nitrogen calculations. [1 packet C; 2 packets F; 1 packet P; 200 kcal]

	Carbon per 100 g	Nitrogen per 100 g
Carbohydrates (C)	37.0	0.00
Fats (F)	79.0	0.00
Proteins (P)	53.0	16.0

PROBLEMS

Each problem may involve more than one physical concept. A problem that requires material from the Enrichment Section, 5.14, is marked by a dagger (†). The numerical answer is given at the end of each problem.

12. A 40.0-kg weight is pulled at constant velocity 10 m/sec over a rough horizontal surface. The coefficient of kinetic friction between the contact surfaces is 0.500. The pulling force is acting upward at an angle of 30° above the surface. How much work is done? [1520 J]

13. A 10.0-kg block is moving with a velocity of 2.00 m/sec on a frictionless horizontal surface. It is brought to rest by compressing a spring which has a spring constant of 4000 N/m. How much is the spring compressed? [10.0 cm]

14. A 1000-kg pile driver is raised 5.00 m above a pile and released. What is the average force exerted by the driver on the pile if the pile is driven 20.0 cm into the ground for each blow? [2.45×10^5 N]

15. There are three frictionless tracks from A to B (Figure 5.15). According to the principle of conservation of energy, the kinetic energy for equal masses sliding down each of the three tracks should be the same. Will all of them arrive at B at the same time if they start at the same time? If not, which one will arrive first? Will this time change if the masses are different? What is the magnitude of the velocity at B? How will it compare with the magnitude of the velocity of a free falling body for the same vertical height h? [no; 3; no; $\sqrt{2gh}$; the same]

16. A 1.00-kg hammer head is traveling 3 m/sec at the instant it strikes the nail. What is the average force acting on the nail if it goes into the wood 0.500 cm? [900 N]

17. A sled with its load has a mass of 60.0 kg and is started in motion by a boy who exerts a constant force of 19.6 N for a distance of 5.00 m. The coefficient of kinetic friction is 0.0200. Find the distance the sled travels after the boy stops pushing. Assume the motion takes place on a horizontal surface. [3.31 m]

18. At point A, a 0.500-kg mass is released and slides to B

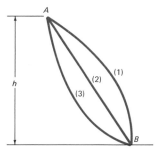

FIGURE 5.15
Problem 15.

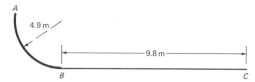

FIGURE 5.16
Problem 18.

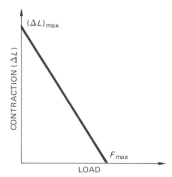

FIGURE 5.17
Problem 24.

over a frictionless track which is a quadrant of a circle with a radius of 4.90 m (Figure 5.16).

a. What is the velocity at B?

b. The track from B to C is rough, and the mass comes to rest in 9.80 m. What is the coefficient of friction of the track BC?   [a. 9.80 m/sec, b. 0.500]

19. Starting from rest, a block of 20.0-kg mass is pushed 20.0 m up a 37° inclined plane by a constant force of 200 N acting parallel to the surface. The coefficient of kinetic friction is 0.300. Find

a. the work done by the acting force

b. the increase in potential energy

c. the work done against friction

d. the increase in kinetic energy   [a. 4000 J; b. 2350 J; c. 940 J; d. 710 J]

20. When a quarterback throws a quick pass, he pushes on the ball with constant horizontal force for 0.100 sec, while his hand moves forward 0.800 m. Assume the mass of the ball is 0.900 kg. Find:

a. the acceleration of the ball

b. the force exerted on the ball

c. the work done on the ball

d. the power developed by the quarterback while throwing the ball   [a. 160 m/sec²; b. 144 N; c. 115 J; d. 1150 watts]

21. A 1000-kg automobile has a speed of 30.0 m/sec on a horizontal road when the engine is developing 35.0 kilowatts. For the same power and friction forces, what will be its speed on an incline of 1.00 m rise in 20.0 m of road?   [21.1 m/sec]

22. A nurse needs to exert a force (parallel to the incline) of 300 N on a wheelchair to move a 75.0-kg patient up a 1.00-m high, 4.00 m long ramp at a uniform rate. What are the TMA and the AMA of the incline, and how much work is done against friction? (*Hint:* by analogy with levers the TMA of the plane is equal to the length of the incline divided by the elevation of plane and $\dot{A}MA$ = (force out/force in) = efficiency × TMA)   [4.00, 2.45, 465 J]

23. A 75.0-kg man chins himself 20 times. The length of each pull is 0.500 m. Assume that no energy goes back into the muscle and that the body muscles are

10.0 percent efficient. How much work is done, and how much chemical energy was supplied to the muscle system?   [7350 J, 73,500 J]

† 24. Given the graph of isotonic (constant-load) contraction for a muscle as shown in Figure 5.17, show that the work done when the load is F is given by

$$\text{work done} = F\,(\Delta L)_{\text{max}} \left(1 - \frac{F}{F_{\text{max}}}\right)$$

where ΔL_{max} is the maximum contraction and F_{max} is the maximum load on the muscle. Using calculus, show that the work is maximum for the load equal to $\frac{1}{2} F_{\text{max}}$.

25. The graphic representation of the force which a muscle can exert in contracting from 8 cm to 2 cm is shown in Figure 5.18. What is the work that is done? (Make a linear approximation.)   [6 J]

† 26. The force to draw a bow is given by $F = 20.0 + 400s$ where F is the force in newtons and s is the draw in meters. How much work is done in drawing this bow 0.500 m?   [60.0 J]

27. In a cyclotron, a beam of protons (mass = m) moves in a circular orbit of radius R. An electromagnet pro-

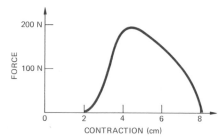

FIGURE 5.18
Problem 25.

vides the force F directed toward the center of revolution. Find the number of revolutions per second and the kinetic energy of each proton in terms of F, R, and m.

$$\left[n = \frac{1}{2\pi} \sqrt{F/mR}, \text{ KE} = \tfrac{1}{2} FR\right]$$

28. A low-friction toy car of mass m is released in the looped track shown in Figure 5.19. The height h is such that the car just makes the loop without falling off at the top. The track is frictionless and the stopping spring at the end of the track has a spring constant k. Find the relationship between h and R, the radius of the loop, for this condition. Find the distance x the spring at the end of the track is compressed in stopping the car. $[h = 2.50R, x = \sqrt{1.33 \ mgh/k}]$

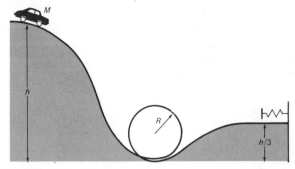

FIGURE 5.19
Problem 28.

GOALS

When you have mastered the contents of this chapter, you will be able to achieve the following goals:

Definitions
Define each of the following terms, and use it in an operational definition:

impulse	elastic collision
impulsive force	inelastic collision
momentum	rocket propulsion

Impulse Problems
Use the relationship between impulse and change in momentum to solve problems.

Conservation of Momentum
Explain the principle of conservation of momentum.

Collision Problems
State the difference in conditions between an elastic impact and an inelastic impact, and use both kinds of conditions to solve problems.

Momentum and Energy Problems
Apply the principles of conservation of momentum and conservation of energy to solve problems.

PREREQUISITES

Before beginning this chapter you should have achieved the goals of Chapter 3, Kinematics, Chapter 4, Forces and Newton's Laws, and Chapter 5, Energy.

6
Momentum and Impulse

6.1 Introduction

The manager of the major league baseball team, which had just won its tenth straight game, was quoted as saying, "Well, we've really got the momentum now. I think we can win the pennant."

After the Judiciary Committee of the House of Representatives had passed three articles of impeachment against President Richard M. Nixon in 1974, a spokesman for the House was reported to have said, "The momentum for impeachment is building up."

If you suddenly decided to take a long trip would your friends say that you were acting on an impulse? In fact, if you are noted for making and acting upon decisions quickly you are probably described by others as an impulsive person.

What are the connotations of these two words, momentum and impulse? How does a baseball team have momentum after ten victories and not after one victory? How does the action of a representative political body have momentum after three articles? Would it also have momentum after one article? And your sudden decision. Suppose you had waited twice as long to make up your mind. Would you still have been acting on an impulse?

The idea of momentum conveyed by the examples above does imply the motion of something and also has a quantitative quality. The idea of impulse conveys the feeling of quickness, of doing something in a short amount of time. How are such ideas related to the use of the definitions of momentum and impulse in physics? Are they similar or contradictory? We hope you will be able to answer that question when you finish this chapter.

6.2 Momentum and Impulse

Suppose we have two blocks moving to the right on a frictionless surface in the same line, with block A moving faster than block B (see Figure 6.1). Block A will eventually come in contact with block B. While they are in contact, block A is exerting a force $\mathbf{F}_{BA}$ on block B, and, in turn, block B is exerting a force $\mathbf{F}_{AB}$ on block A, but $\mathbf{F}_{AB}$ is in the opposite direction of $\mathbf{F}_{BA}$. In accord with Newton's third law of motion, these two forces are equal in magnitude and oppositely directed. One can write $\mathbf{F}_{AB} = -\mathbf{F}_{BA}$. These forces are not constant. Both before and after contact these forces are zero. The magnitude of the net forces varies with time by starting at zero at the instant of contact, (t_1), increasing to some maxi-

FIGURE 6.1
Two blocks moving along a frictionless surface showing condition before impact, during impact, and after impact.

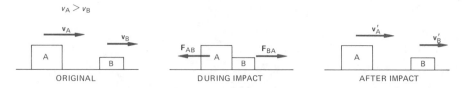

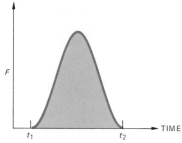

FIGURE 6.2
Representation of a force acting with a variable magnitude over a time period from t_1 to t_2. The impulse of the force is indicated by the area under the curve.

FIGURE 6.3
Wesley Fessler applying an impulsive force to a football. The change in momentum is equal to the impulse applied to the football. Why does a coach recommend "follow-through" in golf, tennis, baseball, and kicking a football? (Harold E. Edgerton, MIT, Cambridge, Mass.)

120 Flash per second photograph of a dub golf shot. Did the driver strike the ground before it hit the ball? Observe the motion of golf tee. Is it rotating? If so, in what direction, and about what? Note that the golf club appears to be bent after hitting the ball. Use some estimates of distances and calculate the velocity of the golf ball. (Harold E. Edgerton, MIT, Cambridge, Mass.)

mum value, and then decreasing again to zero at the instant the blocks lose contact. A plot of the magnitude of $\mathbf{F}_{AB}$ or $\mathbf{F}_{BA}$ as a function of time might be shown as in Figure 6.2. A force of this type, which is a function of time, is called an *impulsive force*. Can you name or give examples of impulsive forces? One example is kicking a football (see Figure 6.3).

The product of the net force and the time interval over which it acts is called the impulse of the force.

This is represented in Figure 6.2 as the area under the curve. Because $\mathbf{F}_{AB}$ and $\mathbf{F}_{BA}$ are equal in magnitude and opposite in direction, and the time of contact is the same for each of the bodies, the impulse of $\mathbf{F}_{AB}$ is equal in magnitude and opposite in direction to the impulse of $\mathbf{F}_{BA}$. If the force is constant, the area under the curve is given by the product of the magnitude of the force times the change in time, $\mathbf{F}(t_2 - t_1)$ where t_1 and t_2 are the times that contact begins and ends, respectively. The impulse of a constant force is given by $\mathbf{F}(t_2 - t_1)$.

The dimensions of impulse in terms of M, L, and T are MLT^{-1}. The units of impulse are newton-second (N-sec) in the SI system. Impulse is a vector quantity. You will recall that the unit of force (newton) is equal to mass times acceleration, force = (mass)(length)/(time)^2. Therefore, the units of impulse are (force)(time) or (mass)(length)/(time). But (length)/(time) is the definition of velocity. So impulse has the same dimensions as the product of mass times velocity. *We call this physical quantity the momentum of a body,*

$$\text{momentum of a body} = \mathbf{p} = m\mathbf{v} \qquad (6.1)$$

Momentum is a vector quantity whose SI units are kg-m/sec.

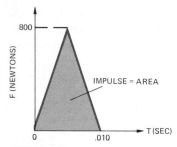

FIGURE 6.4
A triangular impulse.

Momentum and impulse are both vector quantities that are given by the product of a scalar quantity and a vector quantity. The direction of impulse is determined by the direction of the net force, and the direction of momentum is given by the direction of the velocity.

An impulse may be applied to a body to change its momentum. In fact the impulse applied to a body is equal to its change in momentum:

impulse of a force = change in momentum of the body upon which the force acts; this is a result of Newton's third law of motion,

$$\text{impulse} = \Delta \mathbf{p} \tag{6.2}$$

EXAMPLES

1. A body of mass 1.60 kg is acted upon by an impulsive force as shown in Figure 6.4. What is the impulse of the force? What is the change in momentum of the body and the change in velocity?

   The impulse of the force is represented by the triangular area. Thus

   $$\text{impulse} = \tfrac{1}{2}(\text{base of triangle})(\text{height})$$

   $$= \tfrac{1}{2} \times 800 \times .010 = 4.00 \text{ N-sec in the positive direction}$$

   The change in momentum is equal to the impulse, 4.00 N-sec.

   $$\Delta p = m \, \Delta v = 4.00 \text{ N-sec}$$

   $$1.60 \, \Delta v = 4.00 \text{ N-sec}$$

   $$\Delta v = 2.50 \text{ m/sec in the positive direction}$$

2. Short periods of rapid deceleration are characteristic of physical impacts. The current state of knowledge of human tolerances for impacts is not very complete. In fatal collisions a head injury is usually the cause of death. If the head is subjected to an impulse of about 100 N-sec in an abrupt collision, death may ensue. Consider a motorcyclist riding without a helmet to cushion his fall. Assume the mass of his head is 6 kg. If he has an abrupt collision, how slow must he be going to survive?

   $$100 \text{ N-sec} = \Delta p = 6 \text{ kg } \Delta v$$

   $$\Delta v = 16.7 \text{ m/sec, approximately 37 mph!}$$

3. Given the force-time graph for a sprinter's start as shown in Figure 6.5, we wish to find the change in speed between $t = 0.100$ sec and $t = 0.250$ sec for a sprinter with a mass of 75.0 kg.

   The impulse is the area under the curve $ABCD$, or the area of the closed polygon $ABCDEFG$. We can find this area by adding together the area of the triangle BCD, the area of the trapezoid $ABDE$, and the area of the rectangle $AEGF$. For the area of the triangle BCD, we obtain

   $$\text{impulse}_1 = \text{area}_1 = \tfrac{1}{2}(\text{base})(\text{height})$$

   where the base is BD and the height is given by 1900 N − 900 N.

FIGURE 6.5
A force versus time graph for a sprinter's start of a race.

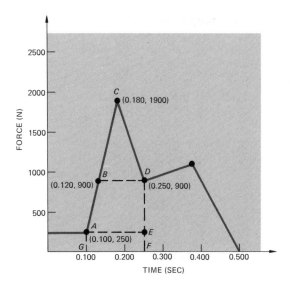

$$\text{impulse}_1 = \tfrac{1}{2}(0.130)(1000)$$

$$I_1 = 65.0 \text{ N-sec}$$

For the area of the trapezoid, we obtain

$$\text{impulse}_2 = \text{area}_2 = \tfrac{1}{2}(\text{height})(\text{base}_1 + \text{base}_2) = \tfrac{1}{2}(650 \text{ N})(0.130 \text{ sec} + 0.150 \text{ sec})$$

$$I_2 = 91.0 \text{ N-sec}$$

For the area of the rectangle *AEFG*, we obtain

$$\text{impulse}_3 = \text{area}_3 = (\text{base})(\text{height}) = (0.150 \text{ sec})(250 \text{ N})$$

$$I_3 = 37.5 \text{ N-sec}$$

$$\text{total impulse} = (65 + 91.0 + 37.5)\text{N-sec} = 193.5 \text{ N-sec}$$

$$\text{change in velocity} = \frac{\text{impulse}}{\text{mass}} = \frac{193 \text{ N-sec}}{75.0 \text{ kg}} = 2.58 \text{ m/sec}$$

4. A baseball with mass 142 g (weight, 5.5 oz) has a velocity of 29.4 m/sec when it is struck by a bat. After impact the ball is traveling with a velocity of 34.4 m/sec in a direction just opposite to the original velocity. What is the impulse imparted to the ball by the bat?

   Let us take the direction of the velocity after impact $\mathbf{v}_f$ as positive. The change in momentum is $m\mathbf{v}_f - m\mathbf{v}_i$ where $\mathbf{v}_i$ is the initial velocity.

   impulse of the bat = change in momentum

   Impulse of the bat = $m(\mathbf{v}_f - \mathbf{v}_i)$ but the direction of $\mathbf{v}_i$ is negative relative to $\mathbf{v}_f$, so in terms of the magnitudes of the velocities $\mathbf{v}_i$ and $\mathbf{v}_f$

$$\text{impulse} = m(v_f + v_i) = 0.142 \times (34.4 + 29.4)$$
$$= 0.142 \times 63.8 = 9.00 \text{ kg-m/sec} = 9.00 \text{ N-sec}$$

6.3 Conservation of Linear Momentum

As we stated earlier from Newton's third law,

impulse of the net $\mathbf{F}_{AB} = -$ impulse of the net $\mathbf{F}_{BA}$ (6.3)

Since the change in momentum is equal to the impulse of the force producing the change in momentum, we can rewrite Equation 6.3 in terms of momentum change, change of momentum of body A $= -$ (change of momentum of body B),

$$\Delta \mathbf{p}_A = -\Delta \mathbf{p}_B$$

$$m_A \Delta \mathbf{v}_A = -m_B \Delta \mathbf{v}_B \tag{6.4}$$

$$m_A \mathbf{v}_{Af} - m_A \mathbf{v}_{Ai} = -(m_B \mathbf{v}_{Bf} - m_B \mathbf{v}_{Bi}) \tag{6.5}$$

where m_A = mass of body A, m_B = mass of body B, $\mathbf{v}_{Ai}$ = initial velocity of body A, $\mathbf{v}_{Af}$ = final velocity of body A, $\mathbf{v}_{Bi}$ = initial velocity of body B, and $\mathbf{v}_{Bf}$ = final velocity of body B.

The above Equation 6.5 can be written with all the initial state terms on one side and the final state terms on the other.

$$m_A \mathbf{v}_{Ai} + m_B \mathbf{v}_{Bi} = m_A \mathbf{v}_{Af} + m_B \mathbf{v}_{Bf} \tag{6.6}$$

The left side of the above equation is the total momentum before collision, and the right side is the total momentum after collision, that is, the momentum before collision is equal to the momentum after collision.

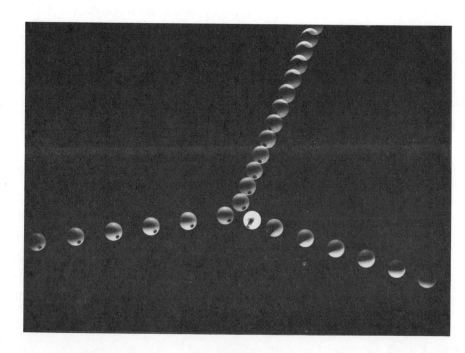

A multiflash photograph of an off-center collision between two balls of equal mass. The dotted ball is the incident ball, and the striped ball is initially at rest. Given that the mass of each ball is 173 g and that there are 30 flashes per second, can you prove that momentum was conserved in the collision? Was the collision perfectly elastic? What is your proof? (Picture from *PSSC Physics*, D.C. Heath & Company, Lexington, Mass., 1965.)

The relationship is called the *principle of conservation of linear momentum.*

The principle applies for any collision. Remember that momentum and impulse are vector quantities. Glancing collisions occur more frequently than head-on collisions, and the situation is complex because of the vector properties of momentum. The total momentum of a system can only be changed by the action of external forces on the system. The internal forces produce opposite and equal changes in momentum that cancel. Thus, we conclude that the *total momentum of an isolated system is constant in magnitude and direction.* This is equivalent to saying that the vector sum of the momenta before collision is equal to the vector sum of momenta after collision,

$$(\Sigma \; m_i \mathbf{v}_i)_{\text{before}} = (\Sigma \; m_t \mathbf{v}_t)_{\text{after}} \tag{6.7}$$

for an isolated system, where Σ indicates the process of adding the momentum of the individual bodies.

EXAMPLES

1. Suppose body A is moving in the x direction with velocity $\mathbf{v}_{Ai}$ and collides with body B which is at rest. Body A is deflected at angle θ above the x axis and body B is deflected on angle ϕ below the x axis after the collision (see Figure 6.6). The x components of momentum are equal before and after the collision,

    $$m_A \mathbf{v}_{Ai} = m_A \mathbf{v}_{Af} \cos \theta + m_B \mathbf{v}_{Bf} \cos \phi \tag{6.8}$$

    The initial y component of momentum is zero, so after the collision the y momentum components must add to equal 0,

    $$0 = m_A \mathbf{v}_{Af} \sin \theta - m_B \mathbf{v}_{Bf} \sin \phi \tag{6.9}$$

2. A body A of mass 2.00 kg and velocity of 3.00 m/sec collides with a body B of mass 3.00 kg that is at rest. After collision, A is moving with a velocity of 2.00 m/sec at an angle of 53° above the original direction. What is the velocity of B after the collision?

    Taking the original direction of $\mathbf{v}_A$ as the x direction, we have for the x components of momentum

    $$\mathbf{p}_i = \mathbf{p}_f$$

    Taking the original direction of $\mathbf{v}_A$ as the x direction, we have for the x components of momentum

    $$2.00(3.00) = 2.00(2.00)(0.600) + 3.00 v_B \cos \phi$$

    $$3.00 v_B \cos \phi = 6.00 - 2.40 = 3.60 \text{ kg-m/sec}$$

    For the y components of momentum, we have

    $$0 = 2.00(2.00)(0.800) - 3.00 v_B \sin \phi$$

    $$3.00 v_B \sin \phi = 3.20 \text{ kg-m/sec}$$

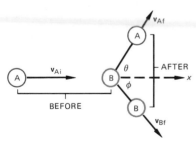

FIGURE 6.6
Body A collides with body B at rest. Body A goes off at an angle of θ and B goes off at an angle of ϕ, each with respect to the original direction of A.

$$\tan \phi = \frac{3.20}{3.60} = 0.889$$

$$\phi = 41.5°$$

Squaring the above equations for v_B and adding them together, we get

$$9.00v_B{}^2 \, (\sin^2 \phi + \cos^2 \phi) = (3.20)^2 + (3.60)^2$$

Because $\sin^2 \phi + \cos^2 \phi = 1$,

$$v_B{}^2 = \frac{(3.20)^2 + (3.60)^2}{9.00} = \frac{10.3 + 12.9}{9.00} = 2.58 \ \text{m}^2/\text{sec}^2$$

$$v_B = 1.61 \ \text{m/sec}$$

6.4 Collisions

In any collision the total momentum is always conserved. However, the same is not always true of kinetic energy. If the kinetic energy of the bodies involved in a collision is the same before and after impact, the collision is said to be a *perfectly elastic collision*. If the bodies stick together after impact, the collision is said to be *completely inelastic*. These represent the two extreme cases. All intermediate cases are possible and are called *partially elastic collisions*.

We have seen many examples of a partially elastic collision. Consider a collision between two automobiles, see Figure 6.7. Momentum is conserved, but the kinetic energy before the collision is greater than the kinetic energy after the collision. Evidence of this is the change in shape of the vehicles as a result of the collision. These "transformations" require an expenditure of energy which comes from the kinetic energy of the system.

The only known perfectly elastic collisions are those between atomic and nuclear particles, and not all of these collisions are perfectly elastic. A collision within your experience that is nearly perfectly elastic is the one between two billiard balls. For a perfectly elastic collision between particles A and B, one can write

FIGURE 6.7
The collision between two automobiles, a partially elastic collision.

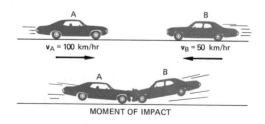

$v_A = 100 \ \text{km/hr}$ $v_B = 50 \ \text{km/hr}$

MOMENT OF IMPACT

TOTAL MOMENTUM IS CONSTANT, BUT
$KE_{initial} > KE_{final}$

$$m_A \mathbf{v}_{Ai} + m_B \mathbf{v}_{Bi} = m_A \mathbf{v}_{Af} + m_B \mathbf{v}_{BF} \qquad \text{conservation of momentum} \qquad (6.6)$$

If $U_1 = U_2$, $mgh_1 = mgh_2$, in Equation 5.10, one can obtain:

$$\tfrac{1}{2} m_A v^2_{Ai} + \tfrac{1}{2} m_B v^2_{Bi}$$
$$= \tfrac{1}{2} m_A v^2_{Af} + \tfrac{1}{2} m_B v^2_{Bf} \qquad \text{conservation of kinetic energy}$$

where the subscripts A and B designate the two particles in collision, the subscript i designates the velocity before collision, and the subscript f designates the velocity after collision. These are two independent equations and can be solved in general to find two unknown quantities, such as the two final velocities.

Let us consider the special case of a two-particle head-on, elastic collision where one body of mass m_B is initially at rest ($\mathbf{v}_{Bi} = 0$). We can rearrange these equations to have all m_A terms on one side and all m_B terms on the other side of the equal sign.

$$m_B \mathbf{v}^2_{Bf} = m_A(\mathbf{v}^2_{Ai} - \mathbf{v}^2_{Af}) \qquad \text{conservation of energy} \qquad (6.10)$$

$$m_B \mathbf{v}_{Bf} = m_A(\mathbf{v}_{Ai} - \mathbf{v}_{Af}) \qquad \text{conservation of momentum} \qquad (6.11)$$

For head-on collisions the velocities are all along the same line of action. We can treat the velocities as positive or negative scalars. Thus we have two equations and two unknowns, and hence a solution is possible. One approach is to eliminate the masses from the two equations by dividing Equation 6.10 by Equation 6.11. The result is given by the following equality between the final velocity of particle B and the sum of the initial and final velocities of particle A,

$$v_{Bf} = v_{Ai} + v_{Af} \qquad (6.12)$$

This expression for v_{Bf} can be substituted into Equation 6.11 to obtain a general relationship between the final and the initial velocities of particle A,

$$v_{Af} = \frac{m_A - m_B}{m_A + m_B} v_{Ai} \qquad (6.13)$$

This relationship is valid for all two particle head-on collisions when one of the objects is initially at rest. We can use this expression to eliminate v_{Af} from Equation 6.11 and obtain a relationship between the initial velocity of particle A and the final velocity of particle B,

$$v_{Bf} = \frac{2m_A}{m_A + m_B} v_{Ai} \qquad (6.14)$$

EXAMPLE OF A PERFECTLY ELASTIC HEAD-ON COLLISION

A neutron (mass = 1.67×10^{-27} kg) collides head on with a nitrogen nucleus at rest (mass = 23.1×10^{-27} kg). The initial velocity of the neutron is 1.50×10^7 m/sec. Find the final velocity of the neutron, and the final velocity of the nitrogen nucleus.

To find the final velocity of the neutron, we can use Equation 6.13.

$$v_{Af} = \frac{m_A - m_B}{m_A + m_B} v_{Ai} = \frac{(1.67 - 23.1) \times 10^{-27} \text{ kg}}{(23.1 + 1.67) \times 10^{-27} \text{ kg}} \times (1.50 \times 10^7 \text{ m/sec})$$

$$v_{Af} = -1.30 \times 10^7 \text{ m/sec}$$

The negative value for v_{Af} in this case, means that the direction of the velocity of the neutron is reversed by the collision.

To find the final velocity of the nitrogen nucleus, we can use Equation 6.14.

$$v_{Bf} = \frac{2m_A}{m_A + m_B} v_{Ai} = \frac{2(1.67 \times 10^{-27} \text{ kg})}{(23.1 + 1.67) \, 10^{-27} \text{ kg}} \times (1.50 \times 10^7 \text{ m/sec})$$

$$v_{Bf} = 2.02 \times 10^6 \text{ m/sec}$$

Let us consider a completely inelastic collision between two objects, A and B. The objects stick together and travel off with a velocity of $\mathbf{v}_f$.

$$m_A \mathbf{v}_{Ai} + m_B \mathbf{v}_{Bi} = (m_A + m_B) \, \mathbf{v}_{Fi} \qquad \text{conservation of momentum} \qquad (6.15)$$

$$\tfrac{1}{2} m_A v^2{}_{Ai} + \tfrac{1}{2} m_B v^2{}_{Bi} - \tfrac{1}{2} (m_A + m_B) v^2{}_f = \text{loss of kinetic energy} \qquad (6.16)$$

EXAMPLE

A 2.00-g rifle bullet is fired into a 2.00 kg block of a ballistic pendulum. The pendulum is 4.00 m long, and it rises 3.60 cm as a result of impact (Figure 6.8). Find the speed of the bullet, and the loss of kinetic energy. Let m_a = mass of bullet, m_b = mass of block, $\mathbf{v}_{ai}$ = initial velocity of bullet, $\mathbf{v}_f$ = velocity of block and bullet, and $\mathbf{v}_{bi}$ = initial velocity of the block = 0.

Upon impact momentum is conserved so

$$m_a v_{ai} = (m_b + m_a) v_f$$

After impact, energy is conserved so

$$\tfrac{1}{2} (m_b + m_a) v_f{}^2 = (m_b + m_a) gh$$

Then $v_f{}^2 = 2gh$.

Putting in the numerical values in the last equation,

$$v_f{}^2 = 2 \times 9.80 \times .0360 = 0.704$$

$$v_f = 0.840 \text{ m/sec}$$

Then from conservation of momentum, with numerical values, we obtain

FIGURE 6.8
A bullet impacts with the wooden bob of a ballistic pendulum.

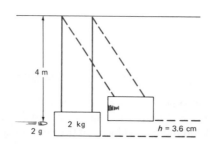

4 m

2 g   2 kg   h = 3.6 cm

$$2v_{ai} = (2000 + 2.00)(0.840)$$

$$v_{ai} = 1001.00 \times 0.840 = 841 \text{ m/sec}$$

$$\text{loss in KE} = \tfrac{1}{2} m_a v_{ai}^2 - \tfrac{1}{2}(m_b + m_a)v_f^2$$

$$= \tfrac{1}{2}(2.00 \times 10^{-3}) \times (841)^2 - \tfrac{1}{2}(0.002)(0.841)^2$$

$$= 707 - 0.71 = 706 \text{ J}$$

6.5 Ballistocardiography

The conservation of momentum law suggests that as the blood is pumped from the heart the body should recoil. The recording of this recoil is called ballistocardiography. We want to look at the physical basis for such measurements.

While measurements are being made, the patient is placed on a "frictionless" platform. Since there are no external forces acting, the total momentum change is zero. We write the following equation for the blood momentum and the momentum for the rest of the body system,

$$m_b \mathbf{v}_b + m_B \mathbf{v}_B = 0 \qquad (6.17)$$

thus $\qquad m_B \mathbf{v}_B = -m_b \mathbf{v}_b \qquad (6.18)$

where subscript b represents blood and subscript B represents the body. Let Δt represent the period of the pump stroke of the heart (time of contraction). We can derive the equation in terms of the distance of body movement. Multiplying both sides of Equation 6.12 by the stroke time Δt and recalling that $v \, \Delta t$ is just the displacement,

$$m_b \mathbf{v}_b \, \Delta t = -m_B \mathbf{v}_B \, \Delta t \quad \text{or} \quad m_b \, \Delta \mathbf{x}_b = -m_B \, \Delta \mathbf{x}_B \qquad (6.19)$$

where $\mathbf{x}_b$ is the blood displacement and $\mathbf{x}_B$ is the body displacement during the time of heart contraction. We have the relationship between the displacement of the body and the displacement of the blood,

$$\Delta \mathbf{x}_B = \frac{m_b}{m_B} \Delta \mathbf{x}_b \qquad (6.20)$$

We can estimate the body displacement by using some typical data. The mass of blood per heart stroke is about 60 g, and we will assume a body mass of 50 kg. A reasonable estimate of the movement of the center of mass of the blood for a contraction can be obtained by using the average velocity of blood flow from the heart 50 cm/sec and a typical contraction time of 0.13 sec. This gives a blood displacement of 6.5 cm. Substituting these values into Equation 6.20 we get

$$\Delta \mathbf{x}_B = \frac{60 \text{ g}}{50 \times 10^3 \text{ g}} (-6.5 \text{ cm}) = -7.8 \times 10^{-3} \text{ cm}$$

This small displacement is detectable with the sensitive displace-

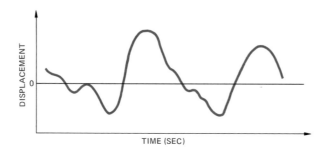

ment transducers available today. A typical ballistocardiogram of displacement versus time is shown in Figure 6.9.

6.6 Rocket Propulsion

Rocket propulsion is a practical example of the conservation of momentum. The momentum of the rocket is initially zero. When the stream of exhaust gases start out the back of the rocket, they have a momentum in the backward direction. For the total momentum of the rocket and fuel system to be conserved, the rocket must be given an equal and opposite momentum.

To simplify the quantitative calculations, let us construct a two-particle model of a rocket. One particle is determined by the rocket ship, the engine, the remaining fuel and the fuel storage tanks. The other particle is the fuel burned and ejected in a given time period (Figure 6.10). We use the conservation of momentum to determine the exit velocity of the fuel $\mathbf{v}_e$, the total thrust exerted by the fuel on the rocket ship, $\mathbf{T}$, and the final velocity of the rocket ship $\mathbf{v}_f$. The initial momentum of the system is given by $(m_r + m_f)\mathbf{v}_i$ where $\mathbf{v}_i$ is the initial velocity of the rocket and the fuel, m_r is the mass of the rocket, m_f is the mass of the fuel. Then the fuel is burned and ejected in a time of t seconds, the ejected

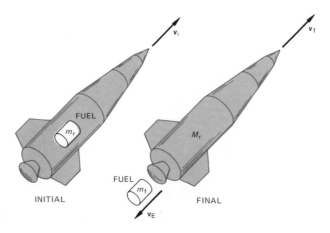

fuel will have an ejection velocity v_e in a direction opposite to v_i. The ejected fuel will exert a thrust upon the rocket, which will attain a final velocity v_f. The final momentum must be equal to the initial momentum,

$$(m_r + m_f)v_i = m_r v_f - m_f v_e \tag{6.21}$$

where the minus sign indicates that v_e is opposite in direction to v_i and v_f, but along the same line, so we dropped the vector notation. The change in momentum of the rocket ship is equal to the impulse imparted to the rocket ship by the ejected gases. Since we know the time of fuel ejection, we can calculate the thrust by dividing the impulse by the time,

net impulse imparted to the rocket ship $= m_r(v_f - v_i)$

$$\text{net thrust} = \frac{\text{impulse}}{\text{time}} = \frac{m_r(v_f - v_i)}{t} \tag{6.22}$$

EXAMPLE

The following data are for a three-stage Saturn rocket launched in 1971 for a trip to the moon. The first-stage fuel burn of liquid oxygen and hydrogen lasted for 165 sec. The initial velocity was zero. The initial total mass of rocket and fuel was 2.91×10^6 kg. The mass of the rocket and its velocity at the end of the burn were 8.36×10^5 kg and 2.32×10^3 m/sec. What is the velocity of the ejected fuel, and what is the average net thrust applied to the rocket ship?

From Equation 6.21 we can write an equation for the velocity of the ejected fuel:

$$v_e = \frac{m_r v_f - (m_r + m_f)v_i}{m_f}$$

$$v_e = \frac{(8.36 \times 16^5 \text{ kg})(2.32 \times 10^3 \text{ m/sec}) - (2.91 \times 10^6) \text{ kg} \times 0 \text{ m/sec}}{(2.91 \times 10^6 - 8.36 \times 10^5) \text{ kg}}$$

$$v_e = 9.37 \times 10^2 \text{ m/sec}$$

The average thrust can be found from Equation 6.22:

$$\text{thrust} = \frac{(8.36 \times 10^5)\text{kg} \times (2.32 \times 10^3 - 0) \text{ m/sec}}{165 \text{ sec}}$$

$$= 1.18 \times 10^7 \text{ N}$$

SUMMARY

Use these questions to evaluate how well you have achieved the goals of this chapter. The answers to these questions are given at the end of this summary with the number of the section where you can find the related content material.

Definitions

1. The impulse of a force is equal to the change of resulting
   a. energy
   b. momentum
   c. potential energy
   d. kinetic energy
   e. all of these
2. The momentum of a car is defined as _____ and is a vector in the direction of _____.
3. Elastic collisions conserve both _____ and _____.
4. Inelastic collisions are characterized by the objects' _____ and conservation of _____.
5. Jet, or rocket, propulsion is possible because the momentum of the vehicle equals the _____.

Impulse Problems

6. If a 10.0-N force acts for 1.00×10^{-2} sec on a 0.100-kg ball at rest the resulting change in velocity is_____.
7. The average force in a head-on collision with a wall of a 0.100-kg ball moving at 10.0 m/sec is 10.0 N. Find the total contact time for the impact with the wall. $\Delta t =$ _____. (Assume elastic collision.)

Conservation of Momentum

8. The condition necessary for conservation of momentum in a given system is that
   a. energy is conserved
   b. one body is at rest
   c. no external force acts
   d. internal forces equal external forces
   e. none of these
9. The superposition principle applied to a system of particles leads to the momentum of the system equal to _____.

Collision Problems

10. The greatest impulse is involved in collisions that are

   a. perfectly inelastic
   b. partially elastic
   c. perfectly elastic
   d. same for all
11. When a 0.1-kg block moving at a speed of 10.0 m/sec collides perfectly inelastically with a 0.400-kg block at rest, the final speed of the blocks is
   a. 33.3 m/sec
   b. 3.33 m/sec
   c. 2.5 m/sec
   d. 2 m/sec
   e. none of these

Momentum and Energy Problems

12. The energy lost in problem 11 is what percent of initial kinetic energy?
13. If the block at rest in problem 11 is attached to a spring with a force constant of 100 N/m, find the distance the spring is compressed by the inelastic collision.
14. In a head-on inelastic collision between two cars, which of the cars will suffer the greater damage—the heavier or lighter vehicle? (Assume both have same momentum before collision.)

Answers

1. b (Section 6.2)
2. $m\mathbf{v}$, $\mathbf{v}$ (Section 6.2)
3. momentum, energy (Section 6.4)
4. sticking together, momentum (Section 6.4)
5. exhaust momentum (Section 6.6)
6. 1 m/sec (Section 6.2)
7. 2×10^{-1} sec (Section 6.2)
8. c (Section 6.3)
9. sum of $m_i \mathbf{v}_i$ (Section 6.3)
10. c (Section 6.4)
11. d (Section 6.4)
12. 80 percent (Section 6.4)
13. 14.1 cm (Section 6.4)
14. lighter car. It has the greatest change in energy (Section 6.4)

ALGORITHMIC PROBLEMS

Listed below are the important equations from this chapter. The problems following the equations will help you to translate words into equations and to solve single concept problems.

Equations

$$\text{impulse} = \mathbf{F}t = \Delta(m\mathbf{v}) = m(\mathbf{v}_f - \mathbf{v}_i) = \Delta\mathbf{p} = \text{change in momentum} \quad (6.2)$$

$$m_A\mathbf{v}_{Ai} + m_B\mathbf{v}_{Bi} = m_A\mathbf{v}_{Af} + m_B\mathbf{v}_{Bf} \quad (6.6)$$

$$(\Sigma m_i\mathbf{v}_i)_{\text{before}} = (\Sigma m_i\mathbf{v}_i)_{\text{after}} \quad (6.7)$$

$$\tfrac{1}{2}m_A\mathbf{v}_{Ai}{}^2 + \tfrac{1}{2}m_B\mathbf{v}_{Bi}{}^2 - \tfrac{1}{2}(m_A + m_B)\mathbf{v}_f{}^2 = \text{loss of KE} \quad (6.16)$$

Problems

1. A tennis ball makes a perfectly elastic collision with a wall and bounces straight back with a speed of 10.0 m/sec. The mass of the ball is 100 g and the time of contact with the wall is 1.00×10^{-2} sec. Find the force acting on the ball exerted by the wall.
2. A force of 100 N acts for 1.00×10^{-3} sec on a 1.00-kg mass. Find the change in momentum of the mass.
3. A boy (mass = 40.0 kg) runs at a speed of 5.00 m/sec and jumps onto a sled (mass = 10.0 kg). Find the speed of the boy and the sled.
4. a. Find the impulse needed to stop a car (mass = 1000 kg) moving at 5.00 m/sec.
   b. Find the force needed to bring this car to rest in 2.00 sec.
5. A railroad car (mass = M) moving at 10.0 m/sec overtakes another car (mass = M) moving in the same direction with a speed of 4.00 m/sec. Find the velocity of the cars after they collide and couple together.
6. Find the fraction of kinetic energy lost in the collision of problem 5. Where did this energy go?

Answers

1. 200 N
2. 0.1 kg-m/sec
3. 4.00 m/sec
4. a. 5000 N-sec

    b. 2500 N
5. 7.00 m/sec
6. 0.155

EXERCISES

These exercises are designed to help you apply the ideas of a section to a physical situation. When appropriate, the numerical answer is given at the end of each exercise.

Section 6.2

1. A 5.00-kg rifle fires a 10.0 g bullet with a speed of 600 m/sec. What is the speed of recoil of the rifle? If the rifle is stopped in 2.00 cm, what is its deceleration and stopping force? [1.20 m/sec, -36.0 m/sec^2, 180 N]
2. A 142-g baseball arrives at a bat with a speed of 40.0 m/sec. It is in contact with the bat for 0.020 seconds, and it leaves the bat going in the opposite direction with a speed of 80.0 m/sec. What is the impulse of the bat on the ball? What is the average force of the bat on the ball? [17.0 N-sec, 850 N]
3. A staple gun fires 10 staples into a wall. Each staple has a mass of 0.400 g and a velocity 80.0 m/sec. What is the impulse acting on the wall? [0.320 N-sec]

Section 6.4

4. Two balls, A (mass 2.00 kg) and B (mass of 2.50 kg),

approach each other head-on with speeds of 6.00 and 10.0 m/sec respectively. Assume an elastic collision.

What is the speed of each ball after the collision? [$v_{Af} = -11.8$ m/sec, $v_{Bf} = 4.22$ m/sec]

PROBLEMS

Each problem may involve more than one physical concept. The answer is given at the end of the problem.

5. A 60.0-g ball A is traveling west with a speed of 1.00 m/sec and collides with ball B (mass 60.0 g) at rest. After the collision ball B moves in a direction 30° west of south. Assume the collision is perfectly elastic. What is the direction of ball A, and the speed of each? [30° north of west, 0.500 m/sec, 0.866 m/sec]

6. An alpha particle of mass 6.66×10^{-27} kg has a velocity of 1.00×10^7 m/sec when it collides head-on with an oxygen nucleus at rest, and with mass four times that of an alpha particle. What is the velocity of each after collision? [4.00×10^6 m/sec, -6.00×10^6 m/sec]

7. An elastic head-on collision occurs between a bombarding mass m and a mass M at rest. Assume the bombarding mass has kinetic energy K. Calculate the kinetic energy of m, K_m, and M, K_M, after impact for five different cases: $m = 0.01M$; $m = 0.1M$; $m = M$; $m = 10M$; $m = 100M$. What conclusions can you reach?

[*Answer*]

m = 0.01M	0.1M	M	10M	100M
K_m/K (99/101)²	(9/11)²	0	(9/11)²	(99/101)²
K_M/K (20/101)²	(40/121)	1	(40/121)	(20/101)²

8. A completely inelastic head-on collision occurs between a bombarding mass m and a target of mass M at rest. Assume the mass m has kinetic energy K. Calculate the kinetic energy after impact for cases: $m = 0.01M$; $m = 0.1M$; $m = M$; $m = 10M$; $m = 100 M$. What conclusion can you reach? [0.00990K, 0.091K, 0.500K, 0.909K, 0.990K]

9. A 1500-kg automobile is going west on Main Street at 15.0 m/sec when it collides with a 4000-kg truck going south on Pine Street at 8.00 m/sec. They become coupled together in the collision. What is the velocity after impact, and how much kinetic energy is lost in the collision? [7.1 m/sec, 35.1° west of south, 1.6×10^5 J]

10. A 10.0-g bullet is fired into a 5.00-kg wood block which is suspended by long cords so that it can swing as a ballistic pendulum.

The impact of the bullet raises block's center of mass 5.00 cm. What was the speed of the bullet? How much kinetic energy is lost? What becomes of the lost kinetic energy? [496 m/sec, 1.23×10^3 J]

11. Two pendulum balls are pulled apart until both are raised 40.0 cm above equilibrium position. When they are released, they collide head-on at the bottom of their swing. If they stick together, find how high they swing and the energy that goes into heating them up after the collision. The masses of the balls are 300 g and 200 g. [1.60 cm, 1.88 J]

12. Three blocks each of mass m are located on a frictionless track as shown in Figure 6.11. After block A is released it makes an inelastic collision with block B and the two of them in turn collide with and stick to block C. Find the final velocity of the three blocks as they move off together. Find the total energy converted to heat in this entire process. [$\frac{2}{3}\sqrt{gh}/2$, $mgh/3$]

13. A neutron of mass 1.67×10^{-27} kg and a velocity of 2.00×10^4 m/sec makes a head-on, completely inelastic, collision with a boron nucleus, mass 17.0×10^{-27} kg, originally at rest. What is the velocity of the new nucleus and its final kinetic energy? What is the loss of kinetic energy in this collision? [1.79×10^3 m/sec, 2.99×10^{-20} J, 3.04×10^{-19} J]

14. An arrow of mass M is shot into a block (mass = $19M$) on the end of a cord 1.00 m in length. The arrow sticks in the block and the combination swings through an arc of 37°. What properties of the system are conserved? Find the speed of the arrow when it hit the block. [momentum, $v_i = 39.6$ m/sec]

15. During the second stage of its launch a Saturn rocket burns 4.52×10^5 kg of fuel in 391 sec. The mass of rocket and fuel at the beginning of the second stage is 6.66×10^5 kg. The average net thrust was 2.32×10^6 N. Use the two particle model to find the velocity

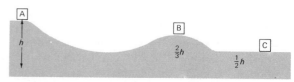

FIGURE 6.11
Problem 12.

of the ejected fuel. What was the final velocity of the rocket if the velocity at the beginning of stage two was 2.73×10^3 m/sec   [$+7.23 \times 10^2$ m/sec in the same direction as v_i, 6.96×10^3 m/sec]

16. If the average force acting on a basketball is equal to its weight when it bounces on the floor and returns to 81 percent of its initial height, H, find the time of contact for the ball on the floor.   [$2.68\sqrt{H/g}$]

17. A rubber raft of mass $3M$ floats past a dock at 1 m/sec toward the east. A girl (mass = $2M$) runs down the dock and jumps into the raft with a speed of 2 m/sec in the north direction. Find the final velocity of the raft and girl and the fraction of kinetic energy lost in this collision.   [$v_E = 0.6$ m/sec, $v_N = 0.8$ m/sec, 54.5 percent KE lost]

GOALS

When you have mastered the contents of this chapter, you will be able to achieve the following goals:

Definitions
Define each of the following terms, and use it in an operational definition:

angular displacement
angular velocity
angular acceleration
uniformly accelerated angular
  motion
torque

center of mass
moment of inertia
rotational kinetic
  energy
angular momentum

Equilibrium
State the conditions for static equilibrium.

Rotational Motion
Write the equations for rotational motion with constant angular acceleration.

Rotational Kinematics
Solve problems for systems with a fixed axis of rotation using the principles of rotational kinematics.

Rotational Dynamics
Solve problems using the principles of rotational dynamics, for systems with fixed axes of rotation, including conservation of energy and conservation of angular momentum.

Equilibrium Problems
Solve problems involving conditions of static equilibrium.

PREREQUISITES

Before beginning this chapter you should be familiar with Chapter 4, Forces and Newton's Laws, and Chapter 5, Energy. The quantitative aspects of rotational motion are very similar to those of kinematics (Chapter 3).

7

Rotational
Motion

7.1 Introduction

Ever since you got up this morning you have been interacting with rotating objects, the clock hands were turning, the door knob was twisted, the water faucet was turned, the door was opened, the automobile or bicycle wheels rolled, and even if you stopped, the earth spun you through space at a high rate of speed. In fact, many people feel that going around in circles is a more common phenomena than going straight.

Since you have had considerable experience with making objects go around in circles, you already know much about rotational motion. How do you characterize the motion of a rotating object? How do you specify how fast it is rotating? Kilometers per hour hardly seems like an appropriate unit since the object really is not going any place. How do you get an object to start rotating? What are the essential features of rotation?

7.2 Angular Motion: Rotational Kinematics

The hands on the face of your watch are excellent examples of angular motion. It is upon the basis of angular displacement of the watch hands that you tell time. In one minute the second hand goes through one complete revolution, an angular displacement of 2π radians (1 revolution or 360°). See Figure 7.1. For the same period of time the minute hand has an angular displacement of one-sixtieth of a revolution, or $\pi/30$ radians. Similarly, the hour hand has an angular displacement of $\pi/360$ radians in one minute. The angle between two positions of the watch hand defines an *angular displacement*. Angular displacement is assigned a direction parallel to the axis of rotation. The direction of the angular displacement can be found by using the *right-hand rule*. Point the fingers of your right hand in the direction of increasing angular displacement θ. Then the thumb of your right hand points in the direction of the angular displacement vector (see Figure 7.2). The fundamental unit for angular displacement is the radian. In Figure 7.1, the direction for the angular displacement of the minute or hour hand is *into* the page.

The *average angular velocity* is defined as the time rate of change of

FIGURE 7.1
The hands of a clock rotate about an axis at the center of the clock face. Each hand has its own constant angular velocity. The angular velocity for the second hand is $2\pi/60$ rad/sec, that of the minute hand is $2\pi/(60 \times 60)$ rad/sec. What is the angular velocity for the hour hand?

FIGURE 7.2
Right-hand rule to determine direction of θ.

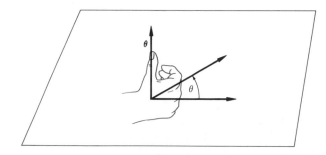

A golf club-ball impact at 1000 flashes per second. Note the rotation of the ball. The rate of rotation is about 125 per second. Can you verify? (Harold E. Edgerton, MIT, Cambridge, Mass.)

angular displacement. Thus,

$$\boldsymbol{\omega}_{\text{ave}} = \frac{\boldsymbol{\theta}_2 - \boldsymbol{\theta}_1}{t_2 - t_1} = \frac{\Delta\boldsymbol{\theta}}{\Delta t} \quad \text{with direction given by a right-hand rule} \qquad (7.1)$$

where $\boldsymbol{\theta}_1$ is the displacement at time t_1, $\boldsymbol{\theta}_2$ is the displacement at a later time t_2, $\Delta\boldsymbol{\theta}$ represents the change in the vector angular displacement in radians $\boldsymbol{\theta}_2 - \boldsymbol{\theta}_1$ in time Δt seconds, $t_2 - t_1$. Angular velocity is a vector quantity and has the units of radians per second (rad/sec). The simplest angular motion occurs if the angular velocity $\boldsymbol{\omega}$ is constant, that is

$$\Delta\boldsymbol{\theta} = \boldsymbol{\omega}\Delta t \qquad (7.2)$$

Many of our appliances have electric motors that start from rest and soon reach a constant angular velocity for continuous operation. During the start-up period there is a change of angular velocity which is called angular acceleration. Angular acceleration is defined as the time rate of change of angular velocity. Thus,

$$\boldsymbol{\alpha}_{\text{ave}} = \frac{\boldsymbol{\omega}_2 - \boldsymbol{\omega}_1}{t_2 - t_1} = \frac{\Delta\boldsymbol{\omega}}{\Delta t} \quad \text{with direction given by that of } \Delta\boldsymbol{\omega} \qquad (7.3)$$

in which $\Delta\boldsymbol{\omega}$ represents the change in angular velocity in radians per second in time Δt in seconds. Angular acceleration is a vector quantity and has units of rad/sec². Instantaneous angular velocity and acceleration are treated in the enrichment section, Section 7.11.

Let us develop the relationship for angular motion in which angular acceleration α is constant. From our definition, the angular acceleration is the change in angular velocity divided by the time in which the change occurs. The magnitude of the angular acceleration is given by the following equation:

$$\alpha = \frac{\omega_f - \omega_0}{t}$$

where ω_f is the magnitude of the angular velocity at time t, and ω_0 is the magnitude of the initial angular velocity. We can solve this equation for the angular velocity at time t,

$$\omega_f = \omega_0 + \alpha t \tag{7.4}$$

The magnitude of the average angular velocity during time t is the sum of the final and initial values of angular velocity divided by two,

$$\omega_{ave} = \frac{\omega_f + \omega_0}{2} \tag{7.5}$$

The magnitude of the change in angular displacement in time t is found by combining Equation 7.1 with Equation 7.4:

$$\Delta\theta = \theta_f - \theta_0 = \omega_{ave}t = \left(\frac{\omega_f + \omega_0}{2}\right)t = \omega_0 t + \tfrac{1}{2}\alpha t^2 \tag{7.6}$$

where the magnitude of the displacement at the times t and zero are given by θ_f and θ_0, respectively.

Another relationship, which is often used, may be obtained by eliminating t in the combination of Equations 7.4 and 7.6:

$$2\alpha(\theta_f - \theta_0) = \omega_f^2 - \omega_0^2 \tag{7.7}$$

If an object starts at the zero position and at rest, $\theta_0 = 0$ and $\omega_0 = 0$, these equations reduce to the following:

$$\omega_f = \alpha t \qquad \theta = \tfrac{1}{2}\alpha t^2 \qquad 2\alpha\theta = \omega_f^2 \tag{7.8}$$

In the chapter on kinematics we discussed the linear velocity and acceleration of a particle moving in a circle. For a rigid body rotating about a fixed axis, every point in the body is moving in a plane perpendicular to the axis of rotation and in a circle whose center is the axis. Let us now develop some relationships between the angular velocity and the angular acceleration of a rotating body and the linear velocities and linear accelerations of points within it. We shall consider only the special case of rotation about a fixed axis. Suppose we have a rigid body rotating about a fixed axis O. See Figure 7.3. The point P represents any point in the body at a distance r from the axis of rotation. The path of P is a circle in the plane of O and P, which is perpendicular to the axis through O. The distance through which P travels, s, is given by the product of the radial distance r and the magnitude of the angle of displacement θ in radians,

$$s = r\theta \tag{7.9}$$

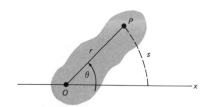

FIGURE 7.3
A rigid body may rotate about a fixed axis so that each part of the body has the same angular displacement and the same angular velocity, but the linear speed of each particle depends upon its distance from the axis of rotation.

As r is constant, an increment Δs is related to an increment of θ as follows:

$$\Delta s = r\Delta\theta \tag{7.10}$$

where $\Delta\theta$ is in radians.

If the body rotates through an angle $\Delta\theta$ in time Δt, the point P moves a distance of Δs in the same time. We can divide both sides of Equation 7.10 by Δt to obtain

$$\frac{\Delta s}{\Delta t} = r\frac{\Delta\theta}{\Delta t} \tag{7.11}$$

But by definition, $\dfrac{\Delta s}{\Delta t} = v_{\text{ave}}$ and $\dfrac{\Delta\theta}{\Delta t} = \omega_{\text{ave}}$ so Equation 7.10 can be re-written as

$$v_{\text{ave}} = r\omega_{\text{ave}} \tag{7.12}$$

By calculus we can show a similar relationship holds for instantaneous linear velocity and instantaneous angular velocity,

$$v_{\text{inst}} = r\omega_{\text{inst}} \tag{7.13}$$

Can you show that an object rolling without slipping has a speed of $r\omega$? If the object is not slipping, then the point of contact and the surface on which it is rolling are at rest with respect to one another. Then the center of the object is a distance r from the point of contact and rotating with an angular velocity of ω with respect to the contact point. Hence, the velocity of the center is given by $r\omega$. If ω is constant, the angular acceleration is zero. Earlier it was shown that a particle moving in a circle at a uniform rate experiences an acceleration toward the center of the circle, called radial acceleration. In the chapter on kinematics we found the magnitude of the radial acceleration to be given by Equation 3.21,

$$a_r = \frac{v^2}{r} \tag{3.21}$$

If we substitute $r\omega$ for v in this expression, we get an expression for the radial acceleration in terms of the angular velocity,

$$a_r = \omega^2 r \tag{7.14}$$

If the angular velocity is not constant, the point P is not rotating at a constant rate but has an acceleration in the direction of motion, tangential to the circle, as well as in the radial direction. This acceleration is called the *tangential acceleration*. To find an expression for the tangential acceleration, we begin with Equation 7.13 and change both v and ω in a time Δt, so $\Delta\omega$ is the change in angular velocity and Δv is the corresponding change in velocity in the time interval Δt,

$$\frac{\Delta v}{\Delta t} = r\frac{\Delta\omega}{\Delta t} \tag{7.15}$$

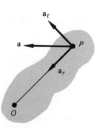

FIGURE 7.4
For a rigid body rotating about a fixed axis at a nonuniform rate, each particle has a tangential and radial acceleration. The total acceleration is the vector sum of the tangential and radial acceleration.

where $\dfrac{\Delta v}{\Delta t}$ is the average tangential acceleration a_t and $\dfrac{\Delta \omega}{\Delta t}$ is the average angular acceleration α. The scalar form of Equation 7.15 is,

$$a_t = r\alpha \tag{7.16}$$

an equation that applies to the motion of an object that rolls without slipping. The resultant acceleration **a** of P is then the vector sum of $\mathbf{a}_r$ and $\mathbf{a}_t$ (see Figure 7.4).

$$\mathbf{a} = \mathbf{a}_r + \mathbf{a}_t \tag{7.17}$$

EXAMPLES

1. Suppose you are riding on a merry-go-round which is making 3.00 revolutions per minute and you are moving in a circle with a radius of 6.00 m. What is your linear speed?

$$\omega \ (\text{rad/sec}) = \frac{\Delta\theta}{\Delta t} = \frac{3.00 \times 2\pi}{60} = \frac{\pi}{10.0} \ \text{rad/sec} \tag{7.1}$$

$$v = \omega r = \frac{\pi}{10}(6.00) = \frac{6.00\pi}{10.0} = 1.88 \ \text{m/sec} \tag{7.12}$$

If the merry-go-round is speeded up to 6.00 rpm in 1.00 minute what is the angular acceleration? What is the tangential acceleration during this minute? What would be the total acceleration when the merry-go-round was going 5.00 rpm?

The angular acceleration can be calculated by using the definition of α:

$$\alpha = \frac{\Delta\omega}{\Delta t} = \frac{12.0\pi/\text{min} - 6.00\pi/\text{min}}{60 \ \text{sec}} \times \frac{1 \ \text{min}}{60 \ \text{sec}}$$

$$= \frac{12.0\pi - 6.00\pi}{60 \times 60} = \frac{6.00\pi}{3600} = \frac{\pi}{600} = 5.24 \times 10^{-3} \ \text{rad/sec}^2 \tag{7.3}$$

The tangential acceleration can be obtained from Equation 7.16:

$$a_t = r\alpha = 6.00 \times \frac{\pi}{600} = \frac{\pi}{100} = 3.14 \times 10^{-2} \ \text{m/sec}^2 \tag{7.16}$$

$$\omega = \frac{(5.00)(2\pi)}{60.0} = \frac{\pi}{6.00} = 5.24 \times 10^{-1} \ \text{rad/sec}$$

$$a_r = \omega^2 r = \frac{\pi^2 \ \text{rad}^2}{(6.00)^2 \ \text{sec}^2}(10.0 \ \text{m}) = \frac{10\pi^2}{36.0} = 2.74 \ \text{m/sec}^2 \tag{7.14}$$

The magnitude of the total acceleration is equal to vector sum of $\mathbf{a}_r$ and $\mathbf{a}_t$,

$$a = \sqrt{\left(\frac{10\pi^2}{36}\right)^2 + \left(\frac{\pi}{100}\right)^2} = 2.74 \ \text{m/sec}^2$$

2. One reads on an electric motor name plate that its operating speed is 1800 rpm. Assume you find that it takes 10.0 sec to reach its operating speed. Find the final angular displacement during start-up. Assume constant angular acceleration.

$$1800 \text{ rpm} = 30.0 \text{ rps} \qquad \omega_f = 30.0 \text{ rps} = 60.0\pi \text{ rad/sec}$$

$$\alpha = \frac{\omega_f - \omega_0}{t} = \frac{(60.0\pi - 0)\text{rad/sec}}{10 \text{ sec}} = 6.00\pi \text{ rad/sec}^2$$

$$\theta = \tfrac{1}{2}\alpha t^2 = \tfrac{1}{2} \times 6.00\pi \text{ rad/sec}^2 \times 100 \text{ sec}^2 = 300\pi \text{ rad}$$

$$\theta = \frac{300\pi}{2\pi} = 150 \text{ revolutions}$$

If you recall your study of kinematics you will notice the great similarity between the concepts we have developed to explain rotational motion and the ones used in Chapter 3 for linear motion. If you mentally bend linear motion into a circular track, it would be the same as rotational motion. In the table below we display the various equations side-by-side to emphasize their similarities.

Linear Kinematics		Rotational Kinematics	
$\mathbf{v_f} = \mathbf{v_0} + \mathbf{a}t$	(3.4)	$\omega_f = \omega_0 + \alpha t$	(7.4)
$\mathbf{s} = \left(\dfrac{\mathbf{v_f} + \mathbf{v_0}}{2}\right)t$	(3.6)	$\theta = \left(\dfrac{\omega_f + \omega_0}{2}\right)t$	(7.6)
$\mathbf{s} = \mathbf{v_0}t + \tfrac{1}{2}\mathbf{a}t^2$	(3.7)	$\theta = \omega_0 t + \tfrac{1}{2}\alpha t^2$	(7.6)
$2\mathbf{a} \cdot \mathbf{s} = \mathbf{v_f}^2 - \mathbf{v_0}^2$	(3.9)	$2\alpha \cdot \theta = \omega_f^2 - \omega_0^2$	(7.7)
		$a_t = \text{tangential acc} = \alpha r$	(7.16)
		$a_r = \text{radial acc} = \omega^2 r$	(7.14)

7.3 Torque

Let us look at the physics of opening a hinged door (see Figure 7.5). As part of life's experiences you have learned to carry out this operation, but can you explain the physics of the operation? You observe that there are hinges on one edge of the door, and there is a latch on the other edge. The type of motion that occurs as a door is opened is rotation about a fixed axis. Assume that the latch is released. You know that you must

FIGURE 7.5
One can open an unlatched door by applying a force not acting through the hinges. For this condition there is a torque acting about the hinges. The magnitude of the torque depends upon the force and the point of application relative to the hinges. Why do you usually apply the force perpendicular to the door and near the edge opposite to the hinges?

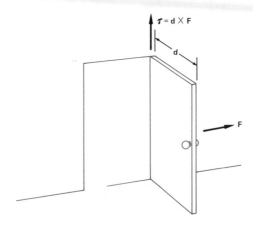

exert a force on the door to open it. Where must the force be applied, and what is its direction? By doing experiments you would find that you cannot open the door by applying the force through the hinges — regardless of the size of the force. This seems to indicate that there is a most desirable position to apply the force. You probably say to apply the force in a direction perpendicular to the door, but not through the axis of the hinge. Again, upon the basis of experience, you would probably apply the force near the edge opposite the hinges. Why? In order to produce a rotational motion you must apply a *torque*. The magnitude of the torque or *moment of force* applied to open the door is given by the product of the force and its lever arm.

The lever arm is equal to the perpendicular distance from the axis of rotation (the hinge of the door) *to the direction of the force.*

The magnitude of the torque of force **F** about the axis of the hinge is given by

$$\tau = Fd \tag{7.18}$$

where τ is the magnitude of the torque, F is the magnitude of the force applied and d is the perpendicular distance from the force to the hinge. The dimensions of torque are ML^2T^{-2}, and the units are newton-meters in the SI system, or dyne-centimeters in the cgs system. The torque of a force is a vector quantity as it has both magnitude and direction. The direction of rotation is either clockwise or counterclockwise. Torque is represented by a vector in which the line of action is through the axis of rotation, in the case of the door it is the hinges. The length of the vector represents the magnitude. Its direction is indicated by clockwise motion as one looks along the axis of rotation. Once again the right-hand algorithm can be used for finding the direction of the torque vector. Point the fingers of your right hand in the direction of the force with your

A string is wrapped around the spindle of a spool and a torque is applied. The direction of rotation when the string is pulled is shown by the arrow. Can one pull on the string so that there will be no rotation? If so, what is the direction of the string?

FIGURE 7.6
A torque is a vector product of two vectors. To determine the direction of the torque vector one uses the right hand. If the fingers of the right hand go around the axis in the direction of rotation, the thumb will then point along the axis in the direction of torque vector.

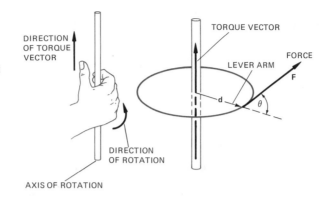

palm open toward the axis of rotation. Then your right thumb points in the direction of the torque (see Figure 7.6). The torque in the figure would be represented by a vector pointing upward with a length proportional to $Fd \sin\theta$.

7.4 Center of Mass and Equilibrium

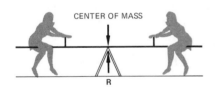

FIGURE 7.7
A teeter-totter with its center of gravity at the balance point.

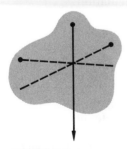

FIGURE 7.8
The intersection of various vertical (plumb-bob) lines establishes the center of gravity.

In your younger days you probably improvised a teeter-totter board. You tried to "balance" it on the axis of support. In its balanced position the torque tending to produce counterclockwise motion is equal to the torque tending to produce clockwise motion. A vertical line (direction of plumb-bob line) through the axis of support will pass through the *center of mass* of the body. The center of mass of a rigid body is its balance point. For the teeter-totter board the support acts upward with a force of **R** which is equal to the weight of the board. See Figure 7.7. We can say the force of gravity (weight) produces zero torque about the center of mass. Thus for a body, you can consider all of its weight as acting at the center of mass. For a uniform body, this coincides with the geometric center. You can determine the center of mass of a body experimentally by finding its point of balance. You can also determine the center of mass of a body such as a plate by hanging the body at different positions and establishing plumb-bob lines for each point of suspension. The point of intersection of the plumb-bob lines determines the center of mass of the body (see Figure 7.8).

In a formal way the center of mass of a system consisting of n particle masses is determined by equating the sum of the mass moments of the particles of a system to the moment of the total mass located at the center of mass for a given axis of rotation. The *moment of a point mass* is defined as the product of the mass times the distance from the point to the axis of rotation,

$$m_1r_1 + m_2r_2 + m_3r_3 + \cdots m_nr_n = \sum_{i=1}^{n} m_ir_i = M\bar{r} \tag{7.19}$$

FIGURE 7.9
Four equal masses are distributed along the horizontal axis. The location of the center of gravity is shown.

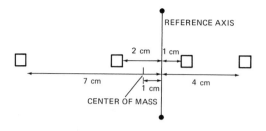

where the symbol $\sum\limits_{i=1}^{n}$ means to add the terms from the first to the nth.

The term m_i is the mass of the ith particle, r_i is the distance of the ith particle from the reference axis, $\bar{r}$ is the location of the center of mass of the system from the reference axis, and M is the total mass of all the n particles in the system:

$$M = \sum_{i=1}^{n} m_i$$

For example, if you are given four equal masses m located as shown in Figure 7.9, you can find the center of the mass by adding

$$m \times 1.00 \text{ cm} + m \times 4.00 \text{ cm} + m \times (-2.00 \text{ cm}) + m \times (-7.00 \text{ cm}) = -4\, m\bar{r}$$

$$\bar{r} = -1.00 \text{ cm}$$

that is, the center of the mass is 1 cm to the left of the reference axis. It is important to note that the choice of axis location does not change the center of mass of a system. Consider the axis through the mass at the far left. One then has

$$0.00\, m + m \times 5.00 \text{ cm} + m \times 8.00 \text{ cm} + m \times 11.0 \text{ cm} = 4.00 m\bar{r}$$

$$24.0 m \text{ cm} = 4.00 m\bar{r}$$

$$\bar{r} = 6.00 \text{ cm}$$

Hence, the center of mass is 6 cm to the right of the far-left mass. This is exactly the same position as indicated in the first calculation.

EXAMPLE

Suppose you have a uniform 4.00–m beam weighing 500 N, and it has a load of 200 N suspended 1.00 m from the left end, one of 300 N at 3.00 m from the left end, and that the system is supported by one hanger (see Figure 7.10).

Where should the hanger be placed? It must be at the center of mass of the system.

Take the left end of the beam as a reference point, then, for moments about O,

$$\bar{r} = \frac{(200 \text{ N})(1.00 \text{ m}) + (500 \text{ N})(2.00 \text{ m}) + (300 \text{ N})(3.00 \text{ m})}{(200 \text{ N} + 500 \text{ N} + 300 \text{ N})} = 2.1 \text{ m}$$

The center of mass of the system is at 2.1 m from the left end.

FIGURE 7.10
A uniform (500 N) beam supporting two masses can be supported in equilibrium by one hanger.

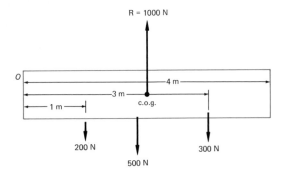

We are now ready to consider the conditions necessary for equilibrium.

A system is said to be in a state of equilibrium if: (1) The sum of the forces acting on the system in any direction is zero (no translational acceleration) (see Chapter 4). (2) The sum of the torques acting on the system (moment of forces) about any axis is zero (no angular acceleration).

EXAMPLES

1. A uniform 4.00-m plank which weighs 200 N is to be used for a teeter-totter board for two girls—one weighing 300 N and one weighing 240 N. Each is to be seated 0.250 m from the end (see Figure 7.11). For equilibrium, where should the fulcrum be placed and how much force must it exert?

   Make a free body diagram. For a condition of equilibrium, **R**, the upward force, must be equal to the total of the downward forces = 240 + 200 + 300 = 740 N. The fulcrum must support 740 N. Consider A as the point about which moments are to be considered. For the second condition of equilibrium to be satisfied, the clockwise torques must be equal to the counterclockwise torques. Then,

   $$(240 \text{ N})(0.250 \text{ m}) + (200 \text{ N})(2.00 \text{ m}) + (300 \text{ N})(3.75 \text{ m}) = 740 \, \bar{r}$$

   $$\bar{r} = \frac{1585}{740} = 2.15 \text{ m}$$

   The fulcrum should be placed at 2.15 m from the end of the plank near the smaller girl.

FIGURE 7.11
Two girls being taken for a ride by a teeter-totter. What conditions must be met for operation?

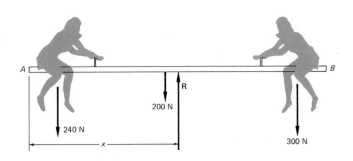

FIGURE 7.12
(a) A diagram of a simple boom crane. (b) The free body diagram of the boom crane. (c) The forces acting at point B on the boom crane.

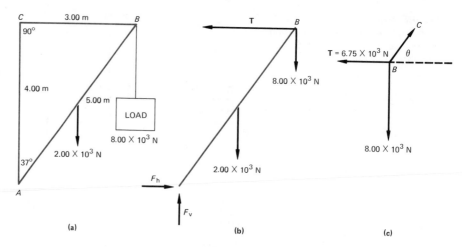

2. A crane is used to support a load. The uniform boom AB weighs 2.00×10^3 N, and when the boom makes an angle of 37° with the vertical, the tie CB is horizontal (see Figure 7.12). Find the tension in the cable, the compression in the boom AB, and the reaction force $\mathbf{F}$ at A. Is the reaction force at A along the boom?

We begin by making a free body diagram of the boom as in Figure 7.12b. The force equations are:

$F_v = 2.00 \times 10^3$ N $+ 8.00 \times 10^3$ N $= 1.00 \times 10^4$ N   for vertical components

$F_h = T$ for horizontal components

Set up a moment equation about A:

$(2.0 \times 10^3$ N$)(1.50$ m$) + (8.00 \times 10^3$ N$)(3.00$ m$) = (4.00$ m$)T$

$$T = \frac{3.00 \times 10^3 \text{N-m} + 2.40 \times 10^4 \text{N-m}}{4.00 \text{ m}} = 6.75 \times 10^3 \text{N}$$

Then

$F_h = 6.75 \times 10^3$ N

$F = \sqrt{F_v{}^2 + F_h{}^2} = \sqrt{(1.00 \times 10^4 \text{ N})^2 + (6.75 \times 10^3 \text{ N})^2}$

$F = 1.21 \times 10^4$ N

$\tan \theta_F = \dfrac{F_v}{F_h} = \dfrac{1.00 \times 10^4}{6.75 \times 10^3} = 1.48$

$\theta_F = 56°$

$\mathbf{F}$ is not along the boom but is at a greater angle with the horizontal. To get the compression C in the boom, let us consider the forces acting at B (see Figure 7.12c).

$C \sin \theta = 8.00 \times 10^3$ N $C \cos \theta = 6.75 \times 10^3$ N

$C = \sqrt{(8.00 \times 10^3 \text{ N})^2 + (6.75 \times 10^3 \text{ N})^2}$

$C = 1.05 \times 10^4$ N

FIGURE 7.13
(a) The human leg and its support systems. (b) The free body diagram of the leg.

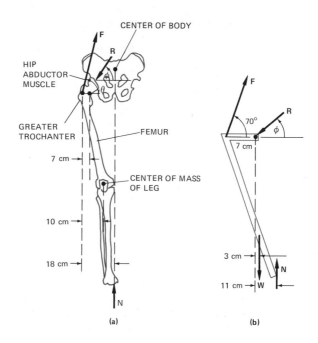

(a)   (b)

3. The human leg provides another example of an equilibrium problem. Let's assume you are standing on your right leg. See Figure 7.13a. Consider the acetabulum and upper end of the femur as a frictionless ball and socket joint. $\mathbf{F}$ is the force acting on the greater trochanter by the hip abductor muscle at an angle θ with the horizontal, and $\mathbf{R}$ is the force the acetabulum exerts on the head of the femur. Given $\theta = 70°$. Find the magnitude of $\mathbf{F}$ and $\mathbf{R}$.

The leg is in equilibrium under the following forces: the normal forces of floor on your foot $\mathbf{N}$, the weight of the leg $\mathbf{W_L}$, the force $\mathbf{F}$, and the force $\mathbf{R}$. Let us draw a free body diagram of the leg. Figure 7.13b.

In order for you to stand on one leg, the force of the floor on your foot must be upward through your center of gravity and equal to your weight. We can write three equations for the equilibrium conditions.

We can resolve $\mathbf{R}$ into its horizontal and vertical components.

$R_h - F \cos 70° = 0$ horizontal components

$N - W_L - R_v + F \sin 70° = 0$ vertical components

As the joint is frictionless and round, the force $\mathbf{R}$ acts through the center of spherical surface and makes an angle ϕ with the horizontal. Set up the moment equation about this point,

$N \times 11.0 \text{ cm} = W_L \times 3.00 \text{ cm} + (F \sin 70°) \times 7.00 \text{ cm}$

These are three equations that can be solved if there are only three unknowns. In order to complete a numerical solution for F and R you need to know your weight and the weight of your leg. Let's assume your weight is 600 N (mass of about 60.0 kg) and that your right leg weight is 15 percent of your total weight, or about 90.0 N. Substituting these values in moment equation we can solve the problem:

$$600 \text{ N} \times 11.0 \text{ cm} = 90.0 \text{ N} \times 3.00 \text{ cm} + 7.00 \text{ cm} \times 0.940 \, F$$

$$F = 962 \text{ N} \qquad \text{the force exerted by the abductor muscle}$$

$$R_h = 962 \times 0.326 = 313 \text{ N}$$

$$R_v = 680 - 90 + 962 \times 0.940 = 1.43 \times 10^3 \text{ N}$$

$$\tan \phi = \frac{R_v}{R_h} = \frac{1.43 \times 10^3 \text{ N}}{313 \text{ N}} = 4.58$$

$$\phi = 77.6°$$

$$R = \sqrt{R_h{}^2 + R_v{}^2} = \sqrt{(313 \text{ N})^2 + (1.43 \times 10^3 \text{ N})^2}$$

$$= 1.47 \times 10^3 \text{ N} = \text{force exerted by the acetabulum.}$$

7.5 Rotational Dynamics

You may recall that in Chapter 4 you found that if an unbalanced force is acting upon a body, an acceleration is produced (Newton's second law of motion). If there is an unbalanced torque acting upon a body, there is an angular acceleration produced. In fact, one can state a law which is parallel to Newton's second law of motion:

If an unbalanced torque is acting upon a body, an angular acceleration is produced that is proportional to the torque and in the direction of the torque.

If you apply an equivalent torque to a different body, you may find that a different value of angular acceleration is produced. The value for α for a given torque depends upon the distribution of the mass of the body relative to the axis of rotation. You can write an equation setting the torque equal to the product of a constant of proportionality times the angular acceleration,

$$\tau = I\alpha \tag{7.20}$$

where I is the *moment of inertia* of a body and its value depends upon the distribution of its mass relative to the axis of rotation. If we have a system made up of point masses $m_1, m_2, \ldots m_n$ as shown in Figure 7.14 the moment of inertia of this system about O is defined by the following equation:

$$I_O = m_1 r_1{}^2 + m_2 r_2{}^2 + m_3 r_3{}^2 + \cdots + m_n r_n{}^2 \tag{7.21}$$

where $r_1, r_2, r_3, \ldots r_n$ are the distances of the various point masses from the axis of rotation, O. If we consider the moment of inertia of a system about a different axis of rotation, we will generally find it to have different moment of inertia. In general, we can write

$$I = \sum_{i=1}^{n} m_i r_i{}^2 \tag{7.22}$$

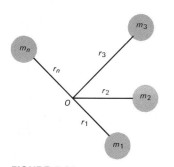

FIGURE 7.14
A number of masses, $m_1, m_2, m_3, \ldots,$ m_n distributed at various distances, $r_1, r_2, r_3, \ldots, r_n,$ from the pivot point O. The property of these masses to resist rotation depends upon their distribution relative to axis of rotation.

FIGURE 7.15
Various regularly shaped uniform
solids and their moments of inertia.

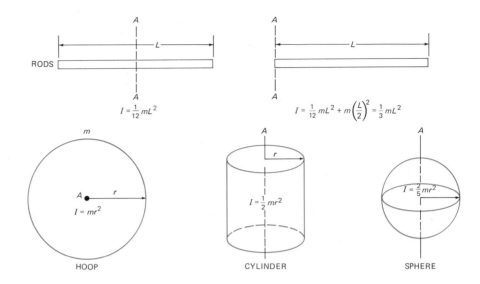

RODS

$I = \frac{1}{12}mL^2$

$I = \frac{1}{12}mL^2 + m\left(\frac{L}{2}\right)^2 = \frac{1}{3}mL^2$

$I = mr^2$

HOOP

$I = \frac{1}{2}mr^2$

CYLINDER

$I = \frac{2}{5}mr^2$

SPHERE

where the symbol sigma

$$\sum_{i=1}^{n}$$

means to add the terms from the first to the *n*th, where each term is given by the product of a point mass times the square of its distance from the axis of rotation.

For a hoop rotating about its geometric center (Figure 7.15), its moment of inertia is mr^2 where *m* is its mass, and *r* is its radius:

$$I = mr^2$$

The moment of inertia has the dimensions of ML^2 and units of kg–m² in the SI system. For solid bodies one uses the methods of calculus to determine the moment of inertia of a body. About the geometric axis the moment of inertia of a solid cylinder is $\frac{1}{2}mr^2$ and for a solid sphere it is $2/5\ mr^2$. However, the moment of inertia has a different value about any other axis. The moment of inertia about the axis parallel to the axis of rotation through the center of mass is given by

$$I = I_G + md^2 \tag{7.23}$$

(known as the parallel axis theorem) where I_G is the moment through the center of mass, I is the moment of inertia about a parallel axis and at a distance of *d* from the center of mass axis.

EXAMPLES

1. Suppose that we have a solid disk of mass 5.00 kg and a radius 0.200 m mounted on frictionless bearings. A constant force of 5.00 N is applied by a

light ribbon wrapped around the rim of the disk. What is the angular acceleration of the disk?

$$I = \tfrac{1}{2} mr^2 = \tfrac{1}{2}(5.00 \text{ kg})(0.200 \text{ m})^2 = 0.100 \text{ kg–m}^2$$

$$\tau = TR = I\alpha$$

$$5.00 \text{ N} \times 0.200 \text{ m} = (0.100 \text{ kg–m})^2 \alpha$$

$$\alpha = 10.0 \text{ rad/sec}^2$$

What mass hanging on the ribbon would produce the same angular acceleration? The tension in the ribbon is to be 5.00 N.

$$T = mg - ma$$

$$a = r\alpha = 0.200 \text{ m} \times 10.0 \text{ rad/sec} = 2 \text{ m/sec}^2$$

$$T = 5.00 \text{ N} = m \times (9.80 - 2.00) \text{ m/sec}^2 = m \times 7.80 \text{ m/sec}^2$$
$$m = 0.640 \text{ kg}$$

2. Recent developments of high-speed flywheels have led to the use of flywheels as energy storage devices. One such use involves using flywheels in each car of a train to store energy during the braking process. This energy increases the speed of the flywheel during braking and then this energy can be used in accelerating the train when it starts up again. Let us consider some typical data. We want to find the moment of inertia of the flywheel when a torque of 17.0π N–m is applied to accelerate it uniformly from 10,200 rpm to 15,000 rpm in 80.0 sec.

$$\alpha = \frac{\Delta\omega}{\Delta t} = \frac{4800 \dfrac{\text{rev}}{\text{min}} \times 2\pi \dfrac{\text{rad}}{\text{rev}} \times \dfrac{1 \text{ min}}{60 \text{ sec}}}{80.0 \text{ sec}} = 2.00\pi \text{ rad/sec}^2$$

$$\text{torque} = I\alpha = I \times 2.00\pi \text{ rad/sec}^2 = 17.0\pi \text{ N–m}$$

$$I = 8.50 \text{ kg–m}^2$$

7.6 Rotational Work

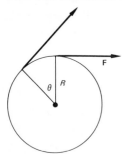

FIGURE 7.16
Work is done in rotation by a torque τ acting through an angle θ. It is illustrated by a force acting perpendicular to the radius of a wheel, $\tau = FR$.

Work is defined as the product of force times the distance through which it acts. If the force acts at a perpendicular distance R from the axis of rotation (Figure 7.16), the distance through which **F** acts is $R\theta$ where θ is the angle of rotation in radians. So we compute the work as the product of the magnitude of the force times $R\theta$, but force times R is equal to the magnitude of the torque τ,

$$\text{work} = FR\theta$$

or in vector form,

$$= \boldsymbol{\tau} \cdot \boldsymbol{\theta} \qquad\qquad (7.24)$$

You notice the similarity between this equation and our previous Equation 5.1 for computing the work done by a force

$$w = \mathbf{F} \cdot \mathbf{s}$$

EXAMPLE

How much work does a girl do in pedaling a bicyle for one minute if the pedal arm is 20.0 cm, a constant force of 10.0 N is exerted on the pedal, and the pedals make 60.0 revolutions per minute? Assume the force is active on each pedal for one-half of the time.

The magnitude of the torque acting to produce rotation is

$\tau = FR = 10.0 \text{ N} \times 0.200 \text{ m} = 2.00 \text{ N-m}$

$\theta = 2\pi \times 60.0 \text{ rpm} = 120\pi \text{ rad/min}$

$w = \tau\theta = 2.00 \text{ N-m} \times 120\pi \text{ rad/min} = 240\pi \text{ joules/min}$

$w = 754 \text{ J in one minute}$

7.7 *Kinetic Energy of Rotation*

If the torque is producing an angular acceleration, there is a change in the kinetic energy of the rotating body. We can derive an expression for the kinetic energy of rotation,

$$\text{Work} = \tau\theta = I\alpha\theta = \text{the change in kinetic energy} \tag{7.25}$$

From the equations of uniform angular acceleration we know that the square of the angular velocity is equal to twice the product of the angular acceleration times the angle,

$$\alpha\theta = \frac{\omega_f^2}{2} - \frac{\omega_0^2}{2}$$

If we substitute this expression for $\alpha\theta$ in Equation 7.25 we obtain another expression for the change in the kinetic energy of rotation:

$$\text{change in KE}_{\text{rotation}} = \tfrac{1}{2}I(\omega_f^2 - \omega_0^2) \tag{7.26}$$

Again, note the parallel to the expression for the change in the kinetic energy of translation, $\text{KE} = \tfrac{1}{2}m(v_f^2 - v_0^2)$.

A rolling solid is an example of a body that has both linear and angular motions simultaneously. Specifically, let us consider the case of the front wheel on a moving bicycle. The wheel rotates about its axle and the axle advances in a forward direction. In such cases it is possible to consider the kinetic energy in two parts:

1. that due to the translation of the center of mass, that is, $\tfrac{1}{2}mv^2$
2. that due to rotation about an axis through the center of mass, that is, $\tfrac{1}{2}I\omega^2$.

Thus the total kinetic energy of a rolling body is given by the sum of the kinetic energy of its translation, plus the rotational kinetic energy.

$$\text{KE}_{\text{total}} = \text{KE}_{\text{translation}} + \text{KE}_{\text{rotation}}$$

$$\text{KE}_{\text{total}} = \tfrac{1}{2}mv^2 + \tfrac{1}{2}I\omega^2 \tag{7.27}$$

Multiple flash photograph (1/30-second intervals) of a moving wrench. The black cross marks its center of mass. Place a ruler on the photograph and note that the positions of the black cross lie essentially in a straight line, and that the distance travelled per interval is constant. The wrench rotates about its center of mass. Forget that the wrench is moving through space and fix your attention upon the rotation of the wrench about the center of mass. What is its angular velocity of rotation? The wrench has both translational and rotational kinetic energy. (Picture from *PSSC Physics*, D.C. Heath and Company, Lexington, Mass., 1965.)

EXAMPLE

Let us consider the relative distribution of the kinetic energy between kinetic energy of translation and kinetic energy of rotation for various objects. Suppose that one has three objects, a hoop, a solid cylinder, and a sphere of equal mass. These are raised to the top of a common incline (Figure 7.17). In this position all of the objects would have the same potential energy (mgh) relative to the foot of the incline. If each body were permitted to roll down the incline (this means there is enough friction to produce rolling and no sliding), and if there is no loss of energy resulting from rolling friction, then the total kinetic energy at the base should be the same for each body and equal to the potential energy at the top.

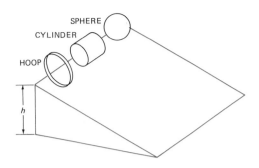

The general equation is then:

$$mgh = \tfrac{1}{2} mv^2 + \tfrac{1}{2} I\omega^2 \tag{7.28}$$

This equation can be rewritten for each object putting in its value for the moment of inertia:

for the sphere $\qquad mgh = \tfrac{1}{2} mv^2 + \tfrac{1}{2} (\tfrac{2}{5} mr^2)\omega^2$

for the cylinder $\qquad mgh = \tfrac{1}{2} mv^2 + \tfrac{1}{2} (\tfrac{1}{2} mr^2)\omega^2$

for the hoop $\qquad mgh = \tfrac{1}{2} mv^2 + \tfrac{1}{2} mr^2\omega^2$

As previously shown for objects rolling without slipping, $v = \omega r$. Note in each case that the velocity at the base is independent of the mass but that it does depend upon the distribution of the mass. That is, it depends upon I. Solving these equations for v, one finds that velocity for each object is given by a different expression:

velocity of sphere $= (\tfrac{10}{7} gh)^{1/2}$

velocity of cylinder $= (\tfrac{4}{3} gh)^{1/2}$

velocity of hoop $= (gh)^{1/2}$

The sphere will require the least time of the three bodies to roll down the plane and the hoop will have the largest percentage of the total kinetic energy in rotational kinetic energy. You should be able to show that for this example half of the total kinetic energy of the hoop is rotational, one-third of the kinetic energy of the solid cylinder is rotational and two-sevenths of the kinetic energy of the sphere is rotational kinetic energy.

7.8 Angular Momentum

Have you watched the performance of a skilled ice skater? You may have seen the skater begin a slow spin maneuver with arms and legs extended (Figure 7.18). Then the skater slowly brings her arms and legs together. As she does this, her rate of spin increases until she has made herself as "thin" as possible and her spin rate has reached a maximum value. Let us discuss the phenomenon in the language of this chapter. At the beginning of her maneuver, the skater with extended limbs has a relatively large amount of inertia (I large) and her angular velocity is small (ω

FIGURE 7.18
An ice skater increases her angular velocity by decreasing her moment of inertia. Her angular momentum remains almost constant.

small). As she changes the position of her body and limbs, her moment of inertia becomes smaller (I small), and her angular velocity becomes large (ω large). If we neglect any friction between the ice skates and the ice, then the skater is an isolated system acted upon only by the vertical force of gravity. If her initial axis of spin is also vertical then the gravitational forces do not apply any torque that will change the rate of rotation of the skater. How then does her spin rate change?

In a way analogous to our definition of linear momentum ($\mathbf{p} = m\mathbf{v}$), we can assume that a rotating body has angular momentum $\mathbf{L}$ which we will define as equal in magnitude to the product of the moment of inertia and the angular velocity,

$$\mathbf{L} = I\boldsymbol{\omega} \tag{7.29}$$

where $\mathbf{L}$ is the angular momentum vector whose direction will be in the direction of the angular velocity vector $\boldsymbol{\omega}$ for simple cases. The direction of $\mathbf{L}$ can be found using a right-hand algorithm (see Figure 7.19).

The line of the vector $\mathbf{L}$ is parallel to the axis of rotation, and the direction is such that one sees clockwise rotation as looking along the axis.

Angular momentum is conserved if there is no unbalanced torque acting on the system. Then the product of the moment of inertia and the angular velocity will remain a conserved quantity of the system,

$$I_f\omega_f = I_i\omega_i \tag{7.30}$$

where the subscripts i and f refer to the initial and final states of the system and I and ω represent the moment of inertia and the angular velocity of the system. A typical example of this conservation law is the skater we discussed above. By reducing her moment of inertia, she increases her angular velocity. She comes out of the spin by increasing her moment of inertia and thus reducing her angular velocity.

FIGURE 7.19
A spinning object has angular momentum **L**. Angular momentum is a vector quantity, and one can use the right-hand rule to determine the direction of the vector.

EXAMPLES

1. Let us consider the case of the skater who goes into a spin. Assume that at the beginning of the spin her moment of inertia is 10.0 kg–m² and the angular

TABLE 7.1
Concepts of Linear and
Rotational Motion

Linear Dynamics		Rotational Dynamics	
Physical Quantity	Rectilinear Motion	Physical Quantity	Rotational Motion About a Fixed Axis
Mass	m	Moment of inertia	I
Force	$\mathbf{F} = m\mathbf{a}$	Torque	$\tau = I\alpha$
Work	$W = Fd$	Work	$W = \tau\theta$
Translational KE	$KE = \frac{1}{2}mv^2$	Rotational KE	$KE = \frac{1}{2}I\omega^2$
Linear momentum	$\mathbf{p} = m\mathbf{v}$	Angular momentum	$\mathbf{L} = I\omega$

velocity is 5.00 rad/sec. If the moment of inertia of the skater is reduced to 2.00 kg–m², what is her final angular velocity?

$$I_i\omega_i = I_f\omega_f$$

$$(10.0 \text{ kg–m}^2)(5.00 \text{ rad/sec}) = (2.00 \text{ kg–m}^2)\omega_f$$

$$\omega_f = 25.0 \text{ rad/sec}$$

2. Assume a high diver is rotating at a slow rate (10.0 rpm) in an outstretched position. The diver goes into a tight tuck position. We want to estimate the new rate of rotation for the tuck position. We will assume that in the initial position we can approximate the diver by a rod rotating about its center of mass (moment of inertia $= \frac{1}{12}ML^2$, where $M =$ mass and $L =$ Length). In the tight tuck position we assume that the diver approximates a disk with a radius of $L/4$. Such a disk (same mass as "rod") has a moment of inertia of $\frac{1}{2}M(L/4)^2 = \frac{1}{32}ML^2$. Since there are no external torques acting on the diver, the angular momentum of the diver must be conserved as expressed in Equation 7.30. Solving for the angular velocity in the tuck position we find:

$$\omega_f = \frac{I_i}{I_f}\omega_i = \frac{\frac{1}{12}ML^2}{\frac{1}{32}ML^2}\omega_i = \frac{32}{12}\omega_i$$

$$\omega_f = 2.67\omega_i = 2.67 \times 10.0 \text{ rpm} = 26.7 \text{ rpm}$$

We have now concluded the definitions of various concepts related to rotational motion. We summarize all these concepts and compare them with their linear motion analogs in Table 7.1.

7.9 Conservation of Energy and Momentum

The principle of conservation of energy applies to bodies with rotational motion as well as to those in translation. Also we have the conservation of angular momentum as well as the conservation of linear momentum in isolated systems. In fact, the conservation of angular momentum may be considered a more general law than the conservation of linear momentum. Linear momentum is conserved for all systems on which no external forces are acting. Angular momentum is conserved for all systems on which there are no external torques acting. It is possible to have forces acting on a system that produce no torque. It is not possible to have a torque produced with no forces acting. Hence, the class of all systems that conserve angular momentum is larger than the class of systems that conserve linear momentum.

Let us summarize the three conservation laws we have discussed in Chapters 5, 6, and 7. For systems upon which no net work is done the total energy of the system is conserved,

total energy = constant

For systems with both kinetic and potential energy, the sum of the kinetic energy and the potential energy remains constant,

KE + PE = constant

For systems near the surface of the earth,

$\frac{1}{2}mv^2 + mgh$ = constant

where m is mass (kg), v is velocity (m/sec), g is gravitational acceleration (9.80 m/sec^2), and h is the distance (m) above a reference line.

For systems upon which no net external forces are acting, the linear momentum **p** is conserved

linear momentum = constant

$\mathbf{p} = m\mathbf{v}$ = constant

where m is the mass of the system and **v** is the velocity of the center of mass of the system. You will notice that this conservation law is a vector law, and so it may be expressed as three independent scalar equations involving each of the three components of the linear momentum, p_x, p_y, and p_z.

For systems upon which no net external torques are acting, the angular momentum **L** is conserved,

angular momentum = constant

$\mathbf{L} = I\boldsymbol{\omega}$ = constant

where I is the momentum of inertia of the system and $\boldsymbol{\omega}$ is the angular velocity of the system. This is also a conservation law that involves vector quantities; so the three components of the angular momentum, L_x, L_y, and L_z, must be constant for a system whose angular momentum remains constant. Notice that angular momentum is conserved for all systems that experience net forces that act through the center of rotation of the system since such forces would not exert any torque upon the system. One example of such a system is the model of our solar system, with the sun as the fixed origin of the rotation of the planets around the sun and which considers only the gravitational force between the sun and each planet. The line of action of the force of attraction between the sun and each planet passes through the sun, the assumed axis of rotation, so the gravitational attraction exerts no torque on the planet. Hence, angular momentum of a planet in orbit around the sun must remain constant.

These three conservation laws can be used to solve problems of particle dynamics in situations where the use of Newton's laws of motion would be quite difficult. Energy, linear momentum, and angular momentum are mental constructs that we can use to explain a wide variety of physical phenomena and to give ourselves a feeling of understanding the universe.

EXAMPLE

A hoop rolls down an inclined plane which has an elevation of 1.00 m. What is its velocity at the foot of the inclined plane? Neglect frictional losses.

At the top of the inclined plane the hoop has potential energy, and at the foot of the plane it has kinetic energy. The potential energy at the top is equal to the kinetic energy at the foot of the plane:

$$PE_{top} = mgh$$

$$KE_{foot} = \tfrac{1}{2}mv^2 + \tfrac{1}{2}I\omega^2$$

It has kinetic energy both of translation and rotation.

The moment of inertia of a hoop is equal to mr^2, giving

$$mgh = \tfrac{1}{2}mv^2 + \tfrac{1}{2}mr^2\omega^2$$

From this one sees that each term is divisible by m, which means the velocity is independent of its mass: $gh = \tfrac{1}{2}v^2 + \tfrac{1}{2}r^2\omega^2$, where $r\omega$ is equal to v if the hoop rolls without slipping; thus one sees the velocity is independent of the radius. Hence, $gh = \tfrac{1}{2}v^2 + \tfrac{1}{2}v^2 = v^2$. This also tells us that the kinetic energy of rotation equals the kinetic energy of translation for the hoop and that

$$v = \sqrt{gh} = \sqrt{9.80 \times 1.00} = \sqrt{9.80} \text{m/sec}$$

If the hoop had slid down a frictionless incline, the velocity would have been

$$v = \sqrt{2gh}$$

7.10 Planetary Motion

The apparent zig-zag motion of the planets across the night sky puzzled human observers of the heavens for many centuries. We now have the mental constructs to discuss planetary motion in a simple way. Consider the sun as a large, massive body that produces a gravitational field. The strength of the field is given by the law of gravitational attraction, $F/m = GM/r^2$ where F is the magnitude of the force of gravitational attraction, G is the universal gravitational constant, M is the mass of the sun, m is the mass of the object in the sun's gravitational field, and r is the distance between the object and the sun. The gravitational force acts toward the sun. It applies no torque to an object moving in a path around the sun. So the angular momentum of objects moving in closed orbits around the sun is constant,

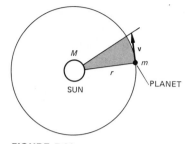

FIGURE 7.20
A planet in a circular orbit around the sun.

$$L = I\omega = (mr^2)(v/r) = mvr = \text{constant} \tag{7.31}$$

Although the planets travel in elliptical orbits with the sun at one focus according to Kepler's first law of planetary motion, let us assume that the planets travel in circular orbits. The result we will derive is also true for elliptical orbits, but the mathematics is simpler for circular orbits. Then we can show that Equation 7.32 is equivalent to Kepler's second law of planetary motion which states;

the line joining the sun and a planet sweeps out equal areas in equal times.

For circular motion v and r are perpendicular to each other and can be drawn as the two sides of a right triangle (see Figure 7.20). The area of such a triangle is proportional to the product of velocity times radius, but vr is proportional to the angular momentum, so

$$\text{area} \propto vr = L/m = \text{constant} \tag{7.32}$$

Kepler's third law can also easily be deduced for circular orbits. This law states that

the ratio of the cube of the radius of the orbit to the square of the time for the planet to make a complete cycle around the sun is a constant for all the planets (see Table 7.2).

The period of time required for a planet to complete one cycle around the sun is the circumference of its orbit divided by its speed of travel.

$$T = \frac{2\pi r}{v} \tag{7.33}$$

where r is the distance of the planet from the sun and v is the speed of the planet. We can obtain an expression for the speed of the planet in its orbit by equating the centripetal force to the gravitational force.

$$\frac{mv^2}{r} = \frac{GMm}{r^2} \tag{7.34}$$

TABLE 7.2
Solar System Data

Planet	Mean Distance to the Sun R(m)	Period of Time for One Cycle T(sec)	R^3/T^2 (m³/sec²)
Mercury	5.79×10^{10}	7.60×10^6	3.36×10^{18}
Venus	1.08×10^{11}	1.94×10^7	3.35×10^{18}
Earth	1.50×10^{11}	3.16×10^7	3.38×10^{18}
Mars	2.28×10^{11}	5.94×10^7	3.36×10^{18}
Jupiter	7.78×10^{11}	3.74×10^8	3.37×10^{18}
Saturn	1.43×10^{12}	9.35×10^8	3.35×10^{18}
Uranus	2.86×10^{12}	2.64×10^9	3.36×10^{18}
Neptune	4.52×10^{12}	5.22×10^9	3.39×10^{18}
Pluto	5.90×10^{12}	7.82×10^9	3.36×10^{18}

It follows that the square of the velocity is inversely proportional to the radius of the orbit.

$$v^2 = \frac{GM}{r} \tag{7.35}$$

This expression for v^2 can be substituted into the square of Equation 7.33 to obtain an equation for the period T in terms of the radius r:

$$T^2 = \frac{4\pi^2 r^2}{v^2} = \frac{4\pi^2 r^2}{\dfrac{GM}{r}} = \frac{4\pi^2 r^3}{GM} \tag{7.36}$$

This equation can be rearranged to show that the ratio r^3/T^2 is a constant, equal in size to the product of the universal gravitational constant times the mass of the sun divided by $4\pi^2$:

$$\frac{r^3}{T^2} = \frac{GM}{4\pi^2} \tag{7.37}$$

Use this expression, the numerical value for G, and the data in Table 7.2 to compute the mass of the sun.

ENRICHMENT
7.11 Instantaneous Angular Velocity and Angular Acceleration

We define average angular velocity as $\omega_{avg} = \Delta\theta/\Delta t$. As Δt approaches 0 ($\Delta t \to 0$), this becomes the instantaneous angular velocity or limiting value of $\Delta\theta/\Delta t$ and is written $d\theta/dt$. Thus,

$$\omega_{inst} = \lim_{\Delta t \to 0} \frac{\Delta\theta}{\Delta t} = \frac{d\theta}{dt}$$

Similarly

$$\alpha_{inst} = \lim_{\Delta t \to 0} \frac{\Delta\omega}{\Delta t} = \frac{d\omega}{dt}$$

We can express the torque produced by a force acting through a lever arm $\mathbf{d}$ and making an angle θ with the force $\mathbf{F}$, as follows (see Figure 7.5):

$$\tau = \mathbf{d} \times \mathbf{F} \tag{7.38}$$

This vector notation carries with it the direction of the right-hand rule. The torque direction is perpendicular to the plane defined by $\mathbf{d}$ and $\mathbf{F}$, and in direction a right-hand threaded screw advances when turned from $\mathbf{d}$ to $\mathbf{F}$. This vector product is commonly called the *cross product* and it has the magnitude of

$$\tau = d \sin\theta \, F$$

where $d \sin\theta$ is the effective lever arm of the applied force.

From Newton's second law we know that

$$\mathbf{F} = \frac{d\mathbf{p}}{dt}$$

and we can write the torque as follows:

$$\boldsymbol{\tau} = \mathbf{d} \times \frac{d\mathbf{p}}{dt}$$

Consider the case of the moving earth around the sun. The gravitational force is parallel to $\mathbf{d}$ and satisfies the centripetal force equation

$$\mathbf{F}_c = \frac{d\mathbf{p}}{dt}$$

Since $\mathbf{F}_c$ is parallel to $\mathbf{d}$ we see that $\boldsymbol{\tau} = 0$. Also $\mathbf{d} \times \mathbf{p} = \mathbf{d} \times m\mathbf{v}$, where the magnitude of this expression is given as:

$$dm\omega d = md^2\omega = I\omega$$

since $I = md^2$ for the earth about the sun at a distance d and ω is the angular velocity of the earth about the sun. The direction of this vector is given by the right-hand rule to be the same as that of $\boldsymbol{\omega}$. Thus we see that $\mathbf{d} \times \mathbf{p}$ is equal to the angular momentum of the earth about the sun. In general we can write the following equation:

$$\mathbf{L} = \mathbf{d} \times \mathbf{p} = I\boldsymbol{\omega} \qquad (7.39)$$

and it follows that

$$\boldsymbol{\tau} = \frac{d\mathbf{L}}{dt} = \frac{d(I\boldsymbol{\omega})}{dt} \qquad (7.40)$$

When the torque is zero, as is the case when $\mathbf{F}$ is parallel to $\mathbf{d}$, then $\mathbf{L}$ is a constant and angular momentum is conserved. When the moment of inertia is constant, we find the following equation holds for the motion:

$$\boldsymbol{\tau} = I\frac{d\boldsymbol{\omega}}{dt} = I\boldsymbol{\alpha}$$

SUMMARY

Use these questions to evaluate how well you have achieved the goals of this chapter. The answers to these questions are given at the end of this summary with the section number where you can find the related content material.

Definitions

1. The angular displacement of the minute hand of a clock between 12:00 and 12:15 in radians is
   a. π

b. $\pi/2$
c. $\pi/3$
d. $\pi/15$
e. $\pi/4$

2. The angular velocity of the minute hand of a clock in radians per second is
   a. $\pi/30$
   b. $\pi/6$
   c. $\pi/1800$
   d. $\pi/3600$
   e. 2π

3. If a turntable reaches a final speed of 120 rpm in 2.00 sec, the angular acceleration of the turntable is
   a. 60 rad/sec^2
   b. 2π rad/sec^2
   c. 1 rad/sec^2
   d. 4π rad/sec^2
   e. π rad/sec^2

4. The torque produced by tangential force, F, applied to the rim of a disk of radius R will be
   a. F/R
   b. F/R^2
   c. $F\sqrt{R}$
   d. FR
   e. R/F

5. The angular acceleration produced by the torque in question 4 will be proportional to
   a. $1/I$
   b. I
   c. I/R^2
   d. I/m
   e. I^2R
   where I is the moment of inertia of the disk.

6. If a free body is suspended on an axis through its center of mass, you can be sure the body will have
   a. clockwise torque
   b. counterclockwise torque
   c. zero weight
   d. zero I
   e. zero torque

7. The rotational kinetic energy of the mass particles in a rotating disk is
   a. the same for all particles
   b. the greatest for those near axis
   c. greatest for those nearest rim
   d. zero
   e. independent of speed of the disk's rotation

8. The angular momentum of a system may be
   a. proportional to ω
   b. proportional to I
   c. directed parallel to $\boldsymbol{\omega}$

d. changed in the direction of applied torque
e. all of these

Equilibrium

9. The conditions necessary for a body to be in static equilibrium are _____.

Rotational Motion and Kinematics

10. An equation that holds for an object of radius R rolling without slipping at a velocity v is given as
    a. $a_t = \alpha R$
    b. $v = \omega R$
    c. $a_c = g$
    d. $a_t = \mu g$
    e. a and b

11. The angular displacement in time t for constant angular acceleration is given by _____.

12. The angular displacement for an object accelerating at 1 rad/sec^2 from 0 to 4 rad/sec is
    a. 4 rad
    b. 8 rad
    c. 16 rad
    d. 2 rad
    e. zero

Rotational Dynamics

13. It is noted that the maximum angular acceleration produced by a given torque applied to a rod with its axis at its center is two times that produced when the axis is at one end. The ratio of the moments of inertia for $\dfrac{I_{center}}{I_{end}}$ is
    a. 2
    b. $\frac{1}{2}$
    c. 4
    d. $\frac{1}{4}$
    e. 8

14. When a chunk of putty of mass M is dropped onto the outer edge (a distance R from the axis) of a turntable, the angular velocity of the turntable is reduced by half. The moment of inertia of the turntable must be _____.

Equilibrium Problems

15. A uniform board is supported by a pivot one-fourth its length from one end. Find the force applied to the opposite end necessary to balance the board. Find the force on the pivot. (M = mass of board and L = length of board.)

Answers

ALGORITHMIC PROBLEMS

Listed below are the important equations from this chapter. The problems following the equations will help you learn to translate words into equations and to solve single concept problems.

Equations

$$\omega_{ave} = \frac{\theta_2 - \theta_1}{t_2 - t_1} = \frac{\Delta \theta}{\Delta t} \tag{7.1}$$

$$\alpha_{ave} = \frac{\omega_2 - \omega_1}{t_2 - t_1} = \frac{\Delta \omega}{\Delta t} \tag{7.3}$$

$$\omega_f = \omega_0 + \alpha t \tag{7.4}$$

$$\omega_{ave} = \frac{\omega_f + \omega_0}{2} \tag{7.5}$$

$$\Delta \theta = \theta_f - \theta_0 = \omega_0 t + \frac{1}{2} \alpha t^2 \tag{7.6}$$

$$2\alpha(\theta_f - \theta_0) = \omega_f^2 - \omega_0^2 \tag{7.7}$$

$$\tau = Fd \tag{7.18}$$

$$\tau = I\alpha \tag{7.20}$$

$$I = \sum_{i=1}^{n} m_i r_i^2 \tag{7.22}$$

$$\text{Work} = \tau \cdot \theta \tag{7.24}$$

$$\text{KE}_{rot} = \frac{1}{2} I \omega^2 \tag{7.26}$$

$$\mathbf{L} = I\boldsymbol{\omega} \tag{7.29}$$

Problems

1. A turntable turns through 10.0 revolutions in 5.00 sec. Find the angular velocity of the turntable in rad/sec.
2. A phonograph turntable is rotating at 45.0 rpm. Find the angle that a point on the rim of the turntable turns through in 12.0 sec.
3. A turntable starting from rest reaches an angular velocity of 10.0 rad/sec in 4.00 sec. Find the angular acceleration of the turntable.

4. If you can exert a 20.0-N force on a wrench 25.0 cm long, find the maximum torque that can be exerted by using this force and wrench.

5. A torque of 10.0 N-m can be applied to a flywheel that has a moment of inertia of 5.00 kg-m^2. Find the angular acceleration of the flywheel that is produced by this torque.

6. Find the rotational kinetic energy of the flywheel in problem 5 if its angular velocity is 100 rad/sec.

Answers

1. 4π rad/sec
2. 9 rev or 18 π rad
3. 2.50 rad/sec^2

4. 5.00 N-m
5. 2.00 rad/sec^2
6. 25,000 J.

EXERCISES

These exercises are designed to help you apply the ideas of a section to physical situations. When appropriate the numerical answer is given at the end of each exercise.

Section 7.2

1. What is the average angular velocity of the earth spinning about its axis? What is the average angular velocity of the earth in its orbit about the sun? [7.27×10^{-5} rad/sec, 1.99×10^{-7} rad/sec]

2. A wheel is fitted with two electrical contacts, separated by 10° for timing circuits. What angular velocity (in rad/sec and rpm) must the wheel have for a time interval between these two contacts for
   a. 0.200 sec?
   b. 0.00400 sec? [a. 0.873 rad/sec, 8.33 rpm; b. 4.36 $\times$ 10^1 rad/sec, 4.17 $\times$ 10^2 rpm]

3. An automobile is traveling on a level road at 80.0 km/hr. The diameter of the wheels is 0.750 m. What is the angular velocity of the wheel about its spindle? [59.2 rad/sec]

4. If the car in exercise 3 stops in a distance of 50.0 m on the highway,
   a. how many revolutions did the wheel make?
   b. what is the average angular acceleration of the wheel in stopping? [a. 21.2 rev; b. 13.1 rad/sec^2]

Section 7.4

5. A dentist grips a pair of forceps exerting a force of 20.0 N on each side at a distance of 10.0 cm from the

pivot. The tooth is gripped at a point of 0.500 cm from the pivot. What is the force exerted on the tooth? [400 N on each side]

6. The radius of the steering wheel of a car is 25.0 cm. The driver exerts a force of 10.0 N tangent to the rim of the steering wheel. What is the torque acting to produce rotation? [2.50 N-m]

7. The weight on the front wheels of an automobile is 900 N, and the weight on the rear wheels is 7200 N. The wheels are 2.80 m apart. Where is the center of gravity? [2.48 m from front]

8. A 4.00-m uniform plank is resting on a pier with 1.50 m extending beyond the edge of the pier. The plank weighs 1400 N. How far can a 50.0-kg student walk beyond the edge of the pier before the plank will tip? [1.43 m]

Section 7.5

9. Find the torque necessary to give a 2.00-kg disk of 10.0 cm radius an angular acceleration of 0.100 rad/sec^2. [1.00×10^{-3} N-m]

10. A rope wrapped around a pulley (disk) of mass M and radius R is given an angular acceleration by an equal mass hanging on the end of the rope. Find the angular acceleration. [2/3 g/R]

Section 7.7

11. Compare the rotational kinetic energies of a hoop and a sphere of equal mass and radius traveling at the same center of mass speed. [KE$_{sphere}$/KE$_{hoop}$ = 0.4]

Section 7.8

12. A tubular ring (mass $= M$) of radius R is filled with a fluid of mass $(M/4)$. If the ring, filled with fluid, spins about an axis through its center at 1.00 rpm, find the spin rate after all the fluid leaks out. [1.25 rpm]

Section 7.9

13. Find the necessary work to roll a ball of mass M, in kilograms, and radius R, in meters, up an incline.

Assume the ball starts from rest and has a speed of v meters/sec at the incline height of H meters. $[MgH + \frac{7}{10} mv^2]$

Section 7.10

14. Using 365 days as Earth's period and its solar orbit radius as 1.00 astronomical unit (AU), find the distance of an asteroid from the sun that has a period of 2920 days. [4.0 AU]

PROBLEMS

Each problem may involve more than one physical concept. The answer is given at the end of the problem. A problem that requires material from the Enrichment Section, 7.11, is marked by a dagger (†).

15. A load of 200 N is supported by a crane as shown in Figure 7.21. The boom AB is uniform and weighs 20.0 N. What is the tension in the tie BC? What is the magnitude and direction of the force exerted by the support at A on the boom? [126 N, 175 N, 55° above horizontal]

16. A man holds a 30–N weight in his hand, keeping his forearm at rest in a horizontal position. In Figure 7.22, E is the elbow, H is the hand, and $\mathbf{F}$ is the force exerted by the muscle at M which makes an angle of 60° with the horizontal. What is the magnitude of $\mathbf{F}$ and the force acting at E? [208 N, 182.5 N down at an angle 55.3° below horizontal]

17. A man carries a uniform pole 2.40 m long, and weighing 50.0 N over his shoulder, holding it horizontally. He pulls down with his hand at the front end 0.600 m from his shoulder. A 20.0-N bundle hangs from a point 0.100 m from the other end. What is the force

exerted by the man's hand and the force exerted against his shoulder? [107 N, 177 N]

18. A uniform ladder 5.00 m long, weighing 100 N leans against a smooth vertical wall. The distance of the foot of the ladder from the wall is 3.00 m. What is the force exerted by the wall on the ladder when an 800-N man stands 1.00 m from the upper end of the ladder? What is the magnitude and direction of the force of the ground on the ladder? What is the coefficient of friction between the ground and ladder for this case? [518 N, 1040 N, 60.1° above the horizontal, 0.576]

19. Find the center of gravity of a unit that is made of two lengths of uniform circular rod (see Figure 7.23).

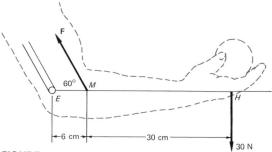

FIGURE 7.22
Problem 16.

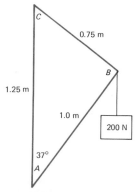

FIGURE 7.21
Problem 15.

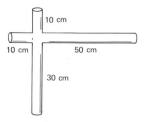

FIGURE 7.23
Problem 19.

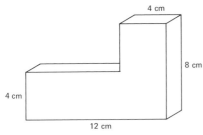

FIGURE 7.24
Problem 20.

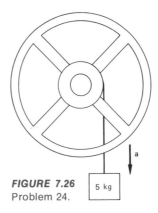

FIGURE 7.26 5 kg
Problem 24.

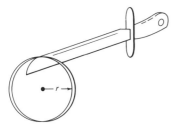

FIGURE 7.25
Problem 21.

[12 cm to right of intersection, 4.0 cm below intersection]

20. Find the center of gravity of a sheet of plastic of uniform thickness with dimensions as shown in Figure 7.24. [7 cm to the right and 3 cm above the lower left corner]

21. A hoop of mass m and radius r is supported by a knife edge (Figure 7.25). If displaced from equilibrium position, it will oscillate about the equilibrium position. What is the moment of inertia of the hoop about the knife edge? What is the torque tending to return it to equilibrium if it is displaced an angle θ from the equilibrium position? [$2Mr^2$, $mgr \sin \theta$]

22. In sharpening a paring knife, a man uses a hand-driven emery wheel which has a handle length of 10.0 cm. Assume a force of 40.0 N is exerted in a direction perpendicular to the handle and that 200 revolutions are made in the grinding process. How much work did he do in sharpening the knife? [5030 J]

23. A solid cylinder is mounted on a frictionless horizontal axis. The cylinder has a radius of 10.0 cm and a mass of 4.00 kg. A constant force of 10.0 N is applied to a cord wrapped around the cylinder. What is the angular acceleration? Would the acceleration be the same if a 10.0-N weight were attached to the cord? [50.0 rad/sec², no]

24. A pulley is mounted on fixed ball bearings (frictionless), and a 5.00-kg mass is hanging from a cord wrapped around the hub which has a radius of 4.00 cm (Figure 7.26). The mass has a downward acceleration of 3.80 m/sec². What is the moment of inertia of the pulley? If the mass of the pulley is 6.00 kg, what is the radius of a hoop which has the same moment of inertia? [.0126 kg m², 0.0458 m]

25. A sphere rolls down an inclined plane that has an elevation of 1.00 m and length of 2.00 m. What is its velocity at the foot of the incline? Assume no frictional losses. How would this velocity compare with the velocity of a body sliding down an equivalent frictionless incline. Compare the kinetic energies of translation and rotation. [3.74 m/sec, 4.43 m/sec, $KE_{rot} = \frac{2}{7} KE_{tot}$, $KE_{trans} = \frac{5}{7} KE_{tot}$]

26. A merry-go-round is moving with an angular velocity of 2.00 rad/sec. Assume the merry-go-round has a moment of inertia of 360 kg-m² and a radius of 2.00 m. A 30.0-kg boy who was riding on the merry-go-round decides to walk on a support toward the center of rotation. What is the angular velocity of the merry-go-round when he reaches 0.500 m from the center? Neglect frictional effects and treat the boy as a point mass. [2.61 rad/sec]

27. Given the schematic of the forearm holding a mass M, shown in Figure 7.27, find the force exerted by the biceps **F** on the forearm for equilibrium if the mass of the forearm is $M/10$. Assume that the center

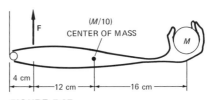

FIGURE 7.27
Problem 27.

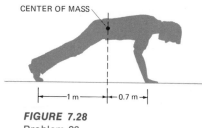

FIGURE 7.28
Problem 28.

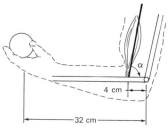

FIGURE 7.30
Problem 30.

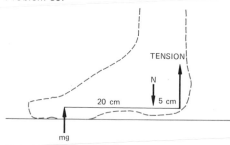

FIGURE 7.31
Problem 31.

of mass of the forearm is 16.0 cm from E. [8.4 Mg N]

28. Figure 7.28 shows a person during push ups. The mass of the person is 80.0 kg. Find the force exerted by each hand and each foot on the floor. [231 N, 161 N]

29. A crouching man on tiptoes puts considerable tension on the Achilles tendon as shown in Figure 7.29. Find the tension on the tendon for a 100-kg man (one-half of his weight is supported by each foot) using the data given. [980 N]

30. The biceps muscle system is shown in Figure 7.30 with typical dimensions. What is the force in the biceps muscle if 2.00 kg is supported by the hand in the horizontal position?
    a. if $\alpha = 80°$?
    b. if $\alpha = 30°$? [a. 160 N; b. 314 N]

31. A 70.0-kg man puts all of his weight on the ball of his right foot. What is the tension of the leg muscle? See Figure 7.31 for relevant data. [2740 N]

32. Suppose that you are standing on your right leg and holding a vertical cane in your left hand which supports $\frac{1}{6}$ of your weight and is 30 cm to the left of the center of mass of your body. Where will your right foot now be placed? See Figure 7.13. With your foot

in the position indicated above, what will be the new value of F? Remember the location of the center of gravity and the length of your leg will remain the same. Compare your values of F with and without a cane. [6 cm to right of your center of mass, 360 N, $F_{without\ cane}/F_{cane} = 2.66$]

33. A clever inventor at the time of Newton decided that he could put carriages in orbit around the earth by releasing them horizontally from the top of a ferris wheel of 30.0-m radius. Find the necessary angular velocity of the ferris wheel to obtain the orbital condition for the carriages. [17.2 rad/sec]

34. There are two different ways you can lift a load of

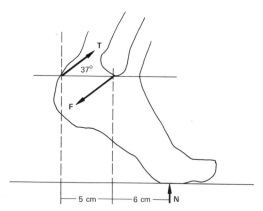

FIGURE 7.29
Problem 29.

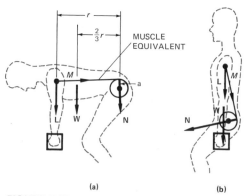

(a) (b)

FIGURE 7.32
Problem 34.

weight L shown in Figure 7.32a and b. Assume the center of mass of your upper trunk is two-thirds of the distance r from your hip pivot to your shoulder, that the load is twice the weight W of your upper trunk, and that the back muscle can be represented by the simple forces **M** and **N** as shown, where r is about $20a$. Compare the forces on your spine in the two cases. [M in part a about 10 times larger than in part b]

35. If a person turns his head through an angle of $60°$ in 0.10 sec, what torque must be applied by the neck muscles? Assume that the head has a mass of 5 kg and is a uniform sphere with a radius of 8 cm. The axis of rotation passes through the center of mass and the motion starts from rest and has constant angular acceleration.   [2.70 N-m]

† 36. The angular displacement of a rotating body is given by $\theta = 4t^2 - 16t + 6$, with θ in radians and t in seconds. What is
   a. the angular velocity at $t = 1$ sec, at $t = 3$ sec?
   b. the angular acceleration?
   c. the time the body is at rest?
   d. the angular displacement when at rest?   [-8 rad/sec, 8 rad/sec^2, 2 sec, 6 rad]

† 37. A 2 kg particle moves through the position $x = 5$ m, at a velocity of 4 m/sec in a direction $37°$ above the $+x$ axis. Find its angular momentum relative to the origin.   [24 kg m^2/sec along the $+z$ axis]

GOALS

When you have mastered the contents of this chapter, you will be able to achieve the following goals:

Definitions
Define each of the following terms, and use it in an operational definition:

fluid	buoyant force
density	streamline flow
specific gravity	viscosity
pressure (absolute and gauge)	

Fluid Laws
State Pascal's law of hydrostatic pressure, Archimedes' principle of buoyancy, Bernoulli's equation for the conservation of energy in a fluid, and the law of conservation of fluid flow.

Fluid Problems
Solve problems making use of the principles of fluids and conservation laws.

Viscous Flow
Use Poiseuille's law of viscous flow to solve numerical problems.

PREREQUISITES

Before you begin this chapter, you should be able to solve problems that use energy concepts (see Chapter 5).

8

Fluid Flow

8.1 Introduction

Fluids play an important part in our everyday life. The basis of our water and air transportation systems is the buoyancy of objects in water and the lift forces of objects moving through the air. These phenomena are results of fluid dynamics.

The circulation of fluids in our bodies plays an essential part in our energy exchange processes. The circulation of atmospheric gases plays a similar part in the energetics of the earth.

Can you give an example of fluid flow used in energy transfer in your present environment? What is the measure of the flow inertia for a fluid? In this chapter you will be introduced to the basic principles of fluid dynamics. Your understanding of these principles will provide you a basis for analyzing and working with fluid systems.

8.2 Fluids

You are familiar with the properties of fluids. How does a fluid contrast to solid matter?

A *fluid* is a substance that flows easily from one location to another. The term thus applies to both liquids and gases. Liquids and gases differ in several ways. A liquid has a fixed volume but takes on the shape of the container up to the limit of its volume.

A gas takes on both the shape and the volume of its container. Another way in which liquids and gases differ is in compressibility: a gas is easily compressed, and a liquid is practically incompressible—at least for our present consideration.

8.3 Density

There are a number of ways in which substances differ from one another. If you have ever lifted a "brick" made of lead or a small bottle filled with mercury your muscles have felt one striking difference between these two substances and other common materials. Small samples of lead or mercury feel massive. That can be a source of muscular surprise. The *density* of a material, assumed to be homogeneous, is defined as the mass per unit volume. So we use as the defining equation:

$$\rho = \frac{m}{V} \tag{8.1}$$

where ρ is the density, and m is the mass of volume V of the material. What are the dimensions of density in terms of M, L, and T? The units are kg/m^3 in the SI system and g/cm^3 in the cgs system. What is the numerical factor required to convert kg/m^3 into gm/cm^3?

TABLE 8.1
Table of Densities and Specific Gravities

Material	Density kg/m³	Specific Gravity
Aluminum	2.7 $\times 10^3$	2.70
Bone	1.85 $\times 10^3$	1.85
Brass	8.6 $\times 10^3$	8.60
Copper	8.9 $\times 10^3$	8.90
Gold	19.3 $\times 10^3$	19.3
Ice	0.92 $\times 10^3$	0.920
Iron	7.8 $\times 10^3$	7.80
Lead	11.3 $\times 10^3$	11.3
Silver	10.5 $\times 10^3$	10.5
Steel	7.8 $\times 10^3$	7.80
Wood (oak)	0.8 $\times 10^3$	0.8
Ethyl alcohol	0.81 $\times 10^3$	0.81
Glycerin	1.26 $\times 10^3$	1.26
Mercury	13.6 $\times 10^3$	13.6
Water	1.00 $\times 10^3$	1.00

A pressure breathing assistor provides intermittent positive-pressure breathing assistance for patients suffering from chronic bronchio-pulmonary diseases such as emphysema, bronchitis, asthma, and other respiratory ailments. (Mine Safety Appliance Company, Pittsburgh, Pa.)

The *specific gravity* of a substance is the ratio of the density of the substance to the density of water. Hence, specific gravity is dimensionless, that is, a pure number, and is independent of the system of measurement that you use. For example, one cubic foot of lead weighs 705 lb and one cubic foot of water weighs 62.4 lb. Calculate the specific gravity of lead from these data and compare it with the value in Table 8.1.

8.4 Force on Fluids

What happens if you push with your finger against the surface of water? Against the surface of ice?

If you investigate the way in which a force acts upon the surface of a liquid and a solid, you find at least one difference. In the case of the solid you can exert a net force in any direction, up, down, or sideways. For an ideal fluid, a fluid with no internal resistance, the net surface force must always be directed at right angles, or normal, to the surface. An ideal fluid at rest cannot sustain a tangential force. The layers of fluid simply slide over one another. An ideal fluid will not maintain a shape of its own. It is the inability of fluids to withstand tangential forces that allows them to flow.

8.5 Pressure

Since a fluid will only sustain a net force perpendicular to its surface, we can speak of a force acting on a fluid in terms of the total force acting on an area of the liquid, or a pressure.

Pressure is defined as the normal force per unit area.

Pulmonary testing unit. The components of this basic unit are a spirometer, a CO_2 absorber-recorder interface, and an X-Y-T recorder. (Searle Cardio Pulmonary Group—CPI, Houston, Texas.)

Pressure is transmitted to the interface boundaries and across any section of a fluid at right angles to the boundary or section. Let us consider a fluid under pressure in contact with a surface. Assume that we have a closed surface that contains a fluid, for example, a balloon filled with water (see Figure 8.1). A small portion of the area of the surface may be represented by ΔA. Let ΔF represent the component of the applied force that is perpendicular to the area ΔA. This perpendicular component is called the normal force.

Then we can define the average pressure P_{ave} acting on that position of the area as $P_{ave} = \Delta F/\Delta A$. The pressure may depend upon the area chosen. This difficulty can be avoided by choosing an area around a point and having the area decrease and approach the point. Thus, the pressure at a point is defined as the ratio of the normal force to the area, as the area is reduced to a very small size. The pressure at a point can be represented mathematically in the following way:

$$P = \lim_{\Delta A \to 0} \frac{\Delta F}{\Delta A} \tag{8.2}$$

The pressure may vary from point to point on a surface. Pressure is a scalar quantity and has the basic units of newtons per square meter (N/m^2) in the SI system, and dynes per square centimeter $(dynes/cm^2)$ in the cgs system. In our daily lives we find many other units of pressure in use (centimeters of mercury, for example), but they are reducible to either of the above. Can you think of other pressure units? What are the units of pressure used by your television weather forecaster to measure atmospheric pressure?

8.6 Pressure of a Liquid in a Column

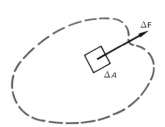

FIGURE 8.1
A closed surface that contains a fluid. ΔA represents the area of a portion of the surface, ΔF is the normal force exerted by the fluid on the surface.

It is a common practice to describe the pressure of the atmosphere by referring to the vertical column of a liquid that the pressure of the atmosphere can support. In this case we are interested in determining the pressure produced by a given height of the liquid in the column shown in Figure 8.2. Let A = area of $GHH'G'$ or $DCC'D'$ and let h = height = $GD = HC = H'C' = G'D'$. For equilibrium, the force acting on the front $GHCD$ is equal in magnitude and opposite in direction to the force acting on the back of the column $G'H'C'D'$. Also the force of $\mathbf{F}_b$ acting on the bottom $DCC'D'$ must be equal in magnitude to the force $\mathbf{F}_t$ acting on the top $GG'H'H$ plus the weight $\mathbf{W}$ of the liquid column:

$$\mathbf{F}_b = \mathbf{F}_t + \mathbf{W}$$

Thus the magnitude of the weight of the liquid column can be expressed as

$$W = F_b - F_t \tag{8.3}$$

The magnitude of weight is the product of mass times gravitational

FIGURE 8.2
A rectangular column of liquid. The liquid in the column has a weight **W**. There is a force **F**$_t$ acting down upon the top of the column and a force **F**$_b$ acting upward on the bottom of the column. The column has a height h equal to the length of the edge GD. The top and bottom of the column have the same area A, which is equal to the area of the rectangles DD'C'C or GG'H'H.

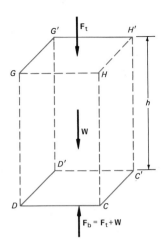

acceleration, or the product of its density and its volume and g, thus

$$W = (\rho V)(g) = (\rho A h)g$$

where g is the magnitude of acceleration due to gravity and the volume of the liquid is expressed as the product of its height h multiplied by the area of its base A. Dividing both sides of Equation 8.3 by the area, (A), this equation becomes

$$\frac{F_b - F_t}{A} = \rho g h$$

where $(F_b - F_t)/A$ is the pressure exerted by the column of liquid of depth h. We can replace the expression $(F_b - F_t)/A$ by the pressure P,

$$P = \rho g h \tag{8.4}$$

This equation provides an explanation for the use of the variety of units for measuring pressure. The height of a liquid column is directly proportional to the pressure supporting the column. Hence, if you know the height of a supported liquid column you can readily compute the pressure. For example, the standard pressure of the atmosphere at sea level will support a column of mercury 76 cm in height. What is standard atmospheric pressure in N/m²?

8.7 Gauge Pressure

If the liquid has a free surface point (Figure 8.3) upon which there is a pressure P_0, then the total or absolute pressure at a point at depth h below the surface is given by the sum of P_0 plus the pressure exerted by the liquid above that point.

$$P_T = P_0 + \rho g h \tag{8.5}$$

For example, the pressure at any depth h of a lake, which has an

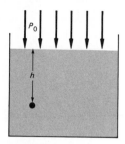

FIGURE 8.3
A liquid in a container with a free surface, upon which a pressure P_0 is acting.

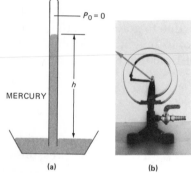

FIGURE 8.4
(a) A mercury barometer. The surface of the mercury in the bowl is open to the pressure of the atmosphere. The pressure on the surface of the mercury in the upper end of the closed tube is zero, except for the vapor pressure of mercury, which can be neglected at room temperature. (b) An aneroid pressure gauge.

atmospheric pressure on its surface, can be calculated by the above relationship where P_0 becomes equal to atmospheric pressure, P_A. Then, $P_T = P_A + \rho gh$.

The difference between P_T and P_A is called the *gauge pressure reading,* and in the case of a liquid the gauge pressure at any depth h becomes equal to ρgh. Usually a gauge pressure reading refers to the pressure above atmospheric pressure. For example, if your automobile tire is inflated to a gauge pressure reading of 2.07×10^5 N/m² (30 pounds per square inch, psi), the absolute or total pressure in the tire is 2.07×10^5 N/m² plus the atmospheric pressure, or about 3.08×10^5 N/m² (45 psi).

EXAMPLES

1. In a mercury barometer, invented by Torricelli in 1643, the pressure above the mercury in the tube is practically zero except for the mercury vapor pressure which can be neglected at room temperature (Figure 8.4). Hence, the atmospheric pressure is equal to

$$P_A = \rho_{Hg} \, gh_{Hg}$$

A barometer reading of 76 cm of Hg is considered standard atmospheric pressure. How long would the tube have to be for a water barometer? Since $\rho_{H_2O}gh_{H_2O} = \rho_{Hg}gh_{Hg}$ and $\rho_{H_2O} = 1$ gm/cm³, we see $h_{H_2O} = (76 \text{ cm})(13.6)$, or, 10.3 meters.

2. If a number of vessels of different shapes are connected as shown in Figure 8.5, it will be found that a liquid poured into the system will rise to the same height in each. That is, the pressure at all points on a horizontal plane such as PP' will have equal pressures in the vessel. This is consistent with Equation 8.5,

$$P_T = P_0 + \rho gh$$

which states that the pressure depends only on the depth below the liquid surface and not upon the shape of the containing vessel.

The device for measuring blood pressure is called a sphygmomanometer. The pressure calibration is in mm of mercury. An inflatable cuff is wrapped around the upper arm, and a stethoscope is placed over the artery in the crease of the elbow below the cuff. Before the cuff is inflated, blood is flowing unimpeded, and there is no sound in the stethoscope. As the cuff is inflated, the blood circulation is gradually shut off, and the doctor or nurse hears the thump-thump through the stethoscope resulting from pumping of blood by the heart. The cuff is inflated

FIGURE 8.5
Differently shaped portions of a vessel containing a liquid. The pressure at any point below the surface of the liquid depends upon the depth of the point below the surface of the liquid.

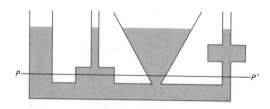

Home use of a sphygmomanometer. Note the height of the mercury column and the use of the stethoscope.

until the listener no longer hears the thump. This means the blood flow has temporarily been shut off by the pressure exerted by the cuff. The pressure is then reduced in the cuff by a controlled release valve, and the observer reads the sphygmomanometer at the exact moment a thump is heard in the stethoscope. This means blood is again flowing in the artery, and the corresponding pressure is called the *systolic pressure*. The pressure in the cuff is further decreased, and the thump disappears at a lower pressure which is also recorded. This is the *diastolic pressure*. There are many factors that influence the blood pressure reading, and there is a variation among individuals. A typical reading for an adult might be about 140/90, meaning a systolic pressure of 140 mm Hg and a diastolic of 90 mm Hg.

The pressures are produced by the heart. The human heart operates as a muscular pump which on contraction can exert a hydrostatic pressure of about 140 mm Hg (systolic) on the blood. On relaxation of the heart there is still tension in the muscle of the left ventricle to maintain a pressure of about 90 mm Hg (diastolic). The pressure exerted by the heart forces the blood out of the left ventricle of the heart into the aorta and through the blood vessel system. The blood is returned to the right atrium of the heart at very nearly zero gauge pressure.

8.8 Pascal's Principle

We have learned that the pressure at a point within a liquid does not depend upon the shape of the vessel, but only upon the depth of the point below the surface of the liquid. We can use this fact to design a mechanical system to multiply our strength. Consider a system like the one diagrammed in Figure 8.6a, which is filled with an incompressible liquid. If the pressure P_0 is increased in any way, such as by inserting a piston on top of the liquid, the pressure at any depth is increased by the same amount. The principle was originally stated by Pascal in 1653 in the following terms:

A pressure applied to a confined liquid is transmitted undiminished to all parts of the liquid and the walls of the containing vessel.

FIGURE 8.6
(a) A closed system filled with an incompressible liquid. The two pistons have different areas. They will support different forces since the pressure exerted by the liquid is the same on both pistons. (b) A laboratory hydraulic press. The scale was reading about 800 lb when the picture was taken. Note the relative size of the two pistons and the mechanical advantage of a lever system. Thus, a relatively small applied force produces a compressive force of 800 lb on the stick of lumber.

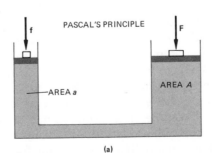

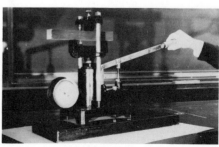

(a)

(b)

The above principle is illustrated by a hydraulic press of which Figure 8.6 is a schematic cross-sectional drawing.

We can use Pascal's principle to derive an expression for the load **F** we can lift on the large piston of area A by applying a force f on the small piston of area a. The magnitude of the pressure we apply to the small piston is given by the ratio of f to a. When this pressure f/a is applied to a liquid such as oil and is transmitted through the connecting pipe to the large piston where the pressure P must be the same. $P = f/a = F/A$ since by definition the pressure P is equal to F/A. We can use this equation to find the size of F,

$$F = f\frac{A}{a} \tag{8.6}$$

A hydraulic press is a force multiplying device with a theoretical mechanical advantage of A/a. What are other examples of Pascal's principle? Use the conservation of energy to show that the distance the load moves equals $(a/A)L$, where L is distance the applied force moves.

Pascal's principle can be used to explain the comfort of water beds. A water filled mattress applies pressure equally to all parts of a body it supports. The use of water mattresses with chronically ill patients can help prevent bed sores. Can you explain why?

8.9 Archimedes' Principle

You have seen an object floating on the surface of a body of water. Perhaps you have tried to lift a rock in water, and you have found that it is easier (that is, it requires a smaller force) to lift the rock in the water than after it is out of the water. Can you explain this phenomenon?

The explanation is a necessary consequence of the laws of fluid mechanics known as Archimedes' principle. If the body is entirely or partially immersed in a fluid at rest, the fluid exerts a pressure at every point of contact. The resultant of all of the pressure forces is called the buoyant force acting on the body. As the body is at rest, and in equilibrium, the horizontal components of the forces cancel each other, and the weight of the body and the resultant buoyant force must have the same line of action. The pressure on the lower side is greater than on the top side; hence the buoyant force acts upward. We can now determine the magnitude of the buoyant force.

Consider a mass surrounded by a fluid (Figure 8.7a) and think of an imaginary boundary isolating an equivalent amount of fluid (Figure 8.7b). We have learned that the pressure of the liquid does *not* depend upon the material of the surface. Hence, the buoyant force on the isolated body of the fluid is the same as that on the mass. So we must determine the force on a given volume of the fluid. The fluid, isolated by the imaginary boundary does not move; so the net force acting on it is zero. This means that the buoyant force is equal to the weight of the

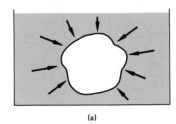 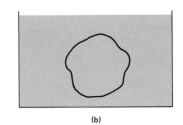

(a) (b)

fluid within the imaginary boundary. Hence, the buoyant force upon the mass is equal to the weight of the displaced liquid.

Archimedes' principle may be stated as follows: *Any object wholly or partially immersed in a fluid is buoyed up by a force equal to the weight of the displaced fluid.*

You can determine the specific gravity of a homogeneous solid by the following experiment (Figure 8.8): the body is weighed first in air and then completely immersed in water. W_a = weight in air, W_w = weight in water, $W_a - W_w$ = buoyant force = weight of displaced liquid. The weight of an object in air is given by the product of its density, its volume, and the acceleration of gravity. Since the solid is completely immersed in the water, the volume of displaced water will be the same as the volume of the solid. Hence, the weight of displaced water, which by Archimedes' principle is equal to the buoyant force $W_a - W_w$, is given by the product of the density of water, the volume of the *solid*, and the acceleration of gravity.

$$W_a - W_w = \rho_w V g$$

where V is the volume of the solid and ρ_w is the density of water. As shown below, the ratio of the weight in air to the buoyant force acting on a completely immersed solid is equal to the specific gravity.

$$\frac{W_a}{W_a - W_w} = \frac{\rho_m V g}{\rho_w V g} = \frac{\rho_m}{\rho_w} = \text{specific gravity of the solid} \qquad (8.7)$$

where ρ_m = density of solid and ρ_w = density of water.

Can you devise an experiment to determine the density of a liquid? There are at least two ways in which this can be done.

W_a — — —

— W_w

H_2O

W_w IS ALWAYS LESS THAN W_a.

Question

The traditional story is that Archimedes had been asked by a ruler of ancient Greece to determine if his newly purchased crown of gold was really made of solid gold. One evening while in his bath, Archimedes is rumored to have shouted, "Eureka, I have found it!" and to have run to the palace, clad only in his bath towel. Explain how Archimedes had found the crown not to be solid gold. Make up some reasonable numbers, and compute an answer for Archimedes.

FIGURE 8.9
(a) A schematic diagram of the flow
of water from the city water mains to
a faucet in a private home. (b) In
streamline flow, the various portions
of the liquid move along
nonintersecting lines.

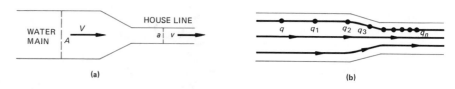

(a)   (b)

8.10 *Fluid Flow*

Consider the flow of water in the city water mains and the water line
when a faucet is opened. Assume a given volume of water is drained and
assume streamline flow, which means that every particle of water that
goes through point q will also follow through $q_1, q_2, q_3, \ldots q_n$, where the
q's represent points along any line of flow (see Figure 8.9). Let the cross-
sectional area of the water main be A and the cross-sectional area of the
line in the house be a. The volume that flows through the main is the
same as the volume which is drained from the faucet and also the same
as that which flows through the house line. The volume that flows
through a pipe in one second must be equal to the cross-sectional area
multiplied by a column length equal to velocity, v^1 in the main, v in the
house. The volume of water that flows in one second is given by the
product of velocity and area. Since the volume of flow is constant everywhere in the system, the velocity-area product must also be a constant,

$$\text{volume of flow per second} = v^1 A = va \tag{8.8}$$

This is an example of a conservation law. Note that the velocity of
flow will be faster through the smaller pipe than the larger pipe.

8.11 *Bernoulli's Theorem*

In our discussion of fluids we have used four physical quantities: pressure P, density of fluid, ρ, the velocity v, and the height h above some
reference level. A relationship between these four quantities was developed by Daniel Bernoulli to describe fluids in motion. Bernoulli's
theorem can be derived by considering an incompressible liquid moving
along a pipe from position 1 to position 2. (See Figure 8.10). In order to
move the liquid, we must exert a force F_1 at position 1. Then the surface

FIGURE 8.10
An incompressible liquid moving
along a bent pipe. The cross-
sectional area and the elevation of
the pipe change from position 1 to
position 2 of the pipe.

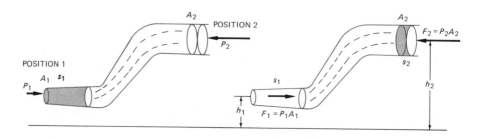

of the liquid near position 1 will move a distance s_1. But there may exist a resisting force F_2 at position 2 where the surface of the liquid will move a distance s_2. We can calculate the net work done on the liquid as we cause it to move from position 1 to 2. The net work will be given by the work we did on the liquid at position 1 minus the work done by the liquid against the resisting force at position 2,

$$\text{net work} = F_1 s_1 - F_2 s_2 \tag{8.9}$$

since work is the product of force times distance. Remember that the volume of liquid that moves near position 1 must be equal to the volume of liquid that moves near position 2 because the liquid is incompressible. So we can calculate the volume of moved liquid by multiplying the cross-sectional area of the pipe times the distance the liquid surface has moved,

$$\text{volume of moved liquid} = V = A_1 s_1 = A_2 s_2 \tag{8.10}$$

where A_1 and A_2 are the cross-sectional areas of the pipe at positions 1 and 2 respectively. We can compute the distances the liquid moved at 1 and 2 from the volume of the moved liquid V and the areas, A_1 and A_2,

$$s_1 = \frac{V}{A_1} \quad \text{and} \quad s_2 = \frac{V}{A_2}$$

Substituting these expressions for s_1 and s_2 into Equation 8.9, we obtain the net work as a function of pressure and volume as follows:

$$\text{net work} = F_1 s_1 - F_2 s_2$$

$$\text{net work} = F_1 \frac{V}{A_1} - F_2 \frac{V}{A_2} \tag{8.9}$$

But the ratio of force to area is the definition of pressure, so F_1/A_1 is equal to the pressure at position 1 and likewise F_2/A_2 is the pressure at position 2. Finally then we find the net work is given by the difference between the pressures at positions 1 and 2 times the volume of liquid that moved,

$$\text{net work} = (P_1 - P_2)V \tag{8.11}$$

From the work-energy theorem we know that the net work that is done on the liquid is equal to the sum of the increase in potential energy and the increase in kinetic energy. As you remember potential energy is energy of position, and kinetic energy depends upon the velocity.

The change in the potential energy between position 1 and position 2 is given by $(mgh_2 - mgh_1)$ where m is the mass of the liquid, g is the acceleration of gravity and h_1 and h_2 are the elevations of positions 1 and 2. Equating the net work to the sum of potential-energy change and kinetic-energy change, we find that net work = change in kinetic energy + change in potential energy:

$$(P_1 - P_2)V = \tfrac{1}{2}(mv_2^2 - mv_1^2) + mgh_2 - mgh_1 \tag{8.12}$$

This equation is a statement of conservation of energy (see Chapters 2 and 4). To obtain an expression that contains the density of the liquid, divide Equation 8.12 by the volume of the moved liquid V. Since the density ρ is given by the ratio of mass to volume, $\rho = m/V$, we obtain

$$(P_1 - P_2) = \tfrac{1}{2}\rho v_2{}^2 - \tfrac{1}{2}\rho v_1{}^2 + \rho g h_2 - \rho g h_1 \qquad (8.13)$$

Rewriting the equation so that all the variables with a subscript 1 are on the left side of the equation and all the subscript 2 variables are on the right, we get

$$P_1 + \rho g h_1 + \tfrac{1}{2}\rho v_1{}^2 = P_2 + \rho g h_2 + \tfrac{1}{2}\rho v_2{}^2 \qquad (8.14)$$

where the subscripts 1 and 2 refer to the two different positions. Bernoulli's equation may be written simply as follows,

$$P + \rho g h + \tfrac{1}{2}\rho v^2 = \text{constant} \qquad (8.15)$$

where P is the absolute pressure (not the gauge pressure), and ρ is the mass density. You will note that each term has the dimensions of pressure.

The total pressure in streamline flow is constant. The total pressure is made up of an applied pressure term (P), an elevation term ($\rho g h$), and a velocity term ($\tfrac{1}{2}\rho v^2$). You can divide each term in Equation 8.15 by the quantity ρg, and then each term is expressed in units of length. The term $P/\rho g$ is called a *pressure head*, h is an *elevation head*, and $\tfrac{1}{2}(v^2/g)$ is a *velocity head*. So the total head in streamline flow is a constant,

pressure head + elevation head + velocity head = constant

$$\frac{P}{\rho g} + h + \tfrac{1}{2}\frac{v^2}{g} = \text{constant} \qquad (8.16)$$

Both Equation 8.15 and Equation 8.16 are equivalent ways of expressing the conservation of energy. Conservation of energy is one of the conservation laws introduced in Chapter 2.

Let us examine Equation 8.15, $P + \rho g h + \tfrac{1}{2}\rho v^2 = \text{constant}$, more carefully. Consider two points in a horizontal flow, $h_1 = h_2$. Then we have $P_1 + \tfrac{1}{2}\rho v_2{}^2 = \text{constant}$. If the velocity at position 1 is greater than the velocity at position 2, the pressure at 1 will have to be less than the pressure at 2 for the equation to remain true. This relationship shows that in a region of higher velocity there is lower pressure.

EXAMPLE

Suppose that you have a spool, a pin, and a card. The pin is pushed through the card and placed in the end of the hollow spool. (See Figure 8.11). (The only purpose of the pin is to keep the card centered.) You try to blow the card from the spool by blowing into the hollow center of the spool. The air escapes between the end of the spool and the card. This produces an area of low pressure, and the atmospheric pressure presses the card to the end of the spool.

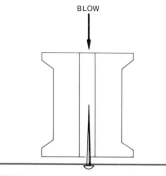

BLOW

FIGURE 8.11
A spool of thread, a card, and a pin can be used to demonstrate the effects predicted by Bernoulli's equation.

FIGURE 8.12
A tank with an orifice 0.2 m above
the bottom of the tank. Water is kept
at a level of 10 m above the bottom.

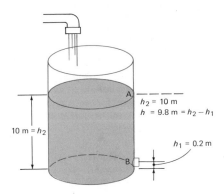

$h_2 = 10$ m
$h = 9.8$ m $= h_2 - h_1$

10 m $= h_2$

$h_1 = 0.2$ m

Bernoulli's equation finds application in almost every aspect of fluid flow. Some applications are:

1. the lift on an air foil
2. the operation of an atomizer
3. a curving baseball or golf ball slice
4. the force pushing two passing trucks together on a highway.

(You have undoubtedly noted this last example in your travels.) You should be able to explain each of these phenomena in terms of Bernoulli's equation.

EXAMPLES

1. Consider water maintained at level h_2 (10 m) in the tank and opening at height h_1 (0.2 m) above the bottom of the tank (see Figure 8.12). Find the velocity of outflow at the opening. The pressure at surface A and surface B are each equal to atmospheric pressure. The velocity at A is 0 if a constant height is maintained. Since h_2 is constant and $v_2 = 0$, Bernoulli's equation becomes

$$\rho g h_2 = \tfrac{1}{2}\rho v_1{}^2 + \rho g h_1$$

We know also that $h_2 - h_1 = h$, so

$$\tfrac{1}{2}\rho v_1{}^2 = \rho g h$$

$$v_1 = \sqrt{2gh} = \sqrt{2(9.80)(9.80)} = 13.8 \text{ m/sec}$$

2. Water flows through the pipe shown in Figure 8.13 at a rate of 120 liters per second. The pressure at position A is 2.00×10^5 N/m². The cross-section of position B is 60.0 cm², and the cross section of position A is 100 cm². What is the velocity at A and at B? What is the pressure at B?

We can let Q be the volume flow rate as follows:

$$Q = Av = 120 \times 10^{-3} \text{ m}^3/\text{sec} = A_1 v_1 = A_2 v_2$$

$$0.120 \text{ m}^3/\text{sec} = 100 \times 10^{-4} \text{ m}^2 \times v_1$$

$$v_1 = \frac{0.12 \text{ m}^3/\text{sec}}{10^{-2} \text{ m}^2} = 12.0 \text{ m/sec}$$

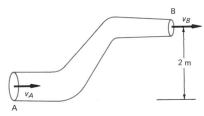

B
v_B

2 m

v_A

A

FIGURE 8.13
A pipe of varying size is used to
transport water.

$$v_2 = \frac{0.120 \text{ m}^3/\text{sec}}{60.0 \times 10^{-4} \text{ m}^2} = 20.0 \text{ m/sec}$$

$$P_1 + \tfrac{1}{2}\rho v_1{}^2 = P_2 + \tfrac{1}{2}\rho v_2{}^2 + \rho g h$$

$$(2.00 \times 10^5 \text{ N/m}^2) + \tfrac{1}{2}(1000 \text{ kg/m}^3)(12.0 \text{ m/sec})^2$$
$$= P_2 + \tfrac{1}{2}(1000 \text{ kg/m}^3)(20.0 \text{ m/sec})^2 + (1000 \text{ kg/m}^3)(9.80 \text{ m/sec}^2)(2.0 \text{ m})$$

$$P_2 = 2.00 \times 10^5 \text{ N/m}^2 + \tfrac{1}{2}(1000)(12)^2 \text{ N/m}^2 - \tfrac{1}{2}(1000)(20) \text{ N/m}^2$$
$$- (2000)(9.8) \text{ N/m}^2$$

$$= (2.00 \times 10^5 + 72 \times 10^3 - 200 \times 10^3 - 19.6 \times 10^3) \text{ N/m}^2$$

$$= 2.00 \times 10^5 \text{ N/m}^2 - 1.476 \times 10^5 \text{ N/m}^2 = 5.24 \times 10^4 \text{ N/m}^2$$

8.12 Poiseuille's Law of Viscous Flow

Bernoulli's equation is applicable for fluid flow cases in which there is no friction. We will now consider the case of viscous flow where friction must be considered. A diagram of a section of a tube in which there is a flowing viscous liquid is shown in Figure 8.14. The dotted line represents a transverse plane. The vectors represent the velocity of the liquid in the tube. An analysis of liquid flow of this type shows that the velocity of flow varies from maximum at center of the tube to minimum at the wall, $v = v(r)$, that is, v is a function of r. The velocity is constant for a given distance r from the center of the tube throughout the length of the tube. The velocity of liquid flow is zero at the wall for nonturbulent flow, and there is a frictional force $\mathbf{F}_f$ that opposes motion since the liquid is viscous. Calculus methods can be used (see Section 8.14) to derive the expression for the velocity of liquid flow as a function of the distance r from the center of the tube, the radius of the tube R, the pressure gradient along the tube $(P_1 - P_2)/L$, and the viscosity of the liquid η,

$$v = \frac{P_1 - P_2}{4\eta L}(R^2 - r^2) \tag{8.17}$$

where L is the length of the tube, $P_1 - P_2$ is the difference in pressure between the ends of the tube, η is the coefficient of viscosity of the liquid, and R is the radius of the tube. The coefficient of viscosity is a relative measure of liquid friction and is equivalent to the ratio of the force per unit area to the change in velocity per unit length perpendicular to the direction of flow. A viscosity of one 10^{-5} N-sec/cm^2 is called a poise. Small viscosities are usually expressed in centipoises (1 centipoise = 10^{-2} poise). The viscosity of fluids is temperature dependent, decreasing with increasing temperature. The coefficient of viscosity for water at normal room temperature is about one centipoise. Olive oil has a viscosity of about 100 times that of water, and the viscosity of castor oil is about 1000 times that of water. The viscosity of whole human blood is about four times the viscosity of water.

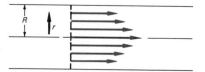

FIGURE 8.14
A section of a tube carrying a viscous liquid flowing in a horizontal direction. The dotted line represents a vertical plane through the tube.

The volume of flow per second (rate of flow) through a tube is given by the product of the average velocity of the liquid flow times the cross-sectional area A of the tube,

$$\text{rate of flow} = \text{average liquid velocity} \times \text{area} = \bar{v}A \tag{8.18}$$

The velocity of liquid flow varies from zero at the tube wall to $R^2(P_1 - P_2)/4\eta L$ at the center of the tube. The average velocity of liquid flow is one-half of the velocity at the center of the tube. Since the area of the tube is given by πR^2, we can obtain an equation that relates the rate of flow in m³/sec of a viscous liquid to the pressure gradient, the radius of the tube, and the viscosity of the liquid,

$$\text{rate of flow (m}^3\text{/sec)} = \bar{v}A$$

$$= \tfrac{1}{2}\left(\frac{P_1 - P_2}{4\eta L}\right) R^2 \pi R^2$$

$$= \frac{\pi}{8\eta}\left(\frac{P_1 - P_2}{L}\right) R^4 \tag{8.19}$$

This equation is the law for viscous flow derived by Poiseuille (pronounced Pwazswee). Poiseuille's law shows that the rate of flow of a viscous liquid is proportional to the pressure gradient $(P_1 - P_2)/L$, inversely proportional to the viscosity of the liquid η, and proportional to the fourth power of the radius R of the tube.

Even though blood vessels are neither circular nor rigid and even though the coefficient of viscosity of blood varies with the pressure gradient, Poiseuille's law is useful in considering the rate of blood flow along the vessels of blood circulation.

EXAMPLES

1. The flow of a liquid through a hypodermic needle is an example of an application of Poiseuille's law of viscous flow. Note that the rate of flow depends upon the fourth power of the radius of the needle and the first power of the pressure gradient (i.e., force exerted on the plunger).
2. Consider the flow of intravenous (I.V.) fluids as administered in a hospital. A typical I.V. arrangement is shown in Figure 8.15. What variables of this system will control the flow of the fluid? Why is the I.V. bottle suspended above the patient? If the I.V. needle is reduced to half its original size, how high would you have to raise the I.V. bottle to keep the flow rate constant?

FIGURE 8.15
Administration of intravenous fluid.

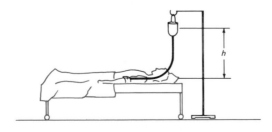

8.13 *Blood Flow in the Human Body*

The blood circulation system in the human body is a closed system. The continuous flow of blood is caused by a circulating pump, the human heart. The heart works to force blood to flow through the human system as shown in Figure 8.16. The pressure of the blood at various parts of the circulatory system is shown in Figure 8.17. The average pressure of the blood as it enters the aorta is 100 mm Hg. The blood pressure has dropped to 97 mm Hg as the blood, leaving the aorta, enters the arteries. The blood pressure at the entrance to the arterioles is 80 mm Hg. The pressure difference across the arterioles is about 55 mm Hg. So the blood pressure at the entrance to the capillaries is only 30 mm Hg. Of that pressure 20 mm Hg is lost in transit through the capillaries. Thus the blood pressure in the veins is not more than 10 mm Hg.

An interesting property of human blood flow is the total quantity of blood that flows through the human system. A typical value for the speed of blood flow through the aorta of an adult is about 35 cm/sec. The aorta, which has a diameter of about 1.8 cm, has a cross-sectional area of about 2.5 cm². The total volume of blood flow is the product of the velocity times the area (see Equation 8.18):

quantity of flow = (35 cm/sec)(2.5 cm²)

$$= 88 \text{ cm}^3/\text{sec}$$

So in a typical human being, about 100 cm³ of blood passes through the system in one second. Of course, during periods of physical exercise the total flow of blood increases. This increase is accomplished by an increase in the blood pressure and by a decrease in the resistance of the circulatory system to the flow of blood. This decrease in resistance is caused by the dilation of the blood vessels, that is, an increase in the diameter of the vessels. Since the flow of blood is proportional to the fourth power of the diameter of the blood vessels, a small increase in the vessel diameters can produce a substantial increase in the rate of blood flow. In a similar way, anything that decreases the effective diameter of the blood vessels will cause the heart to work harder than normal

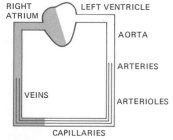

FIGURE 8.16
A schematic diagram of the human circulatory system.

FIGURE 8.17
The blood pressure (mm Hg) in various parts of the circulatory system.

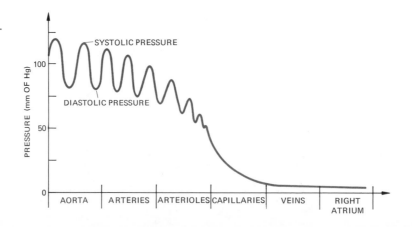

to maintain a constant rate of blood flow. We can calculate the rate at which the human heart does work from the product of the force exerted by the heart F and the velocity of the blood v (see Chapter 5, Equation 5.12),

$$\text{power} = Fv \tag{5.12}$$

Where F is the average force acting on the blood and is given by the blood pressure times the area of the aorta and where v is the velocity of the blood flow and is given by the ratio of the rate of blood flow divided by the area of the aorta. Hence, we can substitute these quantities into Equation 5.12 to show that the power output of a human heart is given by the product of the blood pressure times the rate of blood flow,

$$\text{power} = Fv = \text{pressure} \times \text{volume rate of flow} \tag{8.20}$$

An average blood pressure in a human is about 100 mm Hg (1.3×10^4 N/m^2), and the quantity of flow is on the order of 100 cm^3/sec (1×10^{-4} m^3/sec). So we find that the normal power output of the heart is 1.3 watts, which is about one percent of the total power dissipated by the human body.

ENRICHMENT
8.14 Viscous Flow

When considering the flow of a viscous liquid through a tube, we know that the velocity of the liquid will vary from one point in the tube to another. For example, we know that the viscous liquid will tend to stick to the inside wall of the tube, so the velocity of flow of the liquid will be small near the wall. In order to find a mathematical expression for the flow of a viscous liquid through a tube, let us make the assumption that the velocity of flow is maximum at the center of the tube, that the velocity at the wall is zero, and that the velocity of flow can only be different at different distances from the center of the tube (see Figure 8.14).

The coefficient of viscosity is defined as the ratio of the magnitude of the force required to slide along a unit area of liquid to the velocity gradient in a direction perpendicular to the direction of flow. Let us consider a small cylinder of liquid (radius r) inside of the tube (radius R and length L) as shown in Figure 8.18.

The force acting to slide this small cylinder of radius r in the direction of flow is given by the difference between the pressures acting on the

FIGURE 8.18
A tube of radius R and length L transmitting a viscous liquid from a pressure of P_1 to a lower pressure P_2.

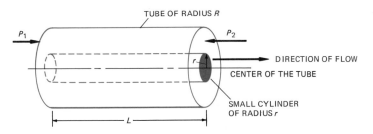

TUBE OF RADIUS R

P_1

P_2

r

DIRECTION OF FLOW

CENTER OF THE TUBE

SMALL CYLINDER OF RADIUS r

L

ends of the cylinder times the area of the end of the cylinder,

$$\text{force} = (P_1 - P_2)\pi r^2 \tag{8.21}$$

The surface area of liquid that is being pushed along by this force is just the outer surface area of the small cylinder,

$$\text{surface area of the cylinder} = 2\pi r L \tag{8.22}$$

Hence, the force acting per unit area to slide the liquid along is given by the ratio of these two equations, Equation 8.21 and Equation 8.22,

$$\text{force per unit area} = \frac{(P_1 - P_2)\pi r^2}{2\pi r L}$$

$$= \frac{(P_1 - P_2)r}{2L} \tag{8.23}$$

We can put this expression into the equation that defines the coefficient of viscosity and obtain an expression that will relate the velocity gradient to the radius of the small cylinder:

$$\eta = \frac{\text{force per unit area}}{\text{velocity gradient}} = \frac{(P_1 - P_2)(r/2L)}{(-dv/dr)} \tag{8.24}$$

where $-dv/dr$ is the velocity gradient in the direction perpendicular to the direction of flow and η is the viscosity. The minus sign in front of dv/dr indicates that the velocity decreases as the value of r increases. In order to find an expression for the velocity of flow as a function of the distance from the center of the tube, we can rearrange the terms in Equation 8.24 to separate the variables v and r. Then we integrate both sides of the equation, as follows:

$$\int dv = -\int \frac{(P_1 - P_2)r}{2\eta L} \, dr \tag{8.25}$$

$$v = -\frac{(P_1 - P_2)}{2\eta L} \frac{r^2}{2} + \text{constant} \tag{8.26}$$

To evaluate the constant in Equation 8.26, we require that the velocity of flow be zero at the wall of the tube where $r = R$:

$$v = 0 = \frac{-(P_1 - P_2)}{4\eta L} r^2 + \text{constant}$$

when $r = R$. Thus the constant has the value

$$\text{constant} = +\frac{(P_1 - P_2)}{4\eta L} R^2$$

and the completed expression for the velocity of flow is the same as presented in Equation 8.17.

$$v = \frac{P_1 - P_2}{4\eta L} (R^2 - r^2) \tag{8.17}$$

FIGURE 8.19
The parabolic distribution of flow velocities for a viscous liquid under conditions of streamline flow.

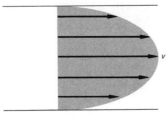

v

FIGURE 8.20
A viscous liquid under conditions of streamline flow contains cylindrical shells moving at constant velocity.

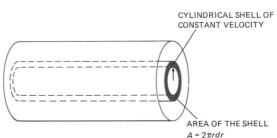

CYLINDRICAL SHELL OF
CONSTANT VELOCITY

AREA OF THE SHELL
$A = 2\pi r dr$

This expression gives a parabolic radial distribution for the flow of a viscous liquid (see Figure 8.19).

We can use the expression for the radial dependence of velocity to derive Poiseuille's law. You may recall that the rate of flow is given by the velocity of flow times the area. Since the liquid flow velocity depends only on the distance of the liquid from the center of the tube, all portions of the liquid at the same distance from the center of the tube are moving with the same velocity. In other words, the liquid moves along as if it were thin cylindrical shells sliding inside of one another (see Figure 8.20). Each cylindrical shell has a different value of flow velocity as given by Equation 8.17. The area of the end of the shell of radius r is the circumference of the shell ($2\pi r$) times the shell thickness (dr),

$$\text{area} = 2\pi r \, dr \tag{8.27}$$

Hence the rate of flow for each shell is given by the product of velocity times area,

$$\text{rate of flow for a shell} = v \times \text{area} \tag{8.28}$$

$$= \left[\frac{P_1 - P_2}{4\eta L} (R^2 - r^2) \right] 2\pi r \, dr \tag{8.29}$$

by substitution from Equations 8.17 and 8.27. The total flow rate for all of the liquid in the tube is the sum of the flow for all the shells; so we integrate Equation 8.29 from $r = 0$ to $r = R$,

$$\text{rate of flow} = 2\pi \left(\frac{P_1 - P_2}{4\eta L} \right) \int_0^R (R^2 - r^2) r \, dr \tag{8.30}$$

$$= \pi \frac{(P_1 - P_2)}{2\eta L} \left[\frac{R^2 r^2}{2} \bigg|_0^R - \frac{r^4}{4} \bigg|_0^R \right] \tag{8.31}$$

$$= \pi \frac{(P_1 - P_2)}{2\eta L} \left(\frac{R^4}{2} - \frac{R^4}{4} \right) \tag{8.19}$$

$$= \frac{\pi}{8\eta} \left(\frac{P_1 - P_2}{L} \right) R^4 \tag{8.19}$$

which is Poiseuille's law of viscous flow.

SUMMARY

Use these questions to evaluate how well you have achieved the goals of this chapter. The answers are given at the end of this summary with the section number where you can find the related content material.

Definitions

Answer questions 1 to 7 from the definitions of the following terms:

fluid
density
specific gravity
absolute pressure

gauge pressure
buoyant force
streamline flow
viscosity

1. A rigid, boxlike object with dimensions of 12 cm × 12 cm × 8 cm has a mass of 3.4 kg. This object is not a fluid because _____.
2. The object in question 1 will _____ (*float/sink*) in water because
   a. it is made out of plastic
   b. it is heavier than air
   c. it is lighter than water
   d. it has a density greater than one
   e. it has a specific gravity greater than one
   f. it has a density less than 10^3 kg/m^3
   g. it has a specific gravity greater than 1 g/cm^3
3. If you add the magnitude of the _____ pressure to the value of the _____ pressure, you will obtain the _____ pressure. Therefore, for the normal systems the _____ pressure is always a smaller number than the _____ pressure.
4. Explain why objects weighed in air and in a vacuum do not weigh the same.
5. If a liquid does exert a buoyant force on an object placed in it, how can you determine if an object will sink of float in the liquid?

6. Compare streamline flow with the assumptions made to treat the viscous flow of a liquid in Section 8.10
7. Viscosity may be considered as an analog to what property of a mechanical system?
   a. potential energy
   b. kinetic energy
   c. actual mechanical advantage
   d. theoretical mechanical advantage
   e. power output
   f. friction
   g. torque
   h. perspicacity
   i. virtuosity

Fluid Laws

8. Group together into sets the letters that represent concepts or words that belong together, and explain the common features of the set. You may put a term in more than one set.

   a. Pascal
   b. Archimedes
   c. Bernoulli
   d. hydrostatics
   e. incompressible fluids
   f. streamline flow
   g. $\rho g h$
   h. specific gravity
   i. conservation of energy
   j. conservation of fluid flow
   k. hydraulic press
   l. buoyancy
   m. golden crown
   n. vA = constant
   o. $P + \rho g h + \frac{1}{2}\rho v^2 = $ constant
   p. $\frac{1}{2}kA^2 + \frac{1}{2}mv^2 = $ constant
   q. a curving baseball
   r. a hot-air balloon

Fluid Problems

9. A hydraulic jack which is designed to lift the head of a hospital bed has a small piston of diameter 1.5 cm which is used to apply pressure to a liquid. The

liquid then applies pressure to a large cylinder of 9.0 cm in diameter. What is the ratio of the force applied by a nurse to the small cylinder to the load lifted by the large cylinder?

a. 6.0   d. 36.0
b. 0.17   e. 13.5
c. 0.028

10. What similarities and differences will you notice if you go swimming in fresh water (specific gravity = 1.00), in the ocean (sp. gr. = 1.03), the Great Salt Lake (sp. gr. = 1.12), and the Dead Sea (sp. gr. = 1.18)?

11. Water is flowing through a closed pipe system, and at one point the velocity of flow is 10.0 m/sec while at a point 30.0 m higher the velocity of flow is 12.5 m/sec. If the pressure at the lower point is 4.00×10^5 N/m² (a) what is the pressure at the upper point, and (b) if the water flow stops, what is pressure at the upper point?

a. 4.00×10^5 N/m², 4.00×10^5 N/m²
b. 4.00×10^5 N/m², 6.94×10^5 N/m²
c. 4.00×10^5 N/m², 1.06×10^5 N/m²
d. 7.78×10^4 N/m², 1.06×10^5 N/m²
e. 7.78×10^4 N/m², 4.00×10^5 N/m²
f. 1.06×10^5 N/m², 7.78×10^4 N/m²

Viscous Flow Problem

12. An elderly heart patient with hardening of the arteries has the effective diameter of his blood vessels reduced by 16 percent. What is the reduction of blood flow at constant pressure, and by what factor must his blood pressure increase if the rate of flow is to remain constant?

a. The flow would decrease by 16 percent; so his pressure must increase by 16 percent.
b. The flow would decrease by a factor of 0.71; so his blood pressure would have to increase by a factor of 1.42.
c. The flow would decrease by 41 percent; so the blood pressure would have to increase by 59 percent.
d. The flow would decrease by a factor of two; so his blood pressure would have to increase by a factor of two.
e. The flow would decrease by 84 percent, and the pressure would increase by 16 percent.

Answers

1. It is rigid (Section 8.2)
2. Sink, e (Section 8.3)
3. atmospheric, gauge, absolute, gauge, absolute (Section 8.7)
4. because the buoyant force exerted upon objects weighed in air reduces their apparent weight (Section 8.9)
5. If its weight is greater than the buoyant force acting on it, the object will sink. Can you rephrase this statement in terms of density or specific gravity using Archimedes principle? (Section 8.9)
6. The viscous flow in Section 8.12 is treated as if it were streamline flow
7. f (Section 8.12)
8. a, d, e, k (Section 8.8)
   b, h, l, m, r (Section 8.9)
   c, f, i, o, q (Section 8.11)
   j, n (Section 8.10)
9. d (Section 8.8)
10. The buoyant force is the same in all cases (i.e., equal to your weight), but less liquid is displaced as you swim in liquid of greater specific gravity. You float with a larger portion of your body above the surface of the liquid (Section 8.9)
11. d (Section 8.11)
12. d (Section 8.12).

ALGORITHMIC PROBLEMS

Listed below are the important equations from this chapter. The problems following the equations will help you learn to translate words into equations and to solve single concept problems.

Equations

$$\rho = \frac{m}{V} \tag{8.1}$$

$$P = \lim_{\Delta A \to 0} \frac{\Delta F}{\Delta A} \tag{8.2}$$

$$P = \rho g h \tag{8.4}$$

$$P_\mathrm{T} = P_0 + \rho g h \tag{8.5}$$

$$F = \frac{fA}{a} \tag{8.6}$$

$$\text{specific gravity} = W_\mathrm{a}/(W_\mathrm{a} - W_\mathrm{w}) \tag{8.7}$$

$$v'A = va \tag{8.8}$$

$$P + \rho g h + \tfrac{1}{2}\rho v^2 = \text{constant} \tag{8.15}$$

$$v = \frac{(P_1 - P_2)}{4\eta L}(R^2 - r^2) \tag{8.17}$$

$$\text{rate of flow} = \frac{\pi}{8\eta}\frac{(P_1 - P_2)}{L} R^4 \tag{8.19}$$

Problems

1. Water moves through a horizontal pipe (radius = 3 cm) with a velocity of 1 m/sec. Find the speed of the water in a section of pipe that has been constricted to a radius of 1.5 cm.
2. Find the pressure difference between the surface and the bottom of a lake 20.0 m deep. The density of water is 1.00×10^3 kg/m³.
3. Water is flowing through a horizontal pipe. The velocity of the flow at point A is 1.0 m/sec, and at point B the velocity is 2.0 m/sec. Find the pressure difference between points A and B.
4. Find the flow rate of olive oil through a pipe 0.5 m long with a radius of 1 cm. The viscosity of the oil is 0.18 N-sec/m². The pressure difference across the pipe is 2×10^4 N/m².
5. A cylinder "weighs" 27 g in air and 17 g in water. Find the specific gravity of the cylinder material.
6. Find the volume of the cylinder in Problem 5.

Answers

1. 4 m/sec
2. 1.96×10^5 N/m²
3. 1.5×10^3 N/m²

4. 9×10^{-4} m³
5. 2.7
6. 10 cm³

EXERCISES

These exercises are designed to help you apply the ideas of a section to physical situations. When appropriate the numerical answer is given at the end of the exercise.

Section 8.3

1. If the earth's atmosphere were uniform with a density of .00129 g/cm³, how high would the atmosphere ex-

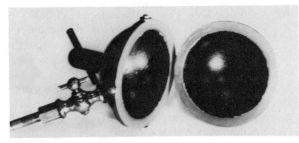

FIGURE 8.21
Magdeburg hemispheres. Exercise 3.

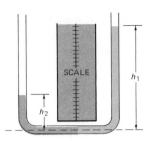

FIGURE 8.22
A U-tube containing two noninteracting liquids.

tend? That is, what is the height of a column of air 1 m^2 in area which has a weight of 1.01×10^5 N? [7.99 km]

Section 8.5

2. Suppose that an airplane is flying at an altitude where the atmospheric pressure is 4.03×10^4 N/m^2 and the airplane has gradually become depressurized, so that the internal pressure is equal to atmospheric pressure. The airplane rapidly descends to an airport where the atmospheric pressure is 1.01×10^5 N/m^2. What is the increased force on the eardrum (area 0.300 cm^2) of a passenger? Perhaps you have experienced this sensation to some degree in airplane travel. [1.82 N]

3. A pair of Magdeburg hemispheres (see Figure 8.21) are 10.0 cm in diameter. The barometric pressure is 75 cm Hg. What force will be required to separate them if the interior is completely evacuated? If the pressure inside is 10 cm Hg? [785 N; 680 N]

Section 8.6

4. Compute the equivalent pressure of 76.0 cm of Hg in dynes/cm^2 and in newtons/m^2. [1.02×10^6 dynes/cm^2, 1.02×10^5 N/m^2]

5. The systolic blood pressure of an individual is 140 mm Hg, and the diastolic blood pressure of the same individual is 90.0 mm Hg. If the aneroid sphygmomanometer is to be calibrated in N/cm^2, what will the equivalent readings be? For constant volume of blood flow compare the work done by this individual's heart with the work done by the heart of an individual who has high blood pressure of 180 mm Hg systolic and 120 mm Hg diastolic. [1.87 N/cm^2, 1.20 N/cm^2, about 1.3]

6. The U-tube shown in Figure 8.22 contains two liquids that do not interact chemically. The density of one is ρ_1 and the density of the second is ρ_2. What is the ratio of height h_1 and h_2? Suppose one liquid is Hg and the other is oil of specific gravity 0.800. If h_{Hg} is 2.00 cm, what is the height of the oil column? If you wished to construct an open tube barometer to indicate small pressure differences would you use a liquid of high or low density? Why? [$h_1/h_2 = \rho_2/\rho_1$, 34.0 cm]

Section 8.7

7. The gauge pressure in an automobile tire is 2.07×10^5 N/m^2. If the wheel supports 4500 N, what is the area of the tire in contact with the road? [2.20×10^{-2} m^2]

Section 8.8

8. The areas of the small and large pistons in a hydraulic press are 1.00 cm^2 and 30.0 cm^2. What force must be applied to the small piston in order to lift a 3600-N load? Through what distance must the applied force act if the load is raised two meters? [120 N, 60 m]

9. The piston under a barber's chair is 4 cm in diameter. If the weight of the chair and its occupant is 250 N, how much pressure is required to raise the chair? If this pressure is produced by means of a plunger 0.500 cm^2 in area, what force must be applied to the plunger? [19.9 N/cm^2, 9.95 N]

Section 8.9

10. A uniform stick of wood, 100 cm long and of density 0.70 g/cm^3, is made to float vertically in water. What length is submerged? How deep would it float in a liquid of density 0.80 g/cm^3? Such a stick could be made into a crude hydrometer. If it were to be used in

liquids in specific gravity of 0.8 to 1.2, how long would the scale be? [70 cm, 87 cm, 29 cm]

11. A 240-kg metal block has a volume of 0.200 m³. The block is suspended by a cord and submerged in oil that has a density of 770 kg/m³. Find
    a. the buoyant force
    b. the density of the metal block
    c. the specific gravity of the oil
    d. the tension in the cord [a. 1.51×10^3 N; b. 1.20×10^3 kg/m³; c. 0.77; d. 840 N]

12. A casting is made of material that has a density of 7.5 g/cm³. In air the casting weighs 1.47 N. When submerged in the water the casting weighs 1.07 N. Is the casting solid, or does it have a cavity? If it has a cavity, what is the volume of the cavity? If there is no cavity, how much should the casting weigh in water? [hollow, 20 cm³, 1.27 N]

13. A cylinder of aluminum is weighed on an equal arm balance. The aluminum is balanced by a 200.0-g brass weight. Assume the density of the brass weight to be 8.00 g/cm³, the density of air to be 1.23×10^{-3} g/cm³, and the density of aluminum to be 2.7 g/cm³. What is the buoyant force on the brass weights? What is the buoyant force on the aluminum cylinder? What is the true mass of the aluminum cylinder? [3.02×10^{-4} N, 8.93×10^{-4} N, $(200 + 6.03 \times 10^{-2})$ g]

Section 8.11

14. Assume that from the human aorta, radius of 1 cm, carrying blood from the heart, the cardiac outflow is 5 liters per minute, and the mean blood pressure is 100 mm Hg. Find the velocity of flow in the aorta and the work done per minute. [26.5 cm/sec, 66 J/min]

15. In a perfume aspirator air is blown across the top of the tube which dips into the perfume. What is the minimum air velocity that will cause the perfume to rise to the top of the tube which is 10 cm long if $\rho_{air} = 1.25 \times 10^{-3}$ g/cm³ and $\rho_{perfume} = 0.9$ g/cm³. [37.5 m/sec]

16. A circular hole 2 cm in diameter is cut in the side of a large vertical pipe 8 m below the water level in the pipe. What is the velocity of outflow and the volume discharged in one minute? [12.5 m/sec, 0.236 m³]

17. It is desired to refuel a plane at the rate of 200 liter/min. The fuel line is an 8.00 cm diameter hose connected to a pump 1.00 m above the ground. A 5.00 cm diameter nozzle delivers gasoline to the plane 5.00 m above the ground. Find the speed of the fuel at the nozzle, the speed of the fuel in the line near the pump, and the pressure difference between the pump and the nozzle. (Take the specific gravity of gasoline to be 0.700.) [170 cm/sec, 66.3 cm/sec., 2.83×10^4 N/m²]

18. A fluid of density 0.800 gm/cm³ is flowing through a 200-cm length of tube that is 1.00 cm in diameter. The flow is found to be 10.0 cm³/sec, and the pressure drop over the length of the tube is 3.25×10^5 dynes/cm². What is the viscosity of the fluid. What are the units? [3.99 poise]

19. Given a hypodermic syringe with an inside diameter of 1.00 cm, and a needle with an inside diameter of 0.500 mm, find the force needed to keep the fluid from coming out of the needle if a 1.00 N force is applied to the plunger. Find the necessary plunger force needed to inject fluid into an artery where the blood pressure is 90.0 mm Hg. [2.50×10^{-3} N, 9.42 N]

Section 8.12

20. Find the pressure necessary to move serum through an intravenous injection tube (radius = 1.00 mm, length = 3.00 cm) at the rate of 1.00 cm³/sec into an artery where the pressure is 100 mm Hg. The viscosity of the serum is 7.00×10^{-3} poise. [1.34 N/cm²]

PROBLEMS

These problems may involve more than one physical concept. The numerical answer is given at the end of each problem.

21. An airplane weighing 5000 N has a wing area of 12 m². What difference of pressure on the two sides of the wing is required to sustain the plane in level flight?

Assume the density ρ of air is 1.2 kg/m³. If the velocity of air relative to the wing is 40 m/sec on the lower side, what is the relative velocity on top of the wing? [420 N/m², 48 m/sec]

22. The large section of the pipe in Figure 8.23 has a cross section of 40.0 cm² and the small pipe's cross section is 10.0 cm². A volume of 30 liters of water is

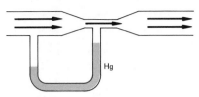

FIGURE 8.23
Liquid flowing in a smoothly tapered tube with two regions of different cross-sectional areas. A U-tube manometer is connected to two parts of the system of different cross-sectional areas. The levels of the mercury in the tube indicate that pressure is lower at the region of higher velocity.

discharged in 5 sec. Find the velocities in both the small and the large pipe. What is the difference in pressure between these portions? What difference in height of a mercury column corresponds to this pressure difference? [150 cm/sec, 600 cm/sec, 1.70×10^5 dynes/cm², 127 mm]

23. Reynolds discovered that the transition from streamline flow (or laminar flow) to turbulent flow (eddy currents form) occurs at a critical relationship among flow velocity, radius of channel, density and viscosity of the fluid. He defined the relationship in terms of the dimensionless number that is now called the Reynolds number Re:

$$Re = (v\rho r)/\eta$$

where v = flow velocity, ρ = density, r = radius, and η = viscosity. If the value of Re is greater than 1000, turbulent flow begins; if Re is less than 1000, laminar, or streamline, flow results. When blood undergoes turbulent flow, vibrations are set up in the blood vessels. Assume that the radius of a blood vessel is 0.800 cm, blood viscosity is 0.0200 poise, and the density of blood is 1.00 g/cm³. Find the velocity that corresponds to the onset of blood vessel wall vibrations [25.0 cm/sec]

24. One method of measuring the viscosity of a fluid is to measure the terminal velocity of a sphere falling through the fluid. The viscous drag force on a sphere is given by Stokes' law:

$$F = 6\pi\eta rv$$

where η = viscosity, r = radius of sphere, v = terminal velocity. Show that the viscosity can be expressed as follows:

$$\eta = \frac{(\rho - \rho')2r^2}{9v} g$$

where ρ = density of sphere, ρ' = density of fluid.

25. The average blood flow velocity in arteries is found to be about 10.0 cm/sec and the average pressure is around 100 mm Hg. Using a density of 1.00 g/cc for blood, compare the energy density due to velocity to that due to pressure. (*Hint:* units of energy density are N/cm².) If a girl is 165 cm tall and has a blood pressure of 100 mm at her heart, find the blood pressure in her feet 100 cm below her heart. [2.32 N/cm²]

26. Use the values of blood pressure given in Section 8.13. Assume that each portion of the blood vessel system has a length of 0.8 m, and take the quantity of blood flow to be equal to 1.00×10^{-4} m³/sec. Use the value of 4.00×10^{-3} N-sec/m² for the viscosity of whole blood. Calculate
    a. the pressure gradients across each element of the blood vessel system, i.e., aorta, arteries, arterioles, capillaries, and veins
    b. the effective radius for each portion of the system
    c. the average velocity of blood flow through each portion of the system [a. 4.99×10^2 N/m³, 2.83×10^3 N/m³, 9.14×10^3 N/m³, 3.33×10^3 N/m³, 1.66×10^3 N/m³; b. 2.13×10^{-2} m, 1.38×10^{-2} m, 1.02×10^{-2} m, 1.33×10^{-2} m, 1.58×10^{-2} m; c. 7.02×10^{-2} m/sec, 1.67×10^{-2} m/sec, 3.00×10^{-2} m/sec, 1.79×10^{-2} m/sec, 1.27×10^{-2} m/sec]

27. Assume that the density of your blood is the same as that of water, that your heart is two-thirds of the way up from your feet to your head, that you are 2.00 m tall, and that your average blood pressure is 1.33×10^4 N/m². In an upright position what is the pressure difference between your head and your feet, between your heart and your feet, and between your heart and your head? [1.96×10^4 N/m², 1.31×10^4 N/m², 6.50×10^3 N/m²]

GOALS

After you have mastered the contents of this chapter you will be able to achieve the following goals:

Transport Equation
Write the quantitative equation for the transport process of a system whose variables are given.

Continuity
State the continuity equation for a system, and explain the flow properties of a system in terms of that equation.

Transport Problems
Use algebraic and graphical methods to solve transport problems for one dimensional systems.

PREREQUISITES

Before beginning this chapter, you should have achieved the goals of Chapter 1, Human Senses, and Chapter 2, Unifying Approaches.

9

Transport Phenomena

9.1 Introduction

During heavy rainstorms puddles of water are formed at various places on the ground. The rainwater flows along the ground from one puddle to another and then, perhaps, into a water drainage ditch. What causes the flow of the water along the ground? Have you noticed that water flows more rapidly along some parts of the ground than others? How do you explain the various rates of water flow over various parts of the surface of the earth?

During the serving of a formal dinner in the palace, the silver serving pieces are placed in piping hot food by the servants. The prince discovers that the handle of the silver gravy ladle is rather warm. How do you explain the warmth of the ladle handle? What causes the flow of heat from the gravy to the handle? Meanwhile, the handle of the prince's china coffee cup remains cool even though his cup contains hot coffee. How do you explain the different rates of heat flow?

Can you take a dry-cell battery, two pieces of wire, and a flashlight bulb and attach them together so that the bulb will light? How do you do it? How do you explain what has happened? Do you use the flow of something to explain this phenomena? If so, what do you say is flowing? What causes the flow to occur?

These are just three examples of a more general phenomena that we observe. The universe in which we live is always changing. Something moves from one part of the universe to another part. These movements are the primary concern of this chapter.

9.2 Temperature Differences and Gradients

In these three examples each of the systems had at least one variable that had different values in various parts of the system. Take the case of the prince and the silver serving pieces of the palace. The temperature varied from one part of the palace to another. In particular the hot food is at a much higher temperature than its surroundings. The silver ladle that was placed in the gravy had different temperatures at its two ends (see Figure 9.1). We can compute the rate at which the temperature changes with distance along the ladle handle by taking the difference in the temperatures between the ends of the ladle and dividing that by the length of the handle to give the rate of change of temperature with distance $= (T_2 - T_1)/L$ where T_2 is the temperature of the gravy, T_1 is the temperature of the cool end of the ladle handle, and L is the length of the handle. This represents an average value for the rate of temperature change over the length of the handle. If the gravy is at a temperature of 82°C and the cool end is at 33°C and the handle is 7 cm long, the rate of change of temperature with distance $= (82 - 33)°/7$ cm $= 7°C/cm$. The temperature of the handle decreases 7° for every centimeter you move away from the gravy. There is no obvious reason why the temperature should change at a constant rate all along the ladle handle. In fact,

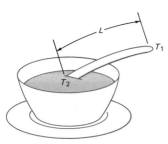

FIGURE 9.1
Temperature difference between the two ends of a metal object—specifically a silver ladle in a gravy bowl. The difference in temperature between the two ends divided by the distance between them is known as temperature gradient. The heat conducted (transported) depends upon the temperature gradient.

FIGURE 9.2
Plot of the temperature vs. distance along the handle of the ladle.

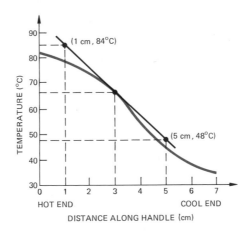

shortly after the ladle is dipped into the hot gravy the temperature as measured along the ladle handle might be represented by the curve in Figure 9.2.

In this case the rate of change of temperature with distance varies along the handle. As you will recall from your experiences with graphs, the rate of change of the y variable with respect to x is the definition of the slope of the curve. The slope is found by taking the ratio of the differences between the ordinates and the abscissa of two points on the straight line tangent to the curve at the point of interest. In Figure 9.2 the value of the slope at the point 3 cm from the hot end is computed to be $(84 - 48)/(1 - 5) = 36/-4 = -9°C/cm$. You can see from this figure that the rate of change of the temperature with distance does not need to be a constant for all values of the distance, but in fact the rate of change of the temperature with distance may have a different value at each point along the handle of the ladle. The rate of change of temperature with distance is called the *temperature gradient*. The magnitude of the *gradient*, as you may recall from Chapter 2, is defined as the rate of change of a variable of a system with respect to distance. Since a variable of a system may change differently in the x and y directions, it is necessary to specify the direction of the gradient of a variable. We will use the following notation where $\text{grad}_x T$ is a symbol for the gradient of the temperature with respect to changes in the distance in the x direction, ΔT is the symbol for the change in temperature $(T_2 - T_1)$, and Δx is the symbol for the change in distance $(x_2 - x_1)$. As you know from your experience in calculating the slopes of curves on graph paper, if the change in x is made sufficiently small, the value of $\Delta T/\Delta x$ will be equal to the slope of the line tangent to the T versus x curve at the point of interest. In Figure 9.2 the $\text{grad}_x T$ at 3 cm is given by the slope of the line tangent to the curve at that point, hence the $\text{grad}_x T$ at 3 cm is equal to $-9°C/cm$.

In summary, the magnitude of the *gradient of a variable* of the system is the ratio of the change in value of that variable to change in distance. If a variable of a system has the units of joules, what will be the units

FIGURE 9.3
Cross section of a hill (a) on a north-south plane and (b) on a west-east plane. (c) From these a contour map can be drawn.

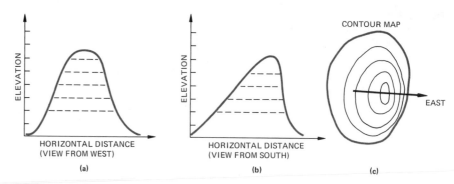

of the gradient of that variable? If a variable has a definite value at each point in the system, then the gradient of that variable will also have a definite value at each point in the system. For most real systems both the values of the variables and of the gradients are smoothly changing quantities over the volume of a system.

EXAMPLE

Let us take the contour map of the hill shown in Figure 9.3 and estimate the components of the gradient of the elevation in given directions for various positions on the hill. Where is the $\text{grad}_E H$ (gradient of height in the easterly direction) the greatest? The least? Zero. What is the relationship between the $\text{grad}_E H$ and $\text{grad}_W H$ (gradient of the height in the westerly direction)? Where is $\text{grad}_S H$ (gradient of the height in the southerly direction) greatest? The least? Zero? Where do you predict that water flowing down the hill will have the greatest rate of flow?

So far we have only considered the temperature of the ladle handle at one particular time. What happens to the temperature of the ladle handle as time goes on?

Questions

1. What does a negative value for the rate of change of temperature with distance indicate?
2. In Figure 9.2 is the rate of change of temperature with distance greater or less at 4 cm than at 3 cm? At 5 cm than at 3 cm?
3. How do you answer these questions quickly by looking at the graph and without making a quantitative calculation?
4. You can imagine that Figure 9.2 represents a one-dimensional contour map of the temperature of the ladle handle. Where does the $\text{grad}_x T$ have its largest value? Smallest? Is it zero anywhere along the ladle handle?

9.3 Flows and Currents

It is natural that when we encounter the continuous motion of something from one part of the universe to another, we should evoke the language and images of those continuously moving objects with which we are most familiar.

So you will find that the subject of transport phenomena is couched in terms that are used in everyday life to describe the motion of water. The actual transport phenomena that we are describing may, in fact, bear little resemblance to the flow of water, yet we will be using terms that we first knew as we poured water from one jar to another in a bathtub, or as we stood beside a puddle of water that fed the trickle of water draining down our driveway. In those situations we remained in one location and watched the quantity of water move past us in time. This quantity of water flow may be called water *current* (quantity of water per second).

9.4 Heat Flow

If we now apply this imagery to the hot gravy system in the palace, we may imagine that the hot gravy contains "heat" which flows along the ladle handle. Then we can define the heat flow as the rate of change of the quantity of heat with respect to time. We will use the symbol I for quantity of flow, and we can write the following equation for heat flow:

$$I \text{ (heat)} = \frac{\text{change in quantity of heat}}{\text{change in time}} = \frac{\Delta H}{\Delta t} \tag{9.1}$$

As a specific example, suppose we have a way of measuring the amount of heat that is contained in a small portion of the ladle handle that is located 3 cm from the gravy. Then the quantity of heat contained in that small portion of ladle handle might vary with time according to the graph in Figure 9.4. We can calculate the heat flow from the slope of the line tangent to the current at the time we chose. For example, let us select the time as $t = 4$ sec, then from the graph we can compute I

FIGURE 9.4
The temperature of a given point on a body that has a different temperature at its two ends changes with time. If time $t = 0$ is the instant that one end is inserted in a higher temperature medium, one can plot the temperature of a given reference point as a function of time. The heat flow through the body depends upon this temperature. A curve for the heat flow as a function of time is shown. The slope of the curve at a given time gives the time rate of heat flow through the reference point.

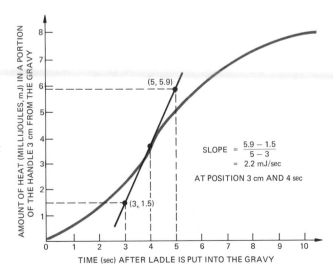

SLOPE $= \dfrac{5.9 - 1.5}{5 - 3}$

$= 2.2$ mJ/sec

AT POSITION 3 cm AND 4 sec

TIME (sec) AFTER LADLE IS PUT INTO THE GRAVY

FIGURE 9.5
Pictorial representation of heat flow.

$$I = \frac{(5.9 - 1.5)\text{mJ}}{(5.0 - 3.0)} = \frac{4.4 \text{ mJ}}{2 \text{ sec}} = 2.2 \text{ mJ/sec} \tag{9.2}$$

where I is the heat flow at a distance of 3 cm from the gravy at the time of 4 sec after the ladle is put into the gravy. This phenomenon may be represented pictorially as in Figure 9.5, where x is the distance along the handle starting from the surface of the gravy and H is the amount of heat contained in the portion of the handle bounded by the solid lines in the figure. The heat flow I in Figure 9.5 is shown as if the change in the quantity of heat in that portion of the handle is completely explained by the flow of heat along the handle.

Questions

5. What are the ways other than along the handle that the heat might escape from that portion of the handle?
6. Does it seem reasonable from your own experience with metal rods held in a fire that the heat flow along the rod is much, much greater than the flow of heat in other directions?

9.5 Water Flow

Let us consider another familiar example of flow, filling a bucket of water from a garden hose. Suppose you have a 19-liter (5-gallon) bucket that you wish to fill with water. The graph of the amount of water in the bucket at any time after you have turned on the hose might be like the one given in Figure 9.6.

In order to use this graph to calculate the flow of water through the hose, what assumptions must you make about water? About the hose? What is the water flow at a time equal to 200 sec? We can answer this question using the slope of the line tangent to the curve in Figure 9.6 at $t = 200$ seconds:

$$I \text{ (water)} = \frac{\text{change in amount of water}}{\text{change in time}} = \frac{\Delta Q}{\Delta t} \tag{9.3}$$

FIGURE 9.6
Graph of the quantity of water in a bucket vs. time of flow.

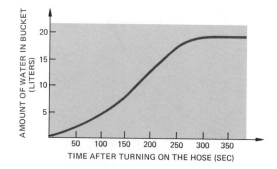

At $t = 200$,

$I = (18.0 - 8.0)$ liters$/(250 - 150)$ sec

$I = 10.0$ liters$/100$ sec $= 0.10$ liters/sec

Questions

7. How can you describe the water flow between 150 and 200 sec?
8. What two explanations can you give for the curve for times greater than 300 sec?

9.6 *Current Density and Continuity*

Is it possible to have a water flow if the amount of water in the bucket does not change with time? If you said yes, you are right. Consider the case of a bucket with a hole in the bottom that allows water to leak out of the bottom at exactly the same rate we are adding water in the top. The total amount of water in the bucket remains the same, but there is a flow of water through the bucket. So we see that our previous definition, while straightforward if the change in amount of time is not zero, does lead us astray in some cases. Let us, therefore, proceed to clear up this messy business with a more precise and more formal definition of a property of the system that we can call the *current density* and represent by the symbol J. Let us take any system and construct an imaginary plane of area A. Then we shall measure the quantity Q of something that passes through that plane in a given time t, and we will define the magnitude of the current density by the ratio

$$J = \frac{Q}{At} \tag{9.4}$$

Hence, J will have the units of the quantity of something divided by the product of area times time, in SI units quantity divided by m²–sec. It is possible that J will depend upon the location of our imaginary plane. So we need to define a direction for J which we will take to be perpendicular to the plane A as shown in Figure 9.7. Then we can indicate the cur-

FIGURE 9.7
A physical flow is in the *x*-direction. To determine the rate of flow an imaginary area A is set up in the *yz* plane. The current density is then the quantity of flow through a unit area per unit of time. That is, $J = Q/At$.

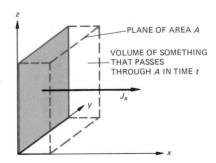

PLANE OF AREA *A*

VOLUME OF SOMETHING THAT PASSES THROUGH *A* IN TIME *t*

J_x

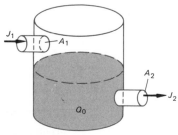

FIGURE 9.8
A liquid container has an inlet tube with area A_1 and an outlet tube of area A_2 and Q_0 quantity of the liquid at $t = 0$. The inlet current density is J_1 and the outlet current density is J_2. The quantity in the container at time t where $t > 0$ is $Q_t = Q_0 + J_1A_1t - J_2A_2t$.

rent density in the x-direction by J_x which will represent the quantity of something that passes through an area A of the yz plane in a time t.

To examine the properties of current density in detail, let us consider a bucket (see Figure 9.8) which has an input tube of area $A_1(m^2)$ and an outflow tube of area A_2. Say the bucket has an amount of water $Q_0(kg)$ in it at time zero and that there is an inflow current density of water of J_1 and an outflow current density of J_2 (kg/m²–sec). From the definition of current density we can show that the amount of water flowing into the bucket in t seconds is given by the product of the input current density times the area of the input tube times the time,

$$\text{inflow} = J_1A_1t \tag{9.5}$$

The amount flowing out of the bucket in t seconds is given by the product of the outflow current density times the area of the outflow tube times the time,

$$\text{outflow} = J_2A_2t \tag{9.6}$$

The net increase in the amount of water in the bucket after t seconds will be the difference between inflow and outflow:

$$\text{net increase} = J_1A_1t - J_2A_2t = (J_1A_1 - J_2A_2)t \tag{9.7}$$

Since at the start there was a quantity Q_0 of water in the bucket, then the total amount of water Q in the bucket at time t is given by the sum of Q_0 and the net increase:

$$Q = Q_0 + (J_1A_1 - J_2A_2)t \tag{9.8}$$

The net change in the amount of water in the bucket in a time of t seconds is given by

$$\frac{\Delta Q}{\Delta t} = \frac{Q - Q_0}{t} = J_1A_1 - J_2A_2 \tag{9.9}$$

We see clearly in this equation that if the net flow is zero, that is, $\Delta Q/\Delta t = 0$, it is not the total current that must be zero, but rather J_1A_1 must be equal to J_2A_2. That is, the amount of water flowing into the bucket must be equal to the amount of water flowing out of the bucket. The amount of water in the bucket remains constant. This is another example of a system in equilibrium as introduced in Chapter 2.

We can use this same intuitive concept to write down a statement in mathematical form for the conservation of matter, which we will call the *continuity equation*. Let us consider a very small imaginary cubical volume of a liquid (Figure 9.9). We have seen above that the time rate of change of the amount of matter in that small volume is given by the difference between the inflow and the outflow,

$$\Delta Q/\Delta t = (J_1 - J_2)a \tag{9.10}$$

where a is the area of one face of our small cube. The difference between the inflow and the outflow is the change in the current density ΔJ. We

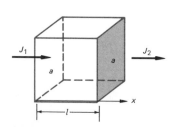

FIGURE 9.9
Volume element for consideration of matter flow.

then rewrite Equation 9.10 as follows,

$$\frac{\Delta Q}{\Delta t} = (\Delta J)a \qquad\qquad (9.11)$$

In order to obtain an expression that is independent of the volume of our imaginary cube, we will divide this equation by the volume of the cube V. The result will be an expression of the change in quantity in a unit volume in a second,

$$\frac{1}{V}\frac{\Delta Q}{\Delta t} = \frac{1}{V}(\Delta J)a \qquad\qquad (9.12)$$

The quantity (mass) of a liquid found in a unit volume is called the density of the liquid, $Q/V = \rho$. The volume of the imaginary cube is equal to the area a times the length of a side of the cube l, $V = al$. We can use these facts to change the form of Equation 9.12,

$$\frac{1}{V}\frac{\Delta Q}{\Delta t} = \frac{\Delta \rho}{\Delta t} = \frac{1}{V}(\Delta J)a = \frac{1}{al}(\Delta J)a = \Delta J/l$$

$$\Delta\rho/\Delta t = \Delta J/l \qquad\qquad (9.13)$$

where $\Delta\rho$ is the change of the density of the liquid. This equation is called the continuity equation.

Hence, the *continuity of matter* can be stated as follows:

The time rate of change of the amount of material in a unit volume of the material is equal to the change of the current density with respect to distance.

In other words, if the amount of the material in a unit volume is changing there must be some net flow of material. We can restate this in a manner very similar to the statement of the conservation laws: If the density of matter in a given volume is not changing, we know that the current density is constant, i.e., J does not change; so the inflow current in the volume is equal to the outflow current from the volume. For incompressible liquids the change of liquid density with time is assumed to be zero, $\Delta\rho/\Delta t = 0$; so the flow of matter is continuous, $\Delta J = 0$.

9.7 *How to Increase the Flow*

What is it that gives rise to the various rates of flow? We already have a vague idea about that. We know that the cool end of a metal rod placed in a fire gets hot faster than the cool end of a metal rod placed in hot water. We have experienced the fact that heat seems to flow away from very hot objects faster than it does from less hot objects. There seems to be some relationship between the rate of heat flow along the rod and the difference in temperature between the two ends of the rod. The greater the temperature difference, the faster heat will flow. Let us, therefore,

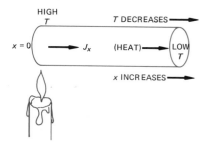

FIGURE 9.10
One end of a rod ($x = 0$) is inserted into a flame. As x increases, the temperature decreases. Is the value of $\Delta T/\Delta x$, the temperature gradient, positive or negative?

make the simplest possible assumption, that is, that the heat current density is directly proportional to the temperature gradient.

$$J_x(\text{heat}) \propto \Delta T/\Delta x \qquad (9.14)$$

Of course, you may say, how can such a simple assumption be justified? You are right, but the test of any assumption is the test of experiment. The question is, does this simple model, with a current density directly proportional to the gradient of a system variable, help in explaining transport phenomena? Does it give us a feeling of understanding current? Does it allow us to make reasonably accurate predictions about the current in systems that we have not yet tested?

We can change the above relationship to an equation by defining a proportionality constant to relate the current to the temperature gradient. In Figure 9.10 you can see that the heat flow is in the positive x direction if the high temperature end of the rod is located at $x = 0$. Therefore, the change in temperature with distance will be negative when the heat flow is positive, and we will have a negative proportionality constant in our equation:

$$J_x(\text{heat}) = -K_H \Delta T/\Delta x \qquad (9.15)$$

where K_H is positive and is called the *coefficient of thermal conductivity*.

So far we have discussed primarily heat and water flow. We have been able to develop some concepts that seem to be useful in understanding the flow of heat from one place to another. We have defined the gradient of a variable as a quantity with both size and direction. The size of the gradient of a variable is given by the ratio of the change in the variable to the change in distance. We have defined a current density as the amount of something that passes through a plane of unit area in 1 sec. We have found that we can write a continuity equation which shows that the time rate of change in the density of matter in some region of space must be equal to the difference between the inflow and outflow of matter in that region. Finally, we have proposed that the current density is directly proportional to the proper gradient, and that the proportionality constant that relates the current to the gradient is negative. Now let us use these concepts to examine some other transport systems.

Questions

9. If the quantity of heat is measured in joules, what are the units for the current density for heat flow in SI units? _____

10. If the temperature is measured in degrees celsius, what are the units for the temperature gradient in SI units? _____

11. What are the units of coefficient of thermal conductivity?

12. If the coefficient of thermal conductivity for a silver gravy ladle is 420 in SI units, what current densities of heat flow can you calculate from Figure 9.2 at distances of 1 cm, 3 cm, and 5 cm from the gravy?

9.8 Diffusion

How does oxygen get into the single cells within the human body where it is essential for life processes? Outside the cell there exists a concentration of oxygen c_o and inside the cell there is a different and lower concentration of oxygen c_c (see Figure 9.11). Hence, there is a gradient in oxygen concentration across the boundary of the single cell. We might expect some flow of oxygen across the boundary. We are now quite prepared to deal with the so-called Fick's law for oxygen diffusion, which is stated as

$$J = -D \frac{\Delta c}{\Delta x} \tag{9.16}$$

where J is the current of oxygen molecules in moles/m^2 sec and c is the concentration of oxygen molecules in moles/m^3 and x is the distance in meters. The proportionality constant D is called the *diffusion constant*.

This equation is simply another form of the relationship we previously developed to relate a current to a gradient. To solve for the rate of diffusion of oxygen into a cell, let us consider a plane surface for the cell boundary and take the diffusion constant for the cell to be D_c and the diffusion constant for the material exterior to the cell to be D_o. Then the current of oxygen is given by (see Figure 9.11),

$$J = \frac{-D_o}{\Delta x}(c_i - c_o) \tag{9.17}$$

in the exterior material and by

$$J = \frac{-D_c}{\Delta x}(c_c - c_i) \tag{9.18}$$

in the cell where c_i is the concentration of oxygen molecules at the cell wall. To solve for the diffusion current of oxygen molecules we simply eliminate the c_i term from these two equations as follows: Solve Equation 9.17 for $c_i = -J\Delta x/D_o + c_o$, substitute that for c_i in Equation 9.18, and solve for J. This results in the following expression:

$$J = \frac{-D_o D_c}{D_o + D_c}\left(\frac{c_c - c_o}{\Delta x}\right) \tag{9.19}$$

If we know the interior and exterior concentration of oxygen molecules, we can use Equation 9.19 to compute the rate of flow of oxygen molecules through the surface of a cell. We can also combine Equation 9.19 with the continuity equation to find the concentration of oxygen molecules in the cell as the time changes. The continuity equation, which assumes that no oxygen molecules are being generated or used inside the cell, may be written from Equation 9.13 as,

$$\frac{\Delta c_c}{\Delta t} = \frac{J}{\Delta x}$$

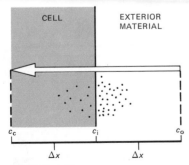

FIGURE 9.11
Diffusion process at the boundary of a cell.

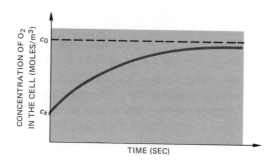

FIGURE 9.12
Concentration of O_2 in a cell as a function of time.

or

$$J = \Delta x \frac{\Delta c_c}{\Delta t}$$

(9.20)

This statement combined with Equation 9.19 enables us to show that the time rate of change of the concentration of oxygen in the cell is directly proportional to the concentration of oxygen in the cell,

$$\frac{\Delta c_c}{\Delta t} \propto c_c - c_0$$

(9.21)

From the section on exponential functions in the mathematics supplement of this book, Equation 9.21 has a mathematical solution given by the following equation,

$$c_c - c_0 = (c_k - c_0)e^{-at}$$

(9.22)

where $a = D_o D_c / (D_o + D_c)\Delta x^2$, c_k is the initial concentration of oxygen inside of the cell, Δx is a characteristic size of the cell and e is the base number for natural logarithms. This result is shown graphically in Figure 9.12.

Questions

13. What are the units of D in the SI units?
14. Check the dimensional consistency of Equation 9.19. Does the current J still have the same units as in the defining Equation 9.16?

SUMMARY

Use these questions to evaluate how well you have achieved the goals of this chapter. The answers to these questions are given at the end of this summary with the number of the section where you can find the related content material.

Transport Equation

1. The quantity of matter transported per unit of time by diffusion is defined as (q = quantity of matter):

   a. $\Delta q / \Delta t$

b. $\Delta q/\Delta x$
c. Δq
d. Δt
e. none of these
2. The matter gradient is defined as
a. $\Delta q/\Delta t$
b. $\Delta q/\Delta x$
c. Δq
d. Δt
e. none of these
3. In diffusion the flow of matter is directly proportional to
a. $\Delta q/\Delta x$
b. $-\Delta q/\Delta x$
c. $\Delta q/\Delta t$
d. $-\Delta q/\Delta t$
e. none of these

Continuity

4. The continuity equation can be derived from the unifying approach involving
a. superposition
b. conservation
c. inertia
d. fields
e. resonance
5. The continuity equation predicts that if more of a quantity flows out of a volume than flows into the volume then
a. physics is violated
b. there is a source inside the volume
c. there is a sink inside the volume

d. the volume is a sphere
e. volume numerically equals area

Transport Problems

6. If the temperature gradient across a window pane is doubled while other variables remain unchanged, the heat current density through the window will
a. quadruple
b. be cut in half
c. double
d. remain the same
e. none of these
7. Which of the following processes have quantity transport proportional to the surface area of the system?
a. thermal conduction
b. water flow
c. diffusion
d. all of these
e. none of these
8. Which of the following ratios is constant for a given window pane?
a. heat current density/temperature gradient
b. temperature change/window thickness
c. change in temperature/change in time
d. all of these
e. none of these

Answers

1. a (Section 9.8)
2. b (Section 9.8)
3. b (Section 9.8)
4. b (Section 9.6)
5. b (Section 9.6)
6. c (Section 9.7)
7. d (Section 9.8)
8. a (Section 9.7)

ALGORITHMIC PROBLEMS

Listed below are the important equations from this chapter. The problems following the equations will help you learn to translate words into equations and solve single concept problems.

Equations

$$I = \frac{\Delta H}{\Delta t} \tag{9.1}$$

$$I = \frac{\Delta Q}{\Delta t} \tag{9.3}$$

$$J = \frac{Q}{At} \tag{9.4}$$

$$\frac{\Delta \rho}{\Delta t} = \frac{\Delta J}{l} \tag{9.13}$$

$$J_x = -K \frac{\Delta T}{\Delta x} \tag{9.15}$$

$$J = -D \frac{\Delta c}{\Delta x} \tag{9.16}$$

Problems

1. If 1.0×10^6 J of heat are conducted through a windowpane per hour, what is the rate of flow of heat?
2. If 12 liters of water flow into a bucket in 30 sec, what is the rate of flow?
3. If the diffusion constant for certain molecules in water is 1.0×10^{-9} m²/sec and concentration gradient is 5.0×10^4 kg/m³ per m across a membrane. Find the diffusion current density across the membrane.
4. The window in a room has a thickness of glass of 2.0 mm. The room temperature is 20.0°C and outside temperature is 0.0°C. The thermal conductivity of glass is 1.0 J/(sec–m–deg). Calculate the rate of heat loss per unit area through the window.
5. During the heating of the air in a hot-air balloon, the density of air inside the balloon changes from 1.29 kg/m³ to 1.09 kg/m³ in 10 min. What is the average net current of air from the balloon per unit volume?

Answers

1. 10^6 J/hr
2. 24 liters/min
3. 5.0×10^{-5} kg/m²-sec

4. 1.00×10^4 J/sec m²
5. 0.02 kg/m³ min

EXERCISES

These exercises are designed to help you apply the ideas of a section to physical situations. Where appropriate the quantitative answer is given at the end of each exercise.

Section 9.2

1. A contour map for Mt. Taum Sauk, the highest mountain in the state of Missouri is shown in Figure 9.13. Compute the height gradients from this map from west to east and from south to north. Sketch how the mountain looks from the West and from the South.

Section 9.6

2. During an iron ore smelting process the quantity of material in a large crucible is continuously monitored. The process engineer notes that the mass of material in the crucible is 4.2×10^3 kg at 10 A.M. and 2.8×10^3 kg at 10:20 A.M. The volume of crucible is 70 m³. (a) What has happened during the time from 10 A.M. to 10:20 A.M.? (b) What is the change in density of all matter in the crucible during this time? (c) What is the average current during this time? [a. 1.4×10^2 kg effused; b. decreases 20 kg/m³; c. 70 kg/min outward.]

3. A service station operator was checking the antifreeze in the radiator of your automobile with a hydrometer. She noticed that the specific gravity of the solution was only 1.05. She added 4 liters of antifreeze of specific gravity 1.80, which just filled the 20-liter radiator. She noticed that the final specific gravity of material in the radiator was 1.30, and she exclaimed, "Your car has a leak in the radiator." Assume the station operator used the continuity equation to arrive at her conclusion. Explain her reasoning. [If no loss, the final specific gravity should have been 1.20.]

FIGURE *9.13*
Exercise 1.

37°36'

37°32'30"
90°45'

90°42'30"

1000   0   1000   2000   3000   4000   5000   6000   7000 FEET

1   5   0   1 KILOMETER

CONTOUR INTERVAL 20 FEET
DATUM IS MEAN SEA LEVEL

Section 9.7

4. An iron rod is placed into a forger's fire. The hot end reaches a temperature of 1000°C. If the rod is 100 cm long and the room temperature is 30°C, what is the temperature gradient along the rod? The thermal conductivity of iron is about 4 J/sec–cm–°C. The diameter of the rod is 2.5 cm. What is the heat flow along the rod? [9.7°C/cm, 190 J/sec]

5. The handle of a silver spoon 5 mm in diameter conducts heat along its handle at the rate of 5 mJ/sec. The thermal conductivity of silver is 4.2 mJ/sec–cm–°C. What is the temperature gradient along the handle? [6.1°C/cm]

Section 9.8

6. The distribution of oxygen molecules in the bloodstream of a patient is given by the expression $c = c_0 + ax^2$ where x is the distance from the heart, c_0 is 3×10^{19} molecules/cm^3, and a is a constant with a value of 3×10^{15} molecules/cm^5.

a. Compute the concentration of oxygen molecules 10 cm, 50 cm, and 100 cm from the heart.

b. Compute the gradient of the oxygen concentration at 10 cm, 50 cm, and 100 cm from the heart.

c. The diffusion constant for oxygen in the blood is 2×10^{-5} cm²/sec. What is the oxygen current flow at 10 cm, 50 cm, and 100 cm from the heart? [a. at 10 cm $c = 3.03 \times 10^{19}$ molecules/cm³; at 50 cm $c =$ 3.75×10^{19} molecules/cm³, at 100 cm $c = 6 \times 10^{19}$ molecules/cm³; b. at 10 cm $\Delta c/\Delta x = 6 \times 10^{16}$ molecules/cm⁴, at 50 cm $\Delta C/\Delta x = 3.0 \times 10^{17}$ molecules/cm⁴, at 100 cm $\Delta c/\Delta x = 6 \times 10^{17}$ molecules/cm⁴; c. at 10 cm $J_x = 1.2 \times 10^{12}$ molecule/sec, at 50 cm $J_x = 6 \times 10^{12}$ molecule/sec, at 100 cm $J_x = 1.2 \times 10^{13}$ molecule/sec]

PROBLEMS

These problems may involve more than one physical concept.

7. The ratio of the diffusion coefficient of KCl to that of NaCl is 1.46. The concentration difference for KCl across a membrane is one-half the equilibrium value. Find the concentration difference for NaCl that will produce the same diffusion rate. [73 percent of the equilibrium concentration.]

8. Show that Equation 9.15 becomes

$$J_h = \frac{-(T_b - T_a)}{x/K_1 + x/K_2}$$

for two layers each of thickness x; K_1 is thermal conductivity of one layer and K_2 is thermal conductivity for the other layer. There is a general equation for any number of layers N as follows:

$$J_h = \frac{-(T_b - T_a)}{\displaystyle\sum_{i=1}^{N} \frac{\Delta x_i}{K_i}}$$

Use this equation to find the amount by which heat loss can be reduced by using good storm windows that provide a dead air space of thickness $5d$ where d is the glass thickness and $K_{air} = \frac{1}{40} K_{glass}$.

9. A styrofoam ice chest is 4400 cm² and 5 cm thick. The chest contains 5 kg of ice (at 0°C). The chest is located in a hot room of 31°C. How long will it take for 1 kg of the ice to melt if the thermal conductivity of styrofoam is 1.6×10^{-4} watts/deg-cm? (*Hint:* 1 g of ice absorbs 336 J of heat energy upon melt-ing.) [The inside area of the box is 4400 cm². The thermal gradient in the styrofoam is 6.2°C/cm. The heat current is 9.9×10^{-4} watts m² or 4.36 watts into the ice chest. So 1 g of ice melts in 1.28 minutes. Then 1 kg of ice melts in 21.4 hours.]

10. A circular pond is 1 m deep and 100 m in radius. Suppose we introduce a bottom feeding fish. The fish respires M g of oxygen per second. Find the necessary diffusion coefficient for oxygen in water if the fish is to survive on oxygen diffusing from the surface where the concentration is $8M$ g of oxygen per cubic meter of water, and the concentration of oxygen at the bottom is $1M$ g per cubic meter. [4.55×10^{-6} m²/sec]

11. Two compartments in a tank are separated by an artificial membrane of 1 mm thickness. In compartment A is a solution with 10 g of glucose per liter, and in compartment B is a solution with 6 g of glucose per liter. The total area of the membrane is 2.0 cm², and the pores constitute 12.5 percent of the total area. The molecular weight of glucose is 180. If the diffusion coefficient $D = 3 \times 10^{-6}$ cm²/sec, what is the current density? What is the original rate of flow through the membrane? [$J = 0.67 \times 10^{-9}$ molecules/cm²-sec; rate $= 1.34 \times 10^{-9}$ molecules/sec]

12. During a blood plasma (specific gravity 1.21) transfer of 1.00 liter, a doctor noticed that the specific gravity of the patient's blood increased from 1.03 to 1.08. (a) What was the initial quantity of blood in the patient? (b) How could a doctor use a plasma transfer to detect internal bleeding in a patient? [3 liters]

GOALS

When you have mastered the contents of this chapter, you will be able to achieve the following goals:

Definitions
Define each of the following terms, and use it in an operational definition:

temperature	mechanical equivalent of heat
thermometer	heat capacity
heat	specific heat
linear expansion	latent heat of fusion
volumetric expansion	latent heat of vaporization
calorie	heat of combustion

Calorimetry
Solve problems in calorimetry.

Gas Laws
Solve problems using the gas laws involving the pressure, volume, and temperature of a confined gas.

PREREQUISITES

Before beginning this chapter you should have achieved the goals of Chapter 5, Energy, and Chapter 9, Transport Phenomena.

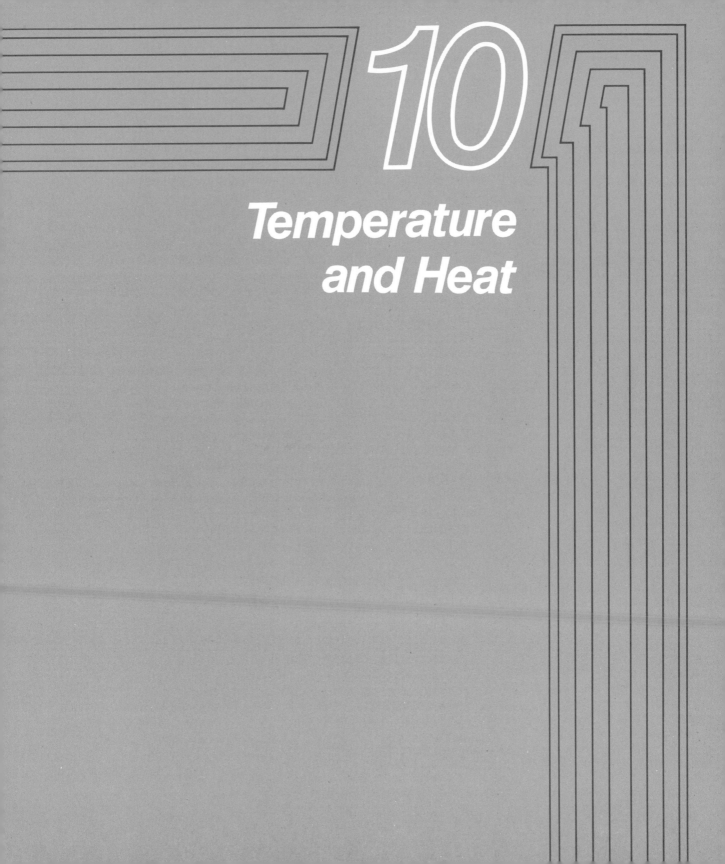

10

Temperature and Heat

10.1 Introduction

Your interactions with your environment provide a variety of experiences that are related to the ideas of temperature and heat. Some of your first autonomous decisions may have been your choices of clothing, choices at least partly influenced by the answer to the question, "What is the temperature going to be today?" You have had many opportunities to influence the temperature of your food. Have you not burned your mouth on something too hot? What methods have you learned to use to cool your food down rapidly so that it is at the proper eating temperature?

Your experiences may have included a number of injunctions that contradict one another. For example, you may think that all water boils at the same temperature, and that water never gets hotter than its boiling temperature, yet the cookbook says to cook the rice in water that is at a "rolling boil." Why? In addition to contradictions, our cultural lore is full of interesting, and perhaps false, statements about temperature and heat: "Hot water placed outside on a very cold winter day will freeze faster than cold water." "A steam burn is worse than a hot water burn." "White clothes are cooler than dark clothes." Furthermore, your firsthand experiences with heat and temperature are not without puzzles. On a hot summer day why would a piece of metal lying on the sidewalk seem hotter to your bare feet than the concrete? Such puzzles helped lead to the present quantitative ideas about temperature and heat. Count Rumford, one of the forefathers of our modern concepts of heat, noted in his personal diary that each time he saw a person burn his mouth on hot apple pie, he wondered what it was about apples that made them seem to retain their heat much longer than other foods.

The point of these observations is that you already have had many encounters with the concepts of temperature and heat. The purpose of this chapter is to discuss these concepts in a way that will give you some quantitative understanding of them and will correct any misunderstandings that you have.

10.2 Temperature

One of the first discriminations you learned was the relative hotness or coldness of an object using the heat sensing organs of your body. However, you also learned that this discrimination was not very reliable. Bath water that feels only slightly warm to your hand may feel rather hot when you sit down in it.

The cookbook says, "Bake in a hot oven for two hours." The novice cook is puzzled by the term "hot" oven, just what is a hot oven anyway? How hot does an oven have to be to be a hot oven? To answer this question we need to define some common measures of hotness that can be used by different people. We use the term *temperature* to speak about the relative hotness or coldness of an object. We then set out to develop some temperature scales that can have universal use. The human heat sensors are transducers that change the heat energy input into electrical signals that are sent to the brain (see Chapter 1). But we know how unreliable those transducers are for use in defining any universal temperatures. Therefore, we must find some other kinds of transducers that have reproducible, quantitative responses to changes in temperature. Such transducers are called *thermometers*. In this context temperature can be operationally defined as the property of a system measured with a thermometer.

A system that has a reproducible linear relationship between temperature and a change in a physical property of the system seems to be a logical choice for a thermometer. A list of some of the kinds of thermometers that are presently in use is given in Table 10.1. None of these transducers is completely satisfactory for all uses and for all ranges of temperature. The best thermometer for a particular use must be selected by taking into account such factors as cost, accuracy, temperature range, and durability.

Once you have selected a transducer system to use for your thermometer, you must establish some way of calibrating it. That is, you need to determine a numbering system that you can use to measure various temperatures. If you wish other people to be able to use your temperature system, you will need to find some events in nature that always occur at the same temperature. You can use these events to serve as reference, or fixed, points on your thermometer.

TABLE 10.1
Temperature Transducers

Physical property that changes with temperature	Type of thermometer	Use
Length	Bimetallic strip	Thermostat (home, car, or radiator)
Volume	Liquid (mercury or alcohol) in glass	Household thermometer
Volume	Gas (at constant pressure)	Rarely used
Pressure	Gas (at constant volume)	Scientific calibration; low-temperature thermometer
Electric potential between two metals	Thermocouple	Furnaces of a few hundred degrees
Electrical resistance of a metal	Resistance	Platinum resistance thermometer high temperature controller
Electrical resistance (semi-conductor)	Thermistor	Digital fever thermometer
Color	Optical pyrometer	Blast furnace
Chemical–color	Thermograph	Crystal temperature
Color	Liquid crystals	Digital thermometer; "mood jewelry"

Based on these ideas, the first accurate thermometer was made in 1641 using alcohol in glass. Then in 1714 the German physicist G. D. Fahrenheit built a mercury-in-glass type thermometer and used three fixed points to calibrate it: the freezing temperature of a mixture of water, ice, and salt which he labeled 0°, the freezing point of pure water labeled 32°, and the temperature of the human body labeled 96°. Please note that this temperature does not agree with current normal body temperature. Although Fahrenheit did not use it to calibrate his thermometer, it was found that water boils at 212° on the Fahrenheit temperature scale at sea level. In 1742 the so-called centigrade scale was developed, which uses the freezing point of pure water (0°) and the boiling point of pure water (100°) as its fixed points with 100 scale divisions between these two points. This scale is attributed to Anders Celsius and is now known as the Celsius temperature scale. Another international temperature scale has also been developed. This scale is based upon a principle of thermodynamics of Lord Kelvin which implies that there exists an absolute minimum temperature. This international scale, called the Kelvin scale, assigns the number zero to this lowest possible temperature. Additional details of the Kelvin scale will be given later in this chapter. The Celsius, Kelvin, and Fahrenheit scale relationships are shown in the following table.

Scale	Absolute zero	Freezing point of water	Boiling point of water
Kelvin (°K)	0°	273.16°	373.16°
Celsius (°C)		0°	100°
Fahrenheit (°F)		32°	212°

Using this table, derive the equations required to convert any temperature from one of these scales to the other two. Fill in the table with the value of absolute zero in °C and in °F. You will find that either the Kelvin scale (°K) or the Celsius scale (°C) is used in this book.

10.3 Changes in Size with Changes in Temperature

You already know that the size of an object changes as its temperature changes. Joints (or cracks) are put in highways, bridges, buildings, and sidewalks to allow for changes in size that occur with the seasonal changes in temperature. As you know, most objects increase in size as their temperature increases. Can you think of any exceptions? Do you know any systems in nature that get larger as their temperatures decrease?

The simplest assumption that we can make about the quantitative changes in the size of an object is to assume that the changes are linearly related to changes in temperature, that is, the changes in length of an object are directly proportional to its temperature changes.

TABLE 10.2
Coefficients of Expansion

Material	per °C	Temperature range (°C)
Linear		
Aluminum	24×10^{-6}	20–100
Brass	19×10^{-6}	0–100
Copper	18×10^{-6}	25–100
Glass (common)	9.0×10^{-6}	20–500
Glass (Pyrex)	3.3×10^{-6}	0–300
Ice	51×10^{-6}	−20 to −1
Steel	11.0×10^{-6}	0–200
Quartz (fused)	0.50×10^{-6}	0–100
Volumetric (at constant pressure)		
Air	3.67×10^{-3}	0–100
Ethyl alcohol	1.01×10^{-3}	30–50
Glycerin	0.49×10^{-3}	5–40
Hydrogen	3.66×10^{-3}	0–200
Mercury	0.18×10^{-3}	0–100
Petroleum	0.90×10^{-3}	20–120
Water	0.21×10^{-3}	at 20°
Water vapor	3.94×10^{-3}	0–200

$$\Delta L \propto \Delta T \tag{10.1}$$

This relationship can be stated as an equality by the use of a proportionality constant α called the *coefficient of linear expansion*. The coefficient of linear expansion is defined as the increase in length per unit length per degree change in temperature, hence

$$\alpha = \Delta L/(L_0 \Delta T) \tag{10.2}$$

where ΔL is the change in length, L_0 is the original length, and ΔT is the change in temperature. The values for the coefficient of linear expansion for various materials are given in Table 10.2. The length of an object at the temperature T is given by

$$L_T = L_0 + \Delta L = L_0 + \alpha L_0 \Delta T \tag{10.3}$$

where ΔT is the difference between the original temperature and the temperature T. It is customary to choose the standard length L_0 to be the length as measured at 0°C. Then Equation 10.3 can be written as

$$L_T = L_0 (1 + \alpha T) \tag{10.4}$$

where T is the temperature in degrees Celsius.

EXAMPLE

If a steel railroad rail of length 10.0 m is laid on a day when the temperature is 0.00°C, how much space should be left between the ends of the adjoining rails for a maximum summer temperature of 45.00°C?

$L_0 = 10.0$ m; $\Delta T = (45.00 - 0.00°) = 45.00°C$

$$\Delta L = (10.0)(11.0 \times 10^{-6}/°C)(45.00°C) = 4.95 \times 10^{-3} \text{ m} = 0.495 \text{ cm}$$

So about one-half centimeter should be left between the ends of the 10-m rails.

10.4 Volumetric Expansion

When a solid or liquid is heated, it expands in all directions. In a manner similar to the definition of the linear expansion coefficient, the coefficient of volumetric expansion can be defined:

$$\beta = \Delta V/V_0 \Delta T \quad \text{or} \quad \Delta V = V_0 \beta \Delta T \tag{10.5}$$

where ΔV is the change in volume V_0 for a temperature change of ΔT. Can you show that the coefficient for the volumetric expansion of a solid is approximately equal to three times the coefficient of linear expansion? Typical values for the volumetric coefficient of expansion of materials are given in the second half of Table 10.2.

Water has many peculiar properties, and one of these is its behavior upon heating. The volume of pure water changes in an unusual way in the range from 0° to 10°C. Starting with a given volume of water at 0° and warming it, you will find that at first the water contracts until a temperature of 4°C is reached. Above this temperature it expands. Water has maximum density at 4°C and about the same density at 0°C and 8°C. This unusual behavior is shown graphically in Figure 10.1.

This behavior of water is of great importance in nature. If water did not change in density this way, large bodies of water would more easily freeze at the bottom. (You will be able to explain this after you have studied convection.) All of the animal life in the lakes or ponds would be destroyed by complete freezing of the water. In Lake Superior the temperature of the water at depth greater than about 74 m is about 4°C during the entire year.

You should be able to explain why a liquid in a glass thermometer using water is useless below about 10°C. Water is not a good liquid for

FIGURE 10.1
Volume-temperature curve for constant mass of water between 0° and 20°C.

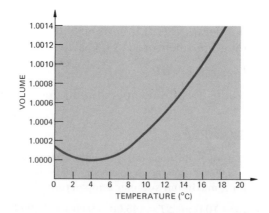

thermometers above 10°C as the rate of expansion is not constant up to 100°C.

EXAMPLES

1. A 100-cm³ Pyrex glass flask is filled with mercury at a temperature of 10°C. The flask is stored in a room that reaches a summer temperature of 40°C. Will the mercury overflow? If so, how much will be lost?

   The approximate coefficient of volumetric expansion for Pyrex is three times its coefficient of linear expansion, $3 \times (3.2 \times 10^{-6})$ or 9.6×10^{-6}. This is much smaller than the volumetric expansion for mercury so the mercury will expand and overflow the flask.

   The overflow will be given by the difference between the increase in the volume of the mercury and the increase in the volume of the flask.

   $$\Delta V_{\text{flask}} = 100 \text{ cm}^3 \times 9.6 \times 10^{-6} \times (40 - 10)°C = 2.9 \times 10^{-2} \text{ cm}^3$$

   $$\Delta V_{\text{Hg}} = 100 \text{ cm}^3 \, (0.18 \times 10^{-3}) \times (40 - 10)°C = 5.4 \times 10^{-1} \text{ cm}^3$$

   $$\text{overflow} = \Delta V_{\text{Hg}} - \Delta V_{\text{flask}} = 0.54 \text{ cm}^3 - 0.029 \text{ cm}^3 = 0.51 \text{ cm}^3$$

2. On a hot summer day gasoline from an underground service station tank is pumped into and completely fills a warm automobile tank. The gasoline is at 20.0°C when it is put into the auto. How much will overflow when it reaches the temperature of the 80.0 liter steel auto tank, which is at 40.0°C? Assume that the coefficient of expansion of gasoline is equal to that of petroleum.

   $$\Delta V_{\text{gasoline}} = (80.0)(0.900 \times 10^{-3})(40.0 - 20.0) = 1.44 \text{ liters}$$

10.5 Bimetallic Switch

A very common use of the thermal expansion properties of materials is in a temperature sensing switch. An on-off switch that responds to changes in temperature has a wide variety of applications, from safety switches in factories to control knobs for electric frying pans. Many of these temperature switches consist of two thin pieces of different metals welded together to form a bimetallic strip (Figure 10.2). Since the two metals have different coefficients of linear expansion, as the temperature changes the two metals will have different lengths, and the

FIGURE 10.2
Bimetallic strip at different temperatures.

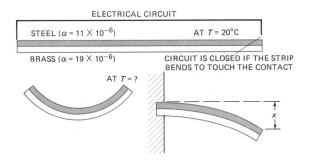

ELECTRICAL CIRCUIT

STEEL ($\alpha = 11 \times 10^{-6}$)   AT $T = 20°C$

BRASS ($\alpha = 19 \times 10^{-6}$)   CIRCUIT IS CLOSED IF THE STRIP
BENDS TO TOUCH THE CONTACT

AT $T = ?$

x

bimetallic strip will bend to accommodate the length difference. Describe the change in curvature of the strip as the temperature is increased and as it is decreased.

For small temperature changes, the bimetallic strip will bend only slightly. List some variables that influence the amount of deflection of the strip, x.

10.6 Heat

In Chapters 1, 2, and 9 you have been introduced to the concepts of energy and heat. The amount of heat energy of an object depends upon the type of material of the object, the quantity of material, and its temperature. It is possible for a given object at low temperature to have more heat energy than another object at a high temperature. Also, an object has thermal inertia toward a change in temperature. You know that if you bring together two objects of different temperatures, they will reach a common temperature that is intermediate between the two initial temperatures. For example, if you have a cup of coffee too hot to drink, you can lower the temperature of the coffee by adding cool water, cream, or milk. You may think of this as a system of mixtures. Qualitatively, you can explain the phenomenon by saying the hot coffee gave up some heat energy, and the cool liquid gained some heat energy. To set up a quantitative equation we need to be able to calculate the transfer of energy involved in a temperature change for a system. In order to do this we must introduce the concept of the amount of energy required to change the temperature of a system. Once again we will fall back upon the simplicity of the linear relationship and assume that the amount of energy ΔQ needed to change the temperature of a system is directly proportional to the temperature change:

$$\Delta Q \propto \Delta T \tag{10.6}$$

We can relate these two quantities by the use of a proportionality constant, which we call the *heat capacity C* of the system. The heat capacity is the amount of heat energy absorbed or liberated from the object for a change in temperature of one degree.

$$\Delta Q = C\Delta T \tag{10.7}$$

where ΔQ is the change in heat energy of the system for a temperature change of ΔT. Since energy has the units of joules in the SI units, C will have the units of J/°C.

You know that it takes longer to heat a large volume of water on a stove than it does to heat a small volume of water. So you already know that somehow the heat capacity is related to the amount of matter in the system. In fact, the heat capacity of an object is directly proportional to the amount of matter of the object. Therefore, to compare the heat ca-

TABLE 10.3
Specific Heat of Some Materials

Material	cal/g-C°	J/g-C°	Temperature range (°C)
Aluminum	0.22	0.92	20–100
Brass	0.094	0.39	20–100
Copper	0.092	0.39	20–100
Glass	0.20	0.84	20–100
Ice	0.50	2.3	−10– 0
Iron, steel	0.11	0.46	20–100
Lead	0.031	0.13	20–100
Mercury	0.033	0.14	20–100
Silver	0.056	0.23	20–100
Water	1.00	4.19	0–100

pacity of different systems, we must take into account their different quantities of matter, or masses. It is useful to define the change in heat energy of a system per degree temperature *per unit of mass*. This property of a system is called its *specific heat, c*, where

$$c = \frac{C}{m} = \frac{\Delta Q}{m\Delta T} \tag{10.8}$$

Table 10.3 gives the value of the specific heat for various materials. In other words, the quantity of heat energy required for a given temperature change of a system is the product of the mass of the matter in the system, the specific heat of the system, and the change in temperature:

$$\Delta Q = mc \, \Delta T \tag{10.9}$$

Unfortunately, the early studies of the properties of heat developed apart from the studies of mechanical energy; so an independent unit for measuring the quantity of heat energy was defined. This quantity of energy is called the calorie and is the amount of energy required to raise the temperature of one gram of water one degree Celsius. This, it turns out, is the same amount of energy as 4.186 joules.

1 calorie = 4.186 joules

This is called the mechanical equivalent of heat.

EXAMPLES

1. An 800-g silver gravy bowl is filled with hot gravy at 100°C. If the silver bowl is originally at room temperature (20°C), how much heat energy is required to warm the bowl up to the temperature of the gravy? From Table 10.3, we see that the specific heat of silver is 0.56×10^{-1} cal/g-°C.

$$Q = 800 \text{ g} \times \frac{0.056 \text{ cal}}{\text{g-C°}} \times (100 - 20)°C$$

$$Q = (64 \times 10^3)(5.6 \times 10^{-2} \text{ cal}) = 3.6 \times 10^3 \text{ calories}$$

2. We can now make a quantitative calculation as a result of conservation of energy: Heat energy lost by hot system = Heat energy gained by cold system.

Assume you have 250 cm³ of coffee in a cup of thermal capacity 20.0 cal/°C and and a silver spoon of mass 80.0 g all at 80.0°C. You desire to have your coffee at 45.0° for drinking. How much water at 10.0°C will you have to add? Assume coffee has the same specific heat as water.

heat energy lost = heat energy lost by coffee + heat energy lost by cup + heat energy lost by spoon

$HEL_{coffee} = 250 \text{ cm}^3 \times 1.00 \text{ g/cm}^3 \times 1.00 \text{ cal/g} \times (80.0 - 45.0)°C = 8750 \text{ cal}$

$HEL_{cup} = 20.0 \text{ cal/°C} \times (80 - 45)°C = 700 \text{ cal}$

$HEL_{spoon} = 80.0 \text{ g} \times 0.056 \text{ cal/gm-°C} \times (80.0 - 45.0)°C = 160 \text{ cal}$

heat energy gained $= m_w \times 1 \times (45.0 - 10.0)°C = 35.0 m_w$

$35.0 m_w = 9610 \text{ cal}$

$m_w = 275 \text{ g of water}$

The physics and math may be correct, but would you want to drink the coffee? What changes would result?

10.7 Changes in the State of Matter

From past experience you know that there are three states of matter: solid, liquid, and gaseous (vapor). A transition from one state of matter to another, liquid to solid, for example, is called a *change of state*. Take the substance H_2O as an example. Its solid state is ice, its liquid state is water, and its gaseous state is steam. Transition from one state to another is accompanied by absorption or liberation of heat energy and usually by a change in volume. Let us now consider a simple experiment in which we take some ice cubes from a refrigerator at a temperature of $-20°C$. We have a thermometer and a heat source, which operates at a constant rate. We will read the temperature at regular intervals of time, plot the temperature as a function of time on a graph, and we will draw a curve through the data points. The curve in Figure 10.3 is a typical curve for an experiment of this kind.

The temperature of the ice will increase steadily from $-20°C$ to $0°C$ (portion *ab* of the curve). When the temperature of $0°C$ is reached there will be no increase in temperature until all ice is melted (*bc*). We now have water, and the temperature will again increase steadily (but at different rate than for ice — why?) until the temperature reaches the boiling point of water (*cd*). The temperature will remain constant at $100°C$ until all of the water is converted into steam (*de*). We must have a closed vessel to finish the transition to $120°C$. The temperature will again increase steadily as heat is added (*ef*).

The above curve is *reversible;* that is, if we remove heat from an equivalent system at the same rate as we heated the system in the original experiment, we will duplicate the above results in the reverse order.

Although water was used in this example, the same type curve would

FIGURE 10.3
Temperature-time curve for a chunk of ice at −20°C being heated to steam at 120°C, with heat being supplied at a constant rate (i.e. constant number of cal/sec)

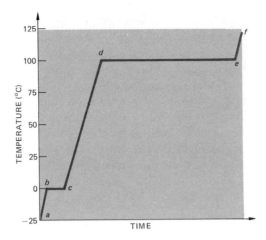

be obtained for many other crystalline substances that have a definite melting point. Amorphous substances such as glass and butter, which soften gradually as the temperature is raised, would not show the constant temperature melting point.

A body has a thermal inertia for a change in its state of matter. Heat energy is required to change an object from one state of matter to another. In other words, heat energy must be added to an object to change it from the solid state to the liquid state, even though the temperature remains constant at the melting point temperature. If there is a change from the liquid state back to the solid state at the melting point, heat energy is liberated. For example, you must add heat energy to convert the ice at 0°C to water at 0°C. And the reverse process occurs when water at 0°C is frozen to ice at 0°C; heat is liberated. This process serves as a source of heat energy. Can you think of an application? The amount of heat required to change a unit mass of material from the solid state to the liquid state at the melting point is called the *latent heat of fusion* L_f. For ice this is 334 joules per gram (J/g) or 79.7 cal/g.

Similarly, for the change from the liquid state to vapor state or vice versa at the boiling point, there is a change of heat energy. In going from the liquid state to the vapor state, heat energy is absorbed, and in the reverse direction, therefore, heat energy is liberated. The process of changing a substance to the vapor state is called *vaporization*. The term vaporization is a general one and covers the processes of evaporation, boiling, and sublimation. *Evaporation* is the process by which the change from a liquid to a gas occurs at the liquid-gas interface. Evaporation lowers the internal energy of the liquid, making evaporation a cooling process. This process is important in maintaining a constant body temperature. If one exercises vigorously, one perspires. The skin is cooled by evaporation. A fever can be reduced by bathing the skin in alcohol. The alcohol evaporates and produces a cooling effect.

Boiling is the change from liquid to gas that may take place anywhere in the entire liquid. Bubbles of vapor are formed throughout the liquid and increase in size as they rise to the surface. This is a process you have surely observed.

Sublimation is the process by which a solid changes directly to the vapor state without going through the liquid state. This process also occurs at the surface. Some examples of sublimation at ordinary room temperatures are the change of iodine crystals, moth balls (naphthalene), and dry ice (solid carbon dioxide) to their respective vapors. Under the proper conditions ice will sublime, as it sometimes does in a freezer.

The heat required to convert one unit of mass of a material from the liquid state to the vapor state at the boiling point temperature is called the *latent heat of vaporization, L_v*. The conversion of 1 g of water at the boiling point of 100°C to steam at the same temperature requires 2260 J, or 540 cal.

The freezing point of a liquid is generally lowered by dissolving some other substance in it. Some examples are antifreeze in water for auto-mobile radiators, salt dissolved in water, sugar dissolved in water. The boiling point is also affected by dissolved substances. It may be either increased or decreased depending on the substance. As an example, a salt solution has a higher boiling point than water and a water-alcohol solution has a lower boiling point than water. Both freezing and boiling points are affected by the external pressure. In general if the volume increases during the change of state, then an increase in the external pressure will raise the temperature of the change of state.

EXAMPLE

Assume you have a 100-g aluminum cup containing 200 g of water and 20.0 g of ice at equilibrium temperature. Dry steam is passed into the mixture and the final temperature is found to be 50.0°C. How much steam is added, and how much water is in the cup at 50.0°C?

The cup and its contents gain an amount of energy calculated as follows:

heat energy gained = heat energy needed to melt ice + heat energy needed to raise temperature of water + heat energy needed to raise temperature of cup

$$= 20 \times 80 + (m_w + m_l) \times 1(50 - 0) + m_c \times c_c \times (50 - 0)$$

$$= 20 \times 80 + (200 + 20)(50 - 0) + 100(.22)(50 - 0)$$

$$= 1{,}600 + 11{,}000 + 1{,}100 = 13{,}700 \text{ cal}$$

This must equal the total heat loss in the system.

heat energy lost = heat energy given up on condensing steam + heat energy given up in cooling water

$$= m_s \times L_v + m_s \times 1.0(100 - 50) = m_s(540 + 50) = 590 m_s$$

energy lost by steam and hot water = energy gained by cup, ice, and cold water

$$590 m_s = 13{,}700 \text{ cal}; \qquad m_s = 23.2 \text{ g}$$

total water in cup at 50°C $= m_i + m_w + m_s = 20 + 200 + 23.2 = 243$ g

10.8 Heat of Combustion

What does it mean for you to speak of a daily diet of 1000 Calories? Let us analyze the physics of the situation. First, the Calorie, which is used in diet discussions, is the *kilocalorie* or the "large calorie." It is equivalent to 1000 of the gram calories we have been using. In speaking of Calories in food, we are talking about the heat energy of the food in terms of complete oxidation of the food. This involves the concept of *heat of combustion*. The heat of combustion of a material is the amount of energy produced by complete oxidation per unit measure of the material. This may be expressed in calories per gram, Calorie per liter, or any other units of energy per quantity. When the complete oxidation of the food you eat in one day releases 1000 kcal of heat energy, you have a daily diet of 1000 Calories. In Table 10.4 are given the approximate values for the heat of combustion of some common foods

On the basis of the principle of energy conservation, in order to lose weight you must consume less food energy per day than your daily energy use. Also in order to gain weight, you must consume more food energy per day than you use daily. From energy considerations only, determine how much daily intake of some common foods you would need for a daily diet of 1000 kcal.

The concept of heat of combustion applies to other fuels as well as to foods. Gasoline has a heat of combustion of about 1150 kcal/100 gm. From the standpoint of economics, we are all interested in the maximum number of calories per dollar. Compare the economics of using butter vs. gasoline as a source of energy.

TABLE 10.4
Energy Value of Some Common Foods

Food	Energy kcal/100 grams
Apples	58
Bread, white	270
Beef, T-bone steak	473
Butter	716
Buttermilk	36
Beans, pinto	349
Chicken, fried	249
Lamb, leg of	313
Milk, whole	65
Potatoes, baked	93
Pork	410
Spinach	23
Sugar, granulated	385

Values taken from *Composition of Foods*, Agriculture Handbook No. 8, USDA, Washington, D.C.

10.9 Gas Laws

As was noted in the discussion of temperature transducers in Section 10.2, gases can be used to build thermometers. In this section we will describe the experiments that can be done to find the quantitative relationships between the pressure, volume, and temperature of a confined gas.

BOYLE'S LAW In the seventeenth century Robert Boyle performed a number of key experiments on gases. You can replicate his results in a simple laboratory experiment. Take a vertical cylinder with a movable piston that fits snugly into it. Trap some air in the cylinder (see Figure 10.4). As the weight applied to the piston is increased, the distance of the piston from the bottom of the cylinder will decrease. In other words, as you increase the force on the piston, the volume of the enclosed air decreases. If you double the total force on the piston, the volume of the confined air will become one-half its original volume. What happens when you change the size of the container? Suppose you double the diameter of your cylinder and piston. What increase in force will be necessary to reduce the volume by a factor of two? Boyle performed such experiments and found that the ratio of the total force applied to the piston divided by the area of the piston was the essential physical variable. This ratio is the pressure on the piston (see also Chapter 8).

$$\text{pressure} = \frac{\text{force}}{\text{area}} \tag{8.2}$$

If the diameter of the piston is doubled, the area is increased by a factor of four since the area of the piston is proportional to the square of the diameter. Hence, for a piston of twice the diameter, the total force must be increased by a factor of four to keep the pressure on the confined gas constant (see Figure 10.4b).

Robert Boyle's results are known as Boyle's law, which can be stated as follows:

The volume of a confined gas whose temperature is held constant will vary in inverse proportion to the pressure on the gas.

$V \propto 1/P$

FIGURE 10.4
Conditions of equal pressure for different size cylinders.

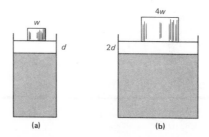

(a)　　　　(b)

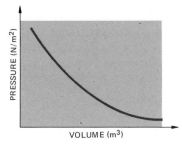

FIGURE 10.5
Pressure-volume curve for a constant mass of gas at constant temperature. Boyle's law: $PV = $ constant

where V is the volume of the confined gas (in m³), and P is the pressure (N/m²). See Figure 10.5.

This inverse proportion is equivalent to stating that the product of the pressure times the volume of a confined gas at constant temperature is a constant,

$$PV = \text{constant} \qquad (10.10)$$

EXAMPLE

A constant mass of gas is confined in a cylinder by a movable piston. The original volume of the gas is 500 cm³ and the pressure is 1.02×10^5 N/m². The gas is compressed to 100 cm³ at constant temperature. What is the new pressure?

$$P_1V_1 = P_2V_2$$

$$(1.02 \times 10^5 \text{ N/m}^2) \times 500 \times 10^{-6} \text{ m}^3 = P_2(100 \times 10^{-6} \text{ m}^3)$$

$$P_2 = 5.10 \times 10^5 \text{ N/m}^2.$$

CHARLES' LAW Almost one hundred years after the work of Robert Boyle, Jacques Charles and Joseph Gay-Lussac were independently studying the properties of gases at constant pressure. They found that the volume of a confined gas changed when the temperature changed. You can also duplicate their experiments with simple laboratory apparatus. Take a small-diameter glass tube closed at one end. Enclose a quantity of gas in the tube by putting a drop of mercury in the tube. Measure the length of the air column in the tube at room temperature. Then place the tube in an ice bath. You will notice that the air column shortens. Next, place the tube in a steam bath. You now will find that the air column lengthens. (see Figure 10.6).

The results of this experiment for the volume of a confined gas at constant pressure show that the volume is a linear function of temperature

FIGURE 10.6
An experimental set-up to demonstrate the variation of the volume of a confined gas with temperature under constant pressure.

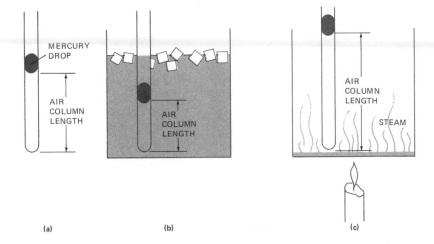

(a)　　　　　(b)　　　　　(c)

FIGURE 10.7

Volume-temperature curve for a constant mass of gas at constant pressure.

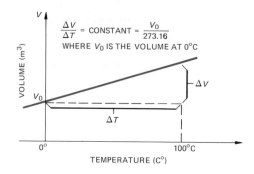

as shown in Figure 10.7. The curve of volume against temperature has a slope of V_0 divided by 273.16 where V_0 is the volume of the confined gas at 0°C. Thus, the change in volume of the gas at constant pressure is directly proportional to the change in temperature. This can be expressed using the idea of a coefficient of volumetric expansion we discussed earlier,

$$\Delta V = V_t - V_0 = V_0 \beta t \tag{10.5}$$

or

$$V_t = V_0 (1 + \beta t) \tag{10.11}$$

where β is the coefficient of volumetric expansion of the gas 1/273.16, t is the temperature in degrees Celsius, and V_t is the volume at the temperature t. You may realize the implication of Equation 10.11: If the temperature becomes −273.16°C, the volume V_t becomes zero.

The law of Charles and Gay-Lussac can be reformulated using the Kelvin (or absolute) temperature scale by replacing the temperature t(in °C) by its equivalent Kelvin reading

$$T(°K) = t(°C) + 273.16$$

$$V_T = V_0 [1 + \beta(T - 273.16)]$$

where $\beta = 1/273.16$.

Thus the volume of a confined gas at constant pressure is directly proportional to its absolute temperature in degrees Kelvin

$$V_T = \frac{V_0}{273.16} T \tag{10.12}$$

This is equivalent to stating that V/T = constant, for constant P.

This equation provides the basis for a gas thermometer with a linear scale, except for temperatures near 0°K. Such a thermometer can be constructed using mercury and a glass tube as described in the experiment above.

EXAMPLE

If 2 liters of a gas at 0°C at constant pressure are cooled to −136.58°C, what is the volume of the gas?

original temperature (°K) = 0°C + 273.16 = 273.16°K

final temperature (°K) = −136.58 + 273.16 = 136.58°K

So,

2 liters/273.16°K = V_{final}/136.58°K

V_{final} = 1 liter

CONSTANT VOLUME   To keep a mass of gas at constant volume one needs a tight vessel made of material with negligible expansion as represented in Figure 10.8a. The application of heat causes the pressure and temperature to increase. The pressures are taken at a number of different temperatures. A pressure-temperature curve is then plotted. See Figure 10.8b. The equation of the curve is given by

$$P_t = P_0(1 + B't)$$

where P_t is pressure at temperature t, P_0 is the pressure at 0°C and B' is the pressure coefficient, which has a numerical value of 1/273.16. The significance of this curve is that we can extend the curve to find the temperature at which $P_t = 0$. The temperature is 0°K, or absolute zero. So we can now say that absolute zero is the temperature at which the pressure of a confined gas is zero. It follows that

the pressure of a confined gas (at constant volume) is directly proportional to its absolute temperature in degrees Kelvin,

$$P_T = \frac{P_0}{273.16} T \qquad\qquad (10.13)$$

EXAMPLE

If the pressure on a constant volume of a gas is doubled, what is the change in the temperature of the gas?

FIGURE 10.8
(a) System for showing how pressure varies as temperature is changed at constant volume. (b) Pressure-temperature curve for a given mass of gas kept at constant volume.

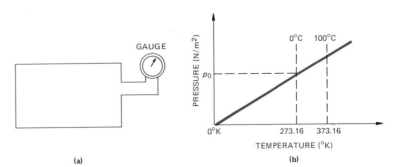

(a)

(b)

The answer is that the absolute temperature in degrees Kelvin will be doubled.

The constant volume gas thermometer is an application of the change of pressure produced by heating or cooling a gas.

Boyle's law and the law of Charles and Gay-Lussac can be combined into a single gas law. Since Boyle's law shows that the product of the pressure and volume of a gas is constant for a constant temperature and the law of Charles and Gay-Lussac shows that the ratio of the volume to the absolute temperature is a constant for constant pressure, these two laws can be combined to show that the product of the absolute pressure times the volume of the confined gas divided by the absolute temperature is a constant,

$$\frac{PV}{T} = \text{constant} \tag{10.14}$$

This expression is equivalent to the following equation:

$$\frac{P_i V_i}{T_i} = \frac{P_f V_f}{T_f} \tag{10.15}$$

where the subscripts i and f refer to the initial and to the final states of the system, and where P and T are the absolute pressure and the temperature, respectively, and where V is the volume of the confined gas.

EXAMPLE

A weather balloon carrying 3 m³ of helium gas at 1 atmosphere of pressure and 27°C is sent aloft to an elevation where the pressure is one-twelfth of an atmosphere and the temperature is a −73°C. What is the volume of the balloon at that elevation? What is the change in the radius of the balloon.

Let us begin by rounding off the Kelvin temperature scale to $T(°K) = t(°C) + 273$.

T_i = initial temperature = 27°C + 273 = 300°K

T_f = final temperature = −73°C + 273 = 200°K

P_i = 1 atmosphere   and P_f = 1/12 atmosphere

$$\frac{(1 \text{ atm})(V_i \text{ m}^3)}{300°K} = \frac{(1/12 \text{ atm}) V_f}{200°K}$$

$8V_i \text{ m}^3 = V_f$

The final volume is eight times the initial volume, and since the volume increases as the cube of the radius ($V = \frac{4}{3}\pi r^3$ for a sphere), the radius will increase by $\sqrt[3]{8}$, or by a factor of 2.

SUMMARY

Use these questions to evaluate how well you have achieved the goals of this chapter. The answers to these questions are given at the end of this summary with the section number where you can find the related content material.

Definitions

Fill in the blanks with correct words.

1. The thermal energy of a system is characterized by its _____ which you can measure using a _____.
2. Many thermal transducers have a positive _____ or increase in length with increase in temperature.
3. Water is unusual in the fact that its _____ is negative from 0°C to 4°C.
4. The study of heat energy developed separately from the study of mechanical energy; so the unit for thermal energy is called a _____ and is equivalent to the amount of energy required to _____.
5. James Prescott Joule carried out a number of experiments which gave a _____ of 1 calorie = 4.186 joules.
6. The amount of energy required to raise the temperature of a system 1°C is called the _____ of the system.
7. The amount of energy required to raise the temperature of 1 kg of material 1°C is the _____ of the material.
8. If a gram of a substance is completely burned in oxygen and gives off 80 calories of energy, then we say its _____ is 80 cal/g.
9. When one gram of ice is melted to water, it absorbs heat energy from its environment in an amount equal to its _____.
10. When one gram of water vapor condenses on the outside of a glass of iced tea, the water has _____ heat energy to its environment in an amount equal to its _____.

Calorimetry

11. Given: average specific heat of ice = 0.50 cal/g-°C; average specific heat of steam = 0.48 cal/g-°C; heat of fusion of ice = 80 cal/g; heat of vaporization of water = 540 cal/g. Calculate the amount of heat needed to heat 20 g of ice from −25°C to steam at 126°C.

Gas Laws

12. As the human diaphragm contracts, it moves downward and expands the volume of the lungs. As a result, the air pressure in the lungs drops from 76.0 cm Hg to 75.8 cm Hg. In a deep breath your lungs will hold 3.5 liters of air at 37°C. What is the volume of cold fresh air at 7°C that is necessary to fill your lungs?

Answers

1. temperature, thermometer (Sections 10.6 and 10.2)
2. linear coefficient of expansion (Section 10.3)
3. volumetric expansion (Section 10.4)
4. calorie, raise the temperature of 1 g of water 1°C (Section 10.6)
5. mechanical equivalent of heat (Section 10.6)
6. heat capacity (Section 10.6)
7. specific heat (Section 10.6)
8. heat of combustion (Section 10.8)
9. latent heat of fusion (Section 10.7)
10. liberated latent heat of vaporization (Section 10.7)
11. 14,900 calories (Section 10.7)
12. 3.15 liters (Section 10.9)

ALGORITHMIC PROBLEMS

Listed below are the important equations from this chapter. The problems following the equations will help you learn to translate words into equations and to solve single concept problems.

Equations

$$°C = \tfrac{5}{9}\,(°F - 32)$$

$$L_T = L_0(1 + \alpha\,\Delta T) \tag{10.4}$$

$$\Delta V = V_0(\beta\,\Delta T) \tag{10.5}$$

$$\Delta Q = mc\,\Delta T \tag{10.9}$$

$$\frac{P_i V_i}{T_i} = \frac{P_f V_f}{T_f} \tag{10.15}$$

Problems

1. The normal temperature of the human body is said to be 98.6°F. What temperature is this on the Celsius scale?
2. If a doctor tells you that you have 6 degrees of fever, what is the equivalent number of Celsius degrees?
3. What is the change in length of a 30–m steel tape in a change of temperature from 0°C to 40°C?
4. What is the percentage change in volume of the mercury in a thermometer for a temperature change from 0°C to 100°C?
5. If you do 1000 J of work to produce heat, how many calories of heat do you produce?
6. How much heat will be required to warm 200 g of ice from −10°C to 0°C.
7. What is the calorie value of a 300-g apple?
8. Suppose you want to consume only milk to get a 1000 Calories/day diet. How much milk would you have to drink per day?
9. An automobile tire, initially with a volume of 0.36 m³, a pressure of 3 atmospheres, and a temperature of 7°C, will increase in volume by 5 percent and in pressure by 15 percent at highway speeds. What is the temperature of the high-speed tire?

Answers

1. 37°C
2. 3.3°C
3. 1.32 cm
4. 1.8 percent
5. 240 cal

6. 1000 cal
7. 174 kcal
8. 1540 g (slightly more than 1.5 liters)
9. 65.1°C

EXERCISES

These exercises are designed to help you apply the ideas of a section to physical situations. When appropriate the numerical answer is given at the end of each exercise.

Section 10.2

1. A human rectal temperature is 102°F. What is this reading on the Celsius scale? A patient is said to have 4 Fahrenheit degrees of fever. What is the equivalent in Celsius degrees? [38.9°C, 2.2°C]
2. At what temperature is the reading on the Celsius and Fahrenheit scales the same? [−40°]

Section 10.3

3. A distance is measured by a steel tape graduated

in millimeters. The tape is correct at 15°C, but a measurement is made on a hot day of 35°C. The length read on the tape was 300.250 m. What was the true distance?   [300.316 m]

Section 10.4

4. Mercury is to be stored in iron spheres which have a capacity of one liter at 10°C. The temperature of the storage room may reach a high of 40°C during the summer. How much mercury can be put in the iron flask at 10°C if no mercury is to overflow at the peak temperature? $\alpha = 12 \times 10^{-6}/°C$. $[1 - 0.0044$ liters$]$
5. The density of mercury is 13.6 g/cm³ at 0°C. What is its density at 200°C? $[13.1$ g/cm³$]$
6. A vertical Pyrex glass tube has an internal area of 10 mm² at 0°C, and it is filled to a height of 80 cm with Hg at this temperature. How long would the tube have to be so that no mercury would overflow if the system is heated to 100°C? Could this system be used as a thermometer? If so, how high would the mercury column be for a temperature of 50°C? Is it a linear system?   [81.4 cm, 80.7 cm, yes]

Section 10.6

7. The British thermal unit (BTU) is a unit of heat energy defined as the amount of heat required to raise the temperature of one pound of water one degree Fahrenheit. How many calories are equivalent to one BTU?   [252]
8. The following data are obtained in an experiment to measure the specific heat of a specimen. An aluminum container of mass 100 g contains 180 g of water at 10°C. A 200-g specimen is taken from a furnace at 200°C and put into the container. The final tempera-

ture of the mixture is 30°C. What is the specific heat of the specimen?   [0.12 cal/g–°C]

Section 10.7

9. If you know that the specific heat of a specimen is 0.98 J/g–°C, what is it in cal/g–°C? What is the heat of fusion of ice in joules/g? What is the heat of vaporization of water in J/g?   [0.234 cal/g–°C, 335 J/g, 2260 J/g]
10. A 200-g copper calorimeter cup contains 400 g of water at 45°C. In the cup 200 g of ice are melted, giving a final temperature of 5°C. Compute the heat of fusion of the ice.   [78.7 cal/g]
11. Calculate the heat required to convert 10 g of ice, originally at temperature of −10°C, to steam at a temperature of 120°C. Use 0.5 cal/g–°C as the specific heat of the steam.   [7355 cal]

Section 10.8

12. A given individual is said to eat 1500 kcal per day. Considering energy only, design a day's diet made up of six of the foods listed in Table 10.4.   [There are many possible answers depending upon food choices.]
13. A continuous-flow calorimeter is used to measure the heat of combustion of a gaseous fuel. For a steady condition water flows at 5000 g/min, and the temperature is raised 10°C. Gas flows at the rate of 4 liter/min. What is the heat of combustion of the gas?   [12,500 cal/liter]
14. A tank contains 2 m³ of nitrogen at an absolute pressure of 150 cm Hg and a temperature of 7°C. What will the pressure be, if the volume is increased to 10 m³ and the temperature is raised 227°C?   [53.6 cm Hg]

PROBLEMS

Each problem may involve more than one physical concept. The numerical answer is given in brackets at the end of the problem.

15. Eight grams of steam at 100°C are delivered by a rubber hose into a 100-g aluminum cup containing 80 g of ice and 200 g of water at 0°C. What is the final state and temperature of the contents of the cup? [All at 0°C, 64 gm of ice melted, 16 gm still ice.]
16. Steam burns are often severe. Compare the energy

liberated if one gram of steam at 100°C is transformed to water at 37°C, and one gram of water is cooled from 100°C to 37°C. If your body absorbed all the heat released, how many times greater is the heat absorbed from the steam than from the water?   [603 cal compared to 63 cal]
17. One hundred square meters of a lake is covered with ice at 0°C. If ice absorbs 0.25 cal/cm²–min, how much ice will melt in one hour? $[188 \times 10^3$ g/hr$]$
18. A bubble of air rises from the bottom of the lake where the pressure is 2.8 atmospheres to the surface

where the pressure is 1 atmosphere. The temperature at the surface is 27°C and 7°C at the bottom. Compare the size of the bubble at the surface with its size at the bottom of the lake?   [$V_t/V_b = 3$]

19. Suppose your coffee maker is rated at 300 watts. How long does it take to boil water for one cup of instant coffee? (Assume one cup contains 250 ml of water at 15°C.)   [nearly 297 sec]

20. If the average loss of heat due to evaporation of perspiration is about 10 kcal/hr, find the rate of water loss in g/hr. Assume the heat of evaporation 575 cal/g.   [17.4 g/hr].

21. Some insects that dive beneath the water surface carry an air bubble with them. Compare the volume of the bubble at a depth of 1 m to the surface volume if the temperature at the surface is 27°C, and the temperature at 1 m is 22°C.   [$0.90 = V_{1\,m}/V_s$]

22. If the human body loses 4 kg of water at 300°K each day due to evaporation, find the body heat loss this represents.   [10.1×10^6 J]

GOALS

When you have mastered the contents of this chapter, you will be able to achieve the following goals:

Definitions
Define the following terms, and use them in an operational definition:
conduction
radiation
convection

Energy Transfer Problems
Solve numerical problems that involve a transfer of energy, temperature gradients, and conduction, convection, or radiation.

Living System Thermal Properties
Explain basic thermal effects in living systems.

PREREQUISITES

Before beginning this chapter, you should have achieved the goals of Chapter 9, Transport Phenomena, and Chapter 10, Temperature and Heat.

11

Thermal Transport

11.1 Introduction

In a time of energy scarcity, all of us become more concerned about the ways we can conserve energy. Various community programs have been started to teach us how to use energy more efficiently. In particular agencies of government have encouraged us to improve the thermal transport properties of our homes. What can we do to reduce the loss of heat from our homes in the winter? Or to prevent the overheating of our homes in the summer? In this chapter we will examine the processes of thermal transport.

11.2 Conduction

We have already discussed the flow of heat through solid objects in Chapter 9. You may recall that we found that the rate at which heat travels along the handle of a silver spoon is related to the temperature gradient (see Section 9.2). The process by which heat energy travels from one part of a solid object to another is called *conduction*. Conduction of heat energy takes place in an object only when different parts of the object have different temperatures. The direction of heat flow is always from the point of higher temperature to that of lower temperature. Conduction involves a transfer of energy within a substance without the material itself being in motion. Conduction is the primary process by which heat energy travels through solids.

The thermal conductivity κ of a material is defined as the flow of heat energy per unit time per unit area per unit temperature gradient, in equation form,

$$\kappa = -\frac{H}{A\left(\dfrac{\Delta T}{\Delta x}\right)}$$

(11.1)

where H is the number of joules of energy transferred in one second, A

TABLE 11.1
Thermal Conductivities of Materials

Metals	κ(J/sec-m-°C)	Various Solids	κ	Fluids (at standard pressure and temp. of 300°K)	
Aluminum	200	Insulating brick	.147	Air*	.024
Brass	100	Clothing (dry)	.040	Helium*	.15
Copper	380	Construction brick	.63	Hydrogen*	.19
Mercury	8.4	Concrete	.84	Neon*	.049
Silver	400	Cork	.042	Oxygen*	.027
Steel	50	Glass	.80	Water	.59
		Glass wool	.042		
		Rock wool	.040		
		Wood	.15		

* The change in κ for gases is of the order of 1 percent per degree near room temperature.

is the area in square meters, and $\Delta T/\Delta x$ is the temperature gradient in degrees Celsius per meter. The negative sign is introduced because H is positive and in the opposite sense to the temperature gradient $\Delta T/\Delta x$. The expression for the flow of heat energy in watts or J/sec is given by

$$H = -\kappa A \frac{\Delta T}{\Delta x} \tag{11.2}$$

For most materials the thermal conductivity is a function of temperature, but the variation with temperature is small for solids and for our purposes may be neglected.

EXAMPLES

1. Compare the heat energy loss by conduction from a brick house of 15 cm wall thickness and from a wooden house of 5-cm wall thickness. Assume the temperature difference between the inside and the outside is 20°C, the house is 6 m × 8 m × 4 m tall, and the heat energy lost through the floor is zero.

   Brick $\quad H_B = -\kappa A \dfrac{\Delta T}{\Delta x}$

   $$= -\frac{(0.63) \text{ watts}}{\text{m–°C}} [2(6.0)(4.0) + 2(8.0)(4.0) + (6.0)(8.0)] \text{m}^2 \times \frac{20°\text{C}}{0.15 \text{ m}}$$

   $$= 1.3 \times 10^4 \text{watts}$$

   Wood $\quad H_w = -\dfrac{(0.15) \text{ watts}}{\text{m°C}} (160 \text{ m}^2) \left(\dfrac{20°\text{C}}{0.05 \text{ m}} \right)$

   $$= 9.6 \times 10^3 \text{watts}$$

2. During the energy crisis of 1973 many homeowners were encouraged to put storm windows on their homes. Does that really make any difference? Compare the heat loss due to conduction for a single window pane and for a storm-window pane with a dead air space between two glass panes, where the dead air space is twice as thick as a pane of glass (see Figure 11.1).

   Let us use the following symbols: T_1 is the inside temperature and T_2 is the outside temperature, so $T_1 > T_2$, κ_1 is thermal conductivity of glass, κ_2 is thermal conductivity of dead air, x_1 is glass thickness, x_2 is air thickness $(= 2x_1)$, and A is the area of window. At equilibrium the heat flow through all interfaces must be equal,

   the heat flow through the inside pane of glass = heat flow through the dead air space = heat flow through the outer pane of glass

   $$-\kappa_1 A \left(\frac{T_1 - T'}{x_1} \right) = -\kappa_2 A \left(\frac{T' - T''}{x_2} \right) = -\kappa_1 A \left(\frac{T'' - T_2}{x_1} \right)$$

   where T' is the temperature at the outside of the inner pane of glass and T'' is the inside of the outer pane of glass. Eliminating T' and T'', we get the following equation for heat transfer through the thermal pane,

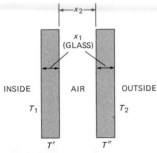

FIGURE 11.1
Double pane window with the spacing between the glass panes twice the thickness of each plate of glass.

$$H = \frac{-A(T_1 - T_2)}{\dfrac{x_1}{\kappa_1} + \dfrac{x_2}{\kappa_2} + \dfrac{x_1}{\kappa_1}} \tag{11.3}$$

This result can be generalized for N layers,

$$H = \frac{-A(T_1 - T_2)}{\sum\limits_{i=1}^{N} \frac{x_i}{\kappa_i}} \tag{11.4}$$

where κ_i is the thermal conductivity of the ith layer of thickness x_i, and $(T_1 - T_2)$ is the temperature difference across the N layers.

For the storm window we have:

$$H_{sw} = \frac{-A(T_1 - T_2)}{\dfrac{2x_1}{\kappa_1} + \dfrac{x_2}{\kappa_2}} = \frac{-A(T_1 - T_2)}{\dfrac{2x_1}{0.84} + \dfrac{2x_1}{0.024}}$$

For a single glass pane we have:

$$H_g = \frac{-A(T_1 - T_2)}{\dfrac{x_1}{\kappa_1}} = \frac{-A(T_1 - T_2)}{\dfrac{x_1}{0.84}}$$

The ratio of storm window heat loss to glass pane heat loss is given by

$$\frac{H_{sw}}{H_g} = \frac{\dfrac{x_1}{.84}}{\dfrac{2x_1}{.84} + \dfrac{2x_1}{.024}} = \frac{1.2}{2.4 + 83} = \frac{1}{71}$$

In this case, the storm window reduces the loss of thermal energy by a factor of 71!

3. A skier has a body surface area of 2.0 m² and wears clothing 1.0 cm thick. The thermal conductivity of the dry clothing is 4.0×10^{-2} watts/m–°K. Assume that the skier's skin temperature is 33°C and the outer clothing surface temperature is 0.0°C. We wish to find the heat loss through the dry clothing and compare it with that through wet clothing (which has a thermal conductivity equal to that of water = 59×10^{-2} watts/m–°K.

$$H_d = -\kappa_d A(T_s - T_o)/x,$$

$$= -4 \times 10^{-2} \times (33 - 0) \times 2/10^{-2} \text{ watts}$$

$$= -260 \text{ watts (heat loss through dry clothing)}$$

$$H_w = -\kappa_w A(T_s - T_o)/x$$

$$= -59 \times 10^{-2} \times (33 - 0) \times 2/10^{-2} \text{ watts}$$

$$= -3900 \text{ watts (heat loss through wet clothing)}$$

This example illustrates why one feels cooler in wet clothing.

11.3 Convection

You have observed a pan of water heating on a stove, and you have probably noticed that the water seemed to be in motion. Also you have observed the trail of the smoke from a lighted cigarette. These are examples of the transfer of heat energy by *convection*, a process in which the heated, or higher energy, matter moves. The transfer of heat by convec-

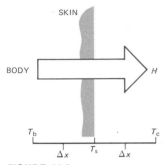

FIGURE 11.2
Heat energy leaving the human body when the body is surrounded with a medium of temperature below body temperature.

tive circulation in a liquid or gas is generally brought about by a difference in density. Liquids and gases, when heated, expand and become less dense. The less dense material is continually displaced by the fluid of greater density. This motion is known as *convection currents,* and heat energy is transferred from one location to another by the motion of the matter with higher internal, or heat, energy.

Convection is an important factor in the design of insulation systems. In order to minimize convection losses, it is necessary to break up open spaces so that the circulation of matter is impeded. For this reason the insulating material used in the walls of a refrigerator or in walls of a house is porous. Insulating materials are not only poor conductors but they contain many small air spaces. No effective long-range convection currents can be set up in the insulating materials.

A detailed mathematical theory of heat convection is quite complicated, but we will discuss the loss of heat energy from the human body by convection in an introductory quantitative way.

Your body has a relatively stable interior temperature, T_b. If you are located in air of temperature T_a which is different from your body temperature there will be a flow of heat by conduction through your skin T_s to the air (Figure 11.2). Using our previously developed equations, we can write statements for the heat current through the skin and through the air as follows:

$$H = \frac{-\kappa_s}{\Delta x}(T_s - T_b)A$$

where κ_s is the coefficient of thermal conductivity of the skin with thickness Δx.

$$H = \frac{-\kappa_a}{\Delta x}(T_a - T_s)A$$

where κ_a is the coefficient of thermal conductivity for still air with thickness Δx.

We can eliminate T_s from these equations as we did in the example of the storm window, and we obtain

$$H = -\frac{\kappa_s \kappa_a}{\kappa_s + \kappa_a} \times \frac{T_b - T_a}{\Delta x}A \tag{11.5}$$

Notice that the heat energy flow across the interface is proportional to the total temperature drop from the interior body temperature to the temperature of the air, $T_b - T_a$. The coefficient of thermal conductivity of still air is small; so the flow of heat energy through perfectly still air is rather low. One of the functions of warm clothes is to provide traps for air to prevent the movement of air near the body and hence to provide good insulation for the body. However, if air is forcibly circulated, or if there is a considerable amount of natural air movement, the rate of energy flow increases because of convection (Figure 11.3).

Right at the skin-air interface, we assume that the flow of heat energy is proportional to the total temperature difference between your skin and

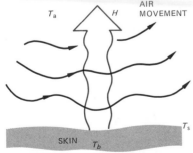

FIGURE 11.3
Heat energy transfer from the skin of an individual when there are convection air currents present.

the air, $T_s - T_a$. We can write an expression for the heat energy flow by forced convection (no stagnant air layer at the surface) as

$$H = -\kappa'(T_s - T_a) \tag{11.6}$$

where κ' is the convection coefficient of the object, which depends upon the nature and area of the interface, the speed of the air movement, the density and viscosity of air, and so on. The value of κ' is usually determined experimentally for a given system.

Under normal indoor conditions, approximately 30 percent of the total heat energy lost from the human body is through convection. As the speed of the air past the body surface increases, as for a bicycle rider, for example, the fraction of heat lost by convection increases.

EXAMPLE

The convection coefficient of a clothed human body at rest is about 2 watts/m²–°C. When the air velocity is 5 m/sec with respect to the body, as for a bicycle rider, the convection coefficient increases to 28 watts/m²–°C. Compare the rates of heat loss in these two cases assuming the same body and environment temperature.

$$H_{rest} = 2(T_b - T_0)$$

$$H_{rider} = 28(T_b - T_0)$$

$$H_{rider}/H_{rest} = 14$$

An empirical rule, known as Newton's law of cooling, was developed to predict the rate at which hot objects are cooled by convection. This rule has the same form as Equation 11.6 and states in mathematical form that the rate of cooling is proportional to the difference between the temperature of the object and the temperature of its surroundings. It can be shown, using the methods of calculus (see Section 11.7), that the temperature of the cooling object can be expressed as an exponential function of the time of cooling,

$$T = T_0 + \Delta T e^{-\kappa' t/C} \tag{11.7}$$

where κ' is the convection coefficient, C is the heat capacity of the object, ΔT is the initial temperature difference between the environment and the object, T_0 is the temperature of the environment, and T is the temperature of the object at a time t.

EXAMPLE

A boiling hot cup of coffee (heat capacity 72 J/°C) in a room of 20°C cools down to a temperature of 60°C in one minute. What is the convection coefficient of this system? How long will it take for the coffee to become lukewarm (48°C)?

To find the convection coefficient, use Equation 11.7 where all the quantities are known except κ',

$$T = T_0 + \Delta T e^{-\kappa t/C}$$

$60 = 20 + (100 - 20)\ e^{-\kappa'(60\ sec)/72\ J/°C}$

$40 = 80\ e^{-0.833\ \kappa'\ sec-°C/J}$

writing 0.50 as $e^{\ln 0.50}$, we obtain,

$0.50 = e^{-0.69} = e^{-0.83\kappa'}$

so

$0.83\ \kappa' = 0.69$ J/sec–°C

$\kappa' = 0.83$ J/sec–°C

Notice that the initial temperature difference of 80° has dropped to a temperature difference of 40° in one minute. It is the property of exponential functions to have constant half-times; that is, the time required for the value of the exponential function to change by a factor of two is constant. Hence, the coffee will have a temperature of 40°C (a temperature difference of 20°) after 2 minutes, 30°C after 3 minutes; 25°C after 4 minutes; 22.5° after 5 minutes, and so on.

11.4 Radiation

You know that heat energy is transferred from the sun to the earth. This method of transfer is called *radiation*. We say that the sun is radiating energy as electromagnetic waves. This type of radiation does not require a material medium for its propagation. Since it is an example of interaction-at-a-distance, we propose that the energy is transferred to the earth from the sun by means of a field (see Chapter 1).

Any object at a temperature above absolute zero emits energy from its surface. This energy is in the form of electromagnetic waves and is called *radiant energy*. The waves travel with the velocity of light. They may be absorbed by the medium in which they travel or by the object which they strike. When this radiant energy is absorbed it becomes heat energy. An object that absorbs this radiant energy, converts the radiant energy into internal, or heat, energy. Experiments have shown that good absorbers are also good radiators. So an ideal radiator is an object that absorbs all of the radiation that falls upon it. It is called an *ideal black body*. No perfect black body is known, but we have some good approximations such as objects whose surfaces are coated with lamp black. Objects with bright shiny surfaces are poor radiators.

The law that expresses the total energy radiated by a black body was originally stated by Stefan on the basis of experimental measurements from a black–body cavity. Later Boltzmann derived the equation from thermodynamic theory. The rate P at which energy is radiated by a black body is given by

$$P = \sigma A T^4 \tag{11.8}$$

where P is expressed in watts, A is the area of the radiator in square

FIGURE 11.4
Intensity of black-body radiation as a function of wavelength for several different temperatures.

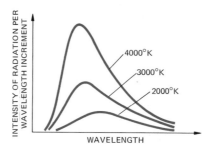

centimeters, and T is the Kelvin temperature, and σ has a value of 5.70 $\times$ 10^{-12} watt/cm^2–$°$K^4. This is known as the Stefan-Boltzmann law. Anything other than a black body will radiate at a lower rate.

The distribution of the energy among the different wavelengths of radiation is characteristic of the temperature of the emitting surface (Figure 11.4). From this distribution you can explain the saying, "White hot is hotter than red hot."

Let us consider a greenhouse. Window glass is transparent for visible radiation but absorbs heat radiation. The glass roof transmits the visible and near infrared radiations from the sun ($T \sim 6000°$K). These rays are converted to heat when they are absorbed by objects in the greenhouse. The absorbing objects are heated and become radiators. Their radiation is characteristic of their temperature (about 360$°$K). Because the glass is opaque to this radiation, this energy does not get out but is reflected back into the greenhouse. So a greenhouse serves as an energy trap.

11.5 Thermal Effects in Living Systems

A living system is constantly interacting thermally with its environment. Environmental equilibrium temperatures on earth are maintained by radiation from the sun. (The solar power reaching the earth is approximately 1.4×10^3 watts/m^2.) Living systems maintain a steady-state equilibrium with their environment by adjusting their metabolic energy production to balance their heat transfer to the environment. The *basal metabolic rate* is the heat production under normal resting conditions. Sleeping organisms operate at levels below basal rates, and active systems operate at higher metabolic rates. Only a limited range of temperatures will support living systems as we currently know them.

Temperature changes can alter the form of proteins in living systems. Such alterations may result in drastic changes in large protein chains. When this occurs the proteins are no longer capable of performing their biological function (such proteins are said to be thermally *denatured*). For many proteins the range of temperatures from 10$°$ to 20$°$C is the range for maximum stability. Much lower or higher temperatures produce denatured proteins.

In all animals the metabolic rate increases as the body temperature increases. In humans a 1°C temperature rise produces about a seven percent increase in the basal metabolic rate. Different systems have maximum efficiency for metabolic processes at different temperatures. Those animals that show a constant body temperature are called *homeotherms* (warm-blooded animals). Animals not possessing the necessary thermoregulation system for a constant body temperature are called *poikilotherms* (cold-blooded animals). The temperature of poikilotherms varies with their environmental temperature. Reptiles are examples of poikilotherms. It seems clear that homeotherms have an evolutionary advantage as they can function independent of significant environmental temperature changes. It has also been suggested that the development and functioning of a complex brain depends on the maintenance of a constant brain temperature.

In humans the core temperature for internal organs is 37°C while the skin temperature is about 33°C. For a human with a surface area of 2 m² the basal metabolic rate (BMR) is about 90 watts. Exercise can increase the metabolic rate by a factor of 10. The conditions affecting the heat transfer to the environment are important factors in determining the metabolic rate. Heat is transferred to the skin from the core of the body by conduction and by forced convection of the blood flowing from the core to the peripheral blood vessels. At high external temperatures or high metabolic rates the human body compensates by increasing its rate of energy loss. This is done by increasing the peripheral blood circulation, dilating the blood vessels (*vasodilation*), and by increasing evaporation cooling. Vasodilation can increase the effective heat transfer by a factor of eight and under optimum conditions sweating can produce a cooling rate of approximately 700 watts.

At low temperatures the human system acts to reduce heat transfer. *Vasoconstriction* reduces blood flow to the surface and thereby reduces heat flow. Energy production is increased for short times by shivering. For longer periods of time, enzyme activity increases metabolic activity in order to maintain equilibrium. The control element for the human body thermoregulator is in the *hypothalamus*, located at the base of the brain. The hypothalamus is connected by the nervous system to the thermal detectors on the body surface. The detective work involved in solving the mystery of the human temperature regulation is an example of success in biophysical research.

Svante A. Arrhenius made one of the most successful theoretical formulations for chemical reaction rates. He suggested that if chemical reactions take place because of collision interactions among molecules, then the rates should be governed by the Maxwell-Boltzmann distribution of energy for molecules in thermal equilibrium at absolute temperature T, a law of classical physics. This distribution can be expressed in equation form as follows: $P \propto e^{-E/kT}$ where P is the probability that a system has energy E when it is in thermal equilibrium at an absolute temperature T and where k is Boltzmann's constant, which has a value of

1.38×10^{-23} J/°K. The reaction rate equation derived by Arrhenius is:

$$\lambda = A_0 e^{-(E_a/kT)} \tag{11.9}$$

where E_a is the activation energy for the reaction, λ is the rate of the reaction, and A_0 is a constant. It has been shown that this same rate equation applies to the activity of biological systems such as fruit flies, guppies, and brine shrimp. The experimental set up consists of a test chamber that is divided into two equal parts separated by a transparent dividing partition in which there is a hole that can be opened. All samples are initially placed in one side of the test chamber, and the hole is opened. The number of samples reaching the second chamber is recorded as a function of time. If the samples are treated as molecules in a kinetic theory model, it can be shown that the following equation results for the number of samples in the second chamber as a function of time:

$$N_2 = \frac{N_0}{2}(1 - e^{-\lambda t})$$

$$\text{or} \quad \ln \frac{N_0}{N_0 - 2N_2} = \lambda t \tag{11.10}$$

where N_0 is the original number of samples placed in side 1, N_2 is the number of samples in side 2 at time t, and λ is the reaction rate (number/sec passing through the hole).

Using this equation and solving for λ, we can then use the Arrhenius equation to determine an activation energy for the activity involved. In a graph of the natural logarithm of λ versus $1/kT$, the slope is equal to the activation energy.[1]

This approach suggests an interesting way to study the effects of such variables as chemicals, light, and magnetic fields on the activity of biological systems. For example consider the question, "Do mosquitoes find their targets through chemical sensing or thermal sensing?" Design an experimental system, and start counting mosquitoes.

EXAMPLE

Natural convection from the human body is found to obey the following heat energy flow equation: $H = -\kappa(T_s - T_a)^{1.25}$ watts/m². We want to compare the natural convection of heat energy loss of a nude person with skin temperature of 33°C in a room at 18°C to the heat loss in the same room maintained at 23°C.

$$H_{18} = -\kappa(33 - 18)^{1.25}$$

$$H_{23} = -\kappa(33 - 23)^{1.25}$$

Thus we have $H/_{18}/H_{23} = (15/10)^{1.25} = 1.66$

The person loses heat 1.66 times faster in a room at 18°C than in the same room at 23°C.

---

[1] O. A. Runquist, C. J. Creswell, J. T. Head-Burgess, *Chemical Principles: A Programed Text*, 2nd ed. (Minneapolis, Minn.: Burgess, 1974), p. 383.

FIGURE 11.5
Cross-section of a microwave oven showing how a microwave pattern coming from a wave guide is broken up so that the radiation is reflected from the sides and bottom of the oven into the food being cooked. (Copyright 1973 by Consumers Union of United States, Inc., Mount Vernon, N.Y. 10550. Reprinted by permission from *Consumers Reports,* April, 1973.)

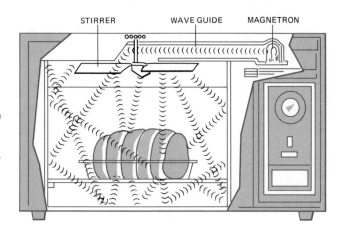

11.6 Microwave Cooking

A recent development in the use of electromagnetic radiation as a source for heat energy is microwave cooking. We offer below two different explanations for how microwaves cook. The first explanation is from a microwave oven manufacturer's cookbook. The second is from an article that evaluated the performance of various microwave ovens.

HOW MICROWAVES COOK

Traditionally, food heats and cooks because of molecular activity caused by a gas flame, burning wood and charcoal, or electricity converted to heat energy. This intense heat must be applied to the bottom of a pan of food or used to surround food in an oven with hot air. If food comes in direct contact with these traditional heat sources, it burns before it cooks through.

Now electrical energy can be converted to microwave energy by means of an electron tube called a magnetron. This tube is inside the microwave oven and sends microwaves directly into food.

Microwaves are classified as electromagnetic waves of a non-ionizing frequency. Microwaves travel directly to food without heating the air or the recommended cooking dishes. The cooking process speeds up because it starts as soon as the oven is turned on.

Microwaves move directly to food because they are attracted to the fat, sugar and liquid or moisture molecules, causing them to vibrate at a fantastically fast rate. This vibration is heat energy. The vibrating molecules bump and rub others, start these molecules vibrating and set up a chain reaction that moves from the outside edges, where microwaves first come in contact with food, toward the center—cooking as it goes. This chain reaction is called conduction. The molecule vibration, or cooking, continues for several minutes after food comes from a microwave oven and is taken into consideration in microwave recipes.

Microwaves' specific attraction for moisture, fat and sugar in molecules, plus the fast molecule vibration rate this causes, results in the amazing speed of microwave cooking.[2]

---

[2] *Variable Power Microwave Cooking,* Litton Systems, Inc. (New York: Van Nostrand Reinhold, 1975), p. 6.

MICROWAVE COOKING "Microwaves reflect from metal objects and pass right through glass, paper and some plastic objects . . . but are absorbed by foods and converted to heat. The waves are not hot; they create heat by agitating molecules in materials that are capable of absorbing them. In addition to food, that includes human flesh and bone. The drawing (Figure 11.5) shows how the microwaves issue from an electronic vacuum tube, called a magnetron, flow through a metal tube, or wave guide, and are dispersed around the oven cavity by a rotating metal paddle, or stirrer. As the microwaves penetrate the food, more energy reaches the interior portion immediately in a microwave oven than in a conventional oven, so that cooking, overall, can be much faster. But the actual time required will depend on the type and quantity of food."[3]

Using the concepts you have studied, evaluate these two explanations of cooking in microwave ovens. Which explanation seems more vivid?

ENRICHMENT
11.7 Newton's Law of Cooling

Let us begin with the equation for the flow of heat energy by forced convection, Equation 11.6

$$H = -\kappa' \ (T - T_0) \tag{11.6}$$

where H is the flow of heat in watts, T is the temperature of the object and T_0 is the temperature of the surroundings. In the notation of calculus the flow of heat is the time derivative of the heat energy of the system, $H = dQ/dt$ where Q is the internal energy of the object and $Q = mcT$ where m is the mass of the object, c is the specific heat of the object and T is the temperature of the object (from Chapter 10). So we can rewrite Equation 11.6 in the calculus notation to obtain the standard form of Newton's law of cooling,

$$mc \ \frac{dT}{dt} = -\kappa' \ (T - T_0) \tag{11.11}$$

Let us use the technique of separation of variables and then integration to find an expression for the temperature of the object as a function of time.

$$\frac{dT}{T - T_0} = -\frac{\kappa' \, dt}{mc}$$

$$\ln \ (T - T_0) = -\frac{\kappa' \, t}{mc} + \text{constant}$$

$$(T - T_0) = (\text{constant}) e^{-\kappa' t/mc} \tag{11.12}$$

[3] *Consumer Reports*, Consumers Union of United States, Inc., 38, Apr. 1973, p. 221.

Since at $t = 0$, the object was ΔT hotter than the environment, we obtain Equation 11.7

$$T = T_0 + \Delta T\, e^{-\kappa' t/mc}$$

Remember $mc = C$ (the heat capacity of body).

SUMMARY

Use these questions to evaluate how well you have achieved the goals of this chapter. The answers to these questions are given at the end of this summary with the section number where you can find the related content material.

Definitions

Write the correct word or phrase in each blank.

1. When heat energy is transferred through a solid object it occurs by a process called _____.
2. Metals have a larger _____ than gases and transfer heat energy by _____.
3. The movement of higher temperature, lower density portions of fluids gives rise to a process of heat energy transfer called _____.
4. In the "zero gravity" of an orbiting space craft it has been found that _____ is no longer an important process of heat transfer because _____.
5. In contrast to the other energy transfer processes _____ can occur in the absence of a material medium.

Energy Transfer Problems

6. How many watts would be conducted through a copper sheet 25 × 25 cm square and 1 mm thick with a temperature difference of 200°C?

Living System Thermal Properties

Write the correct word or phrase in each blank.

7. Living systems that are able to maintain a constant temperature are called _____.
8. Such systems are able to control the rate of their internal _____ by their precise control of _____.
9. The _____ is the control element for the thermoregulation of the human body.

Answers

1. conduction (Section 11.2)
2. coefficient of thermal conductivity, conduction (Section 11.2)
3. convection (Section 11.3)
4. convection, differences in density no longer result in fluid motion (Section 11.3)
5. radiation (Section 11.4)
6. 4.75×10^6 watts (Section 11.2)
7. homeotherms (Section 11.5)
8. chemical reactions, temperature (Section 11.5)
9. hypothalamus (Section 11.5)

ALGORITHMIC PROBLEMS

Listed below are the important equations from this chapter. The problems following the equations will help you learn to translate words into equations and to solve single-concept problems.

Equations

$$H = -\kappa A\, \frac{\Delta T}{\Delta x} \tag{11.2}$$

$$H = \frac{-A(T_1 - T_2)}{\displaystyle\sum_{i=1}^{N} \frac{x_i}{\kappa_i}} \tag{11.4}$$

$$H = -\kappa'(T_s - T_a) \tag{11.6}$$

$$T = T_0 + \Delta T e^{-\kappa' t/C} \tag{11.7}$$

$$P = \sigma A T^4 \tag{11.8}$$

$$\lambda = A_0 e^{-E_a/kT} \tag{11.9}$$

Problems

1. A temperature gradient of 40°C/cm is maintained across an aluminum rod 4 cm² in area. What is the rate of energy transfer by conduction?
2. If the absolute temperature of an object is doubled, by what factor does the rate of energy radiated from the object change?
3. Compare the rate at which energy is radiated by an object at room temperature (27°C) to the rate at which energy is radiated by the object at the temperature of liquid helium (4°K).
4. A hot object of 80°C is placed in a 0°C environment where its energy loss occurs primarily by convection. After one minute the temperature of the object is 60°C; what will the temperature of the object be in 5 minutes? $\kappa'/C = 0.288$/min for this object

Answers

1. 320 watts
2. 16

3. 3.16×10^7
4. 19°C

EXERCISES

These exercises are designed to help you apply the ideas of a section to physical situations. When appropriate the numerical answer is given in brackets at the end of each exercise.

Section 11.2

1. Compare the heat loss by conduction for 5.0 mm of dead air plus 5.0 mm of clothing to that for 1.0 cm of clothing. Assume the temperature difference between the skin and environment to be 40°C. [1.3 times as much heat lost for clothing as for combination]

Section 11.3

2. Under normal forced air convection conditions, the convection coefficient for the human body in air is 13.95 watts/m-°C. If the convective heat current density loss from a body is 60.0 watts/m² for resting, find the temperature difference between the skin and room air. [4.30°C]

Section 11.4

3. If the absolute temperature of a radiating body is increased by 50 percent, how does the rate of radiation increase? [$P_{new}/P_{old} = 5.1$]

PROBLEMS

Each of the following problems may involve more than one physical concept. The numerical answer to the problem is given in brackets at the end of the problem.

4. Compare wood and glass as insulating materials.
5. Three metal rods — copper, aluminum, and brass — each 6 cm long and 2 cm in diameter are placed end

FIGURE 11.6
Heat conducting rod made up of copper, aluminum and brass in series. Problem 5.

to end (Figure 11.6). The free ends of copper and brass are held at temperatures of 100°C and 0°C respectively. Assume that the thermal conductivity of copper is two times that of aluminum, and that of aluminum is two times that of brass. What is the temperature at the copper-aluminum junction and at the aluminum-brass junction when the system reaches equilibrium condition? [$T_{\text{Cu-Al}} = 600/7 \cong 86°C$, $T_{\text{Al-Br}} = 400/7 \cong 57°C$]

6. Analyze the heating system of your home in terms of conduction, convection, and radiation.

7. A house has insulating brick walls of 26 cm thickness and a total area of 240 m². If the difference in temperature between the inside and outside is 20°C, how much natural gas, which has a heat of combustion of 10,000 cal/g, will it take per day to maintain a constant temperature? [5600 g]

8. Given the problem of designing a cold storage container for dry ice using styrofoam of constant thickness, compare the rate of heat flow into spherical, cylindrical ($d = h$), and cubical containers of equal volume when containers are used in identical conditions. [For identical conditions the heat flow will be proportional to the surface area of the container. For a volume of $4\pi/3$ m³, the sphere will have a radius of 1 m and a surface area of 4π m². The cylinder will have a radius of 0.88 m, a height of 1.76 m, and an area of 4.65π m². The cube will have a side length of 1.61 m, and an area of 15.54 m². Hence the heat flow will be the smallest into the spherical container, next smaller into the cylinder, and greatest into the cube in the ratio of 1:1.16:1.24.]

†9. A paramedic team rushes to the scene of an accident and finds a corpse of temperature 26.2°C, five minutes later the corpse temperature is 23.8°C. If the environment temperature is 22.0°C, the normal body temperature is assumed to be 37.0°C, and the rate of heat energy loss assumed to be forced convection cooling,

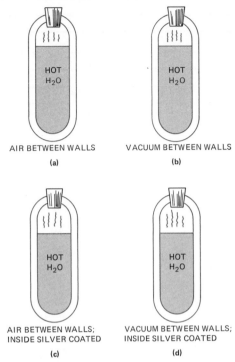

AIR BETWEEN WALLS
(a)

VACUUM BETWEEN WALLS
(b)

AIR BETWEEN WALLS; INSIDE SILVER COATED
(c)

VACUUM BETWEEN WALLS; INSIDE SILVER COATED
(d)

FIGURE 11.7
Four double-walled glass vessels, uncoated and silvered, with air between walls and with a vacuum between walls. Problem 10.

what time elapsed between the death of the person and the arrival of the paramedic team? [7.46 min]

10. Four identical glass, double-walled containers are treated in different ways by evacuating the air from between the walls or not, and by silvering the glass walls or not. Then each container is nearly filled with hot water and a cork stopper is placed in each container. Discuss the rate of heat energy loss from each container (Figure 11.7).

11. Assume a space traveler finds three other planets about the same size as the Earth but with helium, neon, and oxygen atmospheres. Compare the rates of heat conduction from these planets to that from the Earth. What additional assumptions must you make?

GOALS

When you have mastered the contents of this chapter, you will be able to achieve the following goals:

Definitions
Define each of the following terms, and use it in an operational definition:

PV diagram	efficiency of a heat engine
isochoric process	Carnot cycle
isobaric process	refrigerator
isothermal process	coefficient of performance
adiabatic process	of a refrigerator
heat engine	

Laws of Thermodynamics
State three laws of thermodynamics, and explain the operation of a physical system in terms of these laws.

Thermodynamics Problems
Solve problems consistent with the laws of thermodynamics.

PREREQUISITES

Before beginning this chapter you should have achieved the goals of Chapter 5, Energy, and Chapter 10, Temperature and Heat.

12

Thermodynamics

12.1 Introduction

Much of your travel is made possible by a device that we call a heat engine. We can think of a heat engine as being any device which operates by heat energy input, does work, and has a heat energy output. The human body is a heat engine. The refrigerator in your home is a type of heat engine. In this chapter you will be introduced to the fundamental principles of these and other forms of heat engines.

In Chapter 10, it was indicated that when the temperature of an object is changed many different effects may take place—such as change in length or volume, change in resistance, change in pressure, change in electromotive force, change in color, and so on. In such processes there is usually a transfer of heat energy and also a performance of work, a force acting through a displacement. The study of the phenomena that result from energy changes produced by a transfer of heat and performance of work is called *thermodynamics.* These phenomena involve heat energy, temperature, and work. Thermodynamics is the study of the laws that govern thermal processes. When you have completed this study of thermodynamics, you will be able to explain the significance of these laws as they apply to both living and nonliving systems. Thermodynamics concerns itself with a well-defined system, which interacts directly with its surroundings, and by this interaction performs some useful function. Thermodynamics is *not* concerned with internal effects by themselves.

12.2 The Zeroth Law of Thermodynamics

You know that when two equal amounts of ice are added to two identical containers of hot tea to make iced tea, after the ice has melted both containers of hot tea will be at the same final temperature. This is an example of the *zeroth law of thermodynamics.*

The zeroth law of thermodynamics can be stated as follows:

If system A is in thermal equilibrium with system B, and system B is in thermal equilibrium with system C, then A, B, and C are in equilibrium with each other, and they are at the same temperature.

Thermal equilibrium is defined as the condition in which there is no net energy exchange between the systems in equilibrium. The zeroth law is seemingly trivial, but it is the basis of thermometers since it defines operationally temperature as the property of a system that determines its thermal equilibrium with another system. The zeroth law of thermodynamics clearly points out that temperature, as measured by thermometers, involves systems in equilibrium; that is, the temperature being measured by the thermometer in the system must be constant in time, indicating equilibrium, before a measurement is made. In this connection it is important to point out that the thermal inertia of the thermometer determines the time required for the thermometer to reach equilibrium with the system, as well as the effect of the thermometer on the system being monitored. The larger the thermal inertia (mass × specific heat of the thermometer), the longer the response time, and the more energy transferred to or from the thermometer in attaining equilibrium. In this chapter we will use the symbol T to refer to absolute temperature in degrees Kelvin ($0°C = 273°K$).

12.3 External Work

Consider as a system a cylinder containing a working substance (a gas) and closed by a movable piston (Figure 12.1). If the gas pushes the piston out, work is done *by* the system. If an outside agent pushes in the piston, work is done *on* the system. In either case the work is called *external work*.

Work that is done by one part of the system on another part of the same system is called *internal work*. The study of internal work as such is not part of thermodynamics.

Returning to our example system, the cylinder containing gas enclosed by a movable piston, assume the piston has an area A, and a pressure P is acting on the piston. The total force acting on the piston is given by the product of the pressure times area, PA. If the piston is displaced a small distance Δx, then work is done, and the amount of work ΔW is given by the product of the force times the displacement,

$$\Delta W = (\text{force})(\text{displacement}) = (PA)(\Delta x) \qquad (12.1)$$

We notice that the product of area times displacement, $A \Delta x$, is equal to the change in the volume ΔV of the system as the piston moves. Then we can express the amount of work done by the system when it pushes the piston out a small distance Δx by the following equation

$$\Delta W = P \Delta V \qquad (12.2)$$

If the pressure remains constant over a whole series of small displacements, then the total amount of work done is given by the pressure times the change in volume.

$$W = P(V_2 - V_1) \qquad (12.3)$$

where the difference between the final and initial volumes $(V_2 - V_1)$ is the change in volume. If the pressure is not constant, we have to use calculus methods, as we discuss in the Enrichment section of this chapter, Section 12.16.

EXAMPLE

During normal breathing the volume of the lungs increases by about 500 ml in each inspiration. What is the amount of work done on the lungs during inspiration?

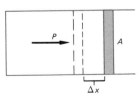

FIGURE 12.1
Cylinder with a movable piston. Work done $= PA \, \Delta x = P \, \Delta V$.

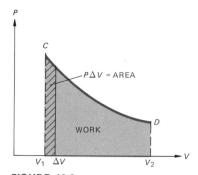

FIGURE 12.2
Geometric representation of work from a P-V diagram. $\Delta W = P\, \Delta V$.

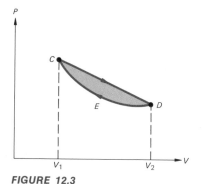

FIGURE 12.3
Graphical representation of the work done during a cycle. The work done in going from one point to another depends upon path. Net work done during the cycle is represented by the enclosed area.

The standard pressure of the atmosphere is about 1.00×10^5 N/m².

work = (pressure)(change in volume)

$$W = (1.00 \times 10^5 \text{ N/m}^2)\left(500 \text{ ml} \times \frac{10^{-6} \text{ m}^3}{1\, ml}\right)$$

$$W = 50 \text{ J}$$

We can interpret the meaning of Equation 12.3 by studying a PV diagram, the graph of the pressure versus the volume of a confined gas (Figure 12.2). If we allow the system to expand, the pressure of the gas will decrease as the volume increased from V_i to V_f. The amount of work done by the gas for a small displacement Δx is given by Equation 12.2 and is shown as the shaded area in the figure since $P\,\Delta V$ is equal to the area of the shaded portion of the graph. This area on the graph has the units of joules. If we add the work done by many of these small displacements as the system expands from V_1 to V_2, we find that the work done by the gas is represented by the area under the curve from C to D between the vertical lines at V_1 and V_2 (shown as the large shaded portion of the graph). If we allow the gas to expand, the system goes from C to D, the gas pushes the piston out, and we say the system does work. By convention we call this positive work. If we compress the gas so that the volume of the system goes from V_2 to V_1 (the pressure changes from D to C), we say work is done on the system. We designate this as negative work.

Let us consider a cyclic system as shown in Figure 12.3. First, the working substance expands from V_1 to V_2 in accordance with line CD. The work done is positive and represented by the area between the curve CD and the volume axis. Then the working substance is compressed from V_2 to V_1 in a process represented by the curve DEC. The work of compression is represented by the area under DEC, DV_2V_1CE, and is negative. For this cycle, the area under CD is greater than the area under DEC. The total amount of work done in one cycle is equal to the enclosed area between CD and DEC. This area represents the *net work* that the system does in one cycle. It is evident that the amount of net work done per cycle depends upon the path that is followed from C to D and back again. That is, it depends upon the processes used during the expansion and compression portions of the cycle.

12.4 The First Law of Thermodynamics

You have probably had the experience of doing mechanical work on a system and thereby changing the temperature of the system. When rapidly pumping up the tires of a bicycle with a hand pump, you may have noticed that both the piston of the pump and the tire become hot. The quantitative relationship between the work done of a system and its thermodynamic properties is expressed by the *first law of thermodynamics*.

The first law of thermodynamics which is a formulation of the conservation of energy states that

the change in the internal energy of a system is equal to the heat added to the system minus the work done by the system.

In equation form this becomes

$$\Delta U = \Delta Q - \Delta W \qquad (12.4)$$

or

$$\Delta Q = \Delta U + \Delta W$$

or

$$\Delta W = \Delta Q - \Delta U$$

where ΔU is the change in internal energy, ΔQ is the change in heat energy, and ΔW is the positive work done by the system. Let us look at this equation carefully. The internal energy of a system *depends only on the state of the system.* For this reason it is called a *state function.* A state function is dependent *only* on the variables defining the state of the system such as the pressure, temperature, and volume for an ideal gas. A state function is independent of the process involved in getting the system to a given state. When a system undergoes a change and moves from one thermodynamic state to another, the change in the internal energy of the system will depend *only* on the final and initial states. The change in internal energy will be the same regardless of the processes involved in changing states. In Chapter 5 we had an analogous case in which the work, or energy change, was independent of path and depended only upon the beginning and ending points. In that situation we defined the system as *conservative,* and we said we were dealing only with conservative forces. In the first law of thermodynamics the change in internal energy depends only upon the change in absolute temperature.

The heat and work terms in Equation 12.4 are *not* state functions; they are dependent on the process involved in the change of state. For example, heat added at constant pressure produces a different change than the same amount of heat added at a constant volume as we shall see later in this chapter. Likewise, the work done on the system depends on the process involved. For example, work at constant temperature produces a different effect than the same amount of work applied at constant pressure. These terms are parallel to work against nonconservative force as discussed in Chapter 5.

The importance of the first law of thermodynamics is that the sum of *the work done* and *the heat added* in moving between two different thermodynamic states *always* produces the same change in the internal energy of the system. The work done on the system may be mechanical, electrical, or chemical. All three of these forms of work occur in living systems.

EXAMPLE

Consider a diet problem. A person undertakes a diet program that provides 2000 kcal per day, and the person expends energy in all forms to a total of 3000 kcal per day. In order to do this much work on this diet, the person must obtain energy stored as internal energy in the body. Loss of body fat will occur as this energy is used. Using the first law of thermodynamics, we see that the heat energy input is represented by the 2000 kcal ($\Delta Q = 2000$ kcal) from the food when burned in the body. The body does work amounting to 3000 kcal ($\Delta W = +3000$ kcal). This work takes many forms, mechanical work in terms of body motion, chemical changes in the muscles, and electrical energy in nerve activity. From the first law it follows that the internal energy must decrease by 1000 kcal per day, $\Delta U = \Delta Q - \Delta W = 2000$ kcal $- 3000$ kcal $= -1000$ kcal. If this energy were stored as adipose (fat) tissue (7500 kcal/kg), then the person would lose about 1 kg in a week.

12.5 Isochoric Processes

An *isochoric process* is one that takes place at a constant volume. An isochoric process is represented on a PV diagram by a vertical line (Figure 12.4). You see that the area under the curve is 0. We know that the pressure times the change in volume is the work done. The volume does not change, so no work is done:

$$\Delta W = P \Delta V$$

where $\Delta V = 0$. Therefore $\Delta W = 0$. For an isochoric process the first law of thermodynamics reduces to an equality between the change in internal energy and the change in heat energy,

$$\Delta U = \Delta Q \tag{12.5}$$

If the heat is added to the system, the internal energy is increased, and the temperature rises. If heat is given up by the system, there is a decrease in internal energy and a decrease in the temperature.

For a confined ideal gas the heat added to the gas during an isochoric process is given by the product of the number of moles of gas, the specific heat of the gas at constant volume, and the change in temperature,

$$\Delta U = \Delta Q = nc_v \Delta T \tag{12.6}$$

FIGURE 12.4
Isochoric process, $\Delta V = 0$. The vertical line *AB* represents the process by which the system goes from state *A* to state *B* without a change in volume.

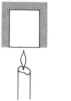
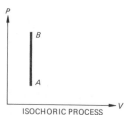

ISOCHORIC PROCESS

TABLE 12.1
Molar Specific Heats of Various Gases
(cal/mole–°K)

	c_v	c_p	$c_p - c_v$	$\gamma = c_p/c_v$
Helium	3.00	4.98	1.98	1.66
Argon	3.00	5.00	2.00	1.67
Nitrogen	4.96	6.95	1.99	1.40
Oxygen	4.96	6.95	1.99	1.40
Air	4.96	6.95	1.99	1.40
Water vapor (steam)	6.74	8.75	2.01	1.31
Carbon dioxide	6.80	8.83	2.03	1.30

where n is the number of moles,° c_v, the molar specific heat at constant volume, is the amount of heat required to raise the temperature of one mole of gas one degree during an isochoric process (see Table 12.1), and ΔT is the change in temperature.

EXAMPLE

Find the amount of heat to be added to raise the pressure of 1.50 moles of helium from 75.0 cm Hg to 100 cm Hg at constant volume. How much is the temperature raised? The original temperature is 27.0°C.

For a gas $P_1 V_1/T_1 = P_2 V_2/T_2$

For this case, $V_1 = V_2$; so

$$\frac{P_1}{T_1} = \frac{P_2}{T_2}$$

$$\frac{75.0 \text{ cm Hg}}{300°\text{K}} = \frac{100 \text{ cm Hg}}{T_2}$$

$$T_2 = 400°\text{K}$$

temperature increase $= 100°\text{K}$

$$\Delta Q = n c_v \Delta T$$

$$c_v = 3.00 \text{ cal/mole–}°\text{K}$$

$$\Delta Q = (1.50 \text{ moles})(3.00 \text{ cal/mole–}°\text{K})(100°\text{K}) = 450 \text{ cal}$$

12.6 Isobaric Processes

An *isobaric process* is one that takes place at a constant pressure and can be represented by a horizontal line in a PV diagram (Figure 12.5). The area under the curve CD is equal to the product of the pressure times the change in volume, $P\Delta V$, and represents the work that is done during the isobaric process. Let us consider a system that is a confined

---

° A mole of any substance is the amount of that substance that contains 6.02×10^{23} molecules and has a mass in grams equal to the molecular mass of the substance in atomic mass units.

FIGURE 12.5

Isobaric process, $\Delta P = 0$. The horizontal line *CD* represents the process by which the system goes from state *C* to state *D* without a change in pressure.

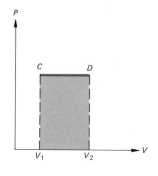

ideal gas. In Section 10.9 on the gas laws, we learned that if the pressure is constant and the volume changes, there is also a change in temperature. A change in temperature means that there has been a change in internal energy. Hence if heat which is added to the system causes work to be done and an increase in internal energy, the first law of thermodynamics for an isobaric process is given by,

$$\Delta Q = \Delta U + \Delta W = \Delta U + P \Delta V \qquad (12.7)$$

As an example, consider the case of changing one gram of water at 100°C into one gram of steam at 100°C. The heat added goes into external work (change in volume) and internal energy.

The heat which is added during an isobaric process is given by

$$\Delta Q = n c_p \Delta T \qquad (12.8)$$

where n is the number of moles, c_p, the molar specific heat at constant pressure, is the amount of heat required to raise the temperature of one mole one degree during an isobaric process, and ΔT is the change in temperature. The change in internal energy, which is independent of the process, is the same as would have occurred by an isochoric process,

$$\Delta U = n c_v \Delta T \qquad (12.6)$$

Now we can rewrite the first law of thermodynamics for an isobaric process,

$$n c_p \Delta T = n c_v \Delta T + P \Delta V \qquad (12.9)$$

This equation shows that the molar specific heat of a gas at constant pressure is always greater than the molar specific heat of a gas at constant volume. By rearranging the above equation, and using the relationship that $P \Delta V = (\text{constant}) \Delta T$ for an ideal gas, we can show that the difference between the two different molar specific heats is a constant number, independent of the kind of gas that is used in the system (see Table 12.1),

$$c_p - c_v = \text{constant} = R \qquad (12.10)$$

where R is the universal gas constant and has a value of 8.31 J/mole-°K.

EXAMPLE

Helium (3.00 moles) originally at 273°K and a volume of 0.067 m³ expands at a constant pressure of 1.01×10^5 N/m² to a volume of 0.134 m³. Find the change in temperature, the change in internal energy, the work done, and the heat added to the system.

From the general gas law $P_1V_1/T_1 = P_2V_2/T_2$. Because $P_1 = P_2$

$$\frac{V_1}{T_1} = \frac{V_2}{T_2}$$

$$\frac{0.067 \; m^3}{273°K} = \frac{0.134 \; m^3}{T_2}$$

$$T_2 = 546°K$$

$$T_2 - T_1 = 273°K$$

The change in internal energy ΔU,

$$\Delta U = n c_v \Delta T = (3.00 \text{ moles})(3.00 \text{ cal/mole-°K})(273°K) = 2460 \text{ cal}$$

The work done ΔW,

$$\Delta W = P \Delta V$$

$$= (1.01 \times 10^5 \text{ N/m}^2)(0.134 - 0.067)\,m^3 = 6.77 \times 10^3 \text{ J}.$$

The heat added ΔQ,

$$\Delta Q = n c_p \Delta T = (3.00 \text{ moles}) \left(\frac{4.98 \text{ cal}}{\text{mole-°K}} \right)(273°K) = 4080 \text{ calories}$$

In order to verify the first law of thermodynamics, you must express all energy quantities in the same units. The work done in calories is work done in joules divided by 4.19 since the mechanical energy equivalent of 1 calorie of heat energy is 4.19 joules (see Chapter 10).

From the first law of thermodynamics, $\Delta W = P \Delta V = \Delta Q - \Delta U$. Now expressing each term in joules, we have

$$\Delta Q - \Delta U = (4080 - 2460) \text{ cal} = 6.79 \times 10^3 \text{ joules}$$

$$\Delta W = P \Delta V = 6.77 \times 10^3 \text{ J}$$

We find that we have a discrepancy of less than one percent, which is not surprising. We may introduce a discrepancy of that size by using only three significant figures.

12.7 Isothermal Processes

An *isothermal process* is one that takes place at a constant temperature. For a system of a confined ideal gas, we find that at constant temperature the pressure and volume of the system are given by Boyle's law (Section 10.9), $PV =$ constant. On the PV diagram an isothermal process for an ideal gas is represented by a hyperbolic curve whose equation is

$$PV = \text{constant}$$

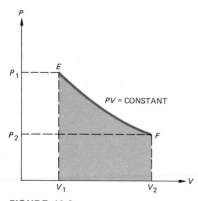

See Figure 12.6. Since the internal energy of a system is a state variable dependent only upon the temperature of the system, for an isothermal process the internal energy of the system remains constant, $\Delta U = 0$. The first law of thermodynamics for an isothermal process states the equality between the work done and the change in heat energy:

$$\Delta Q = \Delta W = P \Delta V \tag{12.11}$$

For an isothermal process the heat added equals the work done by the system and is represented by the shaded area under the curve in Figure 12.6. If the system consists of an ideal gas that expands from a volume of V_1 at a pressure P_1 to a volume of V_2 at a constant temperature, the work done by the gas is given by

$$W = P_1 V_1 \ln \frac{V_2}{V_1} \tag{12.12}$$

(A derivation of this result is given in Section 12.16.)

EXAMPLE

If 50 cal of heat are added to an ideal gas in a cylinder under the condition of a constant temperature, what is the work done by the expanding gas?

The internal energy of an ideal gas depends only on its temperature. In this problem the temperature is constant (i.e., $\Delta U = 0$). Therefore, $\Delta Q = +\Delta W$, the heat input is equal to the work done by the gas. Note that ΔW is positive for work done *by* the gas. Under the constant temperature conditions the work done by the gas is equal to 50 cal, the same as the heat input energy.

12.8 Adiabatic Processes

Any process that takes place very rapidly is essentially adiabatic. An *adiabatic process* occurs if no heat is added to or taken from the system — that is, $\Delta Q = 0$. The curve GH in Figure 12.7 represents an adiabatic process for an ideal gas. The equation for the curve GH is $PV^\gamma = $ constant, where γ is the ratio of the molar specific heats c_p/c_v (Table 12.1). The constant γ has a maximum value of 5/3 for a monoatomic ideal gas, and is always greater than 1. For the adiabatic process in Figure 12.7, the

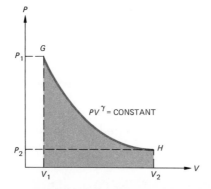

first law of thermodynamics can be expressed by the following equation,

$$\Delta U = -\Delta W = -P\Delta V = nc_v \Delta T = \frac{P_1 V_1 - P_2 V_2}{1 - \gamma} \qquad (12.13)$$

A derivation of this result is given in Section 12.16.

For an adiabatic process if the system does work, there is a decrease in internal energy; if work is done on the system, there is an increase in internal energy — the heating that occurs when you rapidly compress the air in a tire pump, for example.

EXAMPLE

The air in an automobile tire is released. What happens to the temperature of the air as it expands into the atmosphere? Calculate the values of the changes assuming that the original air pressure in the tire was 3.00 atmospheres (atm) and the temperature was 20.0°C.

This expansion is a rapid process and can be approximated by an adiabatic process. The expanding gas does work on the atmosphere so ΔW is positive, and thus ΔU is negative. The internal energy of the gas is reduced; that is, its temperature is lowered. For a unit volume of air (1.00 m^3) in the tire, we can use the expression for an adiabatic process $3(1)^\gamma = 1(V_f)^\gamma$ where the γ for air is 1.40.

$$V_f = \sqrt[1.4]{3} \text{ or } V_f = 3^{1/1.4} = 2.19 \text{ m}^3$$

Now we can use the ideal gas law to find the change in temperature,

$$P_0 V_0 / T_0 = P_f V_f / T_f$$

$$\frac{(3.00)(1.00)}{293} = \frac{(1.00)(2.19)}{T_f}$$

$$T_f = 214°\text{K, or } -59°\text{C}$$

12.9 *The Second Law of Thermodynamics*

The first law of thermodynamics is necessary, but not sufficient, to explain the thermal phenomena that you observe every day. When you mix two glasses of water of different temperatures, what is the final state of the water? Is this final state the only possible state consistent with the first law of thermodynamics? There are an infinite number of ways in which energy could be conserved for this system, but the situation actually occurring is the one in which the final temperature of the water is a definite temperature falling somewhere between the initial temperatures of the two glasses of water. This final state is a result of the second law of thermodynamics.

There are several equivalent ways of stating the second law of thermodynamics. The Clausius statement of the second law is:

Heat energy cannot pass spontaneously from a system at a lower temperature to one at a higher temperature.

Your experiences with hot and cold objects may make this statement of the second law seem trivial. You have never seen an isolated object at room temperature become hot. Why not? We know the total heat energy available in all the objects in the room is enough to warm up one object. But nature does not behave in that fashion. To transfer heat energy from a lower temperature object to a higher temperature we must do work on the system. If the Clausius statement were not true, we would be able to transfer heat energy from inside our homes to the hotter outside air in the summer to air-condition our homes without using any other energy.

The Kelvin-Planck statement of the second law of thermodynamics is that

it is not possible to perform a process which only extracts heat energy from a constant temperature object and performs an equivalent amount of work.

You may think that an isothermal process would violate this statement of the second law because the change in heat energy is equal to the work done by the system, Equation 12.11. However, an isothermal expansion of the system leaves the final system at a different volume than at the beginning of the process. To test this statement of the second law we must return the system to its original volume since the *only* process that can occur is the change of heat energy into work. But it is just this return process that is impossible without changing the way nature performs.

The operational equivalence of the Kelvin-Planck and Clausius statements of the second law is shown by the following example. Suppose the Kelvin-Planck statement can be violated. Then we could construct a heat engine that *only* takes heat energy from a reservoir of heat energy at constant temperature and converts the heat energy into work. We can then use this engine to operate a refrigerator that extracts heat from the inside of the refrigerator and delivers it to a hotter object. The net result of the heat engine-refrigerator combination is the transfer of heat energy from a cold object to a hot object without doing any net work on the total system, a violation of the Clausius statement.

12.10 Heat Engines

The internal combustion gasoline engine and the steam engine are examples of engines that make use of a combustion process to produce motion and to do work. These are both examples of heat engines. A *heat engine* absorbs a quantity of heat energy Q_h from a hot reservoir of heat at temperature T_h, does work W, rejects a quantity of heat energy Q_c to a cold heat reservoir at a temperature T_c, and returns to its original state. For example, your automobile engine burns gasoline at a high temperature T_h from which the engine extracts heat energy Q_h to turn the crank shaft and wheels of the automobile to do work W. The rejected heat energy Q_c is returned, via the exhaust gases and other thermal transport processes, to the atmosphere at a temperature T_c. Part of the work is

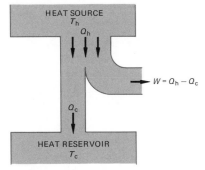

FIGURE 12.8
A heat engine. $T_h > T_c$ and $Q_h > Q_c$.

used to return the engine to its original state, ready to begin the cycle again. Heat engines operate in cycles, using the heat absorbed during each cycle to do work. A schematic diagram of a heat engine is shown in Figure 12.8.

We can combine this cyclic nature of a heat engine with the first law of thermodynamics to develop an expression for the efficiency of a heat engine. Since the initial and final states of a cycle of the heat engine are the same, the internal energy of the engine must be the same at the beginning and end of each cycle, $\Delta U = 0$; then $\Delta W = \Delta Q$, or for the whole cycle

$$W = Q_h - Q_c \qquad (12.14)$$

The *efficiency of a heat engine* is defined as the useful work output divided by the energy input. For the heat engine the energy input is given by the heat energy absorbed from the high temperature reservoir, so

$$\text{efficiency} = \frac{\text{work output}}{\text{energy input}}$$

$$= \frac{W}{Q_h} = \frac{(Q_h - Q_c)}{Q_h}$$

$$= 1 - \frac{Q_c}{Q_h} \qquad (12.15)$$

EXAMPLE

What is the efficiency of a person whose daily diet is equivalent to 4.0 kg of milk and who does useful work at the rate of 50 watts for an eight hour day? Heat of combustion of milk = 650 cal/g.

$$Q_h = 650 \times 4.0 \times 10^3 \times 4.19 = 11 \times 10^6 \text{ J}$$

$$W = \left(50 \text{ J/sec})(8 \text{ hr} \times \frac{60 \text{ min}}{1 \text{ hr}} \times \frac{60 \text{ sec}}{1 \text{ min}}\right) = 1.4 \times 10^6 \text{ J}$$

$$\text{efficiency} = \frac{1.4 \times 10^6 \text{J}}{11 \times 10^6} = .13 = 13 \text{ percent.}$$

Where does all the energy go?

12.11 Carnot's Engine

Both the first and second law of thermodynamics are basic to the study and understanding of heat engines. How do you make the heat engine more efficient? Is there a limit to the efficiency of a heat engine? In 1824 Sadi Carnot, a French engineer, was the first to approach the problem of heat engines from fundamental considerations. Carnot's approach was an entirely theoretical one. He disregarded the mechanical operation and details and considered the fundamentals. In fact, our discussion in the previous section followed Carnot's approach. Carnot also introduced the concept of a reversible cycle which we now know as *Carnot's cycle*.

FIGURE 12.9
Cycle for a Carnot engine and a
possible system for each process.

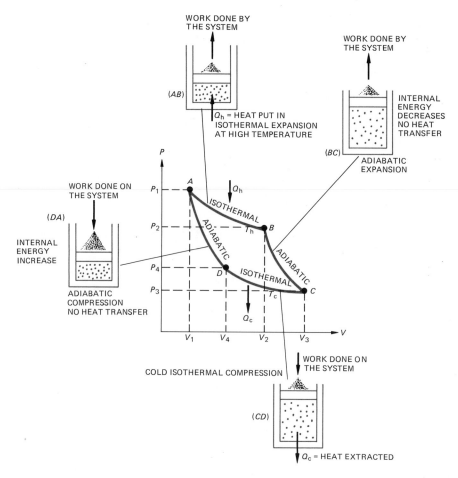

The theoretical Carnot cycle is for an ideal engine that has a cylinder and
piston made of perfect insulating materials and has three separate parts —
one part is a perfect heat conductor and a high temperature heat reservoir
at temperature T_h, one part is a perfect insulator, and the third part is a
perfect conductor and cold heat reservoir at temperature T_c (Figure
12.9). The cylinder was filled with an ideal working substance, such as
an ideal gas. The steps in the Carnot cycle are as follows:

1. The high temperature part T_h is connected to the cylinder and the
   piston moves, giving the isothermal expansion $\Delta Q = \Delta W$ along curve
   AB.
2. The insulating part replaces the heat reservoir T_h, and the gas is further expanded as an adiabatic process, $\Delta Q = 0$. As work is done, the
   temperature of the gas is lowered to temperature T_c along curve BC.
3. The lower temperature T_c reservoir replaces the insulating part. The
   gas is then compressed at a constant temperature T_c. As the gas is
   compressed, the part at T_c must absorb some heat, and the amount
   absorbed is equal to the work of compression (curve CD).

4. Part T_c is replaced by the insulating head, and the gas is adiabatically compressed to the starting point. The work done in compressing the gas increases the internal energy of the gas until it reaches temperature T_h (curve DA).

A summary of the cycle shows that heat Q_h is put into the system in step 1 and that heat Q_c is taken from the system in step 3. The difference between Q_h and Q_c represents the amount of work done during the cycle, $W = Q_h - Q_c$.

Follow the Carnot cycle around the curves on the PV diagram of Figure 12.9. Starting at point A $(P_1V_1T_h)$, step 1 is represented by the curve AB, an isothermal expansion at temperature T_h to P_2V_2. The heat input is Q_h, work is equal to area under the curve AB and bounded by the corners ABV_2V_1. Step 2 is represented by BC, an adiabatic expansion from $P_2V_2T_h$ to $P_3V_3T_c$. The work done is equal to the area under the curve BC and is bounded by the corners BCV_3V_2. There is a decrease in the internal energy so the temperature drops from T_h to T_c. Step 3, represented by curve CD, is an isothermal compression at temperature T_c, heat Q_c is exhausted from the system. Work is done on the system in compressing the ideal gas. It is negative work equal in magnitude to the area under the curve CD and bounded by the corners CV_3V_4D. Step 4, represented by curve DA, is an adiabatic compression from $P_4V_4T_c$ to $P_1V_1T_h$. Work is again done on the system. This is negative work equal in magnitude to the area under the curve DA, which is bounded by DV_4V_1A. This work increases the internal energy. Notice that if all the work done on and by the system is added for one cycle the net work is represented by the area inside of the curves that join the points $ABCD$.

As a result of his theoretical work, Carnot postulated a theorem which we can state in the following way:

No engine working between two heat reservoirs can be more efficient than a Carnot engine operating between those two reservoirs.

In other words the Carnot cycle defines the best possible heat engine permitted by the laws of nature. What is the efficiency of a Carnot engine. We can use Equation 12.15 and our knowledge of isothermal processes to calculate this efficiency in terms of the temperatures of the reservoirs T_h and T_c,

$$\text{efficiency} = 1 - \frac{Q_c}{Q_h} \tag{12.15}$$

For isothermal processes we know that the change in heat energy must be equal to the work done (Equation 12.11), and we have stated that the work done by the system during an isothermal expansion is given by Equation 12.12,

$$Q_h = W = P_1V_1 \ln (V_2/V_1) \tag{12.12}$$

for the expansion AB. For the compression CD, Q_c must be the negative of the work done by the system since the system does negative work, $Q_c = -W_{CD} = -P_3V_3 \ln(V_4/V_3) = P_3V_3 \ln(V_3/V_4)$.

$$\text{efficiency} = 1 - \frac{P_3V_3 \ln(V_3/V_4)}{P_1V_1 \ln(V_2/V_1)} \qquad (12.16)$$

Next we use the ideal gas laws and the relationship for adiabatic processes for a confined gas.

$$\frac{P_1V_1}{T_h} = \frac{P_3V_3}{T_c} \quad \text{so} \quad \frac{P_3V_3}{P_1V_1} = \frac{T_c}{T_h}$$

For the adiabatic processes

$$P_2V_2{}^\gamma = P_3V_3{}^\gamma \quad \text{and} \quad P_1V_1{}^\gamma = P_4V_4{}^\gamma$$

Dividing the first equation by the second, we obtain

$$\frac{P_2V_2{}^\gamma}{P_1V_1{}^\gamma} = \frac{P_3V^\gamma}{P_4V^\gamma}$$

For the isothermal processes

$$\frac{P_2}{P_1} = \frac{V_1}{V_2} \quad \text{and} \quad \frac{P_3}{P_4} = \frac{V_4}{V_3}$$

Then

$$\left(\frac{V_2}{V_1}\right)^{\gamma-1} = \left(\frac{V_3}{V_4}\right)^{\gamma-1} \quad \text{or} \quad \frac{V_2}{V_1} = \frac{V_3}{V_4}$$

Then we can reduce Equation 12.16 to

$$\text{efficiency} = 1 - \frac{T_c}{T_h} \qquad (12.17)$$

where T_c and T_h are the absolute temperatures in Kelvin degrees. The efficiency of the Carnot engine depends only upon the temperatures of the hot and cold heat reservoirs, and no engine can be more efficient than its equivalent Carnot engine.

EXAMPLES

1. A person burns food at a temperature of 37°C and exhausts body heat to an environment of 20°C. What is the maximum efficiency possible for that person?

$$\text{efficiency} = 1 - \frac{T_c}{T_h} = 1 - \frac{293}{310} = 5.48 \times 10^{-2} = 5.48 \text{ percent}$$

2. An automobile engine during combustion produces a source of heat energy with a temperature of 400°C. What is the maximum efficiency of this engine when the outside temperature is 20°C?

$$\text{efficiency} = 1 - \frac{293}{673} = 5.65 \times 10^{-1} = 56.5 \text{ percent}$$

The usual automobile engine has an efficiency of about 20 percent.

12.12 Refrigerators

A refrigerator is, in thermodynamic terms, a heat engine that works in reverse. A *refrigerator* absorbs heat from a low-temperature heat reservoir and exhausts heat to a high-temperature reservoir. To accomplish this task, work must be done on the refrigerator system. Of course, we know from the Clausius statement of the second law that it is not possible to build a refrigerator that requires no work input. From the first law we can write an equation for the heat exhausted to the hot reservoir Q_h as the sum of the work done on the refrigerator W and the heat absorbed from the cold reservoir Q_c,

$$Q_h = W + Q_c \tag{12.18}$$

Refrigerators are characterized by their *coefficient of performance*. The coefficient of performance η of a refrigerator is defined as the ratio of the amount of heat energy absorbed from the cold reservoir to the amount of work done on the refrigerator,

$$\eta = \frac{Q_c}{W} = \frac{Q_c}{Q_h - Q_c} \tag{12.19}$$

Typical refrigerators have performance coefficients of about 5. The larger the performance coefficient, the better is the refrigerator. Is there a maximum possible coefficient of performance for a refrigerator? With our results for the efficiency of a Carnot engine we can transform Equation 12.19 into a relationship between the temperatures of the hot and cold reservoirs. We can combine the information from the equations for the efficiencies of heat engines,

$$\text{efficiency} = 1 - \frac{Q_c}{Q_h} = 1 - \frac{T_c}{T_h}$$

to deduce that

$$\frac{Q_c}{Q_h} = \frac{T_c}{T_h} \tag{12.20}$$

We combine this equation with Equation 12.19 to express the coefficient of performance as the ratio of the cold temperature to the temperature difference between the reservoir temperatures,

$$\eta = \frac{T_c}{T_h - T_c} \tag{12.21}$$

EXAMPLE

A room air conditioner removes heat from a room at 20°C and exhausts it to the outside at 37°C. What is the maximum possible coefficient of performance for this air conditioner?

$$\eta = \frac{293°}{310° - 293°} = 17.2$$

12.13 *Entropy*

One way to characterize the properties of a thermodynamic system is in terms of its orderliness. A system where everything is lined up in rows is highly ordered. A system where everything is happening at random is highly disordered. One form of the second law of thermodynamics is based on the concept of the order of a system. In this form the second law states that *real processes always involve an increase in disorder.* To facilitate the formulation of this law, we define the *entropy* of a system. Entropy is a measure of the disorder of a system, and it is defined in the following equation,

$$\Delta S = \frac{\Delta Q}{T} \tag{12.22}$$

in which ΔS = change in entropy, ΔQ = change in heat energy, and T is the absolute temperature at which process takes place. Therefore, entropy, like internal energy, is a state function. Entropy depends only on the thermodynamic state of the system, and changes in entropy are independent of the changes involved in going from one state to another. Reversible changes are idealized processes. For reversible processes there is no increase in the disorder of the system. Irreversible processes are ones in which there is an increase of entropy and an increase in disorder. Every state of a system has a definite value of entropy. In terms of entropy, the second law of thermodynamics can be written for any isolated system as:

$$\Delta S \geq 0 \tag{12.23}$$

that is, a process that starts in one equilibrium state and ends in another will go in a direction so the entropy of the system plus its environment increases or remains the same. For $\Delta S > 0$, the process is irreversible; for $\Delta S = 0$, the process is reversible (idealized); and for $\Delta S < 0$, the process is impossible.

The entropy of an isolated system never decreases in going from one state to another. It should be noted that the *directedness* of time (arrow of time) is associated with our experience with phenomena that obey the second law of thermodynamics. For example, if you noted a footprint appearing before your eyes on the beach, you would surely conclude that something strange was happening. In fact, you might have the feeling that time was running backward! Our experience has been conditioned by the increase in disorder (the vanishing of a footprint in the sand, for example) associated with the passage of time. Another example in this connection is the phenomena associated with a drop of ink added to a glass of water. The ink drop breaks up and diffuses throughout the water (going from a spatially ordered drop to a randomized distribution of ink particles of considerable disorder). And again, if an ink drop appeared in a glass of water before your eyes, you would certainly have reason to wonder about your sense of time.

12.14 Entropy and Living Systems

Do living systems obey the second law of thermodynamics? Living systems can be characterized as systems that take in energy and create order. These statements certainly are counter to the second law of thermodynamics except that the second law applies only to isolated or closed systems. Upon reflection it becomes apparent that when a living system is isolated, it ceases to live, and then it certainly obeys the second law of thermodynamics. Could you accept the following statement as a definition of life? Any system that takes in energy and creates order is a living system. If this is an unacceptable definition, what would you add?

12.15 The Third Law of Thermodynamics

The third law of thermodynamics is a statement about the changes in physical systems that occur as the temperature approaches absolute zero. In one form, the third law may be written, as $T \to 0$, $S \to 0$, or as the temperature approaches absolute zero, the entropy approaches zero. This gives us some insight into low-temperature phenomena. In particular, the phenomena of superfluidity of liquid helium and superconductivity of aluminum are examples of systems that show increased ordering as absolute zero is approached. In terms of atoms and electrons this means that the systems are in their lowest energy states. This allows exact knowledge of the state of the particles of the system and thus perfect order. This form of the third law of thermodynamics allows us to consider the realm of low temperature research in terms of increased order rather than in terms of energy. Indeed, an important question to ask about a system being considered for low temperature study might well be, "Will increased order make a difference in the behavior of the system?"

ENRICHMENT
12.16 Calculus Derivations of Thermodynamics Equations

Let us derive the equation for an adiabatic expansion of an ideal gas. To obtain this equation we need to obtain a general relationship between the two molar specific heats, c_p and c_v. We can do this using the calculus form of Equation 12.9

$$c_p \, dT = nc_v \, dT + P \, dV \qquad (12.24)$$

for an isobaric process, but for an ideal gas $PV = BT$ where B is a constant.

For a constant pressure process

$$P \, dV = B \, dT \qquad (12.25)$$

Now we combine these two equations to obtain the following general result,

$$nc_p \, dT - nc_v \, dT = B \, dT$$
$$c_p - c_v = B/n \tag{12.26}$$

This result was introduced in Section 12.6, Equation 12.10.

For an adiabatic process we can write the calculus form of Equation 12.13

$$nc_v \, dT + P \, dV = 0 \tag{12.27}$$

For an ideal gas,

$$P \, dV + V \, dP = B \, dT \tag{12.28}$$

Now we solve both equations for dT and set them equal to each other

$$dT = -\frac{P \, dV}{nc_v} = \frac{P \, dV + V \, dP}{B} = dT \tag{12.29}$$

$$0 = (nc_v + B) \, P \, dV + nc_v V \, dP$$

but $B = n(c_p - c_v)$; so

$$0 = (nc_v + nc_p - nc_v)P \, dV + nc_v V \, dP$$

$$0 = c_p P \, dV + c_v V \, dP \tag{12.30}$$

Let $\gamma = c_p/c_v$;

$$0 = \frac{\gamma \, dV}{V} + \frac{dP}{P}$$

Now we integrate this equation

$$\text{constant} = \gamma \ln V + \ln P = \ln V^\gamma + \ln P = \ln PV^\gamma$$

$$PV^\gamma = \text{constant} \tag{12.31}$$

This equation was introduced in Section 12.8.

In Section 12.3 you learned that external work done by a thermodynamic system is $\Delta W = P \, \Delta V$. In calculus notation we can express this relationship as

$$dW = P \, dV \tag{12.32}$$

If we have a curve for a given process plotted on a PV diagram and we wish to determine the work done in going along the curve from A to B, then we integrate this expression,

$$W = \int_{V_A}^{V_B} P \, dV \tag{12.33}$$

If the pressure is a function of volume, $P = f(V)$, then the total work done by the change in the volume of the system from V_A to V_B is,

$$W = \int_{V_A}^{V_B} f(V) \, dV \tag{12.34}$$

EXAMPLES

ISOTHERMAL PROCESS

1. $PV = \text{constant} = C$

$$P = f(V) = \frac{C}{V}$$

$$C = P_A V_A = P_B V_B$$

The work done in an isothermal expansion for an ideal gas is given by

$$W = \int_{V_A}^{B_B} \frac{C}{V}\, dV = C\, \ln V \Big|_{V_A}^{V_B} = C\, \ln \frac{V_B}{V_A} \tag{12.35}$$

This equation was introduced in Section 12.7, Equation 12.12.

ADIABATIC PROCESS

2. $PV^\gamma = \text{constant}$ where $\gamma = c_p/c_v$

$$PV^\gamma = C_1$$

$$P = \frac{C_1}{V^\gamma}$$

$$C_1 = P_A V_A{}^\gamma = P_B V_B{}^\gamma$$

The work done by an adiabatic expansion for an ideal gas is given by

$$W = \int_{V_A}^{V_B} C_1 \frac{dV}{V^\gamma} = \frac{C_1}{1-\gamma}\left(V^{1-\gamma}\right)\Big|_{V_A}^{V_B} = \frac{P_B V_B{}^\gamma\, V_B{}^{1-\gamma}}{1-\gamma} - \frac{P_A V_A{}^\gamma\, V_A{}^{1-\gamma}}{1-\gamma}$$

$$W = \frac{P_B V_B - P_A V_A}{1-\gamma} \tag{12.36}$$

This equation was introduced in Section 12.8, Equation 12.13.

ENTROPY

3. $\Delta S = \Delta/T$, or in calculus notation $dS = dQ/T$.

$$\int_{S_1}^{S_2} dS = \int \frac{dQ}{T}$$

In many processes the change in the quantity of heat dQ is a function of the temperature, $dQ = f(T)\, dT$.
   Thus,

$$S_2 - S_1 = \int_{T_1}^{T_2} \frac{f(T)\, dT}{T}$$

For example for an isochoric process

$$Q = nc_v T, \quad dQ = nc_v\, dT$$

$$S_2 - S_1 = nc_v\, \ln T_2/T_1 \tag{12.37}$$

SUMMARY

Use these questions to evaluate how well you have achieved the goals of this chapter. The answers to these questions are given at the end of this summary with the number of the section where you can find related content material.

Definitions

Write the correct word or phrase in the blank.

1. A _____ is a system that _____ heat from a _____ temperature reservoir, does work upon the universe, and _____ heat to a _____ temperature reservoir, and whose performance is characterized by its _____.

2. A _____ is a system that _____ heat from a _____ temperature reservoir, has work done upon it by an external agent, and _____ heat to a _____ temperature reservoir, and whose performance is characterized by its _____.

3. The _____ is an idealized process that can be used to calculate the greatest possible _____ of a heat engine.

4. A *PV* diagram is a _____ on which the horizontal axis represents the _____ of the system, the vertical axis represents the _____ of the system and the area between a curve and the _____ axis represents the _____.

5. Define each of the following processes, describe the representation of the process on a *PV* diagram and write the first law of thermodynamics equation for the process:
   a. adiabatic
   b. isothermal
   c. isobaric
   d. isochoric

Laws of Thermodynamics

6. State the laws of thermodynamics, and describe the operation of a confined volume of helium in a Carnot engine in terms of these laws.

Thermodynamics Problems

7. A cylinder contains oxygen at a pressure of 2 atm. The volume is 3 liters and temperature is 300°K. The oxy-gen is carried through the following processes:
   a. Heating at constant pressure 500°K
   b. Cooling at constant volume to 250°K
   c. Cooling at constant pressure to 150°K
   d. Heating at constant volume to 300°K.
   Show the processes above in a *PV* diagram giving *P* and *V* at the end of each process. Calculate the net work done by the oxygen.

Answers

1. heat engine, absorbs, high, rejects, low, efficiency

2. refrigerator, absorbs, low, rejects, high, coefficient of performance (Section 12.12)

3. Carnot cycle, efficiency (Section 12.11)

4. graph, volume, pressure, volume axis, work done (Section 12.3)

5. a. constant heat energy, PV^γ = constant curve, $\Delta Q = 0$; $\Delta U = -P \, \Delta V$;
   b. constant temperature, PV = constant, hyperbolic curve, $\Delta U = 0$; $\Delta Q = P \, \Delta V$
   c. constant pressure, a horizontal line, $\Delta Q = \Delta U + P \, \Delta V$ or $nc_p \, \Delta T = nc_v \, \Delta T + P \, \Delta V$
   d. constant volume, a vertical line, $\Delta W = 0$; $\Delta U = \Delta Q = nc_v \, \Delta T$ (Sections 12.5, 12.6, 12.7, 12.8)

6. zeroth law: helium reaches same temperature as high temperature reservoir; first law: the helium expands and $\Delta U = \Delta Q - \Delta W$; second law: the heat absorbed by the helium gas at high temperature is partially converted to work and the rest is rejected at a lower temperature (Section 12.9)

7. See Figure 12.10; net work = area within rectangle = 2 liter–atm (Sections 12.3, 12.5, 12.6)

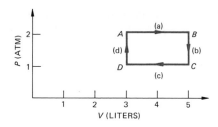

ALGORITHMIC PROBLEMS

Listed below are the important equations from this chapter. The problems following the equations will help you learn to translate words into equations and to solve single-concept problems.

Equations

$$\Delta W = P\,\Delta V \tag{12.2}$$

$$\Delta U = \Delta Q - \Delta W \tag{12.4}$$

$$\Delta U = nc_{v}\,\Delta T \tag{12.6}$$

$$\Delta Q = nc_{p}\,\Delta T \tag{12.8}$$

$$c_{p} - c_{v} = R \tag{12.10}$$

$$W = P_{1}V_{1}\,\ln\,(V_{2}/V_{1}) \tag{12.12}$$

$$\text{efficiency} = \frac{W}{Q_{h}} = 1 - \frac{Q_{c}}{Q_{h}} = 1 - \frac{T_{c}}{T_{h}} \tag{12.15, 12.17}$$

$$\eta = \frac{Q_{c}}{W} = \frac{Q_{c}}{Q_{h} - Q_{c}} = \frac{T_{c}}{T_{h} - T_{c}} \tag{12.19, 12.21}$$

$$\Delta S = \frac{\Delta Q}{T} \tag{12.22}$$

$$\Delta S \geq 0 \tag{12.23}$$

$$PV^{\gamma} = \text{constant} \tag{12.31}$$

$$c_{p}/c_{v} = \gamma \tag{definition}$$

Problems

1. A cylinder is filled with a gas and 2-kg piston of 18 cm diameter closes the cylinder and falls 10 cm in compressing the gas. How much work is done?
2. If 200 cal of heat are added to a system which does 500 J of work, how much is the internal energy of the system changed in this process?
3. It is known that the heat input into an engine to produce 45,000 J of work is 15×10^{4} J. How much heat is lost through the exhaust?
4. What is efficiency of the engine of problem 3?

5. The intake temperature of a Carnot engine is 500°K, and the exhaust temperature is 360°K. What is the efficiency of the engine?

Answers

1. 1.96 J
2. 338 J
3. 105,000 J

4. 30 percent
5. 28 percent

EXERCISES

These exercises are designed to help you apply the ideas of a section to physical situations. When appropriate the numerical answer is given in brackets at the end of each exercise.

Section 12.3

1. A compressed gas is allowed to expand from a volume of 1.0 m³ to 2.5 m³ against the pressure of the atmosphere ($P = 1.02 \times 10^5$ N/m²). What work does the gas do? From where does the energy come to do this work? [1.50×10^5 J, from internal energy or applied heat energy]

Section 12.4

2. One gram of water (1 cm³) becomes 1671 cm³ of steam when boiled at a pressure of 1.013×10^5 N/m². The heat of vaporization at this pressure is 539 cal/g. Compute the external work, the increase in internal energy, and the amount of heat energy added to the system. [1.69×10^2J, 2.09×10^3J, 2.26×10^3J]

3. When a system is taken from A to C via the path ABC, 2000 cal of heat input into the system, and 750 cal of work are done (Figure 12.11). (Ideal gas)

   a. How much heat is put into the system along the path ADC if the work is 250 cal?

   b. When the system returns from C to A along AC the work is 500 cal. Does the system absorb or liberate heat, and how much?

   c. Assume that the internal energy at A is 250 cal and at C is 1500 cal. How much heat is absorbed in process AB and DC? $T_B = 3 T_A$, $T_C = 3 T_D$

   d. What is the change in internal energy for path AC. [a. 1500 cal; b. 1750 cal liberated; c. 500 cal, 1000 cal; d. 1250 cal]

Section 12.5

4. One mole of water vapor in a sealed pressure cooker is heated from 100°C so that the absolute pressure doubles. What is the final temperature? How much heat energy is added to the system? What is the change in the internal energy of the water vapor? [473°C, 2514 cal, 2514 cal]

Section 12.6

5. A confined quantity (9.6 moles) of superheated steam at 427°C drive a piston 30 cm in diameter a distance of 0.80 m at a pressure 5.00×10^5 N/m². The final temperature of the steam is 77°C. What is the work done by the steam on the piston? How much heat energy is *used* during the process? What is the change in the internal energy of the steam? [2.83×10^4 J, 1.23×10^5 J, -9.49×10^4 J]

6. A given quantity of gas at a constant pressure of 10 N/cm² expands from a volume of 10 liters to a volume of 20 liters. How much work is done by

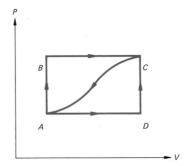

FIGURE 12.11
Exercise 3.

the gas? If the original temperature was 27°C, what was the final temperature? What else would you need to know to calculate the heat added to the system? [1000 J, 600°K]

7. A steel cylinder of cross-sectional area of 20 cm² contains 200 cm³ of mercury. The cylinder is equipped with a tightly fitting piston which supports a load of 30,000 N. The temperature is increased from 15°C to 65°C. Neglecting expansion of the steel cylinder, find
   a. the increase in volume of the mercury
   b. the mechanical work done against the force
   c. the amount of heat added
   d. the change of internal energy   [a. 1.8 cm³; b. 27 J; c. 19,000 J; d. 19,000 J]

Section 12.7

8. Assume that you push down on the plunger of a hand tire pump slowly so that the temperature of the pump remains constant. At the end of your push, the volume of confined air has decreased from 1.05 liters to 0.150 liters. Assume the starting conditions were a pressure of 1.00 atm and a temperature of −13.0°C. What is the final pressure? How much work was done on the system? What is the change in heat energy of the system? What is the change in internal energy of the system? Where did the heat energy go?   [7.0 atm, 2.04 atms–liters, 2.04 atms–liters, 0, the surroundings of the pump]

Section 12.8

9. Assume that you push down the plunger of a hand tire pump so rapidly that no heat escapes from the system during this action. At the end of your push the volume of the confined air has decreased from 1.05 liters to 0.150 liters. Assume the starting conditions were a pressure of 1 atm at a temperature of −13.0°C. What is the final pressure? How much work was done on the system? What is the change in heat energy of the system? What is the change in the internal energy of the system? What is the final temperature of the air?   [15.5 atm, 3.3 atm–liters, 0, +3.3 atm–liter, 304°C]

10. A gas at 27°C at 1 atm is compressed until its volume is one-tenth of its original volume. This compression is done so fast as to be adiabatic ($\gamma = 1.5$).
   a. find the final pressure
   b. find the final temperature assuming the gas is ideal.   [a. 32 atm; b. 960°K]

Section 12.11

11. A Carnot engine whose high temperature reservoir is 127°C takes in 200 cal of heat at this temperature, and it gives up 160 cal to a low-temperature reservoir. What is the efficiency of this engine and the temperature of the exhaust reservoir?   [20 percent, 320°K]

12. A Carnot engine is operating between the two temperatures of 450°K and 300°K. The heat furnished at a high temperature reservoir is 1350 cal. How much work is done by the engine, and how much heat is given out in exhaust?   [450 cal, 900 cal]

Section 12.12

13. A refrigerator requires 450 J of work to exhaust 1350 J of heat to the outside air. How much heat energy is absorbed from the low-temperature reservoir? What is the coefficient of performance for this refrigerator?   [900 J, 2]

14. A room air conditioner can remove 160 J of heat from the room for every 200 J of heat it exhausts to the outside air. What is the work that must be done on the air conditioner to accomplish this? What is the coefficient of performance of the air conditioner?   [40 J, 4]

Section 12.13

15. Given the graph in Figure 12.12 for the molar specific heat of a system at constant pressure near $T = 0$, find the entropy per mole of this material at 4°K.   [1.5×10^{-4} J/°K-mole]

16. Compare the entropy change for 1 mole (16 g) of water going from ice to water at 0°C and from water to steam at 100°C. Does this answer support the idea that greater entropy changes accompany changes of greater disorder?   [$\Delta S_{ws}/\Delta S_{iw} = 4.93$]

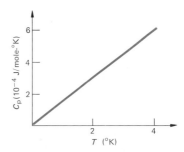

FIGURE 12.12
Exercise 15

PROBLEMS

The following problems may involve more than one physical concept. The numerical answer is given in brackets at the end of the problem.

17. The cycle of a heat engine is described as follows:
    a. start with n moles of gas at $P_0 V_0 T_0$
    b. change to $2 P_0 V_0$ at constant volume
    c. change to $2P_0 2V_0$ at constant pressure
    d. change to $P_0 2V_0$ at constant volume
    e. change to $P_0 V_0$ at constant pressure
       Show this cycle on the PV diagram, and find the temperature for the end of each process. What is the maximum efficiency for these temperatures? [$2T_0$, $4T_0$, $2T_0$, T_0, 75 percent]

18. Assume the gas in problem 17 is an ideal monatomic gas, $\gamma = 5/3$. The specific heat of the gas at constant volume is 3.00 cal/mole–K at $T_0 = 300°$K. Assume you have 0.250 moles of the gas. Find
    a. the heat input
    b. the heat exhaust
    c. the efficiency of the engine   [a. 975 cal; b. 825 cal; c. 15.4 percent]

19. A cylinder contains air at a pressure of 30 N/cm². The original volume at 27°C is 4.0 liters. The air is carried through the following processes:
    a. heating at constant pressure to 227°C
    b. cooling at constant volume to −23°C
    c. cooling at constant pressure to −123°C
    d. heating at constant volume to 27°C
       Show these on a PV diagram, and give the coordinates at the end of each process. Calculate the net work done by the gas.   [a. 30 N/cm², 4 liters; b. 30 N/cm², $6\frac{2}{3}$ liters; c. 15 N/cm², $6\frac{2}{3}$ liters; d. 15 N/cm², 4 liters; $W = 400$ J]

20. Use the data given in problem 19.
    a. How many moles of air were in the cylinder?
    b. Find the heat input during heating at constant pressure to 227°C and heating at constant volume to 27°C given that $c_p = 7.00$ cal/mole–°C and $c_v = 5.00$ cal/mole–°C.
    c. Find the heat liberated during cooling at constant volume to −23°C and cooling at constant pressure to −123°C.
    d. What is efficiency of this device as a heat engine?
    e. What would be the efficiency of a Carnot engine operating in the same temperature range? [a. 0.48 mole; b. $Q_p = 672$ cal, $Q_v = 360$ cal; c. Q_v

= −224 cal, $Q_p = −336$ cal; d. 9 percent; e. 70 percent]

21. Assume that you had a Carnot engine working between the same temperature as an automobile motor. Estimate the temperatures. What efficiency did you get?

22. Given the following data from five actual air conditioners, rank them in the order of efficiency (1 British Thermal Unit (BTU) = 1054.8 joules).
    a. cooling capacity = 24,000 BTU, power = 3600 watts
    b. cooling capacity = 21,000 BTU, power = 2800 watts
    c. cooling capacity = 13,000 BTU, power = 1380 watts
    d. cooling capacity = 10,000 BTU, power = 1375 watts
    e. cooling capacity = 7,800 BTU, power = 850 watts
       [c, e, b, d, a]

23. A racing cyclist is capable of sustaining a power output of 300 watts for extended periods. The associated rate of change of internal energy is measured to be 1400 watts. Find the rate of heat production and the mechanical efficiency (defined as efficiency = P_w/P_u where P_w = power of work or output power and P_u = power of internal energy or input power).   [1700 watts, 21.4 percent]

24. Given the graph in Figure 12.13 for the blood pressure during heart contraction (systole), find the rate of work done by the heart beating 72 beats per minute. $P_2 = 140$ mm (systolic) Hg; $P_1 = 90$ mm (diastolic) Hg; $V_2 − V_1 = 80$ cm³. What percentage of a metabolic rate of 85 kcal/hr is used for this heart action?   [1.47 watts, 1.5 percent]

25. The enthalpy of a system is defined as $H = U + PV$.

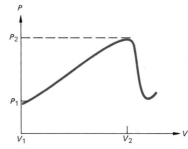

FIGURE 12.13
Problem 24

Thus, $\Delta H = \Delta U + P \, \Delta V + V \, \Delta P$. For constant pressure processes most common for chemical reactions show that ΔH is the heat energy evolved or absorbed.

26. Given the three double containers in Figure 12.14 filled with water at the temperatures shown, which system has the greatest entropy? What conclusion can you make about the entropy and the enthalpy? (See the definition of enthalpy in problem 25.)

27. The Gibb's free energy is defined as $G = H - TS$. Show that the change in the Gibb's free energy is a measure of the energy available for work in an iso-

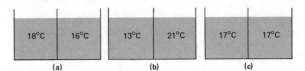

FIGURE 12.14
Problem 26

thermal and isobaric process. (Note that $T \, \Delta S$ is the energy that goes into increasing disorder and measures the unavailable energy in the process.)

GOALS

When you have mastered the contents of this chapter, you will be able to achieve the following goals:

Definitions
Define each of the following terms, and use it in an operational definition:

elastic body   Young's modulus
stress   bulk modulus
strain   modulus of rigidity
elastic limit

Hooke's Law
State Hooke's law.

Stress and Strain
Calculate the strain and stress for various types of deformation.

Elasticity Problems
Solve problems involving the elastic coefficients.

PREREQUISITES

Before you begin this chapter, you should have achieved the goals of Chapter 4, Forces and Newton's Law, Chapter 5, Energy, and Chapter 8, Fluid Flow.

13
Elastic Properties
of Materials

13.1 Introduction

In everyday conversation if someone speaks to you about an elastic body, you probably immediately think of a rubber band. A rubber band yields a great deal to a distorting force, and yet it returns to its original length after the distorting force is removed. Can you think of some biological examples of elastic bodies? In this chapter we will examine the elastic properties of materials.

13.2 Elasticity

Elasticity is a fundamental property of materials. Springs of all kinds are examples of elastic bodies. Let us consider the characteristics of a spring. We find that a spring will respond to distorting force and then return to its original shape after the distorting force is removed. Any material or body can be deformed by an applied force. If it returns to its original shape after the force is removed, it is said to be elastic. Most substances are elastic to some degree. In a technical sense a substance with a high

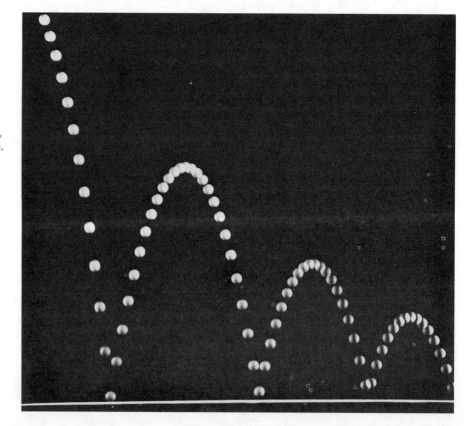

A multiple flash photograph of a bouncing ball. Many physics principles can be studied in this picture—projectile motion, transformation of energy, changes of momentum, elastic properties of material, among others. How would the picture be altered if the ball and surface that it strikes were perfectly elastic? (Picture from *PSSC Physics*, D.C. Heath and Company, Lexington, Mass., 1965.)

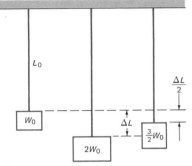

FIGURE 13.1
Strain is proportional to stress. An increase in force of $W_0/2$ produces an elongation of $\Delta L/2$ and an increase in force of W_0 produces an elongation of ΔL.

elasticity is one that requires a large force to produce a distortion—for example, a steel sphere.

In comparing the elasticity of materials there are certain terms we need to define. Suppose that we have a steel wire that is held rigidly at the top end and has a load fastened to the lower end. See Figure 13.1. The wire is then said to be under *stress*. The magnitude of this stress is found by dividing the applied force (the weight in this case) by the cross-sectional area. Thus,

$$\text{stress} = \frac{F}{A} \tag{13.1}$$

The SI units of stress are newtons/meter². If the load is doubled, the wire will be stretched by an amount ΔL. We now introduce another term called *strain*. Strain is a measure of the distortion of an object, and it is defined as the change in a spatial variable divided by the original value of that variable. For example, in Figure 13.1 the variable is length. So,

$$\text{strain} = \frac{\Delta L}{L_0} \tag{13.2}$$

Strain is a dimensionless number. An elastic coefficient is defined as the stress divided by strain. There are three types of distortions that may be produced. These are change in length, change in volume without change in shape and change of shape without change in volume. The only distortion that a fluid resists is volume change. A liquid has a greater ability to resist a change in volume than a gas. Hence, a liquid has a larger value for an elastic coefficient, called *bulk modulus*, than a gas. Solids may have all three types of distortion.

13.3 Hooke's Law

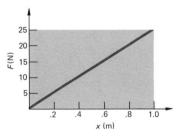

FIGURE 13.2
Hooke's law, $F = kx$, governs the stress-strain relationship within the elastic limits.

In 1676 in his study of the effects of tensile forces, Robert Hooke formulated and stated the law that is still used to define elastic properties of a body. He observed that the increase in length of a stretched body is proportional to applied force F as shown in the experiment above (Figure 13.1).

$$F = kx \tag{13.3}$$

where x is the length increase (m), and k is a proportionality constant or *spring constant* (N/m). This equation is shown graphically in Figure 13.2. Note that k is the slope of the line.

The curve shown in Figure 13.2 applies if the body returns to its original size and shape after the distorting force is removed. If the body does not return to its original condition, it is said to have been distorted beyond its *elastic limit* and takes on a permanent change in length. Hooke's law may be stated in a more general form as follows:

within the elastic limits strain is proportional to the stress, or strain $= C \times$ stress. We will now apply this law to the three different types of distortions we mentioned above.

EXAMPLE

Find the stress on a bone (1 cm in radius and 50 cm long) that supports a mass of 100 kg. Find the strain on the bone if it is compressed 0.15 mm by this weight. Find the proportionality constant C for this bone.

$$\text{stress} = \frac{F}{A} = \frac{(100 \text{ kg}) (9.8 \text{ m/s}^2)}{\pi \times (0.01 \text{ m})^2} = 3.1 \times 10^6 \text{ N/m}^2$$

$$\text{strain} = \frac{\Delta L}{L_0} = \frac{0.15 \times 10^{-3} \text{ m}}{0.5 \text{ m}} = 3.0 \times 10^{-4}$$

Since strain $= C \times$ stress, $C = \text{strain/stress} = 0.96 \times 10^{-10} \text{ m}^2/\text{N}$.

13.4 Young's Modulus

Let us consider a stretched wire. Suppose we have a wire that has a length L and radius r, and suppose a load F is applied to the taut wire to produce an elongation of ΔL. For this case, see Figure 13.3.

$$\text{stress} = \frac{F}{\pi r^2} = \frac{F}{A}$$

$$\text{strain} = \frac{\Delta L}{L} \tag{13.4}$$

From Hooke's law, stress is proportional to strain, or

$$\frac{F}{A} = Y \frac{\Delta L}{L}$$

where Y is the elastic constant of proportionality for a distortion in length and is called *Young's modulus*. Solving for Y, we get

$$Y = \frac{\text{stress}}{\text{strain}} = \frac{F/A}{\Delta L/L} = \frac{FL}{A \, \Delta L} \tag{13.5}$$

The units of Young's modulus are force/unit area, or N/m^2. See Table 13.1 for the value of Young's modulus for a number of solids.

Note that Y has same units as stress. Because neither liquids nor gases will support a linear strain, they have no measured values for Young's modulus.

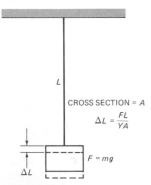

L

CROSS SECTION $= A$

$\Delta L = \frac{FL}{YA}$

$F = mg$

ΔL

FIGURE 13.3
Steel wire subjected to stretching weight *mg*.

EXAMPLES

1. A steel bar 6.00 m long and with rectangular cross section of 5.00 cm $\times$ 2.50 cm supports a mass of 2000 kg. How much is the bar stretched?

TABLE 13.1
Elastic Coefficients of Various Materials

Material	Young's Modulus (10^{10} N/m²)	Modulus of Rigidity (10^{10} N/m²)	Bulk Modulus (10^{10} N/m²)
Aluminum	7.0	2.5	7.0
Bone	0.9–1.3	1.0	–
tensile	1.6	–	1.0
compressive	0.94	–	1.0
Copper	12.0	4.0	13.0
Eggshell	0.006	–	–
Ethanol	–	–	0.09
Granite	5.0	–	4.8
Iron	19.0	6.0	12.0
Mercury	–	–	2.5
Steel	20.0	8.0	16.0
Water	–	–	0.22
Wood (oak)	1.0	1.0	–

From Table 13.1 we find Y for steel is 20.0×10^{10} N/m². Solving for ΔL, we get

$$\Delta L = \frac{FL}{YA} = \frac{(2000)(9.80)(6.00)}{(20.0 \times 10^{10})(.050 \times .025)}$$

$$= 4.70 \times 10^{-4} \text{ m} = 0.47 \text{ mm}$$

2. Your leg bones (cross-sectional area about 9.50 cm²) experience a force of approximately 855 N when you walk. Find the fractional amount your leg bones are compressed by walking. Using $Y = 10^{10}$ N/m² for bone we get:

$$\frac{\Delta L}{L} = \frac{8.55 \times 10^2 \text{ N}}{(9.5 \times 10^{-4} \text{ m}^2)(10^{10} \text{ N/m}^2)} = 9 \times 10^{-5}$$

3. Studies show that for strains less than 0.5 percent bones are elastic. Using values from Table 13.1 we will calculate the elastic limit force for compression and stretch of a humerus 20 cm long and 3 cm² in cross-sectional area.

compression: $\quad F_c = YA \dfrac{\Delta L_c}{L}$

$$= 9.4 \times 10^9 \text{ N/m}^2 + 3.0 \times 10^{-4} \text{ m}^2 \times 5.0 \times 10^{-3}$$

$$F_c = 14000/\text{N}$$

stretch: $\quad F_s = YA \dfrac{\Delta L_s}{L}$

$$= 16 \times 10^9 \text{ N/m}^2 \times 3.0 \times 10^{-4} \text{ m}^2 \times 5.0 \times 10^{-3}$$

$$F_s = 2.4 \times 10^4 \text{ N}$$

SIMPLE EXPERIMENTS

You may be interested in some of the characteristics of a rubber band as an elastic body. You can carry out some simple experiments and record your results. Try stretching a rubber band and observe its change in temperature. You will find that its temperature increases on stretching and decreases upon relaxation. Also have a stretched rubber band in vertical position support a fixed load. Heat the band, and observe how the load moves. You will observe that it rises, indicating a contraction of the rubber band. The above results are exactly opposite to what you would observe for a metal wire. Another experiment you can carry out is one in which you can measure stretch as a function of load. As you increase the load and then decrease the load, you will observe that a rubber band does not follow Hooke's law and that it does not return to the original position after the load is removed. This failure to return to the original position is called an *elastic lag* or *hysteresis*.

13.5 Bulk Modulus

An impulsive force is applied to an elastic sphere. A deformation is produced on impact, but later the ball will return to its spherical shape. Note cracks in the paint. Analyze the transformation in momentum and energy during the time the club head and the ball are in contact. (Harold E. Edgerton, MIT, Cambridge, Mass.)

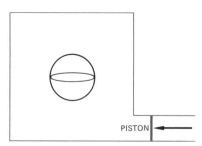

PISTON ←

FIGURE 13.4
Uniform pressure is applied to a body to determine its fractional change in volume as a function of applied pressure.

Suppose that a specimen such as a sphere is placed in a liquid upon which the pressure can be increased by a force applied to the piston (see Figure 13.4). The change in volume of the sphere is a function of the stress applied. The

$$\text{stress} = \frac{F}{A} = \text{pressure applied}$$

$$\text{strain} = \frac{\Delta V}{V} \tag{13.6}$$

From Hooke's law

$$\text{stress} = B \times \text{strain}$$

where B is the constant of proportionality and is called the *bulk modulus.* Then,

$$B = \frac{F/A}{\Delta V/V} = \frac{P}{\Delta V/V} = \frac{PV}{\Delta V} \tag{13.7}$$

Deformation of volume can occur in gases, liquids, and solids. Bulk moduli of liquids and solids are high and of the same order of magnitude. Gases are easiest to compress and hence have the lowest bulk modulus. We often speak of the compressibility of a material which is the reciprocal ($1/B$) of its bulk modulus. The values of bulk moduli of some materials are given in Table 13.1.

EXAMPLE

Find the pressure necessary to change a volume of water by 1.0 percent. Express the pressure in terms of atmospheric pressure units 1 bar $= 10^5$ N/m².

$$P = B \frac{\Delta V}{V} = 2.2 \times 10^9 \text{ N/m}^2 \times 1.0 \times 10^{-2}$$

$$= 2.2 \times 10^7 \text{ N/m}^2 = 2.2 \times 10^2 \text{ bars} = 220 \text{ bars}$$

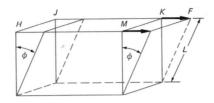

FIGURE 13.5
Applied force F produces shear distortion measured by movement of $HJKL$ plane through a distance $L\phi$. The rigidity modulus η is given by $\eta = \dfrac{F/A}{\phi}$.

13.6 Modulus of Rigidity

The third type of deformation is one of shape but with a constant volume. As an example consider your book resting on a table as you apply a horizontal force to the top cover. You will deform your book in the direction of the applied force. That is, each page tends to slide over the page below it. This may be represented by Figure 13.5. The force F is applied to the plane of $HJKM$ the stress of the force is given by (F/A), where A = area of $HJKM$. Strain ϕ is expressed in radians and given by $\Delta L/L = \phi$. Then by Hooke's law $F/A = \eta\phi$ where η is the constant of proportionality called the shear modulus or the modulus of rigidity, hence

$$\eta = \frac{F/A}{\phi} \qquad (13.8)$$

Note that the radian measure of ϕ (rad) is a ratio of lengths and is therefore unitless.

The moduli of rigidity of some materials are given in Table 13.1.

EXAMPLE

Find the force necessary to produce a shear break of a bone with 3 cm² cross section if the break strain is $6° = 0.10$ rad.

$F = \eta\phi A = 10 \times 10^9 \text{ N/m}^2 \times 10^{-1} \times 3 \times 10^{-4} \text{ m}^2$

$F = 3 \times 10^5 \text{ N}.$

The elastic properties of a material depend upon its molecular structure. Other properties that are closely related to its elastic characteristics and depend also upon molecular structure are ductility, malleability, and hardness.

ENRICHMENT
13.7 Energy of A Hooke's Law System

Energy is stored in a compressed or stretched spring. The work that is done against the elastic forces in deforming a body is a measure of elastic potential energy stored in the deformed body. Consider the

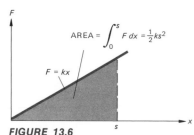

FIGURE 13.6
Work done in compressing or stretching a spring is the area under the F–x graph.

case of a spring that obeys Hooke's law, $F = kx$. The work dW done in stretching a spring a distance dx is given by $dW = kx\, dx$.

$$W = \int_0^s kx\, dx = \tfrac{1}{2} ks^2 \tag{13.9}$$

(see Figure 13.6) in which s is the total stretch, so

$$PE = \tfrac{1}{2} ks^2 \tag{13.10}$$

Similarly, for energy stored in shear we find, where $F = \eta A\phi$ and $ds = L\, d\phi$:

$$W = \int_0^{\phi_f} \eta A\phi \times L\, d\phi = \tfrac{1}{2} \eta AL\phi_f^2 \tag{13.11}$$

We can derive the expression for the coefficient of rigidity for a cylinder as follows: Consider a twisting force applied perpendicular to the radius at the top of the cylinder of length l. If this force produces a twist of the top surface of ϕ (rad), then the strain can be written as $r\phi/l$, at a radius r within the cylinder at its top surface. From the general form of Hooke's law we can write the stress at this top surface as

stress $= \eta$ strain   where η is the modulus of rigidity

Thus

$$\text{stress} = \frac{\eta r \phi}{l}$$

and the torque per unit area at radius r is given by

$$\text{stress} \times r = \frac{\eta r \phi}{l} r$$

If we integrate this torque over the entire area we get

$$\text{torque} = \int_0^R 2\pi r\, dr \times \frac{r\eta r\phi}{l} = \frac{2\pi\eta\phi R^4}{4l} = \frac{\pi\eta R^4 \phi}{2l}$$

Solving for η:

$$\eta = \frac{2l \times \text{torque}}{\pi R^4 \phi} \quad \text{where } R = \text{radius of cylinder}$$

SUMMARY

Use these questions to evaluate how well you have achieved the goals of this chapter. The answers to these questions are given at the end of this summary with the number of the section where you can find related content material.

Definitions

1. Any object can be deformed by applying a force to the object. An object is elastic if _____ after the force is removed. However, if the object has been

deformed beyond its _____, it will remain in a deformed state.

2. The stress on an object is defined as the ratio of the _____ to the _____.

3. The strain of an object is defined as the ratio of the _____ to the _____.

4. The generalized statement of Hooke's law states that _____ is proportional to _____ of the system.

5. Young's modulus is a constant with units of _____, which characterizes the response of a solid to _____ and which has a magnitude equal to _____.

6. The bulk modulus of a material is a constant with units of _____, which characterizes the response of a material to _____.

7. The shear modulus, also called the _____, is a constant with units of _____, which characterize the response of a material to _____ and which has a magnitude equal to _____.

8. A steel rod 2 m long and 0.5 cm² in area stretches 0.24 cm under a tension 12,000 N.
   a. What is the stress of the rod?
   b. What is the strain of the rod?

9. What is the compressibility of water?

Elasticity Problems

10. What is Young's modulus for the steel rod in question 8 above?

11. If the volume of an iron sphere is normally 100 cm³ and the sphere is subjected to a uniform pressure of 10^8 N/m², what is its change in volume?

Answers

1. It returns to its original state, elastic limit (Section 13.3)

2. Applied force, cross-sectional area (Section 13.2)

3. Change in a spatial variable, original value of that variable (Section 13.2)

4. Strain, stress (Section 13.3)

5. N/m², forces applied to change its length, $FL/A\Delta L$ (Section 13.4)

6. N/m², forces applied to change its volume without changing its shape, $FV/A\Delta V$ (Section 13.5)

7. Modulus of rigidity, N/m² force applied to change its shape without changing its volume, $(F/A)\phi$ (Section 13.6)

8. a. 2.4×10^8 N/m²
   b. 12×10^{-4} (Section 13.4)

9. 4.5×10^{-10} m²/N (Section 13.5)

10. 20×10^{10} N/m² (Section 13.4)

11. 6.3×10^{-2} cm³ (Section 13.5)

ALGORITHMIC PROBLEMS

Listed below are the important equations from this chapter. The problems following the equations will help you learn to translate words into equations and to solve single-concept problems.

Equations

$$\text{stress} = \frac{F}{A} \tag{13.1}$$

$$\text{strain} = \frac{\Delta L}{L_0} \quad \text{or} \quad \frac{\Delta V}{V} \quad \text{or} \quad \phi \tag{13.2, 13.6}$$

$$F = kx \tag{13.3}$$

$$Y = \frac{FL}{A\Delta L} \tag{13.5}$$

$$B = \frac{P}{\Delta V/V} \tag{13.7}$$

$$\eta = \frac{F/A}{\phi} \tag{13.8}$$

$$PE = \tfrac{1}{2}ks^2 \tag{13.9}$$

Problems

1. If the stress produced in stretching a wire is 5.00×10^6 N/m² by an applied force of 10.0 N, what is the cross-sectional area of the wire?
2. If the strain for the wire above is 0.100 percent, what is the length of the wire that will have an elongation of 1.00 mm?
3. In an experiment one finds that a force of 160 N produces a stretch of 8.00 cm in a given spring. What is the spring constant of the spring?
4. What is Young's modulus for the wire described in problems 1 and 2?
5. A 5.00-cm cube of gelatin has its upper surface displaced 1.00 cm by a tangential force 0.500 N. What is shear modulus of this substance?
6. What is the potential energy stored in the spring in problem 3 when it is stretched 8 cm?

Answers

1. 2.00×10^{-6} m²
2. 1.00 m
3. 2.00×10^3 N/m

4. 5.0×10^9 N/m²
5. 400 N/m²
6. 6.4 J

EXERCISES

These exercises are designed to help you apply the ideas of a section to physical situations. When appropriate the numerical answer is given in brackets at the end of the exercise.

Section 13.4

1. A wire 0.70 mm in diameter and 2.0 m long was stretched 1.6 mm by a load of 20 N. Find Young's modulus for the wire.   [6.5×10^{10} N/m²]
2. A load of 18×10^4 N is placed upon a vertical steel support of 6.0-m height and with a cross-sectional area of 20 cm². How much is the support compressed by the load?   [2.7×10^{-3} m]

Section 13.5

3. What is the decrease in volume of 2 liters of water if it is subjected to a pressure of 10^{10} N/m²? Compare this decrease with the decrease in 2 liters of mercury under the same pressure. [9×10^{-3}, ratio $H_2O/$ Hg ≅ 11]

4. How much pressure would be required to reduce the volume of a block of aluminum by a factor of one part in one thousand?   [7×10^7 N/m²]

Section 13.6

5. The lower end of a steel wire is 1.0 m long and has a radius of 1.0 mm and is twisted through an angle of 360°. Given $\eta = 2lL/\pi r^4 \phi$, where l = length, L = torque, r = radius, η = shear modulus, and ϕ = twist in rad, what is the torque required?   [0.79 N–m]

PROBLEMS

The following problems may involve more than one physical concept. When appropriate, the numerical answer is given in brackets at the end of the problem.

6. Assume the femur has a diameter of 3.0 cm and a hollow center of 0.8-cm diameter and a length of 50 cm. If it is supporting a load of 600 N, what is the stress in the femur? How much will it be shortened by this load? $Y = 16 \times 10^9$ N/m². [9.1×10^5 N/m²; 2.9×10^{-5} m]

7. Given the density of sea water as 1.03 g/cm³ at the surface, what is its density at a depth where the pressure is 1.00×10^8 N/m²? [1.08 gm/cm³]

8. A steel shaft with a radius of 1.00 cm and a length of 2.00 m transmits 50.0 kilowatts of power at 2400 rpm.
   a. What is the torque?
   b. Through what angle in radians is the shaft twisted? [199 N–m; 0.317 rad]

9. Find the equation for the energy a bone can absorb within its elastic limit in terms of its Young's modulus, cross-sectional area length and compression. $\left[\frac{1}{2}\left(\frac{YA}{L}\right)(\Delta L)^2\right]$

10. Two masses are suspended on a copper and on an iron wire (see Figure 13.7). What is the stress in each

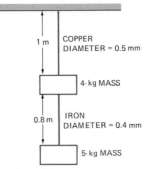

FIGURE 13.7
Problem 10.

wire? What is the elongation for each? What is the elastic PE in each wire? [$(F/A)_{cu} = 4.49 \times 10^8$ N/m²; $(F/A)_{fe} = 3.90 \times 10^8$ N/m²; $(\Delta L)_{cu} = 0.37$ mm; $(\Delta L)_{Fe} = 0.16$ mm; $(PE)_{cu} = 1.65 \times 10^{-1}$ J; $(PE)_{Fe} = 4.01 \times 10^{-2}$ J]]

11. Given Young's modulus for bone $= 1.00 \times 10^{10}$ N/m², find the compression experienced by a leg bone 50 cm long subjected to a load of half the weight a 70-kg person. The cross-sectional area of a leg bone is about 5 cm². [3.4×10^{-5} m]

GOALS

When you have mastered the contents of this chapter, you will be able to achieve the following goals:

Definitions
Define each of the following terms, and use it in an operational definition:

rms velocity	surface tension
adhesion	capillarity
cohesion	osmosis

Molecular Model
Explain the properties of fluids using the molecular model.

Problems
Solve problems that relate the gas laws to the properties of the gas and that involve surface tension, capillarity, and osmosis.

PREREQUISITES

Before beginning this chapter you should have achieved the goals of Chapter 4, Forces and Newton's Laws, Chapter 5, Energy, Chapter 6, Momentum and Impulse, Chapter 10, Temperature and Heat, and Chapter 13, Elastic Properties of Materials.

14

Molecular Model of Matter

14.1 Introduction

In our previous discussions of the properties of materials, we have treated them as if they were continuous media. But if materials are assumed to be a uniform gelatinous substance, how can we explain the existence of three different states of matter, solid, liquid, and gas? In a continuous medium what can be the cause of pressure acting upward on the inside of the top of a container holding a confined gas? Even though all of our previous explanations have been based upon a continuous-media model of matter, they seem to explain natural phenomena properly. However, there are some phenomena that are puzzling in our present framework. We have, for example, no way of explaining the fact that evaporation is a cooling process or the fact that one material can diffuse into another. How is a continuous medium made thin enough to be a gas and also thick enough to be a solid? It is clear that our continuous-media model has some serious limitations. With what model of matter shall we replace it?

14.2 The Molecular Model of Matter

If matter seems not to behave like a uniform gelatin, then perhaps we can describe it as being discontinuous, made up of individual clumps being more or less far apart. In developing our model here is how we shall proceed.

1. We begin by making some hypotheses, or assumptions, or conjectures, to form a model, or mental construct, to explain a natural phenomenon.
2. We use the laws, conservation principles, and mathematics we know with this model to make predictions and develop explanations.
3. We test these predictions and explanations against experimental results.

We cannot show that our model is correct. We can show that the predictions of our model agree with experimental results, but another model that gave the same predictions would be equally good. However, if we can disprove experimentally the predictions that follow from our model, then we know our model is wrong.

You will notice that our model building and testing procedure is a powerful way of showing that a model is wrong. It is less useful in proving a model is correct because it is always possible that if you can make a new prediction from our model it might prove to be wrong. Nevertheless, we tend to find comfort in simple models that we can use to provide us with a feeling of understanding our universe.

On the basis of our experience with materials let us postulate some basic ingredients of our molecular model of matter.

We postulate that all material is composed of small entities called molecules. We may think of molecules as very small, perfectly elastic spheres. The molecules are close together in solids and far apart in gases.

We postulate that molecules are always in motion. This gives us a logical way of explaining the internal energy term in the first law of thermodynamics. It can be considered the total kinetic energy of all the molecules in the material.

We postulate that there are interactions between the molecules. From our experiences and observations of the properties of the three states of matter, we can draw some conclusions about the nature of the interaction between molecules. If you have ever tried to pull a small wire apart, you know that a force is required and that as the diameter of the wire is increased, you will soon find that you cannot pull the wire apart. From this experiment you can predict that on the molecular scale in the wire there is an attractive force. Also, you have observed that very little force is required to separate a liquid into parts. For a gas, it seems that molecules are quite independent from one another. An experiment that suggests a repulsive force comes from attempts to reduce the volume of a solid or liquid. In the chapter on elasticity you learned that large forces are required to compress liquids and solids. These observations suggest that the force between molecules changes as the distance between the molecules change. We expect there exists some equilibrium position where repulsive and attractive forces between molecules cancel. At some very small distances the molecules must repel each other. At larger distances there is an attractive force, and finally at very large distances the molecules feel almost no forces from other molecules. The general shape of the curve that represents the force between molecules as a function of the separation between molecules is shown in Figure 14.1. You will notice that a small displacement from equilibrium results in a force, nearly linear with displacement, which tends to restore the system to equilibrium. At very small separations the repulsive force is strong; so we can treat the molecules as hard, impenetrable particles. Let us apply

FIGURE 14.1
A plot of the force between molecules vs. distance.

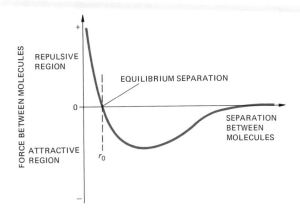

the postulates of the molecular model of matter to explaining the natural phenomena we have studied.

14.3 Kinetic Theory of Gases

We have already studied the properties of gases and used the gas laws to solve many problems in Chapter 10. To use our molecular model to predict the properties of gases, let us make the following assumptions about our idealized gas:

1. The average distance between the molecules is many times greater than the size of the molecules. Therefore we will neglect the volume of the molecules in our calculations.
2. The force of interaction between the molecules is zero except when they collide.
3. All collisions between molecules are perfectly elastic.
4. The idealized gas consists entirely of molecules of mass m which are moving with the same speed v. The molecules are assumed to be moving in completely random directions.

Assume we have N molecules in a rectangular box that has sides of length, l_1, l_2, and l_3. One side is a piston, which has an area of $l_2 \times l_3$ and which is parallel to the yz plane. We want to develop an expression for the force exerted on the piston by the molecules of the gas.

Consider a molecule such as molecule B which is moving with a velocity v (Figure 14.2). This velocity will have the three rectangular velocity components v_x, v_y, and v_z. The magnitude of the velocity will be related to the individual components by the resultant equation from the section on vector addition in Chapter 3,

$$v^2 = v_x{}^2 + v_y{}^2 + v_z{}^2 \tag{14.1}$$

FIGURE 14.2
The pressure exerted by a gas is the total of the pressure exerted by all the molecules of the gas as they strike the walls of the container.

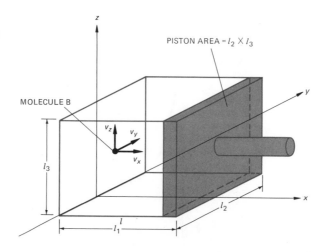

Consider the component v_x as the molecule strikes the face of the piston. The molecule will make an elastic collision with the wall. From Chapter 6 we know that the molecule will rebound with its speed unchanged but moving in the opposite direction. This means that the molecule will have a change of momentum in x direction from mv_x to $-mv_x$. Hence the change of momentum for each collision of molecule B with the piston is $+2mv_x$. From the impulse-momentum equation (Equation 6.2), each collision will impart an impulse of $2mv_x$ to the piston. Impulse is equal to the product of force times the time, so the force on the piston will be equal to the number of impulses the piston receives each second times the magnitude of each impulse.

$$\text{force} = (\text{impulse})(\text{number of impulses per second}) \tag{14.2}$$

If we continue to observe the motion of molecule B, it will not strike the piston again until it has traveled from the back wall and returned to hit the piston, a distance of $2l_1$ in the x direction. The time between impacts of molecule B on the piston is the distance divided by the speed, $2l_1/v_x$. Hence the number of impacts per second of molecule B on the piston is $1/(2l_1/v_x)$, or $v_x/2l_1$. We can substitute these values for impulse and impact rate into Equation 14.2. The force exerted by molecule B on the piston is given by

$$F_\text{B} = (2mv_x)\left(\frac{v_x}{2l_1}\right) = \frac{mv_x^2}{l_1} \tag{14.3}$$

The total force of all N particles acting on the piston is equal to the sum of the contributions of each of the N particles, or to the product of the average force per particle mv_x^2/l_1, multiplied by the number of particles,

$$\text{total force on piston} = N\frac{mv_x^2}{l_1} \tag{14.4}$$

But, of course, with all the molecules moving in random directions we cannot know the x-component of velocity of the molecules. We need to relate this to some property of the whole system. Notice that mv_x^2 is almost proportional to the kinetic energy of a molecule. Let us define the total kinetic energy of the gas as the number of molecules N times the kinetic energy of the individual molecules,

$$\text{total kinetic energy} = N(\text{KE})$$

$$\text{KE} = N(\tfrac{1}{2}mv^2) = \frac{Nmv^2}{2} \tag{14.5}$$

Because the direction of motion of the molecules is random, we can expect the total kinetic energy to be equally divided among the three components of velocity, so

$$mv_x^2 = \tfrac{1}{3}mv^2 \tag{14.6}$$

We can substitute this value into Equation 14.4 to obtain an equation for the force on the piston in terms of the speed of the molecules,

$$\text{total force on piston} = \tfrac{1}{3}N\frac{mv^2}{l_1} \tag{14.7}$$

We can now obtain an expression for the pressure on the piston since the pressure on the piston is equal to the force divided by the area and the area of the piston is $l_2 l_3$,

$$\text{pressure} = \frac{\text{total force}}{l_2 l_3}$$

$$P = \tfrac{1}{3}\frac{Nmv^2}{l_1 l_2 l_3} \tag{14.8}$$

Perhaps as you look at Equation 14.8 it will occur to you that the volume of the container V is equal to the product of the three sides $l_1 l_2 l_3$. We can replace $l_1 l_2 l_3$ in Equation 14.8 by the volume V,

$$P = \tfrac{1}{3}\frac{Nmv^2}{V} \tag{14.9}$$

or

$$PV = \tfrac{1}{3}Nmv^2 \tag{14.10}$$

Look at that expression; PV is a constant of the system by Boyle's law! We can perform some manipulations on Equations 14.9 and 14.10 to obtain some predictions from our model. The density of our gas is given by Nm/V, so Equation 14.9 becomes

$$P = \tfrac{1}{3}\rho v^2 \tag{14.11}$$

where ρ is the density of the gas. For constant velocities the pressure of a confined gas should be proportional to its density. The kinetic energy of a molecule is given by $\tfrac{1}{2}mv^2$; so Equation 14.10 becomes

$$PV = \tfrac{2}{3}N(\text{KE}) \tag{14.12}$$

where KE is the kinetic energy of a molecule.

From the combined gas laws in Chapter 10 you may recall that the (pressure × volume) product is equal to a constant times the absolute temperature. So we can conclude that the kinetic energy of the molecules is related to the temperature. We can write

$$PV = \tfrac{2}{3}N(\text{KE}) = \tfrac{2}{3}N(\tfrac{1}{2}mv^2) = (\text{constant})T \tag{14.13}$$

where T is the Kelvin temperature. The constant is usually expressed as the product of two constants, nR, where n is the number of moles of gas and R is the universal gas constant, which has the value of 8.31 joule/mole–°K. The total number of molecules in our gas N is equal to the number of moles times Avogadro's number, N_0. So we have

$$PV = \tfrac{2}{3} n N_0 (\tfrac{1}{2} m v^2) = nRT \tag{14.14}$$

and the kinetic energy of a molecule

$$KE = \tfrac{1}{2} m v^2 = \tfrac{3}{2} \frac{R}{N_0} T = \tfrac{3}{2} kT \tag{14.15}$$

where k is R/N_0, which has a value of 1.38×10^{-23} joules/molecule–°K and is known as Boltzmann's constant, the gas constant per molecule. Notice that the translational kinetic energy of our gas molecule is proportional to the absolute temperature and independent of the kind of gas.

This kinetic theory of gases seems quite acceptable. Using only the basic principles of mechanics along with the assumptions of our model, we have been able to derive very general results using no mathematics more difficult than algebra. It is rare that we can develop a fairly acceptable theory from facts and relationships learned in an introductory science course. The development of the kinetic theory of gases is certainly an exception.

Now that we know our model can produce interesting results, let us generalize our assumptions a bit. We made the assumption that all of our gas molecules have exactly the same speed v. That is a restrictive assumption not required to obtain the results of Equation 14.15. Instead let us require that the nth molecule have a speed v_N, so that molecule 1 has a speed v_1, molecule 2 has a speed v_2, etc. The system will still have a total kinetic energy given by the sum of the individual kinetic energies,

$$KE_{total} = \tfrac{1}{2} m v_1{}^2 + \tfrac{1}{2} m v_2{}^2 + \cdots + \tfrac{1}{2} m v_N{}^2 \tag{14.16}$$

We can use ratio of the total kinetic energy to the total number of molecules to calculate a value of the average kinetic energy per molecule $\overline{KE}$,

$$\overline{KE} = \frac{KE_{total}}{N} = \frac{(\tfrac{1}{2} m v_1{}^2 + \tfrac{1}{2} m v_2{}^2 + \cdots + \tfrac{1}{2} m v_N{}^2)}{N} \tag{14.17}$$

We can use the value of the average kinetic energy per molecule to obtain a value for the average value of the square of the speeds of the individual molecules $\bar{v}^2$,

$$KE = \tfrac{1}{2} m \bar{v}^2 = \frac{\tfrac{1}{2} m v_1{}^2 + \tfrac{1}{2} m v_2{}^2 + \cdots + \tfrac{1}{2} m v_N{}^2}{N} = \tfrac{3}{2} kT$$

$$\bar{v}^2 = \frac{v_1{}^2 + v_2{}^2 + \cdots + v_N{}^2}{N} = \frac{3kT}{m} \tag{14.18}$$

The square root of $\bar{v}^2$ is called the root mean square (rms) speed. For a collection of molecules with widely different speeds, the root mean square speed v_{rms} is that common value of speed of the molecules in the collection that would give each molecule the average kinetic energy of the collection.

Thus all of our previous results such as Equations 14.10 and 14.15 are generalized to a collection of molecules with a distribution of speeds by replacing the speed in those equations by the root mean square speed.

EXAMPLES

1. Find the rms speed of oxygen molecules at 27°C.

$$\tfrac{1}{2}m\bar{v}^2 = \tfrac{3}{2}kT$$

where $m = 32 \times 1.67 \times 10^{-27}$ kg and $T = 300°$K. Therefore

$$\bar{v}^2 = 3kT/m = \frac{3 \times 1.38 \times 10^{-23} \text{ J/molecule-°K}}{32 \times 1.67 \times 10^{-27} \text{ kg}}$$

$$= 23 \times 10^4 \text{ (m/sec)}^2$$

$$v_{\text{rms}} = 4.82 \times 10^2 \text{ m/sec} = 482 \text{ m/sec}$$

How many times per second will an oxygen molecule in the air transverse your room?

2. According to the kinetic theory of gases the pressure exerted by a gas depends upon three factors. What are they? How can each factor be changed?

From Equation 14.9 we can identify the three factors as:

a. The number of molecules per unit volume (N/V). Changing the total number of molecules N or the volume of the system will change the pressure.

b. The mass of the molecules; so changing the kind of gas would change the pressure.

c. The rms speed of the molecules; so changing the temperature will change the pressure.

3. If the average speed of the molecules of a gas is doubled, what happens to the kinetic energy of the gas?

Since the kinetic energy is proportional to the square of the speed of the molecules, the kinetic energy is four times greater.

14.4 Molecular Model of Liquids

If we change the assumptions we made for the idealized gas, we can use the molecular model to explain the properties of liquids. For liquids the distances between molecules are assumed to be small, and the forces between the neighboring molecules are strong. The force of attraction between like molecules is called the *cohesive force*. The cohesive energy for molecules of a liquid is about equal to the kinetic energy of the molecules.

We can think of a liquid as a large number of very small, slightly sticky glass beads which are continually moving. This model can be used to explain various properties of liquids—for example, their incompressibility, their ability to assume the shape of a container, and their viscous flow.

Molecules at the surface of a liquid will only have cohesive forces acting on them from the interior of the liquid. If a surface molecule has sufficient kinetic energy, it may break away from its neighbors into the gas above the liquid. When this molecule of slightly higher kinetic energy leaves, the average kinetic energy of the liquid is reduced—that is, the temperature of the liquid decreases. This is a molecular explanation of evaporation and its cooling effect.

14.5 *Molecular Model of Solids*

If we imagine the cohesive forces to become even stronger than they are in liquids so that the cohesive energy is much greater than the kinetic energy of the molecules, then we have a model for a solid. In a solid the molecules are constrained in their motion to a small region around their equilibrium locations. The regular spatial patterns of molecules in crystalline solids is the lowest energy state for a system of molecules that interact with strong cohesive forces. Many of the properties of solids can be explained in a qualitative way by this model. The large values of Young's modulus for most crystalline solids is evidence of the strong cohesive forces in solids.

14.6 *Adhesive Forces*

You know that a piece of adhesive tape will adhere to your skin. Why? Within the framework of the molecular model, the tape sticks to your skin because there is an attractive force between the molecules in your skin and the molecules in the coating of the tape.

The attraction of unlike molecules is called adhesion, and the forces of attraction are called adhesive forces.

Other applications of adhesive interactions are: wood and glue, solder and copper or brass, paste and paper, cellophane tape and many materials. You know that the cohesive forces of a solid are greatly weakened if a solid is broken apart. If you wish to put two pieces of wood together, you cannot just press the pieces together so that they will cohere. In order to get them together you must use some adhesive material such as glue.

The relative magnitude of the cohesive and adhesive forces determine the interaction that results at the surface between substances in contact. You have probably observed that if oil is spilled upon a water surface, there is soon an oil film over a large area of water. Try a simple experiment. Take two pieces of a small-diameter glass tube and insert one in mercury and the other in water. Insert the tube into the liquid to some depth and then raise the tube. Observe the surface contour in each case. A diagram of such an experiment is shown in Figure 14.3.

FIGURE 14.3
Examples of adhesive forces between mercury (Hg) and glass and between water and glass. Note the contact angle is determined by the angle made by the liquid surface (tangent) at the point of contact of the three surfaces.

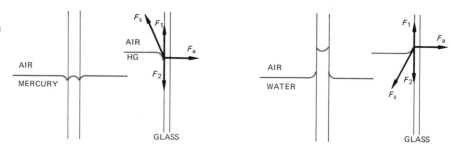

In the mercury you will observe that the surface is convex, and the column of mercury in the tube is actually depressed. In the water the surface is concave, and inside the tube the water surface is higher than the surface in the main container. The surface for mercury is a typical example of the effect produced when the cohesive force is greater than the adhesive force. The water case is typical of situations in which the adhesive force is greater than the cohesive force. At each boundary of surface as indicated in Figure 14.3, there are three substances in contact. There is a surface effect at each surface of separation, and there is a force parallel to each surface. Thus at the point of contact of the three surfaces there are three forces parallel to each of the surfaces—that is, F_1 solid–air surface, F_2 liquid–solid surface, F_S liquid–air surface. In addition there is the fourth force—that of adhesion between the liquid and solid F_A. At the point of contact of the three substances the four forces are in equilibrium. The liquid surface will adjust itself so that this condition is satisfied. If the liquid wets the solid, the angle of contact will be less than 90°, as it is for the water–glass interface. If it does not, the angle of contact will be greater than 90°, as for mercury and glass.

14.7 Surface Tension

Some interesting experiments can be carried out with a soap bubble solution. Suppose that you have a light circular wire frame and tie a double thread across the ring as shown in Figure 14.4a. Then dip the ring in the soap bubble solution so that there is a film over the entire ring. The thread will show no well-defined shape. If you puncture the soap film inside the thread, then the thread takes on the circular shape shown in Figure 14.4b. For the thread to take on this shape there must

FIGURE 14.4
Minimum surface is maintained by surface tension. When a soap film is broken inside the string loop, the outer surface tension pulls it into a circular shape.

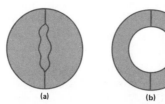

(a)   (b)

TABLE 14.1
Surface Tension

Surface (Liquid-air)	Temperature	Surface Tension (dynes/cm)
Benzene	20	28.9
Ethyl alcohol	0	24.5
Ethyl alcohol	20	22.8
Methyl alcohol	0	24.5
Methyl alcohol	20	22.6
Mercury	20	465
Water	10	74.2
Water	15	73.5
Water	20	72.8
Water	30	71.2

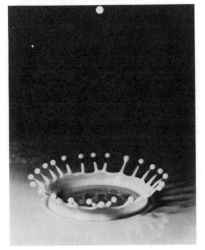

Splash of a milk drop in a crown formation. This illustrates the roll of surface tension in forming drops. (Harold E. Edgerton, MIT, Cambridge, Mass.)

be a force acting on the thread whose direction is perpendicular to the thread. From our molecular model we can explain this surface force as a result of the unbalanced molecular forces that exist at the surface of a liquid. The force per unit length of surface is called the *surface tension* of a liquid. So if you measure the force F acting over a peripheral length L of surface, the surface tension, S, is given by

$$S = \frac{F}{L} \tag{14.19}$$

The unit of surface tension is unit of force over unit of length or, in the SI system, newtons/meter. However, most tables give the value in dynes/cm. Some values of surface tension are given in Table 14.1. (Note that 1 dyne/cm $= 10^{-3}$ N/m.)

14.8 Measurement of Surface Tension

Approximate surface tension measurements can be made in different ways, but accurate measurements are difficult because the surface tension is influenced by impurities and surface dirt, traces of which are extremely hard to remove. An instrument for measuring surface tension is shown in Figure 14.5. A light platinum circular ring is supported on delicate torsion balance. Platinum is used in making the ring as it can be heated until red hot in a flame to remove traces of impurities, particularly greasy materials which are very undesirable substances in these measurements. The ring, held in a horizontal position, is dipped into clean water and is pulled upward by twisting the torsion wire. A film of water clings to both sides of the wire ring. The film from the water becomes vertical as the torsion wire is twisted. The total length of the film then becomes two times the circumference of the ring. The torsion head is twisted until the film breaks. The torsion balance can be calibrated by placing a small known mass on the ring and determining the angle of twist to support the calibrating weight in a horizontal plane. The force F required

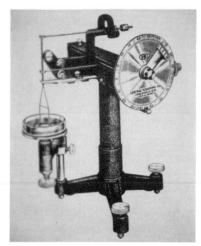

FIGURE 14.5
Tensiometer used to measure surface tension. A wire ring of known length is attached to an arm secured to a torsion wire with a calibrated angular scale.

to break the film is then given by

$$F = W_c \frac{\theta_w}{\theta_c} \qquad (14.20)$$

where W_c is the calibrating weight, θ_c is the angle of twist for calibrating weight, and θ_w is angle of twist for the water. The surface tension S is then given by

$$S = \frac{F}{2(2\pi r)} = \frac{F}{4\pi r} \qquad (14.21)$$

where r is the radius of the ring, and $2\pi r$ is the perimeter of each surface acting on the ring.

The surface tension is dependent on the temperature of the liquid, as shown in Table 14.1, and also on any impurities that may be in the liquid.

Another way to measure surface tension is to use a light wire frame with one movable side (see Figure 14.6). If the frame is dipped into a soap bubble solution so that a film is formed on the inside of the rectangle, there will be a force on the movable side tending to decrease the enclosed area. To hold the movable side of length l in equilibrium a force F will have to be applied to it. This force will be $2l$ times the surface tension S as there will be a film on each side of the wire. If the wire is slowly moved with the force remaining constant, the area increases, and work is done. The work done when the wire is moved a distance x is:

$$W = Fx = 2lSx \qquad (14.22)$$

but the increase in area is lx for each film and for the two films $2lx$, so then

$$W = SA \quad \text{or} \quad S = \frac{W}{A} \qquad (14.23)$$

Thus S, the surface tension, may be regarded as work done per unit area in increasing the film area, and S would be expressed in erg/cm². In the cgs system this is equivalent to dyne/cm.

Any surface that is under tension will always tend to minimize its area for the given boundaries. For example, a droplet tends to take a spherical shape, which makes its surface area a minimum.

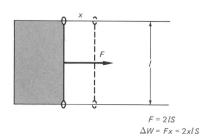

$$F = 2lS$$
$$\Delta W = Fx = 2xlS$$

FIGURE 14.6
Sketch of equipment used to measure surface tension.

14.9 Pressure in Liquid Drops

A drop of liquid has an inside pressure due to the surface tension. Consider a spherical droplet cut into two halves by an imaginary plane. The force holding the halves together results from the surface tension around the circumference of the circle. This force is then $F = 2\pi r S$. The internal pressure balances this force so that, if $\Delta P = P_i - P_o$, where $P_o =$ outside pressure and $P_i =$ inside pressure, then

$$\pi r^2 \, \Delta P = 2\pi r S \quad \text{or} \quad \Delta P = \frac{2S}{r} \tag{14.24}$$

Let us consider a soap bubble. You know that the pressure of the air inside the bubble is greater than that of the air outside because of the process of blowing the bubble. The film of the soap bubble has two sides. The force resulting from surface tension is two times as large in a bubble as in a liquid droplet. Using the same approach as above

$$F = 2(2\pi r) S$$

This force is balanced by pressure so that with $\Delta P = P_i - P_o$

$$\pi r^2 \, \Delta P = 4\pi r S \quad \text{or} \quad \Delta P = \frac{4S}{r} \tag{14.25}$$

EXAMPLE

Compare the pressure difference in a soap bubble with that difference for a water drop of same size. The surface tension of soap solution is one-third that of water.

$$\frac{P_{drop}}{P_{bubble}} = \frac{\dfrac{2S}{r}}{\dfrac{4S/3}{r}} = \frac{6}{4} = 1.5$$

14.10 Capillarity

Earlier it was mentioned that water will rise in a glass tube of small diameter. This effect is caused by surface tension (Figure 14.3) and is called *capillarity*. The upward force is produced by the surface tension and is given by the product of vertical component of the surface tension and the length of surface; namely, the circumference of inside of the tube. So $F_{upward} = (S \cos \theta)(2\pi r)$ where θ is the angle of contact between the liquid surface and the tube and r is the radius of the tube. The equilibrium occurs when the upward force is equal to the weight of the liquid in the tube above the surface of the liquid outside of the tube. The weight of the liquid $= mg = \rho V g$, where ρ is the density of the liquid and V is the volume of the elevated liquid,

$$mg = \rho \pi r^2 h g = (S \cos \theta)(2\pi r)$$

where h is the height of the liquid in the tube above the outside level. We can solve this equation to obtain h:

$$h = \frac{2S \cos \theta}{\rho g r} \tag{14.26}$$

The height h is inversely proportional to the radius of the tube.

EXAMPLE

It is noted that water has a contact angle of $0°$ in a capillary with a radius of 0.20 mm. The water rises 7.4 cm in the capillary; this means $\Delta P = \rho g h$. Find the surface tension of the water.

$$S = \frac{\rho g h r}{2 \cos \theta} = \frac{1.00 \text{ g/cm}^3 \times 980 \text{ cm/sec}^2 \times 7.4 \text{ cm} \times 2.0 \times 10^{-2} \text{ cm}}{2 \times 1.0}$$

$$S = 73 \text{ dyne/cm}$$

14.11 Osmosis

Have you observed what happens if a prune is soaked in water? The dried fruit soaks up water and swells to an enlarged size.

In living systems there are similar cases that involve the movement of water through biological membranes. This transport process is called *osmosis*. The net transport of the host liquid across the membrane is always in the direction that tends to equalize the concentrations of dissolved materials on each side of the membrane. For example, if two compartments of sugar and water are separated by a parchment membrane, water will flow through the membrane from the lower concentration of sugar compartment to the higher concentration of sugar compartment until the two concentrations are equalized or until the pressure in the more concentrated solution prevents further flow. In such osmotic processes the membrane is permeable only to the host liquid, the solvent (Figure 14.7).

By applying pressure in the compartment of the more concentrated solution, the flow of the solvent can be slowed, stopped, or reversed. The applied pressure that is necessary to stop the flow of the solvent from the lower concentration to the higher concentration solution through the membrane is called the *osmotic pressure*. In the 1880s, W. Pfeffer performed experiments to determine the factors that influence

FIGURE 14.7
The principle of osmosis. (a) Originally the sugar solution is contained in the inner container, which is a semipermeable membrane immersed in distilled water. The arrows indicate that the water passes through the membrane but the sugar molecules do not. As water molecules go into the inner tube, the volume of the sugar solution increases, as indicated by the increase in height in the pressure tube. Finally (b) an equilibrium condition is reached in which the number of molecules of water going in each direction through the semipermeable membrane is same. The pressure exerted by the liquid in the small tube is called the osmotic pressure. It is also the pressure that will stop the osmotic process.

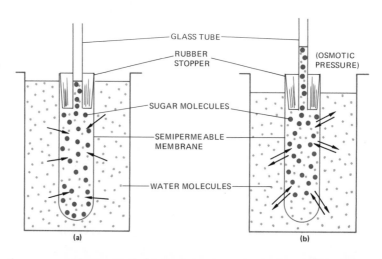

GLASS TUBE

RUBBER STOPPER

(OSMOTIC PRESSURE)

SUGAR MOLECULES

SEMIPERMEABLE MEMBRANE

WATER MOLECULES

(a)   (b)

osmotic pressure. First he carried out a series of experiments at constant temperature and varied the concentration of the dissolved materials, or *solute*. The results indicated that the osmotic pressure was directly proportional to the concentration of solute. In the second series of experiments the concentration of the solute was held fixed and the osmotic pressure was measured as a function of temperature. It was found that the value of the osmotic pressure divided by the Kelvin temperature is constant; that is, the osmotic pressure is directly proportional to the Kelvin temperature. A Dutch chemist named van't Hoff combined these observed relationships to obtain an empirical relationship for the osmotic pressure ΔP_{os},

$$\Delta P_{os} = \Delta c R T \tag{14.27}$$

where Δc is the number of moles of solute per unit volume of solvent, R is the universal gas constant and T is the temperature in degrees Kelvin. The osmotic current density (kg of solvent/m^2–sec) is proportional to the concentration difference across the membrane. This can be expressed in equation form as follows:

$$J_{os} = -k\,\Delta c = -\frac{k}{RT}\,\Delta P_{os} \tag{14.28}$$

where k is the permeability coefficient of the membrane.

EXAMPLE

Given that $k/RT = 6 \times 10^{-10}$ kg/N-sec for a parchment membrane at room temperature. Find the osmotic pressure that will result in 6×10^{-4} kg/m^2-sec of water current density from one side of the parchment to the other side.

$$(\Delta P)_{os} = \left|\frac{J_{os}}{k/RT}\right| = \frac{6 \times 10^{-4}}{6 \times 10^{-10}} = 10^6 \text{ N/m}^2 \cong 10 \text{ atm}$$

SUMMARY

Use these questions to evaluate how well you have achieved the goals of this chapter. The answers to these questions are given at the end of this summary with the number of the section where you can find the related content material.

Definitions

1. The average kinetic energy of the molecules of a confined gas is used to determine the _____, which is equal to _____.

2. Solids have large internal _____ forces, i.e., forces of _____ between _____ molecules.

3. Glues are _____ materials which have large _____ forces, i.e., force of _____ between _____ molecules.

4. The rise of a liquid in a narrow tube is called _____ and is caused by the _____ of the liquid and the _____ between the liquid and the tube.

5. The force that acts upon the _____ of a liquid to make it have the minimum possible surface area is called the _____.

6. The passage of the solvent through a _____ from a region of _____ solute concentration to a region of _____ solute concentration is called _____ .

Molecular Model

7. Use the molecular model to provide an explanation of surface tension (make a sketch).

Problems

8. If you had an open glass tube between a soap bubble of 2-cm radius and one of 8-cm radius, what would happen? Why?
9. A home remedy for the removal of candle wax from clothing is to cover the spot with blotting paper and then pass a hot iron over the paper. Explain.
10. Describe an ideal gas.
11. a. What is the change of momentum per elastic collision per molecule?
    b. How many collisions does each molecule make per second on a given face?
12. Compute the root mean square velocity of oxygen molecules at pressure of 1.01×10^5 N/m² and a temperature of 0°C. The density of oxygen is 1.43 gm/liter.

13. How high will 15°C water rise in a capillary tube of 0.5–mm radius?

Answers

1. rms velocity,

$$\sqrt{\sum_{i=1}^{N} v_i^2 / N}$$

(Section 14.3)
2. cohesive, attraction, like (Section 14.5)
3. adhesive, adhesive, attraction, unlike (Section 14.6)
4. capillarity, surface tension, adhesion (Section 14.10)
5. surface, surface tension (Section 14.7)
6. membrane, low, high, osmosis (Section 14.11)
7. (Section 14.11)
8. small bubble shrinks (Section 14.9)
9. adhesion of liquid candle wax is low to cloth, high to blotting paper (Section 14.6)
10. See the list of assumptions (Sections 14.2, 14.3)
11. a. $2mv$
    b. $v/2l$, where l is the length of the container and v is the velocity parallel to that length (Section 14.3)
12. 460 m/sec (Section 14.3)
13. 3 cm (Section 14.10)

ALGORITHMIC PROBLEMS

Listed below are the important equations from this chapter. The problems following the equations will help you learn to translate words into equations and to solve single-concept problems.

Equations

$$pV = \tfrac{1}{3} Nmv^2 \tag{14.10}$$

$$P = \tfrac{1}{3} \rho v^2 \tag{14.11}$$

$$PV = nRT \tag{14.14}$$

$$KE = \tfrac{3}{2} kT \tag{14.15}$$

$$\bar{v}^2 = \frac{\left(\sum_{i=1}^{N} v_i^2 \right)}{N} = \frac{3kT}{m} \tag{14.18}$$

$$S = \frac{F}{L} \tag{14.19}$$

$$\Delta P = \frac{2S}{r} \tag{14.24}$$

$$\Delta P = \frac{4S}{r} \tag{14.25}$$

$$h = \frac{2S \cos \theta}{\rho g r} \tag{14.26}$$

$$\Delta P_{os} = \Delta c R T \tag{14.27}$$

$$J_{os} = -k \, \Delta c = -\frac{k}{RT}(\Delta P_{os}) \tag{14.28}$$

Problems

1. What force additional to its weight is required to pull a ring of 4.00-cm circumference from a clean water surface at 15°C?
2. What is the pressure within a mercury droplet (20°C) of 4.00-mm diameter?
3. What is the pressure within a soap bubble (20°C) whose radius is 4 cm if the surface tension of the solution is 50 dynes/cm?
4. Assume the angle of contact between the water and glass tube is zero. How high will water rise in a glass tube of radius 0.6 mm if the temperature is 20°C?
5. What is the kinetic energy of a room temperature, 20°C, air molecule?
6. How many moles of air are there in a typical physics classroom (7.2 m × 9.4 m × 3.4 m) at 20°C and at a pressure of 1.01×10^5 N/m²?
7. The density of air at 1 atm pressure and 0°C is 1.29 g/liter. What is the rms speed per molecule?
8. If in problem 7 the pressure remains constant, but the temperature is raised to 100°C, what is the rms speed at 100°C?

Answers

1. 588 dynes
2. 4.65×10^3 dynes/cm² + p_0
3. 50 dynes/cm² + p_0
4. 2.48 cm

5. 6.07×10^{-21} J
6. 9500 moles
7. 485 m/sec
8. 567 m/sec

EXERCISES

These exercises are designed to help you apply the ideas of a section to physical situations. When appropriate, the numerical answer is given in brackets at the end of the exercise.

Section 14.3

1. Ten moving particles have the following speeds (m/sec): 100, 120, 110, 140, 180, 150, 130, 170, 165, 135, find the rms speed for these 10 particles and compare it with the average speed of the particles. [avg = 140 m/s; rms = 142 m/s]

2. If the internal energy U of a gas of N particles (in volume V) is defined as the total kinetic energy of the particles, show that the pressure equals $\frac{2}{3} U/V$.
3. According to the kinetic theory of gases, the pressure exerted by a gas depends upon three factors. With two factors remaining constant, how would the pressure change if
   a. the number of particles per unit volume is increased by a factor of 2?
   b. the gas is changed from hydrogen to oxygen?
   c. the rms speed is doubled? [a. pressure doubled; b. ratio ρ_{O_2}/ρ_{H_2}; c. pressure increased by factor of 4]

Section 14.8

4. When using an apparatus like that shown in Figure 14.6, we find that the equilibrium force required is 500 dynes for a moving wire 2.00 cm long. Find the surface tension of the liquid. [125 dyne/cm]

Section 14.9

5. Using the data from Table 14.1, find the ratio of pressure at 10°C and at 100°C when the surface tension is 59 dyne/cm. [$P_{100}/P_{10} = 0.80$]

Section 14.10

6. Water at 15°C rises in a glass capillary of 0.5-mm radius to a height of 3.00 cm. Find the surface tension of the 15°C water. [73.5 dyne/cm]

Section 14.11

7. Given that $k/RT = 6 \times 10^{-10}$ kg/N-sec for a membrane, find the necessary pressure difference that will produce an osmotic current density of 60 micrograms/cm²-sec of water. [10^6 N/m²]

PROBLEMS

Each problem may involve more than one physical concept. The numerical answer to each problem is given in brackets at the end of the problem.

8. Compute the number of molecules per cubic centimeter in a vacuum tube in which the pressure is 1.00×10^4 mm Hg. Assume temperature is 300°K and molecular mass $= 3.20 \times 10^{-27}$ kg. [3.22×10^{12} molecules/cm³]

9. The horizontal length of wire frame in Figure 14.6 is 5 cm. Compute the value of surface tension from following data:
   a. With clean water a force of 730 dynes was required to pull the frame away from the water.
   b. With water containing a trace of soap the force was 500 dynes. [a. 73 dynes/cm; b. 50 dynes/cm]

10. Mercury is used in a barometer made of glass tubing having an inside diameter of 2 mm. If the mercury stands at 75 cm, what allowance should be made for capillary action, and what should the corrected height be? Angle of contact for mercury is 128°. [add 0.4 cm]

11. Find the ratio of the work necessary to make a soap bubble of radius 1 cm to that required to make a bubble with 2-cm radius. [$W_1/W_2 = 1/4$]

12. Compare the rms velocity of hydrogen and oxygen molecules at 300°K. Both are diatomic molecules; does this affect your answer? [v_{H_2}/v_{O_2} rms = 4]

13. A vertical solid glass rod 0.6 cm in diameter stands partly submerged in water. What is the downward pull on the rod due to the surface tension? [274 dynes]

14. A 0.5-liter flask contains 1.34×10^{22} molecules each of mass 5.31×10^{-23} g with an rms speed 4.50×10^4 cm/sec. Compute the pressure in the flask in atmospheres. [9.61×10^4 N/m² = .96 atm]

15. A needle is 5 cm long. Assuming the needle is not wetted, how heavy can it be and still float on the water? [730 dynes]

16. The surface tension of a soap solution is 40 dynes/cm. How much work must be done in blowing a soap bubble 6 cm in diameter? [905 ergs]

GOALS

When you have mastered the contents of this chapter, you will be able to achieve the following goals:

Definitions
Define each of the following terms, and use it in an operational definition:

period	frequency
simple harmonic motion	restoring force
amplitude	damping
phase angle	

UCM and SHM
Correlate uniform circular motion and simple harmonic motion.

SHM Problems
Solve problems involving simple harmonic motion.

Energy Transformations
Analyze the transfer of energy in simple harmonic motion.

Superposition
Explain the application of the principle of superposition to simple harmonic motion.

Natural Frequencies
Calculate the natural frequencies of solids from their elastic moduli and density values.

PREREQUISITES

Before beginning this chapter you should have achieved the goals of Chapter 5, Energy, Chapter 7, Rotational Motion, and Chapter 13, Elastic Properties of Materials.

15

Simple Harmonic Motion

15.1 Introduction

You are familiar with many examples of repeated motion in your daily life. If an object returns to its original position a number of times, we call its motion repetitive. Typical examples of repetitive motion of the human body are heartbeat and breathing. Many objects move in a repetitive way, a swing, a rocking chair, and a clock pendulum, for example. Probably the first understanding the ancients had of repetitive motion grew out of their observations of the motion of the sun and the phases of the moon.

Strings undergoing repetitive motion are the physical basis of all stringed musical instruments. What are the common properties of these diverse examples of repetitive motion?

In this chapter we will discuss the physical characteristics of repetitive motion, and we will develop techniques that can be used to analyze this motion quantitatively.

15.2 Kinematics of Simple Harmonic Motion

One common characteristic of the motions of the heartbeat, clock pendulum, violin string, and the rotating phonograph turntable is that each motion has a well-defined time interval for each complete cycle of its motion. Any motion that repeats itself with equal time intervals is called *periodic motion*. Its *period* is the time required for one cycle of the motion.

Let us analyze the periodic motion of the turntable of a phonograph. Suppose that we place a marker on a turntable that is rotating about a vertical axis at a uniform rate in a counterclockwise direction when viewed from above. If you observe the motion of the marker in a horizontal plane—that is, viewing the turntable edge-on—the marker will seem to be moving back and forth along a line. The motion you see is the projection of uniform circular motion onto a diameter and is called *simple harmonic motion* (Figure 15.1). To derive the equation for simple harmonic motion, project the motion of the marker upon the diameter AB. The displacement is given relative to the center of the path O and is represented by $x = OC$. From Figure 15.1 we see that $x = R \cos \theta$, where R is the distance of the marker from the axis of rotation. The maximum displacement of the motion is called the *amplitude* of the motion and is represented by the symbol A. The displacement of the marker in a direction parallel to the diameter AOB is then given by the following equation,

$$x = A \cos \theta \tag{15.1}$$

where θ is the angle through which the marker on the turntable has turned. Since we know that the turntable is rotating with a constant angular velocity ω, we recall from Chapter 7 that we can write an ex-

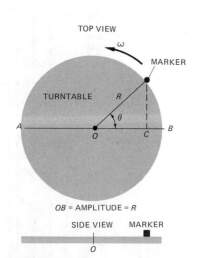

FIGURE 15.1
Simple harmonic motion. Projection of uniform circular motion upon a diameter.

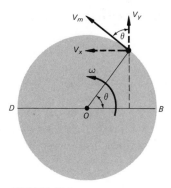

FIGURE 15.2
Velocity in simple harmonic motion.
Projection of the velocity of uniform
circular motion upon a diameter.

pression for the angular displacement θ as the angular speed times the time plus the starting angle,

$$\theta = \omega t + \phi$$

where t is the time of rotation and ϕ, the *phase angle*, is the angular displacement at the beginning, $t = 0$. If we choose the starting position along the line *DOB*, then $\phi = 0$ at $t = 0$. In general, the equation for the x-displacement is given by

$$x = A \cos(\omega t + \phi) \tag{15.2}$$

The velocity of the marker for the position shown in Figure 15.1 is tangential to the circle of motion of the marker and in the direction shown in Figure 15.2. You may recall from Chapter 7 that the magnitude of the velocity is given by

$$v = 2\pi Rn = \omega R = \omega A \tag{7.12}$$

where n is the number of revolutions of the turntable in one second, ω is the angular speed in radians per second, and R is the radius of the circle of motion and is equal to the amplitude of displacement A.

EXAMPLE

What is the velocity of a point on the rim of the standard 12-inch long-playing phonograph record?

$R = 15.2$ cm

$\omega = 33\frac{1}{3}$ rpm $= 5.56 \times 10^{-1}$ rev/sec $= 3.49$ rad/sec.

$v = \omega R = 53.1$ cm/sec

The velocity of the marker in a direction parallel to the line *DOB* is shown by the component v_x in Figure 15.2, where

$$v_x = -v \sin \theta = -\omega A \sin \theta \tag{15.3}$$

The negative sign indicates that the direction of motion is in the negative x-direction. When $\sin \theta$ is positive, the velocity v_x is in the negative x-direction. Notice $\sin \theta$ is always positive for $0 \le \theta \le 180°$; so v_x is negative for those angles and positive for the rest of the motion of one cycle.

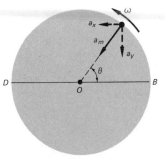

FIGURE 15.3
Acceleration in simple harmonic
motion. Projection of the acceleration
of uniform circular motion upon
a diameter.

The acceleration of the marker for the position shown in Figure 15.1 is perpendicular to the velocity, toward the center of circular motion as shown in Figure 15.3. The magnitude of the acceleration was derived in Chapter 3, $a = v^2/R$ (Equation 3.21). For the case we are considering, we can write the acceleration in terms of the angular speed and the amplitude,

$$a = \frac{v^2}{R} = \frac{\omega^2 R^2}{R} = \omega^2 R = \omega^2 A \tag{15.4}$$

The projection of the acceleration vector onto the line DOB is shown by a_x in Figure 15.3 where

$$a_x = -\omega^2 A \cos\theta \qquad (15.5)$$

The negative sign indicates that the acceleration is in the negative x-direction. The $\cos\theta$ is positive for $-90° \leq \theta \leq 90°$; so a_x is in the negative x-direction for these angles and positive for the rest of each cycle of motion. Notice that from Equation 15.1 we know that $A\cos\theta$ is the x-displacement. The equation for the acceleration can be rewritten in terms of the x-displacement

$$a_x = -\omega^2 x \qquad (15.6)$$

If we substitute the equation for the angular displacement as a function of time, $\theta = \omega t + \phi$, into the equations for x-displacement, x-velocity, and x-acceleration, then the linear displacement, velocity, and acceleration are given in general by the following equations:

$$x = A \cos(\omega t + \phi)$$
$$v_x = -\omega A \sin(\omega t + \phi)$$
$$a_x = -\omega^2 A \cos(\omega t + \phi) \qquad (15.7)$$

Assume $x = A$ at time $t = 0$. Then $\phi = 0$. Then these equations can be represented in graphical form as shown in Figure 15.4.

The curves in Figure 15.4 show that at the time of zero velocity (Figure 15.4a), the acceleration and the displacement are maximum. At a time of maximum velocity (Figure 15.4b), the acceleration and the displacement are zero. For simple harmonic motion the acceleration is proportional to the displacement x and is oppositely directed (Equation 15.6). If the displacement is to the right of the equilibrium position, then the acceleration is to the left, and vice versa. The angular speed ω is a constant, a characteristic of the motion. The angular speed can be expressed in terms of the frequency, or the period, of the motion. In

FIGURE 15.4
Displacement, velocity, and acceleration of simple harmonic motion as functions of time.

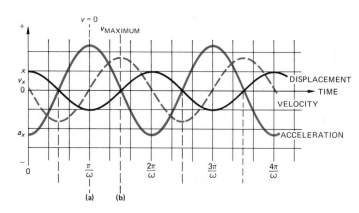

uniform circular motion we defined the number of revolutions per second as the *frequency f.* Then

$$\omega = 2\pi n = 2\pi f \tag{15.8}$$

where *f* is the number of cycles per second (hertz). The period *T* of one vibration is equal to the reciprocal of *f*,

$$T = \frac{1}{f} \tag{15.9}$$

Then the angular speed is proportional to the reciprocal of the period of the motion,

$$\omega = \frac{2\pi}{T} \tag{15.10}$$

where *T* is the period of the motion. By substituting the expression for ω from Equation 15.6 into Equation 15.8 and solving for *f*, we get the relationship between the frequency, the acceleration and the displacement,

$$f = \frac{1}{2\pi} \sqrt{-\frac{a}{x}} = \frac{1}{T} \quad \text{or} \quad T = 2\pi \sqrt{-\frac{x}{a}} \tag{15.11}$$

The value of $-a/x$ is always positive because *a* and *x* are in opposite directions. The above equation can be used to calculate either the frequency (period) or the acceleration or the displacement, if you know the other two variables.

EXAMPLES

1. A butcher throws a cut of beef on spring scales which oscillate about the equilibrium position with a period of $T = 0.500$ sec. The amplitude of the vibration is $A = 2.00$ cm (path length 4.00 cm). Find:
   a. the frequency
   b. the maximum acceleration
   c. the maximum velocity
   d. the acceleration when the displacement is 1.00 cm
   e. the velocity when the displacement is 1.00 cm
   f. the equation of motion as a function of time if the displacement is A at $t = 0$

SOLUTIONS

   a. The frequency $f = 1/T = 1/0.500 = 2.00$ hertz (vibrations/sec) according to Equation 15.9.
   b. By Equation 15.8, the angular velocity is given by

   $$\omega = 2\pi f = 4.00\pi \text{ rad/sec} = 12.6 \text{ rad/sec}$$

   Then by Equation 15.7

   $$a_x = -\omega^2 A \cos(\omega t + \phi)$$

The maximum acceleration occurs when $A \cos(\omega t + \phi)$ is equal to $-A$, or -2.00 for this problem.

$$a_{max} = -(4.00\pi)^2(-2.00) = +32.0\pi^2 \text{ cm/sec}^2 = 316 \text{ cm/sec}^2$$

c. The velocity is given by Equation 15.7:

$$v = -\omega A \sin (\omega t + \phi)$$

The velocity will be maximum when $A \sin(\omega t + \phi)$ is equal to $-A$; so

$$v_{max} = \omega A = 4.00\pi \times 2.00 = 8.00\pi \text{ cm/sec} = 25.1 \text{ cm/sec}$$

d. The acceleration is given by

$$a = -\omega^2 x$$

which is Equation 15.6. When the displacement is 1.00 cm,

$$a = -(4.00\pi)^2 \times 1.00 = -16.0\pi^2 = -158 \text{ cm/sec}^2$$

e. Use Equation 15.2 to find the angular displacement $(\omega t + \phi)$, and then use Equation 15.7 to find the velocity,

$$\cos (\omega t + \phi) = x/A = 0.500$$

So the $\sin (\omega t + \phi) = \sqrt{3}/2 = 0.866$

$$v = -\omega A \sin (\omega t + \phi) = -4.00\pi \times 2.00 \times 0.866$$

Therefore, the velocity when the displacement is 1 cm is

$$v = -6.93\pi \text{ cm/sec} = -21.8 \text{ cm/sec}$$

f. $x = A \cos (\omega t + \phi)$

by Equation 15.2. If $x = A$ at time $t = 0$, then $\phi = 0$. So $x = 2.00 \cos 4.00\pi t$ cm.

2. In a system undergoing simple harmonic motion, the acceleration is -20.0 cm/sec^2 for a displacement of 5.00 cm. What is the frequency and period of motion? What other information is needed to write an equation of motion? Let $x = 5.00$ cm, and $a = -20.0$ cm/sec^2. Then $T = 2\pi \sqrt{-x/a}$ (Equation 15.11). Therefore, the period of motion is $T = 2\pi \sqrt{+5.00/20.0} = 3.14$ sec, and the frequency is $f = 1/T = 1/3.14$ sec $= 0.318$ Hz. If you know the location at any time, then you can determine ϕ and can write the exact equation of the motion.

15.3 *Dynamics of Simple Harmonic Motion*

We have considered simple harmonic motion without regard to the forces that produce such motion. We now want to discover the common characteristics of the forces that produce simple harmonic motion (SHM). In many cases, the recognition of this SHM force not only allows you to predict harmonic motion, but it allows you to predict the frequency of the motion.

What kind of forces produce simple harmonic motion? Look at Equation 15.6. According to this equation, the acceleration in a SHM system must be proportional to the displacement of the system from equilibrium and in the opposite direction. Let us combine this equation with our knowlege of Newton's second law

$$F_x = ma_x \qquad \text{where } a_x = -\omega^2 x \tag{4.1}$$

so $F_x = -m\omega^2 x$

according to Equation 15.6, where the mass m and the angular speed ω are constants of the system. The component of the force F_x in the direction of motion is a *restoring force*, acting opposite in direction to the displacement as indicated by the negative sign. The magnitude of F_x is proportional to the displacement; F_x is a Hooke's law force. Therefore, Hooke's law force systems (Section 13.3) produce simple harmonic motion. If the Hooke's law force equation is written as follows, $F = -kx$ (Equation 13.3) then the force constant k is equal to $m\omega^2$. We can derive equations for the period and frequency of the simple harmonic motion that arises from a Hooke's law force by making use of the equality between the ratio of the force constant to the mass of the system and the square of the angular speed

$$\frac{k}{m} = \omega^2 = (2\pi f)^2 = \left(\frac{2\pi}{T}\right)^2 \tag{15.7}$$

$$\text{frequency} = \frac{1}{2\pi} \sqrt{k/m} \text{ Hz} \qquad \text{from above and} \tag{15.8}$$

$$\text{period} = 2\pi \sqrt{m/k} \text{ sec} \tag{15.9}$$

In summary, all Hooke's law force systems will produce simple harmonic motion. Any system of simple harmonic motion can be used to deduce a Hooke's law force. There are many examples of Hooke's law systems such as spring balances and simple pendulums. In fact, in the enrichment section we will use Taylor's theorem to show that almost *all* equilibrium systems exhibit simple harmonic motion near equilibrium.

EXAMPLE

It is known that a load with a mass of 200 g will stretch a spring 10.0 cm. The spring is then stretched an additional 5.00 cm and released. Find:
a. the spring constant
b. the period of vibration and frequency
c. the maximum acceleration
d. the velocity through equilibrium positions
e. the equation of motion

SOLUTIONS

a. If the force acting $mg = 0.200 \times 9.80$ N and $x = 0.10$ m, then the spring constant is

$$k = \frac{F}{x} = \frac{0.200 \times 9.80}{0.10} = 19.60 \text{ N/m}$$

b. To find the period of vibration let $T = 2\pi \sqrt{m/k}$. Then,

$$T = 2\pi \sqrt{\frac{0.200}{19.6}} = \frac{2\pi}{7\sqrt{2}} = \frac{\sqrt{2}}{7}\pi = 0.634 \text{ sec}$$

The frequency is

$$f = \frac{1}{T} = \frac{1}{0.634} = 1.59 \text{ Hz}$$

c. Given that the amplitude = 5.00 cm,

$$\omega^2 = \frac{k}{m} = \frac{19.6 \text{ N/m}}{0.200 \text{ kg}} = 98.0/\text{sec}^2$$

Then the maximum acceleration is

$$|a_{max}| = |-\omega^2 A| = |-98.0 \times .0500| = 4.90 \text{ m/sec}^2$$

d. To find the velocity through the equilibrium position let $|v_{max}| = A\omega = (0.0500 \text{ m})(\sqrt{98.0/\text{sec}^2}) = 0.495 \text{ m/sec}$

e. $A = 5.00$ cm, $\omega = \sqrt{98}/\text{sec} = 9.90 \text{ sec}^{-1}$, and $\phi = 0$, since $x = A$ at time $t = 0$. Therefore, $x = 5.00 \cos(9.90t)$ cm

15.4 Energy Relationships in Simple Harmonic Motion

You will recall that in previous chapters we have been able to use the concept of conservation of energy to solve mechanical problems. The conservation of energy has been especially useful in problems that involve systems whose total energy is a constant. Perhaps you wonder if energy analyses of SHM systems will yield worthwhile results. Let us consider the energy relationships in simple harmonic motion.

We can show that the potential energy associated with a $-kx$ force is equal to $\frac{1}{2} kx^2$ for a displacement of x. Since potential energy is equal to the work,

$$\text{PE} = W = (F_{ave})(x) = \frac{(kx)x}{2} = \frac{1}{2} kx^2 \tag{15.12}$$

The total energy of the system is equal to the sum of the potential and kinetic energy,

$$\text{Energy} = \text{KE} + \text{PE} = \tfrac{1}{2} mv^2 + \tfrac{1}{2} kx^2 \tag{15.13}$$

where v is the velocity of the moving object whose mass is m. We can use the equations for displacement and velocity as functions of time (Equations 15.2 and 15.7) to write an expression for the total energy as a function of time:

$$E = \tfrac{1}{2} m[-\omega A \sin(\omega t + \phi)]^2 + \tfrac{1}{2} k [A \cos(\omega t + \phi)]^2$$

$$E = \tfrac{1}{2} m\omega^2 A^2 \sin^2(\omega t + \phi) + \tfrac{1}{2} kA^2 \cos^2(\omega t + \phi)$$

where $\omega^2 = k/m$ for Hooke's law system. Therefore,

$$E = \tfrac{1}{2} kA^2 [\sin^2(\omega t + \phi) + \cos^2(\omega t + \phi)]$$

FIGURE 15.5
Plot of the kinetic energy and the potential energy of SHM as a function of displacement. Note that the sum of KE and PE for any displacement is a constant: total energy = KE + PE = constant.

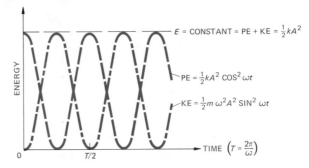

From our knowledge of trigonometry we know that for any angle θ, $\sin^2\theta + \cos^2\theta = 1$,

$$E = \tfrac{1}{2}kA^2 \tag{15.14}$$

The total energy of a SHM system is a constant, independent of time, and equal to one-half of the product of the force constant and the square of the amplitude of oscillation. The relationship between the potential energy and the kinetic energy as a function of time is shown in Figure 15.5. When the potential energy has its maximum value $\tfrac{1}{2}kA^2$, the kinetic energy is zero, and the object is instantaneously at rest at its position of maximum displacement, $x = A$. When the kinetic energy has its maximum value $\tfrac{1}{2}m\omega^2A^2$ or $\tfrac{1}{2}kA^2$, the potential energy is zero, and the object is moving with its maximum velocity at zero displacement. At all other positions neither the potential energy nor kinetic energy is zero. Since the total energy is a constant equal to $\tfrac{1}{2}kA^2$, we can use Equation 15.13 to find an expression for the velocity as a function of displacement,

$$\tfrac{1}{2}kA^2 = \tfrac{1}{2}kx^2 + \tfrac{1}{2}mv^2$$

which we use to solve for velocity at any displacement x,

$$v^2 = \frac{k}{m}(A^2 - x^2) \tag{15.15}$$

Remember that k/m is equal to ω^2; so the magnitude of the velocity is given by

$$v = \omega\sqrt{A^2 - x^2} \tag{15.16}$$

The maximum velocity occurs as the object passes through its equilibrium position, $x = 0$,

$$v_{max} = \pm\omega A \tag{15.17}$$

This is the same result we derived earlier from Equation 15.7.

EXAMPLE

A 0.500-kg mass is vibrating in a system in which the restoring constant is 100 N/m; the amplitude of vibration is 0.200 m. Find

a. the energy of the system
b. the maximum kinetic energy and maximum velocity
c. the PE and KE when $x = 0.100$ m
d. the maximum acceleration
e. the equation of motion if $x = A$ at $t = 0$

SOLUTIONS

a. If $k = 100$ N/m and $A = 0.200$ m, then the total energy of the system $= \frac{1}{2} kA^2$

$E = \frac{1}{2}(100)(0.200)^2 = 2.00$ J

b. Maximum KE $=$ energy of system $= 2.00$ J. To find the maximum velocity let

$KE = \frac{1}{2} mv_{max}^2 = \frac{1}{2}(0.500) \, v_{max}^2 = 2.00$ J

Then $v_{max}^2 = 8.00$; so, $v_{max} = 2.83$ m/sec.

c. The total energy $= \frac{1}{2} kx^2 + \frac{1}{2} mv^2$. At $x = 0.100$,

$PE = \frac{1}{2}(100)(0.100)^2 = 0.500$ J

and

$KE = $ total energy $- PE = 2.00 - 0.500 = 1.50$ J

d. The maximum acceleration $a_{max} = |-\omega^2 A|$ and $\omega^2 = 100/0.500 = 200$ sec^{-2}.

$a_{max} = (200)(0.200) = 40.0$ m/sec^2

e. If $x = A$ and $t = 0$, then $\phi = 0$. Using the equation $x = A(\cos \omega t + \phi)$ where $\omega = \sqrt{k/m} = \sqrt{200} = 14.1$ 1/sec, $x = 0.200 \cos 14.1t$ m.

In summary, the universality and the simplicity of simple harmonic motion is very appealing. The total energy of a SHM system is conserved. From the dynamics of the system, that is, the force constant and mass, all of the properties of the system can be calculated if the maximum displacement is known. From the kinematics of the system, the angular speed, and the amplitude, all the properties of the system can be calculated if the mass is known. Then if the position of the object is known at any time, the system is completely determined. We have summarized the relationships for a SHM system in Table 15.1.

TABLE 15.1
Summary Table for Simple
Harmonic Motion

Physical Quantity	Expressed in Terms of Displacement x	Expressed in Terms of Time t
Displacement (x)		$x = A \cos(\omega t + \phi)$
Velocity (v)	$v = \pm \sqrt{k/m} \sqrt{A^2 - x^2} = \omega \sqrt{A^2 - x^2}$	$v = -\omega A \sin(\omega t + \phi)$
Acceleration (a)	$a = -\left(\dfrac{k}{m}\right) x = -\omega^2 x$	$a = -\omega^2 A \cos(\omega t + \phi)$
Restoring force (F_R)	$F_R = -kx$	$F_R = -kA \cos(\omega t + \phi)$
Potential energy	$PE = \frac{1}{2} kx^2$	$PE = \frac{1}{2} kA^2 \cos^2(\omega t + \phi)$
Kinetic energy	$KE = \frac{1}{2} k(A^2 - x^2)$	$KE = \frac{1}{2} kA^2 \sin^2(\omega t + \phi)$
Frequency (f)	$f = 1/2\pi \left(\sqrt{-\dfrac{a}{x}}\right) = 1/2\pi \left(\sqrt{\dfrac{k}{m}}\right)$	$f = \omega/2\pi$
Period (T)	$T = 2\pi \sqrt{-x/a} = 2\pi \sqrt{m/k}$	$T = 2\pi/\omega$

FIGURE 15.6
Simple harmonic motion showing
damping.

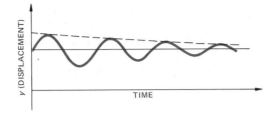

FIGURE 15.7
The degrees of damping: (a) underdamped, (b) critically damped, and
(c) overdamped SHM.

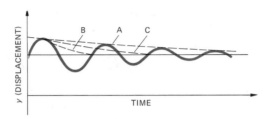

15.5 *Damping, Natural Frequencies, and Resonance*

In real systems simple harmonic motion is subject to energy loss due
to friction. This energy loss process is called *damping*, and it is characterized by decreasing amplitude as shown in Figure 15.6. The degree of
damping may be classed as underdamped (curve A), critically damped
(curve B), or overdamped (curve C) as illustrated in Figure 15.7. The
critically damped case corresponds to the most rapid return to equilibrium for a damped system. Critical damping is the desired situation
for such systems as car suspensions, motor mounts on equipment, and
meter movement in electrical instruments.

A Hooke's law model for materials means that the molecular restoring
forces in the material are linear. We might then expect a material to
oscillate at some frequency if we can displace its molecules. Three common displacements for solids, the longitudinal stretch, the longitudinal
compression, and the shear stress are shown in Figure 15.8. The dis-

FIGURE 15.8
Stretch, compression, and shear
displacements in solids.

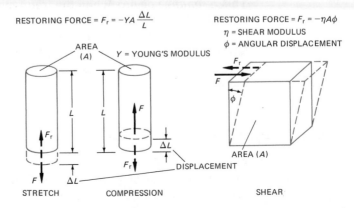

placement forces can be related to the elastic constants of the material (see Chapter 13). For each of these displacements our understanding of SHM predicts a *natural frequency* that is determined by the force constant and the inertial parameter for the displacement involved. We write Equation 15.8 in the following form:

$$\text{natural frequency} = f_0 = \frac{1}{2\pi} \sqrt{\frac{\text{restoring force constant}}{\text{inertial parameter}}} \qquad (15.18)$$

If we excite a compressional vibration in a solid of cross-sectional area A, the natural frequency will be related to Young's modulus and the linear density of the material,

$$f_0 = \frac{1}{2\pi} \sqrt{\frac{\text{Young's modulus}}{(\text{density})(\text{area})}} = \frac{1}{2\pi} \sqrt{\frac{\text{Young's modulus}}{\text{linear density}}} \qquad (15.19)$$

From Tables 8.1 and 13.1, we can obtain the values of density, Young's modulus, and the shear modulus of some materials. From these data we can calculate the natural frequencies of special configurations of these materials.

EXAMPLE

Using the data from Tables 8.1 and 13.1, find the ratio of the natural frequency of bone to that of aluminum for samples of the same geometry.

$$Y_{\text{bone}} \simeq 10^{10} \text{ N/m}^2 \qquad \rho_{\text{bone}} = 1850 \text{ kg/m}^3$$

$$Y_{\text{Al}} = 7.0 \times 10^{10} \text{ N/m}^2 \qquad \rho_{\text{Al}} = 2700 \text{ kg/m}^3$$

Since the geometry of the samples is the same, geometrical factors involved in the inertial parameter in Equation 15.19 will cancel out when we take a ratio. Thus we can write the following equation:

$$f_{\text{Al}}/f_{\text{bone}} = \sqrt{\frac{Y_{\text{bone}}/\rho_{\text{bone}}}{Y_{\text{Al}}/\rho_{\text{Al}}}} \simeq \sqrt{\frac{10^{10} \times 2700}{1850 \times 7.0 \times 10^{10}}} \simeq 0.45$$

When a solid system is subjected to an external periodic force, the Hooke's law model predicts *resonance* just as outlined in Chapter 2. This *resonance* occurs when the frequency of the external force equals a natural frequency of the system. Damping will tend to dissipate the energy into frictional heat and thus decrease the energy that is transferred to the vibration of the solid in resonance. Damping can be used effectively to prevent the destruction of mechanical systems by resonance oscillations.

15.6 Superposition of Simple Harmonic Motions

Any periodic motion, regardless of its complexity, can be reduced to the sum of a number of simple harmonic motions by the application of the superposition principle as discussed in Chapter 2. The resultant dis-

FIGURE 15.9
Superposition of two simple harmonic motions in the same direction.

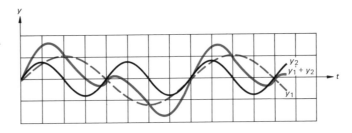

FIGURE 15.10
Blood pressure as function of time.

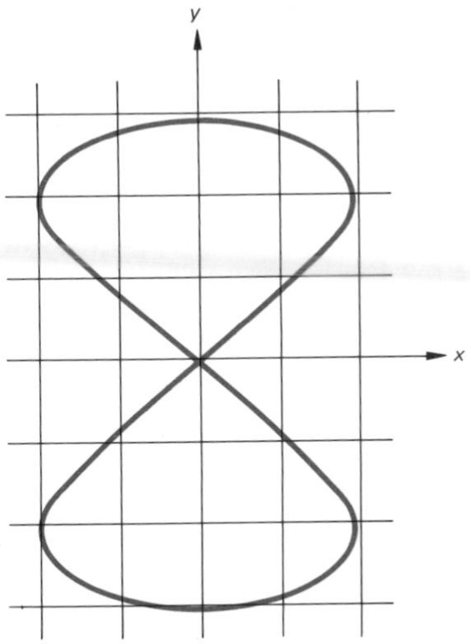

placement of a particle at any time t is the vector sum of the separate displacements of the various natural frequency motions (Figure 15.9). These component motions are the normal modes of the system, and this analysis of complex vibrations in terms of normal modes is an example of Fourier analysis. For example, a person's blood pressure is periodic and may have a tracing like that shown in Figure 15.10. The curve is equivalent to the sum of a number of simple harmonic motions. Using modern computer assisted techniques, we can analyze this curve into component parts. A change in the relative contributions can be used in the diagnosis of various heart conditions. Other examples of periodic phenomena that may be used in medical diagnoses are electrocardiograms, respiration graphs, gastric motility tracings, and ballistocardiograms.

FIGURE 15.11
Lissajou's figure formed by the superposition of two simple harmonic motions at right angles to each other.

EXAMPLES

1. Consider a particle that is subject to two simple harmonic motions given by $y_1 = 6 \sin \pi t$ and $y_2 = 4 \sin 2\pi t$. Plot $y = y_1 + y_2$. What is the displacement of the particle? See Figure 15.9.

2. Find the resultant motion if a body is subjected to the following SHM: $x = 2 \sin 2\pi t$, $y = 3 \sin \pi t$.

    Figure 15.11 shows the resulting motion. This is an example of the patterns called Lissajou's figures that are produced by the superposition of two simple harmonic motions at right angles. These patterns are periodic, and the period is an integral multiple of the periods of the two basic simple harmonic motions.

ENRICHMENT
15.7 Taylor's Theorem

For many systems that have an equilibrium location it is possible to find the potential energy as some function of the displacement of the system from equilibrium,

$$PE = V(x) \tag{15.20}$$

where $V(x)$ is some potential energy function that can be written in terms of the displacement x from equilibrium. From the definition of work and its relationship to force, we can show that the force which gives the potential energy function $V(x)$ is

$$F(x) = -\frac{dV(x)}{dx} \tag{15.21}$$

Taylor's theorem asserts that if $V(x)$ is a continuous and differentiable function, we can write a series of terms that will equal the function if we take enough terms. The form of the Taylor series is shown below,

$$V(x) = a_0 + a_1 x + a_2 x^2 + a_3 x^3 + a_4 x^4 + \cdots + a_n x^n + \cdots \tag{15.22}$$

where the a_n's are constants. We can use this Taylor series to find the form of the force for our system by differentiating the series

$$F(x) = -V'(x) = -a_1 - 2a_2 x - 3a_3 x^2 - 4a_4 x^3 - \cdots - na_n x^{n-1} + \cdots$$

But at equilibrium, all the forces must add to zero, and since $x = 0$ is our equilibrium location: $F(0) = 0 = -a_1$, so $a_1 = 0$.

$$F(x) = -2a_2 x - 3a_3 x^2 - 4a_4 x^3 - \cdots - na_n x^{n-1} \tag{15.23}$$

If we choose x much smaller than 1, then x^2 is much, much smaller than x; and x^3 is much, much, much smaller than x^2, and so forth. Therefore, we can find a small value of x, close to $x =$ zero, the equilibrium location, where the first term of the series for $F(x)$ is much larger than all the other terms,

$$F(x) \simeq -kx$$

a Hooke's law force. Clearly for this to be correct the constant a_2 must be a positive number, a situation that exists for all potential functions near *stable* equilibrium. Think of the scope of this result: all objects in stable equilibrium will perform simple harmonic motion if displaced only slightly from their equilibrium positions. This is a result that applies as well to planets performing small oscillations about their stable orbits around the sun as it does to molecules vibrating about their equilibrium sites in a crystal.

15.8 Simple Pendulum

We want to examine the motion of a simple pendulum, a point mass m suspended by a weightless, frictionless thread. Under what conditions does a simple pendulum exhibit simple harmonic motion? To demonstrate that a system shows SHM, we need to show only that the force that tends to restore the system to equilibrium is of Hooke's law form. Consider the simple pendulum shown in Figure 15.12. For a displacement of θ from equilibrium position, we consider the forces acting on the mass—the weight of the mass and the tension in the string. The tension in the string is always perpendicular to the direction of motion, and the weight of the ball is always vertical. The component of the weight parallel to the direction of motion is the restoring force, and this component is given by $mg \sin \theta$. The restoring force is always directed toward the equilibrium position. However, it is not of the Hooke's law form but rather

$$F = -mg \sin \theta \tag{15.24}$$

Let us recall our result from the previous section. It is only close to equilibrium that we can be sure of finding the Hooke's law force, as the following table of displacements θ shows.

$\theta(°)$	θ (rad)	$\sin \theta$
0.000	0.0000	0.0000
3.000	0.05236	0.05233
6.000	0.1045	0.1047
9.000	0.1571	0.1564
12.000	0.2094	0.2079
15.000	0.2618	0.2588
18.000	0.3142	0.3090
21.000	0.3665	0.3584

Notice that the angle in radians is almost exactly equal to the sine of the angle for small angles; $\sin \theta \cong \theta$ (rad) for $\theta \ll 1$. Then the restoring force for a small angle becomes $F = -mg\theta$. The value of θ is given by x/l, and the restoring force then becomes $F = -mgx/l$. This is of the form $F = -kx$, where $k = mg/l$, which is the correct force relationship for simple harmonic motion. Earlier we found that $k/m = \omega^2$, which reduces to $g/l = \omega^2 = (2\pi f)^2 = (2\pi/T)^2$. Thus, the period for a simple pendulum is given

$$T = 2\pi \sqrt{\frac{l}{g}} \tag{15.25}$$

FIGURE 15.12
A simple pendulum. An analysis of the forces acting is shown. Note that the restoring force is proportional to the displacement for small angles.

In general, if a distortion of an elastic body is produced, and the restoring force or torque is proportional to the magnitude of the change, the system will execute simple harmonic motion. If a shaft is twisted through an angle θ, the restoring torque is of the form $\bar{\tau}=-k'\theta$ where k' is a constant depending upon the shear modulus and geometry of the shaft. From rotational dynamics we learned that $\bar{\tau}=I\alpha$. Equating these values we get

$$I\alpha=-k'\theta \quad \text{and} \quad \alpha=-\frac{k'}{I}\theta$$

and

$$\omega^2=\frac{k'}{I} \tag{15.26}$$

This is the basic relationship for a torsion pendulum and the equation of motion is

$$\theta=\theta_0 \cos(\omega t+0)$$

where

$$\omega=\sqrt{\frac{k'}{I}} \tag{15.27}$$

15.9 Calculus Derivations of SHM Relationships

If the displacement of a particle is represented by the equation $x=A\cos(\omega t+\phi)$, one can find the expression for the instantaneous velocity by finding the derivative relative to t. Thus,

$$v_x=\frac{dx}{dt}=-\omega A \sin(\omega t+\phi)$$

and

$$a_x=\frac{dv_x}{dt^2}=\frac{d^2x}{dt^2}=-\omega^2 A \cos(\omega t+\phi)$$

These are the kinematic equations for simple harmonic motion. The general force equation for simple harmonic motion is

$$m\frac{d^2x}{dt}=-kx \quad \text{or} \quad \frac{d^2x}{dt}+\frac{k}{m}x=0$$

If we let $\omega^2=k/m$, this reduces to

$$\frac{d^2x}{dt^2}+\omega^2x=0$$

Any system that has a force equation equivalent to this form will execute simple harmonic motion.

```
┌─────────────────────────────────────────────────────────┐
│ ┌───────────────────────────────────────────────────┐   │
│ └───────────────────────────────────────────────────┘   │
└─────────────────────────────────────────────────────────┘
```

SUMMARY

Use these questions to evaluate how well you have achieved the goals of this chapter. The answers to these questions are given at the end of this summary with the number of the section where you can find the related content material.

Definitions

Circle the correct answer(s) for each question.

1. Simple harmonic motion has the following characteristic(s):
   a. period
   b. linear restoring force
   c. natural frequency
   d. zero amplitude
   e. none of these

2. The period of a simple pendulum is equal to:
   a. time for one swing
   b. amplitude/velocity
   c. 1/frequency
   d. $2\pi \sqrt{m/k}$
   e. none of these

3. The phase angle of a SHM system where $x = A \cos(\omega t + \phi)$ at $t = 0$ is 90°. The correct form of this SHM as a function of time, is:
   a. $A \cos(\omega t + 90°)$
   b. $A \cos(\omega t + \pi/2)$
   c. $A \cos(\omega t - 90°)$
   d. $A \cos(\omega t - \pi/2)$
   e. $A \cos 90°$

4. Damping of periodic motion always results in:
   a. decreasing amplitude
   b. energy gain in vibration
   c. resonance
   d. energy loss to friction
   e. none of these

UCM and SHM

5. If the frequency of a 0.75-m simple pendulum is 1.5 Hz, the angular frequency on a corresponding reference circle is (in rad/sec):
   a. 1.5π
   b. 0.33π
   c. 3π
   d. 0.5π
   e. 2π

6. The expression for the radial acceleration on the reference circle can be expressed in terms of amplitude A, period T, and maximum speed v_{max} of the corresponding SHM as:
   a. $v_{max}A/T$
   b. v_{max}^2/A
   c. A/T^2
   d. v_{max}/T
   e. none of these

7. If you know the force constant of a spring is 100 N/m, the mass on the spring is 1 kg, and the amplitude is 0.04 m, then the period in seconds of the SHM is:
   a. 20π
   b. $\pi/5$
   c. 5π
   d. $5/\pi$
   e. $\pi/20$

8. For the system described in question 7, the maximum velocity (in m/sec) of SHM is:
   a. 100
   b. 0.04
   c. 0.4
   d. 4.0
   e. 0.004

9. For this same system (question 7) the maximum acceleration (m/sec²) of SHM is:
   a. 4
   b. 0.4
   c. 0.04
   d. 100
   e. 25

10. The minimum speed (m/sec) of SHM for the system of question 7 is
    a. 0.04
    b. 0.4
    c. 0.004
    d. 100
    e. 0

11. The maximum displacement for this system is:
    a. 0.4 m
    b. 0.04 m
    c. 0.004 m
    d. 100 m
    e. 9.8 m

12. The equation for displacement for this system could be written as
    a. $y = 0.04 \cos (10\pi t)$
    b. $y = 0.04 \cos (10t + \phi)$
    c. $y = 0.04 \sin (10t + \phi)$
    d. $y = 0.04 \sin (10\pi t)$

13. For the spring system in question 7 the maximum PE is
    a. 0.08 J
    b. 0.4 J
    c. 0.16 J
    d. 1.6 J
    e. 0.8 J

14. The kinetic energy is equal to the potential energy when the displacement is
    a. 0.04 m
    b. $0.04 \times \sqrt{2}$ m
    c. $0.04/\sqrt{2}$ m
    d. $\sqrt{2}$ m
    e. 0

Superposition Principle

15. If two vibrations of the same frequency are superimposed on a system with equal amplitudes, the maximum resultant amplitude could be
    a. zero
    b. $\sqrt{2}A$
    c. $A/\sqrt{2}$
    d. $2A$
    e. A

16. For the case given in the previous question, the minimum amplitude could be

    a. zero
    b. $\sqrt{2} A$
    c. $A/\sqrt{2}$
    d. $2A$
    e. A

Natural Frequencies

17. The natural frequency of a system is determined in analogy with a spring to be proportional to
    a. $\sqrt{\text{inertia parameter/force constant}}$
    b. $\sqrt{\text{force constant/inertia parameter}}$
    c. amplitude/period
    d. none of these

18. The natural frequency of rod A is $\sqrt{2}$ times the natural frequency of rod B for longitudinal vibrations. If the rods have the same geometry and $\rho_A = 2\rho_B$, then the ratio of Young's moduli for A to B is:
    a. $\sqrt{2}$
    b. 2
    c. 4
    d. $1/\sqrt{2}$
    e. $\frac{1}{2}$

Answers

1. a, b, c (Section 15.2)
2. c, d (Section 15.2)
3. b (Section 15.2)
4. a, d (Section 15.5)
5. c (Section 15.2)
6. b (Sections 15.2, 15.3)
7. b (Section 15.3)
8. c (Section 15.3)
9. a (Section 15.3)
10. e (Section 15.3)
11. b (Section 15.3)
12. b, c (Section 15.3)
13. a (Section 15.3)
14. c (Section 15.3)
15. d (Section 15.6)
16. a (Section 15.6)
17. b (Section 15.5)
18. c (Section 15.5)

ALGORITHMIC PROBLEMS

Listed below are the important equations from this chapter. The problems following the equations will help you learn to translate words into equations and to solve single-concept problems.

Equations

$$F = -kx = -m\omega^2 x \tag{13.3}$$

$$x = A \cos \theta \tag{15.1}$$

$$x = A \cos (\omega t + \phi) \tag{15.2}$$

$$a_x = -\omega^2 x \tag{15.6}$$

$$v_x = -\omega A \sin{(\omega t + \phi)} \tag{15.7}$$

$$a_x = -\omega^2 A \cos{(\omega t + \phi)} \tag{15.7}$$

$$\omega = 2\pi f = 2\pi n, \; f = n = \frac{\omega}{2\pi} = \frac{1}{2\pi}\sqrt{\frac{k}{m}} \tag{15.8}$$

$$T = \frac{1}{f} = \frac{2\pi}{\omega} = 2\pi\sqrt{\frac{m}{k}} \tag{15.9}$$

$$T = 2\pi\sqrt{-x/a} \tag{15.11}$$

$$f = \left(\frac{1}{2\pi}\right)\sqrt{-\frac{a}{x}} \tag{15.11}$$

$$PE = \tfrac{1}{2}kx^2 \tag{15.12}$$

$$E = \tfrac{1}{2}kA^2 = \tfrac{1}{2}mv^2 + \tfrac{1}{2}kx^2 \tag{15.13, 15.14}$$

$$v^2 = \left(\frac{k}{m}\right)(A^2 - x^2) = \omega^2(A^2 - x^2) \tag{15.15}$$

$$v_{\text{max}}{}^2 = \left(\frac{k}{m}\right)A^2 = \omega^2 A^2 \tag{15.16}$$

$$T = 2\pi\sqrt{\frac{l}{g}} \tag{15.25}$$

Problems

1. The force constant of a spring is 10 N/m. Find the period of a 100-g mass on the end of this spring.
2. Find the maximum velocity of the mass in problem 1 if the amplitude of oscillation is 2.0 cm.
3. Find the velocity of the mass in problem 2 when it is 1 cm from its equilibrium position.
4. A mass on the end of a spring is released from a point 2 cm from its equilibrium position. The frequency of oscillation is 4 Hz. Write the equation for the position of the mass as a function of time.
5. The period of a simple pendulum is 2.00 sec. Find the length of the pendulum.
6. Find the maximum energy stored in the spring of problem 1 when it is compressed 2 cm from its equilibrium position.

Answers

1. 0.63 sec
2. 20 cm/sec
3. 17 cm/sec

4. 0.02 cos $(8\pi t)$ m
5. 0.993 m
6. .002 J

EXERCISES

These exercises are designed to help you apply the ideas of a section to physical situations. When appropriate the numerical answer is given in brackets at the end of the exercise.

Section 15.2

1. Show in tabular form the sign of displacement, velocity, and acceleration for each quadrant, i.e., $0 < (\omega t + \phi) < \pi/2$, $\pi/2 < (\omega t + \phi) < \pi$, $\pi < (\omega t + \phi) < 3\pi/2$, $3\pi/2 < (\omega t + \phi) < 2\pi$.

2. A body is executing simple harmonic motion, and the displacement is given by $x = 5 \cos 3\pi t$. Plot the displacement, velocity, and acceleration for two complete periods.

3. If simple harmonic motion has an amplitude of 10 cm, what is the maximum and minimum change in position in one-fourth of the period?   [14 cm, 0 cm]

4. A body is vibrating with simple harmonic motion of amplitude 15 cm and a frequency of 2.0 Hz.
   a. What is its maximum acceleration and maximum velocity?
   b. What is its acceleration and velocity for a displacement of 12 cm?
   c. How long does it take to go from equilibrium position to a displacement of 9.0 cm?   [a. $240\pi^2$ cm/sec^2, 60π cm/sec; b. $-192\pi^2$ cm/sec^2, $\pm 36\pi$ cm/sec; c. 0.051 sec]

5. A particle is moving along the x-axis in accordance with the following:

   $x = 5 \cos (4\pi t + \pi/3)$ cm

   What is the
   a. amplitude of motion?
   b. period of motion?
   c. frequency of motion?
   d. time for $x = 0$?
   e. time for velocity to be 0?
   f. time for acceleration to be a maximum?   [a. 5 cm; b. $\frac{1}{2}$ sec; c. 2 Hz; d. $\frac{1}{24}$ sec, $\frac{7}{24}$ sec, $\frac{13}{24}$ sec; e. $\frac{1}{6}$ sec, $\frac{5}{12}$ sec; f. same as e]

Section 15.3

6. Assume the piston in an automobile engine is executing simple harmonic motion. The length of the stroke (double the amplitude) is 10 cm, and the engine is running at 300 rpm.
   a. What is the acceleration at the end of the stroke?
   b. If the piston has a mass of 0.50 kg, what is the maximum restoring force acting on the piston?
   c. What is the maximum velocity of the piston?
   d. What is the position of the piston as a function of time if it is at the top of the stroke at $t = 0$?   [a. $500\pi^2$ cm/sec^2; b. $2.5\pi^2$ N; c. 50π cm/sec; d. $x = 5 \cos 10\pi t$ cm]

Section 15.4

7. A 2.0-kg mass is attached to a spring with a force constant of 98 N/m. The mass is resting on a frictionless horizontal plane as a horizontal force of 9.8 N is applied to the mass, and it is then released.
   a. What is the amplitude of SHM?
   b. What is its period?
   c. What is the total energy of the SHM?
   d. What is the maximum velocity of the vibrating mass?
   e. What is the PE and KE of the mass when it is 5 cm from equilibrium position?   [a. 0.10 m; b. $2\pi/7$ sec; c. 0.49 J; d. 0.7 m/sec; e. PE = 0.12 J, KE = 0.37 J]

Section 15.5

8. Compare the natural frequencies of longitudinal vibrations for rods of the same geometry made of steel and aluminum. Use values from Table 8.1 and 13.1. [$f_{steel}/f_{Al} = .99$]

Section 15.6

9. A particle is subjected simultaneously to two simple harmonic motions of the same frequency and direction in accordance with the following equations:

   $y_1 = 6 \sin \omega t$ cm

   $y_2 = 8 \sin (\omega t + \pi/3)$ cm

   $\omega = 2\pi$

   Find the amplitude of the resultant motion, and show it in graphical form.   [12.17 cm]

10. A particle is subjected simultaneously to two simple harmonic motions in the same direction in accordance with the following equations:

    $y_1 = 8 \sin 2\pi t$

    $y_2 = 4 \sin 6\pi t$

    Show graphically the resultant path of the particle.

11. A particle is simultaneously subjected to two simple harmonic displacements at right angles to each other. Show the pattern of the particle in the xy plane. The displacements are:

    $x = 5 \cos \pi t$

    $y = 3 \sin (3\pi t)$

PROBLEMS

Each problem may involve more than one physical concept. A problem requiring material from the enrichment section is marked with a dagger †. The answer is given in brackets at the end of the problem.

12. A frame with seats, which weighs 15,680 N, is mounted on springs, and it is executing vertical simple harmonic motion with a frequency of 4 Hz.
    a. What is the force constant of this system of springs?
    b. If four students of total mass 260 kg are seated in the frame, what is the frequency of oscillation? [a. 1.01×10^6 N/m; b. 3.71 Hz]

†13. In a laboratory experiment a student observes that two simple pendulums of different lengths have periods that differ by 10 percent. The student then asks, "What is the percentage difference in length?" What is the answer? Can you make a general statement? [21 percent]

†14. A pendulum clock should have a period of 2 sec to keep accurate time. Observation shows that the clock loses 10 min per day. Assume that the pendulum system behaves as a simple pendulum. What changes should be made and how much? [decrease length of pendulum by 1.38 cm]

†15. The acceleration due to gravity on the moon is about one-fifth that on the earth. What is the period of a simple pendulum on the moon, if it has a period of 2 sec on the earth? [4.5 sec]

16. A spring driven clock has a period of 2 sec on earth. What is the period on the moon? [2 sec]

†17. A simple pendulum can be used to determine experimentally the acceleration due to gravity. What is the acceleration due to gravity at the place where a 1-m simple pendulum has a period of 2 sec? [$g = \pi^2$ m/sec^2]

18. An arrow of length 0.6 m is rotated about a vertical axis through its tail with an angular velocity of π rad/sec. The motion of the tip of the arrow is to be projected upon a diameter of the circle. Find the following:
    a. the amplitude of its motion
    b. its period
    c. the maximum acceleration
    d. the maximum velocity
    e. the acceleration and velocity when the arrow tip is 0.3 m from the center of its projected path. [a. 0.6 m; b. 2 sec; c. 5.92 m/sec^2; d. 0.6π m/sec; e. 2.96 m/sec^2, 1.63 m/sec]

†19. A child's swing has a period of 5.0 sec and an amplitude of 1 m.
    a. What is the angular speed of an imaginary particle moving in a reference circle to represent this vibration?
    b. What is the maximum speed of the swing seat?
    c. What is the length of rope from point of support to the seat? [a. 0.4π rad/sec; b. $4\pi/10$ m/sec; c. 6.21 m]

20. Assume a person's heart is executing SHM with 66 beats per minute. If its amplitude is 3 mm, what is its maximum velocity and acceleration? [2.1 cm/sec, 14.3 cm/sec^2]

†21. A pendulum bob consists of a 1.00-kg ball hung on a string 3m long. If it is drawn back 20 cm from equilibrium position and released, what is its period of motion? What will be its kinetic energy as it passes through the middle of its swing? What are its PE and KE for a displacement of 10 cm? [3.5 sec, 0.685 J, PE = 0.196 J, KE = 0.489 J]

†22. Suppose a force $F = F_0 \cos \omega t$ is applied to a spring-mass system with a natural frequency ω_0. The equation of motion for this system will be

$$F_0 \cos \omega t - kx = m \frac{d^2x}{dt^2}$$

Show that a solution for this equation is:

$$x = \frac{F \cos \omega t}{m(\omega_0{}^2 - \omega^2)}$$

Note that as $\omega \to \omega_0$, the amplitude becomes very large. This is characteristic of undamped resonant oscillations.

GOALS

When you have mastered the contents of this chapter, you will be able to achieve the following goals:

Definitions
Define each of the following terms as it is used in physics, and use the term in an operational definition:

frequency	refraction
wavelength	superposition principle
amplitude	interference
longitudinal wave	diffraction
transverse wave	standing wave
phase	Fourier's theorem
intensity	dispersion
reflection	

Wave Forms
Sketch a longitudinal wave and a transverse wave.

Wave Problems
Solve wave problems involving the relationships that exist between the different characteristics of waves.

Superposition and Fourier's Theorem
Use the superposition principle and Fourier's theorem to explain the wave form of a complex wave.

Standing Waves
Use the superposition principle to explain the formation of standing waves in different situations.

Inverse Square Law
Use the inverse square law to calculate the intensity of a wave emanating from a point source.

PREREQUISITES

Before beginning this chapter you should have achieved the goals of Chapter 5, Energy, Chapter 13, Elastic Properties of Materials, and Chapter 15, Simple Harmonic Motion.

16

Traveling Waves

16.1 Introduction

Have you ever taken the loose end of a rope tied to a fence and given it a shake? Have you ever dropped a stone into a quiet lake? Have you ever shaken the dirt out of a rug? Have you ever given a towel a flip? Each of these is a way of getting energy to travel from one place to another in the form of undulatory, or wave, motion.

Many of your interactions with your environment take place in conjunction with wave motion. Your senses of hearing and sight involve the detection of pressure and electromagnetic traveling waves. In this chapter you are introduced to the quantitative treatment of traveling waves and their associated phenomena.

16.2 Waves

We have previously considered energy transport as it occurs in direct contact interactions with matter. The baseball possesses kinetic energy and makes direct contact with the bat. The gas molecules carry their kinetic energy into a direct contact interaction with the piston of a heat engine. In this chapter we return to the interaction-at-a-distance model and consider wave motion. Energy from a stone dropped into the lake is transported to a floating stick by the water waves generated by the stone. The speaker system of a stereo transfers energy to your ears through the sound waves it generates with its vibrations. The energy we receive from the sun is transported in the form of electromagnetic waves.

A water wave is characterized by its crests and troughs moving past the floating stick. Every form of wave can be characterized by a change in a physical variable that is propagated through space. The wave model calls for energy propagation as a result of a succession of oscillations (crests and troughs) at neighboring points in space. The wave model requires no net transfer of matter for its energy transportation.

For example, in sound, the physical variable undergoing oscillation is the pressure. Sound is referred to as a matter wave, because the vibratory motion of matter, i.e., molecules, is involved in its propagation. Light is referred to as an electromagnetic wave, not a matter wave. The physical variables oscillating in an electromagnetic wave are the electric and magnetic fields. Electromagnetic waves, unlike matter waves, can be propagated through a vacuum.

What are the characteristics of the wave model that distinguish it from the particle, or matter, transport model? The answer lies in the results of the superposition principle when applied to the wave model. You have confronted such results when you hear sound around the corner from the source of the sound. The color of an oil film on water is another example of the results of superimposed waves. The formal designations of these unique superposition results are *interference* and *diffraction*.

16.3 Longitudinal and Transverse Waves

The two different types of waves are illustrated in Figure 16.1. The *longitudinal wave* consists of oscillations parallel to the propagation direction. In a child's toy called a Slinky the wave vibrating parallel to its axis is longitudinal. A *transverse wave* consists of oscillations perpendicular to the direction of propagation. The wave on a vibrating violin string is an example of a transverse wave. Sound and light are understood as *longitudinal* and *transverse* waves respectively.

16.4 Wave Propagation

Consider what happens if you pick up a long rope and start shaking the end of the rope by moving your hand up and down. The shape of the rope for the first few positions of your hand is shown in Figure 16.2, which shows you moving your hand back and forth with an amplitude of motion A and a frequency of one complete cycle every $4t$ seconds. Notice that the waves traveling down the rope have a characteristic wavelength λ.

After you have been shaking the rope for several complete cycles of oscillation, the shape of the rope at some instant in time, t_0, may be given by Figure 16.3a, and the displacement of the rope at some position, x_0, may be given by Figure 16.3b.

Note that the *amplitude* or displacement is repeating itself after each wavelength of distance. The *wavelength* λ of a wave is defined to be the period of the wave in space. Likewise, if you imagine watching one point of the rope you will observe that this point reaches maximum amplitude periodically in time. This time period is the time required for the wave to make one complete oscillation. Thus we have the following relation between period (time) and *frequency* of a wave:

$$f = \frac{1}{\text{period (seconds)}} = f\left(\frac{\text{cycles}}{\text{second}}\right) = f(\text{hertz}) \tag{16.1}$$

We now combine these two periodic aspects of the wave motion to give us a useful wave equation. The distance that the traveling wave

FIGURE 16.1
Longitudinal and transverse waves.

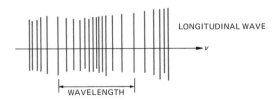

LONGITUDINAL WAVE

WAVELENGTH

WAVELENGTH   TRANSVERSE WAVE

FIGURE 16.2
Propagating a wave train along a rope.

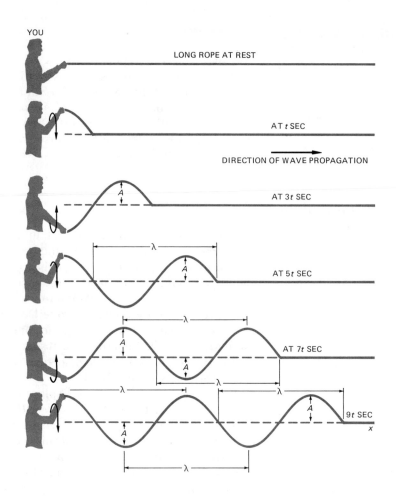

FIGURE 16.3
(a) The shape of the rope shown as vertical displacement for various horizontal locations at time t_0.
(b) The motion of the rope at a fixed location in the horizontal direction x_0 shown as the vertical displacement as a function of time.

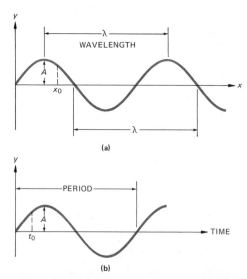

moves during one time period is one wavelength. The speed of the wave is given by the distance traveled divided by the time or

$$v_{wave} = \frac{\lambda}{period} = \lambda f \ (m/sec) \tag{16.2}$$

EXAMPLE

A string has a wave traveling along it with a speed of 50 m/sec at a frequency of 100 Hz. Find the wavelength of the wave. From Equation 16.2 we see that

$$\lambda = \frac{v}{f} = \frac{50 \text{ m/sec}}{100 \text{ Hz}}$$

Thus, $\lambda = 0.50$ m.

16.5 Phase

Assume that your hand which is shaking a rope is executing simple harmonic motion, up and down with a constant amplitude and period. Then an equation for the location of your hand y is given by

$$y = A \ \sin \omega t = A \ \sin 2\pi ft \tag{16.3}$$

where A is the amplitude (m), ω is the angular frequency (rad/sec), and t is the time (sec).

If you study the displacement of a segment of rope located a distance x from the source of the oscillations (your hand), you will notice that the rope is executing the same type of motion as the source except for the phase of the motion, where *phase* is defined as a fraction of a complete cycle from a specified reference, often expressed as an angle. You see in Figure 16.2 that the segment of the rope located a distance of λ from your hand is moving in phase with your hand. In general, the phase difference between two points along the rope is given by $2\pi x/\lambda$. If the wave is traveling in the positive x direction, the phase of the particle at x will be behind the source, and the motion of the particle is given by

$$y = A \ \sin \left(2\pi ft - \frac{2\pi x}{\lambda}\right) \tag{16.4}$$

Show that $y = A \ \sin (2\pi ft + 2\pi x/\lambda)$ is a wave traveling in the negative x direction.

16.6 Energy Transfer

Traveling waves transfer energy from one place to another. For matter waves we can develop an equation for the energy carried by the wave by using our understanding of simple harmonic motion. Waves in matter can be compared to waves propagated along a spring. The elastic properties of the medium play the same role as the spring constant, and the medium has an inertial property that is analogous to the mass on the end of the spring. The energy per cycle of a traveling wave is given by the maximum potential energy of a spring system (Equation 15.14)

$$E = \tfrac{1}{2} kA^2 \tag{15.14}$$

where k is equal to $(2\pi f)^2 m$ and m is the inertial property of the medium,

$$E = \tfrac{1}{2}(2\pi f)^2 mA^2 = 2\pi^2 f^2 mA^2 \tag{16.5}$$

The power transmitted by the wave is equal to the product of the frequency times the energy per cycle. Thus,

$$P = fE = f(2\pi^2 f^2 mA^2) = 2\pi^2 f^3 mA^2 \tag{16.6}$$

The *intensity* of a wave is a measure of the energy transported by the wave through a unit area in one second, or power per unit area. Hence,

$$I = \frac{P}{\text{area}} = \frac{(2\pi^2 f^2 mA^2)}{\text{area}} \cdot \frac{v}{\lambda} = 2\pi^2 f^2 A^2 \left(\frac{m}{\text{area } \lambda}\right) v$$

and this can be expressed as follows for a gas, where

$$\frac{m}{\text{area } \lambda} = \rho$$

the density of the gas. Thus, we can rewrite I in terms of other variables

$$I = 2\pi^2 f^2 A^2 v\rho \tag{16.7}$$

which is an expression for the intensity of a compression (longitudinal) wave in a gas. You will notice that the units for this equation are consistent with our previous definition of power. Also, we can see that the intensity of a wave is proportional to the square of its amplitude from

$$I = \frac{\text{energy transmitted}}{\text{area} \times \text{time}} = 2\pi^2 f^2 A^2 v\rho \propto (\text{amplitude})^2 \tag{16.8}$$

The transfer of energy from one medium to another via waves is maximized when the inertial properties of the two media are matched at their interface. This can be demonstrated by tying two different strings together. The energy transmitted from one string to the next is determined by the linear densities (mass/meter) of the two strings. The more closely the linear densities are matched, the more efficient will be the energy transferred from the first to the second string.

EXAMPLE

A wave of 100 Hz traveling at a speed of 50 m/sec has an amplitude of 2 cm. Find the power transported by the wave in a string of linear density 20×10^{-3} kg/m.

$$P = 2\pi^2 \times (100 \text{ Hz})^2 \left(\frac{m}{\lambda}\right) vA^2 \quad \text{where} \quad \frac{m}{\lambda} = 20 \times 10^{-3} \text{ kg/m}$$

$$= 2\pi^2 \times (100 \text{ Hz})^2 (2 \times 10^{-2} \text{ kg/m}) (50 \text{ m/sec}) (4 \times 10^{-4} \text{ m}^2)$$

$$= 2\pi^2 \times 2 \times 50 \times 10^{-2} \text{ watts}$$

$$= 2\pi^2 \text{ watts}$$

FIGURE 16.4
A pulse is sent down a rope fastened at the other end. The wave is reflected at the fixed end and returns with a 180° phase change.

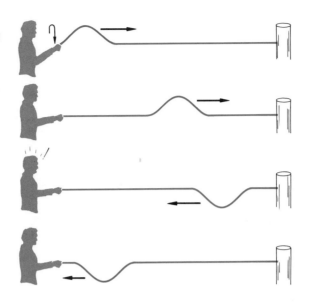

16.7 Reflection and Refraction

If you send a pulse down a string that is tied to a post, you will observe that the pulse comes back to you (Figure 16.4). We say that the pulse is reflected at the post. In this case you will also note that the pulse is inverted upon reflection. This is called a *change in phase*. The incoming pulse is called the incident pulse. The return pulse or wave is known as the reflected pulse. Whenever the inertia parameter increases at an interface, the reflected wave will be one-half wavelength out of phase with the incident wave. For example, at a knot connecting a light rope with a heavy rope the wave traveling from the light to the heavy rope will be reflected exactly out of phase at the knot. A wave going from larger to smaller density undergoes no phase change. *Reflection* is a phenomenon that is common to all waves at the interface between two media. Using the conservation of energy, can you write an equation relating the percent of energy reflected and the percent of energy transmitted at an interface?

A general rule governing reflection is stated as follows: *The angle of incidence equals the angle of reflection.* (Figure 16.5).

$$\angle i = \angle r \tag{16.9}$$

While reflection is a property of wave motion, it does not distinguish between wave motion and the motion of particles. A tennis ball thrown against a wall is also reflected with the incident angle equal to the reflection angle in the idealized case where friction and spin are neglected.

Another common wave phenomenon occurring at an interface is *re-*

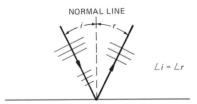

NORMAL LINE

$Li = Lr$

FIGURE 16.5
Reflection of a wave from a plane surface. The angle of incidence equals the angle of reflection, both measured from the normal to the interface.

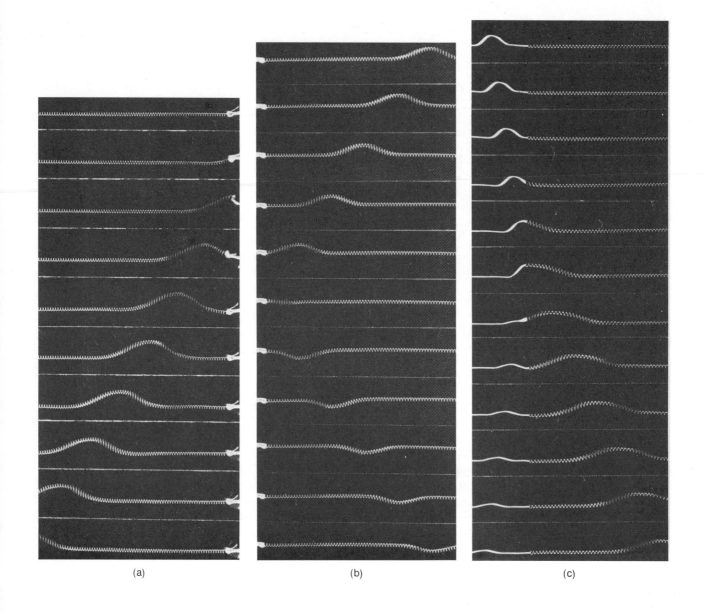

(a)   (b)   (c)

(a) A pulse is generated at the right end of the spring and travels to the left along the spring. Note the displacement of the pulse as a function of time.

(b) A pulse sent down a spring is reflected from a fixed end. The reflected pulse returns in opposite displacement, i.e., 180° out of phase.

(c) A pulse is passing from a heavy spring (left) to a light spring. At the junction the pulse is partially transmitted and partially reflected. Note that the reflected pulse is in same sense as the incident pulse; there is no change in phase. The transmitted pulse continues in the same phase as incident pulse.

(d) A pulse is passing from a light spring (right) to a heavy spring. At the junction the pulse is partially transmitted and partially reflected. Note that the reflected pulse

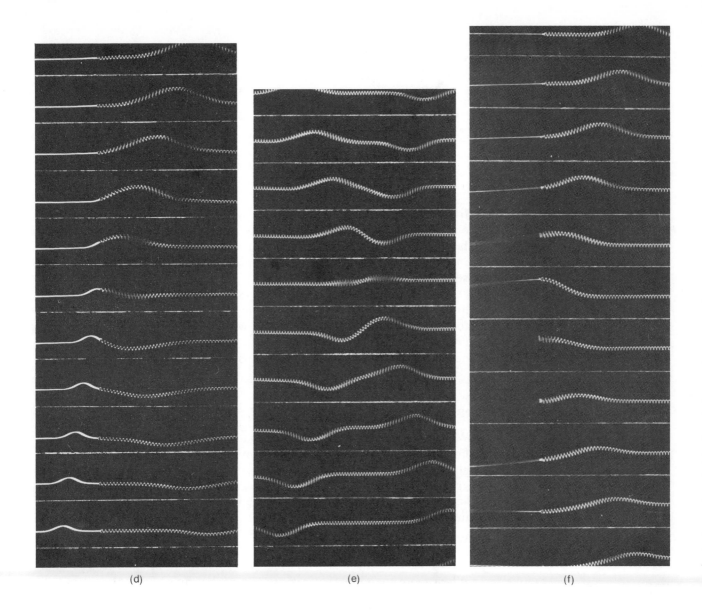

(d)   (e)   (f)

is in the opposite sense to the incident pulse; there is a change in phase of 180°. The transmitted pulse continues in the same phase as incident pulse.

(e) Two pulses are sent down a spring from the ends. Note that when they come together in the opposite sense, they cancel each other. This is an example of the principle of superposition.

(f) A pulse is reflected from the interface between a spring and a light thread. The top pictures show the incident pulse, and the lower pictures show the reflected pulse. Note that there is no change in phase upon reflection. (Picture from *PSSC Physics*, D.C. Heath and Company, Lexington, Mass., 1965.)

FIGURE 16.6
Refraction is the bending of the wave as it passes from one medium to another in which the speed of the wave is different.

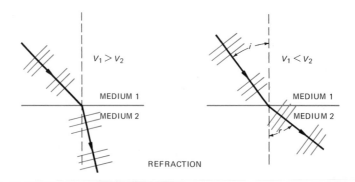

REFRACTION

fraction. Refraction refers to the bending of the wave as it moves from one medium to another. This bending or change in direction at the interface of two media takes place because the wave has different velocities in the two media. List some examples of refraction that you have observed.

The wave in the second medium is bent toward a line perpendicular to the interface when the wave velocity is less in the second medium than in the first medium. When the wave velocity is greater in the second medium than it is in the first medium, it is bent away from this normal line. These two cases are illustrated in Figure 16.6.

16.8 Superposition of Waves

When two or more waves are propagated in the same region of space, we find the *superposition principle* provides us with proper results. The algebraic sum of all amplitudes present at a given point at any instant gives the correct amplitude of the resultant wave.

The superposition principle for two waves on a string is shown in Figure 16.7.

Wave C = Wave A + Wave B

The effect of each wave is independent of the other waves present. When more than one wave is present in the same place, the composite wave displacement is the algebraic sum of the wave displacements at a given place and time. This algebraic sum of individual wave displace-

FIGURE 16.7
A triangular wave and a sine wave added together according to the superposition principle.

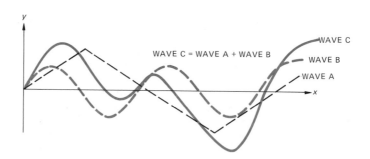

ments depends upon the frequency and phase relations among the waves. Many important consequences of the superposition of waves occur when the waves involved have equal frequencies (or multiples of the same frequency) with fixed phase relationships.

Such conditions result in the unique wave phenomena of interference, diffraction, and standing waves.

16.9 Interference

Suppose that we have two waves of equal frequency and amplitude in the same phase, that is, crest to crest and trough to trough, traveling in the same direction. The displacement of any particle is then the sum of the separate displacements, and the resultant amplitude is twice the amplitude of either. See Figure 16.8a. This is called *constructive interference*. Now, suppose the two waves of the same frequency and amplitude, and traveling in the same direction, are 180° out of phase — that is, crest opposite trough. The net displacement of a particle under these

FIGURE 16.8
Superposition of two waves. (a) When the waves are in phase, the resultant wave shows constructive interference. (b) When they are out of phase by 180° destructive interference occurs. (c) Interference patterns generated in water by two vibrating point sources of water waves. (From *PSSC Physics*, D. C. Heath and Company, Lexington, Mass., 1965.)

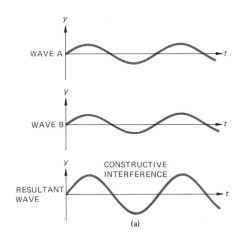

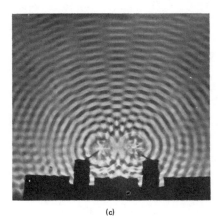

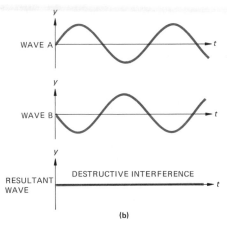

conditions is zero. This condition is called *destructive interference* (Figure 16.8b). We can summarize by saying that *destructive interference* is produced by waves from two coherent sources (phase relationships remain constant) of equal amplitude and frequency, and out of phase by 180° or π radians. In general, we see that two waves can combine to give an amplitude greater than either separate wave or they can combine to give zero amplitude. There is the complete range of possibilities between totally constructive and totally destructive interference. For another case, consider the constructive and destructive case of waves with amplitudes A and $\frac{1}{2}A$. Here the constructive case yields a maximum amplitude of $\frac{3}{2}A$ and the destructive case an amplitude of $\frac{1}{2}A$.

Interference phenomena are unique to waves. Particle models cannot account for experimental observations of interference in nature.

16.10 Diffraction

Diffraction is also uniquely a wave phenomenon. *Diffraction* is the result of the superposition of wave fronts produced when a wave encounters an object. Diffraction patterns are produced by the constructive and destructive interferences of coherent waves. Sound is diffracted when it passes through an open door, resulting in the bending of the sound down a hallway. Light shining through the same door casts a well-defined shadow on the wall. However, when light is incident on a narrow slit, it is bent and shows a diffraction pattern. The rule of thumb for significant diffraction phenomena may be stated as follows: *When a wave encounters an object about the same size as its wavelength, diffraction phenomena can be observed.*

The diffraction of water waves passing through a slit is shown in Figure 16.9. Christian Huygens provided a construction procedure in 1678 that enables us to picture diffraction as follows: each point on the front of a wave serves as a source of secondary waves (traveling with the wave speed in the medium). The superposition of secondary Huygens wavelets determines succeeding waves. In diffraction, Huygens wavelets are held back by obstacles, and the results are the bending of waves around

FIGURE 16.9
Diffraction. Patterns of destructive and constructive interference of waves are produced by waves passing through a slit.

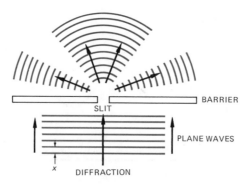

these objects. An opening appears to be a new wave source, and the smaller the opening the more the opening looks like a new point source of waves.

16.11 *Standing Waves*

Standing waves result from the superposition of two traveling waves of *equal frequency* moving in opposite directions (Figure 16.10). You will notice that at time t_2, a time later than t_1, the location of the places where the amplitude is zero has not changed. This fixed location of zero amplitude is called a *node*. The positions of maximum displacement, called *antinodes,* are halfway between adjacent nodes. The appearance of a wave traveling along the string has disappeared. So these waves are called *standing waves.* The energy remains localized in the form of kinetic and potential energies of the antinode regions of the medium. Standing waves result for a particular physical system at its *resonance conditions.* As you recall, resonance occurs when a system oscillates at the natural frequency of the system. Standing wave resonances occur when the wavelengths of the exciting frequencies match those allowed by the physical boundary conditions of the system. For example, a string fixed at both ends must have nodes at each end and such a string has resonant frequencies for the length of the string to be $\lambda/2$, λ, $3/2\,\lambda$, etc., where λ is the wavelength. Only certain frequencies of oscillation can produce standing waves in a given physical system. These frequencies will satisfy the relation $f = v/\lambda$, where v is the velocity of

FIGURE 16.10
Standing waves are created by the superposition of two waves of equal frequency moving in opposite directions in a medium.

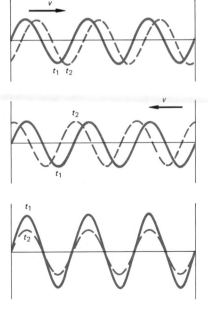

FIGURE 16.11
The normal modes of vibration for
a string with both ends fixed.

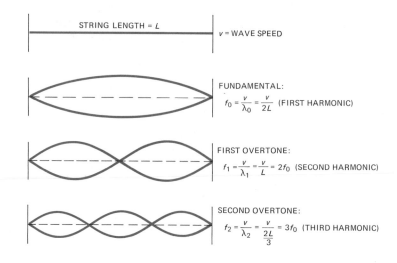

STRING LENGTH = L

v = WAVE SPEED

FUNDAMENTAL:

$f_0 = \dfrac{v}{\lambda_0} = \dfrac{v}{2L}$ (FIRST HARMONIC)

FIRST OVERTONE:

$f_1 = \dfrac{v}{\lambda_1} = \dfrac{v}{L} = 2f_0$ (SECOND HARMONIC)

SECOND OVERTONE:

$f_2 = \dfrac{v}{\lambda_2} = \dfrac{v}{\frac{2L}{3}} = 3f_0$ (THIRD HARMONIC)

waves in the medium and λ is the wavelength which satisfies the
boundary conditions. Fixed boundaries will be nodes and free ends will
be antinodes.

A standing wave represents a *normal mode* of oscillation for a given
system. The lowest normal mode frequency is called the *fundamental
frequency*. All other normal mode frequencies are integral multiples of
the fundamental frequencies of the system and are called overtones or
harmonics. Examples for a string fixed at both ends are shown in Figure
16.11. The first harmonic is the fundamental frequency, the first overtone
is the second harmonic, and so on.

Any possible wave form of a system can be regarded as a superposition of normal modes for the system. The normal mode frequencies are
natural frequencies for resonance of a system. This is the physical basis
for the geometrical shapes of musical instruments such as the sounding
cavities for string instruments and organ pipes.

16.12 Fourier's Theorem

Fourier's theorem provides the mathematical basis for building up any
wave form by the superposition of sine waves of specific wavelengths
and amplitudes. Any wave form, no matter how complex, can be expressed as the sum of sine waves. These sine waves each have different
frequencies and are called the Fourier components of the complex wave.
Wave form analysis, providing Fourier components of wave forms such
as brain waves, has become an important medical application of this
theorem. Sound spectrums or voice prints based on Fourier analysis are
proving to be as dependable as fingerprints in identifying different people. The first four Fourier components of a square wave and their resul-
tants are shown in Figure 16.12.

FIGURE 16.12
Fourier synthesis of a square wave.

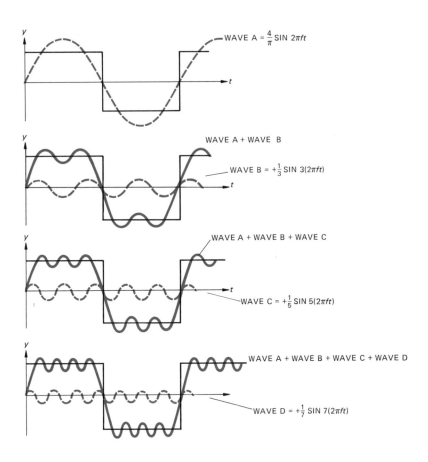

If the change in the wave form is abrupt, the high frequency Fourier components become important in the synthesis of the wave form. The square wave of Figure 16.12 requires a large number of components for its faithful reproduction.

16.13 Dispersion

If the velocity of a wave in a medium is a function of the frequency of the wave, the wave demonstrates *dispersion*. The medium producing this phenomenon is called the dispersive medium. Air is not a dispersive medium for audible sound — all audible frequencies travel through air with the same speed. What would you expect to happen if air were a dispersive medium for sound? Water is a dispersive medium for water waves — low frequency waves are propagated at higher speeds than high frequency water waves. Glass is a dispersive medium for light, and this is the explanation of Newton's famous experiment in which he produced the colored spectrum from sunlight using a glass prism.

16.14 *Waves in Elastic Media*

We have seen that an elastic medium is any material that tends to preserve its length, shape, or volume against external forces. Such a material can be said to have a restoring force that tends to return the material to its original condition after the external force is removed. The restoring force is characteristic of the material and arises from the binding forces between the individual atoms or molecules of the material. As we have seen in Chapter 13, the strength of the restoring force of a material is characterized by a force constant, or restoring modulus. The larger the restoring modulus of a material, the greater is the tendency of the material to resist changes in its length, shape, or volume. You may recall that Young's modulus, Y, is the measure of a material's tendency to maintain its length against external forces,

$$Y = \frac{F/A}{\Delta L/L} \tag{13.5}$$

where F is the magnitude of the external force, A is the area over which the force is acting, ΔL is the change in length, and L is the original length. In a similar way, we defined the bulk modulus B, the tendency of an object to maintain its volume,

$$B = \frac{F/A}{\Delta V/V} = \frac{P}{\Delta V/V} \tag{13.7}$$

where F is the magnitude of the external force, A is the area over which the force is acting, ΔV is the change in volume and V is the original volume. Some typical values for Young's modulus and for the bulk modulus are shown in Table 13.1. The general form of the equation for the velocity of a wave in an elastic medium is as follows:

$$\text{velocity} = \sqrt{\frac{\text{restoring factor}}{\text{inertial variable}}}$$

Specifically, we have the following applications: (all in SI units)

WAVE ON A STRING

$$v = \sqrt{\frac{T}{\mu}} \qquad \begin{array}{l} \text{where } T = \text{tension} \\ \mu = \text{mass per unit length} \end{array} \tag{16.10a}$$

COMPRESSIONAL WAVES (SOUND) IN A FLUID

$$v = \sqrt{\frac{B}{\rho}} \qquad \begin{array}{l} \text{where } B = \text{bulk modulus of fluid} \\ \rho = \text{density of fluid} \end{array} \tag{16.10b}$$

COMPRESSIONAL WAVES (SOUND) IN A SOLID

$$v = \sqrt{\frac{Y}{\rho}} \qquad \begin{array}{l} \text{where } Y = \text{Young's modulus of the solid} \\ \rho = \text{density of the solid} \end{array} \tag{16.10c}$$

EXAMPLES

1. A string 2.00 m long has a mass of 40.0 g. Find the wavelength of the waves on the string if the tension is 50.0 N and the frequency is 200 Hz.

$$v = f\lambda$$

$$\lambda = \frac{v}{f} = \frac{\sqrt{T/\mu}}{f} = \frac{\sqrt{50.0 \text{ N}/0.020 \text{ kg/m}}}{200 \text{ Hz}} = \frac{50.0}{200} \text{ m} = 25.0 \text{ cm}$$

2. The measurement of the velocity of sound is used to determine the Young's modulus of the solid. Given the velocity of sound to be 5000 m/sec in a solid with a density $= 8.90$ g/cm³, find Young's modulus for this solid.

$$Y = v^2\rho = (5000)^2 \text{ (m/sec)}^2 (8.90 \times 10^{-3} \text{ kg}/10^{-6} \text{ m}^3)$$

$$= 22.3 \times 10^{10} \text{ N/m}^2$$

3. A string is vibrating in its fundamental mode with a length of 0.500 m when under a tension of 180 N. The linear density of the string is 2.00×10^{-3} kg/m. Find the fundamental frequency of this string.

$$f_0 = \frac{v}{\lambda_0} = \frac{v}{2L} = \frac{\sqrt{T/\mu}}{2L} = \frac{\sqrt{180/(2.00 \times 10^{-3})}^{\frac{1}{2}}}{2 \times 0.5} = 3.00 \times 10^2 \text{ Hz}$$

16.15 *Intensity of a Wave Emanating from a Point Source*

If you are located a great distance from the source of a wave, no matter how large the source, the source appears to be small and we can represent it by a point in space. For example, even the largest of stars in the sky appears as a small point source to us. Consider a source that emits energy at a constant rate, that is, the energy emitted per unit of time remains constant. It follows from Equation 16.8 that the intensity of the wave times the area through which the wave is transported remains a constant. For an isotropic source, the wave spreads out uniformly in all directions. The area through which the wave passes as it leaves the source will be the surface area of a sphere which is centered on the point source and which surrounds the source of the wave. As the distance the wave travels from the source increases, the area through which the emitted energy passes increases. In fact, the area through which the energy passes increases as the square of the distance from the source. We can use this fact to derive the *inverse square law* for the intensity of a wave emitted from a point source as follows:

energy emitted = intensity × area × time

$$E \quad = \quad I \quad \quad A \quad \quad t \quad = I(4\pi r^2)t$$

where the surface area of a sphere of radius r is $4\pi r^2$.

$$\text{intensity} = \frac{E}{4\pi r^2 \, t} \qquad\qquad (16.11)$$

If the time rate of energy emitted from a small isotropic source is constant ($E/t=$ constant), the intensity is inversely proportional to the square of the distance from the source. Notice that in Equations 16.11 and 16.8 the SI units of intensity are joules per second per square meter or watts per square meter

$$\text{intensity at point } A \propto \frac{1}{r^2} \tag{16.12}$$

where r is the distance from the source to point A.

EXAMPLE

Suppose you are trying to read a road map on a dark sidewalk and the map is 48 meters from the only nearby street light. How far will you have to move to have the intensity of the light shining on the map increase by four times?

We can solve this problem using proportional reasoning.

$$\text{intensity at 48 m} \propto \frac{1}{48^2}$$

$$\text{intensity at new location} = 4\times \text{(intensity at 48 m)} \propto \frac{1}{r^2}$$

Now we divide the first proportional relationship by the second one.

$$\frac{\text{intensity at 48 m}}{\text{intensity at new location}} = \frac{1}{4} = \frac{(1/48^2)}{(1/r^2)} = \frac{r^2}{48^2}$$

Taking the square root of both sides, then

$$\frac{1}{2} = \frac{r}{48} \qquad r = 24 \text{ m}$$

You would have move to a position where the map is 24 m from the street light.

SUMMARY

Use these questions to evaluate how well you have achieved the goals of this chapter. The answers to these questions are given at the end of the summary with the number of the section where you can find related content material.

Definitions

1. The _____ of a wave is its period in space, and its appropriate SI units are _____.
2. The _____ of a wave is its period in time, and its units are _____.
3. The ratio of amplitudes for two waves is 1:4. Find the ratio of intensities of these waves if their frequencies are the same.

Wave Forms

4. Transverse waves are not possible in fluids because _____.

Problems

5. Given two waves of equal frequency but 90° out of phase, find the resultant wave amplitude if $A_2 = \frac{1}{2}A$.

Superposition

6. Given the waves shown in Figure 16.13, use the superposition principle to sketch the resultant wave.

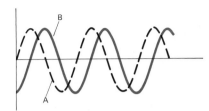

FIGURE 16.13
Question 6. Two waves moving to the right at the same speed.

Standing Waves

7. Compare the fundamental frequencies of a string when both ends are fixed and when one end is free.

Inverse Square Law

8. To reduce by half the intensity of a sound wave reaching you from a point source, you have to change your distance from the source by how much?

Answers

1. wavelength, m (Section 16.4)
2. frequency = 1/period, Hertz (cycle/sec) (Section 16.4)
3. $I_1/I_2 \propto A_1^2/A_2^2 = 1/16$ (Section 16.6)
4. fluids have no transverse restoring force (Section 16.14)
5. $\sqrt{5}\,A/2$ (Section 16.8)
6. see Figure 16.7 (Section 16.8)
7. $f(\text{fixed})/f(\text{free}) = 2$ (Section 16.11)
8. increase the distance by 40 percent (Section 16.15)

ALGORITHMIC PROBLEMS

Listed below are the important equations from this chapter. The problems following the equations will help you learn to translate words into equations and to solve single concept problems.

Equations

$$v = \lambda f \tag{16.2}$$

$$y = A \sin 2\pi ft \tag{16.3}$$

$$y = A \sin(2\pi ft - 2\pi \frac{x}{\lambda}) \tag{16.4}$$

$$E = \tfrac{1}{2}kA^2 = 2\pi^2 f^2 mA^2 \tag{16.5}$$

$$P = fE = 2\pi^2 f^3 mA^2 \tag{16.6}$$

$$I = P/_{\text{area}} = 2\pi f^2 A^2 v\rho \tag{16.7}$$

$$I \propto A^2 \tag{16.8}$$

$$v = \sqrt{\frac{T}{\mu}} \tag{16.10a}$$

$$v = \sqrt{\frac{B}{\rho}} \tag{16.10b}$$

$$v = \sqrt{\frac{Y}{\rho}} \tag{16.10c}$$

$$I \propto \frac{1}{r^2} \tag{16.12}$$

Problems

1. A reference particle is vibrating with an amplitude of 10 cm and a frequency of 100 Hz. Write the equation for the displacement of the particle as a function of time.
2. If a second particle is 100 cm from the reference particle of problem 1 in a transmitting medium in which the velocity of propagation is 300 m/sec, write the equation of displacement of the second particle as a function of time.
3. If the particle in problem 1 has a mass of 1 g, what is the energy per cycle?
4. What is the power transmitted by the wave described in the first three problems?
5. Two similar waves differing only in amplitude are being propagated through a medium. One wave has an amplitude of 6 cm and the other has an amplitude of 10 cm. Compare the intensities of the two waves.
6. What is the speed of a transverse wave in a cord that is 200 cm long and has a mass of 10 g. The tension is 5×10^{-2} N.
7. The speed of sound in water is about 1440 m/sec. What is the coefficient of bulk modulus (volume of elasticity) of water?
8. The elasticity modulus of a certain metal is found to be 16×10^8 N/m²; its density is 7800 kg/m³. What is the speed of a compressional wave in it?
9. If the solar intensity at the earth is 1.35×10^3 watts/m², what is the solar intensity at the planet Mercury, whose distance from the sun is 0.387 times the distance from the sun to the earth?

Answers

1. $y = 10 \sin (200\pi t)$ cm
2. $y = 10 \sin 2\pi[100t - (x/3)]$ cm
3. 2×10^7 ergs $= 2$ J
4. 2×10^9 ergs/sec $= 2 \times 10^2$ watts
5. ratio $= 36/100$

6. 3.1×10^2 cm/sec
7. 2.07×10^9 N/m²
8. 450 m/sec
9. 9.01×10^3 watts/m²

EXERCISES

These exercises are designed to help you apply the ideas of a section to physics situations. When appropriate the numerical answer is given in brackets at the end of the exercise.

Section 16.5

1. Given $y_1 = 2 \sin (0.628x - 314t)$ m,
   a. find the amplitude, frequency, and wavelength of the wave
   b. determine whether the wave is transverse or longitudinal
   c. find the speed of the wave [a. 2 m, 50 Hz, 10 cm; b. transverse; c. 500 m/sec]
2. A piano string emits a sound with a frequency of 341 Hz. The velocity of sound through air at this particular time is 340 m/sec. What is the wavelength of the sound wave in air? If the observer is 4.00 m from the source, what is the phase difference between the source and the wave at the observer? [1 m, 8π]
3. A string is attached to a vibrating source and the displacement of the source is given by $y = 2 \sin 60\pi t$ cm. The velocity of wave in the string is 100 cm/sec.
   a. Sketch the form of the string for the first 10 cm at time $t = 0.5$ sec.
   b. Sketch the displacement of a particle 5 cm from the source for 0.1 sec. after the wave reaches the particle.
4. A traveling transverse wave is described by the equation $y = 2 \sin (512\pi t + \pi x/20)$ cm.

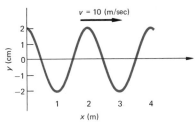

FIGURE 16.14
Exercise 6.

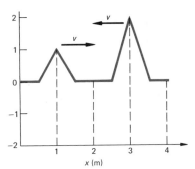

FIGURE 16.15
Exercise 10.

a. What are the amplitude, frequency, and wavelength of the wave?
b. What is the velocity of the wave?
c. What are the displacement and velocity of a particle at $x = 30$ cm and $t = 0.4$ sec? [a. 2 cm, 256 Hz, 40 cm; b. -102 m/sec; c. $y = 2 \sin 54° = 1.6$ cm; velocity $= 18.4$ m/sec]

5. A sinusoidal wave travels along a string. If the time required for a given particle to travel from maximum displacement to 0 displacement is 0.10 sec, what are the period and the frequency? If the wavelength is 1.2 m, what is the velocity of propagation? [0.4 sec, 2.5 Hz, 3 m/sec]

6. Given the wave in Figure 16.14 with a speed of 10 m/sec, sketch the wave 0.15 sec after the figure shown.

Section 16.6

7. Find the ratio of the intensities of two compressional waves of equal amplitude traveling in a gas if one has a frequency of 500 Hz and the other 1000 Hz. $[I(500 \text{ Hz})/I(1000 \text{ Hz}) = 1/4]$

8. If two waves of the same frequency are found to have an intensity ratio of $I_1/I_2 = 0.16$ in air, find the ratio of their amplitudes. $[A_1/A_2 = 0.4]$

Section 16.8

9. Suppose a wave given by $y_2 = 1 \sin(0.628x - 314t + 90°)$ cm is traveling in the same medium as wave y_1 from exercise 1. Find the resultant wave using the superposition principle. $[y = \sqrt{5} \sin(0.628x - 314t + 26.50°)$ cm]

10. Two triangular waves are traveling on a string as shown at $t = 0$ sec in Figure 16.15; the speed of each is 0.5 m/sec. Sketch the shape of the string at $t = 1.0$ sec, 2.0 sec, and 3.0 sec.

Section 16.11

11. The frequency of the fundamental tone of a violin string is 256 Hz. What are frequencies of second and third harmonics? [512 Hz, 768 Hz]

Section 16.14

12. Find the tension necessary to make the fundamental frequency of a 0.5 m string ($\rho = 2 \times 10^{-3}$ kg/m) equal to 150 Hz. [45 N]

Section 16.15

13. An ultraviolet lamp is designed to provide a typical fair-skinned person with a mild sunburn in 10 minutes if the person is 3.0 m from the lamp. How long should the person stay under the lamp when it is at a distance of 5.0 m? [28 minutes]

PROBLEMS

Each of the following problems may involve more than one physical concept. Numerical answers are given in brackets at the end of the problem.

14. The relative index of refraction of two media is defined as ratio of the wave velocities in the media and is related to the angles of incidence and refraction by

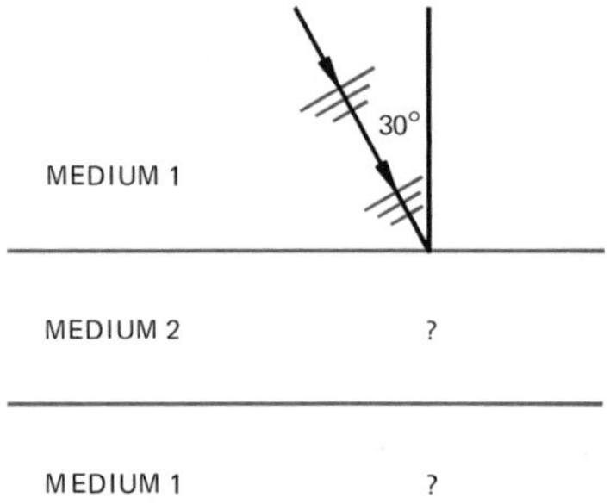

MEDIUM 1

30°

MEDIUM 2   ?

MEDIUM 1   ?

FIGURE 16.16
Problem 15.

Snell's law:

$$n = \frac{v_1}{v_2} = \frac{\sin i}{\sin r}$$

where i = angle of incidence and r = angle of defraction (see Figure 16.16). If the velocity in medium 2 is three-fourths that in medium 1 and the angle of incidence is 30°, what is the angle of refraction? [$\sin r = \frac{3}{8}$, $r = 22°$]

15. Trace a wave through the system shown in Figure 16.16 if the index of refraction is $\frac{3}{2}$. (See problem 14.) If medium 2 is 10 cm thick, what is the lateral displacement of the ray in passing through medium 2? [1.94 cm]

16. A string fastened at both ends is resonant with wavelengths of 0.16 m and 0.20 m. What is the minimum possible string length? [0.4 m]

17. A wire is vibrating in its first overtone mode with a frequency of 250 Hz. If the tension on the wire is 250 N and its linear density is 10^{-3} kg/m, find the length of the wire. [2.0 m]

18. A stretched string 75 cm long is vibrating in its third overtone at a frequency of 300 Hz. Find the velocity of the traveling waves that produce this standing wave by superposition. [113 m/sec]

GOALS

When you have mastered the contents of this chapter, you will be able to achieve the following goals:

Definitions
Define each of the following terms and use it in an operational definition:

sound wave
intensity of sound wave
amplitude
frequency
wavelength
acoustic impedance
decibel

infrasonic
ultrasonic
distortion
standing wave
beats
Doppler effect

Sound Wave Form
Graph the amplitude of a sound wave in pressure versus space or time coordinates.

Sound Level
Calculate the sound level in decibels (dB).

Sound Intensity
Solve sound intensity and sound amplitude problems.

Sound Application
Explain the use of sound in selected medical applications.

Hearing System
Describe the properties of the human sound detection system.

Doppler Effect
Solve Doppler-effect and standing-wave problems.

PREREQUISITES

Before beginning this chapter you should have achieved the goals of Chapter 5, Energy, Chapter 10, Temperature and Heat, Chapter 15, Simple Harmonic Motion, and Chapter 16, Traveling Waves, respectively.

17

Sound and the Human Ear

17.1 Introduction

How does sound originate? Do you know what gives rise to sounds? Hold a ruler tight against your desk and strike the end that extends in the air. What happens? What can you hear? How can you change what you hear? If you put your head behind a book and strike the ruler again, can you still hear something? What does that tell you about the ability of sound to travel around objects?

At sometime or other you probably have heard the old philosophical question—if a tree falls in the forest and no one is there, does the tree crashing to the ground make a sound? If you know the physical principles of sound, you can answer this question logically.

This chapter is intended to introduce you to the basic physical principles of sound and of your ear. Sound is becoming increasingly useful in medical applications. Your understanding of sound will not only enable you to appreciate the properties of your ear as a sound detector, but may enable you to understand the design of a sound system for making medical diagnoses.

17.2 Sound

The term *sound* has two distinct uses. The psychologist and physiologist think of sound as a sensation due to the stimulation of the auditory nervous centers. To the physicist it is a form of vibrational energy that produces the audio sensation. Sound is modeled as longitudinal wave motion which is transmitted by an elastic medium.

A sound system has three components. First, there is a sound source consisting of a vibrating object, for example, a speaker. Second, there is the medium through which the sound is transmitted, air in a room, for example. And finally, there is the detector, the human ear. Sound is an example of interaction-at-a-distance. The vibrating source interacts with a detector some distance away.

For the familiar case of a vibrating object in air, we can think of the vibrating source as sweeping out a small volume of space and causing fluctuations in the density (or pressure) of the air in the vicinity of the source. These density or pressure fluctuations are then transmitted to the detector as traveling waves in the air.

We can define sound as a variation in pressure set up by a vibrating source. Then sound is seen as waves propagated through matter and characterized by a periodic variation of pressure (or density) in a medium. These periodic pressure variations occur in the direction of the wave propagation. Thus, sound is defined as a longitudinal wave as discussed in Chapter 16. A sound wave diagram is shown in Figure 17.1. The mathematical formula for the pressure in this sound wave being propagated in the x-direction is given by Equation 17.1.

FIGURE 17.1
Diagrammatic representation of a
sound wave, showing positions of
compression and rarefraction. The
curve shows the variations from
normal pressure within a typical wave.

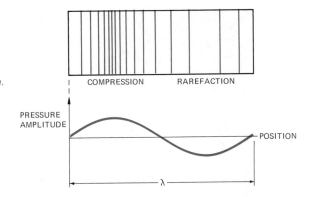

$$P = P_0 \sin\left(2\pi ft - \frac{2\pi x}{\lambda}\right) \tag{17.1}$$

P_0 = pressure amplitude (N/m²) at source

x = position (m) or distance from source

f = frequency (Hz) = number of complete pressure cycles per second

λ = wavelength (m) = distance between adjacent equal pressure amplitudes

t = time in seconds

This equation expresses the periodicity of the pressure in both space (x) and time (t). You can see this periodicity, if you examine the equation with time constant or with position constant.

EXAMPLES

1. Consider the following as an equation for a typical sound wave:

$$P = (2 \times 10^{-2} \text{ N/m}^2) \sin\left(1000\pi t - \frac{\pi x}{0.34}\right)$$

Graph P against x for two different times, $t_1 = 0.8 \times 10^{-3}$ sec and $t_2 = 10^{-3}$ sec. What is the wavelength of this wave? What is its speed? Which direction is it traveling? See Figure 17.2.

FIGURE 17.2
The plot of variation of pressure as
a function of x (position) for a given
time t, i.e., t = a constant value.

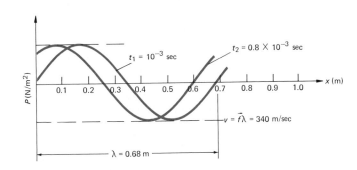

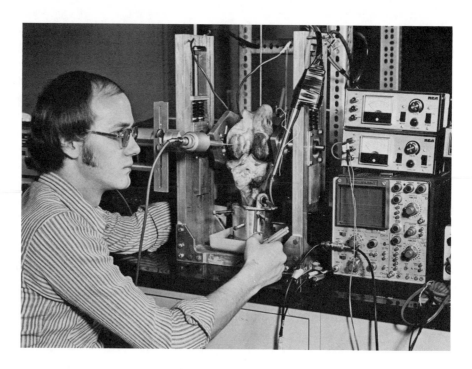

The photograph shows Larry Zavodney with apparatus and a cow's knee joint which he and Mamerto Chu are using for *in vitro* studies of sounds produced in joints. Their work was described in the June 1976 issue of *Physics Today* on page 20.

Knee-joint problems of various kinds seem to affect almost everyone sooner or later, highly priced football players and ordinary mortals alike. A group at the University of Akron have been listening to the clicking and scraping noises of defective knees, and they reported to the 91st Meeting of the Acoustical Society of America, in Washington last April, on what they had heard. The photograph shows Larry Zavodney, a graduate student in mechanical engineering, with a cow's knee joint that he and Mamerto L. Chu are using for *in vitro* studies of these sounds.

A cartilage surface of this joint has been artificially roughened to simulate a pathological condition—say, arthritis—and the noise produced on flexing the joint is picked up by condenser microphone. After amplification, conversion to digital form, and computer-aided statistical analysis, the signal shows unique acoustical characteristics for various types of knee-joint diseases. Zavodney and Chu claim that this acoustical "signature" may become useful as a rough diagnosis of the various diseases that can affect joints.

For a parallel study of the progress of joint diseases in living specimens, Zavodney and Chu, joined by Richard Mostardi and Ivan Gradisar, are doing a similar experiment with rabbits. Here the Akron team create an artificial "disease" by surgery on a live specimen and follow its progress by listening to the sounds generated, at different times, by the flexing joint, just as in the *in vitro* experiments on cow's knees. Finally, surgical examination enables cartilage roughness at an advanced stage of the disease to be measured. Preliminary results reported to the ASA meeting suggest that there is a close connection between the degree of roughness of the diseased cartilage and the acoustical signature detected. The goal is a technique for diagnosis of conditions such as arthritis at earlier stages than is currently possible.

(Photograph courtesy of Dr. Mamerto L. Chu and the Department of Mechanical Engineering, University of Akron.)

FIGURE 17.3
The variation of pressure as a
function of time for a given value of x.

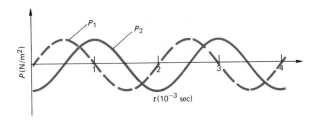

2. Draw a graph P versus t for two different values of x, $x_1 = 0$, and $x_2 = 0.17$ m as shown in Figure 17.3. For $x_1 = 0$,

$$P_1 = 2 \times 10^{-2} \sin 1000\pi t$$

For $x_2 = 0.17$ m

$$P_2 = 2 \times 10^{-2} \sin (1000\pi t - \pi/2)$$

What is the period of oscillation for this wave? $T = 2 \times 10^{-3}$ sec. Can you compare it to the frequency of some known sound? $f = 500$ Hz. (*Hint:* middle C on a piano has a frequency of 256 Hz.) Could this sound be heard? Recall that the magnitude of the velocity, the speed, of a wave is given by $f\lambda = v$ (Equation 16.2). Find the speed of the sound for this example.

In the first case we have a sinusoidal wave in space (x-direction), and in the second case for a given x, the pressure varies sinusoidally in time as the wave goes by a given position. Such a wave transports energy from one place to another in the direction of the wave propagation.

In Chapter 16 we learned that the velocity of a compressional wave such as sound in a solid is given by

$$v_s = \sqrt{\frac{Y}{\rho}}$$

where Y is Young's modulus for the solid and ρ is the density of the solid.

$$v_s = \sqrt{\frac{B}{\rho}} \qquad \text{for a fluid}$$

where B is the bulk modulus and ρ is the density of the fluid. And

$$v_s = \sqrt{\frac{\gamma P_0}{\rho}} \qquad \text{for a gas}$$

where γ = ratio of c_p/c_v, P_0 = undisturbed gas pressure, and ρ = the density of the gas.

The velocity and the wavelength change in such a way that $v_s = f\lambda$ is always satisfied. When a sound wave passes from one medium to another, for example, from air to water, the frequency of the wave remains constant. In other words, the frequency is a property of the source and not of the transmitting medium. On the other hand, the velocity is a property of the elastic medium in which the wave is traveling. The wavelength results from the combination of these two properties.

Consider the problem of comparing the velocity of sound in hydrogen and deuterium. What are reasonable assumptions to make for these two gases? Since they are isotopes, it should be a good approximation to assume the same elastic constant (bulk modulus) for each. The velocity of sound is then primarily determined by the density of the gases according to the previous equation. Since the mass of a deuterium molecule is twice that of a hydrogen molecule, we find the magnitudes of the velocity of sound in the two gases are related as

$$v_s(H_2) = v_s(D_2) \times \sqrt{2}$$

This result agrees with the actual measured values to within 2 percent.

17.3 Sound Intensity

The intensity of a traveling wave is defined as the energy per unit time (that is, power) transported perpendicularly across a surface of unit area (Figure 17.4). The units of wave intensity in the SI are watts/meter². You should verify these as the correct units from the definition of intensity.

In Chapter 16 on traveling waves, you were given an equation for the intensity of a traveling wave (Equation 16.7). The equation for the *intensity of a sound* wave follows by analogy. In the medium of density ρ, we have the following expression for the sound intensity.

$$I = \tfrac{1}{2}(A\omega)^2 \rho v_s \text{ (watt/m}^2) \tag{17.2}$$

where v_s = magnitude of the velocity of sound (m/sec), A = amplitude of motion (m), and $\omega = 2\pi f$ (rad/sec).

The intensity of a sound wave can also be expressed in terms of the pressure amplitude P_0, of the wave as follows:

$$I = \frac{1}{2}\frac{P_0^2}{(\rho v_s)} \tag{17.3}$$

where ρv_s is the *acoustical impedance* of the medium, Z_a.

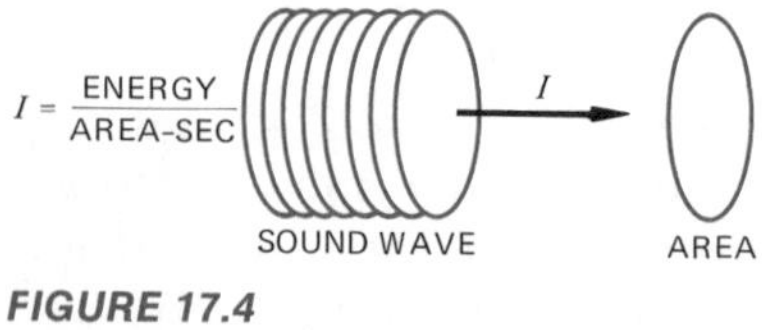

EXAMPLE

Normal conversation has an intensity of about 1.00×10^{-6} watts/m². The velocity of sound at room temperature is 345 m/sec and density of air in the room is 1.26 kg/m³. What is the pressure amplitude?

$$I = \frac{1}{2}\frac{P_0^2}{(\rho v_s)}$$

$$P_0^2 = 2I\rho v_s$$

$$P_0 = \sqrt{2I\rho v_s}$$

$$= \sqrt{2 \times 10^{-6} \times 1.26 \times 345} = 29.5 \times 10^{-3} \text{ N/m}^2$$

Normal atmospheric pressure is about 10^5 N/m².

FIGURE 17.4
The intensity of a traveling wave is the energy going through a unit area in one second. This is equivalent to power per unit area.

The transmission and reflection of sound energy at an interface between two media is determined by the acoustic impedances of the media involved.

Let Z_A be the acoustic impedance of first medium and Z_B, the acoustic impedance of second medium. For sound traveling perpendicular to the interface from medium A to medium B, it can be shown that the ratio of reflected intensity I_r to incident intensity I_0 is given by:

$$\frac{I_r}{I_0} = \frac{(Z_A - Z_B)^2}{(Z_A + Z_B)^2} \tag{17.4}$$

and the ratio of transmitted intensity I_t to incident intensity is given by:

$$\frac{I_t}{I_0} = \frac{4Z_A Z_B}{(Z_A + Z_B)^2} \tag{17.5}$$

Note if the impedances match ($Z_A = Z_B$), no energy is reflected at the interface and $I_t = I_0$. This is another example of maximum energy transfer when inertial properties match at the interface of two systems.

EXAMPLE

An application of ultrasound is its use in brain surgery. Let us determine the energy transmitted from water to bone and compare it with energy transmitted from air to bone. $Z_a(\text{water}) = 1.43 \times 10^6$ (kg/m²-sec); $Z_a(\text{bone}) = 6 \times 10^6$ (kg/m²-sec); $Z_a(\text{air}) = 430$ kg/m²-sec. Thus, we have $I(\text{bone})/I(\text{water}) = 4 \times 1.43 \times 10^6 \times 6 \times 10^6/(7.43 \times 10^6)^2 = 0.62$, or 62 percent of the energy is transmitted from the water to the bone. For $I(\text{bone})/I(\text{air}) = 4 \times 430 \times 6 \times 10^6/(6.00043 \times 10^6)^2 = 0.29 \times 10^{-3}$, or .03 percent of the energy is transmitted from air to the bone. This illustrates why water is used to match an ultrasonic transducer to bone or brain tissue.

17.4 Threshold of Hearing

We can calculate the amplitude of motion for the minimum intensity detected by a normal human ear from Equation 17.2. The minimum threshold of hearing occurs at a frequency of approximately 3500 Hz and has an intensity value of 1.00×10^{-12} watt/m² for the typical human being. The density of air is 1.25 kg/m³, and $v_s = 345$ m/sec at 300°K. Thus the amplitude of motion is given by:

$$I = 1.00 \times 10^{-12} \text{ watts/m}^2 = \frac{A^2}{2}(3500 \times 2\pi)^2 \times 1.25 \times 345$$

$$A^2 = \frac{2.00 \times 10^{-12}}{1.25 \times 345. \times 12.25 \times 10^6 \times 4 \times \pi^2}$$

$$= \frac{2.00 \times 10^{-18}}{1.25 \times 345. \times 12.25 \times 4 \times \pi^2}$$

$$= 9.60 \times 10^{-24}$$

$$A = 3.10 \times 10^{-12} \text{ m}$$

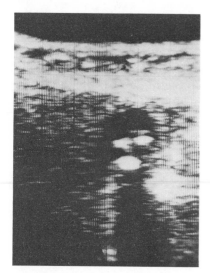

An ultrasonic scan showing multiple gall stones. This is a longitudinal scan through the right lobe of the liver, the finely distributed liver structure. The lumen of the gall bladder shows as a black reflection. [free area, with 4 light reflections at the gall stones, dorsally the sound shadow zone.] (Siemens Corporation.)

Note that this amplitude of motion for the threshold of hearing is of the order of magnitude of the diameter of an atom (10^{-10}m). By using proportional reasoning, calculate the amplitude of an intensity of 1 watt/m^2 (this corresponds to the threshold of pain) at the same frequency.

The range of intensities that the human ear can successfully detect varies from 10^{-12} watt/m^2 to 1 watt/m^2. Because of this wide range of intensities, a logarithmic scale is used for defining sound level. The sound level in *decibels* (dB) is defined as follows:

$$\text{sound level (dB)} = 10 \log_{10} (I/I_0) \qquad (17.6)$$

where I is the sound intensity and I_0 is the standard intensity of 10^{-12} watt/m^2.

Ordinary conversation is approximately 60 dB. This corresponds to $\log I/I_0 = 60/10$, or $I = I_0 \times 10^6 = 10^{-6}$ watts/m^2.

EXAMPLES

1. Table 17.1 lists some sound levels as measured at the ear. Complete the table using a frequency of 2000 Hz.
2. Suppose that a snowmobile has a sound level of 80 dB at 30 m from the machine. If the maximum noise limit is 77 dB, how much must the intensity be reduced to meet this requirement, and what is the intensity at this noise limit? Using Equation 17.6 we have

$$80 \text{ dB} = 10 \log I/10^{-12} \text{ watt/m}^2$$

Thus,

$$I = 10^{-12} \times 10^8 = 10^{-4} \text{ watt/m}^2$$

For 77 dB,

$$I = 10^{7.7}/10^8 \times 10^{-4} \text{ watt/m}^2$$

TABLE 17.1
Sound Levels

Type of Sound	Sound Level at Ear (dB)	Amplitude (m)	Intensity (watts/m^2)
Threshold of hearing (youth)	0	3×10^{-12}	10^{-12}
Threshold of hearing (age 50)	10		
Rustle of leaves	20		
Average room	40		
Normal conversation	60		
Street traffic	70		
Symphony orchestra (full blast)	90		
Rock band, riveter	100		
Threshold of discomfort	120		
Threshold of pain (30 m behind a jet airplane at takeoff)	130		

Thus for 77 dB,

$$I = 10^{-4.3} \text{ watt/m}^2$$

a reduction of 0.5×10^{-4} watt/m².

Notice that doubling the intensity increases the sound level by 3 dB.

17.5 Infrasonics and Ultrasonics

Recall the equation for traveling waves, $v = f\lambda$, that relates the magnitude of the velocity of the wave (v) to the frequency (f) and wavelength (λ). In phenomena involving sound, the frequency range is usually divided into three different regions, frequencies below 20 Hz are referred to as *infrasonic*, between 20 Hz and 20,000 Hz is the *audio frequency range* for the human ear, and frequencies above 20,000 Hz are referred to as *ultrasonic*. The effect of infrasound on living systems is not well understood, and it is a current area of research.

There are several applications of ultrasonics in medical science. One such application is found in neurosonic surgery. Ultrasonic waves are focused on a particular part of the brain, destroying only the tissue at the focal point without interferring with the normal activity of other parts of the brain. This application has proved to be effective in treating Parkinson's disease, which seems to have well-localized centers in the brain. What are the particular properties of ultrasound that make it useful for this kind of application? Try to design a model of interaction for the tissue and the ultrasonic wave.

17.6 Superposition Principle and Sound

Thus far we have considered pure sinusoidal sound waves. Such waves are sometimes referred to as pure tones. Using the mathematical formulation known as Fourier analysis (see Section 16.12), it is possible to synthesize any wave shape by adding up a series of sinusoidal waves of the proper amplitudes, frequencies, and phases. Most natural sounds are complex in shape and involve many of these Fourier components, or pure sound waves. If the amplitude of the wave changes rapidly, the number of different frequency components necessary to synthesize the sound is large. For example, a square wave must have more high-frequency components than a wave with slowly varying amplitude. The relative amplitudes of the harmonics making up a sawtooth wave and a square wave of the same frequency are illustrated in Figure 17.5a. Such a component analysis of a sound into its harmonics is called a sound spectrum. The sound spectrum for a clarinet and a french horn playing the same note are shown in Figure 17.5b. Your voice can be analyzed when you say certain words, but your voice will have considerable variability. By using computers to make the Fourier analysis of voice

FIGURE 17.5
The relative ratios of the fundamental
and the overtones at a given constant
frequency are shown for a saw tooth
wave, a square wave, a clarinet, and
a French horn.

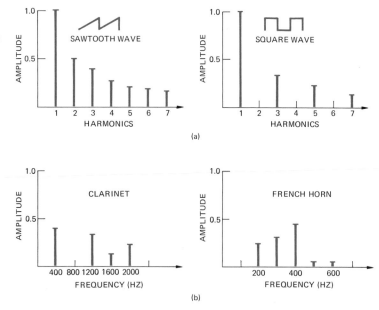

FIGURE 17.5
The relative ratios of the fundamental
and the overtones at a given constant
frequency are shown for a saw tooth
wave, a square wave, a clarinet, and
a French horn.

prints, it is possible to detect changes that are due to the emotional state
of the speaker. Such voice print analysis systems may become the "lie
detectors" of the future. It has already been demonstrated that each of
us has a voice print as unique as our fingerprints, and thus such prints
can be used for identification.

17.7 Distortion

Not all sound systems satisfy the linear approximation necessary for
the superposition principle to be valid. When this linear condition is
not satisfied by any part of an acoustical system, *distortion* is introduced.
Distortion may take the form of new components added to the sound
wave. Distortion is frequently encountered when working with large
amplitude sound waves. Large amplitude waves may force the source,
medium, or detector to operate outside the region where the deformation of a system is proportional to the force producing the deformation,
that is, outside the region of Hooke's law behavior for the system.

17.8 Standing Waves

We will now consider other applications of the superposition principle.
First, we will apply this principle to the addition of waves in a medium
within a specific volume with well-defined boundaries. When a sound
wave meets such a boundary, it undergoes reflection. The determining
factor for the amplitude and phase shift of the reflected wave is the
acoustical impedance of the boundary layer. The acoustic impedance is

defined as the density of the medium times the velocity of sound in the medium,

$$Z_a = \rho v_s \tag{17.7}$$

You will recall that this parameter appeared in the expression for the intensity of a sound wave. Under certain boundary conditions the addition of reflected and incident waves in a defined volume yields a *standing wave* pattern. The sound in this volume is not described as a traveling wave, but instead we find an equation such as the following:

$$P = 2P_0 \sin kx \cos \omega t \tag{17.8}$$

which results from the addition of $P_1 = P_0 \sin(\omega t + kx)$ and $P_2 = P_0 \sin(\omega t - kx)$. You will recall that standing waves are possible only for the wavelengths that satisfy the conditions of nodes at fixed ends and antinodes at free ends. For both ends fixed, $L = n(\lambda/2)$ where $n = 1$, $2, \ldots$. For one end free, $L = m(\lambda/4)$, where $m = 1, 3, 5, \ldots$ (odd integers only).

An example of such a standing wave is the motion of a string fixed at each end and plucked in its center. Such standing waves are the basis of sound production for many musical instruments. The lowest frequency standing wave of a system is called the fundamental or first harmonic of the system. Other possible frequencies of standing waves for a system involve integral multiples of the fundamental frequency. If all the boundaries of a system are the same, the enclosed system will support standing waves of all integral harmonics. If all boundaries are not the same, that is, with the pipe opened at one end and closed at the other, only odd harmonics will be supported by the system.

Some possible representations for standing waves for open and closed pipes are shown in Figure 17.6. Please note that for an open pipe, $L = n(\lambda/2)$ where n is an integer and for a closed pipe, $L = n(\lambda/4)$ where n is an odd integer.

FIGURE 17.6
Standing waves in (a) an open and (b) a closed tube for the fundamental and the two overtones.

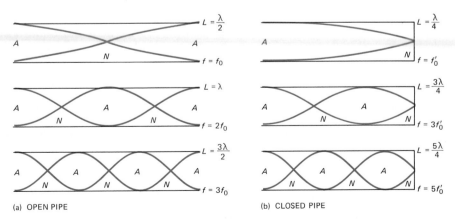

(a) OPEN PIPE   (b) CLOSED PIPE

A = LOCATION OF MAXIMUM AMPLITUDE OF VIBRATION = ANTINODE.
N = LOCATION OF MINIMUM AMPLITUDE OF VIBRATION = NODE.
$f = v/\lambda$

FIGURE 17.7
Two waves of different frequencies
add to produce a beat frequency
equal to the difference of the
frequency of the two waves.

ENVELOPE
FREQUENCY = $f_1 - f_2$

17.9 Beat Frequency

Another result of the application of the superposition principle is the production of beats when two pure sound waves are generated together. *Beats* are the pulsation of the amplitude of the sum of two waves with slightly different frequencies. *The beat frequency is equal to the difference between the frequencies of the two waves involved.* An example of such a beat phenomenon is shown in Figure 17.7. Beats are heard as periodic variation in the loudness of the sound. The human ear can detect beats to about seven beats per second. Higher beat frequencies do not yield distinct beats, and the sound is characterized by dissonance. See the enrichment section of this chapter for a mathematical treatment of beats.

17.10 The Doppler Effect

The *Doppler effect* involves the change in frequency of a wave due to the relative motion of the source and observer. First, consider the case where the sound source is approaching an observer at rest. Let v designate the velocity of the source toward the observer and v_s the magnitude of the velocity of sound in air. The motion of the source toward the observer means that more waves will be crowded into each meter than would be present if both source and observer were at rest. This means that the apparent wavelength is decreased (see Figure 17.8).

FIGURE 17.8
The Doppler effect.

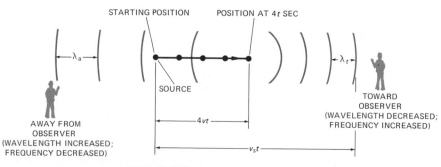

STARTING POSITION   POSITION AT $4t$ SEC

λ_a

λ_t

SOURCE

TOWARD
OBSERVER
(WAVELENGTH DECREASED;
FREQUENCY INCREASED)

$4vt$

$v_s t$

AWAY FROM
OBSERVER
(WAVELENGTH INCREASED;
FREQUENCY DECREASED)

v_s = SPEED OF SOUND

Since during one cycle the source will move v/f, this wavelength will be reduced by v/f, so the apparent wavelength is

$$\lambda' = \lambda - \frac{v}{f} = \frac{v_s}{f} - \frac{v}{f} = \frac{v_s - v}{f} \tag{17.9}$$

and the frequency of the approaching source will appear to be

$$f' = \frac{\text{velocity}}{\text{wavelength}} = \frac{v_s}{(v_s - v)/f}$$

or

$$f' = \frac{fv_s}{v_s - v}$$

If the source moves away from the observer at the same speed, the same argument yields a frequency $f' = fv_s/(v_s + v)$. If the observer moves toward a source at rest with a speed of v_0, the apparent velocity of the sound waves relative to the observer is $v_s + v_0$ and, thus, the apparent frequency of the sound is given by the ratio of the velocity of the wave to its wavelength,

$$f' = \frac{v_s + v_0}{\lambda} = \frac{f(v_s + v_0)}{v_s} \tag{17.10}$$

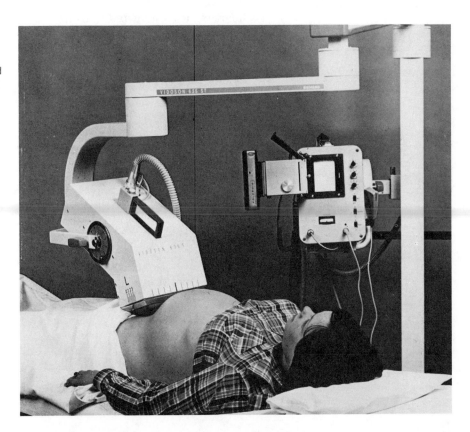

Ultrasonic imagining unit allows a pregnant mother to observe her child on the monitor. In obstetrics and gynecology the unit may be used for the assessment of embryonic and fetal movements. In addition the ultrasonic automatic scanner has many uses in internal medicine, pediatrics, cardiology, urology, and radiology. (Courtesy of Siemens.)

If the observer moves away from the source at the same speed, the result is a frequency of

$$f' = \frac{f(v_s - v_0)}{v_s}$$

We can combine these results into an equation for the relative motion of approach and an equation for the relative motion of separation.

For relative approach: $\qquad f' = \dfrac{f(v_s + v_0)}{(v_s - v)}$ (17.11)

For relative separation: $\qquad f' = \dfrac{f(v_s - v_0)}{(v_s + v)}$

where f = frequency with no relative motion, v_s = speed of sound, v_0 = speed of observer, and v = speed of source.

The Doppler effect is used in medicine by measuring the frequency shift of an ultrasonic wave that is reflected off a moving organ inside the body. Such a technique can be used to study the heart beat of the fetus early in pregnancy. A small ultrasonic transducer is placed on the mother's abdomen. Some of the ultrasound waves are reflected back by the beating fetus' heart. The fetal heart serves as a moving reflector and provides a Doppler shift of the ultrasound when received. When the heart is moving away the frequency is lower and when the heart moves toward the receiver the frequency of the detected signal is increased.

EXAMPLE

An observer is standing near the edge of a state highway, watching an automobile traveling on the highway at 20 m/sec. The automobile approaches the observer, and then recedes from him. The natural frequency of the automobile's horn is 540 Hz. If the horn is sounded on the approach and on the separation from the observer, what frequency does the observer hear? On this particular day the velocity of sound in air is 340 m/sec.

On the automobile's approach, the horn's sound seems to have a frequency greater than the natural frequency.

$$f' = f \frac{v_s}{v_s - v}$$

where the velocity of sound in air = v_s = 340 m/sec. And the velocity of the car (source) = v = 20 m/sec.

$$f' = 540 \times \frac{340}{340 - 20} = 540 \times \frac{340}{320} = 574 \text{ Hz}$$

As the automobile moves away, the frequency of the horn's sound is heard at less than the natural frequency:

$$f' = f \frac{v_s}{v_s + v} = 540 \times \frac{340}{340 + 20} = 510 \text{ Hz}$$

Now suppose the car is stationary and you are in a car moving along the highway under the conditions stated above. As you approach the horn, you hear a frequency greater than the natural frequency.

$$f' = f\frac{v_s + v}{v_s} = 540 \times \frac{340 + 20}{340}$$

$$= 573 \text{ Hz}$$

Moving away from the horn you hear a frequency less than the natural frequency:

$$f' = f\frac{v_s - v}{v_s} = 540\left(\frac{340 - 20}{340}\right)$$

$$= 508 \text{ Hz}$$

17.11 Sound Detectors

We will now consider some common sources and detectors of sound. Two important characteristics of a sound source or detector are efficiency and fidelity. The efficiency of a sound source is a measure of its effectiveness in converting the power supplied to the source into power of the generated sound wave. The efficiency is dependent upon the coupling of the source to the medium of propagation as well as the transduction efficiency of the source. The fidelity of a sound source is a measure of how faithfully the source reproduces the frequencies supplied to it. The speakers of electronic sound reproduction systems have efficiencies and fidelities that are dependent partly on their size. A single speaker is not equally efficient at both low and high frequencies. A good high-fidelity sound system requires different size speakers for different frequency ranges. (Can you figure out the relation between speaker size and frequency for best sound reproduction?)

When considering a sound detector we are interested in its sensitivity to different frequencies, which is sometimes referred to as its frequency response. There are several different kinds of microphones available. The common ones are the condenser, the crystal, and the electromagnetic, or dynamic, microphones. Microphones have varying sensitivities and display frequency responses that vary over the audible frequency range (20 Hz to 20,000 Hz).

17.12 The Human Ear

The human ear is a superb sound transducer coupled to a computer center, the human brain. Figure 17.9 shows a block diagram of the human sound detection system.

The ear has the following typical specifications: frequency response from 20 Hz to 20,000 Hz, with nonuniform sensitivity in this range; sound level response from 0 dB to 120 dB (see Figure 17.10); and thresh-

FIGURE 17.9
Block diagram of the human system for detecting sound waves.

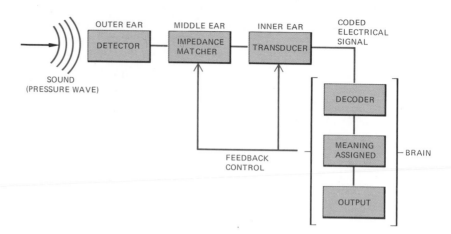

FIGURE 17.10
Plot the threshold of hearing for the "normal" ear.

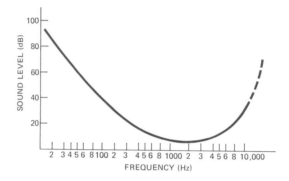

old amplitude response for molecular vibrations with amplitudes less than the size of an atom. The ear is a detector that transduces the incoming pressure wave information into an electrical signal that is transmitted to the brain via nerve impulses. A cross section of the ear is illustrated in Figure 17.11.

For our analysis of the ear we consider three main parts, the external canal, the middle ear, and the inner ear. The external canal can be thought of as a closed-end organ pipe. The closed end is formed by the tympanic membrane, or ear drum, that separates the external canal from the middle ear. Do you know that your minimum threshold of hearing occurs for sounds of frequencies near 3.5 kHz? Compare this frequency with the fundamental frequency of a 2.5-cm outer ear canal.

The middle ear cavity contains three small bones forming a lever system attached to the ear drum. The purpose of the middle ear is to transmit the vibrations in the air due to sound stimulus to the fluid inside of the inner ear. The middle ear mechanism is a nonlinear coupling system. At high sound levels the middle ear produces harmonics of the frequencies exciting it. One of the main functions of the middle ear is that of impedance matching between the sound input and the inner ear. The middle ear has been studied in detail, and it has been shown that the middle ear provides a pressure amplification of 17 at low sound

FIGURE 17.11
Diagram of the human ear.

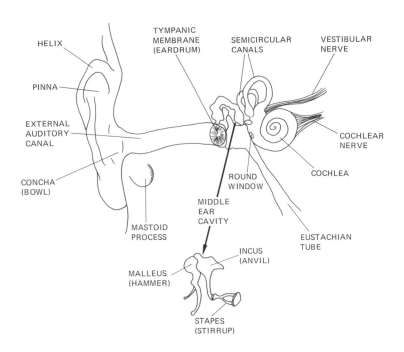

levels. At high sound levels the middle ear action resembles that of an automatic volume control on a radio. The middle ear reduces the energy fed into the inner ear through a feedback control system that attempts to maintain a constant input to the inner ear.

The inner ear consists primarily of the *cochlea*, a spiral canal inside the bone of the skull. The cochlea is divided into three channels by fibrous membranes. These channels are filled with fluid. The fluid transmits the sound energy from the middle ear. The basilar membrane contains the endings of the auditory nerve. These nerve cells have hairlike projections that are anchored in the tectorial membrane. The basilar membrane has a varying thickness as it proceeds toward the apex of the spiral. Relative motion between the tectorial and basilar membranes produces an electrical potential in the nerve cell. The varying thickness of the basilar membrane causes different frequencies to stimulate maximum signals at different locations along the canal. The sound information is coded into an electrical signal that is transmitted to the brain. The signal has all of the frequency and amplitude information that is available from the input sound stimulus. The brain, through a process as yet not well understood, interprets this message and supplies the necessary feedback controls and output action orders. There is still much to learn about the entire hearing process. Some of the questions yet to be resolved are: How is the phase information transduced and analyzed? What determines the limit of the ears' ability to hear beat frequencies? How is it possible to ignore loud noises and detect much lower sound levels? If you are intrigued by such questions, you may wish to pursue studies in this area of biophysics and psychoacoustics.

FIGURE 17.12
Schematic sketch of method to produce an ultrasonic hologram. The interference pattern produced by the reference beam and the object beam is produced on a liquid surface. An optical system is needed to photograph this pattern. If one wishes to produce a visual image, the camera is replaced by a TV camera and a TV monitor.

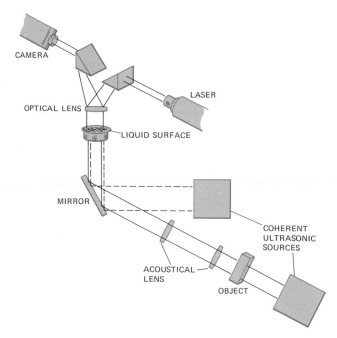

17.13 Medical Applications of Acoustic Holography

A hologram is a photographic record of an interference pattern generated by a set of waves and a set of reference waves. One can make holograms using sound waves provided that the interference pattern can be recorded, Figure 17.12. Many parts of the human body react to high-frequency sound waves in a much different manner than they do to other forms of waves such as x rays. The nonbony parts of the human body are essentially transparent to x rays. However, these parts of the body absorb and reflect the sound waves in different ways. Hence the application of holographic techniques are being developed. This form of sound imaging provides a differentiation between the various types of tissues. It is being predicted that the detection of tumors in the human body will be one of the greatest contributions of acoustical holography to medicine.

ENRICHMENT
17.14 Velocity of Sound in an Ideal Gas

The equation for the velocity of sound in a gas as a function of the elastic modulus and the density of the gas can be derived as follows: The elastic modulus is the bulk modulus of the gas, B, which is defined as

$$B = -\frac{\Delta P}{\Delta V} V \tag{17.12}$$

where ΔP is the pressure change producing volume change ΔV to an initial volume V.

In calculus notation this becomes

$$B = -V \frac{dP}{dV} \tag{17.13}$$

It can be shown that the sound wave passes through a gas in an essentially adiabatic process. How would you show that this is so? The gas law for adiabatic processes is

$$P = (\text{constant})V^{-\gamma}$$

where $\gamma = C_p/C_v$ (Chapter 12).
   Thus

$$\frac{dP}{dV} = (\text{constant})(-\gamma V^{-(\gamma+1)})$$

Then

$$B = -V(\text{constant})(-\gamma V^{-(\gamma+1)}) = \gamma(\text{constant})V^{-\gamma}$$

Therefore, $B = \gamma P$.
   Thus, the expression for the speed of sound in an ideal gas becomes

$$v_s = \sqrt{\frac{\gamma P}{\rho}} = \sqrt{\frac{\gamma RT}{m}}$$

where R is ideal gas constant and m is the molecular mass of the gas. While this derivation is based on the ideal gas law, it gives good results for many gases at room temperature.

17.15 Derivation of Beat Frequencies

The production of beats can be analyzed mathematically as follows: Consider two waves given by

$$y_1 = A \sin\left(\frac{2\pi x}{\lambda_1} - 2\pi f_1 t\right) \quad \text{and} \quad y_2 = A \sin\left(\frac{2\pi x}{\lambda_2} - 2\pi f_2 t\right)$$

The superposition principles give

$$y = y_1 + y_2 = 2A \sin\left[\left(\frac{2\pi}{\lambda_1} + \frac{2\pi}{\lambda_2}\right)x - \left(\frac{2\pi f_1 + 2\pi f_2}{2}\right)t\right]$$

$$\times \cos\left[\left(\frac{2\pi}{\lambda_1} - \frac{2\pi}{\lambda_2}\right)x - \left(\frac{2\pi f_1 - 2\pi f_2}{2}\right)t\right]$$

since $\sin\theta_1 + \sin\theta_2 = 2 \sin\frac{1}{2}(\theta_1 + \theta_2) \cos\frac{1}{2}(\theta_1 - \theta_2)$. This result can be written as follows noting that $v = f_1\lambda_1 = f_2\lambda_2$.

$$y = 2A \sin\left[\frac{2\pi(f_1 + f_2)(x/v - t)}{2}\right] \cos\left[\frac{2\pi(f_1 - f_2)(x/v - t)}{2}\right]$$

This is the equation of a product of two waves, one of frequency

$f_c = (f_1 + f_2)/2$ and the other $f_m = (f_1 - f_2)/2$. The first is the *carrier frequency* of the waveform and the latter is the *modulation frequency*. A beat occurs each time the amplitude as governed by the modulation frequency is a maximum—that is whenever

$$\cos\left[2\pi \frac{(f_1 - f_2)(x/v - t)}{2}\right] = +1 \text{ or } -1$$

Since this happens two times each cycle, the beat frequency $= f_1 - f_2$.

SUMMARY

Use these questions to evaluate how well you have achieved the goals of this chapter. The answers to these questions are given at the end of the summary with the number of the section where you can find related content material.

Definitions

1. The amplitude of a sound wave is measured in what SI units?
2. The pulsations in wave amplitude are called_____ and arise from _____.
3. The sound level of a sound is 40 dB; this means the intensity of the sound is _____ watts/m².
4. Infrasound has frequency _____ than audio range.
5. Ultrasonic waves have frequencies _____ than audio range.
6. Standing waves are set up in organ pipes whenever the driving frequency produces _____.
7. Matching acoustic impedance at boundaries assures _____.
8. The Doppler effect is characterized by a change in _____ due to relative _____. If the observer and source approach each other, the frequency is _____; if the source and observer retreat from each other, the frequency _____.

Waveforms

9. Graph the amplitude of the sound wave $P = 30 \sin 2\pi (440t - x/.77)$ N/m².
   a. as a function of time for $x = 0$
   b. as function of x for $t = 1/440$ sec

Sound Level and Intensity

10. What is the intensity of a sound wave whose pressure amplitude is 2 N/m²? What is the intensity level in dB? See Example on page 376.
11. What is the difference between intensity levels in dB if the intensity of one wave is twice that of another?

Applications

12. Name possible applications of sound in dental or medical professions.

Hearing System

13. What are the essential parts of the human sound detecting system?

Doppler Effect

14. A train is moving toward a listener with a speed of 20 m/sec.
    a. If the whistle of the train has a frequency of 330 Hz, what frequency does the listener hear?
    b. After the train passes the listener, what frequency does he hear?

Standing Waves

15. An open pipe resonates at 400 Hz (fundamental). At what frequency will it resonate when one end is plugged?

Answers

1. N/m² (Section 17.2)
2. beats, superposition of two waves of nearly the same frequency (Section 17.9)
3. 10^{-8} watts/m² (Section 17.4)
4. lower (Section 17.5)
5. greater (Section 17.5)
6. wavelengths satisfying boundary conditions (Section 17.8)
7. maximum power transfer (Section 17.3)
8. frequency, motion of observer and source, increased, decreased (Section 17.10)
9. a. sine curve with amplitude of 30 N/m² and period of 1/440 sec, i.e., $y = 30 \sin 2\pi$ $(440t)$ N/m²
   b. sine curve of amplitude of 30 N/m² and wavelength of 0.77 m. $P = 30 \sin 2\pi$ $[1 - x/.77)]$ (Section 17.2)
10. 0.01 watts/m², 100 dB (Section 17.4)
11. 3 dB (Section 17.4)
12. ultrasonic cleaning for cleaning instruments, ultrasonic drilling or cleaning of teeth, Doppler shift, heart beat embryo, acoustical hologram (Sections 17.5, 17.6, 17.13)
13. See Figure 17.11 (Section 17.12)
14. a. 351 Hz
    b. 312 Hz (Section 17.10)
15. 800 Hz (Section 17.8)

ALGORITHMIC PROBLEMS

Listed below are the important equations from this chapter. The problems following the equations will help you learn to translate words into equations and to solve single-concept problems.

Equations

$$P = P_0 \sin \left(2\pi ft - \frac{2\pi x}{\lambda} \right) \tag{17.1}$$

$$I = \frac{1}{2} (A\omega)^2 \, \rho v_s \frac{\text{watts}}{\text{m}^2} = \frac{1}{2} \frac{P_0^2}{(\rho v_s)} \tag{17.2, 17.3}$$

$$\text{sound level (dB)} = 10 \log_{10} \frac{I}{I_0} \tag{17.6}$$

$$f' = \frac{f(v_s + v_0)}{(v_s - v)} \tag{17.11}$$

$$f' = \frac{f(v_s - v_0)}{(v_s + v)} \tag{17.11}$$

for closed pipe: $L = n\dfrac{\lambda}{4}$ where n is an odd integer

for open pipe: $L = n\dfrac{\lambda}{2}$ where n is an integer

Problems

1. Fill in the blanks in Table 17.1 in Example 1 of Section 17.4.
2. An automobile has a horn with a frequency of 510 Hz, and the velocity of sound in air is 340 m/sec. What frequency does the observer hear if
   a. the car is approaching the observer with velocity of 20 m/sec when the horn is sounded

b. the car is receding from the observer with a velocity 20 m/sec at the time the horn is sounded

c. the observer is approaching the parked car with a velocity of 20 m/sec when the horn is blown

d. the observer is leaving the parked car with a velocity of 20 m/sec when the horn is blown

3. A closed organ pipe has a length of 44 cm. The velocity of sound in air is 342 m/sec at the temperature of the pipe.

a. What is wavelength of the fundamental note?

b. What is the frequency of the fundamental note?

c. What is the wavelength of the first overtone?

d. What is the frequency of the second overtone?

4. An open organ pipe has a length of 1 m. The velocity of sound in air at the temperature of the pipe is 340 m/sec.

a. What is the wavelength of the fundamental note?

b. What is the frequency of the fundamental note?

c. What is the wavelength of the second overtone?

d. What is the frequency of the first overtone?

Answers

2. a. 542 HZ

   b. 482 HZ

   c. 540 Hz

   d. 480 Hz

3. a. 176 cm

   b. 194 Hz

   c. 59 cm

   d. 970 Hz

4. a. 2m

   b. 170 Hz

   c. 2/3 m

   d. 340 Hz

EXERCISES

These exercises are designed to help you apply the ideas of a section to physical situations. When appropriate the numerical answer is given in brackets at the end of the exercise.

Section 17.2

1. Using appropriate values of B and ρ for air, calculate the velocity of sound in air. Assume $\gamma = 1.4$ for air. [330 m/sec]

2. Using the value of v_s from problem 1, determine the pressure amplitude associated with a 1000 Hz sound wave of 10^{-12} watt/m^2 intensity. (This corresponds to a barely audible sound.) Also determine the pressure amplitude for an intensity of 1 watt/m^2. [2.9×10^{-5} N/m^2, 2.9×10 N/m^2]

Section 17.3

3. Compare the acoustic impedance of water and air. If you are able to drive a sound source with the same amplitude in air as in water, what is the ratio of the intensity of the sound generated in air and water by such a source? The velocity of sound in water is 1400 m/sec. [$I_w/I_a = 1/3.20 \times 10^3$]

4. Compute the intensity of a compressional wave at 0°C and 760 mm Hg, if its frequency is 512 Hz and its amplitude is 2.0×10^{-3} cm. The speed of the wave in air is 331 m/sec. [8.6×10^{-1} watts/m^2]

Section 17.4

5. If a 60-dB sound level is reduced to 1/10 its inten-

sity, the new sound level will be what in decibels? [50 dB]

6. If one trumpet will produce a sound of 60 dB at the back row of the auditorium,
   a. How many dB will two trumpets produce?
   b. How many dB will four trumpets produce?
   c. How many dB will ten trumpets produce?
   d. How many dB will 100 trumpets produce?
   e. How many trumpets will be required to produce the discomfort threshold of 120 dB at the back row of the auditorium? [a. 63 dB; b. 66 dB; c. 70 dB; d. 80 dB; e. 10^6]

7. One person talking to you from a distance of one meter produces a sound of 40 dB. How many people will it take simultaneously talking to you from 1 m to produce a sound of 80 dB? (This is sometimes called a stampede.) [10^4]

Section 17.8

8. Find the fundamental frequency and the first three overtones of a 0.5-m closed-end organ pipe. Assume the velocity of sound is 340 m/sec. How many overtones could the average student hear? [170 Hz, 510 Hz, 850 Hz, 1190 Hz, about 59]

Section 17.10

9. At the Exploratorium Science Exhibit in San Francisco there is a whistle mounted on a movable belt. By turning a crank you can cause the whistle (assume a natural frequency of 540 Hz) to move first away from you and then toward you. If you can make the whistle move at a speed of 34 m/sec, what frequencies will you hear? [receding 491 Hz, approaching 600 Hz]

PROBLEMS

Each problem may involve more than one physical concept. The numerical answer to the problem is given in brackets at the end of the problem. A problem that requires material from the enrichment section is marked by a dagger.

10. What is the amplitude of vibration of air molecules for a 40-dB sound with a frequency of 2000 Hz? The air has an acoustic impedance of about 440 kg/m²–sec. [5.3×10^{-10} m]

11. The Super Sound Rock Group of Rolla, Missouri, uses two 150-watt, 15-inch (19-cm) radius speakers to put out sweet sounds to their admiring audiences. Assume that 75 percent of the power transmitted to the speakers is actually converted into sound energy.
    a. What is the intensity of sound at the opening of one speaker (due to that speaker alone) when it is being operated at full power of 150 watts? How many dB is that?
    b. Assuming that the intensity is inversely proportional to the square of the distance you are from the speakers, if the speakers are located together and are being operated at full power, how far away from them will you have to stand to have the sound be less than the pain of threshold of 120 dB? [a. 1000 watts/m², 150 dB; b. 45 m]

12. If the sound level increases from 50 dB to 70 dB, find the ratio of the amplitudes of the second sound level to the first. [10]

13. A 2000-Hz sound can just be distinguished from another of the same frequency if it is about 1.00 dB louder than the other. (This is known as one JND = just noticeable difference.) Find:
    a. the ratio of the JND sound intensities
    b. the pressure amplitude ratio associated with this JND pair [a. 1.26; b. 1.12]

14. It is well known that the high noise background of an industrial society causes your threshold of hearing to increase. Suppose that your hearing threshold is being increased by 0.33 dB per year, and that your present threshold is 10 dB.
    a. What will your hearing threshold be in 60 years?
    b. How much will the intensity of a barely audible sound have to be increased for you to be able to hear it in 60 years? Assume that you can presently hear a sound intensity 10^{-11} watts/m².
    c. How much will the pressure amplitude of a barely audible sound have to be increased for you to hear it in 60 years? [a. 30 dB; b. 100 times, 10^{-9} watts/m; c. 10 times]

15. As you are hitchhiking along the highway a passing motorist honks his horn (frequency f_0) at you in a 60 km/hr speed zone. You note that his horn changes its apparent pitch by $(1/10)f_0$ as he goes past ($v_s = 350$ m/sec).
    a. Does the pitch appear to increase or decrease?
    b. How fast is the motorist going? Should he be given

a speeding ticket? [a. $\Delta f = f_0/10$, decrease in pitch; b. $v = 17.5$ m/sec or 63 km/hr; give him a ticket]

16. Estimating the size of your external ear canal, and using the previously developed ideas for a closed-end pipe, calculate the resonant frequency for your ear in air. The velocity of sound in water is about 5 times that in air. How does the resonant frequency of your ear canal change when your ears are full of water? What are you able to hear when you are swimming under water? [~3400 Hz, 5]

17. The fundamental frequency of the A string of a violin is 440 Hz. The vibrating portion of the string is 37.0 cm long and has a mass of 1.28 g. With what tension must it be stretched to be at the correct pitch? [367 N]

18. If the ear is assumed to be resonating as a closed pipe in problem 16, find the length of the ear canal if 2000 Hz is its fundamental frequency. [4.3 cm]

19. What gas (in the ideal gas approximation) would give the greatest ratio of the velocity of sound between any two of its isotopes? [hydrogen]

20. A recent report indicates that ultrasonic Doppler shift equipment is being used by police to determine the speed of cars on freeways. Suppose that a 1.00 MHz sound was used, and a Doppler shift results in a detected signal of 0.88 MHz signal. What was the speed of the car as it receded from the policeman? (Be careful here. Remember that the source at 1.0 MHz is with the policeman at rest, and the speeding car reflects this sound as it moves away from the observer.) [22 m/sec, or 78 km/hr, or 49 mph]

†21. A beat frequency of 6 beats/second is heard between two sources, one of which is at rest and the other moving toward the observer. If the frequency of the source at rest is 1000 Hz, find the speed of the moving source. [2m/sec]

22. An obstetrician uses a 22,000.00 Hz ultrasonic sound wave to detect the fetal heart motion. If she obtains Doppler frequencies of 22,002.84 and 21,997.16 on her detector, what is the velocity of the motion of the fetus' heart? The speed of sound in the body tissue is 1550 m/sec. See suggestion in problem 20. [0.1 m/sec]

†23. Two whistles, Y and Z, each have a frequency of 500 Hz. Y is stationary, and Z is moving to the left, away from Y, with a velocity of 50 m/sec. An observer is between the two whistles and moving to the right at a velocity of 30 m/sec. What is the frequency of each whistle as heard by the observer? How many beats would he hear? Assume the velocity of sound in air to be 340 m/sec. [544 Hz, 398 Hz, 146 beats/sec which cannot be heard as beats]

24. Show that for speeds in which $v \ll \mu$ where $v =$ velocity of source and $\mu =$ velocity of wave, the equation for Doppler shift may be expressed in an approximate form as $\Delta\lambda/\lambda = v/\mu$ where $\Delta\lambda$ is the shift in wavelength.

25. Derive an equation that could be used to determine the bulk modulus of an unknown gas by comparing compressional wave intensity with that in a known gas using the same sound source. $[B_x/B_s = \sqrt{I_x/I_s}]$

GOALS

When you have mastered the content of this chapter, you will be able to achieve the following goals:

Definitions
Define each of the following terms and use it in an operational definition:

light ray	optical axis
object distance	converging optical elements
image distance	diverging optical elements
index of refraction	real image
reflection coefficient	virtual image
internal reflection	magnification
focal point	aberrations—chromatic and spherical

Ray Diagrams
Draw ray diagrams for some common optical systems.

Lens and Mirror Equations
Apply the basic equations for lenses and mirrors to optical systems with one or two components.

Optical Devices
Explain, using physical principles, the operation of a reading glass, camera, microscope, and fiber optics.

PREREQUISITES

Before beginning this chapter you should have achieved the goals of Chapter 16, Traveling Waves.

18

Optical Elements

18.1 Introduction

You observe many different kinds of interactions between light and objects around you. The bowl of a spoon may serve as a mirror. You will note that at arm's length your face appears upside down in the spoon. As you move the spoon toward your face there is a point where the image appears to be right side up. You have probably discovered that a swimming pool or lake appears to be shallower than it actually is.

How do eyeglasses help the wearer to see more clearly? Why do some large panes of window glass distort images of objects viewed through the glass? Photographic slides must be inserted backward and upside down in the projector for correct projection on the screen. Why?

These are common examples of reflection and refraction of light.

Much of the information we receive from our environment involves light and its interaction with optical elements such as lenses and mirrors. Instruments such as telescopes, microscopes, and cameras have contributed much to our understanding of natural phenomena.

In this chapter we shall explore the physical principles of reflection and refraction that form the basis of lenses and mirrors and their applications.

18.2 Straight-Line Wave Propagation and Ray Diagrams

Can you see around corners? Can you hear around corners? If sound and light are both wave phenomena, why do they behave differently? Do sound waves cast sound "shadows"? What is the characteristic of light waves that makes seeing around corners very difficult, but enables you to make shadow pictures on the wall?

In this chapter we will be studying optical phenomena that we can understand by using a model of straight-line propagation of light waves.

FIGURE 18.1

Straight-line propagation of light is illustrated by the shadow cast by a corner (a) or by your hand in front of a light (b).

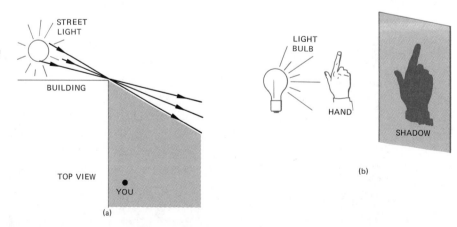

We find this model of the straight-line wave motion of light works well in a uniform medium such as air, glass, or water at constant temperature and pressure.

We assume in this model that the path of a beam of light is a straight line in a uniform medium. Hence we cannot see around corners. By placing a hand between a light source and the wall and stopping some of the light waves at that point, we can create shadow pictures. See Figure 18.1. Although it is possible to change the direction of the light paths at the interface between media in this model, the light path coming to the interface and the light path leaving the interface are treated as straight lines. It is possible to represent light paths on a diagram by straight lines. This representation is known as a *ray*. A *ray diagram* is a pictorial representation of an optical system in which lines (rays) are used to show some selected paths of light.

18.3 Reflection at Plane Surfaces

A young girl is watching the reflection of the setting sun on a small lake. Use a ray diagram to illustrate what she sees. (*Hint:* What do you know about the angles of incidence and reflection?) How many images can be seen in two mirrors at 90° to one another?

Our model predicts that the reflection of waves is governed by the law of reflection which states the angle of incidence equals the angle of reflection, as stated in Equation 18.1 and illustrated in Figure 18.2.

$$\angle i = \angle r \tag{18.1}$$

It is customary to measure these angles from the line perpendicular (called the *normal line*) to the reflecting surface.

A ray diagram for an object in front of a plane mirror is shown in Figure 18.2b. Rays striking the mirror along the normal from a point on the object reflect back along this normal line. Other rays passing through a point on the mirror are reflected at angles satisfying the $\angle i = \angle r$ condition as shown.

FIGURE 18.2
Ray diagrams illustrate the law of reflection (a) where the angle of incidence *i* equals the angle of reflection *r*. The image appears as far behind the mirror as the object is in front of the mirror (b).

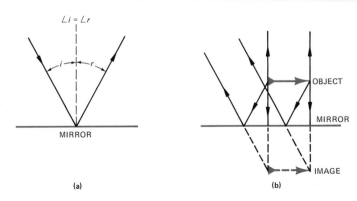

The intersection of the extension of any two rays from the same object point gives the image position for this object point.

A carefully drawn diagram shows that the distance of the image from the mirror, *image distance* behind the plane mirror is equal to the distance of the object from the mirror *object distance* in front of the mirror. Such an image is called a *virtual image* because the light appears to be diverging from this image but it does not actually pass through the image position.

As was pointed out in Chapter 16, waves undergo reflection at the interface between two media. The mirror, as we have treated it, is the perfect reflector, reflecting all of the incident wave energy. At other kinds of surfaces, the fraction of energy reflected depends on the angle of incidence and on the index of refraction of the two media defining the interface. The index of refraction of a material is defined in Equation 18.2. The *index of refraction* plays the analogous role of system inertia for the energy transfer across the interface between two media.

$$\text{index of refraction} = n = \frac{\text{velocity of light in vacuum}}{\text{velocity of light in medium}} \qquad (18.2)$$

One of the accomplishments of Maxwell's electromagnetic theory was the theoretical prediction that related the index of refraction to the reflection coefficient for a given interface and angle of incidence. For example, the *reflection coefficient* or fraction of light reflected R at normal incidence ($i = 0°$) is given by Equation 18.3.

$$R = \frac{(n_2 - n_1)^2}{(n_2 + n_1)^2} \qquad (18.3)$$

where n_1 is the index of refraction of incident medium and n_2 is the index of refraction of second medium. From this equation and from the conservation of energy, we can see that if the reflected energy approaches zero, the maximum energy is transmitted into the second medium. From Equation 18.3 we see that this condition is satisfied when the two indices of refraction are equal. Again we find the maximum energy is transferred from one system to another when the inertial parameters are matched at the boundary. This condition for zero reflection can be used to determine the index of refraction for transparent materials. The unknown samples are immersed in liquids with known indices of refraction that are not solvents for the sample. If the index of refraction of the liquid matches that of the sample, no light will be reflected at the surfaces of the sample, and the sample will disappear in the liquid.

EXAMPLE

We want to find the fraction of light reflected at normal incidence at the interface between the human eye lens ($n = 1.40$) and the vitreous humor ($n = 1.33$).

$$R = \frac{(1.33 - 1.40)^2}{(1.33 + 1.40)^2} = \frac{(-.07)^2}{(2.73)^2} = 6.6 \times 10^{-4}$$

We see that very little light is lost due to reflection at this interface in the human eye.

18.4 Refraction

In addition to reflection at the interface between two media, there is a change in direction of the transmitted light ray in the second medium. This bending of light as it passes from one medium to another is called refraction and it is governed by Snell's law as expressed in Equation 18.4.

$$n_1 \sin \theta_1 = n_2 \sin \theta_2 \tag{18.4}$$

where n_1 is the index of refraction of the medium of the incident ray, θ_1 is the angle of incidence, θ_2 is the angle of refraction, and n_2 is the index of refraction of the medium of the refracted ray. The refraction of light at the interface between two media is illustrated in Figure 18.3. The in-

FIGURE 18.3
Reflection and refraction of light at an interface between air and glass. Note that the angle of incidence is equal to angle of reflection and that the angle of refraction is less than the angle of incidence. The index of refraction for the glass is given by $n = \sin i / \sin r$; for this case $n = 1.52$. (Pictures from *PSSC Physics,* D.C. Heath and Company, Lexington, Mass., 1965.)

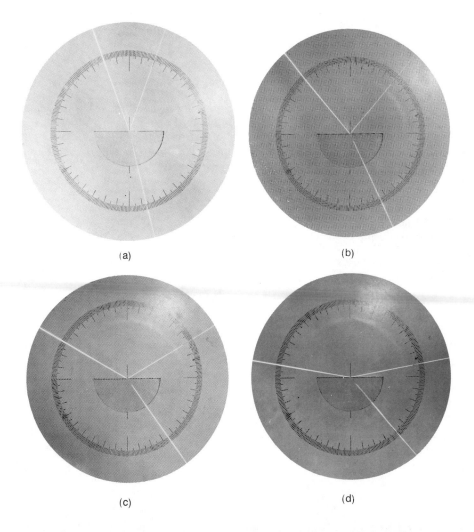

(a)

(b)

(c)

(d)

TABLE 18.1
Indices of Refraction of Common Materials

Substance	Index of Refraction
Acetone	1.359
Methyl alcohol	1.329
Ethyl alcohol	1.362
Benzene	1.501
Carbon disulfide	1.628
Turpentine	1.472
Water	1.333
Crown glass	1.517
Light flint glass	1.575
Heavy flint glass	1.650
Rock salt (NaCl)	1.544
Quartz	1.544
Polystyrene	1.591
Ice	1.309

dices of refraction of some materials measured relative to air are given in Table 18.1. All measurements were made with yellow light (wavelength = 589 nm) at 20°C.

EXAMPLES

1. Locate the image of a rock as seen in water. Consider two rays, one which is normal to the surface and one with an angle of incidence other than 0°. See Figure 18.4. For ray A the path of light from A to the flying observer is $AA'BO_1$. For ray B the path of light from A to the standing observer is ACO_2. However, the standing observer interprets a straight line path, as equivalent to $A'CO_2$. The apparent position of the rock A is a distance BA' below the surface.

FIGURE 18.4
The apparent depth of an object at a depth of BA below the surface of water appears to be BA'. Apparent depth equals actual depth $\times \dfrac{n_1}{n_2}$.

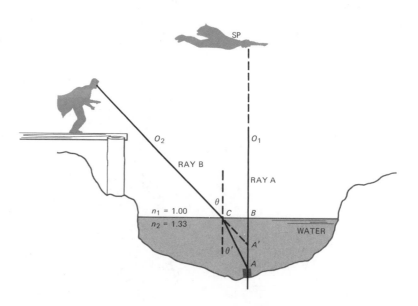

For near normal incident viewing (that is, $CB \ll AB$) we have

$$\tan \theta' = \frac{CB}{BA}$$

$$\tan \theta = \frac{CB}{BA'}$$

For small angles $\sin \theta' \simeq \tan \theta'$ and $\sin \theta \simeq \tan \theta$. From Snell's law

$$n_2 \sin \theta' = n_1 \sin \theta$$

Therefore,

$$\frac{\tan \theta'}{\tan \theta} = \frac{CB/BA}{CB/BA'} = \frac{BA'}{BA} = \frac{n_1}{n_2} = \frac{1}{n_2}$$

Behavior of light incident on a pair of parallel faces of a glass slab. The incident beam is from the left. At the top surface there is both reflection and refraction. For the reflected beam the angle of incidence is equal to the angle of reflection. The refracted beam changes direction at the interface in accord with Snell's law ($\sin i / \sin r = n$). At the other face part of the beam is reflected, and part is refracted. The laws of reflection are obeyed by the reflected portion of the beam, and the laws of refraction are obeyed by the refracted beam. The reflected beam from the lower face is refracted at the upper face. Note that the beams emerging from the top surface are parallel to the incident beam. Compare the intensity of the beams. Does the intensity of the incident appear to be equal to the sum of the other three? (Picture from *PSSC Physics,* D.C. Heath and Company, Lexington, Mass., 1965.)

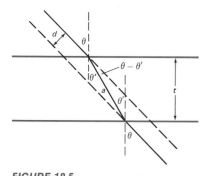

FIGURE 18.5
The displacement d of a near normal incident ray ($\cos \theta \sim 1$) is found for a slab of thickness t with index of refraction n and parallel sides.

if $n_1 = 1$, or,

$$\frac{BA}{BA'} = n_2$$

The apparent depth is reduced in comparison to the actual depth by a factor of the index of refraction of the medium, when viewed from a medium such as air, which has an index of refraction equal to one.

2. In everyday life we often look through media with parallel sides. What is the effect? See Figure 18.5. You note that a parallel shift occurs. This is not significant for window panes and aquarium sides, but in high-powered microscopy it becomes important, and a calibrated tilted glass plate may be used for measurement through a microscope. For small values of θ in the equation for d, the displacement of the ray is determined as follows:

$$a \cos \theta' = t \qquad d = a \sin (\theta - \theta')$$

from the geometry of the two right triangles. Now we substitute the expression $t/\cos \theta'$ for a in the equation for d and expand $\sin(\theta - \theta')$ using a standard trigonometric identity,

$$d = \frac{t}{\cos \theta'} (\sin \theta \cos \theta' - \cos \theta \sin \theta')$$

$$d = t (\sin \theta - \cos \theta \tan \theta')$$

For small angles, $\cos \theta \simeq 1$ and $\tan \theta' \simeq \sin \theta'$. From Snell's law we have

$$\theta \simeq \sin \theta = n \sin \theta' \quad \text{or} \quad \theta \simeq n\theta'$$

Therefore,

$$d \simeq t\left(\theta - \frac{\theta}{n}\right) = t\theta\left(\frac{n-1}{n}\right)$$

For glass, viewing at near normal incidence, the sideways displacement of a ray d is about one-third the thickness of the glass times the angle of incidence.

18.5 Internal Reflection

The path of light from medium 2 to medium 1 is the reverse of the path from medium 1 to medium 2 up to a given angle, which is known as the critical angle. For incident angles greater than the critical angle, total *internal reflection* occurs; that is, no light passes out of the medium (Figure 18.6). To determine the critical angle we set $\theta = 90°$, so that $\sin \theta = 1$. Then

$$\sin \theta_c = \frac{n_1}{n_2} = \frac{1}{n_{21}} \tag{18.5}$$

where n_{21} is the ratio of the index of refraction of medium 2 to the index of refraction of medium 1.

For water to air the critical angle is $\sin \theta_c = \frac{3}{4}$.

$$\theta_{c_{\text{water}}} \simeq 48.6°$$

FIGURE 18.6
Total internal reflection occurs when the angle of incidence θ is greater than the critical angle θ_c where $\sin \theta_c = n_1/n_2$.

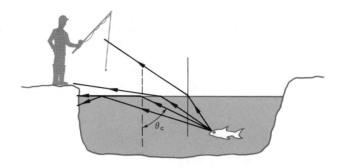

and for crown glass to air (approximate)

$$\sin \theta_c = \tfrac{2}{3}$$

$$\theta_{c_{\text{glass}}} \simeq 42°$$

There are many applications of internal reflection in optical instruments. Consider an isosceles right angle prism with 45° angles. Figure 18.7a shows the light path has a 90° change in direction. The light path for a 180° change in direction is shown in Figure 18.7b. Can you name an optical instrument in which these devices are used? Please note that the first arrangement is equivalent to a single mirror, and that the second is equivalent to two mirrors at right angles. However, the glass prism is more efficient, since the reflection coefficient for a mirror is always less than 100 percent. Use a ray diagram to test the statement "If you wish to see yourself as other people see you, look into a prism (Figure 18.7b) or two mirrors at 90° to each other."

Fiber optics is another application of internal reflection. Light is incident upon one end of a transparent fiber, and the total internal reflection keeps the light inside to the emergent end. One can think of the system as a light pipe. One very important use of the light pipe has been the study of the blood flow in animals and humans. The light pipe is used to illuminate a section to be studied so that the phenomenon can be photographed (Figure 18.7c).

Glass fibers are used in bundles to convey light in bronchoscopes, gastroduodenoscopes, and other such instruments used for examinations of the internal parts of humans. The individual fibers in these bundles are about 10 μm in diameter.

FIGURE 18.7
Total internal reflection is widely used in optical instruments to reflect light in desired directions. Prisms like those in (a) and (b) are used in binoculars, and (c) is a modern light pipe.

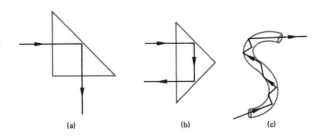

(a)   (b)   (c)

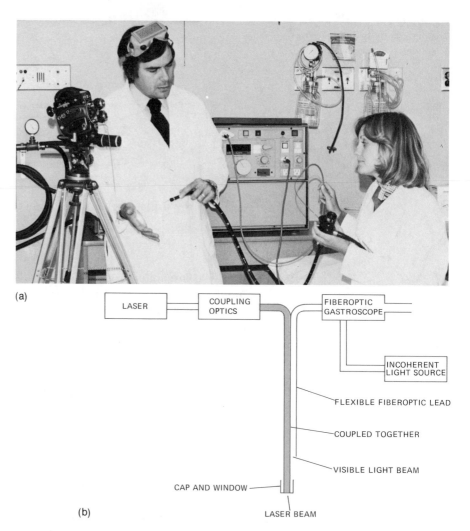

(a)

(b)

(a) Coupled fiberoptic argon-ion laser gastroduodenoscope. The endoscope can see inside the stomach. Note the reflection from center of palm of hand of Dr. Dwyer. This unit is used for the phototherapy of bleeding gastric lesions. (Photo by Alan Goldstein, courtesy of Dr. Richard Dwyer and Harbor General Hospital Multi-Media Department, Torrance, California.)

(b) Block diagram of a coupled argon-ion retinal photocoagulator and flexible fiberoptic endoscope.

An argon laser is connected to a flexible fiberoptic system to deliver laser light into the upper gastrointestinal tract. This flexible fiberoptic system is connected to another independent fiberoptic system connected to a noncoherent visible light source with endoscope.

For more detailed information see the following two references.

Endoscopic Argon-Ion Laser Phototherapy of Bleeding Gastric Lesions by Yullin, Dwyer, Craig, Bass & Cherlow, *Archives of Surgery*, Vol. 111, p. 750 (1976)
Gastric Hemostasis by Laser Phototherapy in Man by Dwyer, Haverback, Bass & Cherlow, *Journal of the American Medical Association*, Vol. 236, No. 12, Fig. 1383 (1976).

FIGURE 18.8
The focal length *f* of (a) a concave mirror and (b) a convex mirror.

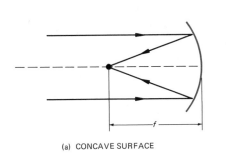

(a) CONCAVE SURFACE

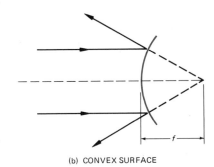

(b) CONVEX SURFACE

18.6 *Reflection from Spherical Mirrors*

The use of curved surface mirrors makes it possible to converge or diverge light rays upon reflection. They produce images that may be either real or virtual and either enlarged or reduced compared to the object. The most common curved mirror surfaces are spherical in shape. The spherical surface has an imaginary symmetry axis which is called the *optical axis* of the surface.

The *focal point* of any optical element is the point on the optical axis where rays parallel to the optical axis are focused. The focal points for a concave and convex mirror respectively are shown in Figures 18.8a and b. The distance along the optical axis from the mirror to the focal point is called the *focal length f.*

You may use the basic principle of reflection and a ray diagram to solve spherical mirror problems. It should be noted that spherical mirrors exhibit *spherical abberation* for rays some distance away from the optical axis of the mirror. This means that sharp images are only possible for rays near the optical axis of the mirror. Spherical aberration is illustrated in Figure 18.9.

There are three rays that can be used to locate images formed by a mirror and that are easy to draw (Figure 18.10). These are

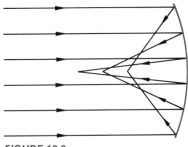

FIGURE 18.9
Spherical aberration for a concave mirror.

FIGURE 18.10
Ray diagram for a concave mirror. Three ray paths for locating the image are given.

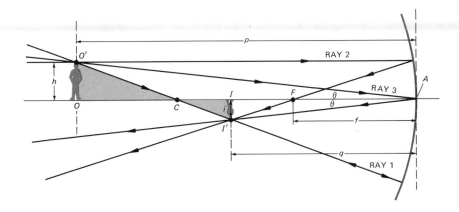

1. Any ray that passes through the center of curvature C. It has an incident angle of $0°$ since its path is perpendicular to the mirror at the point of incidence and is reflected back along the same path. This ray is normal to the mirror surface at the point of reflection.
2. An incident ray that is parallel to the optical axis OA of the mirror. Such a ray is reflected to intersect the optical axis at the focal point F. Any ray that passes through F will be reflected parallel to the optical axis. The distance of the point F from the surface of the mirror is the focal length f.
3. An incident ray that strikes the mirror at the optical axis and makes an angle θ with the axis is reflected at the same angle θ on the opposite side of the axis.

If these three rays are drawn from a point of the object O' they will intersect at a point I' of the image as shown in Figure 18.10. In any case, only two of these three rays are needed to locate an image. So we select the two that are most convenient for the particular diagram we are drawing.

EXAMPLE

Let us use ray diagrams to locate images in the following cases in Figure 18.11:

a. a concave mirror with object distance greater than the focal length
b. a concave mirror with object distance less than the focal length
c. a convex mirror

If the image formed by a mirror can be displayed on a screen, it is called a *real image*. Which of the previous three images we sketched are real? How can you tell from our ray diagram which images are real? All

FIGURE 18.11
Ray diagrams for (a) a concave mirror with the object at a distance greater than the focal length, (b) a concave mirror with the object at a distance less than f, and (c) a convex mirror. An object outside of concave mirror center of curvature c gives a real, inverted, and reduced image. An object inside focal length of concave mirror gives a virtual, upright and enlarged image. An object outside of focal length of convex mirror gives a virtual, upright, and reduced image.

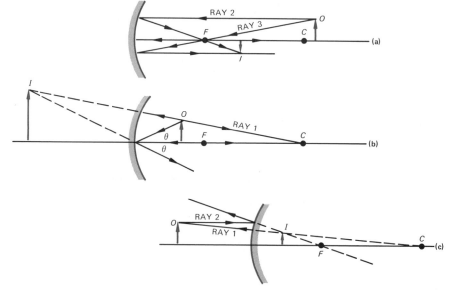

other images are called virtual images. What are their characteristics? The image formed in Figure 18.11a is real, inverted, and reduced. The image in Figure 18.11b is virtual, upright, and enlarged. The image in Figure 18.11c is virtual, upright, and reduced.

Although graphical methods are instructive in depicting the properties of mirrors, we can obtain algebraic relationships between the object distance p, image distance q, and focal length f. Consider the case in Figure 18.10, but let us require that the size of the mirror be small relative to the radius of curvature.

By definition ray 3 makes equal angles with the optical axis OA. What can you deduce about the two triangles $O'OA$ and $I'IA$? These two right triangles are similar triangles. What is the ratio of h to i? The sides of these two triangles are of equal ratios: $h/p = \tan \theta = i/q$. Can you derive a relationship between the ratio h/i and p and q?

Magnification M is defined as the negative of the ratio of the image size to the object size,

$$M = -\frac{\text{size of image}}{\text{size of object}} = -\frac{i}{h} = -\frac{q \tan \theta}{p \tan \theta}$$

$$M = -\frac{q}{p} \tag{18.6}$$

where the minus sign indicates that the image is inverted and real for positive values of q and p. From similar triangles $OO'C$ and $II'C$ you can show that the ratio of object size to image size is equal to the ratio OC/CI, but $OC = p - CA$ and $CI = CA - q$, where CA is equal to the radius of the curvature R of the mirror.

$$\frac{h}{i} = \frac{OC}{CI} = \frac{p - CA}{CA - q} = \frac{p - R}{R - q}$$

But we found before that

$$\frac{h}{i} = \frac{p}{q} = \frac{p - R}{R - q}$$

That is,

$$pR - pq = pq - qR$$

By dividing each term by pqR we can show that this expression is equivalent to

$$\frac{1}{p} + \frac{1}{q} = \frac{2}{R} \tag{18.7}$$

If we let p become very large, the incident rays are parallel to each other and to the optical axis and they are reflected to intersect the optical axis at F, the focal point. Since for very large p the reciprocal $1/p$ approaches zero, q becomes $R/2$. Thus the focal length is equal to one-half the radius of the curvature, and we can rewrite Equation 18.7 in terms of the focal length:

$$\frac{1}{p} + \frac{1}{q} = \frac{1}{f} \tag{18.8}$$

In order to use Equation 18.8 for all spherical mirrors, it is convenient to adopt the following convention of signs:

p is positive for a real object
q is positive for a real image
f is positive for a concave mirror
p is negative for a virtual object and is possible only for systems with at least two optical elements in which light converges toward the virtual object on the back side of the mirror
q is negative for a virtual image
f is negative for a convex mirror

EXAMPLES

1. A concave mirror has a 50.0-cm radius of curvature, that is, a focal length of 25.0 cm. The object distance is 60.0 cm. Where is the image? How would you describe the image?

    $R = +50.0$ cm, $f = +25.0$ cm, and $p = +60.0$ cm.

    Using $1/p + 1/q = 1/f$, we can substitute:

    $$\frac{1}{60.0} + \frac{1}{q} = \frac{1}{25.0}$$

    By rearranging,

    $$\frac{1}{q} = \frac{1}{25.0} - \frac{1}{60.0} = \frac{60.0 - 25.0}{1500}$$

    Therefore, $q = 1500/35.0 = +42.9$ cm. The magnification thus is

    $$M = -\frac{q}{p} = -\frac{42.9}{60.0} = -0.714$$

    The image is real, inverted, reduced. Draw a ray diagram of this problem in Figure 18.12.

2. Using the same concave mirror as in example 1 and an object distance of 40.0 cm, where is the image? How would you describe the image?

    $f = +25.0$ cm and $p = +40.0$ cm.

    $$\frac{1}{40.0} + \frac{1}{q} = \frac{1}{25.0}$$

    $$\frac{1}{q} = \frac{1}{25.0} - \frac{1}{40.0} = \frac{40.0 - 25.0}{1000}$$

    $$q = \frac{1000}{15.0} = +66.7 \text{ cm}$$

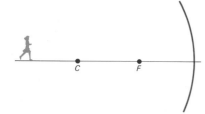

FIGURE 18.12
An object at 60 cm in front of concave mirror with 50-cm radius of curvature.

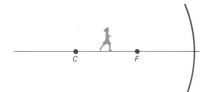

FIGURE 18.13
An object 40 cm in front of concave mirror with 50-cm radius of curvature.

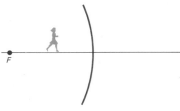

FIGURE 18.14
An object 20 cm in front of concave mirror with 50-cm radius of curvature.

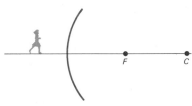

FIGURE 18.15
An object 20 cm in front of convex mirror with 50-cm radius of curvature.

$$M = -\frac{q}{p} = -\frac{66.7}{40} = -1.67$$

The image is real, inverted, and enlarged. Draw a ray diagram of this problem in Figure 18.13.

3. Again using the mirror in example 1 with an object distance of 20.0 cm, where is the image? Describe the image.

$$f = +25.0 \text{ cm and } p = +20.0 \text{ cm.}$$

$$\frac{1}{20.0} + \frac{1}{q} = \frac{1}{25.0}$$

$$\frac{1}{q} = \frac{1}{25.0} - \frac{1}{20.0} = \frac{20.0 - 25.0}{500}$$

$$q = -100 \text{ cm}$$

$$M = -\frac{-100}{20.0} = +5.0$$

The image is virtual, erect, and enlarged. Draw a ray diagram of this problem in Figure 18.14.

4. For a *convex* mirror of radius of curvature of 50.0 cm, $f = -25.0$ cm. The object is 20.0 cm from the mirror. Where is the image? Describe the image.

$$f = -25.0 \text{ cm and } p = +20.0 \text{ cm.}$$

$$\frac{1}{20.0} + \frac{1}{q} = -\frac{1}{25.0}$$

$$\frac{1}{q} = -\frac{1}{25.0} - \frac{1}{20.0} = \frac{-20.0 - 25.0}{500}$$

$$q = -\frac{500}{45.0} = -11.1 \text{ cm}$$

$$M = -\frac{-11.1}{20.0} = +\frac{11.1}{20.0} = +0.556$$

The image is virtual, erect, and reduced. Draw a ray diagram of this problem in Figure 18.15.

18.7 Lenses

The primary optical elements in a microscope are lenses. We can define a lens as two curved surfaces with a common axis separated by a refractive medium between them. The refractive medium should be transparent to light rays. Glass is the most common material used as the refractive medium; other materials used are plastics and quartz. The lens is immersed in some medium, and in our case we will consider a glass lens in air. How are the properties of a glass lens changed when it is used under water?

FIGURE 18.16
Cross-sections of different types of
converging and diverging lens.

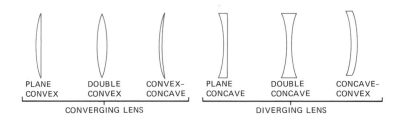

The two surfaces of a lens may be concave, convex, planar, or a combination of them. A lens that is thicker at the center than at the edge is said to be convex, or converging. One that is thinner at the center than at the edge is said to be concave, or diverging (Figure 18.16). We will treat only the properties of thin lenses whose thicknesses are much less than their diameters. When this is the case we can assume all of the refraction occurs at the center of the lens.

The terms *converging* and *diverging* indicate what happens to parallel beams of light incident upon the lenses. For a convergent lens the emergent beam converges toward the axis and for a divergent lens it diverges from the axis.

In a fashion similar to the definition of a focal point F of a mirror, we can define the focal point of a lens as the point on the optical axis of the lens where incident parallel rays are, or appear to be, focused (Figure 18.17). The distance from the lens to the focal point is called the focal length f.

A ray diagram can be made for a thin lens to locate an image formed by the lens. There are three incident rays whose refracted directions are easiest to draw. They are as follows:

1. An incident ray parallel to the optical axis; it is refracted through the focal point.
2. An incident ray through the center of the lens. It passes undeviated because the angle of incidence is equal to 0°.
3. An incident ray through the focal point. It is refracted to a direction parallel to the axis. These three rays are shown in Figure 18.18.

We can use a ray diagram to develop the algebraic relationship be-

FIGURE 18.17
Focal length illustrations for a
convergent (convex) and a divergent
(concave) lens.

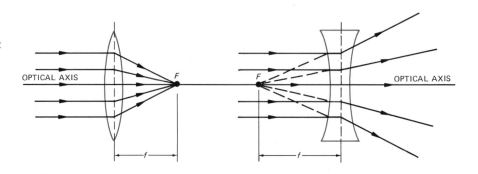

FIGURE 18.18
Three rays used to locate the image
in a ray diagram involving a thin
lens.

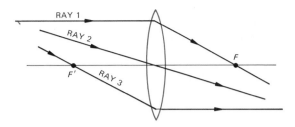

tween the object distance p, the image distance q, and the focal length f.
Can we find two similar triangles that contain h and i in Figure 18.19?
Yes, triangle $O'OA$ contains h and is similar to $I'IA$ which contains i.
How is the h/i ratio related to OA/IA, or p/q? From the properties of
similar triangles $h/i = OA/IA = p/q$.

The magnification is defined, as for mirrors, as the negative of the
ratio of the size of the image to the size of the object,

$$M = -\frac{i}{h} = -\frac{q}{p} \tag{18.9}$$

where the minus sign indicates that a real image is inverted relative to
the object.

Another pair of similar triangles is $OO'F'$ and ADF'. The ratio of the
sides of these two triangles must be equal,

$$\frac{h}{OF'} = \frac{AD}{AF'}$$

But AD is equal to i; so

$$\frac{h}{i} = \frac{OF'}{AF'} = \frac{p-f}{f} \tag{18.10}$$

The third pair of similar triangles is ABF and $II'F$. The ratio of the
sides of these two triangles must be equal,

$$\frac{AB}{AF} = \frac{i}{FI}$$

FIGURE 18.19
The image formation of a single
convex lens.

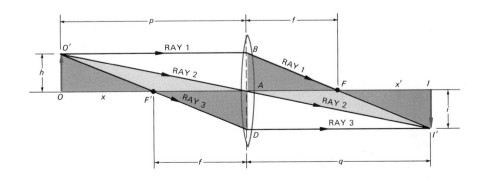

But AB is equal to h, so

$$\frac{h}{i} = \frac{AF}{FI} = \frac{f}{q - f} \tag{18.11}$$

Now we can set Equation 18.10 equal to Equation 8.11

$$\frac{h}{i} = \frac{p - f}{f} = \frac{f}{q - f} \tag{18.12}$$

If we let $(p - f) = x$, and $(q - f) = x'$ (please consider the physical meaning of x and x'), then $xx' = f^2$. This is a useful lens relationship and can be used to solve many problems, but it does not represent the goal for which we started.

We can transform Equation 18.12 into a form where the three variables are separated by multiplying Equation 18.12 by $(q - f)$ and dividing the result by qpf to obtain,

$$\frac{1}{f} = \frac{1}{p} + \frac{1}{q} \tag{18.13}$$

The above relationship was derived for a simple converging lens. However, it holds for any thin lens if you use the following sign convention:

p is positive for real objects

p is negative for virtual objects, which are only possible for systems with at least two optical elements in which light converges toward a virtual object on the back side of the lens

q is positive for real images

q is negative for virtual images

f is positive for converging (convex) lenses

f is negative for diverging (concave) lenses

EXAMPLES

1. If an object is 30.0 cm in front of a converging lens of 20.0-cm focal length, what is the position of the image? Describe the image. Draw the ray diagram in Figure 18.20.

Let $p = 30.0$ and $f = 20.0$.

$$\frac{1}{p} + \frac{1}{q} = \frac{1}{f}$$

so

$$\frac{1}{30.0} + \frac{1}{q} = \frac{1}{20.0}$$

$$\frac{1}{q} = \frac{1}{20.0} - \frac{1}{30.0} = \frac{30.0 - 20.0}{600}$$

Therefore, the distance to the image is $q = 60.0$ cm.

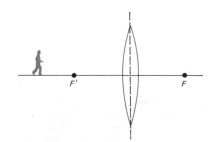

FIGURE 18.20
An object 30 cm in front of converging lens with a 20-cm focal length.

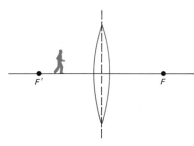

FIGURE 18.21
An object 15 cm in front of converging lens with a 20-cm focal length.

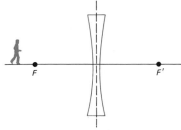

FIGURE 18.22
An object 30 cm in front of diverging lens with a 20-cm focal length.

$$\text{magnification} = -\frac{q}{p} = -\frac{60.0}{30.0} = -2.00$$

The image is real, inverted, and enlarged.

2. If an object is 15.0 cm in front of a converging lens of 20.0-cm focal length, what is the position of the image? Draw the ray diagram in Figure 18.21.

Let $p = 15.0$ and $f = 20.0$

$$\frac{1}{p} + \frac{1}{q} = \frac{1}{f}$$

So

$$\frac{1}{15.0} + \frac{1}{q} = \frac{1}{20.0}$$

$$\frac{1}{q} = \frac{1}{20.0} - \frac{1}{15.0} = \frac{15.0 - 20.0}{300} = -\frac{5.0}{300}$$

Therefore, the position of the image is $q = -60$.

$$\text{magnification} = -\frac{q}{p} = -\frac{-60}{15.0} = 4.0$$

The image is enlarged, upright, and virtual.

3. If an object is 30.0 cm in front of a diverging lens of 20.0-cm focal length, what is the position of the image? Describe the image. Draw the ray diagram in Figure 18.22.

Let $p = 30.0$ cm, and $f = -20.0$ cm.

$$\frac{1}{p} + \frac{1}{q} = \frac{1}{f}$$

So

$$\frac{1}{30.0} + \frac{1}{q} = \frac{1}{-20}$$

$$\frac{1}{q} = \frac{-1}{20.0} - \frac{1}{30.0} = \frac{-20.0 - 30.0}{600}$$

Therefore the image distance is

$$q = \frac{600}{-50.0} = -12.0$$

$$\text{magnification} = -\frac{q}{p} = -\frac{-12.0}{30.0} = 0.400$$

The image is reduced, upright, and virtual.

18.8 Magnifier or Reading Glass

A magnifier or reading glass is a single converging lens. The object is placed between the focal point and the lens. Thus an upright, enlarged, virtual image is formed. The eye sees the virtual image. The apparent

FIGURE 18.23
The angular magnification of a
converging lens with an object inside
its focal length.

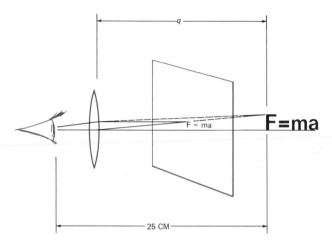

size of an object is determined by the size of the retinal image. If the
eye is unaided, the size of the retinal image depends upon the angle
subtended by the object at the eye. This suggests that all you need to do
to magnify an object is to bring it closer to your eye. Your eye, however,
cannot focus on objects closer than the *near point*. The near point is
defined as the point of the closest distance for distinct vision. This distance varies from one individual to another, but it is usually considered
to be about 25 cm. With a converging lens it is possible to produce an
enlarged virtual image which will increase the angle subtended on the
retina. To obtain maximum magnification locate the object between the
lens and its focal point so the virtual image is about 25 cm from the
lens when the lens is held close to your eye (Figure 18.23) We can use
Equations 18.9 and 18.13 to calculate the magnifying power of this lens.
For the situation we have described above the image distance is given
by -25 cm, and the *angular magnification* is given by

$$M = \frac{-q}{p} = \frac{25 \text{ cm}}{p}$$

where $1/p$ can be computed from Equation 18.13 and is equal to $(1/25 +
1/f)$ or $(f + 25$ cm$)/25f$. Then we can find the angular magnification M' in
terms of the focal length of the lens,

$$M' = \frac{25 \text{ cm}}{p} = 25 \text{ cm} \times \frac{f + 25 \text{ cm}}{f \times (25 \text{ cm})} = 1 + \frac{25(\text{cm})}{f(\text{cm})} \tag{18.14}$$

Questions

1. A typical reading glass has a focal length of 10 cm. What is its magnification?
   [3.5×]
2. The typical ocular magnifier of medical doctors has a magnification of 8. What
   is its focal length?   [3.6 cm]

FIGURE 18.24
Schematic diagram of a simple camera.

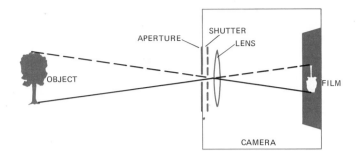

18.9 Cameras

The camera is essentially a converging lens, a light tight box, and a photo-sensitized detector (photographic film). The object distance is greater than the focal length, in general many times greater, and a real, inverted, reduced, image is formed on the sensitized detector. The camera has an additional feature called the *aperture* whereby the area of the lens that transmits the incident radiation is controlled (See Figure 18.24.)

The camera aperture is indicated by the f-number which is the ratio of the focal length of the lens to the diameter of the aperture; e.g., f/2.8 designates a focal length 2.8 times the diameter of the lens aperture. The intensity of light is defined as the incident energy per unit area per second. Then the total energy that enters the lens is equal to the product of the intensity I, area of aperture A, and the exposure time t. The energy per unit area that strikes the photographic surface determines the photo-chemical activity upon development of the photographic film. For the desired photographic result this means that IAt is a constant.

Questions

What is the relation between f-number and exposure time for a constant total energy input? Typical f-numbers are 1.4, 2.8, 4, 5.6, 8, 11, 16, and 22. What is the ratio of the energy per unit area between consecutive f-numbers?

18.10 Optical Systems of More than One Element

A number of optical instruments are made of a combination of single optical elements. For the analysis of a complex optical system you can treat each element alone and make use of the concept that the image formed by the first element becomes the object for the second element, the image formed by the second element becomes the object for the third element, and so on.

FIGURE 18.25
Two lens problem. The superposition principle applies to yield the image of the first lens as the object of the second lens.

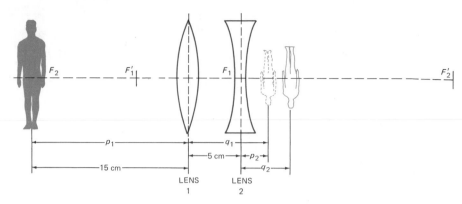

EXAMPLE

An object is 15 cm in front of a converging lens of focal length 5.00 cm. A diverging lens of focal length 20.0 cm is placed 5.00 cm beyond the converging lens. Find the position of the image and describe the image (Figure 18.25).

For the first lens: (all distances are in cm)

$$\frac{1}{p_1} + \frac{1}{q_1} = \frac{1}{f_1}$$

$$\frac{1}{15.0} + \frac{1}{q_1} = \frac{1}{5.00}$$

$$\frac{1}{q_1} = \frac{1}{5.00} - \frac{1}{15.0} = \frac{3.00 - 1.00}{15.0} = \frac{2.00}{15.0}$$

$$q_1 = 7.50$$

For the second lens:

$$\frac{1}{p_2} + \frac{1}{q_2} = \frac{1}{f_2}$$

$$|p_2| = |7.50 - 5.00| = 2.50$$

and the object treated is a virtual object because it is on the back side of the lens, $p_2 = -2.50$, f_2 is -20.0.

$$\frac{1}{-2.50} + \frac{1}{q_2} = \frac{1}{-20.0}$$

$$\frac{1}{q_2} = \frac{1}{-20.0} + \frac{1}{2.50} = \frac{8.00 - 1.00}{20.0}$$

$$q_2 = \frac{20.0}{7.00} = 2.86$$

The image for the system is real and inverted. The magnification is the product of the magnification of lens 1 and the magnification of lens 2. Thus

$$M_t = M_1 M_2 = \frac{-q_1}{p_1} \times \frac{-q_2}{p_2} = \frac{-7.50}{15.0} \times \frac{-2.86}{-2.50} = \frac{-21.4}{37.5} = -0.571$$

18.11 The Compound Microscope

The microscope is a most important instrument for the medical professions. The invention of the microscope began the modern era of bacteriology. In fact, much of your understanding of illness may be the result of the findings of microscopic observations that have made their way into the common folk understanding of medicine for our society.

The microscope is a compound optical instrument. The way to develop an understanding of it is to use the superposition concept; that is, we can study the individual components and then analyze their properties when they are used together in the microscope. You have learned the basic principles of single lenses. You can use these principles to develop an explanation of the operation of a microscope.

The microscope is an optical instrument in which a combination of lenses is used to produce a greatly enlarged image for your eye. What kind of image is it, real or virtual? How do you photograph a microscope image?

The first (objective) lens is a short focal length lens, and the object distance is only slightly larger than the focal length. A real image is formed by the objective lens just within the focal length of the second lens (the eyepiece). The final image formed by the eyepiece is virtual. See Figure 18.26. The total magnification M_t of the compound micro-

FIGURE 18.26
Schematic diagram of a compound microscope with image at 25 cm.

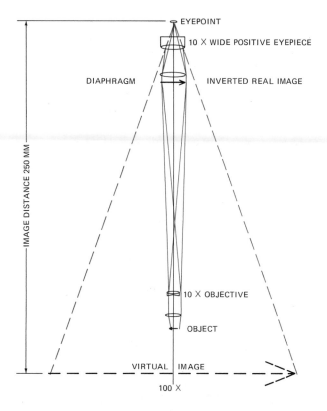

scope is the product of the magnification of the objective and the magnification of the eyepiece,

$$M_t = M_0 M_e$$

$$M_t = -\frac{q_1}{p_1}\left(\frac{25}{f_e} + 1\right) \tag{18.15}$$

For practical considerations, q_1 is approximately the length of the microscope tube, normally set at a standard value of 16.0 cm, and p_1 is approximately equal to f_0. A working expression for the magnification becomes

$$M_t = \frac{-400}{f_0 f_e} \tag{18.16}$$

where f_0 and f_e are the focal lengths of the objective lens and the eyepiece lens in centimeters.

EXAMPLE

A microscope has an objective lens with a focal length of 1.00 cm and an eyepiece lens with a focal length of 2.00 cm. If the objective has a magnification of 10.0 and the final image is at 25.0 cm from the eyepiece, find the total magnification and the length of the microscope tube. (All distances are in centimeters.)

First we can use the expression for the magnification of the objective to find the location of the objective lens image.

$$M_0 = -10.0 = -\frac{q_1}{p_1} \qquad q_1 = 10.0 p_1$$

$$\frac{1}{p_1} + \frac{1}{10.0 p_1} = \frac{1}{1.00} \qquad \frac{10.0 + 1}{10.0 p_1} = 1.00$$

$$p_1 = 1.10 \text{ cm} \quad \text{and} \quad q_1 = 11.0 \text{ cm}$$

Second, we can use the location of the eyepiece image to find the location of the eyepiece object. For $q_2 = -25.0$ cm

$$\frac{1}{p_2} + -\frac{1}{25.0} = \frac{1}{2.00}$$

$$\frac{1}{p} = \frac{1}{2.00} + \frac{1}{25.0} = \frac{25.0 + 2.00}{50.0}$$

$$p_2 = \frac{50.0}{27.0} = 1.85 \text{ cm}$$

length of the tube $= q_1 + p_2 = 12.85$ cm

$$\text{magnification}_t = M_0 M_e = (-10.0)\left(\frac{25.0}{f} + 1\right)$$

$$= (-10.0)\left(\frac{25.0}{2.00} + 1\right) = -135$$

18.12 *Lens Distortions*

There are two kinds of distortion that are common for single-lens systems. *Spherical aberration* is the term used to describe the distortion caused by rays which pass through the outer edges of the lens being focused at different points than rays near the optical axis of the lens. This distortion is common for all lenses. It is minimized by using apertures with lenses to limit the effective lens diameter and to reduce the spherical aberration. *Chromatic aberration* is the term used to describe the distortion caused by the fact that different wavelengths (colors) of light are focused at different places on the optical axis. This means that white light will produce a spread of colored images along the optical axis. This distortion can be reduced by using multiple lens systems and thin film coatings to make optical elements that are called *achromatic systems*. These systems are designed to minimize chromatic aberration in visible light applications.

Spherical lens distortion. The lens is placed so that the page of the telephone directory is slightly below the focal region. Note the difference in focal length for different parts of the lens. (Picture from *PSSC Physics*, D.C. Heath and Company, Lexington, Mass., 1965.)

ENRICHMENT
18.13 Depth of Field and Focus

The *depth of field* of an optical system is the range of object distances that produces an image that meets an acceptable sharpness of image criteria. The associated variation of the image distance is referred to as the *depth of focus*. The aperture size or diaphragm controls the depth of field. Let us consider the case for the eye. For a given amount of de-focusing or blurring there is a corresponding distance between the image and the retina. The diameter of the blur circle (also known as the circle of least confusion) for a point object is proportional to the diameter of the opening in the eye lens as shown in Figure 18.27.

$$\text{depth of focus for the eye} = q_2 - q_1 \tag{18.17}$$

$$\text{depth of field for the eye} = p_1 - p_2 \tag{18.18}$$

If an object at p_o is in focus on the retina at a distance l from the eye lens, it can be shown that the focal length of the eye lens is given by

$$f_{\text{eye}} = \frac{p_o l}{p_o + l} \tag{18.19}$$

FIGURE 18.27
Depth of focus diagrams.

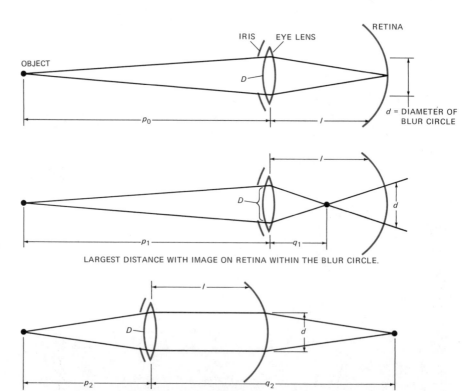

LARGEST DISTANCE WITH IMAGE ON RETINA WITHIN THE BLUR CIRCLE.

SHORTEST DISTANCE WITH IMAGE ON RETINA WITHIN THE BLUR CIRCLE.

where l is length of the eyeball, that is, the distance from the eye lens to the retina.

With the eye still focused on an object p_0 meters from the eye lens, what is the distance p_1 to the farthest object whose image lies within the blur circle of the eye? (See Figure 18.27b.)

From similar triangles,

$$\frac{q_1}{D} = \frac{l - q_1}{d}$$

$$q_1 = \frac{Dl}{D + d} \tag{18.20}$$

where d is the diameter of the blur circle and D is the diameter of the iris.

Using Equation 18.13 for lenses, we can find the object distance, p_1,

$$p_1 = \frac{p_0}{1 - (dp_0/lD)} \tag{18.21}$$

What is the shortest distance to an object that will still be in focus? (See Figure 18.27c.)

From similar triangles

$$\frac{D}{q_2} = \frac{d}{q_2 - l}$$

$$q_2 = \frac{Dl}{D - d} \tag{18.22}$$

Once again the lens equation, Equation 18.13 can be used to obtain an expression for p_2,

$$p_2 = \frac{p_0}{1 + (dp_0/lD)} \tag{18.23}$$

Since the diameter of the blur circle for the eye d is much smaller than the diameter of the iris, the depth of focus of the eye is well approximated by,

$$\text{depth of focus} = q_2 - q_1 = \frac{2dl}{D} \tag{18.24}$$

and the depth of field by,

$$\text{depth of field} = p_1 - p_2 = p_0 \frac{2dp_0}{lD} \tag{18.25}$$

Note that for a given blur circle size (constant d), the depth of focus is increased when D, the diameter of the iris, is decreased. The increase in depth of focus corresponds to an increased depth of field. Your depth of focus is decreased in dim light because the D increases. The angle of resolution of the human eye (d/l) is approximately 10^{-3} radians.

EXAMPLE

What is the depth of field for a human eye at its near point (25.0 cm) in bright room light where the iris diameter is 2.00 mm?

$$\text{depth of field} = (25.0 \text{ cm})2(10^{-3})\frac{25.0 \text{ cm}}{0.200 \text{ cm}} = 6.25 \text{ cm}$$

Let us consider the problem of determining the depth of field for a camera. The blur circle is determined by the properties of the photographic film. In general, the blur circle of the film is smaller than the minimum angle of resolution of the human eye. If a photograph is held at your near point (25 cm), what is the maximum size of a blur circle you will see on the film?

The equations for the depth of field of a camera are similar to those derived above for the human eye. Since the angle of resolution of the eye ($10^{-3} = d/l$) determines the blur circle for viewing a film, this value for the d/l ratio can be substituted into Equation 18.21 and 18.23:

$$p_1 = \frac{p_0}{1 - (p_0/1000D)} \tag{18.26}$$

$$p_2 = \frac{p_0}{1 + (p_0/1000D)} \tag{18.27}$$

where D is the diameter of the camera aperature. If $p_0 = 1000D$, then p_1 moves to infinity. This value of p_0 is called the *hyperfocal distance* for the camera. If $p_0 = 1000D$, p_2 is $500D$. For this case the depth of field $(p_1 - p_2)$ is from $500D$ to infinity.

EXAMPLE

Given a camera with a 55-mm focal-length lens. If the lens is stopped to f/11, then $D = 55/11$ mm. The hyperfocal distance is $1000D$, or 5 m, and the depth of field ranges from 2.5 m to infinity.

SUMMARY

Use these questions to evaluate how well you have achieved the goals of this chapter. The answers to these questions are given at the end of this summary with the number of the section where you can find the related content material.

Definitions

1. The focal point of an optical element is defined as

the point of
a. center of curvature
b. image formation
c. object position
d. parallel light focus

2. The numerical value of the magnification of a lens is equal to
a. image distance/object distance
b. image size/object size

c. object distance/image distance
d. object size/image size
e. none of these

3. A _____ image can be focused on a screen, and a _____ image is the apparent source of diverging light.

4. A _____ lens has a negative focal length, and a _____ lens has a positive focal length.

5. Internal reflection occurs only when the light originates in a medium of refractive index n_2 which is in an external medium of refractive index n_1 that satisfies the condition
   a. $n_1 < n_2$
   b. $n_1 > n_2$
   c. $n_1 = n_2$
   d. any of these
   e. none of these

6. The reflection coefficient for a glass-air interface is equal to
   a. transmission coefficient
   b. $1 -$ transmission coefficient
   c. one
   d. zero
   e. none of these

7. Spherical aberration in a lens results in multiple images along the optical axis because of different focal lengths for
   a. different colors
   b. different materials
   c. off-axis rays
   d. source positions
   e. none of these

Ray Diagrams

8. Sketch a ray diagram for an object located a distance $3f/2$ in front of a positive lens of focal length f.

Lens and Mirror Equations

9. If an object is $2f$ in front of a lens with focal length f, the magnification of the image will be
   a. 2
   b. -2
   c. 1
   d. -1
   e. -4

10. A diverging lens will always give an image that is
    a. virtual, reduced
    b. virtual, enlarged
    c. real, reduced

    d. real, enlarged
    e. none of these

11. If a mirror gives an enlarged, virtual, image, you know the mirror must be
    a. plane
    b. concave
    c. convex
    d. none of these

Optical Devices

12. A reading glass gives a virtual enlarged image. It must be a
    a. diverging lens
    b. converging lens
    c. prism
    d. none of these

13. The object for a reading glass must be placed
    a. between f and $2f$
    b. outside $2f$
    c. inside f
    d. none of these

14. In normal viewing a compound microscope has a _____ lens producing a real image and a _____ lens giving a virtual image.

15. The f number of a camera is the ratio
    a. aperture/focal length
    b. focal length/aperture
    c. shutter speed/focal length
    d. focal length/shutter speed

16. Light pipes using fiber optics are based on
    a. long focal length mirrors
    b. short focal length lenses
    c. vacuum transmission
    d. internal reflection
    e. none of these

Answers

1. d (Section 18.6)
2. a, b (Section 18.7)
3. real, virtual (Section 18.6)
4. diverging, converging (Section 18.7)
5. a (Section 18.5)
6. b (Section 18.3)
7. c (Section 18.12)
8. see (Section 18.7)
9. d (Section 18.7)
10. a (Section 18.7)
11. b (Section 18.6)
12. b (Section 18.8)
13. c (Section 18.8)
14. objective, eyepiece, (Section 18.11)
15. b (Section 18.9)
16. d (Section 18.5)

ALGORITHMIC PROBLEMS

Listed below are the important equations from this chapter. The problems following the equations will help you learn to translate words into equations and to solve single-concept problems.

Equations

$$\angle i = \angle r \tag{18.1}$$

$$R = \frac{(n_2 - n_1)^2}{(n_2 + n_1)^2} \tag{18.3}$$

$$n_1 \sin \theta_1 = n_2 \sin \theta_2 \tag{18.4}$$

$$\sin \theta_c = \frac{1}{n_{21}} \tag{18.5}$$

$$M = -\frac{q}{p} \tag{18.6, 18.9}$$

$$\frac{1}{p} + \frac{1}{q} = \frac{2}{R} = \frac{1}{f} \tag{18.7, 18.8}$$

$$\frac{1}{f} = \frac{1}{p} + \frac{1}{q} \tag{18.13}$$

$$M' = 1 + \frac{25}{f} \tag{18.14}$$

$$M_t = -\frac{q_1}{p_1}\left(1 + \frac{25}{f_e}\right) \tag{18.15}$$

$$M_t = -\frac{400}{f_o f_e} \tag{18.16}$$

$$q_2 - q_1 = \frac{2dl}{D} \tag{18.24}$$

$$p_1 - p_2 = p_o \frac{2dp_o}{lD} \tag{18.25}$$

Problems

1. What is the index of refraction of carbon disulfide relative to water?
2. What is the critical angle for quartz in air?
3. The magnification of a compact mirror is 1.20. Where is the image if the object distance is 10.0 cm? Interpret.
4. What is the focal length of a reading glass which has a magnification of 6?
5. A microscope has a total magnification of 500 and the eyepiece has a focal length of 2.00 cm. What is the ratio of image distance to the object distance for the objective?
6. If the distance between the objective lens and the eyepiece in the microscope of problem 5 is 16.0 cm, what is the focal length of the objective?

7. Find the percentage of light reflected at normal incidence on an air-water interface. ($n_{water} = 1.33$)

Answers

1. 1.22
2. arcsin $= 0.648$ or $40°$
3. $q = -12$ cm, virtual, in front of the mirror, upright image

4. 5 cm
5. 500/13.5
6. 0.37 cm
7. 2 percent

EXERCISES

These exercises are designed to help you apply the ideas of a section to physical situations. When appropriate the numerical answer is given in brackets at the end of each exercise.

Section 18.3

1. Find the minimum length of a plane mirror in which a man 2 m tall can see both his feet and the top of his head. Where would he stand relative to the mirror? [1 m, independent of object distance]
2. Assuming that there is no absorption at the air-glass interface, find the percentage of light transmitted into the glass at normal incidence. (Use $n = 1.5$ for glass.) [$T = 96$ percent]
3. Find the index of refraction of a piece of clear plastic if it is found that 9.00 percent of the normal incident light is reflected at the air-plastic interface. [$n = 1.86$]

Section 18.4

4. A plate glass window is 0.250 cm thick and has an index of refraction of 1.50. How much is the line of sight displaced by the window when the angle of incidence is 6°? [8.3×10^{-3} cm]
5. A monochromatic beam of light is incident at 50° on one face of an equilaterial prism whose index of refraction is 1.60. Draw a ray diagram, and calculate
   a. the angle of refraction at the first surface
   b. the angle of incidence at the second surface
   c. the angle of emergence at the second surface
   d. the total angle of deviation   [a. $28°26'$, b. $31°24'$; c. $56°28'$; d. $46°28'$]

Section 18.5

6. A skin diver, 5 m beneath the surface of the ocean, stops work long enough to watch the sun set. At what angle does he look if the index of refraction of the ocean water is 1.38?   [$46.4°$ from vertical]

Section 18.6

7. It is desired to design a make-up or shaving mirror which has a magnification of 1.25. What should be the focal length of the mirror? (Assume the image is at 25 cm.) Is it concave or convex? Draw a ray diagram.   [concave, $f = 100$ cm]
8. A 2.0-cm object is 40 cm from a concave mirror which has a radius of curvature of 50 cm. Where is the image formed? Make a ray diagram, and describe the image. [+67 cm]
9. A 4.00-cm high lighted candle is 3.00 m from a convex spherical reflector of 30-cm diameter. Where is the image formed, and can it be screened? Draw a ray diagram.   [−7.3 cm]

Section 18.7

10. A slide projector for 35.0 mm slides produces an image with a linear magnification of 40.0. The focal length of the projection lens is 30.0 cm. What are the object and image distances for this projection system?   [30.8 cm, 1232 cm]
11. The dimensions of a lantern slide are 7.62 cm × 10.2 cm (3 inches × 4 inches). If we wish to project an image of the slide enlarged to 1.83 m × 2.44 m

(6 ft × 8 ft) on a screen 9.14 m (30 ft) from the projection lens, find
a. the focal length of the lens
b. the position of the slide with respect to the lens
[a. 36.6 cm (1.2 ft); b. 38.1 cm]

Section 18.8

12. Scientific supply stores sell small plastic "bug boxes" for children. The bug box is a clear plastic cube 5.00 cm on a side with a lens in the top. Determine the focal length and magnification of this lens. [6.25 cm, 5]

Section 18.9

13. If you have a camera with an f/2.8 lens of 8-cm focal length, what is the diameter of the lens? With this lens the proper exposure time for a given scene is 1/100 sec at setting f/2.8. What would be the exposure time for f/5.6 setting?   [2.8 cm, 0.04 sec]

Section 18.11

14. The focal length of the objective of a microscope is 12 mm. An object is placed at 13 mm from the objective. The magnification of the eyepiece is 5. Parallel rays are emerging from the eyepiece. Where is the image of the objective formed, and what is the overall magnification?   [156 mm, 60]

15. A microscope has an objective ($f = 4.0$ mm) and an eyepiece ($f = 25$ mm). If the eyepiece is adjusted for normal viewing (image at 25 cm), find the actual magnification of the microscope when it is focused on an object 5.0 mm from the objective lens. Find the separation of the objective and eyepiece.   [$M = -44$, $l = 4.3$ cm]

PROBLEMS

Each of the following problems may involve more than one physical concept. A problem requiring material from the enrichment section is marked with a dagger †. The answer is given in brackets at the end of each problem.

16. A local boutique owner comes to you with the following questions:
    a. "I would like to install a mirror in my shop that makes all of the customers look 10 percent thinner. Is that possible?"
    b. "If so, what kind of mirror would I need to install and what are its properties?" (That means radius of curvature and focal length, etc. You may need to make a few back-of-the-envelope calculations and assumptions for which physicists are famous, if you are to give your friend a numerical answer.)   [For virtual, upright image of magnification of +0.90, a convex mirror of focal length 18 m would suffice for persons standing 2 m in front of the mirror]

17. A parallel beam of light which is composed of two wavelengths is incident upon the prism as described in exercise 5. The index of refraction of one wavelength is 1.60 and of the other is 1.62. What is the angle of separation of the two emergent rays? This spread is called dispersion. The emergent rays for each wavelength are parallel and incident upon a lens of 25-cm focal length. What is the separation of images formed by this lens?   [1.9°, 0.8 cm]

18. A telephoto lens combination consists of a converging lens of focal length +20 cm and a diverging lens of focal length −5.0 cm, the separation between the lenses is 17.5 cm. What should be the position of the film to photograph an object 100 m in front of the first lens?   [5.0 cm from second lens]

19. A compound microscope has an objective lens of focal length 1.0 cm and an eyepiece of focal length 2.0 cm. The object is an insect wing 0.20 mm in diameter on a slide 1.10 cm from the objective lens.
    a. How far apart are the lenses when the lenses are adjusted for the image at infinity?
    b. What is the angle subtended by the image of the magnified insect wing?
    c. It is desirable to focus the image of the bug on a screen 100 cm from the eyepiece. How far from the original position must the eyepiece be moved?
    d. What is the linear size of the image of the insect wing on the screen?   [a. 13 cm; b. 57.3°; c. 0.04 cm; d. 98 mm]

20. The schematic diagram for an opthalmoscope is shown in Figure 18.28. The opthalmoscope is an instrument for viewing the retina of the human eye. The primary element of the opthalmoscope is a partially silvered mirror, which reflects part of the light incident upon its surface and allows the other part of the light to pass through the mirror. Draw several light rays to show how the opthalmoscope permits the doctor to see the patient's retina.

FIGURE 18.28
Problem 20.

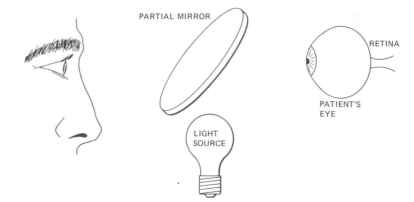

PARTIAL MIRROR

RETINA

PATIENT'S
EYE

LIGHT
SOURCE

†21. Derive an exact equation for the depth of field for the human eye. Using $d/l = 10^{-3}$, plot the depth of field as a function of distance for various iris sizes. [$p_1 - p_2 = 2dDlp_0^2/(D^2l^2 - d^2p_0^2)$]

22. A novelty company sells "eyeglasses for college administrators" to see "everything clearly." These eyeglasses consist of small pinholes in opaque lenses. Explain the physics of these lenses.

23. It is found that 92 percent of the normal incident light is transmitted through a plate of glass in air. Assuming normal incidence and no absorption losses, find the index of refraction of the glass. (You need only to consider first reflections at each interface. Can you show that this is a good approximation?) [$n = 1.5$]

24. A compound microscope consists of an objective of 1.5-cm focal length and an eyepiece of 5.0-cm focal length. The lenses are 17.5 cm apart. If the user adjusts the microscope for the image at infinity, find the distance the object must be from the objective, and determine the overall magnification for this use. [1.7 cm, 37]

†25. As you leave a lighted room, the light intensity decreases by a factor of 10. Assume the iris of the eye expands to keep the light intensity on the retina constant. By what factor does your depth of field change? [decreases by a factor of $1/\sqrt{10}$]

GOALS

When you have mastered the contents of this chapter, you will be able to achieve the following goals:

Definitions
Define each of the following terms, and use it in an operational definition:

interference	optical path length
diffraction	optical activity
dispersion	coherent source
polarization	noncoherent source
birefringence	

Application of Wave Properties
Explain the physical basis for:

thin film colors	polarimetry
diffraction grating spectrometry	holography
resolving power of optical systems	

Problems Involving Wave Properties
Solve problems involving interference, diffraction, and polarization.

Optical Activity
Design an experimental system capable of measuring the optical activity of a solution.

Lasers
Compare the laser with other light sources in terms of their optical characteristics.

PREREQUISITES

Before you begin this chapter you should have achieved the goals of Chapter 16, Traveling Waves, and Chapter 18, Optical Elements.

19

Wave Properties of Light

19.1 Introduction

Can you list some of your everyday experiences that are based on the wave properties of light? The colors observed in a soap bubble or an oil film on water are some examples. What wave property of light is involved in these observations? Polaroid sun glasses are designed to reduce glare from reflected light. What wave property of light is used in these glasses? The laser is a new light source that has unique properties; can you name some of these properties?

In this chapter we will explore the experimental basis of the wave phenomena of light. We will discuss the questions mentioned here and point out some applications of the wave nature of light.

19.2 Light Waves

Light is another example of an interaction-at-a-distance. Once again, as we discussed in Chapter 2, we postulate that the source of light is the source of a field that fills the space between the source and the receiver. We know of properties of light that can be explained as interference or diffraction phenomena. Hence, we construct a model of light as a wave. We shall begin by introducing you to the properties of our wave model of light, and then we shall discuss how this model helps in understanding the various properties of light. In our model light is a transverse electromagnetic wave. Light waves consist of electric and magnetic fields perpendicular to each other oscillating with the wave frequency at right angles to the direction of propagation. The speed of light in a vacuum c is the maximum possible speed for the transmission of energy. ($c \approx 3.0 \times 10^8$ m/sec) The speed of light in matter is less than c. The index of refraction of a material is c divided by the speed of light in the material. Since our model for light satisfies the wave equation we have the following relationships:

$$c = f\lambda \qquad n = \frac{c}{v} = \frac{f\lambda_0}{f\lambda} = \frac{\lambda_0}{\lambda} \tag{19.1}$$

TABLE 19.1
Index of Refraction of Glasses Relative to Air as a Function of Wavelength

	Wavelength in Nanometers (nm)							
	361	434	486	589	656	768	1200	2000
Variety of Glass	Ultraviolet	Purple	Blue	Yellow	Red		Infrared	
Zinc crown	1.539	1.528	1.523	1.517	1.514	1.511	1.505	1.497
High dispersion crown	1.546	1.533	1.527	1.520	1.517	1.514	1.507	1.497
Light flint	1.614	1.594	1.585	1.575	1.571	1.567	1.559	1.549
Heavy flint	1.705	1.675	1.664	1.650	1.644	1.638	1.628	1.617
Heaviest flint	—	1.945	1.919	1.890	1.879	1.867	1.848	1.832

where λ is the wavelength in medium, n is the index of refraction of the medium, f is the frequency of the light wave, and λ_0 is its wavelength in a vacuum. The traditional units for light wavelengths are angstroms (Å), $1\ \text{Å} = 10^{-10}$ m, and millimicrons (mμ), $1\ \text{m}\mu = 10^{-9}$ m. However in the approved SI, nanometers ($1\ \text{nm} = 10^{-9}$ m) are the proper unit to use. The index of refraction for various types of glass is shown in Table 19.1 as a function of the wavelength. The change in the refractive index of a glass with wavelength is called *dispersion* and shows that light of different wavelengths travels with different speeds in glass. Such glasses are called *dispersive media*.

19.3 Interference

As we pointed out in our discussion of traveling waves, the superposition principle applied to the wave model produces unique interference phenomena. For waves on a string we found a complete range of possibilities ranging from total constructive interference to total destructive interference. The determining factors were the amplitudes and phase relationships among the waves present at a given place and time. It was assumed that the waves were of the same frequency with constant phase differences between any two waves. If the waves had been of different frequencies, the phase difference between two waves would be continuously varying in time (except for the special case where the frequencies were all multiples of a common frequency). What are the necessary conditions for interference phenomena in light waves? What examples of light interference have you observed in natural phenomena? What results when two or more beams of light are present in the same region of space? The answer to this question depends on the nature of the light beams. If the beams are from different sources (or even different parts of a large single source), the resultant energy at any point is the sum of the energies produced by the individual beams. Such independent sources are called *noncoherent* sources. The law that governs noncoherent sources is called *photometric summation*. If the beams are *coherent* (that is, the beams have a constant phase relationship during observation), the results of the two or more beams are the summation of the amplitudes of the individual beams. When the superposition principle is applied to coherent light beams, the light intensity varies from place to place, giving rise to intensity maxima and minima called *interference fringes*. The maxima occur where the waves are in phase. The superposition principle leads to *constructive interference* with the resultant amplitude equal to the sum of the in-phase amplitudes. The minima (*destructive interference*) occur in places where the waves are exactly out of phase (phase differences of one-half wavelength). A graphic example of constructive and destructive interference due to coherent sinusoidal waves of unequal amplitude is shown in Figure 19.1. What would you expect if the

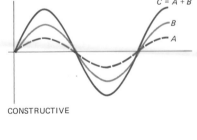

CONSTRUCTIVE

DESTRUCTIVE

FIGURE 19.1
Summation of two waves. (a) In phase (constructive interference) and (b) out of phase (destructive interference).

FIGURE 19.2
(a) Diagram of experimental set up for double-slit interference. (b) line drawing for double-slit interference showing the path difference Δ as a function of the angle θ. The two slits have a width w and are a distance d apart. The viewing screen is a distance L from the slits. The distance along the screen is x.

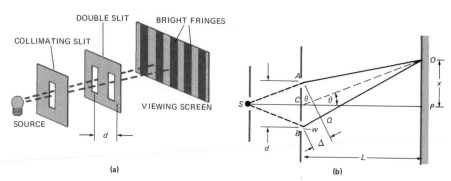

amplitudes were equal in these two cases? How can you reconcile interference fringes with the principle of conservation of energy?

If you wanted to produce interference fringes, how would you do it? Thomas Young was the first to record such an experiment, and it was accepted as the crucial experiment in support of the wave model for light. The essentials of Young's experimental set up are shown in Figure 19.2.

A narrow slit in front of the source provides a coherent plane wave that is incident upon two very narrow slits. Each of these slits acts as a new source, but these two sources are coherent, spatially separated sources. Every point on the screen represents a particular *optical path length* (defined as the path from each slit to the screen along each straight line). Since the slits are coherent sources with zero phase difference, we should expect constructive interference where the optical path difference between the two slits is an integral number of waves.

$$\Delta = m\lambda \qquad (19.2)$$

where Δ is the optical path difference and $m = 0, 1, \ldots$. From the geometry of the double-slit apparatus (Figure 19.2b), we see that this relation can be expressed as:

path difference $= \Delta = d \sin \theta$

where $d =$ slit separation.

Thus, constructive interference occurs when

$$\Delta = m\lambda = d \sin \theta \qquad (19.3)$$

where $m = 0, 1, 2, \ldots$ and is called the *order number*.

The zeroth order bright fringe is symmetrically centered with respect to the slits. There is a first-order maximum on each side of the central bright fringe at an angle θ given by $\sin \theta = \lambda/d$. For small angles, $\sin \theta \simeq \tan \theta$, and $\sin \theta \simeq x/L = m\lambda/d$ for bright fringes.

EXAMPLE

Given a pair of slits separated by 0.2 mm, illuminated by green light (500 nm) in a coherent parallel beam, find the separation of the two first-order green fringes on a screen 1 m from the slits.

FIGURE 19.3
Intensity pattern for double-slit
interference.

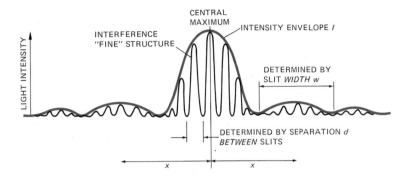

The waves from the two slits must be in phase at positions of maximum intensity. See Figure 19.3. For $m = 1$ (first order),

$$d = 2 \times 10^{-4} \text{ m} \qquad \lambda_o = 5 \times 10^{-7} \text{ m}$$

$$\sin \theta = \frac{\lambda}{d}$$

$$\sin \theta = \frac{5 \times 10^{-7}}{2 \times 10^{-4}} = 2.5 \times 10^{-3}$$

For small angles,

$$\frac{x}{L} = \tan \theta \simeq \sin \theta \simeq \theta$$

when θ is in radians.

We want to find two times the distance between central bright and the first maxima on either side of the central bright fringe,

$$2x = 2 \times 10^{2} \text{ cm} \times 2.5 \times 10^{-3}$$

$$2x = 5 \times 10^{-1} \text{ cm} = 5 \text{ mm}$$

19.4 Effective Optical Path Lengths

Young's experiments show that it is possible to set up interference fringes by introducing optical path differences between coherent wave trains. In Young's experiments the path differences were in air, and we did not correct for the difference of the wavelength in a vacuum and the wavelength in air. (If $n = 1.0003$ for air, what is the error in neglecting this wavelength change?)

Effective optical path lengths in materials involve the wavelength of white light in the medium. For example, the path differences for constructive interference must be an integral number of waves in the given material. Since the wavelength of light in a material with an index of refraction n is given by its wavelength in a vacuum λ_0 divided by n, then a thickness t of this material is equivalent to nt/λ_0 wavelengths. This means that the effective *optical path length* for a sample of thickness t with an index of refraction n is nt:

optical path length $= nt$ (19.4)

EXAMPLE

Compare the effective optical path of 10^{-6} m in a vacuum, air and water.

10^{-6} m (vacuum) = 1.0003×10^{-6} m (air) = 1.3333×10^{-6} m (water)

19.5 Thin-Film Interference Patterns

Thin-film interference patterns are the result of the superposition of waves reflected from the front surface and the back surface of the film. The phase difference between these front and back surface reflected waves determines the nature of the interference pattern for a given film system. An important factor that must be included in the analysis is that light reflected at an interface where $n_2 > n_1$ (the second medium has a greater index of refraction than the first medium) undergoes a one-half wavelength phase shift (π radians) upon reflection. Light reflected from an interface where $n_1 > n_2$ undergoes no phase shift.

Let us consider an oil film ($n = 1.50$) floating on water ($n = 1.33$). When such a film is viewed in white light, a series of colored fringes will be observed. We will calculate the minimum thickness of the oil film for constructive interference of purple (400 nm) and red (700 nm) light. There is a one-half wavelength shift at the air-oil interface and no phase shift at the oil-water interface (Figure 19.4). Thus for constructive interference between the first and second surface reflections, we must have a minimum optical path of one-half wavelength in the oil. Note that any optical path length equal to an odd multiple of one-half wavelength in oil will produce constructive interference for a given wavelength, and an even multiple of one-half wavelength will produce destructive interference. The effective optical path difference between the front and back surface reflections is $2tn_{\text{oil}}$, where t is the film thickness. Light reflected from the second surface travels a distance of $2t$ in the oil. Thus the condition for constructive interference can be expressed as

$$2tn = \frac{m}{2}\lambda_0 \tag{19.5}$$

for $m = 1, 3, 5, \ldots$ (constructive) and $m = 2, 4, 6, \ldots$ (destructive), where t = film thickness and n is the index of refraction of the film, or

$$t(\text{violet}) = \frac{\lambda_0}{4n} = \frac{400 \text{ nm}}{6} = 66.7 \text{ nm}$$

$$t(\text{red}) = \frac{700 \text{ nm}}{6} = 116 \text{ nm}$$

Consider the similar problem for an oil film suspended on a wire frame in air.

The colors of some insect wings are due to such interference phenomena. (What information about insect wings could you obtain by studying these color fringes? What experiments would you do?)

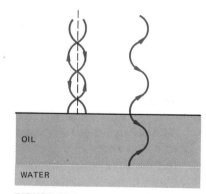

FIGURE 19.4
The difference in phase shift on reflection at an air-oil interface (from rarer to denser medium) and an oil-water interface (from denser to rarer medium).

OIL

WATER

19.6 The Interference Microscope

The interference microscope is designed to convert the phase difference introduced by the different optical path length through the specimen and through the surrounding fluid into a difference in intensity that can be detected by the observer. This is done by taking a parallel coherent beam that passes only through the air and making it π radians ($\frac{1}{2}$ wavelength) out of phase with the beam passing through the fluid alone. These two beams produce destructive interference on a view screen. Any light passing through the specimen will introduce an additional phase shift. The phase shift $\Delta\theta$ is given by

$$\Delta\theta = 2\pi \; (n_s - n_f) \; \frac{t}{\lambda_0} \tag{19.6}$$

where n_s = specimen index and n_f = fluid index. At places where $\Delta\theta = 2\pi$ there will be constructive interference and as the specimen thickness varies, intensity variations occur that make the specimen visible on the screen.

19.7 Diffraction

The bending of waves around objects is common to all wave motion. This wave phenomenon is called *diffraction*. Diffraction patterns result from the interference of waves that travel different distances around objects or through apertures. Consider the diffraction due to a single slit as shown in Figure 19.5a.

To derive the conditions for *destructive* interference we divide the

FIGURE 19.5
(a) Diffraction due to a single slit.
(b) The path difference Δ from the two zones of a single slit of width w.

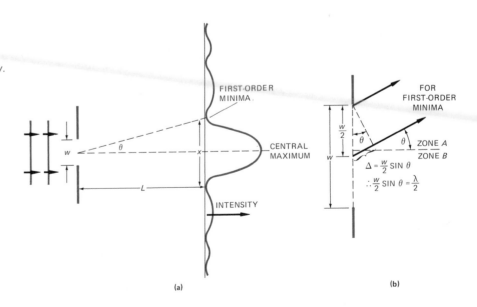

Single-slit diffraction pattern with laser source of light. The irregularities in the photograph were produced by dust particles on an indentation in the edges of the slit. This photo was taken by undergraduate students. (Courtesy of Dr. Richard Anderson.)

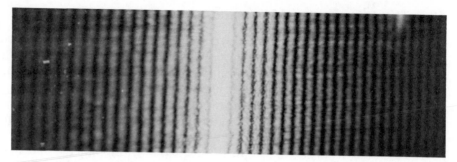

slit of width w into two equal zones (see Figure 19.5b). For destructive interference the waves from zone A cancel the waves from zone B in pairs. The path difference for these cancelling pairs results in the following condition for first-order minima: $(w/2) \sin \theta = \lambda/2$ for the first-order minima on either side of the central bright band.

Higher-order minima are given by path differences equal to odd multiples of one-half wavelength. This general condition for single-slit diffraction minima can be expressed as

$$w \sin \theta = m\lambda \tag{19.7}$$

where $m = 1, 2, 3, \ldots$ is the order of the minimum and w is the width of the slit.

EXAMPLE

Find the separation of the two second-order minima for red (600 nm) parallel light incident on a slit of 0.100 mm width which is 1.00 m from the viewing screen.

$$w \sin \theta = 2\lambda$$

$$\sin \theta = 2 \times 6.00 \times 10^{-7} \text{ m}/10^{-4} = 12.0 \times 10^{-3}$$

Let x_2 be the distance along the screen from the central maximum to the second order minimum, for small angles

$$\sin \theta \simeq \frac{x_2}{L}$$

so

$$x_2 = 12.0 \times 10^{-3} \text{ m}$$

and the separation between the two second-order minima is $2x_2 = 24 \times 10^{-3} \text{ m} = 0.024$ m.

19.8 Diffraction Grating

The diffraction grating consists of a number of close, uniformly spaced, diffracting elements (either transmitting slits or reflecting grooves). Such gratings are used in spectrometers to measure the wavelengths of

FIGURE 19.6
A diffraction grating, a screen with
many fine slits equally spaced a
distance d apart.

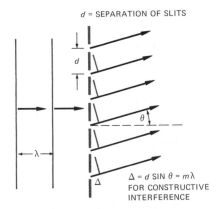

d = SEPARATION OF SLITS

$\Delta = d$ SIN $\theta = m\lambda$
FOR CONSTRUCTIVE
INTERFERENCE

spectra. The diffraction pattern produced by a grating is the result of the
interference of the waves from the different diffracting elements. A diagram of a diffracting grating is shown in Figure 19.6. *Constructive* interference occurs when the waves from adjacent elements have path differences of an integral number of wavelengths between the grating and
screen. The equation expressing this condition is:

$$d \sin \theta = m\lambda \tag{19.3}$$

where $m = 1,2,3, \ldots$ and is the order of the diffraction maxima, where
d is the distance between the slits and is considered to be much larger
than the width of each slit.

EXAMPLE

Find the angular spread of the first-order visible spectrum from a grating with
a number of lines per unit length N of 10,000 lines/cm. The visible spectrum
range is from 400 nm (purple) to 700 nm (red).

$d = 1/N = 10^{-4}$ cm

$\sin \theta$ (purple) $= 4 \times 10^{-5}$ cm/10^{-4} cm $= 0.4$

θ(purple) $= 23.6°$

$\sin \theta$(red) $= 7 \times 10^{-5}$ cm/10^{-4} cm $= 0.7$

θ (red) $= 44.6°$

$\Delta\theta = \theta$(red) $- \theta$(purple) $= 21°$ angular spread of first-order visible spectrum

19.9 Resolution Factors

The ability of a grating to separate two wavelengths increases as the
order increases. However, there are problems concerning intensity and
orders that overlap in high-order spectra. The intensity of the diffraction
decreases with increasing order. In addition, the overlap for a given

grating produces a higher-order, shorter-wavelength line between lower-order, longer-wavelength lines in the spectra under observation. For example, third-order 400-nm light will be at same angle as second-order 600-nm light. Thus, the use of higher orders to improve resolution involves a compromise to solve the resolution problem. We have ignored the single slit patterns that are superimposed for the actual grating. Our treatment has assumed that the individual slits have widths approximately equal to the wavelength of light.

The limiting factor in the resolving power of an optical instrument is the diffraction pattern produced by the system. The apertures of optical instruments produce diffraction patterns of point sources. The resolving power of an optical system is measured by the minimum angle subtended by two *just resolvable sources* of light. The "just resolvable" criterion we use is called the Rayleigh criterion. Lord Rayleigh suggested that two images are just resolvable if the central maximum of one source is found at the first minimum of the other source. This condition leads to the following equation for the resolving power of circular apertures:

$$\sin \theta_R = \frac{1.22\lambda}{D} \tag{19.8}$$

where λ = wavelength, D = diameter of aperture, and θ_R is the angle between two just resolvable sources. We see that the resolving power may be increased by using shorter-wavelength light and/or larger apertures.

The resolving power of a telescope is illustrated in Figure 19.7. The images of just resolvable stars are displayed on a screen in the focal plane of the telescope eyepiece. The stars, separated by a distance s, are a distance L from the objective lens of the telescope. The angle subtended by these stars is equal to the Rayleigh criterion angle for light of wavelength λ and an objective of diameter D.

$$\theta_R \simeq \sin \theta_R = \frac{1.22\lambda}{D} = \frac{s}{L}$$

The distance x between the just resolvable images at the focal length of the eyepiece is given by

$$x = f_e\theta_R = f_e \frac{1.22\lambda}{D}$$

FIGURE 19.7

Resolution of a telescope. The angle between two just resolvable objects is given by θ_R.

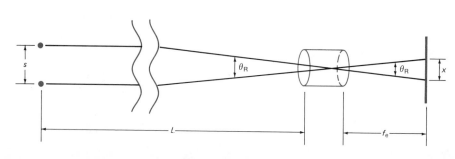

FIGURE 19.8
Intensity plot of Rayleigh's criteria
for resolution.

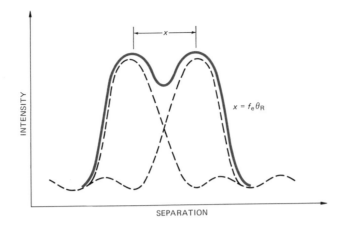

where f_e is the focal length of the eyepiece of the telescope. The intensity pattern for such resolvable images is shown in Figure 19.8.

19.10 Polarization

Wave amplitude oscillations in a plane perpendicular to the direction of propagation are possible only for transverse waves. The orientations of the wave amplitude vibrations in a plane perpendicular to the direction of propagation are called the directions of polarization of the wave (Figure 19.9). The transverse nature of light, with its electromagnetic field oscillations perpendicular to the direction of propagation, makes it possible to observe light polarization phenomena. In general, light sources produce light waves that are randomly polarized. This means the electric field vector (sometimes called the *light vector*) in the electromagnetic wave is vibrating with a random orientation in a plane perpendicular to the direction of propagation (Figure 19.9). Why do you think most natural sources produce unpolarized light?

There are three types of polarizations possible for the electric field vector in xy plane: plane polarized, elliptically polarized, and circularly polarized. These three polarizations are illustrated in Figure 19.10. The plane polarized wave can be resolved into x- and y-components vibrating in phase with each other. The elliptical polarization can be re-

FIGURE 19.9
Graphical representation of ordinary
light showing the electric and magnetic wave amplitudes in planes
perpendicular to the direction of the
propagation of the light.

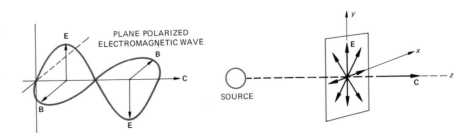

FIGURE 19.10
Representation of (a) plane, (b) elliptical, and (c) circular polarized light. (d) The combination of two simple harmonic vibrations perpendicular to each other shown for various phase differences.

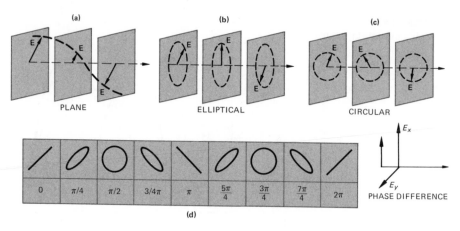

solved into x- and y-components not in phase with each other. The circular polarization can be resolved into x- and y-components 90° out of phase with each other. Circular polarization is said to be right handed if the electric field vector rotates in the clockwise direction as you look at the wave as it travels toward you. Left-handed circularly polarized light corresponds to the E vector rotation in the counter-clockwise direction as the wave travels toward you. We can write the following rules for combining two simple harmonic vibrations of same frequency and amplitude perpendicular to each other:

1. When the phase difference is an even multiple of π, the result is plane polarization at 45° to both original vibrators.
2. When the phase difference is an odd multiple of π, plane polarization results at 90° to those of case 1.
3. When phase differences are odd multiples of $\pi/2$, circular polarization results.
4. All other phase differences produce elliptical polarizations (see Figure 19.10d).

Polarized light can be produced by double-refracting crystals such as tourmaline or other polarizing materials which have an optical axis. Incident light will have all vibrations except those along the optical axis of the polarizing material heavily damped. That is, light other than that with vibrations parallel to the axis is absorbed. In this way incident unpolarized light will be plane polarized when it leaves such a polarizer. The amplitude of a plane polarized beam making an angle θ with respect to the optical axis of the polarizer will be proportional to the component parallel to the other polarizing optical axis, that is, amplitude $\propto \cos \theta$. The intensity of the light wave is proportional to its amplitude squared, thus

$$I = I_0 \cos^2 \theta \qquad (19.10)$$

where I_0 is the intensity for $\theta = 0°$. By using two polarizers you can set up a system in which you can continuously change the output intensity by varying the angle between the two polarizers. See Figure 19.11.

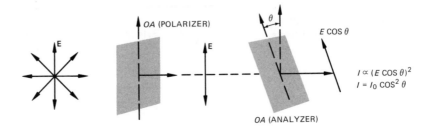

Other ways to produce polarized light are by the reflection of light
from a dielectric surface or by scattering light from small particles. The
boundary value problem of electromagnetic waves incident upon a
dielectric surface yields solutions showing that the reflected light has
various amounts of polarization. There is a critical angle of incidence,
called Brewster's angle, that results in a completely plane polarized
reflected wave. The condition for this critical angle is that the angle of
incidence plus the angle of refraction equals 90°, $i + r = 90°$. When this
condition is imposed on Snell's law, we get the equation

$$n_1 \sin i = n_2 \sin r$$

But

$$\sin (90° - i) = \cos i$$

Thus,

$$\frac{\sin i}{\cos i} = \tan \theta_B = n_{21} \tag{19.11}$$

where n_{21} is the relative index of refraction of the two media and equals
n_2/n_1, and θ_B is the Brewster angle. This is known as Brewster's law, and
at the Brewster angle of incidence, the reflected polarized light will have
its electric field plane polarized parallel to the plane of the reflecting
surface. Polarization by reflection at the Brewster angle is illustrated in
Figure 19.12.

One effect of polarization due to reflection is to produce enhanced
polarization of sunlight over a water surface. It is not surprising to find
evidence that some water insects have sensitivity to polarized light as
this may be useful in survival in their natural environment.

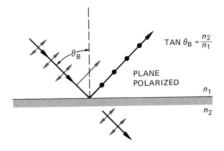

TABLE 19.2
Index of Refraction of Rock Salt, Sylvite, Calcite, Fluorite and Quartz

Wavelength (nm)	Rock Salt	Sylvite (KCl)	Fluorite	Calcite, Ordinary Ray	Calcite, Extraordinary Ray	Quartz, Ordinary Ray	Quartz, Extraordinary Ray
185.	1.893	1.827	—	—	—	1.676	1.690
198.	—	—	1.496	—	1.578	1.651	1.664
340.	—	—	—	1.701	1.506	1.567	1.577
589.	1.544	1.490	1.434	1.658	1.486	1.544	1.553
760.	—	—	1.431		1.483	1.539	1.548
884.	1.534	1.481	1.430				
1179.	1.530	1.478	1.428				
1229.	—	—	—	1.639	1.479		
2324.	—	—	—	—	1.474	1.516	
2357.	1.526	1.475	1.421				
3536.	1.523	1.473	1.414				
5893.	1.516	1.469	1.387				
8840.	1.502	1.461	1.331				

Light scattered from small particles is also selectively plane polarized. You can detect this polarization in sky light. It is thought that bees use this light polarization in their navigation from hive to food and back. Polarization by scattering occurs because light is a transverse wave, and thus the particles will selectively scatter light polarized perpendicularly to the direction of travel of the incident light. The particles of dust, ice, and salt crystals in the atmosphere produce partial polarization of skylight due to scattering. This same scattering, called Rayleigh scattering, also produces the blue color of the sky. The intensity of the scattered light is proportional to $1/\lambda^4$, and thus the blue region of the solar spectrum is scattered most, producing the blue sky. Can you use this information to explain why a sunset is usually red? See Figure 19.13.

There are many materials that have different indices of refraction for different planes of polarization. This property of having different velocities for the propagation of light of different polarizations is called

FIGURE 19.13
Polarization by scattering.

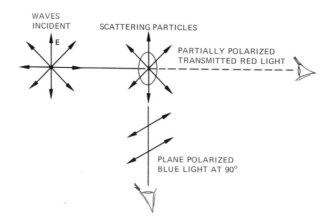

double refraction, or *birefrigence*. Calcite, for example, is a birefrigent crystal. For these uniaxial materials the two different refractive indices are called *ordinary* and *extraordinary*, and each is slightly wavelength dependent (Table 19.2).

EXAMPLE

Given the index of refraction for an ordinary beam to be 1.320 and the index of refraction of 1.330 for the extraordinary beam in a biological specimen (for a 400 nm light), find the phase difference between these beams after passing through a specimen 1μ thick. $t = 10^{-6}$ m, $\lambda_0 = 400$ nm $= 4 \times 10^{-7}$ m

$$\text{phase difference} = \frac{2\pi \,(\text{optical path difference})}{\lambda}$$

$$\text{phase difference} = \frac{2\pi \,(n_e - n_0)t}{\lambda_0}$$

$$= \frac{2\pi \times 10^{-6} \text{ m} \times (1.330 - 1.320)}{4 \times 10^{-7} \text{ m}} = \frac{\pi}{20}$$

Recall that optical path length $= tn$ since $\lambda = \lambda_0/n$.

19.11 Applications of Polarization

Since many biological specimens are birefringent, the phase difference introduced in the specimen can be used in designing special microscopes. One such instrument is the polarizing microscope which is used to increase the contrast between different parts of the specimen under study. In some cases the actual thickness of the various parts can be determined. The polarizing microscope makes use of polarized light transmitted through the specimen. An analyzer is placed between the objective lens and the eyepiece of the microscope. This analyzer is crossed (at 90°) with the polarizer. If the object is homogenous throughout, the field of view will be dark. A different thickness, or orientation, of the sample will produce a varying light intensity, and thus make sharp contrasts visible within the sample. By using a compensating wedge of a known birefringent material (usually quartz), this contrast can be enhanced for thin samples. A calibrated wedge makes it possible to measure specimen thickness by cancelling out the phase shift of the specimen with a known thickness of wedge.

19.12 Optically Active Materials

Some substances (especially in living systems) have the property which causes the plane of polarization of light to be rotated as it passes through them. Such materials are called *optically active* materials. There are both right- and left-handed substances that rotate the plane of polarization in opposite directions. Sugar is found in both forms, right-handed *dextrose* and left-handed *levulose*. The optical activity is demonstrated by solutions of these materials. The optical activity of a solution is proportional to the concentration of optically active molecules in the solution and the length of the light path through the solution. The angle of rotation of plane of polarization θ is given by

$$\theta = KLc \tag{19.12}$$

where $K =$ proportionality constant for a given substance, $L =$ path length, and $c =$ concentration of optically active material.

This equation can be used with appropriate instruments such as a polarimeter or saccharimeter to measure unknown concentrations of optically active substances. Polarized light is passed through the sample and the angle through which the analyzer must be rotated to achieve extinction is measured.

EXAMPLE

A concentration of dextrose of 1.00 g/cm³ produces a rotation of 5.3°/cm of path. If light passing through 10.0 cm of an unknown dextrose solution is rotated through 8.3 degrees, find the concentration of the unknown. If $5.3° = K \times 1$ (cm) $\times 1.00$ (g/cm³), and $8.3° = K \times 10.0$ (cm) $\times c$ (g/cm³) then

$$\frac{5.3}{8.3} = \frac{1.00 \ (\text{gm/cm}^3)}{10.0 \ c} \quad \text{or} \quad c = \frac{8.30}{53.0} = 0.156 \ \text{g/cm}^3$$

19.13 Lasers and Laser Applications

In general, the light from a single light source consists of a series of random wave trains. Each wave train lasts for about 10^{-9} sec, and thus a wave train is approximately 30 cm long. Since these wave trains are uncorrelated with each other in phase, they are noncoherent. Light from different sources is also noncoherent. The development of the laser has provided a high intensity, monochromatic, highly directional, coherent source of light.

The laser produces a continuous train of waves. The actual wave train from a laser is approximately 30 m long. The laser beam is highly correlated in space and time. Laser is an acronym for *l*ight *a*mplified by the *s*timulated *e*mission of *r*adiation. There are many different kinds of lasers available today, but the physical basis of each is essentially the same. We will outline the basic physics involved in the operation of the laser in the chapter on atomic physics (Chapter 28).

There are many applications of lasers because of their unique properties as light sources. Their high intensity and extreme directionality has made them ideal for alignment and communication applications. The laser beam can be focused by optical lenses. It is possible to produce power concentrations greater than 10^9 watts/cm³ at the focal point of a

Xenon ion laser. An electrical discharge in xenon gas filled tube (the long glass tube) produces a laser beam made up of four separate wavelengths. The beam is incident (at left) on a diffraction grating which disperses the light. Note the four monochromatic beams coming from the grating. The wavelengths of the beams shown are at 539, 535, 526, and 495 nanometers. The intensity of the beam shown is equivalent to a 1500-watt light bulb except that the laser beam concentrates all its energy in a single direction and within very small wavelength intervals. (Courtesy of Dr. Laird D. Schearer, University of Missouri—Rolla.)

Laser ophthalmology. Photocoagulation of retinal hemorrhages is performed with an argon laser. (Courtesy of Dr. Felix N. Sabates, Department of Ophthalmology, University of Missouri–Kansas City.)

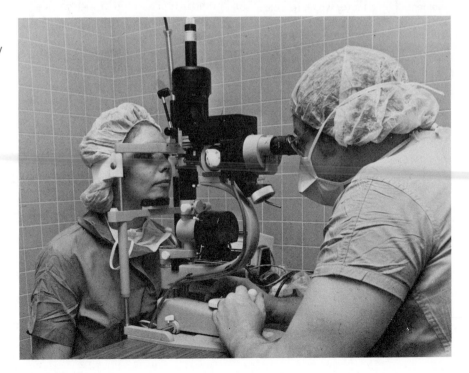

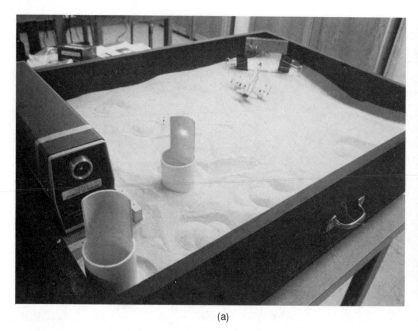

(a)

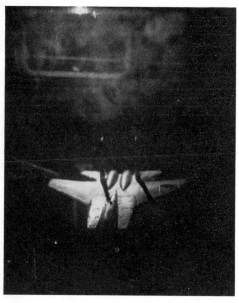

(b)

Sandbox holography. (a) Set up for single beam transmission hologram. The light source is a 3-milliwatt helium-neon laser. The laser beam is incident upon a mirror which reflects part of the beam directly to the photographic film, and another part of the beam is reflected from the object (in this case a toy airplane) to the film. The film is placed between the glass plates held together by spring clamps. (b) A hologram taken with this set up. (Courtesy of Joseph Ferry and Dr. Richard Anderson, University of Missouri—Rolla.)

lens. Such power levels enable lasers to be used for cutting materials (for example, in laser band saws) and drilling small holes with great precision. This well-focused high power beam has been used in medicine for such things as welding detached retinas and bloodless surgery using the self-cauterizing property of the laser beam. Recently the laser has been used to treat patients suffering from diabetic retinopathy. This condition arises in diabetics when tiny blood vessels deteriorate and new vessels grow on the surface of the retina. When these vessels hemorrhage into the normally clear vitreous humor, vision is severely impaired. Scar tissue may also detach the retina from the back of the eye. A fine laser beam focused on the weakened blood vessels can produce coagulation and the proliferating new vessels can be destroyed. The incidence of vision loss has been cut by 60 percent through use of the laser in this treatment.

A similar application of the laser is the control of hemorrhaging in the gastrointestinal tract. A fiber optic bundle can be inserted through the mouth into the stomach. With the bleeding sites visible, the laser can be discharged through the bundle. The intense laser beam can bring on coagulation and cease hemorrhaging within ten minutes.

A new use of the laser is being investigated in dentistry. New tooth filling materials that can be cured by laser radiation are being tested. It is hoped that a material that can be welded to the tooth with a pulse of laser radiation will be found.

One of the most promising uses in medicine involves the laser in holography. The hologram is a three-dimensional photograph of the exposed object. This photograph can be used to reconstruct a three dimen-

FIGURE 19.14
Set up for formation of a hologram.

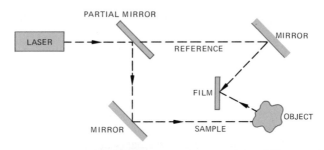

Sandbox set up for double-beam transmission holograms. The light source is a 3-milliwatt helium-neon laser. The laser beam is reflected by a mirror upon a beam splitter. One beam goes to plane *mirror—aperture to film.* The other beam goes through the aperture to the plane mirror to the object (in this case a frog) and is reflected to the film. The film holder is a pair of plate glass flats held together by spring clamps. See Figure 19.14 for diagram. The cat has no part in this optical set up. It is just another object to replace the frog. (Courtesy of Joseph Ferry and Dr. Richard A. Anderson, University of Missouri–Rolla.)

sional image of the original object. A diagram of a set up used for producing holograms is illustrated in Figure 19.14.

The partially silvered mirror produces two coherent beams, a reference beam and a sample beam. The sample beam is reflected from the object onto the film where it interferes with the reference beam producing an interference pattern on the film. This interference pattern contains all of the three-dimensional information available from the object. When a laser beam strikes this record at the angle of the reference beam, it produces a three-dimensional image of the original object. Much research has been devoted to holography, and it offers great potential in many applications. A three-dimensional hologram television picture would be most useful for producing three-dimensional images of organs inside of the body.

19.14 *Light Interactions with Living Systems*

The interactions of light and living matter can be classified as ionizing interactions and nonionizing interactions. The ionizing radiation consists of short wavelength radiation including ultraviolet, x-rays, and gamma rays. Of these we will consider only the ultraviolet radiation in this chapter.

Ultraviolet radiation is important in photochemical reactions. The ultraviolet light provides needed activation energy that makes certain reactions possible. Photosynthesis in plants is a most important ultraviolet induced photochemical reaction.

Light can act as a catalyst in biochemical reactions. The synthesis of vitamin D in our bodies is an example of a photocatalytic reaction. These reactions are the basis of much ultraviolet therapy. Some people also experience reactions like allergies to ultraviolet light. Some drugs (especially antibiotics) can induce an allergic sensitivity to ultraviolet light. Such drugs combined with ultraviolet light treatments are being developed to treat such chronic skin diseases as psoriasis. Prolonged exposure to ultraviolet light (like sun bathing) can damage skin pigments. Proteins and pigments in the skin absorb maximally at about 280 nm. Fortunately much of the sun's ultraviolet radiation is absorbed by the ozone in the earth's atmosphere. There is currently concern that the propellent gas in aerosol cans may be causing depletion of the earth's ozone layer. An intensive research effort is underway to investigate this newly discovered environmental threat.

Ultraviolet light is used in treating jaundice in newborn babies. The condition develops because of poor liver functioning in newly born babies. It has been found that ultraviolet light can be used as a treatment until the baby matures and its liver functions properly. Ultraviolet radiation is also used to kill bacteria. Special lamps with high ultraviolet output are sold as bactericidal lamps.

Recent studies have shown a dramatic effect of ultraviolet light combined with a dye (hematoporphyrin) that accumulates preferentially in malignant tissue. Studies with rats have given kill rates of 75 to 90 percent for malignant brain tumors. This interaction, like others between ultraviolet light and tissue, is not completely understood. There remains much to learn about the interaction of ultraviolet radiation and living matter.

Infrared light is a form of nonionizing radiation. Infrared radiation is associated with the heating effects in molecular systems. The absorption of infrared light increases the energy of molecules in the absorbing material. When infrared radiation penetrates tissue, the associated warming effect is the basis of infrared therapy.

The visible spectrum (400-700 nm) is that part of the solar spectrum that is used in human vision. This important electromagnetic interaction with our environment is the subject of a separate chapter (Chapter 20). Another effect associated with visible light is the entrainment of the individual's daily temperature cycle. Changing the light cycle for an individual changes the phase of the body temperature cycle.

Photobiology offers an exciting area for research by biologists, chemists, and physicists. There are many unanswered questions concerning the interaction of light with living matter.

EXAMPLE

It is known that approximately 14×10^{-13} joule is required to kill a single bacterium. Find the kill rate $\Delta N / \Delta t$ for a 20 watt uv bacterial light if 20 percent of its output is lethal for bacteria.

$$\text{effective killing power} = 0.2 \times 20 \text{ J/sec} = \frac{\Delta N}{\Delta t} \times 14 \times 10^{-13} \text{ J/bacterium}$$

Thus

$$\frac{\Delta N}{\Delta t} = 2.9 \times 10^{12} \text{ bacteria/sec}$$

SUMMARY

Use these questions to evaluate how well you have achieved the goals of this chapter. The answers to these questions are given at the end of the summary with the number of the section where you can find related content material.

Definitions

1. Assign the correct term to each of the following physical phenomenon:
   a. color of a soap bubble
   b. solar spectrum produced with a prism
   c. intensity pattern of light after passing through a pin hole
   d. two beams of laser light used in holography
   e. light characteristic when reflected from a dielectric surface
   f. the rotation of the plane of polarization by sugar solutions

Applications of Wave Properties

2. Thin-film colors involve the superposition of reflected waves. Phase differences involved arise from _____ and _____.
3. A diffraction grating spectrometer designed for high resolving power should have a _____ number of lines/mm.
4. The resolving power of a telescope is improved by increasing the size of _____ used.
5. Holography results in a two dimensional _____ on photographic film that produces a _____ when viewed with laser light under the proper conditions.

Problems Involving Wave Properties

6. If a soap film appears bright when viewed in 600 nm light, find its minimum thickness if $n = \frac{3}{2}$.
7. A diffraction grating with N lines per meter is L meters from a screen. For wavelength λ find the separation between first- and second-order fringes on the screen. Assume $L \gg d$.
8. Light reflected from a surface at $53°$ is noted to be plane polarized.
   a. If you look at this reflected light through Polaroid sun glasses find the fraction of this light reaching your eyes if you tilt your head $45°$ from the vertical.
   b. Find the index of refraction of the dielectric material.

Optical Activity

9. Sketch a system that could be used to measure optical activity.

Lasers

10. List three important characteristics of laser light.

Answers

1. a. interference
      (Section 19.3)
   b. dispersion
      (Section 19.2)
   c. diffraction
      (Section 19.7)
   d. coherent
      (Section 19.13)
   e. polarization
      (Section 19.10)
   f. optical activity
      (Section 19.12)

2. reflection, optical path differences (Section 19.5)
3. large (Section 19.8)
4. objective lens (Section 19.9)

5. interference pattern, three-dimensional image (Section 19.13)
6. 100 nm (Section 19.5)
7. *LN*λ (Section 19.8)

8. 50 percent, $n = 1.33$ (Section 19.10)
9. source polarizer, sample, analyzer, detector (Section 19.12)

10. coherence, mono-chromicity, direc-tionality, intensity (Section 19.13)

ALGORITHMIC PROBLEMS

Listed below are the important equations from this chapter. The problems following the equations will help you learn to translate words into equations and to solve single concept problems.

Equations

$$n = \frac{\lambda_0}{\lambda} \qquad n = \frac{c}{v} \tag{19.1}$$

$$d \sin \theta = m\lambda \qquad m = 1, 2, 3 \dots \tag{19.3}$$

$$\text{optical path length} = nt \tag{19.4}$$

$$2nt = \frac{m}{2} \lambda_0 \qquad m = 1, 3, 5, \dots, \text{constructive}, \qquad m = 2, 4, 6, \dots, \text{destructive} \tag{19.5}$$

$$\Delta\theta = 2\pi (n_s - n_f) \frac{t}{\lambda_0} \tag{19.6}$$

$$\sin \theta_R = 1.22 \frac{\lambda}{D} \tag{19.8}$$

$$I = I_0 \cos^2 \theta \tag{19.10}$$

$$\tan \theta_B = n_{21} \tag{19.11}$$

$$\theta = KLc \tag{19.12}$$

Problems

1. The wavelength of one of the lines in the emission spectrum of sodium is 589 nm in a vacuum. What is its wavelength in heavy flint glass?

2. A coherent parallel beam of the green light (546 nm) of mercury is incident upon a pair of slits. The separation of the first-order interference pattern is 2 mm from the central image on a screen 1 meter from the plane of the slits. What is the distance between the slits? (*Hint:* $\sin \theta \approx \theta$.)

3. What is minimum thickness of an oil film on water that will give destructive interference for 546-nm light by reflection from the surfaces of the film? n = 1.50 for oil.

4. What is the sine of the angle of diffraction for the second-order maxima for the 546 nm light of the mercury spectrum incident upon a diffraction grating with 5000 lines/cm?

5. What is Brewster's angle for water?   n = 1.33 for water.
6. A polarizer and an analyzer are set for maximum intensity of transmission. If the analyzer is turned through 37°, what is the new intensity of transmission?
7. An optically active material of a given concentration c_0 and path length 10 cm produces a rotation of 10° of the plane of polarization. What would be the concentration of the same material that would produce the same angle of rotation for an optical path length of 15 cm?

Answers

1. 357 nm
2. 0.273 mm
3. 182 nm
4. 0.546

5. 53°
6. 0.64 I_0
7. 2/3 c_0

EXERCISES

These exercises are designed to help you apply the ideas from one section to physical situations. When appropriate the numerical answer is given in brackets at the end of the exercise.

Section 19.2

1. Plot the index of refraction as a function of wavelength for higher dispersion crown glass and for heavy flint glass. What is the physical meaning of the slope of the curve? Compare the slopes at 400 and 600 nm. What is the index of refraction of each for the 546-nm green light from mercury?   [~ 1.523 at 546 nm, ~ 1.656 at 546 nm]
2. If the wavelength of the green line of mercury is 546 nm in a vacuum, what is it in water? In heavy flint glass?   [410 nm, 331 nm]

Section 19.3

3. Two narrow slits are spaced 0.25 mm apart and are 60 cm from a screen. What is the distance between the second and third bright lines of the inference pattern if the source is the 546 nm light from mercury? [0.13 cm]
4. Given a double slit with a separation of 0.2 mm, find the separation of consecutive bright fringes on a screen 1 m from the slits for red (600 nm) parallel light incident on the slits.   [0.3 cm]

Section 19.5

5. What is the minimum thickness of the film of a soap bubble with a refractive index of 1.33 if the film shows constructive interference for the reflection of the yellow sodium light (589 nm) at normal incidence in air?   [110 nm]
6. What is the minimum thickness of a plastic film (index of refraction 1.4) on your eye glasses which will give destructive inference for the reflection of light of wavelength 560 nm?   [200 nm]

Section 19.8

7. What is the wavelength of a line which is diffracted 20° in the first order for normal incidence upon a transmission grating? What is the second-order diffraction angle for this wavelength? Assume the grating has a ruling of 6000 lines/cm.   [570 nm, 43.2°]
8. For orders greater than one, there is an overlap of orders in the visible spectrum from a diffraction grating. What is the basic relationship that shows this? What third-order line coincides with the second-order line of 589-nm light?   [393 nm]

Section 19.9

9. Find the resolving power of the 508-cm Mount Palomar telescope. Use 550 nm for the wavelength of light. Find the separation of just resolvable objects near Jupiter (6.5×10^6 km from earth).   [1.32×10^{-7} rad, 0.86 km]

PROBLEMS

Each of the following problems may involve more than one physical concept. When appropriate, the answer is given in brackets at the end of the problem.

10. For a double refracting crystal such as calcite, the geometry may be such that only one beam of light is transmitted while the other is internally reflected at the surface of a 60° crystal. What are the largest and smallest incident angles that can be used to separate the two beams by this method? Assume a sodium source ($\lambda = 589$ nm). See Figure 19.15. [26.9°, 39.8°]

11. A beam of light from a sodium arc is incident upon a 60° heavy flint prism at the proper angle to give the minimum angle of deviation. (This condition calls for ray parallel to the prism base inside the prism; see Figure 19.16.) What is the value of this angle? [51.2°]

12. A piece of quartz crystal has two parallel sides 1 mm apart. A beam of yellow sodium light ($\lambda = 589$) is at normal incidence on one of the parallel sides of the crystal. What is the minimum change in phase difference between the ordinary and extraordinary rays in going through the crystal? [96 rad]

13. How does the diffraction angle for a given wavelength line vary as a function of slit separation for a given order? How does the diffraction between two wavelengths, say 546 nm and 589 nm, compare in first and second and in second and third orders? What does this suggest to you in trying to resolve two lines of wavelengths close together? This solution may not be practical. Why? Remember in nature you do not get something for nothing!

14. What is the maximum diffraction order of a red light

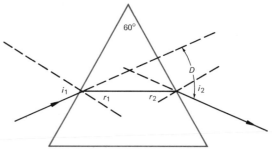

FIGURE 19.16
Problem 11.

of 656 nm that you can get with a grating of 6000 lines/cm? 600 lines/cm? [Second order, twenty-fifth order]

15. A beam of light is incident upon a salt solution (1.36), and the reflected beam is completely polarized. What is the angle of refraction of the beam? [36.3°]

16. How can you combine two beams of plane polarized light to produce a beam of circularly polarized light?

17. A soap film ($n = 1.33$) is displayed in a vertical wire loop. It is noticed that the film has a dark reflection band at the top just before it breaks. If green light (500 nm) is used, find the thickness of the film at the position of the first bright band. [94 nm]

18. A thin film of water ($n = 1.33$) is floating on glycerine ($n = 1.47$). Find the minimum thickness of water that will produce constructive interference for reflected red (600 nm) light. [226 nm]

19. Find the minimum thickness of an oil film ($n = 1.50$) on water that will give constructive interference for reflected red (600 nm) light. [100 nm]

20. Compare the effect of temperature on thin-film interference due to thermal expansion and variation of index of refraction n. For benzene $\Delta n/\Delta t = -6 \times 10^{-4}$ /°C and $\Delta(\text{vol})/\Delta t = \alpha V_0 = (.24 \times 10^{-3} \ V_0/°C)$

21. A reflected green light (500 nm) is completely plane polarized at an angle of incidence of 53° on an insect wing. There is also a dark band in this reflected green light at the thinnest part of the wing. Find this minimum thickness, assuming it is due to destructive interference for the green light when viewed in air. [188 nm]

22. For objects that are not self illuminating the criterion for resolution is given in terms of the radius of the first dark fringe of the diffraction of a circular

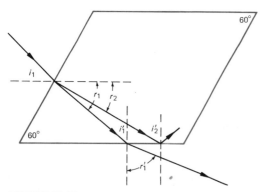

FIGURE 19.15
Problem 10.

aperture. The diffraction fringe radius r is given by $r = 0.61 \ \lambda/n \ \sin i$, where i is the angle subtended by the aperture at the object, λ is the wavelength and n is the index of refraction of the object space. Two objects are said to be resolved when the separation of images is equal to the diffraction fringe radius.

    a. Find the percent improvement in resolving power of a microscope that is obtained by using an oil immersion lens system with $n = 1.5$.

    b. Find the separation of just resolvable objects in green light (500 nm) if $n \sin i$ (called the numerical aperture) is one.   [33 percent, 305 nm]

23. A standard sugar solution ($1 \ g/cm^3$) is found to rotate the plane of polarization of light by 5.4° per cm of path length. A sugar sample of unknown concentration rotates the plane of polarization through 10.6° in a 5 cm long sample tube. Find the concentration of the sample.   [0.4 g/cm^3]

24. The Brewster angle for a specimen is 55° for 500-nm light. If the specimen is 0.001 mm thick, find the phase shift introduced by this specimen as compared with an equal thickness of water.   [72°]

25. A thin wedge of air is formed between a sheet of glass 5 cm long and a horizontal glass plate. One end of the sheet of glass is in contact with a glass plate. The other end is supported by a thin metal film 0.05 mm thick. The horizontal plate is illuminated from above with light 589 nm. How many dark interference fringes are observed per cm in the reflected light? [17 fringes/cm]

26. Given a single slit of width $D = k\lambda$, where $k = $ con-

stant, find the angle of the first-order minimum for each of the following values of k:

    a. 1; b. 10; c. 100; d. 1000.   [a. 90°; b. 5.7°; c. 0.57°; d. 0.0057°]

27. Given a diffraction grating with 5000 lines per cm, find the diffraction angles for bluish-purple (400 nm) light in the first and second order. At which order does the 400 nm light overlap the reddish-orange light (600 nm)?   [11.5°, 23.6°, third]

28. A helium-neon laser source produces a second-order spectrum for light of wavelength 632.8 nm at an angle of 30° using a certain diffraction grating.

    a. Find the angle for the first-order sodium yellow ($\lambda = 589$ nm).

    b. Assume the grating is a reflection grating of 1-m focal length. Find the second-order separation in the focal plane for the sodium doublet lines of λ equal to 589.0 and 589.6 nm.   [a. 13.5°; b. 0.535 mm]

29. The axes of a polarizer and an analyzer are oriented at 60° to each other.

    a. If polarized light of intensity I is incident on the analyzer system, find the intensity of the transmitted light.

    b. If the incident light (intensity I) is plane polarized at an angle of 30° with respect to the polarizer axis, find the intensity of the transmitted light.   [a. $0.25I$; b. $0.188I$]

30. What is the minimum thickness of a water film on glass that will give destructive interference for 546-nm light by reflection from surfaces of the film? [102.4 nm]

GOALS

When you have mastered the contents of this chapter, you will be able to achieve the following goals:

Human Vision
Characterize the physical parameters that are significant in human vision.

Visual Defects
Explain the causes and corrections for the visual defects of myopia, hypermetropia, presbyopia and astigmatism.

Characteristics of Vision
Define the following terms:

visual acuity   accommodation
scotopic vision   dichromat
photopic vision

Corrective Lenses
Solve problems involving visual defects and their corrections using lenses.

PREREQUISITES

Before beginning this chapter you should have achieved the goals of Chapter 18, Optical Elements, and Chapter 19, Wave Properties of Light.

20

Human Vision

20.1 Introduction

From your own experience, what descriptive statements can you make concerning the following aspects of your vision: sensitivity for low light levels, response to different colors, ability to discriminate between two objects at a distance, and the reliability of visual depth perception?

The human eye is our main source of information about our external environment. It responds to light and supplies messages to the brain. In this chapter we will explore the human visual system and the physical principles behind its operation.

20.2 The Powerful Eye

In this chapter we will examine one of the most amazing optical systems you will ever encounter. The human eye is an optical system of unsurpassed versatility. Its sensitivity is so great that it apparently detects single packets of light, yet visual information is processed only when sufficient numbers of these packets are detected coincidently to produce a visual pattern. This same system is capable of handling intensities more than 10^9 times the minimum energy threshold. Let us look at the physics behind the performance of the human eye.

The eye is analogous to a camera in its basic operation (Figure 20.1). There are three basic parts of the eye: a light focusing system, an automatic aperture system, and a light sensitive detection system. In the light focusing system, the cornea and the lens account for the refractive power of the eye. Most of this power is due to the curvature of the cornea, but the muscles controlling the curvature of the lens provide the flexible focusing power of the eye. These eye, or *ciliary*, muscles permit the normal eye to focus on the retina light from objects very distant, essentially parallel rays of light, as well as light from objects as close as the normal near point of 25 cm from the eye. Measure your near point.

FIGURE 20.1
Diagram of the human eye.

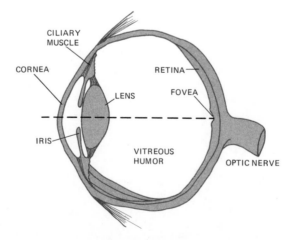

Move this page toward your eyes until the images of the letters begin to blur. The distance from your eye to the page at the nearest point of clear vision is your near-point vision. You have greater eye accommmodation than normal if your near-point distance is less than 25 cm. The power of a lens is a measure of its ability to bend rays of light. A lens of high power has a small focal length. The numerical value for the power of a lens in *diopters* is computed by taking the reciprocal of the focal length of a lens when the focal length is measured in meters. For example, a lens of 5-cm focal length has a power of 20 diopters. The human eye has a variable focal length so the eye can change its power to focus on objects at various distances from the eye. Let us calculate the accommodation ability of a typical human eye. If the distance from the eye lens to the retina is given by l (about 3 cm for a human eye), the power of the eye lens for distant objects, symbolized by P infinity P_∞, can be calculated by using the thin lens formula from Chapter 18,

$$\frac{1}{p} + \frac{1}{q} = \frac{1}{f} \tag{18.13}$$

where p is the distance to the object, q is the distance to the image, and f is the focal length of the lens. For distant objects the reciprocal of the object distance $1/p$ is approximately zero, so the distance from the eye lens to the image on the retina is equal to the focal length of the eye. Therefore, the power for the eye lens for distant objects P_∞,

$$P_\infty = \frac{1}{f} = \frac{1}{l} \simeq 30 \text{ diopters} \tag{20.1}$$

where l is the distance from the eye lens to the retina.

The power of the eye for near objects, P_n, can be similarly calculated,

$$P_n = \frac{1}{f} = \frac{1}{p} + \frac{1}{q} \tag{20.2}$$

where p is the distance to an object at your near point (say, 25 cm) and q is the distance from the eye lens to the retina l,

$$P_n = \frac{1}{\text{near point}} + \frac{1}{l} = \frac{1}{0.25} + \frac{1}{l} = 4 + P_\infty \simeq 34 \text{ diopters} \tag{20.3}$$

Hence, the change in the power of a human eye lens to accommodate objects at various distances from the eye is about 4 diopters. Design a simple experiment using a meter stick to measure the ability of your eyes to accommodate objects at various distances from your eye. If you wear eyeglasses or contact lenses, perform this experiment both with and without your glasses or lenses.

The automatic aperture system involves a negative feedback system through the brain to the iris (Figure 20.1). The system automatically adjusts the opening size of the iris in response to the ambient level of light intensity. In response to high intensity light, the iris stops down the opening to minimize damage to the detector system and to provide the proper intensity levels to the detecting system. The response to low light levels, referred to as dark adaption, takes up to 45 minutes. It provides the eye with its maximum entrance pupil size and allows the eye to assume its most sensitive light gathering configuration.

The light sensitive detecting system of the eye consists of the light sensitive receptors distributed over the back of the eyeball. This light sensitive membrane is called the retina. There are two kinds of photosensitive cells present in the retina. These cells are *rods* and *cones*, so named because of their shape. The rods are more sensitive to light and distinguish light from dark when the light level is very low. Vision at low levels of light intensity is called *scotopic vision*. The cones are less sensitive to light, but are capable of resolving light into components giving details of images and information involving the color of the received light. The rods and cones are distributed with densities greater than 100,000/mm² on the retina, the greatest concentration of cones occurring in the region called the *fovea*.

20.3 The Defective Eye

The most common defects of human vision involve the light refracting system of the eye. Normal vision provides clear vision for letters that subtend *three minutes of arc* at a distance of *6.10 m (20 ft) from the observer*. At this distance, defective eyes may produce images that are too blurred to read. If a person requires letters at 6.10 m (20 ft) that normal vision reads at 12.2m (40 ft) the person is said to have 20/40 vision. The numerator expresses the distance between observer and the letters; the denominator is the distance at which the normal eye can read these letters. Normal vision is expressed as 20/20 vision.

EXAMPLE

Find the size of the letters the normal eye can read at 6.10 m (20 ft), if the criterion is that they subtend 3 minutes of arc at the eye. Using this information what size object is required at 10 m?

angle (rad) = object size/distance
1 min = 2.91×10^{-4} rad
3 min = $8.73 \times 10^{-4} = x/6.10$ m

$x = 0.533$ cm

At 10 m,

$x = 10$ m $\times 8.73 \times 10^{-4} = 8.73 \times 10^{-3} = 0.87$ cm.

20.4 Correction of Defective Sight

Many of the defects of the human eye are correctable. *Myopia* is the term for nearsightedness. Myopia results when parallel light is focused in front of the retina. This means that there exists a farthest point of clear

FIGURE 20.2
Myopia. The near-sighted eye forms the image in front of the retina. This condition is corrected with a negative lens that has a focal length equal to the far point distance.

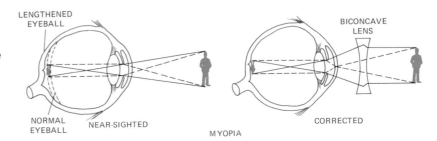

vision, usually only a few meters from the eye. Only objects closer to the eye than the farthest distance of clear vision can be properly focused on the retina. This defect is most frequently associated with an elongated eyeball. The correction for myopia is the use of a diverging lens with a focal length equal to the far point distance, the distance of farthest clear vision. This negative spectacle lens makes incoming parallel light appear to diverge from the far point (Figure 20.2).

Farsightedness (*hypermetropia*) due to a short eyeball, results in parallel light being focused behind the retina (Figure 20.3). The same result caused by the hardening of the lens with aging is called *presbyopia*. These two conditions mean that the near-point distance for a person is increased beyond 25 cm. The correction for each of these defects is the use of a converging lens, which produces for an object at the normal near point (25 cm) a virtual image at the actual near point of the eye. This means that the focal length of the correcting lens is given by the equation $1/f = 1/25 - 1/x$, where x is the actual near point of the eye (all distances in cm).

The defect of *astigmatism* is the formation of a distorted image on the retina. The usual cause of this defect is the nonuniform curvature of the cornea. This defect can be corrected by using a lens that is ground to compensate for the cornea defect. Often this correction is obtained through the use of an appropriately shaped cylindrical lens.

EXAMPLES

1. A man has a far point of 2 m. Find the focal length of the corrective lens for seeing distant objects, and find the near point of this eye with corrective lens if the man has the normal near point without glasses.

FIGURE 20.3
Hypermetropia. The far-sighted eye forms the image behind the retina. This condition is corrected by using a converging lens that places the image of an object placed at 25 cm (the normal near point) at the far-sighted eye near point.

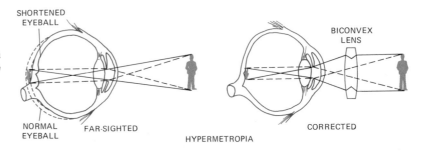

The lens needed must have a focal length of -2 m so that an object at 2 meters will have an image at infinity. The near point with this lens is x, where $1/x - 1/25 = 1/f$ (in cm). Thus x equals $200/7$ cm.

2. A person has a near point of 50 cm. Find the corrective lens required to enable this person to read at the normal distance of 25 cm.

Using the equation $1/f = 1/25 - 1/50$, we find that the corrective lens should have a focal length of $+50$ cm.

20.5 The Receptive Eye

As stated earlier research has provided evidence that individual receptor cells on the retina can be activated by single packets (photons) of light. Fortunately, such phenomena are not processed by the brain; the noise signals by such random activity would be most distracting. The research shows that of the order of 100 photons per second will excite a localized set of rods of about 100 in number causing visual perception to occur. The basis of this coincident detection system is still not clear, and it continues to be a current area of research.

The wavelength, or color, sensitivity of the human eye is also a subject of present-day research. It has been established that the cone cells in the retina play the central role in color, or *photopic*, vision. The cones are most closely packed in the fovea region of the retina; so when you direct your visual attention to an object your eye muscles rotate your eyeballs so as to place the focused image of that object on the fovea of the retina.

It has been found that there are four colors whose apparent color does not change as the intensity of the light focused upon the retina is increased above the threshold required to stimulate the cone cells. These "pure" colors are red, yellow, green, and blue, and for a typical observer these pure colors will have dominant wavelengths of 630 nm, 570 nm, 510 nm, and 470 nm, respectively. On the other hand, it is possible for a person to perceive yellow from a proper mixture of blue and red light, without any 570-nm light present. So a human being is capable of synthesizing colors in a way that is not clearly understood at the present time.

In an attempt to refine our understanding of human vision, direct measurements of the response of the retinal cones to various exciting lights have been made. Three types of cone cells have been identified, each particularly sensitive to a different range of the spectrum. However, it is clear that these cones do not send nerve signals directly to the brain. The cone signals pass through a complex coding system before transmission to the brain where the visual messages are translated into a person's visual perception.

It has long been known that humans see objects differently in dim light and in bright light. We now know that the rod cells in the retina are most sensitive to low intensity of light, and these rod cells are most closely packed in the periphery of the retina, rather than in the fovea. To see objects in dim lighting you should look slightly to one side of the

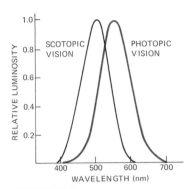

FIGURE 20.4
The relative sensitivity for the dark-adapted eye (scotopic vision) and the light-adapted eye (photopic vision).

object, not directly at it, so that the light entering the eye falls not on the fovea but on the periphery of the retina where the rods are more concentrated. The rods and the cones also differ in their wavelength response to light. The rods are more sensitive to blue light than are the cones. The relative luminosity curves for *scotopic* (dim light) and photopic vision are shown in Figure 20.4. *Relative luminosity* is defined as the reciprocal of the normalized threshold intensities, obtained by dividing each wavelength threshold by the minimum threshold for the entire system. (Can you suggest an analogous analysis for the human ear?)

The range of wavelengths that make up the visible spectrum (380–770 nm) is determined by the absorption of ultraviolet by the lens and by the insensitivity of the receptors in the eye to wavelengths greater than 770 nm.

There are people who cannot discriminate colors. The degree of color-blindness varies among the population. There is evidence that some color blind persons have only two of the necessary pigments (thus they are called *dichromates*) for full color vision. Those with no color vision are called *monochromates*.

20.6 The Perceptive Eye

The visual acuity of the eye is defined as the minimum angular separation between two objects that can be resolved by the eye. Several factors can set this limit, the spacing of receptor cells with independent nerve paths, diffraction effects (Rayleigh criterion), and lens system defects. The closest spacing of receptors in the human eye is that of cones in the fovea, where the center to center distance of adjacent cones is from 2 to 5 microns. For an eye with 2 cm between cornea and retina this corresponds to about 10^{-4} rad for the angle subtended by two stimulated cells separated by one unstimulated cell. This criterion determines the resolving power of the human eye.

EXAMPLE

Find the distance from the eye for two objects 1 cm apart that subtend an angle of 10^{-4} rad

$$\frac{s}{r} = \theta$$

$$\frac{1 \text{ cm}}{r} = 10^{-4}$$

$$r = 100 \text{ m}$$

Using the Rayleigh criterion and appropriate values for wavelength and for the aperture of the eye, find the limit for visual acuity, and compare it with the 10^{-4} rad value calculated from receptor separation. Does this comparison support the idea that the visual acuity of the eye is essentially independent of color?

FIGURE 20.5
A schematic diagram illustrating depth perception with binocular vision. The different viewing angles are associated with distances through learning experiences.

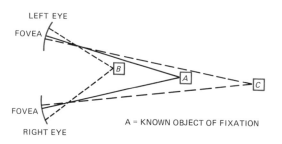

20.7 Binocular Vision

Depth perception in human vision is primarily a result of binocular vision. Each eye produces a slightly different image on their respective retina due to the different angle the object subtends at each eyeball. The brain learns to interpret these slightly different images in terms of distance from the eye. The experimental data suggest that depth perception can be improved with practice. Experience is used to correlate the actual size of the object with the size of the image on the retina and the actual distance to the object. The following cues are important in depth perception:

1. viewing angle differences for each eyeball (Figure 20.5)
2. muscle tension for convergence upon a point of fixation
3. angle subtended by known object at a known distance
4. color contrast interpreted in terms of light scattering which relates to distance
5. relation of light and shadows
6. overlapping contours
7. parallax (the apparent movement of objects in the visual field with observer motion)
8. context of object and environment

ENRICHMENT
20.8 Derivation of the Power of Lenses Used Close to the Eye

To derive the power of a lens to be used close to the eye, let the eye lens be replaced by a thin convex lens of fixed focal length f_e in meters. Then the power of the eye will be given by P_e where

$$P_e = \frac{1}{f_e} \tag{20.4}$$

Next take another lens of focal length f_1 in meters with a power of $P_1 = 1/f_1$ and place it close to the first lens. If an object is placed several meters away from the lens we can find the image distance q_1 by applying the thin-lens equation (Equation 18.13),

$$\frac{1}{q_1} = \frac{1}{f_1} - \frac{1}{p} = \frac{p - f_1}{p f_1}$$

Then   (20.5)

$$q_1 = \frac{p f_1}{p - f_1}$$

This image distance q_1 becomes the virtual (negative) object distance for the eye lens so that the final location of the image in the eye is given by q_e:

$$\frac{1}{q_e} = \frac{1}{f_e} - \frac{1}{p_e} = \frac{1}{f_e} + \frac{1}{q_1}$$

where $p_e (= -q_1)$ is the object distance to the eye lens.

But the power of the two lens system,

$$P_s = \frac{1}{f_s} = \frac{1}{p} + \frac{1}{q_e} = \frac{1}{f_1} - \frac{1}{q_1} + \frac{1}{f_e} + \frac{1}{q_1}$$

                                                                                             (20.6)

$$P_s = \frac{1}{f_1} + \frac{1}{f_e} = P_1 + P_e$$

Hence the *power of a lens-eye system is the sum of the glass lens power plus the eye lens power.*

EXAMPLE

A man has a near point of 50 cm from his eyes and a far point at infinity. What is the useful accommodation power of his eyes? Consider the lens of the eye as a simple lens.

$$\frac{1}{f_e} = \frac{1}{p} + \frac{1}{q}$$

For the far point

$$\frac{1}{f_\infty} = \frac{1}{\infty} + \frac{1}{q_\infty} = \frac{1}{q_\infty}$$

Now $\dfrac{1}{f} = P_\infty = \dfrac{1}{q_\infty}$

$$P_n = \frac{1}{0.50} + P_\infty = 2.0 + \frac{1}{q_\infty} = \frac{1}{f_n}$$

accommodation power $= P - P_\infty = 2.0$ diopters

What lens should be prescribed for this individual?

For the normal eye the near point should be 25 cm. This means that the image formed by the lens must be at least 50 cm from the eye lens.

$$\frac{1}{p} + \frac{1}{q} = \frac{1}{f}$$

$$\frac{1}{25} - \frac{1}{50} = \frac{1}{f}$$

$$\frac{1}{f} - \frac{2-1}{50} = \frac{1}{50 \text{ cm}}$$

$$P = \frac{1}{0.50 \text{ m}} = 2.0 \text{ diopters}$$

A general rule for an auxiliary lens for the eye is that the image formed by this lens must be between the near and far point of the unaided eye. One then uses the principles of a multiple lens system to prescribe the corrective lens.

SUMMARY

Use these questions to evaluate how well you have achieved the goals of this chapter. The answers to these questions are given at the end of this summary with the number of the section where you can find related content material.

Human Vision

1. For each of the following optical properties write a statement to relate it to human vision: light intensity, light wavelength, lens power, and resolving power.

Visual Defect

2. A person who suffers from hypermetropia is unable to clearly see objects that are _____ the eye lens, has the _____ point distance _____ in size, and can wear corrective eyeglasses of _____ power.
3. A person who suffers from myopia is unable to clearly see objects that are _____ the eye lens, has the _____ point distance _____ in size, and can wear corrective eyeglasses of _____ power.

Characteristics of Vision

Match each term below with the lettered items.

4. visual acuity
5. scotopic vision
6. photopic vision
7. visual accommodation
8. dichromat
9. binocular vision

a. looking at photographs
b. two-colored rug
c. contact-lens holder
d. perspicacity
e. depth perception
f. color perception
g. low intensity
h. carrots
i. color blindness
j. blue sensitive
k. red sensitive
l. fovea region
m. 0.01 rad
n. reciprocal of normalized intensities
o. lens power
p. iris aperture size
q. diopters
r. cones
s. rods
t. 20/20
u. red, yellow, blue, green
v. Rayleigh criterion
w. eye muscle tension
x. perspiration

Corrective Lenses

10. A young person with a near point of 20 cm, is not able to see an object clearly if it is more than 5 m from his eye. What kind of lens should you prescribe to correct his vision?
    a. +5 diopters   d. −0.2 diopters
    b. −5 diopters   e. −2.0 diopters
    c. +0.2 diopters

Answers

1. Section 20.5, Section 20.2, Section 20.6
2. near, near, increased, positive (Section 20.4)
3. far from, far, decreased, negative (Section 20.4)
4. d, l, m, p, t, v (Section 20.6)
5. g, j, s (Section 20.5)

6. f, i, k, l, r, u (Section 20.5)

7. o, q (Section 20.2)

8. f, i (Section 20.5)

9. e, w (Section 20.7)

10. d (Section 20.2)

ALGORITHMIC PROBLEMS

Listed below are the important equations from this chapter. The problems following the equations will help you learn to translate words into equations and to solve single concept problems.

Equations

$$\frac{1}{p} + \frac{1}{q} = \frac{1}{f} \tag{18.13}$$

$$P = \frac{1}{f(\text{m})} \tag{20.1}$$

$$P_\infty = \frac{1}{l(\text{m})} \tag{20.1}$$

$$P_n = \frac{1}{\text{near point}} + \frac{1}{l} \tag{20.3}$$

Problems

1. At what distance should a person with 20/20 vision be able to read a letter 1 cm high?
2. What is the power of the lens for each example in Section 20.4, Correction of Defective Sight?

Answers

1. 11.5 m

2. −0.5 diopters, +2.0 diopters

EXERCISES

These exercises are designed to help you apply the ideas of a section to physical situations. When appropriate, the numerical answer is given in brackets at the end of the exercise.

Section 20.2

1. A girl can adjust the power of her eye's lens-cornea system between limits of +60.0 diopters and +65.0 diopters. With the lens relaxed she can see distant objects such as stars clearly.

a. Find the girl's near point.

b. How far is her retina from her eye lens? [20.0 cm, 1.67 cm]

Section 20.4

2. A myopic student wears contact lenses of −8.00 diopter power in order to see distant objects. The distance from the eye lens to the retina is 2.00 cm.

a. Find the power of the student's eye when relaxed.

b. If the student can give an extra +4.00 diopters to

his eye lens with ciliary muscles, find the near point of his eyes without glasses. [58.8 diopters, 8.33 cm]

3. A nearsighted person wears glasses of $f = -50.0$ cm. Through these glasses he has perfect distant vision, but he cannot focus clearly on the objects closer than 25.0 cm. How close can he focus if he removes his glasses? What is his far point? [16.7 cm, 50.0 cm]

4. A pair of bifocal lenses have components with focal lengths of 50 cm and -250 cm. What are the near and far points for the wearer of these bifocal lenses? [50 cm and 250 cm]

PROBLEMS

Each of the following problems may involve more than one physical concept. When appropriate, the answer is given in brackets at the end of the problem.

5. Typical vision data for a female subject is as follows:

	Near Point	Far Point
Without glasses	13 cm	39 cm
With glasses	20 cm	∞

What is the subject's power of accommodation without glasses? With glasses, what is the power of accommodation of the optical system? Explain these results. What is the power of the glasses lens? Does this person suffer from myopia, hyperopia, or presbyopia? [5.1 diopters, -2.6 diopters, myopia]

6. An automobile is approaching you on a straight road on a pitch black night. The lights on the car are 1.5 m apart. Assume the aperture of your eye is 1 cm and the effective wavelength of the light is 550 nm.

   a. Using the Rayleigh criterion, how far will the automobile be from you when you can just be sure that you are seeing an automobile and not a motorcycle?

   b. Using the spacing of receptor cells in the eye, how far will the automobile be from you when you can just be sure that you are seeing an automobile and not a motorcycle?

   c. According to NASA sources the minimum separation acuity for the average astronaut is 0.4 minutes of arc. How far will the automobile be from an astronaut before he can be sure that it is not a motorcycle? [2.2×10^4 m, 1.5×10^4 m, 1.3×10^4 m]

7. A photon of visible light has energy of about 3.6×10^{-19} J. The absolute luminance threshold for the dark-adapted hyman eye is 10^{-5} mL (milli-Lambert $= 1.47 \times 10^{-6}$ watts/cm²). How many photons per second per cm² are required for minimum human vision? [4.1×10^7 photons/cm²–sec]

8. In the fovea portion of the eye there are 136 thousand cones per square millimeter, and the lower absolute threshold of illumination is 10^{-3} mL. Using the average energy of a visible photon of light as 3.6×10^{-19} J, how many photons per second are required to provoke a visual response from a cone? [300 photons/cone–sec]

9. On a horizontal angle of 20° from the fovea there are 158 thousand rods per square millimeter, and the lower absolute threshold of illumination is 10^{-5} mL. Using an average energy of visible light photons as 3.6×10^{-19} J, how many photons per second are required to provoke a visual response from a rod? [2.58 photons/rod–sec]

10. Given the cone and rod average distribution per unit area in problems 8 and 9, what is the minimum acuity angle for vision using only cones? Using only rods? The distance from the cornea to the retina is 25 mm. [2.16×10^{-4} rad, 2×10^{-4} rad]

11. Design an experiment to measure relative luminosity for the eye as shown in Figure 20.4.

12. An older person has a near point of 75 cm. Find the correct lenses to prescribe for this person. [bifocals, clear glass and $+2.67$ diopters]

GOALS

When you have mastered the contents of this chapter, you will be able to achieve the following goals:

Definitions
Define each of the following terms, and use it in an operational definition:

dielectric constant   equipotential surfaces
electrical field   dipole
potential gradient   capacitance of a capacitor
potential difference

Coulomb's Law
Apply the basic model of an electrostatic field, and use Coulomb's law to calculate the force on one, two, or three given point charges.

Potential Gradient
Apply the gradient to electrical field phenomena.

Moving Charged Particles
Explain the motion of charged particles in an electric field.

Capacitance
Solve capacitance and capacitor problems, including the use of a capacitor as a means of storing electrical energy.

Applications of Electrostatics
List a number of applications of electrostatic principles to daily living and to medical equipment.

PREREQUISITES

Before beginning this chapter you should have achieved the goals of Chapter 2, Unifying Approaches, Chapter 4, Forces and Newton's Laws, Chapter 5, Energy, and Chapter 9, Transport Phenomena.

21

Electrical Properties of Matter

21.1 Introduction

You have observed and experienced many phenomena that are examples of the electrical nature of matter. During the winter you have experienced a shock by walking over a rug and then touching a metal object or by sliding across your automobile seat and touching the door handle. You have seen bolts of lightning. Clothes you take out of a clothes dryer often cling together due to static electricity. The focusing and imaging system of a television picture tube makes use of the force on electrons moving in an electric field. Your heart is an electric device that keeps its steady beat with electric synchronization.

These examples all involve the electrical nature of matter and electric interactions. In this chapter, the electric field model will be introduced and used to explain electric phenomena.

21.2 Electrical Forces

A dry glass rod after being rubbed with silk will pick up small bits of paper. A dry rubber rod, after being rubbed with cat's fur, will pick up small bits of paper and will attract the rubbed glass rod. An analysis of many experiments such as these has lead us to postulate that there are two kinds of electricity which we call positive and negative (Figure 21.1). Conventionally, positive electricity is defined as the electricity that appears on a glass rod when it is rubbed with silk, and negative electricity

FIGURE 21.1
Two different kinds of charge are produced in the illustrated electrostatic experiment. The glass rod rubbed with silk becomes positively charged as the silk becomes negatively charged. The rubber rod becomes negatively charged as the fur becomes positively charged.

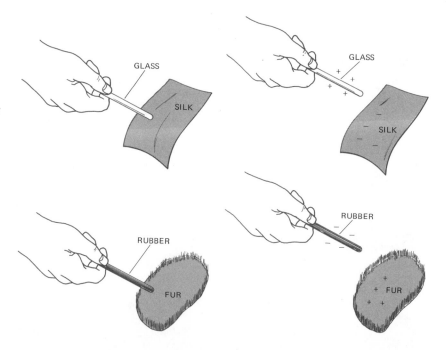

is defined as the electricity that appears on a hard rubber rod rubbed with cat's fur. If an object has equal amounts of positive electricity and negative electricity, it is said to be neutral. If an object has an excess of negative electricity, it is said to be negatively charged. If an object has a deficiency of negative electricity, it is said to be positively charged.

In 1909 Robert A. Millikan performed a series of experiments on charged oil drops. He reported that these drops always contained an amount of electrical charge that was an integer times a fundamental constant:

$$\text{total charge} = Ne \tag{21.1}$$

where N is an integer, $0, \pm 1, \pm 2, \pm 3, \ldots$ and e is the magnitude of fundamental charge, which we call the electronic charge. The best measurements of the size of the electronic charge now indicates that the unit of electric charge in the SI units, which is called a coulomb (C), is equal to a very large number of electron charges.

$$1 \text{ coulomb} = 1 \text{ C} = 6,241,450,300,000,000,000 \, e$$

$$1 \text{ C} = 6.24 \times 10^{18} e \tag{21.2}$$

or

$$e = 1.60 \times 10^{-19} \text{ C} \tag{21.3}$$

In the eighteenth century, experiments like those illustrated in Figure 21.2 had shown that like charges repel each other and unlike charges attract each other. The inverse square law for electrical forces was verified by the experiments of Charles Coulomb. According to Coulomb's law of electrical interaction, the magnitude of the electrical force between two point charges is proportional to the product of the charges and inversely proportional to the square of the distance between the charges,

$$F \propto \frac{q_1 q_2}{r^2} \tag{21.4}$$

where q_1 and q_2 are the amounts of electric charge of the two point charges and where r is the distance between the point charges. This expression may be written as a vector equation as follows:

$$\mathbf{F}_{12} = \frac{k q_1 q_2}{r^2} \hat{\mathbf{r}}_{12} \tag{21.5}$$

where $\mathbf{F}_{12}$ is the force acting on charge 2 because of the presence of charge q_1 and $\hat{\mathbf{r}}_{12}$ is a vector of unit magnitude which points in the direction from charge 1 to charge 2 (Figure 21.3). If q_1 and q_2 are of like sign, Equation 21.5 has the force between the charge acting so as to push the two charges apart (repulsion). See Figure 21.3a and b. If the charges are of opposite sign, the force between the charges acts to push the charges together (attraction) as in Figure 21.3c and d. The value of the proportionality constant k which appears in Equation 21.5 depends

FIGURE 21.2
Photographs of electric field patterns formed by grass seeds in an insulating liquid. (a) A single charged electrode, (b) Two electrodes with opposite charges, (c) Two electrodes with same kind of charge, (d) A single charged metal plate, (e) Two parallel plates, no electric field between them, (f) a pair of oppositely charged parallel plates. (Pictures from *PSSC Physics,* D.C. Heath and Co., Lexington, Mass., 1965.)

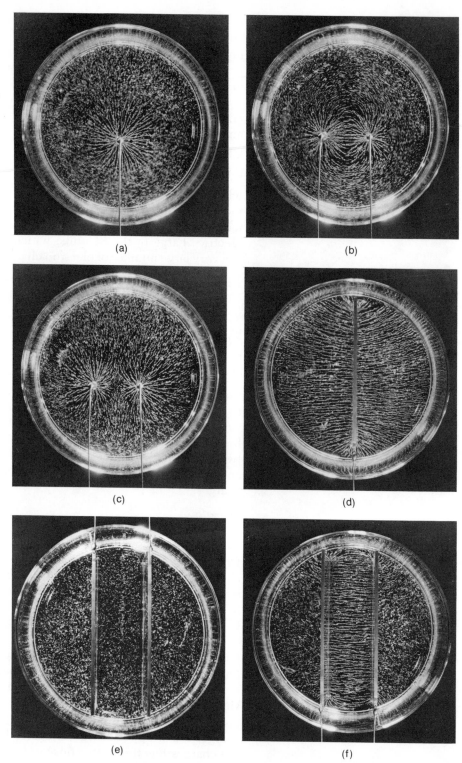

(a)

(b)

(c)

(d)

(e)

(f)

FIGURE 21.3
Directions and magnitudes of forces
between charges for repulsion and
attraction.

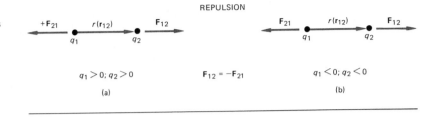

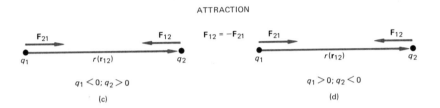

upon the system of units that is being used and upon the medium in
which the two charges are located. If the point charges are located in a
vacuum and the force is measured in newtons, the charges in coulombs
and the distance between the charges in meters, then k has the following
value and units:

$$k = \frac{1}{4\pi\epsilon_0} = 8.99 \times 10^9 \ \frac{\text{N--m}^2}{\text{C}^2} \simeq 9.0 \times 10^9 \ \text{N--m}^2/\text{C}^2 \qquad (21.6)$$

where ϵ_0 is the *permittivity of free space* and has the numerical value of
$8.85 \times 10^{-12} \ \text{C}^2/\text{N--m}^2$. For problems in this book you may use the value of
k at $9.00 \times 10^9 \ \text{N--m}^2/\text{C}^2$ in SI units. If the charges are located in a medium that has a *dielectric constant* ϵ, then Equation 21.5 must be replaced by the following equation,

$$\mathbf{F}_{12} = \frac{kq_1q_2}{\epsilon r^2}\ \hat{\mathbf{r}}_{12} \qquad (21.7)$$

The dielectric constant accounts for the contribution of the molecular
charges in the medium in the determination of the electric field inside
such materials. For many gases the charge distribution of the molecules
is symmetrical around the center of the molecule. For gases with this
spherical charge symmetry, ϵ is about 1. Materials which show a charge
distribution that is not symmetric but is instead characterized by the
separation of its negative and positive charge centers, are called *polar*
materials. Polar molecules are electrically neutral but their nonsymmetric charge distribution results in dielectric constants that are much
greater than 1. A simple example of such a distribution would be equal
volumes of positive and negative spherical charges with their centers
slightly displaced from each other.

Since electrical charges interact with each other across some distance, they represent an example of systems that interact-at-a-distance.

An electric charge moving on one side of the room seems to cause changes in the motion of an electrical charge on the other side of the room. We then construct the model of an electric force field that fills all the space in the vicinity of an electrical charge. Since there are electrical charges in all molecules, the universe is more or less filled with electric fields. We construct electric fields that obey linear relationships. We can use the principle of superposition to solve problems that involve the interaction-at-a-distance of many different charges. In such cases we solve the problem for each individual charge as if the others did not exist, and then add up all of the individual results. Hence, in any region where an electric force is acting upon an electric charge at rest, we say there exists an electric field. The direction of the field is defined as the direction of the force that would act upon a positive charge at that location in space. The magnitude of the field at that point is given by the force divided by the charge. The electric field $\mathbf{E}$ at a point is given by

$$\mathbf{E} = \mathbf{F}/q \tag{21.8}$$

where $\mathbf{F}$ is the force acting on a charge q at that location. The electric field is a vector quantity defined as the force upon a unit charge. The SI units for $\mathbf{E}$ are newtons per coulomb. In the region where there is a single positive point charge q, the electric field is given by,

$$\mathbf{E} = k \frac{q}{r^2} \hat{\mathbf{r}} \tag{21.9}$$

where $\hat{\mathbf{r}}$ is a vector of unit magnitude that points in a radial direction away from the charge q. If there are several point charges in the region, the resultant electric field is the vector sum of the separate fields resulting from each individual charge.

EXAMPLE

In the diagram shown in Figure 21.4, determine the field at B resulting from a charge of $+1.6 \times 10^{-12}$ C at A, 4 cm north of B, and a charge of -1.8×10^{-12} C 3 cm east of B.

The field at B resulting from the charge at A is a repulsive force whose magnitude is given by,

$$E_A = k \frac{q_A}{r_A^2} = 9 \times 10^9 \times \frac{1.6 \times 10^{-12}}{(0.04)^2} = 9 \text{ N/C}$$

and which is directed south. The field at B resulting from the charge at C is an attractive field whose magnitude is given by,

$$E_C = k \frac{q_C}{r_C^2} = \frac{9 \times 10^9 \times 1.8 \times 10^{-12}}{(0.03)^2} = 18 \text{ N/C}$$

directed to the east.

$$|\mathbf{E}_B| = (9^2 + 18^2)^{1/2} = 20.1 \text{ N/C}$$

$$\tan \theta = \frac{9}{18} = \frac{1}{2}$$

FIGURE 21.4
Determination of electric fields.

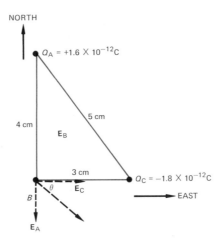

$\theta = 26.6°$ south of east.

A charge of $+3$ C located at point B would experience a force of 60.3 N acting 26.6° south of east. A charge of -2 C at point B would experience a force of 40.2 N acting 26.6° north of west.

21.3 *Electrical Forces Acting on Moving Charges*

Now consider the behavior of a charged particle free to move in an electric field. We first investigate the case of a charged body in a uniform electric field. Suppose we have charge q in a field of **E**. The force acting on the charge is q times **E**,

$$\mathbf{F} = q\mathbf{E} \tag{21.10}$$

in the direction of **E**. In an earlier chapter on Newton's laws (Chapter 4), we learned that a constant force acting on a free body produces an acceleration given by Newton's second law,

$$\mathbf{a} = \frac{\mathbf{F}}{m} = \frac{q\mathbf{E}}{m} \tag{21.11}$$

where, in this case, the force is given by $q\mathbf{E}$. We can apply the equations for uniformly accelerated motion to this situation. So, if we start from rest at time $t = 0$, at any later time t the velocity is given by the acceleration multiplied by the time,

$$\mathbf{v} = \mathbf{a}t = \frac{q\mathbf{E}}{m}\, t \tag{21.12}$$

The displacement **y** is given by one-half the acceleration times the square of the time,

$$\mathbf{y} = \tfrac{1}{2}\mathbf{a}t^2 = \tfrac{1}{2}\frac{q\mathbf{E}}{m}t^2 \tag{21.13}$$

The square of the velocity is proportional to the product of the magnitudes of the field and the displacement,

$$v^2 = 2|\mathbf{a}| \, |\mathbf{y}| = \frac{2q|\mathbf{E}| \, |\mathbf{y}|}{m} \qquad (21.14)$$

The kinetic energy acquired after moving a distance y is given by the work done in moving the charge, or the product of the force times the distance,

$$\text{KE} = \tfrac{1}{2} m v^2 = q|\mathbf{E}| \, |\mathbf{y}| \qquad (21.15)$$

Thus, we see that the motion of charged bodies can be influenced by the application of electric fields. There are many applications of this principle.

EXAMPLES

1. An instrument that makes use of this phenomenon is the oscilloscope. Consider the case of a beam of particles of charge $-e$ moving with a velocity v in a horizontal direction. They enter a vertical electric field $\mathbf{E}$ in the upward direction. What happens to the particles? What is the shape of their path in the electric field?
2. *Electrophoresis* is a process that is used to separate different types of molecules. It is very useful in separating proteins and amino acids for further analysis. The electrophoresis process is based on the different motions of charged particles (ions) in certain solvents (called buffers) under the influence of applied electric fields. Different molecular constituents in the solution migrate at different speeds. For example, when blood is subjected to this process, serum albumin and three globular proteins are separated according to their speed of migration in the applied electric field. What properties of the constituents would be involved in determining the speed of migration in an electric field?

21.4 Electrical Potential

Consider two points, A and B, in an electric field. A test charge q is moved from A to B, and we measure the work done in moving q from A to B without any acceleration. The electric potential difference between A and B, V_{AB}, is defined as the work done per unit charge. So,

$$V_{AB} = V_B - V_A = \frac{W}{q} \quad \text{or} \quad W = qV_{AB} \qquad (21.16)$$

W may be positive, zero, or negative. If W is positive, B is said to be at a higher potential than A, and an external agent is required to move a positive charge q from A to B (Figure 21.5a). If W is 0, A and B are at the same potential (moving a charge at right angles to the electric field lines, for example, as in Figure 21.5b). If W is negative, B is said to be at a lower potential than A and the work is done by the electric field (moving

FIGURE 21.5
The relationship between electric field, work and potential.

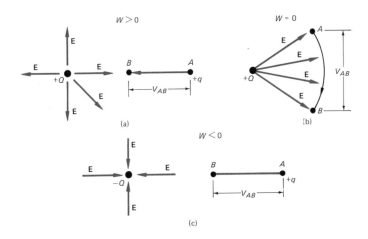

a positive charge from A to a point B closer to a negative charge as in Figure 21.5c). In the SI units, W is in joules, q in coulombs, and the potential is expressed in joules per coulomb, which is called *volts* (V). One volt equals an energy of one joule expended on one coulomb. The difference in potential between the two points is given by Equation 21.16. If one selects point A as a reference point and assigns it a value of zero potential, then any other potential, such as that at any point P, is given by

$$V_P = \frac{W_P}{q} \qquad (21.16)$$

where W_P represents the work required to take charge q from A to the point P under consideration. Usually the potential is taken to be zero at some point an infinite distance away.

An electric charge of positive q will gain kinetic energy if it is released at an electric potential V and moves to zero potential. The work done on the charge is $W = Vq$. Since work is done on the charge, it will gain kinetic energy,

$$KE = \tfrac{1}{2}mv^2 = qV \qquad (21.17)$$

The velocity of a charged particle, which starts from rest and undergoes a change of electric potential V is given by

$$v = \sqrt{2qV/m} \qquad (21.18)$$

Since the electric field was developed as a linear model for interaction-at-a-distance, the principle of superposition holds for potential, and you can find the resultant potential at a given point due to the separate individual charges by adding the individual potentials,

$$V_{total} = V_1 + V_2 + V_3 \cdots + V_n$$

It can be shown that the potential due to a point charge at a distance r from the charge q is given by

$$V = k\frac{q}{r} \tag{21.19a}$$

(See Equation 21.40.) Hence the potential at a given point produced by a number of point charges is given by

$$V = k\left(\frac{q_1}{r_1} + \frac{q_2}{r_2} + \frac{q_3}{r_3} \cdots + \frac{q_n}{r_n}\right) \tag{21.19b}$$

EXAMPLE

Find the potential at B resulting from a charge of $+1.6 \times 10^{-12}$ C at A, 4 cm north of B, and a charge of -1.8×10^{-12} C at C, 3 cm east of B. See Figure 21.4. The potential at B due to the charge A is

$$V_A = k\frac{q_A}{r_A} = 9 \times 10^9 \times \frac{1.6 \times 10^{-12}}{4 \times 10^{-2}} = 0.36 \text{ V}$$

The potential at B due to charge C is

$$V_C = k\frac{q_C}{r_C} = 9 \times 10^9 \frac{-1.8 \times 10^{-12}}{3 \times 10^{-2}} = -0.54 \text{ V}$$

V at $B = V_A + V_C = 0.36 + (-0.54) = -0.18$ V where the potential at infinity is taken as zero.

If no work is required to move a charge from one point to another, the two points are at the same potential and are said to be on an *equipotential surface*. Let us consider the case of two large, parallel, flat conducting plates separated by a distance d with plate A charged positively, and plate B charged negatively. The field is directed from plate A to plate B, perpendicular to the plates, and is uniform (Figure 21.6). The work required to take a positive charge q from plate B to plate A is

$$w = FD = +(-qEd) \tag{21.20}$$

because the force F must be equal but opposite to the electric field qE in order to move the charge from plate B to plate A. We can also express the work in terms of the potential difference between plates B and A,

$$w = (V_A - V_B)q \tag{21.21}$$

We can set Equations 21.20 and 21.21 equal to each other and solve for E

$$E = -\frac{V_A - V_B}{d} \tag{21.22}$$

The magnitude of the electric field E is the ratio of the change in potential $(V_A - V_B)$ to the change in distance d. In other words, the *electric field is a potential gradient*, and the negative sign indicates that the electric field is oppositely directed to the potential gradient (that is, **E** is directed from high to low potential). In the case of parallel plates one notes that the direction of the electric field is perpendicular to the equipotential surfaces. This result was derived for a special case; however, it

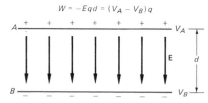

$W = -Eqd = (V_A - V_B)q$

FIGURE 21.6

The work required to take a unit positive charge from the negatively charged plate to the positively charged plate is illustrated for the case of the uniform electric field between oppositely charged plates.

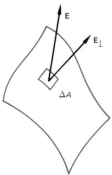

FIGURE 21.7
The relationship between flux and electric field. $\Delta\phi_E = E_\perp \Delta A$

can be shown that the electric field is always equal to the negative of the potential gradient:

$$E = -\frac{\Delta V}{\Delta s} \quad \text{(directed} \perp \text{to surfaces of constant potential)} \qquad (21.23)$$

where ΔV is the potential difference over the distance Δs. Also, the direction of the electric field is always perpendicular to the equipotential surfaces, directed from high to low potential.

Questions

Show that the electric field must always be perpendicular to the equipotential surfaces by assuming that **E** is not perpendicular to an equipotential surface and showing that this leads to a contradiction for an equipotential surface.

21.5 Gauss' Law

Another formulation of the inverse square law for electric fields due to point charges is known as Gauss' law. First we must define the electric flux ϕ_E. To compute the electric flux through some small area ΔA, multiply the component of the electric field that is perpendicular to that area times the area ΔA; that is, $\Delta\phi_E = E_\perp \Delta A$ (Figure 21.7). For a closed surface, Gauss' law states that the total flux passing through the area of that closed surface will be equal to the total electric charge enclosed divided by the permittivity of free space, ϵ_0,

$$\phi_\epsilon = q_{total}/\epsilon_0 \qquad (21.24)$$

where ϕ_E is the total flux through the surface and q_{total} is the total charge enclosed by the surface. The total flux, the sum of all the small fluxes $\Delta\phi_E$ that pass through all the small areas ΔA needed to make up the total enclosed surface,

$$\phi_E = \Sigma\Delta\phi_E$$

Therefore

$$\phi_E = \Sigma\,(E_\perp \Delta A) \qquad (21.25)$$

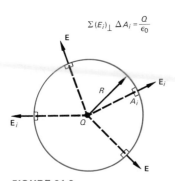

FIGURE 21.8
Gauss' law applied to a point charge. The electric field flux through any spherical surface with the charge Q at its center is equal to Q/ϵ_0. All points on the sphere are equidistance from the charge, and thus the field is constant over the sphere and is directed radially outward.

EXAMPLES

1. Consider an isolated point charge. We know by symmetry that the electric field must be radial and depend only upon the distance R from the point charge. So we choose a spherical surface which has an area $4\pi R^2$ (Figure 21.8). Thus the electric flux is

$$\phi_E = E(4\pi R^2)$$

and

$$\phi_E = E(4\pi R^2) = q/\epsilon_0$$

from Gauss' law.

FIGURE 21.9
Gauss' law applied to a line of charge. The electric field flux through any cylindrical surface with the line charge along its axis is equal to the total line Q/ϵ_0. All points on the cylindrical surface are the same distance from the line charge and thus the field is constant over the surface and directed radially outward.

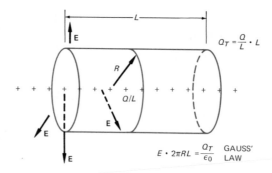

Solving for the value of the electric field we find:

$$E = \frac{q}{4\pi R^2 \epsilon_0} \qquad \text{(directed radially outward)} \qquad (21.26)$$

This is Coulomb's law for a point charge q. Gauss' law is primarily useful when dealing with problems where we know from symmetry that the electric field is constant over a properly chosen surface around the given charge distribution. Gauss' law shows that the field outside any *spherical distribution of charge* is identical to that of a single point charge q_{total} located at the center of the spherical distribution.

2. Let us apply Gauss' law to a long cylinder of positive charge neglecting end effects. (This is usually stated by saying the length is much greater than the radius of cylindrical charge.)

Again the symmetry of the problem is such that we know there can only be an electric field pointing radially outward from the center of the cylinder. Therefore, if we pick a cylindrical surface surrounding the line charge (see Figure 21.9), the field is perpendicular to the surface and constant over the whole cylindrical area. Then from Gauss' law we get:

$$\phi_E = E \text{ (area of cylindrical surface)} = q_{total}/\epsilon.$$

$$E(2\pi RL) = q_{total}/\epsilon_0$$

$$E = \frac{1}{2\pi\epsilon_0} \frac{q_{total}/L}{R} \qquad (21.27)$$

where q_{total}/L is the linear charge density inside the cylinder. We see that a long line of charge produces a field that is proportional to $1/R$ as compared with $1/R^2$ for point or uniformly spherically distributed charges.

3. In the case of electrostatic equilibrium the electric field inside a conductor must be zero. If this were not the case, the free charge of the conductor would move under the influence of the electric field. Consequently, conductors are equipotential surfaces for electrostatic phenomena. We can use Gauss' law to show that the field inside an empty cavity in a conductor must be zero. Imagine a closed surface inside the cavity. From Gauss' law it follows that the field everywhere on the surface must be zero since there is no charge inside the surface. Thus, everywhere inside the cavity the electric field is zero. This result is completely independent of any charge on or outside the conductor. Such a space becomes an electrostatic shield, as motion of charge outside the conductor will have no affect on apparatus inside the cavity (Figure 21.10).

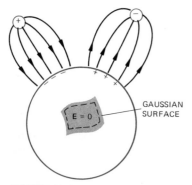

FIGURE 21.10
The field inside a hollow conductor is zero everywhere, and thus this space is shielded from effects due to charges outside of the conductor. This is the basis of electrostatic shielding that is used for many biological experimental set ups.

21.6 Electric Dipoles

Recall that polar materials result from neutral charge distributions that have their positive and negative charge centers separated by a distance we will call s. Such charge distributions are called *electric dipoles* (Figure 21.11). The product of the charge and the distance between charges qs gives the magnitude of the electric dipole and its vector is directed from the negative charge to the positive charge. It can be shown that the electric field due to a dipole is proportional to $1/r^3$ where r is the distance from the dipole to the field point and $r \gg s$. Substances with large molecular electric dipoles have high dielectric constants, for example, for water, $\epsilon = 80.36$ at 20°C.

21.7 Electrostatic Applications

The basic principles of electrostatic phenomena given above are sufficient to provide qualitative understanding of the electrical activity of the heart. One can ignore the electrical conductivity of the body and use Coulomb's law in order to show that the electrocardiogram (ECG) measures the potential distribution of a set of dipoles within the heart.*

The dielectric constant of a material is a measure of the internal electric field for the material. An external electric field applied to a material with a high dielectric constant aligns the molecular dipoles inside the material producing a high internal electric field. The external field does work on the molecular dipoles, and this work is stored as potential energy of orientation within the material. The potential energy is a function of temperature. Can you propose a model to predict its qualitative temperature dependence? When the external field is removed, the molecular dipoles will gradually return to random orientations (what effect will temperature have on this process?) and produce very small depolarization currents in the materials.

The dielectric constant of water is ~80. This high value is typical of biological cells and tissues. The interactions between such dielectric materials and externally applied fields are important areas of study in biophysics.

The effectiveness of microwave ovens is a result of the resonant absorption of electromagnetic energy by molecular dipoles in the food being cooked in the oven. The frequency of these resonances is in the microwave region of the electromagnetic spectrum. Recent research shows that the mapping of absorption of microwaves by human tissue may be very useful in early detection of cancer.

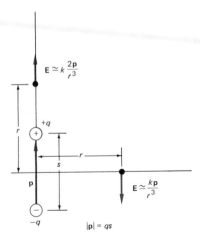

FIGURE 21.11
The electric dipole, $\mathbf{p} = q\mathbf{s}$ directed from positive to negative charge.

* For a complete discussion of electrostatic phenomena as applied to electrocardiograms, see R. K. Hobbie, "The Electrocardiogram as an Example of Electrostatics," *American Journal of Physics* **41** (June 1973): 824–831.

21.8 Capacitance

Electrical energy can be stored in a device called a *capacitor*. A capacitor can be made of two conducting plates separated by a nonconductor (dielectric). In an uncharged condition both plates are neutral. In a charged condition one plate is positively charged, and the other is negatively charged; that is, the two plates are at different potentials. The ratio of the charge on plates to the potential difference between the plates is defined as the capacitance.

$$C = \frac{q}{V} \tag{21.28}$$

where q is the charge in coulombs, V is the potential in volts, and C is the capacitance in farads. The farad is a very large unit of capacitance, and the capacitance of typical devices is given in micro (10^{-6}) farads or pico (10^{-12}) farads.

The capacitance of a capacitor depends upon the area of the plates, the thickness, the properties of the dielectric, and the geometric configuration. For a parallel plate capacitor

$$C = \frac{\epsilon \epsilon_0 A}{d} \tag{21.29}$$

where C is in farads, ϵ is the dielectric constant, A is the area in square meters, d is the thickness in meters, and ϵ_0 is the permittivity of free space $(8.85 \times 10^{-12} \text{ F/m})$. Note that the charge stored is proportional to the dielectric constant for a constant applied voltage.

EXAMPLE

What is the capacitance of an air capacitor whose parallel plates have an area of 1.00 cm² and are spaced 1.00 mm apart?

$\epsilon_{air} = 1.00$

$C = 1 \times 1.00 \times 10^{-4} \times 8.85 \times 10^{-12} / 1.00 \times 10^{-3}$

$= 8.85 \times 10^{-13} \text{ F} = 0.885 \text{ pF}$

21.9 Combinations of Capacitors

In a variety of applications it is desirable to use a combination of capacitors. We can connect two charged capacitors together in two ways. We can connect the two oppositely charged plates or the two similarly charged plates.

When the unlike plates of charged capacitors are connected together, the arrangement is called a *series* connection (Figure 21.12). The charges will move from one capacitor to the other until the positive

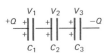

FIGURE 21.12
Capacitors in series. The charges on the capacitors are equal, and the total voltage across capacitances is the sum of the individual voltages.

charge on one plate is equal to the negative charge on the plate connected to it,

$$q_1 = q_2 = q_3 = Q = \text{total charge in the combination} \tag{21.30}$$

In this configuration the potential across the total combination is the sum of the individual potentials,

$$V = V_1 + V_2 + V_3 \tag{21.31}$$

but the potential across a capacitor is defined as the ratio of the charge to the capacitance. We can make use of that definition to derive an expression for the effective capacitance of the three used in series,

$$\frac{Q}{C} = \frac{q_1}{C_1} + \frac{q_2}{C_2} + \frac{q_3}{C_3} \tag{21.32}$$

By making use of Equation 21.30, we find that the effective capacitance C of a series configuration of capacitors is calculated by taking the reciprocal of the sum of the reciprocals of the individual capacitances. For three capacitors, $C_1, C_2,$ and C_3,

$$\frac{1}{C} = \frac{1}{C_1} + \frac{1}{C_2} + \frac{1}{C_3} \tag{21.33}$$

The configuration in which the like plates of charged capacitors are connected together is called a *parallel* connection (Figure 21.13). In this configuration the charges rearrange themselves until the potential V is the same across all of the capacitors

$$V = V_1 = V_2 = V_3 \tag{21.34}$$

and the total charge in the combination is the sum of the individual charges,

$$Q = q_1 + q_2 + q_3 \tag{21.35}$$

We can use the definition of the charge on a capacitor, CV, to derive an equation for the effective capacitance for a number of capacitors connected in parallel,

$$CV = C_1V_1 + C_2V_2 + C_3V_3 = (C_1 + C_2 + C_3)V$$
$$C = C_1 + C_2 + C_3 \tag{21.36}$$

The effective capacitance of a number of capacitors connected in parallel is the sum of the individual capacitances.

The charged capacitor has the ability to do work, and thus possesses energy. The energy, which is stored in the dielectric, can be dissipated by connecting the two plates by a conductor. If we do so a redistribution of charges occurs. The energy stored in a charged capacitor can be calculated by finding the increment of work required to charge a capacitor from zero to a small charge Δq,

$$\Delta w = V \Delta q \tag{21.37}$$

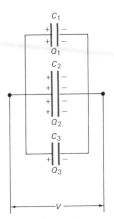

FIGURE 21.13
Capacitors in parallel. The voltages across capacitors are equal, and the total charge is the sum of the individual charges.

So the total energy stored in the capacitor will be equal to the total work required to build up the charge to Q,

$$w = V_{ave}Q \qquad (21.38)$$

where $V_{ave} = \frac{1}{2}Q/C$.

$$w = \frac{1}{2}\frac{Q^2}{C} = \frac{1}{2}CV^2 = \frac{1}{2}QV \qquad (21.39)$$

where C is in farads, Q is in coulombs, V is in volts, and w is in joules.

The capacitor is an important component of many electrical instruments and appliances. One example is the direct current defibrillator. *Defibrillation* is the application of an electric shock to the heart to stop the rapid, uncoordinated contractions of the heart muscle called fibrillation. In a defibrillator, the capacitor is usually of the order of 10 to 20 μF, the charging potential may be of the order of a few thousand volts, and the energy delivered to the patient may be as much as 400 J. A capacitor is also used in the electrical circuit of an electrocardiograph. The capacitor is used to block any direct current component of the ECG signal. A third application is in a heart pacemaker. A pacemaker is an electronic device which provides a regular periodic electrical stimulus to the heart to make regular the rhythmic performance of the heart. The energy requirement for a pacemaker is many orders of magnitude smaller than that for a defibrillator.

ENRICHMENT
21.10 Calculus Derivations of Electrostatic Relationships

In Chapter 5 on work and energy you learned that work is given by

$$\Delta w = F \cos \theta \Delta s \quad \text{or} \quad w = \int_a^b F \cos \theta \; ds$$

where ds is the incremental element of displacement. Let us apply this to an electric field. The field is radial from a charge q (Figure 21.14). The line ab represents some arbitrary path between these two points. Consider some element of path where $\mathbf{E}$ makes an angle with the path. The work done to take unit charge along length ds is then $E \cos \theta \; ds = dw$. The integral of the portion

$$\int_a^b E \cos \theta \; ds$$

is the integral along the line from a to b. For the field in the vicinity of a point charge,

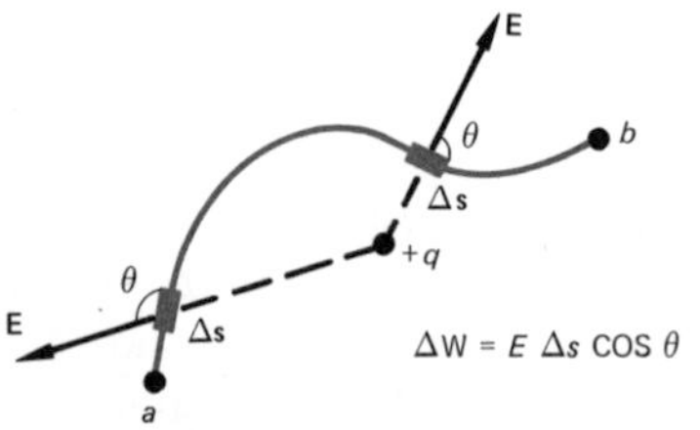

$$|\mathbf{E}| = \left(\frac{1}{4\pi\epsilon_0}\right)\left(\frac{q}{r^2}\right)$$

FIGURE 21.14
The work per unit charge done in moving in the field of a point charge.

$\Delta W = E \; \Delta s \; \cos \theta$

$$ds \cos \theta = dr$$

where θ is the angle between ds and dr.

Hence the work is equal to the integral of electric intensity along the line ab,

$$w = \int_a^b E \cos \theta \, ds = \int_{r_a}^{r_b} \frac{1}{4\pi\epsilon_0} \left(\frac{q}{r^2}\right) dr = \frac{1}{4\pi\epsilon_0} q \int_{r_a}^{r_b} \frac{dr}{r^2}$$

$$w = kq \left(\frac{1}{r_b} - \frac{1}{r_a}\right) = V_b - V_a \text{ for a unit charge}$$

where

$$V_a = \frac{kq}{r_a} \qquad V_b = \frac{kq}{r} \tag{21.40}$$

If we choose $r_a = \infty$ as reference point and $V_a = 0$, then in general the potential due to a point q of a point r meters away is

$$V = \frac{1}{(4\pi\epsilon_0)} \frac{q}{(r)} \tag{21.41}$$

To show that the work done in moving a unit charge around any closed path in an electrostatic field is zero, we proceed as follows: If a point charge q' is moved in an electric field, the force is $q'E$. The work for moving q' a distance ds is then

$$dw = (q'E \cos \theta)(ds)$$

Again,

$$E = \left(\frac{1}{4\pi\epsilon_0}\right)\left(\frac{q}{r^2}\right)$$

$$dw = \left(\frac{1}{4\pi\epsilon_0}\right)\left(\frac{q}{r^2}\right) q' dr$$

but $ds \cos \theta = dr$. Then,

$$w = \frac{qq'}{4\pi\epsilon_0} \int_{r_a}^{r_a} \frac{dr}{r^2} = \frac{qq'}{4\pi\epsilon_0} \left(\frac{1}{r_a} - \frac{1}{r_a}\right) = 0$$

where r_a is the starting and ending point of the loop.

In charging a capacitor, charge is transferred from a lower potential to a higher potential. This requires an expenditure of energy. The work done to transfer charge dq is $dw = V_A dq$. But, $V = q/C$; so

$$dw = \frac{q}{C} dq$$

$$w = \frac{1}{C} \int_0^Q q \, dq = \frac{1}{2} \frac{Q^2}{C}$$

Because $Q = CV$,

$$w = \frac{1}{2}\frac{Q^2}{C} = \frac{1}{2}CV^2 = \frac{1}{2}QV$$

This result was introduced in Section 21.9 in an intuitive way.

SUMMARY

Use these questions to evaluate how well you have achieved the goals of this chapter. The answers to these questions are given at the end of the summary with the number of the section where you can find related content material.

Definitions

1. The electric field model for electrostatics is analogous to the gravitational field model. Give the analogous electrostatic term for the given gravitational term:
   source of field: *mass*, _____; distance dependence: $1/r^2$, _____.
2. The dielectric constant for a material is defined by which of the following ratios (E_e = external electric field in free space, E_i = internal electric field)
   a. E_e/E_i
   b. E_i/E_e
   c. q_i/q_e
   d. q_e/q_i
3. The relationship between the potential gradient and electric field can be written as gradient of $V(r)$ equals
   a. energy
   b. **E**
   c. $-\mathbf{E}$
   d. q_{total}
   e. none of these
4. Equipotential surfaces are satisfied by the following
   a. conductors
   b. $\mathbf{E} = 0$ surfaces
   c. potential gradient $= 0$
   d. charges at rest
   e. all of these
5. The unit for potential differences is the volt; this is equivalent to:
   a. joule-second
   b. watt–second
   c. joule/second
   d. joule/coulomb

   e. joule–coulomb
6. The electric dipole consists of a neutral charge distribution with a separation between the centers of the positive and negative charge. The electric field of the dipole is proportional to (s = charge separation)
   a. s^2
   b. $1/s^2$
   c. s
   d. $1/s^3$
   e. $1/s$
7. The capacitor is a device that stores energy in its electric field. This energy depends on the capacitor's
   a. geometry
   b. applied voltage
   c. total charge
   d. dielectric
   e. all of these
8. Capacitance has units of farads. One farad is equal to
   a. volt/coulomb
   b. coulomb/volt
   c. volt $\times$ coulomb
   d. volt/meter
   e. none of these
9. By Coulomb's law the force on a unit positive charge half way between identical charges q (in terms of the force F due to one charge) is
   a. $2F$
   b. F
   c. zero
   d. $F/2$
   e. $4F$
10. The magnitude of the force on a unit positive charge half way between $+q$ and $-q$ charge separated by a distance $2d$ is given as:
    a. $k(q/d^2)$
    b. $2(kq/d^2)$
    c. $kq/2d^2$
    d. zero
11. The direction of the force in question 10 is toward
    a. $+q$

b. origin

c. -q

d. undetermined

Potential Gradient

12. The constant electric field in a region is known to be 10 N/C. The potential difference between two points 0.5 m apart is
    a. 10 V
    b. 5 V
    c. 20 V
    d. 0.05 V
    e. cannot tell from these data

13. The direction of the electric field is opposite the direction of the potential gradient where the potential gradient is:
    a. minimum
    b. zero
    c. maximum
    d. unknown
    e. any value

Moving Charged Particles

14. The energy gained by a particle of charge $+q$ and mass m accelerating through a potential difference V is given by
    a. qV
    b. $(qV)^{1/2}$
    c. $2qV^{1/2}/m$
    d. V/q
    e. q/m

15. The energy gained by a charged particle passing through a potential difference V is independent of the particle's
    a. charge

b. mass

c. path length

d. initial speed

e. none of these

Capacitance

16. The energy stored by a capacitor can be written in terms of C (capacitance), V (voltage), and q (charge) as
    a. $\frac{1}{2}qV$
    b. $\frac{1}{2}CV^2$
    c. $\frac{1}{2}(q^2/C)$
    d. none of these
    e. $\frac{1}{2}q^2C$

Applications of Electrostatics

17. Gauss' law suggests that shielding equipment from electric fields can be accomplished by putting it inside
    a. a cavity in dielectric
    b. a vacuum
    c. a cavity in conductors
    d. none of these

Answers

1. charge, $1/r^2$ (Section 21.2)
2. a (Section 21.2)
3. c (Section 21.4)
4. a,c (Section 21.4)
5. d (Section 21.4)
6. c (Section 21.6)
7. e (Sections 21.8 and 21.9)
8. b (Section 21.7)
9. c (Section 21.2)
10. b (Section 21.2)
11. c (Section 21.2)
12. b (Section 21.4)
13. e (Section 21.4)
14. a (Section 21.4)
15. b,c,d (Section 21.4)
16. a,c,b, (Section 21.9)
17. c (Section 21.5)

ALGORITHMIC PROBLEMS

Listed below are the important equations from this chapter. The problems following the equations will help you learn to translate words into equations and to solve single-concept problems.

Equations

$$\mathbf{F} = k\frac{q_1 q_2}{r^2}\,\hat{\mathbf{r}}_{12} \tag{21.5}$$

$$k \simeq 9.00 \times 10^9 \text{ N-m}^2/\text{C}^2 \tag{21.6}$$

$$\mathbf{E} = \frac{\mathbf{F}}{q} = k\frac{q}{r^2}\hat{\mathbf{r}}$$

(21.8, 21.9)

$$KE = \tfrac{1}{2}mv^2 = q|\mathbf{E}||\mathbf{y}|$$

(21.15)

$$W = qV$$

(21.16)

$$KE = qV$$

(21.17)

$$E = -\frac{\Delta V}{\Delta s} \qquad \text{(directed perpendicular to an equipotential surface)}$$

(21.23)

$$\phi_E = \Sigma(E_\perp \, \Delta A)$$

(21.25)

$$C = \frac{q}{V}$$

(21.28)

$$C = \frac{\epsilon\epsilon_0 A}{d}$$

(21.29)

$$\frac{1}{C} = \frac{1}{C_1} + \frac{1}{C_2} + \frac{1}{C_3} \qquad \text{(series)}$$

(21.33)

$$C = C_1 + C_2 + C_3 \qquad \text{(parallel)}$$

(21.36)

$$\text{Energy} = \frac{1}{2}QV = \frac{1}{2}CV^2 = \frac{1}{2}\frac{Q^2}{C}$$

(21.39)

$$V = \frac{kq}{r}$$

(21.19a)

Problems

1. What is the electrostatic force between two ions in vacuum if one has a charge of 1.6×10^{-19} C and the other has a charge of 3.2×10^{-19} C and the charges are separated by a distance of 4×10^{-10} m?
2. What is the kinetic energy of an ion that has a charge of 1.6×10^{-19} C and is accelerated through a potential of 1.0×10^6 V?
3. Two parallel plates separated by 1.0 mm in a vacuum have a potential difference of 1000 V. What is the electric field of the capacitor?
4. A 2-μF capacitor is charged to a potential difference of 100 V. What is the charge on the capacitor?
5. What is the energy stored in the charged capacitor of problem 4?
6. A 2.0-μF and a 4.0-μF capacitor are connected in series. What is the capacitance of an equivalent single capacitor?

Answers

1. 2.88×10^{-9} N
2. 1.6×10^{-13} J
3. $E = 1.0 \times 10^6$ N/C from positive to negative plate; $\mathbf{E}$ is a vector quantity perpendicular to the plates

4. 200 μC
5. 10^{-2} J
6. 1.3 μF

EXERCISES

These exercises are designed to help you apply the ideas of a section to physical situations. When appropriate, the numerical answer is given in brackets at the end of the exercise.

Section 21.2

1. Two point charges $+4.00$ C and -2.00 C are located along the x-axis at the origin and at 20 cm respectively. Sketch the electric field in the region of these charges. What is the field far, far away from the origin? Locate all the points on the x-axis where the electric field is zero. $[2k/r^2$ N/C; $x = 68.3$ cm]

Section 21.3

2. A hydrogen ion of charge $+e$ with a mass of 1.67×10^{-27} kg is initially at rest in an electric field of 1.00×10^{-6} N/C. What is the velocity of the ion after 5 sec? How far has it traveled? What is its kinetic energy? $[v = 479$ m/sec; $s = 12$ m; $\Delta K = 19.2 \times 10^{-23}$ J]

Section 21.4

3. For the charges and locations given in exercise 1, find the location of zero potential points on the x-axis. What is the potential far, far away from the origin? $[40$ cm, 13.3 cm, $\sim 2k/r$ J/C]
4. In atomic and molecular experiments a unit of energy called an *electron volt* (eV) is used. One electron volt is the energy gained by a charge e as it changes its electric potential by one volt. Calculate the value of an electron volt in joules. $[1$ eV $= 1.6 \times 10^{-19}$ J]

5. In a given medical x-ray tube, electrons are accelerated through a potential of 10,000 V. How much energy does the accelerated electron have? What is the velocity of the electron after acceleration? $[1.6 \times 10^{-15}$ J or 10^4 eV, 5.9×10^7 m/sec]

Section 21.5

6. Given a hollow spherical conducting shell with a $+Q$ charge at the center of its inner cavity. Show that the charge on the inner surface of the conductor is $-Q$ and that a charge of $+Q$ is on the outer surface.
7. Show that if a copper ball is given a charge Q, the entire charge resides on its outer surface.

Section 21.8

8. A parallel plate capacitor consists of two flat plates 20 cm square separated by a dielectric 0.2 mm thick.
   a. Find the capacitance if the dielectric is air.
   b. Find capacitance if the dielectric is mica (dielectric constant = 6). $[a. C = 17.8 \times 10^{-10}$ F; b. 106.2×10^{-10} F]

Section 21.9

9. If one has three capacitors with capacitances of 0.5, 1.0, and 2.0 μF, what capacitance can be produced by connecting these in various parallel and series combinations? $[$all series: 0.29 μF; all parallel: 3.5 μF; six other arrangements: 2.33 μF; 1.40 μF; 1.17 μF; 0.86 μF; 0.71 μF; 0.43 μF$]$

PROBLEMS

The following problems may involve more than one physical concept. Where appropriate, the numerical answer is given in brackets at the end of the problem.

10. Find the force of attraction between an electron and a proton (hydrogen nucleus) at a distance of 5.3×10^{-11} m. How does this compare with their gravitational attraction? $[82 \times 10^{-9}$ N, 3.6×10^{-47} N]
11. Given that the dielectric constant of sodium chloride is 6.12, find the force of attraction between a Na^+ ion and Cl^- ion in a salt crystal if the separation is 2.8×10^{-10} m. Calculate the energy of the ionic bond in

a vacuum and in water ($\epsilon = 80$). $[0.48 \times 10^{-9}$ N; $e_{vac} = 8.2 \times 10^{-19}$ J, $e_{water} = e_{vac}/80]$
12. A square $ABCD$ is 10.0 cm on a side. A charge of 2.00×10^{-10} C is placed at B, and a charge of -3.00×10^{-10} C is placed at C. Find:
   a. the field at D
   b. the potential at D
      $[a.$ 216 N/C down to right making angle of $17°$ with CD; b. -14.3 volt$]$
13. If a charge of 5.00×10^{-10} C is moved from D in problem 12 to the center of the square, how much work is done? $[7.55 \times 10^{-10}$ J$]$

14. Given that the capacitance of 1.0 cm² of a cell membrane is 1.0 μF, find the number of ions necessary to charge the membrane to 70 mV (resting potential for an axon). Assume that the ions are singly charged with $q = 1.6 \times 10^{-19}$ C.   [44×10^{10} ions]

15. An air capacitor (0.1 μF) is charged with 20 μC of charge. The separation of the plates is 1 mm.
    a. Find the electric field between the plates of the capacitor.
    b. Find the energy needed to charge the capacitor as given.   [a. $E = 2 \times 10^5$ V/m; b. $w = 2 \times 10^{-3}$ J]

16. The electrical potential difference between the inside and the outside of a heart muscle is about 90 mV. If the wall of each cell is an insulating layer of thickness of 5.0×10^{-9} m, what is the electric field in the cell membrane? Compare this with other electrical fields, such as when an electrical breakdown occurs in air (10^4 V/cm).   [18×10^6 V/m, 3×10^6 V/m]

17. A direct current defibrillator has a maximum energy output of 400 J. The capacitance of the capacitor is 20 μF. What is the maximum charging potential required? What is the electrical charge impulse sent through the body for maximum charging potential? [6300 V, 0.13 C]

18. If the energy for each impulse of a pacemaker, 2.4×10^{-4} J, is stored in a capacitor charged to a potential of 6.0 V, what is the capacitance of the capacitor, and what is the charge of each impulse? If 70 impulses are given per minute, how much energy is needed per day?   [13.3 μF, 80 μC, 24 J]

19. Specially designed capacitors called ion chambers are used to detect ionizing radiation. The incoming radiation produces ion pairs in the capacitor. These pairs tend to neutralize the charged capacitor. The battery attached to the capacitor recharges the capacitor. Outline a method that could be used to determine the energy deposited in the capacitor.

20. A 3.0-μF and 6.0-μF capacitor are connected in series to a 120-V source of potential. The capacitors are disconnected and reconnected with positive plates together and negative plates together. What is:
    a. The original charge on each capacitor?
    b. The initial potential difference for each capacitor?
    c. The final charge on each capacitor?
    d. The final potential difference for each?
    e. The change in energy of the charged capacitors? How do you account for the difference? [a. $Q_3 = Q_6 = 240$ μC; b. $V_3 = 80$ V, $V_6 = 40$ V; c. $Q_3 = 160$ μC, $Q_6 = 320$ μC; d. $V_3 = V_6 = 53$ V; e. 1600 μJ]

21. A 2.00-μF and 3.00-μF capacitor are connected in parallel across a 100-V line. They are disconnected and reconnected with the positive plate of each capacitor connected to the negative plate of the other. What is the final charge on each capacitor and potential difference across each? What is the change in energy? Explain the difference.   [$Q_2 = 40$ μC, $Q_3 = 60$ μC, $V_2 = V_3 = 20$ V; $\Delta E = 24.0 \times 10^{-3}$ J]

GOALS

When you have mastered the contents of this chapter, you will be able to achieve the following goals:

Definitions
Define each of the following terms, and use it in an operational definition:

ampere	electromotive force (emf)
electrical conductivity	Seebeck effect
ohm	Peltier effect
resistivity	piezoelectric effect

Resistors
Determine an equivalent resistance for a series or parallel combination of resistors.

Ohm's Law
Solve problems using the relationship among resistance, potential difference, and electric current—that is, apply Ohm's law to simple circuits.

Power Loss
Solve problems for instantaneous power in resistive elements obeying Ohm's law.

DC Circuits
Analyze direct-current circuits consisting of resistances and sources of emfs, and find the currents, terminal potential difference of sources of emfs, potential drops, and power developed in circuit elements.

DC Instruments
Explain the basic principle of operation of direct-current instruments: potentiometer, Wheatstone bridge, and thermocouple.

Bioelectricity
Describe some application of electricity in human physiology.

PREREQUISITES

Before you begin this chapter, you need to have mastered the basic concepts of Chapter 5, Energy, Chapter 11, Thermal Transport, and Chapter 21, Electrical Properties of Matter.

22

Basic Electrical Measurements

22.1 Introduction

You have probably interacted with the various dials and meters on the dashboard of an automobile. One may be a gauge (Figure 22.1) or a light marked ALT or Generator or AMP. It is an electrical device designed to provide you with some feedback about the performance of the electrical system of your automobile.

Look around you and note how many different electrical devices there are in your room. How much electrical energy do you use during a day? How could you measure it?

This chapter is intended to introduce you to basic electrical measurements. Electrical measurements are the basis of much of the instrumentation now used in the life sciences. A knowledge of electrical measurements and circuits will enable you to understand the electrical characteristics of your body.

22.2 Electrical Charges in Motion

If you put wire, a battery, and a light bulb together in an appropriate way, the bulb will light. When you disconnect a wire from the battery the bulb stops giving off light. When the light bulb is lit the wires, bulb, and battery form an electrical "circuit." Once again we will evoke the language of continuous flow to explain what is happening in an electrical circuit.

In our description of the motion of electricity from one location to another, we use the mental images that we have developed as a result of our experiences with the flow of water. Let us begin our quantitative analysis of electrical charge flow with a definition of current density J. The magnitude of the current density is given by the ratio of the quantity of electric charge Q that passes through an area A in a time t,

$$J = \frac{Q}{At} \tag{22.1}$$

The SI units for electrical current density are coulombs per (meter)^2 per second. The total current, or flow of electric charge, I through the area A is given by

$$I = JA \tag{22.2}$$

where SI units for total current are called *amperes* (A) and are equivalent to the number of coulombs per second (C/sec).

When an electric field is applied to many materials, a current results as "free" charge carriers move under the influence of the applied field. The ability of material to support electric current is determined by its atomic structure and electron energy levels. In the transport of charge in an electric circuit, the relationship between the current density and its driving force, the potential gradient, is given by

FIGURE 22.1
An indicator in your automobile indicating the performance of its electrical system.

$$J = -\sigma \frac{\Delta V}{\Delta l} \qquad (22.3)$$

where σ is the proportionality conductivity constant between the electric current density and the potential gradient given by $\Delta V / \Delta l$, the potential difference across Δl. (You should be able to rewrite this equation in terms of the applied electric field.) The units of σ, the electrical conductivity, are given by

$$\frac{amperes/meters^2}{volt/meter} \quad \text{or} \quad ohm^{-1} \, meter^{-1}$$

The gradient of the electric potential can be approximated by the total electric potential divided by the length of the circuit.

$$\frac{\Delta V}{\Delta l} \approx \frac{V}{l} \qquad (22.4)$$

where V is the electric potential measured in volts or joules/coulomb.

For a piece of material with cross-sectional area A, the magnitude of the total current is

$$I = JA = A\sigma \frac{V}{l} \qquad (22.5)$$

Then,

$$V = I \frac{l}{\sigma A} \qquad V = \frac{lI}{\sigma A} = IR \qquad \text{(Ohm's law)} \qquad (22.6)$$

where the *resistance* of a material R is defined in terms of its specific resistance or resistivity, length, and cross-sectional area.

$$R = \frac{l}{A\sigma} = \rho \frac{l}{A} \qquad (22.7)$$

where ρ is the resistivity and equals $1/\sigma$ and R is measured in ohms. What are the units of ρ? Ohm's law in its most common form is given in Equation 22.6. This equation states that *the current through a sample is equal to the voltage difference across it divided by its resistance.* Materials that readily conduct current are called *ohmic conductors.*

Questions

1. If the charge of one electron is 1.6×10^{-19} C, how many electrons must pass through an area of 1 cm^2 in one second for the current density to be 1 ampere/m^2?
2. How many electrons per second are needed for a current of one ampere?
3. What conditions must exist for the approximation given in Equation 22.4 to be exact?
4. Should a good conductor have a large or small value for its resistivity?

22.3 *Sources of Electric Energy*

The question now arises what are the possible sources for generating electric current in materials? Recall the definition of electric potential difference. The electric potential difference between two points is the measure of the energy needed to move a unit charge (one coulomb) from one point to the other. A 1.5-V battery supplies 1.5 J of energy to transport 1 C of electric charge from the positive terminal to the negative terminal of the battery. The battery is a source of electrical energy because it expends some other form of energy (in this case, chemical energy) to separate the positive (+) and negative (−) charges. We can generalize from this idea to all sources of electric potential energy. A source of electric potential energy is a charge separation produced by expending some other energy. Thermocouples, solar batteries, generators, crystal phonograph cartridges, nerves, and static electricity all produce this charge separation. Sometimes it is desirable to eliminate sources of electric potential, for example, static electricity in a hospital operating room which contains ether vapors.

The maximum potential difference between the terminals of a source is called the electromotive force or emf $\mathcal{E}$. The emf voltage is the ideal voltage available when no current is flowing through the source. All sources have some internal resistance r. Thus when a current flows through the source, there is a potential difference across the internal resistance which must be subtracted from the emf to obtain the terminal voltage for a particular current. This situation is expressed in the equation

$$\Delta V_{\text{terminal}} = \mathcal{E} - Ir \tag{22.8}$$

EXAMPLE

A flashlight battery has an emf of 1.5 V and an internal resistance of 2 ohms (Ω). If a current of 0.2 amps (A) flows when this battery is used in a flashlight, what is the terminal voltage of the battery?

$$\Delta V_{\text{terminal}} = \mathcal{E} - Ir = 1.5 \text{ V} - (0.2 \text{ A})(2 \text{ } \Omega)$$
$$= 1.1 \text{ V}$$

This example indicates what happens to most batteries when they "run down." The emf remains constant, but the internal resistance of the battery increases. So the terminal voltage available decreases. Using this information, suggest a procedure that could be followed to determine whether a battery is "good" or "bad."

Questions

5. How might you minimize electrostatic potentials?

22.4 Electric Circuits: Ohm's Law and Joule's Law

In order to sustain an electric current the material through which the electric charge flows must form a closed circuit. This requirement results from the conservation of charge. Charge cannot be created or destroyed, and the continuity concept suggests that as much charge must enter a circuit element as leaves it in each unit of time. In order to study electrical circuits we will use a symbolic circuit diagram. The conventional symbols used are:

$\dfrac{+}{} |{-} =$ battery

$\sim\!\!\wedge\!\!\wedge\!\!\wedge\!\!\sim$ = resistance element

$\diagup$ = switch

$-\!\!(A)\!\!-$ = ammeter (measures current)

$-\!\!(V)\!\!-$ = voltmeter (measures electric potential difference)

Consider the circuit in Figure 22.2. Since charge is conserved, we see that the current through each element in this *series* circuit is the same. Likewise, since energy is conserved, the sum of the individual potential differences across the elements must equal the emf of the source in the circuit. We can summarize these two facts in the following equation form:

$$\mathscr{E} = Ir + IR_1 + IR_2$$

or

$$\mathscr{E} = I(r + R_1 + R_2) = IR_{\text{eff}}$$

The effective resistance is equal to the sum of separate resistances.
From this relation we can calculate the current in the circuit.

$$I = \frac{12.0 \text{ V}}{(0.10 + 2.0 + 3.9) \ \Omega}$$

$$I = \frac{12.0 \text{ V}}{6.0 \ \Omega} = 2.0 \text{ A}$$

The terminal voltage scross the battery in this circuit is

$$\mathscr{E} - Ir = 12.0 \text{ V} - (2.00 \text{ A} \times 0.10 \ \Omega) = 11.8 \text{ V}$$

FIGURE 22.2
A DC series circuit.

$$\mathscr{E} = 12 \text{ V}$$
$$R_1 = 2.0 \ \Omega$$
$$R_2 = 3.9 \ \Omega$$
$$r = 0.1 \ \Omega = \text{INTERNAL RESISTANCE OF BATTERY}$$

SERIES CIRCUIT

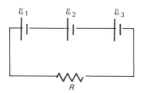

FIGURE 22.3
Three sources of emf connected in series. $\mathcal{E}_{total} = \mathcal{E}_1 + \mathcal{E}_2 + \mathcal{E}_3$.

Note that the effective resistance of resistors in series is the sum of the individual resistances. This example has demonstrated some important approaches that can be generalized.

1. Elements in series carry the same current.
2. The sum of the potential differences in a series circuit must equal the *net* applied emf in the circuit. For example, in Figure 22.3

$$\mathcal{E}_{tot} = \mathcal{E}_1 + \mathcal{E}_2 + \mathcal{E}_3 = \sum_i IR_i \qquad (22.9)$$

3. The effective resistance for n resistances in series is equal to the sum of the individual resistances:

$$R_{eff} = R_1 + R_2 + \cdots + R_n \qquad (22.10)$$

4. Conservation of charge means that electric charge entering a junction must be equal to the sum of electric charge leaving the junction. In Figure 22.4

$$I = i_1 + i_2 + i_3 \qquad (22.11)$$

Equations 22.9 and 22.11 are equivalent to Kirchoff's laws which are discussed in Section 22.14.

EXAMPLE

What is the effective resistance of five 100–Ω resistors in series?
By Equation 22.10 it is the sum of the individual resistances or 500 Ω.

Elements in a circuit can also be connected in parallel. A parallel combination is illustrated in Figures 22.4 and 22.5.
For each of the parallel branches (see Figure 22.4), the potential difference across each branch is the same, because the energy required to move a charge from one position in a circuit to another is independent of the path between the two positions.

$$V = i_1 R_1 = i_2 R_2 = i_3 R_3 \qquad (22.12)$$

Suppose we try to find an effective resistance for this network of branches such that

$$V = IR_{eff} \qquad (22.13)$$

FIGURE 22.4
Current through parallel resistances at a junction in a DC circuit $I = i_1 + i_2 + i_3$.

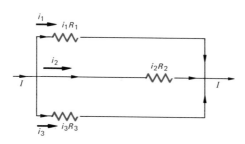

FIGURE 22.5
A series-parallel circuit. The patient and nurse form a parallel resistance electrical current path.

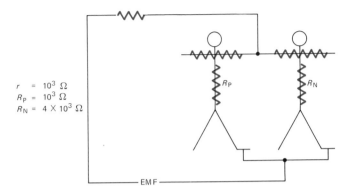

$$r = 10^3 \ \Omega$$
$$R_P = 10^3 \ \Omega$$
$$R_N = 4 \times 10^3 \ \Omega$$

where R_{eff} is the effective resistance of the three parallel resistances. We know the total current I is given by the sum of the three branch currents,

$$I = i_1 + i_2 + i_3 \tag{22.14}$$

Using Ohm's law we can replace each current by the ratio of the potential difference to resistance,

$$\frac{V}{R_{\mathrm{eff}}} = \frac{V}{R_1} + \frac{V}{R_2} + \frac{V}{R_3} \tag{22.15}$$

We can divide by the potential difference V to obtain an expression for the effective resistance,

$$\frac{1}{R_{\mathrm{eff}}} = \frac{1}{R_1} + \frac{1}{R_2} + \frac{1}{R_3} \tag{22.16}$$

EXAMPLE

The situation shown in Figure 22.5 may actually happen when a ground lead breaks on a piece of electrical equipment. The important factor is the current through the patient and the nurse. A current of 5 milliamps (mA) is a painful shock, and a current of 160 mA can cause fibrillation (the rapid, uncoordinated series of contractions) of the heart muscle. The potential difference across all elements in parallel is the same (see Figure 22.5),

$$V_{\mathrm{p}} = V_{\mathrm{n}} \quad \text{or} \quad I_{\mathrm{p}} R_{\mathrm{p}} = I_{\mathrm{n}} R_{\mathrm{n}}$$

The total current must be equal to the sum of the currents through the patient and the nurse,

$$I = I_{\mathrm{p}} + I_{\mathrm{n}}$$

When we apply Ohm's law to this circuit, we can express the emf in the circuit as the sum of two terms, the potential difference across the load r and the potential difference across either the patient or the nurse,

$$\mathscr{E} = Ir + I_{\mathrm{p}} R_{\mathrm{p}} = Ir + I_{\mathrm{n}} R_{\mathrm{n}} \tag{22.17}$$

where R_{p} and R_{n} are the respective resistances of the patient and the nurse. To solve for an expression for the current through the patient, set the two expressions for the current through the nurse equal to each other:

$$I_{\mathrm{n}} = I - I_{\mathrm{p}} = \frac{I_{\mathrm{p}}}{(R_{\mathrm{n}}/R_{\mathrm{p}})} \tag{22.18}$$

Solving for the current through the patient we obtain

$$I_{\mathrm{p}} = I \frac{R_{\mathrm{n}}}{R_{\mathrm{p}} + R_{\mathrm{n}}} \tag{22.19}$$

The expression for the total emf in the circuit is found by substituting Equation 22.19 in Equation 22.17 to eliminate I_{p}:

$$\mathscr{E} = Ir + \frac{IR_{\mathrm{p}}R_{\mathrm{n}}}{R_{\mathrm{p}} + R_{\mathrm{n}}} \tag{22.20}$$

A resistance of $R_{\mathrm{p}}R_{\mathrm{n}}/(R_{\mathrm{p}} + R_{\mathrm{n}})$ ohms is equivalent to the resistors R_{n} and R_{p} in parallel.

This example illustrates the following generalizations for parallel circuits:

1. Elements in parallel have equal potential differences.
2. The current into a parallel network equals the sum of the current in parallel branches.
3. The effective resistance of n resistors in parallel is found by

$$\frac{1}{R_{\mathrm{eff}}} = \frac{1}{R_1} + \frac{1}{R_2} + \cdots + \frac{1}{R_n} \tag{22.21}$$

EXAMPLE

What is the effective resistance of five 100-Ω resistors in parallel?

The effective resistance is the reciprocal of the sum of the reciprocals. The sum of the reciprocals is $5(1/100\ \Omega) = 5/100\ \Omega$. The reciprocal of that is 20 Ω.

Let us consider the energy dissipated in an electrical circuit. Suppose we construct a model of a solid conductor that consists of vibrating massive atoms surrounded by a gas of small charge carriers. We can think of the energy dissipated as the energy lost by means of inelastic collisions between the moving charges and the atoms. The energy lost by the charges will go to increase the vibrational kinetic energy of the atoms. The internal energy of the conductor will be increased. So the energy dissipated results in the heating of all resistive elements in a circuit.

The energy expended per unit time is the power. A potential source supplies some energy for each charge and the current is a measure of the time rate of flow of charges. Hence the product of the potential times the current is the time rate of energy supplied, or power,

$$\text{power} = V\left(\frac{\text{joules}}{\text{coulomb}}\right)I\left(\frac{\text{coulomb}}{\text{second}}\right) = VI\left(\frac{\text{joules}}{\text{second}}\right) = \frac{\text{energy}}{\text{time}} \tag{22.22}$$

The watt which is equal to joule/second is the basic unit of power. For the power dissipated in a resistor we can use Ohm's law to transform the above equation for power into another form. Let us substitute the product of the current times the resistance for the potential difference in Equation 22.22 to obtain Joule's Law:

$$P = VI = (IR)I = I^2R \qquad (22.23)$$

From the conservation of energy, it follows that the power supplied by the sources in a circuit must be equal to the total power dissipated in the circuit.

The total energy dissipated in an electrical circuit is calculated as the product of the dissipated power times the length of time the circuit is operated. Since the electrical power is measured in watts or kilowatts, the electrical energy used is often measured in watt hours or kilowatt hours, energy used = (power) × (time).

EXAMPLE

What must be the minimum power rating of each resistor in the diagram of a series-parallel circuit shown in Figure 22.6? (Surpassing the power rating results in overheating and "burning out" of resistors.)

First we find the total current in the circuit. This is also the current through R_3 where

$$R_{eff} = \frac{R_1 R_2}{R_1 + R_2}$$

Thus

$$\mathscr{E} = \frac{I R_1 R_2}{R_2 + R_1} + I R_3$$

$$= \frac{I(12.0 \times 4.00)}{16.0} + 9.0I = 12.0I = 12.0 \text{ V}$$

Therefore, $I = 1$ A.

The power rating of R_3 must be $I^2 R_3 = I^2 \times 9 = 9.00$ watts.

The power rating of R_2 is $P_2 = V_2^2/R_2 = (1.00 \times 3)^2/4 = 9.00/4.00 = 2.25$ watts. Likewise, the power rating of R_1 must be V_1^2/R_1. Therefore, $P_1 = (1.00 \times 3)^2/12 = 9/12 = 0.75$ watts. The power supplied by the battery must be $\mathscr{E}I = 12$ watts. Does this check with the total power dissipated in the circuit? What generalization can you make for such a series-parallel circuit? What resistor determines the maximum current in the circuit?

Questions

6. How would you use Equation 22.6 to show that the effective resistance of any two resistors in parallel is equal to the product of their resistances divided by the sum of their resistances.

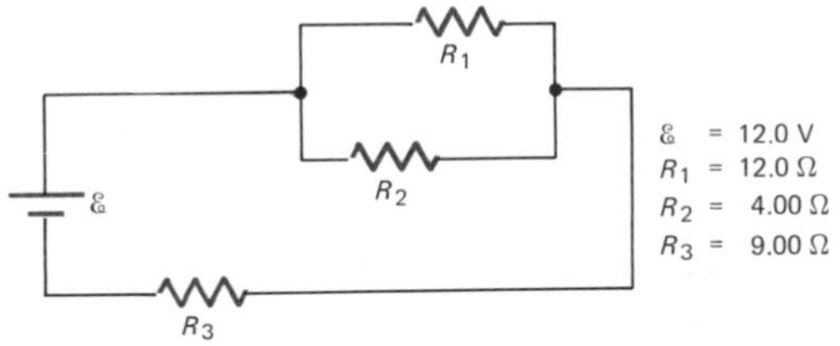

FIGURE 22.6
Circuit for example problem on a series-parallel circuit.

7. What is the expression for power dissipated in terms of the potential difference and the resistance?
8. What experimental design can you develop to use an electrical circuit to verify the conservation of energy?

22.5 Galvanometers

The basic meter for electrical measurements is a sensitive current measuring galvanometer. Some galvanometers are capable of measuring currents as small as 10^{-9} A. The internal resistance of such a sensitive galvanometer may be higher than 1000 Ω.

By adding the appropriate resistances to the basic galvanometer it is possible to convert a galvanometer to an ammeter used for measuring current or a voltmeter used for measuring potential difference. To make an ammeter it is necessary to shunt most of the current around the galvanometer by hooking a low resistance in parallel with the meter. A typical ammeter circuit is shown in Figure 22.7. To make a voltmeter from a galvanometer it is necessary to increase the potential difference across the galvanometer by adding a high resistance in series with the galvanometer. A typical voltmeter circuit is shown in Figure 22.8.

Let us summarize some important facts concerning ammeters and voltmeters.

1. Ammeters are always connected in series with the circuit element under study.
2. Voltmeters are always connected in parallel with the circuit element under study.
3. Since the connection of meters in a circuit changes the circuit, it is necessary to take into consideration the resistances of the meters when they are used in circuits.

EXAMPLES

1. Consider a galvanometer that reads 1.00×10^{-4} A for a full-scale deflection with a resistance of 100 Ω. Find the shunt resistance needed to convert this galvanometer into a 0.100-ampere full-scale ammeter.

   The shunt resistance R_s will carry the current in excess of the maximum galvanometer current, so the shunt current I_s is given by the total current I minus the galvanometer current I_g,

   $$I_s = I - I_g = (10^{-1} - 10^{-4})\ \text{A}$$

   $$I_s = 9.99 \times 10^{-2}\ \text{A}$$

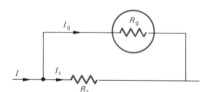

FIGURE 22.7
Circuit diagram for an ammeter.

FIGURE 22.8
Circuit diagram for a voltmeter.

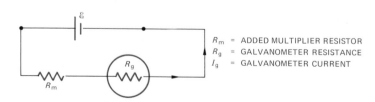

R_m = ADDED MULTIPLIER RESISTOR
R_g = GALVANOMETER RESISTANCE
I_g = GALVANOMETER CURRENT

FIGURE 22.9
Circuits for measuring resistance by
ammeter-voltmeter method.

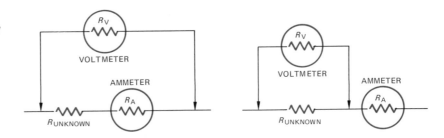

Now, since the galvanometer and the shunt are in parallel, the potential difference across them will be equal; so by Ohm's law we write the following:

$$I_g R_g = I_s R_s$$

or

$$R_s = \frac{I_g R_g}{I_s} = \frac{1.00 \times 10^{-4} \times 100}{(9.99 \times 10^{-2})\ \Omega} = 1.00 \times 10^{-1}\ \Omega$$

$$R_s \simeq 10^{-1}\ \Omega.$$

2. Using the same galvanometer (1.00×10^{-4} A full-scale with 100 Ω resistance), find the value of R_m to make a 1.00-V full-scale meter.

The excess potential difference must occur across the large resistor in series with the galvanometer. Since the resistor and the galvanometer are in series they will both carry the same current, namely the maximum allowed by the characteristics of the galvanometer, or 10^{-4} A in this case. The potential difference (1 V) will be equal to the total potential difference across the resistor and the galvanometer,

$$1.00\ V = 1 \times 10^{-4}\ (R_m + 100)$$

$$R_m = 1 \times 10^4 - 100 = 9900\ \Omega$$

Questions

9. What is the voltage across a 1000-Ω galvanometer measuring a current of 10^{-9} A?

10. A perfect ammeter (one not affecting the circuit under study) would have high or low resistance?

11. A perfect voltmeter (one not drawing any current) would have high or low resistance?

12. Suppose you want to determine an unknown resistance by using the ammeter and voltmeter readings in Ohm's law. Can you determine the error made by such a calculation for each of the circuits shown in Figure 22.9?

22.6 Potentiometers

The *potentiometer* is an instrument whose effect upon the circuit approaches the capability of a perfect voltmeter. The potentiometer draws no current from the circuit measured. The circuit for a typical potentiometer is shown in Figure 22.10.

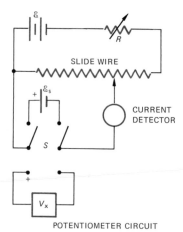

SLIDE WIRE

CURRENT DETECTOR

POTENTIOMETER CIRCUIT

FIGURE 22.10
A simple potentiometer circuit.

The operation of the potentiometer is as follows: An external power supply $\mathcal{E}$ with a terminal voltage larger than the standard emf cell or the unknown voltage is required. The working current supplied by the power supply $\mathcal{E}$ must be stable. First, the switch S is closed so that the standard emf cell is in the circuit. The potentiometer slide wire is set to the emf of the standard cell, and the calibration resistor R is adjusted until the current detector indicates that there is no current through the standard cell. This condition can be expressed in equation form as

$$IL_s = \mathcal{E}_s \tag{22.24}$$

where L_s is the resistance of the slide wire when it is adjusted to read the known value of the standard emf $\mathcal{E}_s$ and I is the current in the slide wire.

Second, to measure an unknown voltage the switch is changed to introduce the unknown potential source V_x into the circuit. The slide-wire contact is then adjusted until the current detector again reads zero. This condition also means the same standardizing current is set up in the slide wire as when the standard cell was in the circuit. (*Note:* the resistor R is only altered when standardizing the potentiometer.)

There is no current through the unknown; this condition is expressed by the equation

$$IL_x = V_x \tag{22.25}$$

where L_x is the resistance of slide wire corresponding to voltage reading V_x. If the wire is uniform, its resistance is directly proportional to its length. The unknown voltage is usually read from the slide wire which can be calibrated in volts. If the wire is not calibrated, we can find V_x by noting that the last two equations lead to the following ratio equation

$$V_x = \mathcal{E}_s \frac{L_x}{L_s} \tag{22.26}$$

An important characteristic of the potentiometer is that it draws no current from the unknown source when it is correctly balanced. This means that it reads the emf of a source, and it is not affected by the internal resistance of the unknown potential source. The potentiometer circuit is very useful. In many laboratories a chart recorder is used to convert electrical signals into a graph. The chart paper is driven at constant speed under the tip of a pen that is attached to the potentiometer slidewire. As the potential signal changes, the pen moves to provide a graph of the changing electrical properties of the system under study.

22.7 The Wheatstone Bridge

The Wheatstone bridge is a circuit used to measure resistance. A typical Wheatstone bridge is shown in Figure 22.11. The bridge is balanced by adjusting variable resistors R_2 and R_3 until the current detector be-

FIGURE 22.11
Wheatstone bridge circuit.

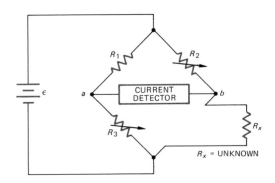

tween a and b reads zero. This balanced condition satisfies the following equations:

$$i_a R_1 = i_b R_2 \qquad (22.27)$$

$$i_a R_3 = i_b R_x \qquad (22.28)$$

$$R_x = \frac{R_2}{R_1} R_3 \qquad (22.29)$$

Hence, the value of the unknown resistance R_x can be calculated.

Questions

13. How can you justify Equations 22.27 and 22.28?
14. Can you show that the effective voltage across the current detector V_{ab}, before balance is:

$$V_{ab} = \frac{\mathscr{E}}{R_1 + R_3} R_1 - \frac{\mathscr{E}}{R_2 + R_x} R_2$$

Figure 22.12 shows the effective resistance seen by the current detector:

$$R_{ab} = \frac{R_1 R_3}{R_1 + R_3} + \frac{R_2 R_x}{R_2 + R_x}$$

22.8 The Seebeck Effect

The Seebeck effect is the physical basis for the operation of the thermal transducers known as thermocouples. T. J. Seebeck discovered that an emf is generated when the junction of two dissimilar metals is heated or

FIGURE 22.12
Equivalent resistance circuit.

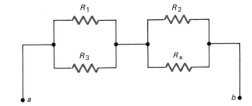

FIGURE 22.13
Two circuits for use of a thermo-couple to measure a given temperature.

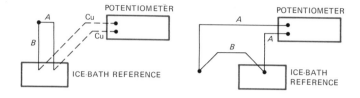

cooled. Seebeck also discovered the thermoelectric series of metals (bismuth, nickel, cobalt, palladium, platinum, copper, manganese, titanium, mercury, lead, tin, chromium, rhodium, iridium, gold, silver, zinc, tungsten, cadmium, iron, antimony) which determines the magnitude and polarity of the emf generated. In general, the further apart the two metals are in the series, the greater will be the emf, and the positive metal at the hot junction will be the one coming earlier in the series as it is listed above. Two different thermocouple arrangements are illustrated in Figure 22.13. One junction of metals A and B is the probing junction while the other junction is referenced at 0°C. The emf produced is proportional to the temperature differences over various ranges of temperatures depending upon the metals used in the thermocouple.

22.9 The Peltier Effect

The Peltier effect is the reverse of the Seebeck effect. An electric current through a junction of dissimilar metals will cause the junction to liberate or absorb heat energy. A reversal of the current will reverse the effect. The quantity of heat liberated or absorbed is directly proportional to the quantity of charge passing through the junction. The Peltier effect can be used to cool small samples under a microscope. A semiconductor device known as a *frigistor* can be incorporated into the microscope stage. By simple control of the current through the frigistor, the specimen can be cooled or heated.

22.10 The Piezoelectric Effect

When certain crystals are mechanically deformed, electric charges are separated along particular crystal axes. This charge separation results in a potential difference across specific faces of the crystal. This effect is known as the *piezoelectric effect*. A small strain on such a crystal will produce a linear output voltage. The common crystal phonograph pickup and crystal microphone are devices that use the piezoelectric effect. Piezoelectric crystals typically have high resistances, and this means that when they are used as transducers for small displacements or forces, they must be monitored with very little current drain.

The piezoelectric effect is reversible—an applied voltage across the crystal will produce mechanical deformation of the crystal. This effect is the basis for crystal ultrasonic generators. A high-frequency voltage is applied at the natural frequency of the crystal. The crystal vibrates at the applied frequency, and its mechanical vibrations generate sound waves in the medium around it. Piezoelectric transducers have many biomedical applications. They are used to measure heart sounds, muscle pull, respiration, and pulse rates.

Recent research has shown that human bone is piezoelectric. It seems, for example, that as you walk across a room the bending forces acting on your leg bones generate changing electrical voltages across your leg bones. These electrical voltages serve an important function in the system that controls the growth and strength of your leg bones. In some cases of forced physical inactivity, repeated electrical pulses have been used to simulate the effects of walking and to enable the leg bones to maintain their size and strength.*

Questions

15. What kind of monitoring instrument would be best to use with a piezoelectric crystal?
16. The use of a periodic voltage applied at the natural frequency of a piezoelectric crystal is an example of what physical phenomena?

22.11 Electrical Applications in Human Physiology

The human body has electrical characteristics that have proven to be important in the health sciences. The electrical characteristics of your body serve to indicate the state of your health and the condition of your various parts. Recent research using human signals and appropriate transducers has led to some important applications of biofeedback.

Let us first consider the electrical signals that are detectable at the surface of the human body. There are important voltage signals that can be detected by surface electrodes. These include: ECG (electrocardiogram), EMG (electromyogram), and EEG (electroencephalogram).

The ECG measures the electrical activity of the heart. The heart has a changing electric dipole moment as it pumps. The voltages at the ECG electrodes are caused by the electric dipole fields of the beating heart. This dipole moves over the heart and synchronizes the heartbeat. A typical ECG wave form is shown in Figure 22.14 along with a schematic diagram of the heart.

The sinoatrial node, or pacemaker, initiates the ECG wave and serves to synchronize the heart contraction. *P* represents this atrial changing dipole signal. The *QRS* wave represents the ventricular changing di-

---

* See R. O. Becker, "Boosting Our Healing Potential," *Science Year, the World Book Science Annual, 1975.* Chicago: Field Enterprises Educational Corp., 1974. See also R. K. Hobbie, "The Electrocardiogram as an Example of Electro-statics," *American Journal of Physics* **41** (June 1973): 824–831.

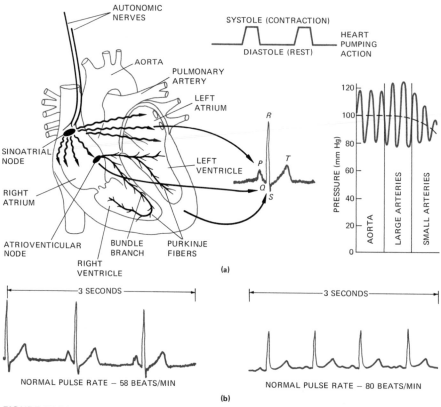

FIGURE 22.14

(a) Diagram of human heart. (b) Normal ECG tracing, plotting of millivolts vs. time. The time scale is 3 sec between the short vertical marks. From this plot one can determine the pulse rate. Count the number of complete cycles over a 3-sec span, and multiply that number by 20. On the second normal trace, one counts 4 during this interval; hence the pulse rate is 80.

The *P*, *Q*, *R*, *S*, and *T* waves are the parts of a normal heart cycle.

pole, and *T* is an indication of ventricular dipole restoration. The actual placement of electrodes determines the observed ECG wave form. Sometimes different parts of the heart beat independently. This condition is called *fibrillation*. An external electrical pulse occurring just before the *T* wave can produce fibrillation.

In emergency cases where the heart has stopped, an electrical pulse is applied to start it again. In cases where the heart is unable to maintain its own synchronization, a pacemaker device may be implanted in the patient. Recent developments in this area include power sources with life times greater than twenty years. The pacemaker circuitry is relatively simple and provides an electric synchronizing pulse at a regular rate. Several abnormal ECG's are shown in Figure 22.15.

The electroencephalogram (EEG) is the term used for the electrical potentials measured on the surface of the head. These potentials are

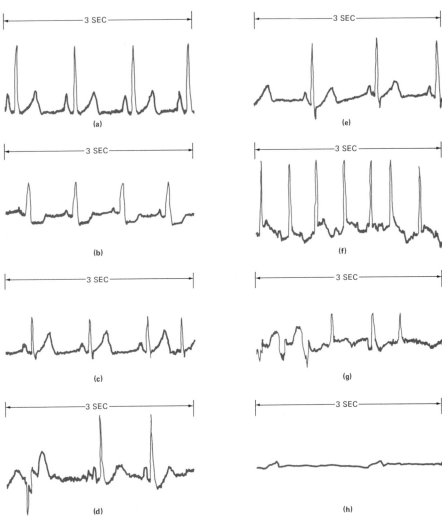

FIGURE 22.15
Abnormal ECG tracings. (a) High *P* wave is an indication of pulmonary heart disease.
(b) *S–T* wave depression indicates ischemic heart disease. (c) Atopic atrial beat.
(d) Ventricular ectopic beat; no rhythm. (e) Premature ventricular heart contraction.
(f) Atrial fibrillation with rapid ventricular rate. (g) The four nonregular pulses
in the center indicate ventricular fibrillation, a serious situation unless soon stabilized.
(h) Terminal case; a dying heart trace.

associated with the activity of the brain. The characteristic form of the
EEG patterns is useful in the diagnosis and treatment of epilepsy. Brain
injuries can also be diagnosed from EEG patterns. There are four different patterns that make up a normal EEG. Alpha waves have frequencies
of from 8 to 13 Hz and seem to be correlated with eye activity. Typical
alpha wave amplitudes are the order of 10-20 μV. Beta waves are from
14–50 Hz and have a higher amplitude (50–100 μV). They are always present in adults. Delta waves (0.5 to 4 Hz) and theta waves (5 to 7 Hz) are

also found in the EEG. The use of computers in EEG analysis coupled with the electrode stimulation of the brain may provide needed information to understand the relationships between these waves and the activity of various parts of the brain. Biofeedback experiments have shown that it is possible for an individual to alter his or her EEG pattern willfully. This may prove to be medically significant for treatment of central nervous system problems.

The electromyogram (EMG) potentials are associated with muscle activity. Electrodes are placed on the skin near the muscle under study. The voltage between the muscle electrode, and a neutral electrode is measured. The EMG is useful in diagnosing diseases affecting muscles and the nerve cells that control body movements.

Surface electrodes can also be used to measure the resistance changes of the skin. These studies are referred to as bioimpedance studies and are usually made with alternating currents. Of particular interest is the galvanic skin response (GSR), a basic component of the polygraph lie detector. The galvanic skin response is thought to be caused by a change in the resistance of the skin with the action of the sweat glands. The emotional condition of the subject causes changes in sweat production, and thus the GSR serves as an index for the subject's emotional state. An experienced operator can use the GSR very effectively as a lie detector.

22.12 Elements of Neuroelectricity

Nerve transmission involves the propagation of a voltage signal (action potential) along the nerve. The nerve consists of a bundle of nerve fibers, or *axons*. Axons are part of single nerve cells, the *neurons*. The conduction of the electric impulse along an axon travels at much slower speeds than electrical currents in conductors. The speed of transmission of the action potential depends on the diameter of the nerve fiber involved. Fibers of 1 micron (μm) transmit impulses at 1 m/sec or faster. The nerve transmission speed also increases with a rise in temperature.

The basis of the bioelectrical phenomena associated with nerve transmission is the separation of charge across a membrane. The charge separation is called polarization. This polarization in the nerve is typified by a low concentration of potassium ions (K^+) inside the nerve cell and a high concentration of sodium ions (Na^+) outside of the nerve cell. The inside of the cell is negative with respect to the outside of the cell. The potential difference between the outside and inside is about 90 mV. When the nerve is stimulated, the membrane depolarizes, generating the action potential that is propagated along the nerve. The stimulation may be mechanical, thermal, or electrical in nature. A typical action potential is shown in Figure 22.16. After this spike potential passes there is a recovery period during which the sodium pump restores the axon to readiness for the next impulse. The recovery time for this process is in milliseconds. The nerve can transmit impulses in both directions, but the

FIGURE 22.16
Nerve action potential.

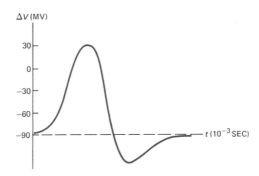

connection junctions (synapses) between neurons transmit signals only in one direction.*

The use of microelectrodes and modern electronic instrumentation has helped us gain a much better understanding of bioelectric phenomena.

22.13 Electrical Thresholds and Effects

The electric current is the major cause of damage to a person subject to an electrical shock. The facts determining the effect of a current passing through the body are the current path, the current magnitude, and the direction of the current.

A current path through the heart is the most serious. Using Ohm's law you can see that for a given voltage the current will depend on the body's resistance at the points of contact between the person and the potential source. The typical resistance between two points on the body ranges from 10,000 to greater than 100,000 Ω.

A 1-mA current is perceptible. A 5-mA current is painful. A 10-mA current will cause muscle contractions. For currents greater than 15 mA, voluntary muscle control is lost. Ventricular fibrillation may be produced by 70-mA currents if the current duration is one second. Longer durations can prove fatal. It is known that currents as small as 200 μA can cause fibrillation if the electrodes make direct contact with the heart.

Recent laboratory research and clinical practice have substantiated the following electromagnetic effects: low level DC currents stimulate bone growth and multi-tissue regenerative growth and influence nerve activity and function.† The interaction of electromagnetic forces and living systems needs more study. The basic interactions are not well understood even though there is growing use of electromagnetic effects in clinical medicine. There are many other research areas to be explored, such as the relationship between biological clocks and periodic geo-

* For more details of the operation of these synaptic junctions, see Bernard Katz, *Nerve, Muscle and Synapse*. New York: McGraw-Hill Book Company, 1966.
† Becker, op. cit.

physical fields, the effects of microwaves on living systems, the modification of behavior through the use of electromagnetic fields, and the possible connection between the earth's magnetic field reversal and the extinction of species.

Questions

17. What can you conclude about the dangers of static electricity based upon the content of this section?
18. If the damage due to electric shock is assumed to be caused by the energy dissipated, what arguments can you give to support the following relationship:

$$\text{damage} \propto \frac{(\text{current})^2}{\text{area of contact}} \times \text{duration of shock}$$

22.14 Network Circuits

Complex electrical network circuits cannot easily be solved by methods used in the previous series or parallel examples. Kirchhoff has pointed out two conditions for methods to use to solve circuit networks. These are:

1. At any point in an electric circuit where two or more conductors are joined, the currents into the junction equal the currents out of the junction. This is another way of saying we have conservation of charge. If you call the currents toward the junction positive and those away from the junction negative, then you can say $\Sigma I = 0$.
2. Around any closed loop in an electric circuit, the sum of IR drops is equal to the applied emf. If one goes around the loop in the direction the positive charge is flowing, the potential drop is positive, and if one goes against the direction of the current, the drop is negative. If the positive charge is flowing through the battery from negative to positive, the emf is positive and if the positive charge is flowing through the battery from positive to negative the emf is negative. Then $\Sigma \mathscr{E} = \Sigma IR$.

EXAMPLE

Find the current through each resistor and the power supplied by each battery in Figure 22.17. All resistances are in ohms. Assume the direction of currents is as indicated by the arrows. Apply Kirchhoff's law,

$$i_3 = i_1 + i_2 \quad \text{(condition 1)}$$

Loop 1: $\mathscr{E}_1 - \mathscr{E}_2 = i_1(2+3) + (i_3)10 = 5i_1 + 10i_1 + 10i_2 \quad \text{(condition 2)}$

$$(10-5)\,V = 15i_1 + 10i_2$$

$$5\,V = 15i_1 + 10i_2$$

FIGURE 22.17
Analysis of a circuit network using
Kirchhoff's principles.

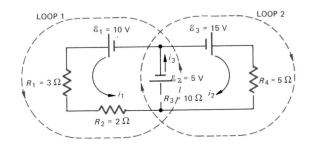

Loop 2: $\mathscr{E}_3 - \mathscr{E}_2 = i_2(5) + (i_3)10 = 5i_2 + 10i_2 + 10i_1$ (condition 2)

$$15 \text{ V} - 5 \text{ V} = 10 \text{ V} = 10i_1 + 15i_2$$

We now have two equations with two unknowns to solve. These are (dividing both equations by 5).

$$1 \text{ V} = 3i_1 + 2i_2$$

$$2 \text{ V} = 2i_1 + 3i_2$$

Multiplying the first by 2 and the second by 3, we obtain

$$2 \text{ V} = 6i_1 + 4i_2$$

$$6 \text{ V} = 6i_1 + 9i_2$$

Eliminating i_1, we have

$$-4 \text{ V} = -5i_2$$

$$i_2 = 4/5 \text{ A}$$

Substituting this value in the top equation for i_2,

$$3i_1 + 8/5 \text{ V} = 1 \text{ V}$$

$$i_1 = -1/5 \text{ A}$$

The same set of equations can be solved by determinants:

$$i_1 = \frac{\begin{vmatrix} 5 & 10 \\ 10 & 15 \end{vmatrix}}{\begin{vmatrix} 15 & 10 \\ 10 & 15 \end{vmatrix}} = \frac{75 - 100}{225 - 100} = -\frac{1}{5} \text{ A}$$

$$i_2 = \frac{\begin{vmatrix} 15 & 5 \\ 10 & 10 \end{vmatrix}}{\begin{vmatrix} 15 & 10 \\ 10 & 15 \end{vmatrix}} = \frac{150 - 50}{225 - 100} = \frac{100}{125} = \frac{4}{5} \text{ A}$$

The current through R_1 and R_2 is $\frac{1}{5}$ A in the direction opposite the arrow since i_1 is negative. The current through R_4 is $\frac{4}{5}$ A in direction of i_2 as shown.

Power is supplied to $\mathscr{E}_1$—that is, the battery is being charged—at the rate

$$\mathscr{E}_1 i_1 = 10 \text{ V} \times -\tfrac{1}{5} \text{ A} = -2 \text{ watts}$$

power supplied to $\mathscr{E}_2 = -5 \text{ V} \times \tfrac{3}{5} \text{ A} = -3 \text{ watts}$

power supplied by $\mathscr{E}_3 = 15 \text{ V} \times \frac{4}{5} \text{ A} = 12$ watts

total power dissipated in resistances:

$P_1 = i_1^2 R_1 = \frac{1}{25} \times 3 = \frac{3}{25}$

$P_2 = i_1^2 R_2 = \frac{1}{25} \times 2 = \frac{2}{25}$

$P_3 = (i_2 + i_1)^2 R_3 = \frac{9}{25} \times 10 = \frac{90}{25}$

$P_4 = i_2^2 R_4 = \frac{16}{25} \times 5 = \frac{80}{25}$

$P_{\text{total}} = \frac{175}{25} = 7$ watts

ENRICHMENT
22.15 Maximum Power Transferred to a Resistive Load

We wish to study the condition necessary for maximum power transfer from a DC power source to a resistive load. Consider the circuit shown in Figure 22.18.

The current drawn from the energy source is given by the equation $I = \mathscr{E}/(r + R)$, where r is the internal resistance of the source. The power supplied to the resistor R is $I^2 R = [\mathscr{E}/(r + R)]^2 R$. We wish to find the value of the resistance R that will make this power a maximum. This value of R can be found by the maximization technique learned in calculus. We differentiate the power expression with respect to the resistance R, and set this derivative equal to zero. What is the reason for this approach? Explain this procedure by drawing a graph of the power versus the resistance R. If

$$P = \frac{\mathscr{E}^2}{(r + R)^2} R$$

then

$$\frac{dP}{dR} = \mathscr{E}^2 \left[\frac{1}{(r + R)^2} - \frac{2R}{(r + R)^3} \right]$$

Setting this derivative equal to zero we have:

$$\mathscr{E}^2 \frac{(r + R - 2R)}{(r + R)^3} = 0$$

Thus $R = r$ is the condition for maximum power transfer from a power source to a resistive load. This is another example of the unifying principle that the maximum transfer of power occurs when the inertial properties of the interacting systems are equal (see Chapter 2).

FIGURE 22.18
DC series circuit.

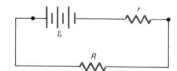

SUMMARY

Use these questions to evaluate how well you have achieved the goals of this chapter. The answers to these questions are given at the end of the summary with the number of the Section where you can find related content material.

Definitions

1. The _____ is the unit for measuring the flow of electrical charge and is equal to _____ per _____.
2. The _____ is the unit for measuring the electric potential and is equal to _____ per _____.
3. The _____ is the unit for measuring the electric resistance of a circuit element and is equal to _____ per _____.
4. The specific resistance of a substance has the units of _____ and is called _____.
5. The _____ of a substance is a constant of proportionality between the electrical current density and the electric potential gradient; it is the reciprocal of the _____, and is measured in _____.
6. The _____ names the property of the junction of two dissimilar metals which, when heated, produces a _____.
7. The maximum potential difference between the terminals of an electrical energy source is called the _____.
8. When a human bone is flexed, it produces a _____ which results from the property of human bone called _____.

Resistors

9. Given three resistors of resistances, 4.0 Ω, 3.0 Ω, and 6.0 Ω, find the resistance of:
   a. the three in series
   b. the three in parallel
   c. the 4.0-Ω resistor in series with the 3.0-Ω resistor and the 6.0-Ω resistor in parallel

Ohm's Law

10. A 100-V battery has an internal resistance of 5.0 Ω. A voltmeter with resistance of 500 Ω is connected across the terminals of the battery. What voltage will the voltmeter show, and what is the current through the voltmeter?

Power Loss

11. A battery with an emf of 3.00 V and internal resistance of 0.20 Ω is connected to a lamp which is carrying a current of 0.75 A. What is the resistance of the lamp and the power that is being dissipated in it?

DC Circuits

12. In the circuit shown in Figure 22.19, what is the current in the battery $\mathscr{E}_2$? What is the resistance R?

DC Instruments

13. a. Why is the use of a potentiometer and a Wheatstone bridge called a "null" method?
    b. If a thermocouple is used with a potentiometer, what is detected?

Bioelectricity

14. Name and characterize at least three types of electrical measurements of the human body that can be made.
15. The transmission of nerve pulses involves the generation of an _____ which travels along as axon at a speed of about _____, much _____ than typical speeds for _____ in conductors.
16. Electrical shocks cause damage to the human body because of the effects of the _____.

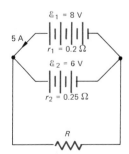

$\mathscr{E}_1 = 8$ V
5 A
$r_1 = 0.2\ \Omega$
$\mathscr{E}_2 = 6$ V
$r_2 = 0.25\ \Omega$
R

FIGURE 22.19
Question 12.

Answers

1. ampere, coulomb, second (Section 22.2)
2. volt, joule, coulomb (Section 22.2)
3. ohm, volt, ampere (Section 22.2)
4. ohm–meter, resistivity (Section 22.2)
5. electrical conductivity, resistivity, ohm^{-1} m^{-1} (Section 22.2)
6. Seebeck effect, potential difference (Section 22.8)

7. electromotive force (emf) (Section 22.2)
8. potential difference, piezoelectricity (Section 22.10)
9. a. 13.0 Ω
   b. 1.33 Ω
   c. 6 Ω (Section 22.4)
10. 99.0 V, 0.198 A (Section 22.4)
11. 3.8 Ω, 2.1 watts (Section 22.4)
12. 4 A, 7 Ω (Section 22.14)

13. a. when balanced, no current flows through a detector (Sections 22.6 and 22.7)
    b. a potential difference related to the temperature of the bimetallic junction (Section 22.8)
14. ECG (heart potentials), EMG (muscle potentials), EEG (brain waves), GSR (skin resistance) (Section 22.11)
15. action potential, 1 m/sec, slower, electrical conduction (Section 22.12)
16. electric current (Section 22.13)

ALGORITHMIC PROBLEMS

Listed below are the important equations from this chapter. The problems following the equations will help you learn to translate words into equations and to solve single-concept problems.

Equations

$$J = \frac{Q}{At} \tag{22.1}$$

$$I = JA \tag{22.2}$$

$$J = -\sigma \frac{\Delta V}{\Delta l} \tag{22.3}$$

$$V = \frac{l}{\sigma A} \quad I = IR \tag{22.6}$$

$$R = \rho \frac{l}{A} \tag{22.7}$$

$$\mathscr{E} = \mathscr{E}_1 + \mathscr{E}_2 + \mathscr{E}_3 \tag{22.9}$$

$$R_{\text{eff}} = R_1 + R_2 + R_3 + \cdots + R_n \tag{22.10}$$

$$I = i_1 + i_2 + i_3 \tag{22.11}$$

$$\frac{1}{R_{\text{eff}}} = \frac{1}{R_1} + \frac{1}{R_2} + \frac{1}{R_3} \tag{22.16}$$

$$\text{power} = VI \tag{22.22}$$

$$P = I^2 R \tag{22.23}$$

$$V_x = E_s \frac{L_x}{L_s} \tag{22.26}$$

$$R_x = \frac{R_2}{R_1} R_3 \qquad\qquad (22.29)$$

Problems

1. If 100,000 C of electricity flows through a conductor of 1.00 mm² cross-sectional area in 10.0 minutes, what is the current density?
2. The resistivity of copper is 1.7×10^{-8} Ω-m. What is the resistance of 1 kilometer of wire which has an area of 1 mm²?
3. Aluminum has a resistivity of 2.8×10^{-8} Ω-m, and copper has a resistivity of 1.7×10^{-8} Ω-m. How must the areas of copper and aluminum wires compare if the resistance of a given line is to be the same?
4. If one has three resistances of 2.0, 3.0, 4.0 Ω.
   a. What is the maximum resistance using all three resistors, and how are they connected?
   b. What is the minimum resistance for the three, and how are they connected?
5. A portable radio is operated by three 1.5–V dry cells. What is the maximum potential that could be applied to any element in the radio?
6. Heat is developed in a resistor at the rate of 10 watts when the current is 3.0 A. What is the resistance?
7. Is the filament resistance higher in a 500-watt or a 100-watt lamp if each operates on a 110-V line?
8. In a potentiometer system with a standard cell of emf of 1.08 V, the galvanometer shows zero deflection for a balanced point of 100 cm. What is the emf of a cell which balances at 150 cm?

Answers

1. 1.67×10^8 A/m²
2. 17.0 Ω
3. $\dfrac{A_{Cu}}{A_{Al}} = 0.61$
4. a. 9.0 Ω series
   b. 0.92 Ω, parallel

5. 4.5 V
6. $R = 1.11$ Ω
7. 100-watt lamp
8. 1.62 V.

EXERCISES

These exercises are designed to help you apply the ideas of a section to physical situations. Where appropriate the numerical answer is given in brackets at the end of each exercise.

Section 22.2

1. The manganese used in a standard resistance coil has a resistivity of 43.0×10^{-6} Ω–cm. How long would the wire for a 1.00-Ω coil be if the diameter of the wire is 0.100 mm? What diameter manganese wire would have a resistance of 5.00 Ω? [1.83 cm, 3.31×10^{-2} cm]
2. A piece of uniform wire of a given material 10.0 m long and 1.00 mm in diameter has a resistance of 1.00 Ω. What would be the resistance of a wire of the same material 50.0 m long and 0.500 mm in diameter? [20 Ω]

Section 22.4

3. Find the effective resistance of the circuit in Figure 22.20. [12.5 Ω]
4. A thermistor changes its resistance from R to R/2 in the circuit in Figure 22.21. Find the change in voltage across the thermistor corresponding to this resistance change. Assume $R_0 = R$. [$\Delta V = E/6$]
5. For the circuit in Figure 22.22, find the reading of

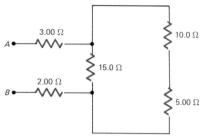

FIGURE 22.20
Exercise 3.

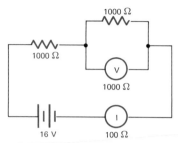

FIGURE 22.22
Exercise 5.

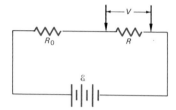

FIGURE 22.21
Exercise 4.

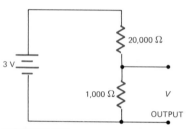

FIGURE 22.23
Exercise 6.

the ammeter (I) and the voltmeter (V). Find the power supplied by the battery. The resistance of the voltmeter is 1000 Ω; that of the ammeter is 100 Ω. [0.01 A, 5 V, 0.16 watts]

6. The output of the voltage divider shown in Figure 22.23 is measured by voltmeters with internal resistances of 1,000 Ω and 20,000 Ω respectively. Find the reading of each voltmeter. [$V_{1,000} = 0.073$ V, $V_{20,000} = 0.15$ V]

7. A lamp of resistance of 240 Ω and a variable resistance R are connected in series to a 120-V source. What is the power dissipation in the lamp if $R = 0$, if $R = 240$ Ω? What is the value of R if the power dissipated in the lamp is 50 watts? [60 watts, 15 watts, 22.8 Ω]

Section 22.5

8. It is desired to make an ammeter from a meter that has 100 μA full-scale current when a potential of 10 mV is across it. Find the necessary resistance and sketch the circuit for this ammeter if it is 1.0 A at full scale. [$R_{\text{meter}} = 100\ \Omega, R_{\text{shunt}} = 0.01\ \Omega$]

9. A galvanometer can detect a maximum current of i A. A shunt resistor of R Ω converts this galvanometer to an ammeter of I A full scale.
   a. Find the resistance of the galvanometer in terms of given data.

b. Find the maximum power rating of the galvonometer. [a. $R_g = R(I - i)/i$; b. $P_{\text{max}} = (I - i)iR$]

10. A meter movement has an internal resistance of 400 Ω and a full-scale current of 20.0 μA.
   a. Sketch a circuit that can be used to convert this meter into a voltmeter reading 12.0 V full scale. Determine all resistance values necessary for this conversion.
   b. If this voltmeter is to be used with an ammeter (1.00 A full scale) made from the same kind of meter to measure a resistance of approximately 1,000,000 Ω, find the shunt resistance for the ammeter and sketch the circuit that would give best results. [a. series resistance 599,600 Ω; b. parallel resistance $8 \times 10^{-3}\ \Omega$]

Section 22.6

11. In the circuit in Figure 22.24, what value of R will give no voltage across the detector? What is this circuit? [960 Ω]

12. The emf of a battery is measured by means of a slide-wire potentiometer. A standard cell of emf 1.018 V gives a zero reading when 18.0 cm are intercepted by the potentiometer slide. The battery intercepts 27.0 cm for zero balance. Find the emf of the battery. [emf = 1.53 V]

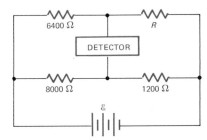

FIGURE 22.24
Exercise 11.

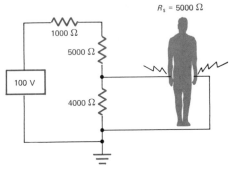

FIGURE 22.27
Exercise 14.

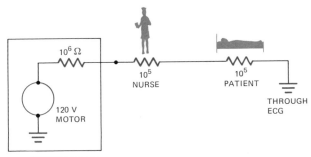

FIGURE 22.25
Exercise 13a.

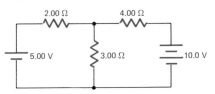

FIGURE 22.28
Exercise 15.

the current through the patient. [a. 100 μA; b. 133 μA]

14. An instrument provides the potential shock hazard as shown in Figure 22.27. Find an equivalent circuit and the shock current when the subject makes contact. [5.41×10^{-3} A]

Section 22.14

15. Given the circuit in Figure 22.28, find the current through each resistance. [$I_2 = 0.19$ A, $I_4 = 1.35$ A, $I_3 = 1.54$ A]

16. In the circuit in Figure 22.29, find the current through the 6 V and $\mathscr{E}_1$ cells, the value of $\mathscr{E}_1$, and the potential difference V_{ab}. [$V_{ab} = 12$ V, $\mathscr{E}_1 = 6$ V, $I = 0.5$ A to the right; $I_{3.5} = 1.5$ A to the right]

Section 22.13

13. The following examples, similar to actual electrical hazards found in hospitals have been pointed out by R. K. Hobbie (University of Minnesota).
    a. The ground lead breaks on a motorized bed. The nurse touches the bed and the patient (with ECG electrodes in place) at the same time. The circuit diagram is shown in Figure 22.25. Find the current through the patient.
    b. A vacuum cleaner bumps an ECG monitor leaking 1.0 A to ground through 0.08 Ω of wire. The complete circuit is shown in Figure 22.26. Find

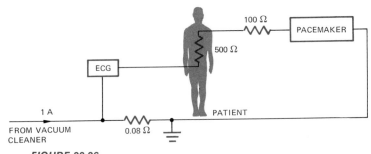

FIGURE 22.26
Exercise 13b.

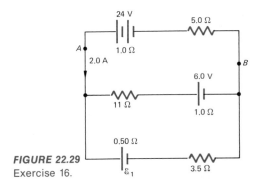

FIGURE 22.29
Exercise 16.

PROBLEMS

Each of these following problems may involve more than one physical concept. Where appropriate, the numerical answer is given in brackets at the end of the problem.

17. Three resistors (50 Ω each) are wired as shown in Figure 22.30. Each resistor is rated for 0.5 watt. Find the maximum voltage that can be put across this circuit.   [7.5 V]

18. An electrical heater has a rating of 1500 watts when connected to a 120-V line. What current does it draw, what is its resistance? What is the cost of operating the heater for ten hours at nine cents per kilowatt hour?   [12.5 amp, 9.6 ohm, $1.35]

19. Nichrome wire is to be used in the construction of an electric furnace. The resistivity of nichrome is 100×10^{-6} Ω–cm. If the furnace is to have a resistance at room temperature of 10.0 Ω, what length of 0.500 mm radius wire is needed? If connected to a 120-V line, how many joules should the furnace produce in 10 minutes? (1 cal = 4.18 J)   [785 cm, 8.59×10^5 J]

20. A three-way light bulb can be constructed with two filaments, R_1 and R, connected to leads a, b, and c (see Figure 22.31). By means of a switch 120 V can be placed across ac, ab, or bc. When the bulb is connected to ac, the rating is 50 watts, and the rating is 75 watts when the bulb is connected to ab. What is the resistance of R_1 and R, and what is the power

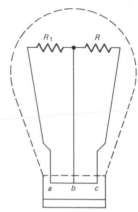

FIGURE 22.31
Problem 20.

rating when the bulb is connected across bc?   [$R = 96$ Ω, $R_1 = 192$ Ω, $P = 150$ watts]

21. A storage battery has an emf of 12.4 V and an internal resistance of 0.20 Ω. It is charged at the rate of 15.0 A for eight hours. How much energy goes into charging the battery? What percentage of this energy is lost as heat? [1.49 kwh, 24 percent of the energy goes to heat]

22. When a car battery is being charged at 15.0 A, the terminal potential is 12.9 V, and when it is being discharged at 10.0 A, the terminal potential is 11.4 V. What is the emf and the internal resistance of the car battery?   [12 V, 0.06 Ω]

23. Most conductors show a change in resistance as their temperature changes. Platinum has a thermal coefficient of resistance of 3.60×10^{-3} (deg^{-1}). If a platinum resistance thermometer has a resistance of 8.00 Ω at 20.0°C, what is the temperature at which it has a resistance of 20 Ω?   [437°C]

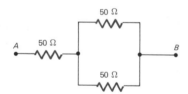

FIGURE 22.30
Problem 17.

GOALS

When you have mastered the content of this chapter, you will be able to achieve the following goals:

Definitions
Define each of the following terms, and use it in an operational definition:

magnetic field   current sensitivity
magnetic forces   ferromagnetism
ampere

Biot-Savart Law
Apply the basic relationship between current and its associated magnetic field.

Magnetic Forces on Moving Particles
Explain the motion of a charged particle in a uniform magnetic field.

Magnetic Interactions
Discuss the interaction of magnetic fields.

Electric and Magnetic Fields
Explain the difference between the behavior of charged particles in electric and magnetic fields.

Magnetic Field Applications
Explain such applications as: electromagnetic pump, focusing of charged particles by a magnetic field, DC electric meters, motors, and the Hall effect.

PREREQUISITES

Before beginning this chapter you should have achieved the goals of Chapter 4, Forces and Newton's Laws, Chapter 21, Electrical Properties of Matter, and Chapter 22, Basic Electrical Measurements.

23

Magnetism

23.1 Introduction

Magnetism is a concept introduced in physics to help you understand one of the fundamental interactions in nature, the interaction between moving charges. Like the gravitational force and the electrostatic force, the magnetic force is an interaction-at-a-distance.

Can you list five different ways in which magnetism has played a part in your life today? How strong is the earth's magnetic field? Are there interactions between living systems and the earth's magnetic field?

In this chapter we will discuss the basis of the magnetic field model for the interaction between moving charges and explore the close relationship between electricity and magnetism.

23.2 The Magnetic Field Model

As early as 600 years B.C. it was known that naturally occurring lodestone would attract iron. This material was found in the country of Magnesia, and the name magnet was applied to individual specimens. It was found that when this material was suspended in such a way that it was free to rotate, it would align itself in an approximate north-south direction (Figure 23.1a). By studying the interaction between two specimens, it was found one pair of ends attracted each other and another pair of ends repelled each other (Figure 23.1b). The laws of interaction were similar to those of electrostatics. The relative strengths of the magnets were expressed in terms of pole strengths. The entire field of magnetostatics can be developed in a parallel manner to that for electrostatics. However, there is one great physical difference. To date there have been only a few disputed reports of detection of a magnetic monopole, which would be analogous to the single electric charge.

We shall approach our study of magnetism from the standpoint of magnetic effects of an electric current and the interaction between magnetic fields. We may say that in any region in which a compass needle (a magnet) takes a definite direction, there exists a magnetic field. (If there were no magnetic field, the compass needle would take

FIGURE 23.1
(a) Direction a freely moving magnet takes. (b) The force relationships between magnetic poles.

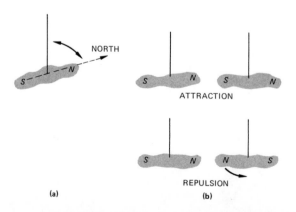

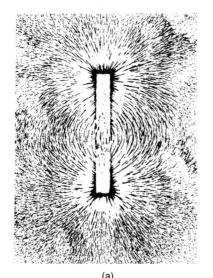

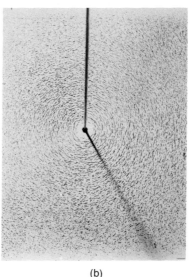

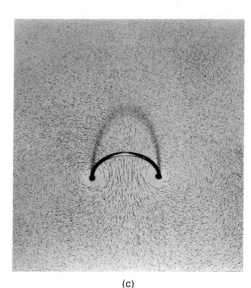

(a)

(b)

(c)

Iron filings in a magnetic field. (a) Magnetic field pattern for a bar magnet. (b) Magnetic field pattern produced by an electric current in a long straight wire in a plane perpendicular to the wire. (c) Magnetic field pattern produced by an electric current in a single circular coil in a plane perpendicular to the plane of the coil. (Picture from *PSSC Physics*, D.C. Heath and Co., Lexington, Mass., 1965.)

any random direction.) On the earth a compass takes a definite direction as a result of the earth's magnetic field; the earth itself behaves magnetically as a bar magnet. The compass needle aligns itself parallel to the magnetic field at the point of suspension. The end of the compass needle that points in a northerly direction is called the *north pole*, and the other end is called the *south pole*.

A *dip needle* is a magnetized needle that is mounted so that it can rotate freely in a vertical plane. When a dip needle is placed in a north-south plane, the needle points in the direction of the earth's magnetic field. The dip angle is measured from the horizontal position. For example, the dip angle at Washington, D.C., is 71°.

In 1873 James Clark Maxwell published his famous theory of electricity and magnetism. In this work the magnetic field is ascribed to interactions involving moving charges. Moving charges are seen to be the sources of all magnetic phenomena. The magnetic field is introduced into physics to explain the interaction between moving charges, which is more complex than the Coulomb interaction between electric charges at rest. Maxwell's theory predicts that whenever there are moving charges, there are both electric and magnetic fields that can be used to explain observed physical phenomena. Another prediction of Maxwell's theory is that accelerating charges generate electromagnetic waves that travel with the speed of light.

23.3 Properties of the Magnetic Force

Assume that a magnetic field exists in the region of consideration and that there are no electric or gravitational forces acting on the charged particles. We can explore the properties of the magnetic force, and hence the magnetic field, by directing a beam of charged particles into this

FIGURE 23.2
Direction of force acting on a moving charge in a magnetic field (a) parallel to the field and (b) perpendicular to the field. (c) The magnitude of the force depends upon the magnitude of the charge.

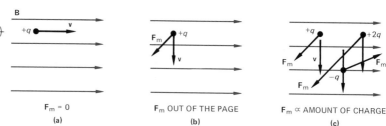

$F_m = 0$
(a)

F_m OUT OF THE PAGE
(b)

$F_m \propto$ AMOUNT OF CHARGE
(c)

region. The paths of the charged particles gives us information about the magnetic force $\mathbf{F}_m$ which is acting upon them. The experimental facts are:

1. There exists one particular orientation in which $\mathbf{F}_m$ is zero. This means that the charged particle goes in this direction at constant velocity. The line of the magnetic field, $\mathbf{B}$, is this line, which is the same direction defined by the compass needle approach (Figure 23.2a).

2. For other orientations the magnetic force $\mathbf{F}_m$ is always perpendicular to $\mathbf{v}$ the velocity of the particle (Figure 23.2b).

3. The magnitude of the magnetic force is proportional to the charge q and is in the opposite direction for positive and negative charges (Figure 23.2c).

4. The magnitude of the magnetic force is directly proportional to the component of the velocity perpendicular to the magnetic field; that is,

$$F_m \propto \sin \theta$$

where θ is the angle between the velocity and the direction of the magnetic field $\mathbf{B}$.

From properties 3 and 4 we can define the magnitude of the magnetic induction field $\mathbf{B}$,

$$B = \frac{F_m}{qv \sin \theta} \tag{23.1}$$

The SI units for B are webers/m² (wb/m²) or *tesla* (T) where

$$1 \text{ tesla} = \text{weber/m}^2 = \frac{N}{(C)(m/sec)}$$

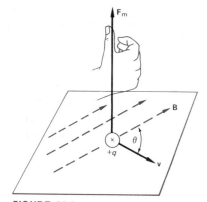

FIGURE 23.3
The magnetic force on a charge moving in a magnetic field. The direction given is that of a right-hand rule.

5. The magnetic force is always at right angles to the plane of the velocity line and line of the magnetic induction. Experiment shows that the direction relationship is as shown in Figure 23.3 where $\mathbf{v}$ is the velocity of the positive particle. A right-hand rule determines the direction of magnetic force. The fingers point in the order of the multiplication (in this case $\mathbf{v}$ into $\mathbf{B}$), and the thumb will point in the direction of the product. The product is a vector and is perpendicular to the plane of $\mathbf{v}$ and $\mathbf{B}$. This is an operational procedure and definition, and is known as a vector product. It is normally written in the form

$$\mathbf{F}_m = q\mathbf{v} \times \mathbf{B} \qquad F_m = qvB \sin \theta \qquad\qquad (23.2)$$

EXAMPLES

1. An electron of charge $-e$ (-1.60×10^{-19} C) is traveling east at 3.00×10^6 m/sec in the magnetic field of the earth, 0.563×10^{-4} tesla (T) north. What is the direction and magnitude of the force on the electron?

    We note the $\mathbf{v}$ and $\mathbf{B}$ are perpendicular to each other so $\sin \theta = 1$. Rotate fingers of your right hand from east to north and your thumb will point up, so the force on a positive charge would be up away from the earth, but an electron is negative so the force is directed down.

    $$F_m = qvB = (1.60 \times 10^{-19} \text{ C})(3.00 \times 10^6 \text{ m/sec})(0.563 \times 10^{-4} \text{ T})$$

    $$F_m = 2.70 \times 10^{-17} \text{ N}$$

2. A particle with a charge of 1.00 C is traveling with a velocity of 3.00 m/sec in a magnetic field of 7.00×10^{-4} T. The particle experiences a force of 1.50×10^{-3} N. What is the angle between the direction of motion and the magnetic field?

    $$\sin \theta = \frac{F_m}{qvB} = \frac{1.50 \times 10^{-3} \text{ N}}{(1.00 \text{ C})(3.00 \text{ m/sec})(7.00 \times 10^{-4} \text{ T})} = \frac{1.50 \times 10^{-3}}{2.10 \times 10^{-3}} = 0.714$$

    $$\theta = 45.6°$$

The quantity which we have been calling magnetic induction B, is also referred to as magnetic induction field, or magnetic flux density. The SI unit, weber/m² or tesla, is equal to 10^4 gauss. The gauss (G) is the cgs unit for magnetic induction. To give you a feeling for the size of the units of magnetic fields, the magnetic field of the earth is of the order of 0.5×10^{-4} W/m², a field of the order of 1 W/m² will pull steel rulers and screw drivers from pockets, and the maximum producible constant fields are of the order of 20 Wb/m².

23.4 *Magnetic Effects of Electrical Currents*

In this chapter we are interested in exploring the magnetic effects of moving charges. To show the magnetic effect of a current, which was first discovered in 1820 by Hans Christian Oersted, we can do the following experiment: place a wire carrying an electric current in the vicinity of a compass needle. Note the deflection of the needle both with current on and off. Reverse the direction of current, that is reverse the connections to the battery, and again note the deflections. We find that the compass needle is deflected by the presence of a magnetic field produced by the current in the conductor. The direction of the magnetic field at a point is defined as being in the direction of the force on a north magnetic pole at that point. The relationship between the direction of the magnetic field and of the direction of the electric current is given by a right-hand rule: If you place the fingers of your right hand around the current carrying conductor with your thumb pointing in the direc-

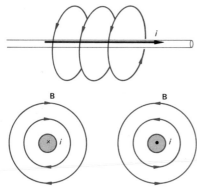

FIGURE 23.4
The magnetic field lines around long current filaments. If the thumb of right hand is pointed in the direction of the current, the fingers curl in direction of the magnetic field. (Remember the convention for current directions is that of positive charge flow.)

tion of the current (from positive terminal of battery to negative terminal in the external circuit), your fingers will encircle the conductor in the direction of the magnetic field. The direction of the magnetic induction for a long, straight conductor perpendicular to the page is shown in Figure 23.4.

The symbol $\otimes$ represents a vector (direction of current) into the page. It may be considered the tail of an arrow. The symbol $\odot$ represents a vector (direction of current) out of this page. It may be considered the point of an arrow.

According to the Biot-Savart law, the magnitude of the magnetic induction resulting from a current in a circuit element Δl at a point a distance r and direction θ from Δl is given by (see Figure 23.5):

$$B = \left(\frac{\mu_0}{4\pi}\right)\left(\frac{i\Delta l \, \sin\theta}{r^2}\right) \quad \begin{array}{l}\text{(direction of } \mathbf{B} \text{ given by the right-hand} \\ \text{rule)}\end{array} \qquad (23.3)$$

or in vector product form:

$$\mathbf{B} = \left(\frac{\mu_0}{4\pi}\right)\left(\frac{i\Delta l \times \hat{\mathbf{r}}}{r^2}\right) \qquad \text{where } \hat{\mathbf{r}} \text{ is a unit vector}$$

one unit in length in the direction from the increment of length to the point of interest, Δl is the vector in the direction of the current flowing through the increment of the conductor of length Δl, i is the current in the small element Δl (in the plane of the page) which establishes a magnetic induction contribution at point P, and μ_0 is the permeability constant. In SI units, i is in amperes, Δl is in meters, r is in meters, $\mathbf{B}$ is in webers per square meter, and μ_0 has the value of $4\pi \times 10^{-7}$ weber per ampere-meter.

In order to develop the quantitative value of the magnetic induction due to a current in a conductor, one generally must use calculus methods. For special geometric shapes we can find $\mathbf{B}$ directly from Equation 23.3. Let us apply Equation 23.3 to the case of the magnetic induction in the plane and at the center of a single circular turn of a conductor. For this case $r =$ radius of the coil and is constant, Δl becomes $2\pi r$, and $\sin\theta = 1$. thus

$$B = \frac{\mu_0 I}{2r} \qquad (23.4)$$

FIGURE 23.5
The magnetic field due to a current i in an element Δl. Check the direction with a right-hand rule.

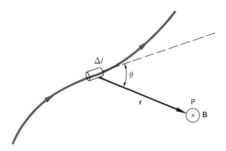

FIGURE 23.6
(a) Side view of the magnetic field around a long current filament.
(b) The magnetic field due to a long solenoid. A right-hand rule applied to each turn gives **B** direction.

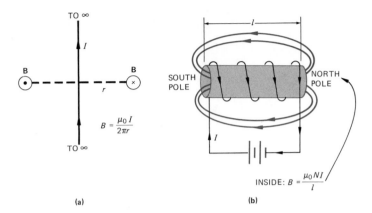

(a)   (b)

For N turns, the value for B is N times larger. For a long straight conductor the value of B is given by:

$$B = \frac{\mu_0 I}{2\pi r} \tag{23.5}$$

(see Figure 23.6a) where r is the distance from the conductor. Equation 23.4 and Equation 23.5 can be derived from Equation 23.3 by calculus methods (see Section 23.13). For a solenoid the value of B within the solenoid is given by:

$$B = \frac{N\mu_0 I}{l} \tag{23.6}$$

(see Figure 23.6) where N is the number of turns in solenoid and l is the length of the solenoid.

23.5 Motion of Charged Particles in A Uniform Magnetic Field

The fundamental relation for the magnetic force on a moving charge is given by Equation 23.2:

$$\mathbf{F}_m = q\mathbf{v} \times \mathbf{B} \tag{23.2}$$

As stated earlier $\mathbf{F}_m$ is always perpendicular to the velocity which means that there is no component of the force along the direction of motion, and no work is done by the magnetic force. With the velocity at right angles to the force we have the conditions for uniform circular motion (see Chapter 4). If a charged particle (charge $= q$, mass $= m$) is projected with velocity $\mathbf{v}$ at right angles to a uniform magnetic field $\mathbf{B}$, the path of the particle will be a circle (Figure 23.7a). Because $\sin \theta = 1$, we then have:

$$Bqv = \frac{mv^2}{r} \tag{23.7}$$

The paths of charged particles in a magnetic field are marked on the lower pole face of the switchyard magnet of the University of Colorado Nuclear Physics Laboratory. The beam from the cyclotron enters the chamber through the tube near the lower left-hand corner. The magnetic field is perpendicular to the pole faces (the upper face was removed when picture was taken). By means of the magnetic field the beam may be directed upward from the lower pole. Then positively charged particles are bent to the right. (Apply the right-hand rule.) (Photo by Jack Groft. In Dr. Albert A. Bartlett, *Physics in Pictures:* the paths of charged particles in a magnetic field, *The Physics Teacher*, Vol. 13, No. 8. p. 499 (1975).

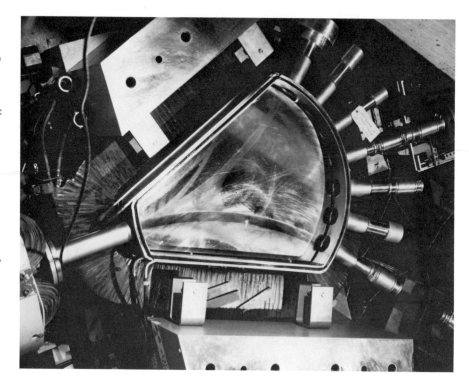

and

$$r = \frac{mv}{Bq} \tag{23.8}$$

Thus we note that $Bqr = mv = p$, the momentum of the particle; so the radius of the path of the particle is proportional to its momentum.

EXAMPLE

A proton is projected with a velocity of 1.00×10^6 m/sec perpendicular to a magnetic field of 0.100 Wb/m². What is the radius of its path?

FIGURE 23.7
(a) The magnetic force on a positive charge moving at right angles to a uniform magnetic field. Verify the force direction with a right-hand rule. (b) Path of a charged particle in a magnetic field, if the particle enters the field at angle so that it has a component of velocity in the direction of the field.

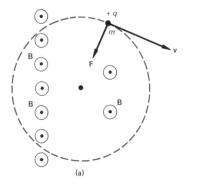

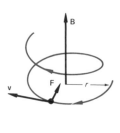

$$m_p = 1.67 \times 10^{-27} \text{ kg} \qquad\qquad B = 0.100 \text{ Wb/m}^2$$

$$q = 1.60 \times 10^{-19} \text{ C}$$

$$r = \frac{1.67 \times 10^{-27} \times 10^6}{0.160 \times 10^{-19}} = 0.104 \text{ m}$$

If the charged particle, which is projected into a uniform magnetic field, has a component of velocity parallel to the direction of **B**, the particle will have a spiral path (Figure 23.7b) around the **B** direction.

23.6 *Applications of the Principles of Charged Particles Moving in a Magnetic Field*

There are many applications involving charged particles moving in a magnetic field. Some important present-day instruments are electron microscopes, mass spectrometers, beta-ray spectrometers, cyclotrons, and betatrons.

Another application, which is important in some medical cases, is *electromagnetic pumping*. The basic electromagnetic pump is shown in Figure 23.8. A fluid which is a good conductor of electricity is placed in a closed system, circulated by means of an electromagnetic pump. A magnetic field is established over a portion of the closed system, and an electric current is passed through the fluid at right angles to the magnetic field. The portion of the fluid carrying the current then experiences a force whose direction is given by the right-hand rule. The fluid moves through the system as each element of fluid reaches this section of the system and experiences this force. The fluid moves through the system as long as the magnetic field and the electric current are maintained.

The electromagnetic pump has been used to circulate blood in an artificial heart machine. An external magnetic field is created and the ions in the blood carry the electric current. Thus the pumping action can be achieved. The electromagnetic pump is preferable to the mechanical pump because a mechanical pump has moving parts that may damage the blood cells, but the electromagnetic pump has no moving parts and does not injure the blood cells.

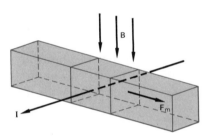

FIGURE 23.8
Electromagnetic pumping. A conducting fluid carries current in direction of **I** perpendicular to the applied magnetic field, **B**. The magnetic pumping force direction is given by a right-hand rule.

23.7 *Interaction Between Magnetic Fields*

There is a simple experiment that we can do to show the interaction of magnetic fields:

We place a conductor between the poles of a strong magnet perpendicular to the field. A current is sent through the conductor, and we notice that there is a force acting upon the conductor when it carries a current and that there is no force when there is no current. We also notice that the direction of the force depends upon the direction of the magnetic field and the direction of current in the conductor.

FIGURE 23.9
A right-hand rule is applied to give the direction of the force on a current at right angles to an external magnetic field.

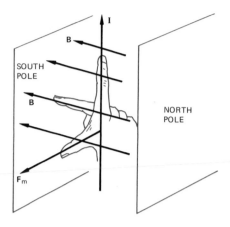

The interaction of magnetic fields is shown in Figure 23.9. Where **B** is the direction of the magnetic field due to the magnet, the current in the conductor is up in the figure, and the direction of the force is out of the page. Let us look at the example in Figure 23.9 in terms of interacting magnetic fields. The magnetic field produced by the current in the conductor is clockwise when looking in the direction of the current. In back of the conductor the current **B** field is in the same direction as applied **B**, and in front of the conductor the field from the current is in the opposite direction to the applied **B**. We can say that the total field is increased behind the conductor and decreased in front of it, and the direction of the force is from the stronger to the weaker field. This may be shown by use of a right-hand rule by directing the thumb, first finger, and middle finger so that they are mutually perpendicular to each other with the thumb representing direction of force, the first finger, the direction of current, and the middle finger the direction of magnetic field of the magnet (Figure 23.9).

Because a magnetic field exerts a force on a moving charge, we expect it to exert a force on the moving charges constituting a current in a conductor. Consider the current to be carried by free electrons which have a drift speed v_d to the left (Figure 23.10). This means the conventional current is to the right, with the magnetic field into the page. The direction of the magnetic force is toward the top of the page. The current carrying conductor experiences an upward force.

FIGURE 23.10
The magnetic force on electrons moving in a magnetic field. Note that this is an example of a left-hand rule for negative charge.

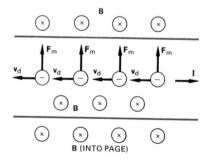

23.8 *Measuring Magnetic Fields*

Let us develop a quantitative expression for the magnetic force on a wire. We have learned that the magnitude of the force on a charged particle is: $F_e = Bqv \sin \theta$. The force on a length L is then said to be the product of the number of particles in length L and the force on each. The number of free charged particles is the number per unit volume times the volume, that is, nAL, where n is the number of conduction particles per unit volume, A is the cross-sectional area of the conductor, and L is the length. Then, the total force magnitude is just nAL times the force on an individual charged particle,

$$F_m = nALBqv \sin \theta \tag{23.9}$$

but the total current is given by the product of the number of charge carriers per unit volume times their charge, their speed, and the cross-sectional area of the conductor,

$$I = nAqv \tag{23.10}$$

If the wire is perpendicular to the magnetic field, the force becomes:

$$F = BIL \tag{23.11}$$

One newton is the magnetic force on a 1 meter conductor, perpendicular to a magnetic induction of 1 Wb/m² and carrying a current of 1 A. By combining Equations 23.5 and 23.11 one can arrive at the following definition of the ampere:

One ampere is that unvarying current, which, if present in each of two parallel conductors of infinite length, and one meter apart in empty space, causes each conductor to experience a force of exactly 2×10^{-7} newtons per meter of length.

23.9 *Force and Torque on a Loop Conductor*

In many basic galvanometers a rectangular loop (width b, length a) of wire is suspended in a uniform magnetic induction. If a line perpendicular to the plane of the coil makes an angle θ with the field **B** when current passes through the coil (see Figure 23.11a and 23.11b), there is a force upon each side of the coil in accordance with Equation 23.11. However, the forces on side AB and side CD cancel. The forces upon sides BC and DA are oppositely directed and exert a torque on the coil. The magnitude of the force is $F_m = BIb$. For N turns the force is N times as large.

The torque about the axis of suspension is given by the force times the perpendicular distances

$$\tau = 2F_m d = 2NBb \frac{a \sin \theta}{2} = NBIA \sin \theta \tag{23.12}$$

FIGURE 23.11
(a) The magnetic force on a current carrying loop in a uniform magnetic field. (b) A side view of the loop in the magnetic field.

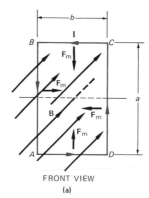
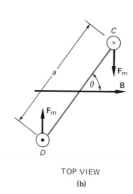

FRONT VIEW
(a)

TOP VIEW
(b)

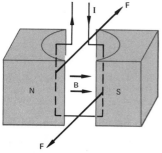

FIGURE 23.12
Meter coil in a magnetic field. The torque produced by current in the coil produces an angular deflection which is opposed by the suspension system.

for a coil of N turns, and where the area of the coil A is equal to ab. In the case of a galvanometer, ammeter, or voltmeter the deflecting torque given above is opposed by a restoring torque which depends upon the suspending system. This restoring torque is of the form

$$\tau_r = k\phi \tag{23.13}$$

where ϕ is the angle of deflection in radians and k is a constant of the suspension system. The coil takes an equilibrium position in which the torque produced by the current in the magnetic induction is equal to the restoring torque:

$$BIA \sin\theta = k\phi \tag{23.14}$$

In many instruments the construction is such that the magnetic field always points in a radial direction, that is, θ is 90°. Then (see Figure 23.12) Equation 23.14 reduces to,

$$I = \frac{k}{NBA}\phi \tag{23.15}$$

Model of a DC electric motor. Note the index mark showing the split in commutator ring. The brush traces are shown on the lower part of the commutator. The coil rotates in the magnetic field produced by the two electromagnets.

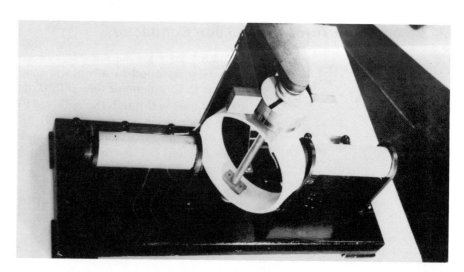

where N = number of turns. For a given system k/NBA is a constant, and one has

$$I = K\phi \tag{23.16}$$

where K is called the current sensitivity of the instrument.

The basic principle of an electric motor is the same as that of the loop in a magnetic field except the motor is constructed so that the torque is in the same direction at all times. With a torque applied to a coil or armature in a constant direction, rotation is produced. A rotating body has the ability to do work. We can convert electrical energy into mechanical energy. We can define any device that has an input of electrical energy and an output of mechanical energy as an electric motor.

23.10 The Hall Effect

One application of the forces on the moving charges in a conductor in a magnetic field is in the Hall effect. In this application a flat conductor is placed perpendicular to a magnetic field. If we consider the moving charge to be positive, then the diagram is as shown in Figure 23.13.

We expect an excess of negative charges on the back edge and an excess of positive charge on the front edge. This charge separation results in an electric force in the conductor. At equilibrium, the force resulting from the magnetic field is equal to the electric force due to the charge separation. This separation of charge produces an electric potential difference between the edges which is called the Hall voltage. In Figure 23.13 we would expect the front to be higher potential. If the moving charge carriers are negative, then the front of the conductor will be negative or at a lower potential. Hence the sign of the Hall potential indicates the sign of the charge carriers. Let us derive an expression for the magnitude of the Hall potential.

At equilibrium the electric force magnitude will be equal to the magnetic force magnitude,

$$F_e = F_m$$

where the electric force is given by q, the magnitude of the charge on the moving charges, times the electric field E_H for Hall field, and where the magnetic force magnitude is given by Bqv_d where v_d is the drift speed of the charges in the conductors. (This magnetic force is in the opposite direction from the electric force.)

FIGURE 23.13

The Hall effect. Positive charges will be deflected to the front for positive charge motion to the right, and electrons will be deflected to the front for negative charge motion to the left when the **B** vector points upward.

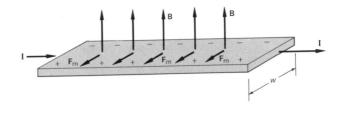

$$qE_H = Bqv_d \tag{23.17}$$

If the width of the strip is w, then $E_H = V_H/w$ where V_H is the Hall voltage across the strip, and we can write the expression for the Hall voltage in terms of the magnetic field

$$V_H = Bwv_d \tag{23.18}$$

But the drift velocity of the charges is difficult to measure; so let us recall that the total current I is given by the product of the number of charge carriers per unit volume n times their charges times their velocity v_d times the area of the conductor A:

$$I = nqvA$$

where A is equal to the product of w and d, that is, the thickness in the direction of B. We can substitute this expression in Equation 23.18 to eliminate v_d,

$$V_H = \frac{BwI}{nAq} = \frac{BI}{nqd} \tag{23.19}$$

In the above equation all quantities except n, can be measured in the laboratory, enabling us to calculate the value of n. For monovalent metallic conductors the value of n is nearly the same as the atom density. Once n is known for a given material, then a Hall effect probe can be used to measure magnetic fields.

23.11 Electron Microscope

Earlier you learned that the path of a charged particle could be altered by application of either an electric field or a magnetic field. In Chapter 18 you studied an optical microscope. A microscope which uses a beam of electrons instead of a ray of light is called an electron microscope. The comparison of the two are shown in Figure 23.14.

In the electron microscope the lenses for focusing the beam of electrons may be either electrostatic or magnetic. In the United States the magnetic focusing lenses are more commonly used. Although there are some parallel features between optical and electron microscopes, there are also significant differences. Some of these differences include:

1. The magnetic lens of an electron microscope does not have a fixed focal length as does an optical lens. The focal length of the magnetic lens depends upon the strength of the field, which is altered by changes in the current in the coil.
2. In an electron microscope the magnification is not changed by using different lenses, as is the case in the optical unit, but by changing the focal lengths.
3. The depth of field in the optical microscope can be observed directly

FIGURE 23.14

Schematic diagram showing the analogy between a light microscope and an electron microscope.

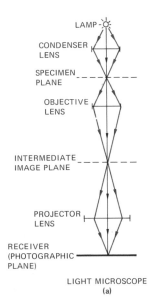

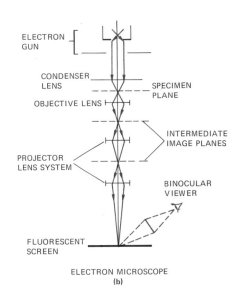

LIGHT MICROSCOPE
(a)

ELECTRON MICROSCOPE
(b)

by altering the focus control. The visual observation in electron microscopes is not so acute.

4. The image formation is different in the two systems. In the optical unit image formation results from a differential absorption of the light incident by the specimens. For the electron microscope variations in intensity of the image depends upon the scattering of the electrons by the individual elements of the specimen.

One great advantage of the electron microscope over an optical unit is the much greater magnification and higher resolving power. Some electron microscopes have a magnification of 200,000 and are capable of separating objects of a few angstrom units (10^{-10} m) in size.

There were two difficulties with the electron microscope: only dead specimens could be used, and the images formed were entirely two-dimensional. The first difficulty results from the low penetrating power of the electrons. The thickness of water in a living cell is a great barrier to the electrons, and impairs the image. In order to study structures in three dimensions, generally a large number of sections were required. This meant each section must be photographed, and from a number of photographs a three-dimensional model of the object constructed. However, the recent development of the *scanning electron microscope* has allowed research scientists to obtain some striking three-dimensional photographs of a variety of specimens.

In biological work one is interested in obtaining contrast in the microscope image. The difficulties outlined above make this difficult in electron microscopes. Nevertheless, the electron microscope is an important instrument for life scientists and the medical profession. If you wish to study more fully the contributions of the electron microscope to various

FIGURE 23.15
A simple form of a mass spectrometer. Particles enter the main chamber with a constant velocity due to velocity selector. The mass/charge ratio determines the radius of the particle path to the detecting film.

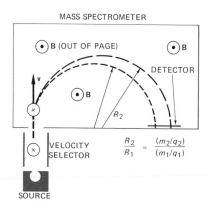

areas, you should consult a bibliography of electron microscopy. The literature contains many references and observations about the structure of bacteria, cells, tissues, and viruses.

The optical microscope was developed over a period of about two centuries. The first commercial electron microscope was completed less than fifty years ago. Much progress has been made since that time, and as the recent development of the scanning electron microscope indicates, we can anticipate more improvements in the future.

23.12 The Mass Spectrometer

The mass spectrometer is an instrument designed to measure and compare masses of individual atoms. The atoms to be analyzed are first ionized and passed through an accelerator and a velocity selector before they enter a uniform magnetic field that bends them into circular paths. A schematic diagram of a mass spectrometer is shown in Figure 23.15. All of the particles enter into the main chamber with the same velocity. The magnetic field **B** in this chamber is directed perpendicular to the plane of the path of the particles. The equation for the motion of the particles was discussed in Section 23.5, where we found the radius of the circular path of the particles to be proportional to their momentum mv,

$$r = \frac{mv}{qB} \tag{23.8}$$

If all the particles have the same velocity, particles of different ratios of mass to charge m/q are detected at different radii in the mass spectrometer.

23.13 Magnetic Materials and Biomagnetism

Since the discovery of magnetic monopoles (point sources for magnetic fields) has not been completely verified, we may assume that the sources of magnetic fields inside materials are atomic current loops (sometimes

FIGURE 23.16
The magnetization of soft iron using the field of a solenoid. Saturation occurs when all of the magnetic domains are aligned with the external field.

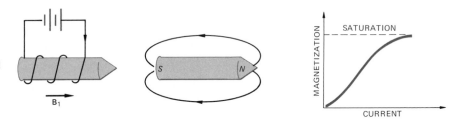

called magnetic dipoles). Materials are classified as *ferromagnetic, paramagnetic,* or *diamagnetic* depending on the nature of the atomic current loops within the material. Ferromagnetic materials (iron, nickel, and cobalt at room temperature) have permanent atomic magnetic dipoles that align with external magnetic fields. This cooperative interaction produces very strong internal magnetic fields within ferromagnetic materials.

One model for ferromagnetism is based on electron current loops in the atoms of these elements. The electronic magnetic dipoles combine to form macroscopic magnetic regions (*domains*) that align with external magnetic fields to produce unusually large magnetization. Ferromagnetic materials and their alloys are the basic materials for our permanent magnets and electromagnet cores.

The magnetic induction of magnetic material inside a solenoid is determined by its permeability K_m

$$B = K_m \mu_0 \frac{NI}{l}$$

so materials of large permeability, such as iron with a K_m ranging from a few hundred to ten thousand can greatly increase the magnetic field strength of a solenoid. You can make an electromagnet by using a piece of soft iron, a coil of wire, and a dry cell (Figure 23.16). The physical basis for this electromagnet is as follows. The external field due to the current in the coil interacts with the electronic magnetic dipoles in the domains of the iron. This interaction causes the domains to align their magnetic fields with the direction of the external field. This alignment process continues with increasing external field until all the domains are aligned and saturation is reached. The iron is then said to be completely magnetized. If the current is turned off, the domains remain aligned giving the iron a permanent magnetic field. This magnetic field can be reduced by heating the iron (allowing the domains to randomize) or by demagnetizing the iron with a field in the opposite direction. Ferromagnetism involves an interesting application of quantum physics and electromagnetism.

Paramagnetic materials such as liquid oxygen have molecular magnetic dipoles which tend to align with external magnetic fields. Paramagnetism is much weaker than ferromagnetism, but it can be used to study molecular electronic structure, as well as reaction rates for certain biochemical processes (enzyme catalyzed reactions, for example).

Most organic molecules are diamagnetic, that is they have no per-

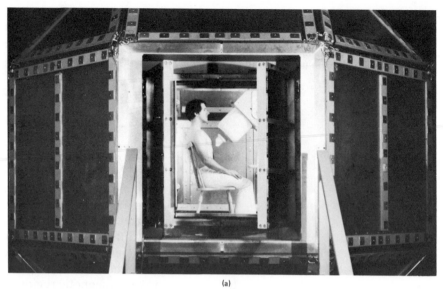

(a)

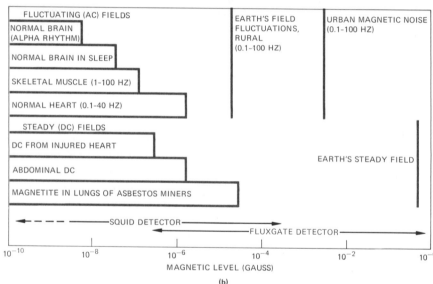

(b)

Magnetic fields of the human body. (a) The magnetic field of a human subject's heart is measured by a superconducting magnetometer inside the MIT shielded room. The shielding is provided by a five-layered wall with the addition of negative-feedback loops. The magnetometer is capable of detecting fields of the nanogauss range (10^{-13} tesla). Fields produced by a normal human heart are of the microgauss range (10^{-10} tesla). in accord with table shown. (b) Levels of the fields around the body and in the background. The strongest fluctuating field is over the lower part of the heart; the strongest steady fields, in welders and asbestos miners, are produced by particles of iron oxide in the lungs. The bottom of the diagram indicates the sensitivity of the two detectors, fluxgate and SQUID, for a 1 Hz bandwidth; the broken line shows that the limit moves to the right with greater bandwidth.

Details were published in an article by David Cohen in *Physics Today*, August 1975, page 34. (Courtesy of Dr. David Cohen.)

manent magnetic dipoles. When they are placed in a magnetic field, atomic current loops are set up that produce magnetic fields that oppose the external field.

In biological materials we need only to note that one place we might expect to find significant magnetic interactions is in systems that include ferromagnetic atoms. Recent research suggests that birds are capable of using the earth's magnetic field for navigation. The basis for such a magnetic sense is not yet known. The entire area of biomagnetism still lacks firm cases of magnetic phenomena. Recent sophisticated methods and instrumentation promise more insight into possible biomagnetic phenomena.

ENRICHMENT
23.14 Calculus Derivations Using the Biot-Savart Law

We will make use of the Biot-Savart Law, Equation 23.3, to calculate the magnetic induction of a long, straight conductor at point P (see Figure 23.17).

$$|dB| = \frac{\mu_0}{4\pi} I \frac{dx}{s^2} \sin \theta \tag{23.9}$$

The induction **dB** set up at P by any other element of the conductor is parallel to the vector shown. The resultant is the algebraic sum or the integral of the dB. It becomes simpler if we let θ be the independent variable,

$$s = r \csc \theta$$

$$x = r \cot \theta$$

$$dx = -r \csc^2 \theta \, d\theta$$

FIGURE 23.17
Diagram for developing the value of magnetic induction from Biot-Savart law by calculus methods.

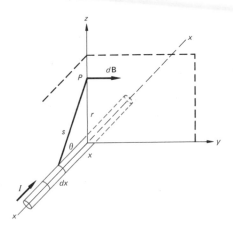

FIGURE 23.18
The magnetic field due to a current
loop at a point on its axis.

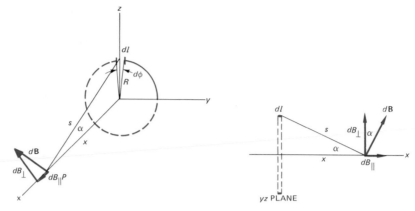

Substituting,

$$dB = -\frac{\mu_0}{4\pi}\frac{I}{r}\sin\theta\, d\theta$$

$$B = -\left(\frac{\mu_0}{4\pi}\right)\left(\frac{I}{r}\right)\int_\pi^0 \sin\theta\, d\theta = -\frac{\mu_0}{4\pi}\frac{I}{r}(-\cos\theta)\Big|_\pi^0$$

$$= \left(\frac{\mu_0}{4\pi}\right)\left(\frac{2I}{r}\right) = \frac{\mu_0}{2\pi r}I \tag{23.5}$$

We will now apply the Biot-Savart law to calculate the magnetic induction of current in a single turn of wire (Figure 23.18). Many devices use multiples of this configuration. A coil of one turn with radius R is in the yz plane with center at the origin. The increment dB is resolved into two components $db_\perp$ and $dB_\parallel$. For a point on the x axis all $dB_\perp$ cancel out, and B total is the sum of $dB_\parallel$ contributions.

$$dB_\parallel = dB\,\sin\alpha$$

$$dB_\parallel = \frac{\mu_0}{4\pi}I\,dl\frac{\sin\alpha}{s^2}\sin\theta$$

As $\sin\theta = \sin 90° = 1$,

$$s^2 = R^2 + x^2$$

$$dl = R\, d\phi$$

$$\sin\alpha = \frac{R}{\sqrt{R^2 + x^2}}$$

$$B = \int dB = \frac{\mu_0}{4\pi}I\int_0^{2\pi}\frac{R}{(R^2 + x^2)^{1/2}}R\, d\phi$$

$$B = \frac{\mu_0 I R^2}{2(R^2 + x^2)^{3/2}}$$

Note that if $x = 0$, this reduces to

$$B = \frac{\mu_0}{2R}I \tag{23.4}$$

SUMMARY

Use these questions to evaluate how well you have achieved the goals of this chapter. The answers to these questions are given at the end of this summary with the number of the section where you can find related content material.

Definitions

1. The ampere is defined in terms of the force between two parallel wires; it is the specified
   a. force/meter
   b. current
   c. separation of wires
   d. area of each wire
2. The magnetic field is introduced in physics because it
   a. is analogous to the electric field in all interactions
   b. is analogous to the gravitational field in all interactions
   c. can be used in explaining interactions of currents
   d. was defined in the Bible
   e. explains the origin of magnetic monopoles
3. The force between parallel currents in the same direction is
   a. repulsive
   b. zero
   c. parallel to current
   d. attractive
   e. none of these
4. The deflection of a galvanometer coil is due to
   a. suspension system heating
   b. gravitational attraction due to current
   c. diamagnetism
   d. paramagnetism
   e. torque on a loop conductor in a B field
5. Ferromagnetism is based on atomic current loops. It is characterized by internal magnetic fields that are:
   a. weak
   b. nearly zero
   c. independent of external fields
   d. dependent on external fields
   e. strong

Biot-Savart Law

6. For a current flowing into the paper, $\oplus I$, its magnetic field will be directed as shown.

e. none of these
7. For a long straight wire the magnetic field varies with the distance r from the wire and it is proportional to
   a. r
   b. $1/r^2$
   c. r^2
   d. $\sqrt{r}$
   e. $1/r$

Magnetic Forces on Moving Particles

8. The direction of the force on a charged particle moving with velocity v in a magnetic field B is
   a. perpendicular to v but not to B
   b. parallel to v but not to B
   c. perpendicular to v and B
   d. parallel to B but not to v
   e. in the plane of v and B
9. The magnetic force on a charged particle moving through a uniform magnetic field B always produces
   a. no change in energy of the particle
   b. no change in direction of the particle motion
   c. no centripetal acceleration for the particle
   d. none of these

Magnetic Interactions

10. When you consider the interaction of moving charges in terms of the interactions of their magnetic fields, the force on a moving charge is directed
    a. toward strongest B region
    b. away from strongest B region
    c. parallel to strongest B
    d. parallel to weakest B

Electric and Magnetic Fields

11. Negatively charged particles moving in a region containing perpendicular **B** and **E** fields will be accelerated
    a. parallel to **B**
    b. parallel to **E**
    c. parallel to −**E**
    d. parallel to −**B**
    e. perpendicular to **B**
12. If a positive charge at rest is the same distance from a 1 C point charge as it is from a bar magnet producing a 1 Wb/m² field, the ratio of the magnetic force to electric force will be
    a. 1
    b. ∞
    c. undetermined
    d. zero
    e. 1.6×10^{-9}
13. Based on their relative strengths, which field would you expect to be the most important in natural processes
    a. gravitational
    b. electric
    c. magnetic

Magnetic Field Applications

14. Applications of magnetic fields *always* involve
    a. permanent magnets
    b. moving charges
    c. monopoles
    d. charges at rest
    e. Hall effect
15. The Hall effect probe on the Viking I spacecraft on Mars enables NASA to measure Mar's
    a. atmospheric currents
    b. atmospheric storms
    c. water abundance
    d. magnetic field
    e. gravitational field

Answers

1. b (Section 23.8)
2. c (Section 23.3)
3. d (Section 23.7)
4. e (Section 23.9)
5. d, e (Section 23.13)
6. a (Section 23.4)
7. e (Section 23.4)
8. c (Section 23.3)
9. a (Section 23.3)
10. b (Section 23.7)
11. c, e (Section 23.3)
12. d (Section 23.3)
13. b (Section 23.3)
14. b (Section 23.2)
15. d (Sections 23.7, 23.10)

ALGORITHMIC PROBLEMS

Listed below are the important equations from this chapter. The problems following the equations will help you learn to translate words into equations and to solve single-concept problems.

Equations

$$B = \frac{F_m}{qv \sin \theta} \tag{23.1}$$

$$\mathbf{F}_m = q\mathbf{v} \times \mathbf{B} \tag{23.2}$$

$$B = \left(\frac{\mu_0}{4\pi}\right)\left(\frac{i\Delta l \sin \theta}{r^2}\right) \qquad \mu_0 = 4\pi \times 10^{-7} \text{ webers/A-m} \tag{23.3}$$

$$B = \frac{\mu_0 I}{2r} \tag{23.4}$$

$$B = \frac{\mu_0 I}{2\pi r} \tag{23.5}$$

$$B = \frac{N\mu_0 I}{l} \tag{23.6}$$

$$r = \frac{mv}{Bq} \tag{23.8}$$

$$F_m = BIL \tag{23.11}$$

$$\tau = NBIA \sin\theta \tag{23.12}$$

$$I = K\phi \tag{23.16}$$

$$V_H = \frac{BwI}{nAq} \tag{23.19}$$

$$\tan(\text{dip angle}) = \frac{B_v}{B_H}$$

Problems

1. At the equator the earth's magnetic field is about horizontal and directed from south to north with a strength of 0.5×10^{-4} webers/m². Find the force direction and magnitude on a 30-m horizontal wire and carrying a current of 40 A from east to west.
2. If an electron is shot with a speed of 1.00×10^6 m/sec upward into the earth's magnetic field that has a horizontal component directed north of 1.60×10^{-5} Wb/m² (tesla). What is the magnitude and direction of the force acting on the electron?
3. Two long, straight, parallel wires are 4 cm apart and carry a current of 5 A each in the same direction. What is the value of B midway between the two?
4. What is the value of B midway between the wires of problem 3 if the directions of currents are opposite.
5. A solenoid is 1.0 m long; its length is many times greater than its diameter. It has 4000 turns and carries a current of 5.0 A. What is B at the center of the solenoid?
6. It is found that a certain galvanometer has a deflection of 30° for a current of 10 A. What is the value of the torsion constant?
7. The dip angle is 71° at Washington, D.C., and the vertical component of the magnetic field **B** is 0.543×10^{-4} Wb/m². What is the magnitude of the earth's magnetic induction there?

Answers

1. 6×10^{-2} N down
2. 2.56×10^{-18} N east
3. zero
4. 10^{-4} Wb/m² or 10^{-4} T.

5. 2.5×10^{-2} T
6. 19.0 A/rad or 00.33 A/degree
7. $.574 \times 10^{-4}$ T

EXERCISES

These exercises are designed to help you apply the ideas of a section to physical situations. When appropriate the numerical answer is given in brackets at the end of the exercise.

Section 23.3

1. What quantity is the symbol **B** used to represent?

What are its units in the SI system? Give the SI equivalent of these units in meters, kilograms, seconds, and amperes.
2. Make a sketch of a positive charge traveling at right angles to a uniform magnetic field. Show the force that acts upon the charge and the subsequent path of the charge.
3. Make a sketch of a negative charge traveling at right

angles to a uniform magnetic field. Show the force that acts upon the negative charge and the subsequent path of the charge.

Section 23.4

4. Find the field produced by a current of 10 A at a distance of 5 cm from a very long conductor. [4×10^{-5} Wb/m²]
5. Calculate the field produced at the center of a 200-turn flat coil of 10-cm radius for a current of 5.00 A. [6.3×10^{-3} Wb/m²]
6. What is the magnetic field at the center of an air core solenoid of 1000 turns, 50.0 cm long, and carrying a current of 10.0 A? If this coil is wrapped upon an iron core with permeability of 500, what is the field? [25.1×10^{-3} Wb/m², 12.6 Wb/m²]
7. Make a sketch of a long, straight, current carrying wire, and show the magnetic field in its vicinity.
8. A solenoid 1.0 m long with 3.0×10^{3} turns carries a current of 6.0 A. Calculate the magnetic induction at the center of the solenoid if it has an air core. [.023 Wb/m²]

Section 23.5

9. An electron of charge 1.6×10^{-19} C moves with a speed of 8.0×10^{7} m/sec in a magnetic field of 0.50 (in SI units).
   a. What is the force on the electron?
   b. What is the radius of its path? ($m_e = 9.1 \times 10^{-31}$ kg)
   c. Make a sketch showing the electron, the field, and its direction of curvature. (*Hint:* The electron has a negative charge. Assume directions of the motion of the electron and the magnetic field.) [a. 6.4×10^{-12} N; b. 9.1×10^{-4} m]

Section 23.8

10. A current of 5 A is directed up in a vertical wire in a uniform magnetic field of 0.020 Wb/m² directed north. Find the magnitude and direction of the force on a 6-cm section of the vertical wire. [.006 N west]
11. A straight wire 20 cm long and carrying a current of

15 A is placed in a field where the magnetic induction is 0.50 in SI units.
   a. If the wire and the field are perpendicular to each other, find the force on the wire.
   b. Sketch the wire, the field, and the direction of the force upon the wire. (Assume whichever directions you wish for the current and the magnetic induction.) [a. 1.5 N]

Section 23.9

12. A rectangular galvanometer coil is suspended in a uniform magnetic field of 0.10 Wb/m². The coil is 1.0 cm wide and 4.0 cm long, has 200 turns, and carries a current of 1.0×10^{-8} A for a deflection of 30°. What is the torsion constant of the suspending system? [$7.6 \times <\neq^{-11}$ N-m/red]

Section 23.10

13. A current of 150 A is flowing in a silver ribbon in the x-direction. The thickness of the ribbon is 1.00 mm in the y-direction and 2.00 cm in the z-direction. The magnetic field of 1.20 Wb/m² is in the y-direction. Assume there are 7.40×10^{28} free electrons per cubic meter. Find:
   a. the drift velocity of the electrons
   b. the direction and magnitude of the field due to the Hall effect
   c. the Hall potential [a. 6.3×10^{-4} m/sec; b. 7.6×10^{-4} V/m; c. 1.52×10^{-5} V, the + side is in the z-direction]

Section 23.12

14. A mass spectrometer can be used to separate the isotopes of chlorine, ^{35}Cl and ^{37}Cl. These ions have masses of 58.45×10^{-27} kg and 61.79×10^{-27} kg respectively. Assume the chlorine ions enter a common slit after being accelerated through a potential of 10,000 V. At the slit the ions enter perpendicularly a magnetic field of 1.00 Wb/m², and they are turned through an angle of 180° to the detector. What is the separation between the ^{35}Cl and ^{37}Cl at the detector? [.48 cm]

PROBLEMS

These problems may involve more than one physical concept. When appropriate, the numerical answer is given in brackets at the end of the problem.

15. A velocity selector for accelerated sodium ions can be made by passing the ions through the combination of an electrostatic and a magnetic field. Can you de-

sign such an instrument? For what velocity of Na^+ is your instrument designed?

16. A proton of mass 1.67×10^{-27} kg moves with a constant velocity in the earth's magnetic field at the equator of 1.00×10^{-4} Wb/m². Find the magnitude and direction of the proton's velocity that will cause the magnetic force to just cancel the weight of the proton mg. ($g = 9.80$ m/sec²) [1.02×10^{-3} m/sec east]

17. In a DC motor the magnetic field is shaped such that the long side (12 cm) of the armature winding is always moving perpendicularly to the 0.50 Wb/m² field. What torque will the motor deliver if the 500 turn armature carries a current of 5.0 A and has a radius of 5.0 cm? If the armature revolves at 1800 rpm, what should be the output rating of the motor? [15 N-m, 2800 watts]

18. A long, straight conductor which has a linear density of 5.5×10^{-3} kg/m is supported in mid-air by a magnetic field. If the current in the conductor is 15 A, what is the strength of the magnetic field? [3.6×10^{-3} Wb/m²]

19. A 1-m length of wire with a resistance of 12 Ω is subjected to a voltage of 120 V and experiences an upward force of 2 N. What is the magnitude and direction of the magnetic field in the vicinity of the wire? (Draw a sketch.) [$B = 0.2$ Wb/m²]

20. Two long, parallel wires are originally 2 cm apart with a current of 20 A in the same direction in each. What is the original force of attraction between these? Plot a curve for the force of attraction as a function of distance between the wires. [4×10^{-3} N/m]

21. A long, iron-cored solenoid which has 5.00×10^{3} turns per meter carries a current of 5.00 A. The iron has a permeability of 3000.
    a. What is the value of the magnetic field at the center of the solenoid? $\mu_0 = 4\pi \times 10^{-7}$ Wb/A-m
    b. What happens to the magnetic field if the iron core is removed? What is its final value? [a. 94.2 Wb/m²; b. 3.14×10^{-2} Wb/m²]

22. a. If the electrons in a TV tube are accelerated through 8.20 kV, what is their velocity? $m_e = 9.1 \times 10^{-31}$ kg.
    b. If these electrons travel 20 cm in a southerly direction in the earth's magnetic field, what happens to them?
    c. Calculate the magnitude of the deflection if the vertical component of the earth's magnetic field is 4.55×10^{-5} in SI units. [a. 5.36×10^{7} m/sec; b. deflected to west; c. 3.9×10^{-16} N, $d = 29.8 \times 10^{-4}$ m]

23. Make a sketch of the previous problem showing the motion of electrons in a TV tube. Show the magnetic field direction, the electron direction, the direction of positive current flow, and the directions of any forces on the electrons.

24. At the magnetic equator of the earth the field is north-south with only a horizontal component equal to 5.6×10^{-5} in the SI system. It is possible for a current carrying wire of linear density 0.028 kg/m to be prevented from falling. How? What must the direction of the wire be? What current must the wire carry? (*Hint*: take $g = 9.8$ m/sec² at the magnetic equator.) [4.9×10^{3} A, E $-$ W E]

25. A blood flow meter is made to operate on the principle of the Hall effect. The blood flow through a magnetic field of 2 WB/m² generates a transverse voltage of 600 microvolts across electrodes separated by 1 mm in the blood stream. Find the velocity of the blood perpendicular to the magnetic field. [0.3 m/sec]

26. Find the force on a stream of singly ionized particles (with speed of 4.0×10^{6} m/sec) that are moving parallel to a wire carrying a current of 10 A. The distance between the ion stream and the center of wire is 2.0 cm. [$F = 6.4 \times 10^{-17}$ N]

27. Find the magnitude and direction of the force on an electron moving radially outward from a wire carrying a current I. Assume the velocity of the electron is v at a radial distance r from the center of the wire. [$F = \mu_0 I e v / 2\pi r$]

28. A long (length l) thin (width b) rectangular loop carrying a current I is at a distance a from a long wire (in the plane of the loop) carrying current $2I$. Find the net force on the loop.

$$\left[\frac{\mu_0 2I^2 l}{2\pi} \left(\frac{1}{a} - \frac{1}{a+b} \right) \text{ toward the wire} \right]$$

GOALS

When you have mastered the content of this chapter, you will be able to achieve the following goals:

Definitions
Define each of the following terms, and use it in an operational definition:
Faraday's induction law   Lenz's law
self-inductance   henry

Applications of Induction
Explain the physical basis for the operation of each of the following:
AC generator
electromagnetic damping
search coil

Faraday's Law Problems
Solve problems involving Faraday's induction law.

Lenz's Law
Predict the correct directions for induced current flow in electromagnetic induction phenomena.

Inductors
Solve problems involving combinations of inductors.

PREREQUISITES

Before beginning this chapter, you should have achieved the goals of Chapter 22, Basic Electrical Measurements, and Chapter 23, Magnetism.

24

Electromagnetic Induction

24.1 Introduction

What is the physical basis for the static or noise produced in a radio when certain appliances are used in its vicinity? What is the principle supporting the operation of the ferrite rod antenna in a radio? How does the power company generate AC voltage? The answers to all of these questions involve electromagnetic induction. In this chapter you will study the physical phenomena associated with electromagnetic induction.

24.2 Induced EMF: Faraday's Law

Let us consider a simple experiment. The equipment needed consists of a coil of wire, a galvanometer and a bar magnet. The coil of wire is connected to the galvanometer. We observe the impulsive deflection of the galvanometer under a series of conditions. First, if we move the magnet towards the coil of wire, we see a deflection of the galvanometer. The deflection means that there is a current in the galvanometer coil circuit, hence an induced emf. Second, if we move the magnet away from the coil, we see another deflection of the galvanometer. We observe in these cases that the deflections are in opposite directions. Third, if we reverse the ends of the magnet and repeat the experiment described above, we observe that the galvanometer deflection for approach is opposite to that in the first case and likewise on taking the magnet away. What conclusions can we reach from this series of experiments? Now let us try a set of experiments in which the magnet remains fixed and the coil is moved toward and away from the magnet—first toward one end of the magnet and then toward the other end. We will find that when the coil and a given end of the magnet are approaching each other the direction of deflection is the same regardless of which is moving. Also the direction of deflection is the same when the coil and a given end are receding from each other, regardless of which is moving.

We can get additional information relative to the time involved in the process. If we move either the magnet or the coil faster, the deflection is increased. The results of these experiments indicate that an emf is induced in a coil of wire whenever there is a change in the magnetic flux interlinking it, and that there is a definite relationship between the direction of the field and the direction of motion and between the magnitude of the induced emf and how rapidly we change the magnetic flux. On the basis of experiments such as these, Michael Faraday, one of the greatest experimental investigators in the history of science, formulated his now famous law for electromagnetic induction. Faraday envisioned the magnetic field associated with a magnet to be ethereal tentacles that grasp the iron filings and pull them toward the magnet. His induction law expressed the following idea: When the number of tentacles cutting through the area capping a circuit changes in time, there will be an

FIGURE 24.1

The magnetic flux through a circuit in a uniform magnetic field equals the area of the circuit multiplied by the component of the field that is perpendicular to the plane of the circuit.

induced electric field in the circuit and thus an induced voltage and current. We can put this induction law in a useful quantitative form. We define the magnetic flux as the magnetic induction field magnitude B, times the area of the circuit that is perpendicular to $\mathbf{B}$ (Figure 24.1). The magnetic flux ϕ can be written as,

$$\phi = BA \cos \theta \tag{24.1}$$

where θ is the angle between the magnetic induction $\mathbf{B}$ and the line perpendicular to the area A. Faraday's law states that whenever the magnetic flux through an area changes, there will be an emf induced in the perimeter of the area (or equivalently there will be an electric field induced at the perimeter of the area). We can express this in the following equation:

$$\mathscr{E} \text{ (volts)} = -\frac{\Delta\phi}{\Delta t} \text{ (webers/second)} \tag{24.2}$$

where $\mathscr{E}$ is the electromotive force in volts and $\Delta\phi$ is the change of magnetic flux in webers in a change of time Δt in seconds.

The negative sign in the equation is due to Lenz's law which states that the induced voltage is in such a direction as to oppose its cause; that is, the magnetic field due to the induced current *opposes the changing magnetic flux*. The induced current has an associated magnetic flux which tries to maintain the status quo. This is another example of inertia in a physical system. Various examples of the direction of induced emf are shown in Figure 24.2. If B increases into the paper, the field produced by the induced current would oppose this increase. Hence the current would be in a counterclockwise direction (Figure 24.2a). If B

FIGURE 24.2

The induced current flows in a direction such that its magnetic field opposes the flux change that produces it. In (a) an increasing flux into the paper produces a counterclockwise induced current, which has a magnetic field out of the paper. Apply this principle (Lenz's law) to verify the induced current directions in (b), (c), (d), and (e).

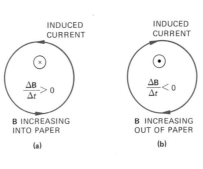

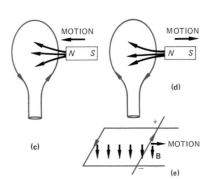

increases in a direction out of the paper the direction of the induced current would oppose this; it would be in a clockwise direction (Figure 24.2b). If the north pole of a bar magnet moves into the coil, the induced current would be in a direction to oppose this motion as is indicated in the diagram (Figure 24.2c). If the motion is in the opposite direction, the induced current will also be in the opposite direction (Figure 24.2d). When the area increases, the induced current is such that its **B** field opposes the applied **B** field passing through the increasing area (Figure 24.2e).

24.3 Applications of Faraday's Law

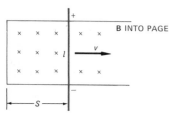

FIGURE 24.3
A simple generator. When the rod is moved to the right as shown an emf will be induced across its length equal to Blv. The top end of the rod will be positive.

There are many different ways to produce induced currents. Faraday's law indicates that a changing field **B**, a changing area, or a changing angle between **B** and the circuit can each produce an induced emf. We now consider some specific quantitative examples.

Consider a U-shaped conductor in a plane perpendicular to a magnetic field with induction **B**. A rod of length l is moving to the right with a velocity v (Figure 24.3). According to Faraday's law:

$$\mathscr{E} = -\frac{\Delta\phi}{\Delta t} = \frac{Bls}{t} = Blv \qquad (24.3)$$

where B is the magnitude of the magnetic induction in tesla, l is the length of the rod, s is the distance in meters through which the rod moves in t seconds, and v is the magnitude of the velocity of the rod.

EXAMPLE

Find the induced emf in a 0.50-m rod moving with velocity of 2 m/sec perpendicular to a field of 2×10^{-4} Wb/m², or tesla. Then,

$$\mathscr{E} = 2 \times 10^{-4} \times 2 \times 0.50 = 2 \times 10^{-4} \text{ V}$$

Suppose a coil of wire of 1-cm radius is removed from a field of 0.1 Wb/m² in 0.01 sec. The induced emf in the coil will be:

$$\mathscr{E} = \frac{-0.1 \text{ Wb/m}^2 \times \pi(.01)^2\text{m}^2}{10^{-2} \text{ sec}} = -\pi \times 10^{-3}\text{V} = -\pi \text{ mV}$$

If the coil has 100 turns of wire, the flux is increased by a factor of 100: $\phi = NBA = 100\,BA$. Such a coil can be used as a search coil to measure an unknown magnetic field (Figure 24.4).

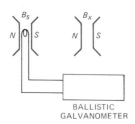

BALLISTIC
GALVANOMETER

FIGURE 24.4
A search coil and a ballistic galvanometer can be used to measure magnetic fields. Removal of the coil from magnetic field B_s produces a deflection, θ_s, that is proportional to the flux change NAB_s, where N is the number of turns in the coil and A is the coil area. When the coil is removed from the unknown field, B_x, the deflection is proportional to the flux change NAB_x. Thus $B_x = (\theta_x/\theta_s)\,B_s$.

The key to this measurement is the ballistic galvanometer which measures total charge flowing through it. The angular deflection θ_x of the galvanometer pointer is a constant times the total charge passing through it, Q. The search coil is removed from an unknown field producing an angular deflection, θ_x,

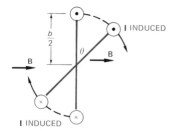

FIGURE 24.5
End view of a single rectangular coil rotating in a constant magnetic field.

$$\theta_x = kQ_x = k(I_x\Delta t) = k\frac{(\mathscr{E}_x)\Delta t}{R} = k\frac{(A\ \Delta B_x)\ N}{R}$$

where R is the resistance of coil, A is the area of coil, k is the galvanometer sensitivity constant, N is the number of turns in the coil, and B_x is the unknown magnetic field. The search coil is then removed from a standard field B_s giving a deflection of θ_s,

$$\theta_s = k\frac{A\ \Delta B_s\ N}{R} \tag{24.4}$$

In both cases the change in B is from maximum field to zero, $\Delta B_x = B_x$ and $\Delta B_s = B_s$. Thus we have the unknown field B_x in terms of known quantities,

$$B_x = \frac{\theta_x}{\theta_s}\ B_s \tag{24.5}$$

(What would be the effect of increasing the number of turns in the search coil in this measurement?)

Another important application of electromagnetic induction is a rotating coil in a magnetic field. Suppose that we have a rectangular coil free to rotate about a horizontal axis which is perpendicular to the magnetic induction B in horizontal direction. Thus the coil whose area is length l times width b (Figure 24.5) is rotating in a uniform field so that the lower side is coming up and the top side is going down for the position shown. The direction of the current in the coil is shown in this position. In fact, in this position each length side of the coil is moving perpendicularly to the magnetic flux and the induced emf is maximum. When the coil is turned through 90° each side will be moving in a direction parallel to the magnetic flux and the induced emf will be zero. If we measure the angle θ that the plane of the coil makes with a plane perpendicular to the field, the velocity of the coil wire across the field is $v\sin\theta$, where $\theta = \omega t$. The velocity of the coil is $2\pi rn$ where r is equal to one-half the width of the coil, n is the number of revolutions per second, and $2\pi n = \omega$, the angular velocity. Earlier we learned the induced emf by a conductor with length l moving perpendicularly to magnetic field is given by $\mathscr{E} = Blv$. For a rotating coil of one turn this reduces to

$$\mathscr{E} = 2\left(Bl2\pi\frac{b}{2}\ n\ \sin\omega t\right)$$

as the coil has two sides. (See Figure 24.6 for a side view of the coil.) This becomes N times as large if the coil has N turns. The product of lb is the area A of the coil and $2\pi n$ is the angular velocity. Then the induced emf in a rotating coil is given by

$$\mathscr{E} = BAN\omega\ \sin\omega t \tag{24.6}$$

$$\mathscr{E}_{max} = BAN\omega \tag{24.7}$$

So

$$\mathscr{E} = \mathscr{E}_{max}\ \sin\omega t \tag{24.8}$$

FIGURE 24.6
Side view of a coil rotating in a
constant magnetic field with the
direction of the induced
current shown.

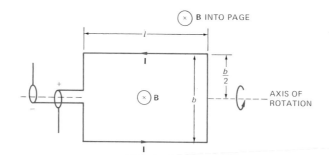

FIGURE 24.7
A rotating coil in a magnetic field
becomes an AC generator.

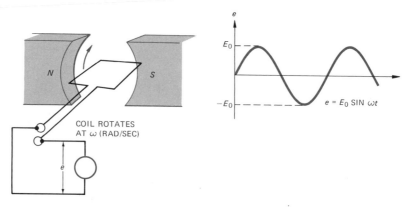

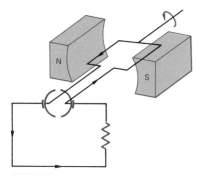

FIGURE 24.8
A single plane coil with commutator
contacts serves as a DC generator.

The induced emf of a rotating coil is shown in the Figure 24.7. Note that this emf is a sine curve and its direction changes. The electric current flowing in a conductor as a result of this emf is called an *alternating current*. This example gives the principle of operation of an AC generator. Each end of the coil is connected to a given slip ring which is connected by a brush to an external circuit. The alternating current has a frequency equal to the number of revolutions per second.

What can we do to make a DC generator? When the side of the coil is moving across the field in a given direction the direction of the induced emf is the same. So we need a device such that the side of the coil which is moving downward across the field is always connected to a given brush and the side of the coil which is moving upward across the field is always connected to another brush. This is done by replacing the slip rings by a *commutator*, which for the loop in one plane is nearly semicircular (Figure 24.8). A brush changes its contact from

FIGURE 24.9
Plot of the induced EMF for a single
loop DC generator.

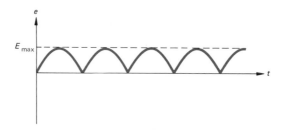

FIGURE 24.10
Plot of the induced EMF for two
perpendicular loops as a DC
generator.

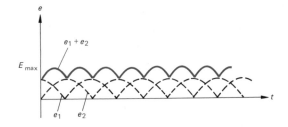

one side of the coil to the other when the side of the coil is moving
parallel to the magnetic field, that is, at zero induced emf. A plot of
induced emf as a function of time is shown in Figure 24.9, in which
$E_{max} = BAN\omega$. This is a pulsating unidirectional current called *direct
current* (DC).

By adding more planes of coils and more segments to the commutator
one can approach a constant direct current generator. For a four-segment
commutator, one with a two-plane armature, the representation of the
induced emf would appear as in Figure 24.10.

24.4 Electromagnetic Damping

Electromagnetic damping is used in analytical balances and galvanometers to reduce oscillations around equilibrium positions. In the case
of the balance, a thin conducting plate moves with the pans. This plate
moves between the poles of a permanent magnet. As the plate moves
into the magnetic field, induced currents are set up in the plate. These
currents provide an induced magnetic field that opposes the permanent
magnetic field, and thus a repelling force is acting on the plate. As the

FIGURE 24.11
Electromagnetic damping results
when the magnetic field of the
induced current opposes the external magnetic field.

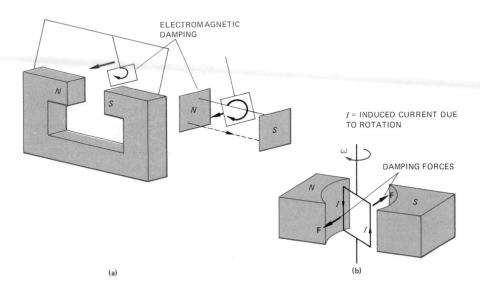

plate leaves the permanent magnet, its induced currents change direction and give rise to an induced field that is now attracted by the permanent magnet. In both entering and leaving the permanent magnet field the motion is damped by the induced field interaction (Figure 24.11a).

In the case of the moving-coil galvanometer the motion of the coil through the permanent magnet field induces a current through the coil. This induced current in turn interacts with the permanent magnetic field and produces a torque that opposes the rotation of the coil (Figure 24.11b).

24.5 Self-Inductance

A single coil of many turns has inductive properties due to the magnetic interaction of adjacent loops in the coil. If the loops are wound in the same direction the total effect of this interaction is an induced back emf that opposes the initial current change. This *self-inductance* is a form of inertia resisting current change in the coil. The self-induced emf $\mathscr{E}$ can be expressed in equation form as follows:

$$\mathscr{E} = -L\frac{\Delta i}{\Delta t} \tag{24.9}$$

where L is the self-inductance and $\Delta i/\Delta t$ is the current change per unit time in the coil. If $\mathscr{E}$ is measured in volts and $\Delta i/\Delta t$ in amperes per second, the inductance is given in henries, 1 H = 1 V–sec/A. The negative sign is consistent with Lenz's law and conservation of energy.

EXAMPLE

If the current changes in an air core inductor having a self-inductance of 25×10^{-5} H from zero to 1.0 A in 0.10 sec, what is the magnitude and direction of the self-induced emf?

$$\mathscr{E} = -L\frac{\Delta i}{\Delta t} = (-25 \times 10^{-5})\frac{1}{0.10} = -2.5 \times 10^{-3} \text{ V}$$

As the current is increasing, the direction of the induced emf opposes the growing current; hence, the negative sign. The effect of an inductor in a circuit is to oppose any change in current. Thus, inductance is analogous to inertia in a mechanical system.

To minimize self-inductance in wire-wound resistors, the loops are wound back on themselves. This noninductive winding tends to cancel out the inductive interactions of the loops.

There are consequences of self-inductance that are useful to know. Electric motors have large coils as part of their armature. When a motor is turned on, the changing current in these coils produces a back emf which tends to limit the current build up. Once operating speed is reached in the motor, the induced back emf limits the current to a value less than would be expected based on the resistance of the coil. When working with an electromagnet, it is good practice to reduce the current to zero before turning off the power supply to the magnet. The coils of an electromagnet have very high self-inductance; if the switch is opened with charge still flowing, the sudden collapse of the magnetic field will generate very large voltages. These large induced voltages can arc through the insulation in the windings or arc across the switch causing considerable damage.

24.6 Inductors

An inductor is a device that stores energy in the magnetic field set up by the current through a coil. By analogy the inductor plays the same role for magnetic field energy that the capacitor does for the electric field energy. We can derive an equation for the energy stored in an inductor as follows: the work done in moving charge q against the back emf is $\mathscr{E}q$ or in equation form,

$$w = \mathscr{E}q = -L\frac{\Delta i}{\Delta t} \times i\Delta t$$

since $q = i\,\Delta t$. If we start with the current at zero and the current increases at a constant rate to the final value I, we can use the average value of $I/2$ to get the work done. We then get the following equation for the work done in building up the magnetic field:

$$w = LI\left(\frac{I}{2}\right) = \tfrac{1}{2}LI^2 = \text{energy stored in the field} \tag{24.10}$$

This represents the work done in building up a current of I in the self-inductor and establishing the magnetic field. It also represents the energy that is stored in the magnetic field and that becomes available if the current is reduced to zero. This is shown by the flash on opening a switch in a DC circuit in which there is an inductor.

EXAMPLE

The back emf induced in a coil when the current changes from 100 mA to zero in 1 msec is 1 V. Find the self-inductance of the coil and the energy dissipated when the current goes to zero.

$$L = \frac{\mathscr{E}_b}{\Delta i/\Delta t} = \frac{1\text{ V}}{100\text{ A/sec}} = 10 \text{ millihenries (mH)}$$

$$w = \tfrac{1}{2}LI^2 = \tfrac{1}{2} \times 10 \times 10^{-3} \times 0.1^2 = 0.5 \times 10^{-4} \text{ J}$$

24.7 *Combinations of Inductors*

There are some circumstances in which we may find it useful to connect inductors together into some configuration. As with resistors and capacitors we can connect them together in series (Figure 24.12a), or in parallel (Figure 24.12b). Let us derive expressions for the equivalent inductance for both of these situations.

When the inductors are connected in series (Figure 24.12a) the current is the same through all of the inductors and the potential is the sum of the individual emfs,

$$V = \mathcal{E}_1 + \mathcal{E}_2 + \mathcal{E}_2$$

$$V = -L_1 \frac{\Delta i_1}{\Delta t} - L_2 \frac{\Delta i_2}{\Delta t} - L_3 \frac{\Delta i_3}{\Delta t} = -L_{\text{eff}} \frac{\Delta I}{\Delta t}$$

where L_1, L_2, and L_3 are the inductances and where ΔI is the change in the current in the system in a time Δt. For a series configuration, all of the currents are the same:

$$\Delta i_1 = \Delta i_2 = \Delta i_3 = \Delta I$$

$$-V = (+L_1 + L_2 + L_3) \frac{\Delta I}{\Delta t} = L_{\text{eff}} \frac{\Delta I}{\Delta t}$$

$$L_1 + L_2 + L_3 = L_{\text{eff}} \tag{24.11}$$

When inductors are connected in series, their effective inductance is the sum of the individual inductances.

When the inductors are connected in parallel, the potential across all of the inductors is the same, but the current through the system is divided into components so that the total current is the sum of the individual components, $I = \Delta i_1 + \Delta i_2 + \Delta i_3$,

$$V = -L_1 \frac{\Delta i_1}{\Delta t} = -L_2 \frac{\Delta i_2}{\Delta t} = -L_3 \frac{\Delta i_3}{\Delta t} = -L_{\text{eff}} \frac{\Delta I}{\Delta t}$$

FIGURE 24.12
(a) Inductances in series, and (b) inductances in parallel.

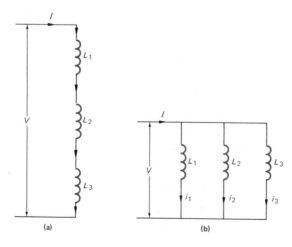

(a)   (b)

Then solving for, $\Delta i_1/\Delta t$, $\Delta i_2/\Delta t$, and $\Delta i_3/\Delta t$ we have

$$\frac{\Delta I}{\Delta t} = \frac{\Delta i_1}{\Delta t} + \frac{\Delta i_2}{\Delta t} + \frac{\Delta i_3}{\Delta t}$$

$$= \frac{L_{\text{eff}}}{L_1}\frac{\Delta I}{\Delta t} + \frac{L_{\text{eff}}}{L_2}\frac{\Delta I}{\Delta t} + \frac{L_{\text{eff}}}{L_3}\frac{\Delta I}{\Delta I}$$

Thus,

$$\frac{1}{L_{\text{eff}}} = \frac{1}{L_1} + \frac{1}{L_2} + \frac{1}{L_3} \tag{24.12}$$

When inductors are connected in parallel, their effective inductance is the reciprocal of the sum of the reciprocals of the individual inductance.

EXAMPLE

Inductors of 3 H, 5 H, and 10 H are connected

a. in series
b. in parallel
c. with the 3-H inductor in series with a parallel connection of the 5-H and 10-H inductors. What is the effective inductance of each configuration?

a. *Series*

$$L_{\text{eff}} = 3\text{ H} + 5\text{ H} + 10\text{ H} = 18\text{ H}$$

b. *Parallel*

$$\frac{1}{L_{\text{eff}}} = \frac{1}{3\text{ H}} + \frac{1}{5\text{ H}} + \frac{1}{10\text{ H}} = \frac{10}{30} + \frac{6}{30} + \frac{3}{30} = \frac{19}{30}$$

$$L_{\text{eff}} = \frac{30}{19} \approx 1.6\text{ H}$$

c. *Combination*

$$L_{\text{eff}} = 3\text{ H} + \left(\frac{5\text{ H} \times 10\text{ H}}{5\text{ H} + 10\text{ H}}\right) = \frac{95}{15} \approx 6.3\text{ H}$$

24.8 *Applications of Electromagnetic Inductance*

In summary, we find electromagnetic induction phenomena when there is a changing magnetic flux. Static on a radio results from the changing fields produced by appliances or lightning. The ferrite used in the antenna of a radio serves to increase the flux through the antenna coil. Self-induction is a form of electrical inertia against changing current in coils, and thus is an important factor when studying alternating-current circuits. *Mutual inductance* measures the coupling between circuits that are not physically in contact. An AC transformer is a device that operates on the principle of mutual inductance. As you look around you, how many examples of electromagnetic induction do you see?

ENRICHMENT
24.9 Conservation of Energy

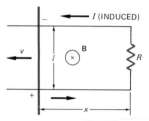

FIGURE 24.13
Simple moving-rod generator. The power expended in moving the rod at a constant speed v equals the electric power generated.

We consider the simple generator shown in Figure 24.13. A conducting rod slides without friction at a constant velocity over a conducting track. The entire circuit formed by the track and rod is in a constant magnetic field **B** (into the paper). The total resistance of the circuit is R (Ω). We wish to determine the emf induced across the rod due to the flux change through the circuit, and we want to compare the electrical power supplied to the circuit with the mechanical power needed to keep the rod moving at a constant speed.

The induced emf is given by Faraday's law as follows:

$$\mathcal{E} = -\frac{\Delta\phi}{\Delta t} = -\frac{d\int B\,dA}{dt} = -\frac{d(B\int dA)}{dt} = -\frac{d(Blx)}{dt}$$

where $\int dA = lx$, since $v = dx/dt$ (velocity of the rod). The induced current in the circuit is given by

$$I = \frac{\mathcal{E}}{R} = \frac{Blv}{R}$$

and the power supplied is

$$P = I^2 R$$

$$P = \left(\frac{Blv}{R}\right)^2 R = \frac{(Blv)^2}{R}$$

The power needed to keep the rod moving at a constant speed can be determined as follows: The force on the rod due to the induced current is given by $F = BIL$ (in opposite direction to v). In order to move the rod at a constant speed an equal force must be applied in the direction of v. The power supplied by this mechanical force must be

$$P = \frac{dw}{dt} = \frac{F\,dx}{dt} = Fv$$

$$= BIlv = Blv\left(\frac{\mathcal{E}}{R}\right) = \frac{(Blv)^2}{R}$$

The power supplied to move the rod at constant speed is equal to the electrical power supplied to the circuit. This should reaffirm your faith in the conservation of energy! (Note that we have neglected friction. How would friction affect this problem?)

ENRICHMENT
24.10 Calculus Derivation for Current through an Inductor

Many of the examples cited above can be treated with more rigor using calculus methods. Faraday's law becomes

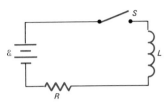

FIGURE 24.14
A resistor and inductor in series with a battery.

$$\mathscr{E} = -\frac{d\phi}{dt} \tag{24.13}$$

In many cases $\phi = f(t)$ so

$$\mathscr{E} = -\frac{d}{dt} f(t)$$

Also, $d\phi = B \, dA$:

$$\frac{d\phi}{dt} = B \frac{dA}{dt}$$

for constant field.

For self-inductance,

$$\mathscr{E} = -L \frac{di}{dt} \tag{24.14}$$

Energy in a magnetic field,

$$E = -\Sigma \, L \frac{di}{dt} i \, dt = \int_0^I Li \, di = \tfrac{1}{2}LI^2 \tag{24.15}$$

Let us consider a DC circuit with a resistor R and an inductor L in series with an emf $\mathscr{E}$ as shown in Figure 24.14. Ohm's law becomes upon closing switch,

$$\mathscr{E} - L \frac{di}{dt} = iR \tag{24.16}$$

in which $\mathscr{E}$ is the emf of the battery, $L(di/dt)$ is the emf induced in the inductor and is in the opposite sense to $\mathscr{E}$, and iR is the potential drop resulting from the resistance in the circuit. In this equation, i is the instantaneous current and is equal to zero at $t = 0$ and becomes equal to I or $\mathscr{E}/R$ for steady state. The instantaneous current at any time is

$$i = \frac{\mathscr{E}}{R} (1 - e^{-Rt/L}) \tag{24.17}$$

Show that this is a solution of the equation above by either of the following methods:

a. Showing that this value of i satisfies Equation 24.16.
b. By direct solution of the differential equation.

Notice that the ratio L/R is a measure of the time response of this circuit. It is called the time constant of this circuit and is the time in seconds required for the current to reach 63 percent $(1 - e^{-1})$ of its equilibrium value.

EXAMPLE

Given $L = 10$ H and $R = 693$ Ω, find the time it takes for the current to reach half its equilibrium value.

Substituting the given values in Equation 24.17, $i = \mathscr{E}/2R$, since $\mathscr{E}/R$ is equilibrium current.

$$i = \frac{\mathscr{E}}{2R} = \frac{\mathscr{E}}{R}[1 - \exp(-693t/10)]$$

$$\tfrac{1}{2} = 1 - \exp(-693t/10)$$

$$\exp(-693t/10) = \tfrac{1}{2}$$

$(693/10)\,t$ is the exponent to which e, the base of the natural logarithms, must be raised to get 2.

Since $e^{.693} = 2$,

$$\frac{693}{10}t = 0.693$$

$$t = 0.01 \text{ sec}$$

ENRICHMENT
24.11 Calculus Derivation for Current through a Capacitor

In Chapter 21 we considered the properties of capacitors. A capacitor which consists of an insulator between two conducting plates will not support a steady flow of electric charge. However, during the charging and discharging processes charges will flow in a circuit that includes a capacitor. Let us consider a circuit with a resistor R (Ω), in series with a capacitor C (farads), and an emf $\mathscr{E}$. Ohm's law becomes, upon the closing of the switch.

$$\mathscr{E} = iR + \frac{q}{C} \tag{24.18}$$

where iR is the potential difference across the resistor and q/C is the potential difference across the capacitor.

$$i = \frac{dq}{dt}$$

$$\mathscr{E} = R\frac{dq}{dt} + \frac{q}{C} \tag{24.19}$$

The instantaneous charge on the capacitor at any time t is

$$q = \mathscr{E}C(1 - e^{-t/RC}) \tag{24.20}$$

where the charge is zero at $t = 0$. The current in the circuit as a function of time is found to be

$$i = \frac{dq}{dt} = \frac{\mathscr{E}}{R}e^{-t/RC} \tag{24.21}$$

Notice that the product RC is a measure of the time response of the circuit. It is called the time constant of the circuit and is the time in seconds required for the current to reach $0.37(e^{-1})$ times its equilibrium value.

EXAMPLE

Given $C = 10\mu\text{F}$ and $R = 1.44$ MΩ, find the time it takes for the current to reach one-half its equilibrium value. Substitute the given values in Equation 22.33.

$$\frac{\mathscr{E}}{2R} = \frac{\mathscr{E}}{R} e^{-t/1.44}$$

$$e^{-t/1.44} = 1/2 = e^{-.693}$$

$$.0693t = .693$$
$$t = 10 \text{ sec}$$

SUMMARY

Use these questions to evaluate how well you have achieved the goals of this chapter. The answers to these questions are given at the end of this summary with the number of the section where you can find related content material.

Definitions

1. The magnetic flux can be found by multiplying the _____ by the _____.
2. List the properties of the magnetic flux that can influence the emf induced in a wire in the region where the flux exists.
3. The direction of the induced emf is determined by _____ law which is another statement of the _____ properties of an electromagnetic system.
4. The _____ law equates the magnitude of the _____ to the _____ rate of change of the _____.
5. In a coiled wire the neighboring coils are linked by _____ when a current is flowing in the coil, and if the _____ changes with _____, a back emf will be _____ giving rise to the effect that is called _____.
6. The henry is the unit of measurement for _____.

Applications of Induction

7. Without the aid of the text, explain
   a. the principles and operation of an AC generator
   b. the principles and operation of a DC generator
   c. the principles and operation of a search coil

Faraday's Law

8. The coefficient of self-induction of a coil is 0.015 henries (H). If the current changes from 0.20 A to 1.0 A in 0.010 sec, what is the emf of self-induction?
9. What is the maximum induced emf in a 400-turn coil of area 400 cm² revolving at 30 rps about an axis in the plane of coil and perpendicular to a field of 0.050 tesla (T)?

Lenz's Law

10. A flip coil is originally in a horizontal plane and is free to rotate about a north-south axis in the Northern Hemisphere. It is suddenly rotated through 90°. What is the direction of the self-induced emf in coil? Draw a sketch of the system, and show all related directions.

Inductors

11. If you have one 4-H and one 12-H inductor, what inductances can you achieve by using them together? What configurations would you use to obtain these inductances?

Answers

1. area, B component perpendicular to area $BA \cos \theta$ (Section 24.2)
2. direction of flux, change in flux, time used to change the flux, direction of the wire (Section 24.2)

3. Lenz's, inertial (Section 24.2)
4. Faraday induction, induced emf, time, magnetic flux (Section 24.2)
5. magnetic flux, current, time, induced,

self-inductance (Section 24.5)
6. inductance (Section 24.5)
7. a. (Section 24.3)
   b. (Section 24.3)
   c. (Section 24.3)
8. 1.2 V (Section 24.5)

9. 150 V (Section 24.3)
10. Sections 24.1 and 24.2)
11. series, 16 H; parallel, 3 H (Section 24.7)

ALGORITHMIC PROBLEMS

Listed below are the important equations from this chapter. The problems following the equations will help you learn to translate words into equations and to solve single-concept problems.

Equations

$$\mathscr{E} = -\frac{\Delta\phi}{\Delta t} \tag{24.2}$$

$$\mathscr{E} = Blv \tag{24.3}$$

$$B_x = \frac{\theta_x}{\theta_s} B_s \tag{24.5}$$

$$\mathscr{E} = \omega BAN \sin \omega t \tag{24.6}$$

$$\mathscr{E}_{max} = BAN\omega \tag{24.7}$$

$$\mathscr{E} = \mathscr{E}_{max} \sin \omega t \tag{24.8}$$

$$\mathscr{E} = -L\frac{\Delta i}{\Delta t} \tag{24.9}$$

$$w = \tfrac{1}{2}LI^2 \tag{24.10}$$

$$L_{eff} = L_1 + L_2 + L_3 \quad \text{(series)} \tag{24.11}$$

$$\frac{1}{L_{eff}} = \frac{1}{L_1} + \frac{1}{L_2} + \frac{1}{L_3} \quad \text{(parallel)} \tag{24.12}$$

Problems

1. What is the maximum flux through a coil that has an area of 400 cm² and a value of 0.500 T (weber/m²) for B?
2. The total flux within a coil of one turn is reduced from 0.0400 Wb to 0 in 0.0100 sec. What is the induced emf?
3. What is the direction of an induced current in an east-west conductor moving north horizontally in a vertical magnetic field pointing upward?
4. An emf of 2.00 V is induced in an inductor when the current is changing at the rate of 8.00 A/sec. What is the self-inductance of the inductor?
5. What is the energy required to establish a magnetic field by a 5.00 H inductor carrying a current of 10.0 A?

Answers

1. 0.020 Wb
2. 4.00 V
3. east

4. 0.250 H
5. 250 J

EXERCISES

These exercises are designed to help you apply the ideas of a section to physical situations. When appropriate, numerical answers are given in brackets at the end of the exercise.

Section 24.2

1. The north pole of a bar magnet is inserted in a flat coil (closed) of 20 turns resting on a table. If the flux changes from 10^{-4} Wb to 10^{-3} Wb in 0.25 sec, what is the induced emf? What is the direction of the induced current in the coil as you look down upon the coil — Clockwise or counterclockwise? [0.07 V, counterclockwise]

2. A search coil has 60 turns, each of an area of 5 cm². It is moved from a magnetic field of 0.5 Wb/m² to one of 0.0 Wb/m² in 0.15 msec. What is the induced emf? [100 V]

3. A coil with a radius of 2.00 cm and 20.0 turns is removed from a magnetic field of 0.300 T in 4.00 msec. Find the emf induced in the coil.   [1.88 V]

4. A coil of wire in the plane of the paper has a permanent magnet dropped (north pole down) through it. Draw a sketch showing the direction of the induced current in the coil when the north pole is just above the coil and when the north pole is just below the coil.

Section 24.3

5. Suppose a student decides that he wants to make a simple generator on his automobile by using a straight horizontal conductor moving in the earth's field. The conductor is 1.5 m long. How much emf will be induced between the ends of the conductor if he drives north at 25 m/sec, and the vertical intensity of the earth's field is 7.0×10^{-5} T. Which end is at the higher potential? Is this a practical generator?   [2.6×10^{-3} V, west end has higher potential]

6. A rectangular coil of 100 turns has dimensions of 10 cm by 15 cm. It rotates about an axis through the midpoint of the short sides. The axis of rotation is perpendicular to the direction of the magnetic field of strength 0.50 T, and it is rotating at 600 rpm.
   a. When is the emf induced in the coil a maximum?
   b. Is the induced emf ever zero? If so, when?
   c. What is the maximum induced emf?   [a. when $\phi = 0$ and $\Delta\phi/\Delta t$ is max; b. yes: when ϕ is max, and $\Delta\phi/\Delta t = 0$; c. 47 V]

7. If the coil in exercise 3 is rotated in the field at 100 rad/sec, find the maximum emf induced in the coil. [75 V]

Section 24.7

8. Suppose you have a collection of three inductors of 2.00 H, 4.00 H, and 8.00 H, how many different values of inductance can you make using all of these inductors in the same circuit?   [all series 14.00 H, all parallel 1.14 H, 2 H + 8 H in series in parallel with 4 H = 2.86 H, 2 H + 4 H in series in parallel with 8 H = 3.43 H, 4 H + 8 H in series in parallel with 2 H = 1.71 H, 2 H and 4 H in parallel in series with 8 H = 9.33 H, 8 H and 2 H in parallel in series with 4 H = 5.60 H, 8 H and 4 H in parallel in series with 2 H = 4.67 H]

PROBLEMS

These problems may involve more than one physical concept. Problems requiring the material from the enrichment section are marked with a dagger †. The answer is given in brackets at the end of each problem.

9. An experiment is carried out to determine the flux density between the poles of an electromagnet. A 50-turn coil with 20Ω resistance and enclosing an area of 2.0 cm² is placed perpendicularly to the field.

When it is quickly withdrawn from the field, a calibrated detector shows a charge of 8.0×10^{-4} C flowing in the circuit. What was the flux density? [1.6 T]

10. You are given 100 cm of #16 gauge copper wire (cross-sectional area 0.0214 cm²) with a resistance of 0.010 Ω. It is formed into a circular loop and placed in a plane perpendicular to a magnetic field which is changing at the rate of 1.0×10^{-2} Wb/m²/sec. How much energy is converted into heat per second? [6.33×10^{-5} J/sec]

11. An earth inductor is a large coil which rotates through 180°. The coil is connected to a ballistic galvanometer which gives a deflection of 15.0 cm when cutting the vertical component of the earth's magnetic field and a deflection of 6.0 cm when cutting the horizontal component. What is the angle of dip at this position? What deflection should you get if the coil is placed so that the earth's total field is cut, and how would you place the coil? [68°12′, 16.2 cm]

12. Given two identical coils, one directly above the other, answer the following questions for this system.

   a. A clockwise current is started in the top coil. What is the direction of the induced current in the bottom coil?

   b. A clockwise current is flowing in the bottom coil, and the top coil falls onto it. What is the direction of the induced current in the top coil?

   c. A clockwise current is flowing in the top coil, and the bottom coil is rotated through 90°. What is the direction of the induced current? (Give a sketch for each answer.) [a. counterclockwise; b. counterclockwise; c. clockwise]

13. An airplane with a wing span of L meters is flying south with a speed of v meters per second in the earth's magnetic field (vertical component is 5×10^{-5} T). Find the induced emf across the wing of the plane, and give its polarity. (*Hint:* the induced electric field force just balances the magnetic force on the charge carriers moving with the wing, i.e., $qE = qvB$) [$5 \times 10^{-5}\, Lv$ V]

†14. The flux within a circular coil of wire changes in accordance with $\phi = 2 + 3 \sin 120\pi t$ Wb. What is the general expression for the induced emf in the coil? What is the maximum induced emf? What can you say about the direction of the induced emf? [$360\pi \cos(120\pi t)$, 360π V]

†15. Given a long rectangular loop of length l and width w, derive an expression for the self-inductance per meter of the loop assuming $w \gg r$, where r is the radius of the wire. $\left[L = \dfrac{\mu_0}{\pi} \ln\left(\dfrac{w - r}{r} \right) \right]$

†16. A circuit is made up of a battery $\mathscr{E}$, switch, inductor L, and resistance R connected in series. What is the rate of increase of the current at any time t after the switch is closed. [$\mathscr{E}/L \exp(-Rt/L)$]

†17. The time constant of a circuit containing only resistance R and inductance L is equal to L/R and is defined at the time required to get within $1/e$ of final value. Find the time constant for the circuit in the example problem in Section 24.10. Plot curve for i versus t for closing the switch and also upon opening the switch. Interpret the meaning. [$\tau = 1/69.3$ sec]

†18. A circuit is made up of a battery E, capacitor C, and resistor R connected in series. What is the rate of decrease of the current at any time t after the switch is closed? [$(E/R^2 C)\, e^{-t/RC}$]

†19. Assume the circuit in problem 18 has had the switch closed for a time much greater than RC. What is the current? What is the charge of the capacitor? The switch is suddenly opened at a time $t = 0$, and at the same time the resistor is placed in series with the charged capacitor (possible with the proper switch) to form a closed circuit. What is: (i) the charge on the capacitor as a function of time? (ii) the current in the circuit at any time t? (iii) the total amount of energy dissipated in the resistor? [0, $\mathscr{E}C$, $\mathscr{E}Ce^{-t/RC}$, $(-\mathscr{E}/R)e^{-t/RC}$, $\mathscr{E}^2 C/2$]

GOALS

When you have mastered the contents of this chapter, you will be able to achieve the following goals:

Definitions
Define each of the following terms and use it in an operational definition:

effective values of current and voltage

reactance

impedance

power factor

resonance

Q-factor

AC Circuits
Solve alternating-current problems involving resistance, inductance, and capacitance in a series circuit.

Phasor Diagrams
Draw phasor diagrams for alternating current circuits.

Transformer
Explain the operation of the transformer.

AC Measurements
Describe the use of alternating currents in physiological measurements.

PREREQUISITES

Before you begin this chapter you should have achieved the goals of Chapter 22, Basic Electrical Measurements, and Chapter 24, Electromagnetic Induction.

25

Alternating Currents

25.1 Introduction

Most of your experience with electricity has probably been with alternating current (AC) circuits. Do you know the difference between AC and direct current (DC) electricity? At some time or another you have probably read the printing on the end of a light bulb. Usually a light bulb has printed on it that it is for use in a 120-volt AC circuit. What was the manufacturer trying to indicate? Could you use a 120-V DC bulb in your household sockets? What, if anything, would happen? Why is AC the dominant form of electrical energy in use? More than 90 percent of electrical energy is used as AC electricity.

If you owned an electric train in your childhood, you may recall that you used a transformer to reduce household voltage to the voltage required by the electric train. The transformer is a device that plays a most important role in the use of electrical energy. Do you know what its functions are in transmission of electric power? What is the role of the transformer in the coupling of two AC circuits? These are a few of the questions we will discuss in this chapter.

25.2 Nomenclature Used for Alternating Currents

In Chapter 24, we learned that a simple AC generator can be made by rotating a coil of conducting wire in a constant magnetic field. The voltage produced by such a generator has the form shown in Figure 25.1. This sinusoidal voltage *alternates* polarity + to − and produces an alternating current in a circuit connected to such an AC source. This alternating polarity is in contrast to the unidirectional nature of DC current.

In the United States the standard frequency in home and industrial use is 60 cycles/second (Hz). This means that there is a reversal of the direction of the current every 1/120 second. Radio broadcast frequencies are of the order of 10^6 Hz. Some microwave devices have frequencies of the order of 10^{10} Hz. So the AC current in common use is of relatively low frequency.

If you have two equal resistors and in one there is an AC current of 1 ampere maximum and in the other a DC current of one ampere, would you expect the two resistors to produce the same heat? Let us consider

FIGURE 25.1
Alternating emf from an AC generator.

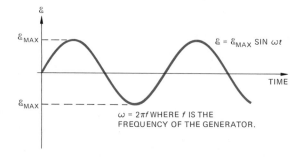

the situation to see what we should expect. In the resistor with the DC current, the current is constant. In the resistor with AC current, the current varies from 0 to 1 A in one direction and the same in the opposite direction. You have learned that the heat produced depends upon the square of current for a given resistor. Hence, you would have been correct if you had said that more heat is produced by the 1–A DC current. For a varying current we would expect the heat produced to be proportional to the average value of the square of the current. The square root of this quantity is called *effective current*. That is, it is the current that would produce the same heating effect as one ampere of DC current. Let us find the relationship between the effective AC current and maximum AC current. The AC current at any instant of time can be expressed as a sinusoidal function of time,

$$i = i_{max} \sin \omega t \tag{25.1}$$

where i_{max} is the maximum value of the current, ω is the angular frequency of the AC, and t is the time in seconds. The square of the current will be proportional to the power dissipated in the resistor; so

$$i^2 = i^2_{max} \sin^2 \omega t \tag{25.2}$$

We can draw a graph of the square of the current versus the time (Figure 25.2). The total heating effect will be proportional to the area under the curve. You see that the curve between π and 2π radians is exactly the same as that from 0 to π radians. So we will need to consider only half of the total cycle. The curve for the square of the DC current is a horizontal line represented by AC. The problem is then to find the altitude for the rectangle with the base of π which has the same area as under the $i^2_{max} \sin^2 \omega t$ curve. By inspection you might say that the area under AB is about equal to the area under the $i^2_{max} \sin^2 \omega t$ curve if $A_1 = A_2$ and $A_3 = A_4$. This is true if OC is equal to $i^2_{max}/2$. You can show this by measuring the two areas.

Let us use the symbol I for the effective current. We can see from Figure 25.2 that the effective current squared is equal to one-half of i_{max} squared,

$$I^2 = \tfrac{1}{2} i^2_{max} \tag{25.3}$$

FIGURE 25.2
Graph of $(i_{max} \sin \omega t)^2$ as a function of ωt. The mean value of the current squared is the DC value that produces equal joule heating as the sinusoidal current over one period.

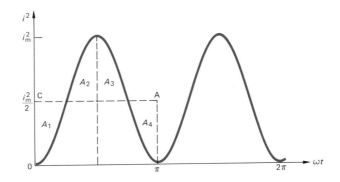

It follows, taking the square root of both sides of Equation 25.3 that the effective current is equal to the maximum AC current divided by the square root of two,

$$I = \frac{i_{max}}{\sqrt{2}} = 0.707 \ i_{max} \qquad (25.4)$$

The effective current is called the root mean square current. Similarly we will use $\mathscr{E}$ to represent the effective emf of an AC source and V to represent the effective voltage drop in an AC circuit:

$$\mathscr{E} = \mathscr{E}_{rms} = \frac{\mathscr{E}_{max}}{\sqrt{2}} = 0.707 \ \mathscr{E}_{max} \qquad (25.5)$$

$$V = V_{rms} = \frac{V_{max}}{\sqrt{2}} = 0.707 \ V_{max} \qquad (25.6)$$

The AC meters which you use measure the V_{rms} and I_{rms} values of an AC circuit. If the AC line in your home is said to be 110 V, that is the effective value of the voltage. The maximum or peak voltage would be $110 \times \sqrt{2}$ or 155 V.

25.3 Phase Relations of Current and Voltage in AC Circuits

In an AC circuit containing only resistance, the instantaneous voltage and current are always in phase. This means they are both zero at the same time and reach their maximum value at the same time. This is shown in Figure 25.3.

In Chapter 24 you learned that whenever the current is changing in an inductive coil, an emf is induced. This induced emf depends upon the induction of the coil and on the time rate of change of current. In a coil, continuously changing AC current produces an alternating emf from self-induction. According to Lenz's law this induced emf is opposite to the applied emf. If you neglect the resistance of the coil, the emf of the source will just be equal to the emf of self-induction in the coil. The induced emf will cause the current in the coil to lag behind the applied emf by one-quarter of a cycle. We say that the current in an inductor lags the voltage by 90°, or that the phase of the voltage across the inductor leads the phase of the current by 90°. See Figure 25.4.

FIGURE 25.3
The voltage $\mathscr{E}$ and current I are in phase in a pure resistance AC circuit as shown.

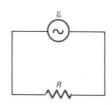

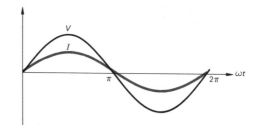

FIGURE 25.4
The voltage leads the current by 90°
in a pure inductance AC circuit as
shown.

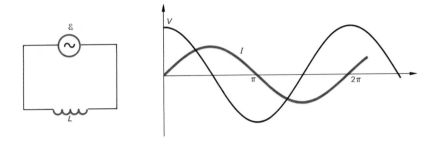

The inductance in the circuit not only causes the current to lag the emf but it also reduces the current to a smaller value than it would have if there were no inductance present. The voltage drop V across an inductance L is given by

$$V = I\omega L = 2\pi f L I \tag{25.7}$$

where f is the frequency in cycles per second and L is the inductance in the circuit in henries. It follows that the AC current in an inductance is given by the voltage drop across the inductance divided by $2\pi f$ times the inductance,

$$I = \frac{V}{2\pi f L} \tag{25.8}$$

The factor $2\pi f L$ is called the inductive reactance of the circuit, represented by X_L, and is measured in ohms if L is in henries and f is the frequency (in Hz),

$$X_L = 2\pi f L \tag{25.9}$$

An inductive element in an AC circuit acts as an inertia element and impedes the alternating current. We also note that this effect depends directly upon the frequency. Hence for high frequency an inductor exhibits a large inertia and thus greatly impedes a high-frequency alternating current.

If you connect a capacitor to an AC source the plates of the capacitor become charged alternately positive and negative, according to the surge of charges back and forth in the connecting wires. So, even though there can be no constant DC current through a capacitor, we can say that there can be an alternating current through a capacitor. A capacitor does present an impedance to an alternating current which is called *capacitive reactance*. We shall represent the capacitive reactance by X_C. In a way analogous to the definition of capacitance as the ratio of voltage to charge, we shall define the capacitive reactance as the ratio of the voltage drop across the capacitor to the current through the capacitor,

$$X_C = \frac{V}{I}, \text{ using rms values for voltage and current.} \tag{25.10}$$

FIGURE 25.5

The voltage lags the current by 90° in a pure capacitance AC circuit as shown.

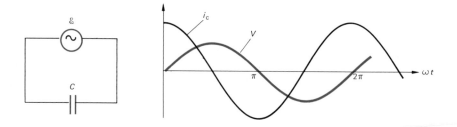

where

$$X_C = \frac{1}{2\pi f C} \text{ ohms} \tag{25.11}$$

and C is the capacitance in farads and f is the frequency in cycles per second (Hz). Note that the capacitive reactance decreases as the frequency increases. A capacitor has infinite impedance for DC sources. The impedance of a capacitor in an AC circuit decreases as the frequency increases.

In a circuit with only a capacitor the current leads the voltage by one-quarter of a cycle (see Figure 25.5). We say that phase of the voltage across a capacitor lags the current by 90°.

25.4 AC Series Circuits

Let us consider a series circuit including a resistor and an inductor (see Figure 25.6). The source is 120 V, 60 cycle. Suppose you find that an AC voltmeter connected between the points N and O reads 90.0 V, connected between O and P reads 79.4 V and connected between N and P it reads 120 V. You see that the voltage drop across NO added to the voltage drop across OP is greater than 120 V. Is this possible? Can the sum of two voltage drops across series elements be greater than the applied voltage? The answer is yes.

As we indicated in the previous section, the voltage drops across and the currents in various AC circuit elements may not be in phase with each other. To correctly add the voltage drops across AC circuit elements, these phase differences must be taken into account. To enable you to perform these additions correctly, we will introduce you to the graphical technique called *phasor diagrams*. The currents and voltages will be represented by vectors in the diagrams. In a series circuit the current is the same in all circuit elements; so we will choose the phasor of the current to be the x-axis (see Figure 25.7). Since the voltage drop across a resistor is in phase with the current through it, the voltage drop across a resistor, V_R, in magnitude equal to IR, will also be a phasor along the x-axis. The voltage drop across an inductor leads the current by 90° so the voltage drop across the inductor V_L in magnitude equal to IX_L or $I\omega L$, is shown as a phasor at 90° to the x-axis and pointing in the

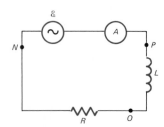

FIGURE 25.6

A resistor and inductor in series with an AC generator.

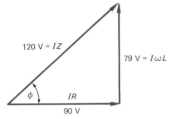

FIGURE 25.7
A phasor diagram for the circuit shown in Figure 25.6. The voltage leads the current for the resistance-inductance series circuit.

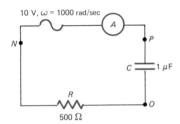

FIGURE 25.8
A resistor and capacitor in series with an AC generator.

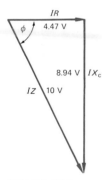

FIGURE 25.9
The phasor diagram for the resistance and capacitance circuit of Figure 25.8. The voltage lags the current in all resistance-capacitance series circuits.

positive y-direction. These two voltages can then be added using the rules of vector addition. The total applied voltage is represented by the hypotenuse of the right triangle formed by V_R and V_L. To preserve the simple structure of Ohm's law for use with AC circuits, we will define a quantity called the *total impedance* of the AC circuit, Z. The impedance Z is given by the number that must be multiplied by the current I to obtain the magnitude of the applied voltage. The impedance also has a phase angle, namely the angle between the current and the applied voltage. Remembering the form of multiplication for AC circuits involves magnitude and phase angle, we can write an AC circuit form of Ohm's law,

$$\mathscr{E} = IZ \tag{25.12}$$

where $\mathscr{E}$ is the applied emf, I is the current, and Z is the impedance. For a series circuit containing a resistor and an inductor we can show the Z is calculated by the following equation:

$$Z = \sqrt{R^2 + X_L^2} = \sqrt{R^2 + (\omega L)^2} \tag{25.13}$$

The phase angle ϕ is given by the angle whose tangent is the ratio of X_L to R,

$$\tan \phi = \frac{X_L}{R} = \frac{\omega L}{R} \tag{25.14}$$

For a series circuit the ratio X_L/R is the same as the ratio V_L/V_R. From the Figure 25.7, we can calculate the phase angle for that example,

$$\tan \phi = \frac{X_L}{R} = \frac{V_L}{V_R} = \frac{79.4}{90.0} = 0.882 \tag{25.15}$$

so $\phi = 41.42°$. For this example the phase angle represents the angle by which the applied voltage leads the current.

Let us now consider an AC circuit that contains only a resistor and a capacitor (see Figure 25.8). Consider a 10-V source with $\omega = 1000$ rad/sec in series with $1 \mu F$ capacitor and a 500Ω resistor. If you measure the voltage across the resistor (NO) and find it to be 4.47 V and across the capacitor OP and find it to be 8.94 V, how can you explain it? The two voltage drops seems to have a total greater than the 10 V of applied emf. We now return to our phasor diagram. Since we know that the voltage across a capacitor *lags* behind the current through it, we will indicate the voltage drop across the capacitor V_C, equal in magnitude to IX_C, by a phasor at 90° to the current through the resistor and to V_R pointing in the negative y-direction (see Figure 25.9). As before the total applied emf will be represented by the hypotenuse of the triangle but this time by one whose sides are V_R and V_C. We can again define an impedance and phase angle for the circuit,

$$Z = \sqrt{R^2 + X_C^2} = \sqrt{R^2 + \left(\frac{1}{\omega C}\right)^2} \tag{25.16}$$

$$\tan \phi = \frac{-X_C}{R} = -\frac{1}{\omega RC} = -\frac{V_C}{V_R} \tag{25.17}$$

where the negative phase angle indicates that the voltage lags behind the current. For the circuit shown in Figure 25.8, we obtain the following results:

$$V_R = 500I = 4.47 \text{ V}$$

$$X_C = \frac{1}{\omega C} = \frac{1}{1000 \times 1 \times 10^{-6}} = 1000 \ \Omega$$

$$V_C = IX_C = 8.94 \text{ V}$$

$$\tan \phi = \frac{-8.94}{4.47} = -2.00 \qquad \phi = -63.4°$$

which indicates that the applied 10 V lags behind the current by 63.4°.

Finally, let us consider a circuit consisting of a resistor, capacitor, and inductor in series with an AC generator (see Figure 25.10a). The current in such a circuit determines the total impedance of the circuit. From our discussion of inductance and capacitance, we know that these elements exhibit voltages 180° out of phase with each other, shown by vectors pointing in opposite y-directions on a phasor diagram as shown in Figure 25.10b. For this circuit we can find the applied voltage using the same procedures we used in the above two cases,

$$V_{\text{applied}} = \sqrt{V_R{}^2 + (V_L - V_C)^2} \tag{25.18}$$

We find that the impedance and phase angle are defined as before

$$Z = \sqrt{R^2 + (X_L - X_C)^2} = \sqrt{R^2 + \left(\omega L - \frac{1}{\omega C}\right)^2}, \tag{25.19}$$

$$\tan \phi = \frac{X_L - X_C}{R} = \frac{\omega L - (1/\omega C)}{R} \tag{25.20}$$

The power dissipated in the circuit is accounted for entirely by the resistor in the circuit. Ideal inductors and capacitors only store and ex-

FIGURE 25.10
An AC generator connected in series with a resistor, inductor and capacitor. The phasor diagram for the circuit illustrates the resultant phase angle between the voltage and current in the circuit.

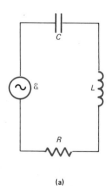

(a)

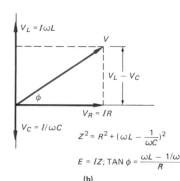

(b)

FIGURE 25.11
The series RLC AC circuit has an effective AC impedance that determines the magnitude of the current in the circuit.

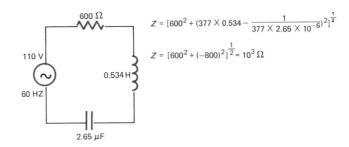

$$Z = [600^2 + (377 \times 0.534 - \frac{1}{377 \times 2.65 \times 10^{-6}})^2]^{\frac{1}{2}}$$

$$Z = [600^2 + (-800)^2]^{\frac{1}{2}} = 10^3 \ \Omega$$

change magnetic and electrical field energies. The power dissipated in the circuit is given by the product of the square of the current times the resistance,

$$P = I^2 R = I^2 Z \cos \phi \tag{25.21}$$

since the resistance is equal to the impedance times the cosine of the phase angle between the voltage and the current. This factor, $\cos \phi$. is known as the power factor of the circuit.

EXAMPLE

Given a 110-V (60-Hz) AC source in a circuit consisting of a 2.65 μF capacitor, a 0.534-H coil, and 600-Ω resistance (see Figure 25.11), find the current in the circuit, the phase angle between current and voltage (which leads?), the power factor, and power dissipated in the circuit.

$$\omega = 2\pi f = 377 \ \text{rad/sec}$$

$$Z = \sqrt{600^2 + \left[(377)(.534) - \frac{10^6}{(377)(2.65)} \right]^2}$$

$$= 1000 \ \Omega$$

$$I = \frac{V}{Z} = 0.11 \ \text{A}$$

$$\tan \phi = \frac{-800}{600} = -1.33 \qquad \phi = -53°$$

The current leads the voltage by 53°.

power factor $= \cos \phi = 3/5 = 0.6$

power $= I^2 R = (0.11)^2 (600) = 7.26$ watts

25.5 Resonance

We have noted that the voltages across a capacitor and inductor are exactly out of phase with each other. Consider the special case when these voltages are equal in magnitude in a series *RLC* circuit. By inspecting the equations for the impedance of an AC series circuit, Equation 25.19, we see that this condition will occur when X_C and X_L are equal,

$$\omega L = \frac{1}{\omega C} \tag{25.22}$$

This condition is called resonance. So the angular frequency for which resonance occurs is given by,

$$\omega_0 = \frac{1}{\sqrt{LC}} \tag{25.23}$$

This is called the *natural frequency* of the circuit and is symbolized by ω_0.

As an example a radio station is broadcasting on a given frequency. This frequency is determined by the inductance and capacitance in its output circuit. In order for you to receive the signal from this station, you must tune the receiving circuit of your radio so that it has the same natural frequency as the broadcasting station. That is the product of L and C will have a fixed value. In order to do this one could theoretically change either L or C, or both. In practice the tuning of a radio receiver is done by changing only the capacitance.

Questions

1. What is the maximum current in a resonant circuit?
2. It should be noted that resonant circuits can exhibit voltages across L or C that are much larger than the AC source voltage. Does this violate the conservation of energy? Explain how this can be true.

25.6 Electrical Oscillations

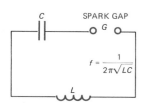

FIGURE 25.12

An AC resonant circuit results when the voltages across the inductor and capacitor are equal. Since these voltages are exactly out of phase, the effective impedance for such a circuit is equal to the total resistance in the circuit. The condition for resonance is $x_L = x_C$ or $f = 1/(2\pi\sqrt{LC})$. The current is a maximum for resonant conditions.

In our study of mechanical vibratory motion we found that most objects can be set into vibration about their equilibrium points. If a slight displacement of the object from equilibrium is produced, then a restoring force may cause the object to oscillate until friction damps out the motion.

Electric oscillations can be set up in an electrical circuit if analogous conditions are met. Earlier in this chapter it was pointed out that an inductor acts as inertia in impeding the building up of the flow of charge in the circuit. The accumulation of charge in the capacitor plates produces a restoring force on the electrons. The resistance in the circuit produces heat and is analogous to friction in the mechanical system.

The frequency of the mechanical vibrations depended upon the inertia and the restoring force constant. In the electrical circuit we will see that the frequency of the electrical oscillations depend upon the inductance and capacitance in the circuit. In the circuit shown in Figure 25.12, a capacitor C and inductance L are connected in series with spark gap G. The spark gap G will have a high resistance until a spark jumps across it and a low resistance after the spark. The capacitor is charged until the potential across the gap is sufficiently high to produce a spark. The capacitor will then discharge. The effect of the inductor is to oppose the buildup of charge so the current does not stop at zero but the capacitor is charged in the opposite direction. It then discharges again the current reversing in the circuit. Originally, the energy of the system was stored in the charged capacitor. Upon complete discharge, the energy goes into the magnetic field of inductor and into heat produced by the resistance of the circuit. The circuit continues to oscillate until all of the energy originally stored in the capacitor has been converted to heat. If there is a source of energy of input to compensate for the heat losses through resistance, the system will continue to oscillate. The frequency of oscillation depends upon the values of L and C and corresponds to the natural frequency of the circuit, which occurs for $X_L = X_C$ and

$$\omega_0 = \frac{1}{\sqrt{LC}} \tag{25.23}$$

the same equation as used to calculate the resonance frequency.

25.7 Q-Factor

The *quality factor* of a resonance is the measure of its sharpness. The *Q-factor* of a resonant circuit is illustrated in Figure 25.13. If the current through an AC circuit is plotted as a function of the angular frequency ω, then the current is maximum at the natural resonance frequency ω_0. The height of the current peak compared to its width is a measure of the sharpness of resonance. The quality factor for an AC circuit is defined as the ratio of the reactance at resonance to the resistance,

$$Q = \frac{\omega_0 L}{R} = \frac{\omega_0}{\Delta\omega} \tag{25.24}$$

where ω_0 is the resonance frequency.

A high Q circuit has low resistance and low fractional energy loss per cycle. A radio receiver with a high Q tuning circuit will have good discrimination between radio signals of nearly the same frequency.

FIGURE 25.13
The current as a function of angular frequency ω shows a maximum at resonance $\omega_0 = 1/\sqrt{LC}$. The Q-factor measures the sharpness of the resonance; Q-factor $= \omega_0/\Delta\omega$.

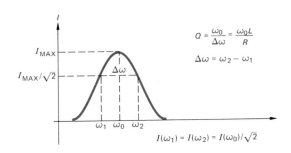

EXAMPLE

It is desirable to make a resonant circuit for EEG waves at 50 Hz. Given an inductor of 2.4 H and 100 Ω resistance, find the capacitance necessary for resonance and the Q-factor for the circuit.

$$\omega_0 = 2\pi f_0 = 2\pi(50) = 100\pi \text{ rad/sec} = \frac{1}{\sqrt{LC}}$$

Thus,

$$C = \frac{1}{(100\pi)^2 L} = \frac{10^{-4}}{\pi^2(2.4)} = 4.15 \ \mu F$$

$$Q = \frac{\omega_0 L}{R} = \frac{100\pi \times 2.4}{100} = 7.6$$

Questions

3. Show that the Q-factor is proportional to the ratio of the energy stored to the energy dissipated in the circuit per cycle?

25.8 The Transformer and Its Applications

The transformer is a most important AC device. In its simple form a transformer consists of two coils wound around a common ferromagnetic core. A changing current in one coil (the primary coil) produces a changing magnetic flux through the second (secondary) coil and thereby produces an emf in the secondary coil. There is also a back emf in the primary coil almost equal to the applied emf. This back emf, $\mathscr{E}_b$, equals the rate of total flux change,

$$\mathscr{E}_b = -N_p \frac{\Delta\phi}{\Delta t} \tag{25.25}$$

where N_p is the number of turns in the primary coil and $\Delta\phi/\Delta t$ is the time rate of change in magnetic flux through each turn.

The emf $\mathscr{E}_s$ induced in the secondary coil likewise is given by

$$\mathscr{E}_s = -N_s \frac{\Delta\phi}{\Delta t} \tag{25.26}$$

where N_s is the number of turns in the secondary. The ratio of the primary emf to the secondary emf is given by the ratio of Equations 25.25 and 25.26, since the primary coil emf is almost equal to the back emf, $\mathscr{E}_b$,

$$\mathscr{E}_p/\mathscr{E}_s = \frac{-N_p \ \Delta\phi/\Delta t}{-N_s \ \Delta\phi/\Delta t} = \frac{N_p}{N_s} \tag{25.27}$$

A transformer with $N_s > N_p$ is called a step-up transformer (stepping up voltage), while a transformer with $N_s < N_p$ is a step-down transformer. Transformers can be made with 99 percent efficiency. From conservation of energy, the power input should almost equal the power output,

power input $\simeq$ power output

Thus,

$$I_p \mathscr{E}_p = I_s \mathscr{E}_s$$

$$\frac{I_s}{I_p} = \frac{\mathscr{E}_p}{\mathscr{E}_s} = \frac{N_p}{N_s} \tag{25.28}$$

Transformers make it possible to transmit AC electric power over high voltage (120,000 V) transmission lines, reducing the I^2R energy loss in the power line. When the power reaches the desired destination, the voltage can be stepped down (to 240 or 120 V) for use. A step-down transformer can be used to produce high currents at low voltages as is common in electric welders.

EXAMPLE

What is the maximum current available in a step-down transformer with $\mathscr{E}_p = 120$ V, $I_p = 5$ A, and $\mathscr{E}_s = 6$ V?

Since $\mathscr{E}_p I_p = \mathscr{E}_s I_s$,

$$I_s = \frac{120 \text{ V} \times 5 \text{ A}}{6 \text{ V}} = 100 \text{ A}$$

Another important use of transformers is that of impedance matching. Maximum power will be transferred from an AC source to a load of the same impedance magnitude with a 180° phase difference between their reactance components. This means the impedance of the load should equal the impedance of the source. If the impedance of a load Z_L equals the secondary impedance Z_s, then we can show that the load impedance is equal to a constant times the impedance of the primary,

$$Z_L = Z_s = \frac{\mathscr{E}_s}{I_s} = \left(\frac{N_s}{N_p}\right)^2 \frac{\mathscr{E}_p}{I_p} = \left(\frac{N_s}{N_p}\right)^2 Z_p \tag{25.29}$$

A transformer can be used to match impedance between sources and loads. A common example of this use of transformers is the coupling of audio-speaker systems (low impedance 8 or 16 Ω) with audio amplifiers (high impedance output).

EXAMPLE

It is desired to match an 8-Ω sound speaker to the audio amplifier from the ECG set up that has a 10,000 Ω output impedance. Find the turn ratio needed for the transformer to be used in this impedance match.

$$8 \ \Omega = 10^4 \left(\frac{N_s}{N_p}\right)^2$$

Therefore,

$$\frac{N_p}{N_s} = \frac{10^2}{\sqrt{8}} = 36$$

25.9 Alternating Current Applications

Direct-current measurements may cause polarization voltages in ionic conduction. These polarization voltages are generated by the charge separation produced by the electric field. To minimize the effects of polarization voltages, it is desirable to use AC measurements. One such measurement procedure involves the use of an AC impedance bridge, analogous to the DC Wheatstone bridge, to measure bioimpedances. For resistance measurements the equations are the same as those for the Wheatstone bridge. Such a bridge diagram is shown in Figure 25.14. Versatile AC bridges use reactive elements (capacitors and inductors) in the branches and are capable of measuring both resistive and reactive components of an unknown impedance.

It has been found that electrode implantation can be aided by measuring the impedance between the inserted electrode and a reference electrode. For example, different tissue layers have different impedance values, and boundary layers are discerned by impedance changes as the electrode is moved into its desired position.

Alternating-current impedance measurements are also used for physiological measurements. Impedance plethysmography involves measuring changing impedance across the chest associated with breathing and the pulsating blood flow. (Plethysmography is the study of blood volume changes within an organ.)

Most of the electrical signals generated by the human body are AC in nature. Some of these signals, their frequency, and amplitudes are shown in Table 25.1. (It should be noted that while these signals are periodic they are *not* sinusoidal.)

FIGURE 25.14
An AC bridge circuit is analogous to the DC Wheatstone bridge. The AC bridge produces less heating in unknown resistors and avoids problems of polarization in liquid or ionic systems.

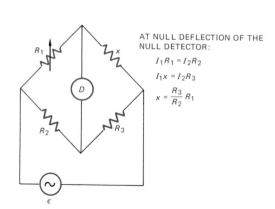

AT NULL DEFLECTION OF THE NULL DETECTOR:

$$I_1 R_1 = I_2 R_2$$
$$I_1 x = I_2 R_3$$
$$x = \frac{R_3}{R_2} R_1$$

TABLE 25.1
Frequency and Amplitudes of Some
Electrical Signals Generated in the
Human Body

Signal	Frequency (Hz)	Amplitudes (order of magnitude)
EEG α	8–13	20 μV
EEG β	14–50	10 μV
EEG δ	0.5–4	50 μV
EEG θ	5–7	50 μV
ECG beats/min	1–1.5	1 mV
Eye blink potentials	1–3	0.5 mV

Electrical hazards are associated with leakage currents of electrical equipment. These currents result from pathways to ground that are not intended. Most frequently this path results from capacitive coupling between the high voltage side of the power line and ground. (One wire of the powerline is grounded. The third wire of modern equipment grounds the case of the equipment.) Maximum leakage currents of electrical equipment are subject to federal regulations and should be less than 1 mA. For capacitive leakage (a pair of conducting wires have such capacitive coupling), the effective impedance for a pathway is given by $Z_C = 1/\omega C$ for 1 mA current at 120 V

$$Z = \frac{120 \text{ V}}{10^{-3} \text{ A}} = 1.2 \times 10^5 \ \Omega = \frac{1}{\omega C}$$

If $\omega = 2\pi \ (60)$, then $C = 0.22 \ \mu$F.

This is a rather large "leakage" capacitance, a more typical value might be 10^{-9} F or 1000 picofarads.

Currents as small as 20 μA have been known to cause fibrillation when internal electrical contacts (such as implanted electrodes or catheters) are used. A current of this size would result for a capacitance leakage given by

$$Z = \frac{120 \text{ V}}{20 \times 10^{-6} \text{ A}} = 6 \times 10^6 \ \Omega = \frac{1}{\omega C}$$

If $\omega = 2\pi(60)$, then $C = 440 \times 10^{-12}$ F.

This is a small leakage capacitance and points out the care that must be taken when internal electrical contacts are present.

Another possible electrical hazard results when ground connections break or equipment is improperly grounded. In these cases, a human body may provide the pathway to ground when the instrument or piece of equipment "floats" at a potential above ground. This commonly occurs when three-wire equipment is used without proper ground connection. In these cases, the subject provides resistive coupling to ground. The value of this resistive coupling is determined by the nature of the electrical contacts to the high voltage side and to the ground side of the AC power. Moisture at the contact points reduces the resistance and increases danger of electrical shock.

ENRICHMENT
25.10 Calculation of Effective Current

The instantaneous power dissipated in a resistor in an AC circuit is $P = i^2R = (I_0 \sin \omega t)^2 R$.

We want to determine the average power dissipated. This can be found by averaging over a period τ as follows:

$$P_{\text{ave}} = \frac{1}{\tau} \int_0^\tau i^2 R \, dt = \frac{1}{\tau} \int_0^\tau I_0^2 R \sin^2 \omega t \, dt$$

$$P_{\text{ave}} = \frac{I_0 2R}{\tau} \int_0^\tau \left(\frac{1 - \cos 2\omega t}{2} \right) dt$$

$$P_{\text{ave}} = \frac{I_0 2R}{2\tau} \left(t - \frac{\sin 2\omega t}{2\omega} \right) \Big|_0^\tau \quad \text{where } \tau = \frac{2\pi}{\omega}$$

$$P_{\text{ave}} = \frac{I_0^2 R}{2} = I^2 R$$

So

$$I = \frac{I_0}{\sqrt{2}}$$

Similarly, using the instantaneous voltage $P = V^2/R$, again the average power is

$$P_{\text{ave}} = \frac{1}{\tau} \int_0^\tau \left(\frac{V_0^2 \sin^2 \omega t}{R} \right) dt = \left[\int_0^\tau \left(\frac{1 - \cos 2\omega t}{2} \right) dt \right] \frac{V_0^2}{R\tau}$$

$$P_{\text{ave}} = \frac{V_0^2 \tau}{R\tau 2} = \frac{V_0^2}{2R} = \frac{V^2}{R}$$

Thus,

$$V = \frac{V_0}{\sqrt{2}}$$

and

$$P_{\text{ave}} = VI = \frac{V_0 I_0}{2}$$

The effective values of the current and voltage in an AC circuit are also referred to as the root-mean-square (rms) values. Most AC meters are calibrated to give rms values. These rms values produce the same heating as equal dc values of voltage and current.

ENRICHMENT
25.11 Calculation of Reactances

We can determine the reactance of a capacitor and an inductor as follows: The voltage across a capacitor is defined as charge (coulombs) divided by capacitance.

$$\mathcal{V}_C = \frac{Q}{C} = \int_0^t \frac{I\,dt}{C} = \int_0^t \frac{I_0}{C} \sin \omega t\, dt = \frac{I_0 \cos \omega t}{\omega C}$$

We see that the voltage across a capacitor is 90° out of phase with the AC current, that is, in this case the voltage lags the current. Also, if we define reactance such that

$$\mathcal{V}_C = X_C I_0 \cos \omega t$$

then $X_C = 1/\omega C$. The voltage across an inductor (based on Faraday's law of induction) is

$$\mathcal{V}_L = -L\frac{di}{dt}$$

$$\mathcal{V}_L = -L\frac{d}{dt}\,(I_0 \sin \omega t) = -\omega L I_0 \cos \omega t$$

$$= \omega L I_0 (-\cos \omega t) = I_0 X_L\,(-\cos \omega t)$$

Again we see the voltage is 90° out of phase with the AC current; in this case the voltage leads the current. The inductive reactance X_L equals ωL.

SUMMARY

Use these questions to evaluate how well you have achieved the goals of this chapter. The answers are given at the end of this summary with the number of the section where you can find related content material.

Definitions

1. The values for AC current and AC voltage that result in the same average power per cycle as the DC values are called the _____ or _____ values of current and voltage and they are equal to _____ times the peak current and voltage values respectively.
2. The sharpness of the _____ of an AC circuit is measured by the _____ which is equal to the ratio of the inductive _____ to the resistance when the frequency is equal to ω_0, the _____ frequency.
3. For an AC circuit the ratio of the voltage to the current is called the _____ which has both a _____ and a _____.
4. In general the product of voltage and current for an AC circuit is not equal to the power, but that product

must be reduced by a _____ factor, equal to _____, where ϕ is the angle between the AC current and the AC voltage.

AC Circuits

5. A coil of wire has a resistance of 30.0 Ω and an inductance of 0.100 H. Find
   a. its inductive reactance if connected to a 60-cycle line
   b. its impedance
   c. What would the current be if the coil were connected to a 120-V DC line?
   d. What would the current be if it were to a 120-V AC, 60-cycle line?
6. A 120-Ω rheostat and a 15 μF capacitor are connected in a series circuit to 120-V, 60-cycle emf.
   a. What is the reactance of the capacitor?
   b. What is the total impedance of the circuit?
   c. What is the current through the circuit?
   d. What is the voltage drop across each circuit element?

Phasor Diagrams

7. Draw a phasor diagram for problems 5 and 6 above and determine the phase angle for each.

Transformer

8. If a 110-V line is connected to the primary of a step-up transformer, it delivers 2 amps on the secondary coil. The ratio of turns on the two windings is 25. Assume no losses in the transformer. Find
   a. the secondary voltage
   b. the primary current

AC Measurements

9. Electrical equipment may be hazardous because of _____ .

10. List at least three periodic electrical signals generated by the human body, and give their typical frequencies and amplitudes.

Answers

1. effective, rms, 0.707 (Section 25.2)
2. resonance, Q-factor, reactance, resonance (Section 25.5 and 25.7)
3. impedance, magnitude, phase angle (Section 25.4)
4. power, $\cos \phi$ (Section 25.4)
5. a. 37.7 Ω
   b. 48.2 Ω
   c. 4 A
   d. 2.49 A (Section 25.3 and 25.4)
6. a. 177 Ω
   b. 214 Ω
   c. 0.561 A
   d. $V_C = 99.3$ V, $V_R = 67.3$ V (Section 25.3 and 25.4)
7. prob. 5, $\phi = 51.5°$ prob. 6, $\phi = -55.9°$ (Section 25.4)
8. a. 2750 V
   b. 50 A (Section 25.8)
9. leakage currents (Section 25.9)
10. ECQ, ~ 1 Hz, ~ 1 mV; EEG α, ~ 10 Hz, ~ 20 μV; EEG θ, ~ 5 Hz, ~ 50 μV (Section 25.9)

ALGORITHMIC PROBLEMS

Listed below are the important equations from this chapter. The problems following the equations will help you learn to translate words into equations and to solve-single concept problems.

Equations

$$i = i_{max} \sin \omega t \tag{25.1}$$

$$I = \frac{i_{max}}{\sqrt{2}} \tag{25.4}$$

$$V = \frac{V_{max}}{\sqrt{2}} \tag{25.6}$$

$$X_L = 2\pi f L \tag{25.9}$$

$$X_C = \frac{1}{2\pi f C} \tag{25.11}$$

$$Z = \sqrt{R^2 + (X_L - X_C)^2} = \sqrt{R^2 + [1/\omega L - (1/\omega C)]^2} \tag{25.19}$$

$$\tan \phi = \frac{X_L - X_C}{R} = \frac{\omega L - 1/\omega C}{R} \tag{25.20}$$

$$P = I^2 Z \cos \phi \tag{25.21}$$

$$\omega_0 = \frac{1}{\sqrt{LC}} \tag{25.23}$$

$$Q = \frac{\omega_0 L}{R} = \frac{\omega_0}{\Delta\omega} \qquad\qquad (25.24)$$

$$\frac{\mathscr{E}_p}{\mathscr{E}_s} = \frac{N_p}{N_s} = \frac{I_s}{I_p} \qquad\qquad (25.28)$$

Problems

1. If the effective voltage to an electric stove is 208 V, what is the peak voltage?
2. Compare the inductive reactance of a 1-H inductance on a 25-cycle source and a 60-cycle source.
3. Compare the capacitive reactance of a 2 μF capacitor connected to a 25-cycle source and a 60-cycle source.
4. What is the impedance of a circuit which has a resistance of 30 Ω and an inductive reactance of 40 Ω?
5. What is the phase angle for a circuit that has an inductance reactance of 30 Ω, a capacitive reactance of 20 Ω and 20 Ω resistance?
6. What is the capacitance needed in a circuit to produce resonance in a 60-cycle circuit having an inductance of 1 H?
7. What is the ratio of primary turns to secondary turns in a transformer which is designed to operate a 6-V bell system when connected to a 114-V line?

Answers

1. 293 V
2. $X_L(f = 60) = 2.4X_L(f = 25)$
3. $X_c(f = 25) = 2.4X_c(f = 60)$
4. 50 Ω

5. $\phi = 26.6°$, voltage leads current
6. 7.04 μF
7. 19:1

EXERCISES

These exercises are designed to help you apply the ideas of a section to physical situations. When appropriate, the numerical answer is given in brackets at the end of the exercise.

Section 25.2

1. What does a 120-V, 60-cycle source mean? What peak voltage must insulation stand for this source? What is its effective voltage? What is its average voltage? Compare it with a 120-V DC source. [170 V, 120 V, 0 V]

Section 25.3

2. What is the reactance of a 2.00-μF capacitance at a frequency of 1, 60, 440, 10^6 Hz? What does this indicate? [7.96×10^4 Ω, 1.3×10^3 Ω, 181 Ω, 7.96×10^{-2} Ω]

3. What is the reactance of a 2-H inductor at frequency of 1, 60, 440, 10^6? What does this indicate? [1.26×10^1, 7.54, $\times 10^2$, 5.53×10^3, 1.26×10^7]

Section 25.4

4. What is the total impedance at 60 cycles of the resistor, capacitor and inductor shown in Figure 25.15? [1290 Ω]

5. In Figure 25.15, what is the current, and what are the voltages V_{ab}, V_{bc}, V_{cd}, V_{ac}, and V_{bd}, if a 120-V, 60-cycle source is connected across ad? [0.093 A; $V_{ab} = 4.65$

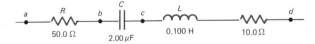

FIGURE 25.15
Exercises 4, 5, and 6.

V; $V_{bc} = 123$ V; $V_{cd} = 3.63$ V; $V_{ac} = 123$ V; $V_{bd} = 119$ V]

6. Using data from problems 4 and 5, what is the power loss for the entire circuit and for each component? What is the phase angle? [$P_{tot} = 0.52$ watt; $P_R = 0.43$ watt; $P_C = 0$; $P_L = 0.09$ watt; phase angle $= 89.96°$]

7. A resistance of 100 Ω, an inductance of 75.0 mH, and a capacitor of 4.0 μF are connected in series with a generator (100 volts at 2500 rad/sec). Find
   a. the current in the circuit
   b. the voltage across each circuit component
   c. the power dissipated in the circuit
   d. the phase angle between the current and the voltage in the circuit.
   Draw an appropriate phasor diagram. [a. 0.75 A; b. $V_R = V_C = 75$ V; $V_L = 141$ V; c. 56 watt; d. 41.2°]

8. An AC series circuit has $R = 300$ Ω, $L = 0.90$ H, and $C = 2.0$ μF with a generator of 50 V and $\omega = 1000$ rad/sec. Find
   a. the current in the circuit
   b. the voltage across each of R, L, and C
   c. the phase angle between voltage and current in the circuit.
   Draw a phasor diagram.
   d. the power dissipated in the circuit [a. 0.10 A; b. 30 V, 90 V, 50 V; c. 53°; d. 3 watts]

Section 25.5

9. Assume your radio has an inductance of 18 mH in its receiving circuit. To what capacitance must you turn your radio dial to receive your favorite radio programs broadcast at 1490 kilocycles? [0.63 $\mu\mu$F]

Section 25.7

10. Assume the receiving circuit of your radio has a resis-

tance of 1800 Ω. What is the Q-factor of the circuit? What is its ratio of response to the signal from a nearby station at 1540 kc when you have it adjusted to receive your favorite 1490-kc program? Do you judge this to be a high-quality or low-quality radio? [93.6, $S_{1540}/S_{1490} = \frac{1}{2}$, low quality]

Section 25.8

11. A step-up transformer has a turns ratio of 200:1, and 100 V are applied to the primary side of this transformer.
    a. Find the secondary output voltage.
    b. If the secondary current is 100 mA, find the primary current.
    c. Find the power output of the transformer. [a. 2 $\times 10^4$ V; b. 20 A; c. 2 $\times 10^3$ watts]

12. It is desired to operate a 4-V lamp by using a transformer on a 120-V supply. Find the ratio of the secondary current to the primary current in the operation of this transformer. [30]

13. The internal resistance of an AC source is 100 Ω. Find the turns ratio of a transformer that could be used to match this source to a 25-Ω load with maximum power transferred to the load. [2]

Section 25.9

14. A patient undergoes fibrillation while being catheterized. A current of 5×10^{-5} A is produced by leakage potential of 120 V (60 Hz) through capacitive coupling. Find the value of this capacitance. [1.1 $\times 10^{-9}$ F]

15. One potential hazard of a microwave oven is its capacitive coupled leakage current. For an oven operating at 100 MHz (1 MHz $= 10^6$ Hz) at 120 V, find the leakage current if the capacitive coupling is 10^{-12} F. [7.5×10^{-2} A]

PROBLEMS

These problems may involve more than one physical concept. When appropriate, the answer is given in brackets at the end of the problem.

16. A rectangular coil of 100 turns, which has dimensions of 10 cm by 15 cm, rotates at 300 rpm about an axis through midpoint of the short sides. The axis of rotation is perpendicular to the direction of the mag-

netic field of strength 0.5 Wb/m². Plot the induced emf for two complete revolutions of the coil. Choose a position for $t = 0$. [$\mathscr{E} = 7.5\pi \sin 10\pi t$]

17. Assume you have available a 50-Ω resistor, an inductor with resistance of 10 Ω and inductance of 0.10 H, a 1.0-μF capacitor, and two sources of electric energy, a DC source of 120 V and an AC source 60-cycle 120 V. What currents would you get if you

connected two of these components in series with a source?

	RC	RL	LC
DC	0.0 A	2.0 A	0.0 A
AC	0.045 A	1.7 A	0.046 A

18. A 120-V, 60-cycle source is dangerous. It has been estimated that the maximum safe current is 15 mA. At this frequency it is thought that a current of 70 mA for one second could be lethal, that is, it could produce ventricular fibrillation. What is the impedance of the body for these currents? How much energy would be expended in the electrocution? [8000 Ω; 1700 Ω, 8.4 J]

19. Given a resistance of 100 Ω, an inductor of 250 mH, and a capacitor of 1.00 μF in a series with a 10.0-V variable-frequency generator, find
    a. the resonant frequency for the circuit
    b. the voltage across each of R, L, and C at resonance
    c. the power supplied by the generator at resonance
    d. the Q-factor of the circuit.   [a. 318 Hz; b. 10 V, 50 V, 50 V; c. 1 watt; d. 5]

20. A circuit consists of a 500 Ω resistor, an inductor of 100 mH, and a variable capacitor connected in series with a 100-V generator operating at 100 Hz. Find
    a. the value of C that produces resonance in the circuit
    b. the voltage across L, R, and C at resonance
    c. the Q-factor for this circuit
    d. the power dissipated in the circuit.   [a. 25.3 μF; b. 12.6 V, 100 V, 12.6 V; c. 0.126; d. 20 watt]

21. Inductive coupling is much less common in leakage currents than capacitive coupling, but two adjacent conductors 10 m in length have inductive coupling as well as capacitive coupling. Find the leakage current for a 10^{-5} H leakage inductance with a 1.20 μV (60 Hz) leakage potential. (Is the neglect of resistance of the inductance justified? $R_e = 1.6 \times 10^{-3}$ Ω/m)   [3.18 $\times 10^{-4}$ A]

22. For an inductive leakage of 1.00×10^{-5} H in series with a capacitive leakage of 1.00×10^{-9} F,
    a. find the resonant frequency
    b. If 10.0 V leakage voltage results at this resonant frequency, find the leakage current if the resistance (due to dielectric loss) is 1.00×10^7 Ω.
    c. Find the voltage across the capacitance and inductance in this problem.   [a. 1.6×10^6 Hz; b. 10^{-6} A; c. 10^{-4} V]

23. Find the power loss in a transmission line whose resistance is 2 Ω, if 50 kilowatts are delivered by the line
    a. at 50,000 V
    b. at 5000 V
    c. What kind of transformer would you need at the input end if the voltage at the generator is 2500 V? [a. 2 watts, b. 200 watts; c. for 50 kv, $N_p/N_s = 1/20$, for 5 kv, 1/2]

GOALS

When you have mastered the content of this chapter, you will be able to achieve the following goals:

Definitions
Define each of the following terms and use each term in an operational definition:

feedback
linearity
amplification
frequency response

interference noise
signal-to-noise ratio
stability

Electronic Devices
Explain the basis of operation and potential use for each of the following:

diode rectifier
transistor amplifier
operational amplifier

differential amplifier
oscilloscope

Oscilloscopes
Evaluate the specifications provided for commercially produced oscilloscopes

PREREQUISITES

Before beginning this chapter you should have achieved the goals of Chapter 21, Electrical Properties of Matter, Chapter 22, Basic Electrical Measurements, Chapter 23, Magnetism, and Chapter 25, Alternating Currents.

26

Bioelectronics and Instrumentation

26.1 Introduction

Many of the advances that have been made in the life sciences in recent years are the result of the use of electronics in modern bioinstrumentation. Electronic devices called transducers can convert forms of energy that are not detectable by human senses into easily detected and recorded information. Electronic amplifiers make possible the study of heart potentials, muscle potentials, nerve action potentials, and brain waves in physiology laboratories. It is not necessary to be able to design your own instruments, but it is important to understand the basis of operation of modern electronic instrumentation so that you can use it intelligently. It is helpful to know the limitations of your instruments and thereby make sure that you use them correctly. As you improve your understanding of the basic operation of electronics, you will be able to increase your use of electronic instrumentation and you will discover many different approaches to the study of life science problems.

The advent of solid state electronics has greatly improved instrumentation. Technology has developed the compact integrated circuits (IC) which incorporate many components into complete, single-unit, complex circuits. The IC is the building block of modern electronic instrumentation. The discrete elements such as *diodes* and *transistors* may still be used, but if the circuit has wide enough applicability, it is likely to be produced as an IC.

26.2 Instrumentation Characteristics

Assume we are interested in studying the response of a human eye to light flashes (see Figure 26.1). The stimulus in this case is a light, and the transducers might be a television camera monitoring the subject's eyes and a microphone monitoring the subject's heart rate. The feedback control can provide random light flashes with possible variations in light intensity and color. In more sophisticated systems the feedback control and recording parts of the system can be handled by a minicomputer (or any other versatile electronic control system). The signal

FIGURE 26.1

A schematic block diagram illustrating an experimental system for measuring human response to a physical stimulus.

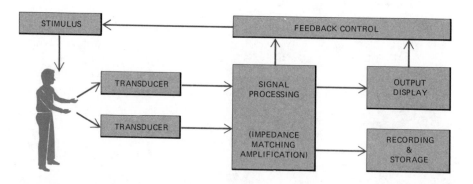

processing would include matching the impedance of the transducer to the impedance of the amplifier as well as shaping of the signal to assure *linearity* of output signal with input signal. By linearity we mean that the output signal is a linear function of the input signal:

$$\text{output signal} = (\text{constant}_1)(\text{input signal}) + \text{constant}_2 \qquad (26.1)$$

The linearity criteria is very important if quantitative relationships are sought between input and output signals.

In many cases the signal we obtain from our experimental system is small. In fact, it may be too small to measure with our recording instrument. Our signal must be made larger, or *amplified*, before we record it. We can use an electronic instrument called an amplifier for this purpose. The ratio of the output signal to input signal is called the *amplification factor*. We can choose to amplify the current, voltage, or power of our experimental signal.

Considerable effort is spent in making linear amplifiers for wide ranges of *frequency responses* and amplitudes of input signals. Again the matching of the transducer with the amplifier is important in order to maximize the measuring capability of the system. The *frequency response* of either the transducer or the amplifier refers to the range of frequencies to which the system responds without distortion.

In all present-day bioscience research laboratories there are a large number and variety of electronic devices. The air is literally full of weak electromagnetic waves generated as a by-product of electronic instrumentation. These electromagnetic waves can cause random signals which interfere with our detection systems. This random signal interference is called *interference noise*. The interference noise is due to coupling of environmental energy sources into the experimental system.

One of the important sources of interference noise is the 60-Hz noise picked up by electromagnetic induction from all of the AC equipment in the laboratory. Shielded coaxial cable should be used, and in some cases it might be necessary to isolate the subject in a shielded screen room in order to minimize the noise picked up by the electrodes attached to the subject. Interference noise must be reduced before amplification because the amplifier will amplify the noise along with the signal under study.

One way to characterize the quality of our experimental system is to compute the ratio of the average amplitude of the signal we are measuring to the average amplitude of the interference noise. We call this the *signal-to-noise ratio* for the system, and it should be greater than 10. In life science applications the signals are usually so small that interference noise must be kept at a minimum in order to maintain a signal-to-noise ratio that allows adequate measurements.

Finally, it is desirable to have the system as stable as possible. The *stability* of the electronics is a measure of the system's ability to return to equilibrium after an input disturbance. The stability of a system can often be improved by taking a small portion of the output signal, changing its polarity and feeding it back into the input side of the system. This is called negative feedback (Section 2.6) or feedback control.

26.3 Diodes

Modern solid state electronics is based on the physical properties of semiconductor materials such as germanium and silicon. Devices such as thermistors, diodes, transistors, solar cells, and integrated circuits are examples of semiconductor devices. Solid state electronics is based on the ability to vary the conductivity of the semiconductor materials by varying the number of current carriers per unit volume. Besides electrons (in n-type semiconductors), it is possible to make semiconductors with holes, or positive charge carriers (in p-type semiconductors).

Semiconductor diodes are made by joining a n-type and a p-type semiconductor. When the p-type region has a higher potential than the n-type region, the current is maintained as electrons fall into holes at the junction with an equal number of holes being created as electrons are pulled from the p-type region. This configuration is a *forward biased diode* that has low impedance to current. When the polarity is reversed, the electrons are pulled out of the n-type region and the holes are pulled out of the p-type region, leaving the junction void of current carriers. This situation corresponds to very high impedance to current and this is the *reversed biased diode* that blocks current flow. (See Figure 26.2 for the current characteristics of a diode.) Thus the diode can function as an on-off current switch which is controlled by the polarity of the voltage applied to the diode.

When a current source of alternating polarity is connected to a diode as shown in Figure 26.3, the result is an output voltage of a single polarity corresponding to the forward biased polarity of the diode. The diode is said to have rectified the current. Such a circuit is called a *rectifier circuit*.

Notice that the arrowhead part of the symbol for a diode points in the direction for positive current. The primary use of diodes is as rectifiers. Another use of semiconductor diodes in life science research is that of temperature transducer. The voltage drop across a forward biased diode

FIGURE 26.2
A typical current versus voltage characteristic curve for a solid-state diode.

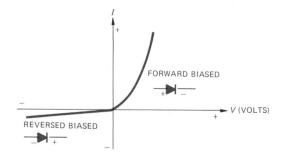

FIGURE 26.3

The solid-state diode in an AC circuit produces a pulsating DC voltage since it only allows current when it is forward biased.

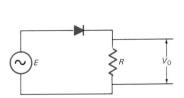

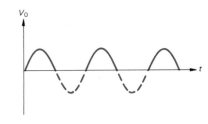

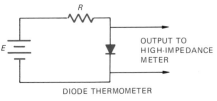

OUTPUT TO HIGH-IMPEDANCE METER

DIODE THERMOMETER

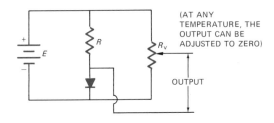

(AT ANY TEMPERATURE, THE OUTPUT CAN BE ADJUSTED TO ZERO)

OUTPUT

FIGURE 26.4

Two circuits showing the use of a solid-state diode as a temperature transducer. The voltage across the diode is proportional to its temperature.

is quite sensitive to temperature. This sensitivity is linear, being about -2.5 mV/°C for silicon diodes. Two circuits for diode temperature transducers are illustrated in Figure 26.4. The best results are obtained with very small diodes which have a small thermal inertia and with a diode current of approximately 100 μA

26.4 Transistors

The transistor is a three terminal semiconductor (silicon or germanium) device. The three terminals are called emitter, base, and collector, and transistors are designated as either pnp or npn types. This notation refers to the nature of the majority carriers in the emitter, base, and collector respectively. The p represents a majority of positive carriers (holes) and the n represents a majority of negative carriers (electrons). In the schematic representation of a transistor an arrow is shown at the emitter to indicate the direction of positive current (Figure 26.5).

In a circuit with two input wires and two output wires, a three-terminal device such as a transistor can be installed in three different ways.

FIGURE 26.5

Schematic diagrams of npn and pnp transistors. The arrow on the emitter terminal indicates the direction of positive current flow.

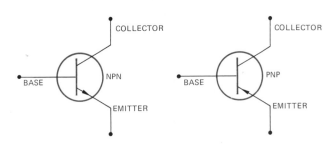

FIGURE 26.6

Three transistor amplifier configurations for a pnp transistor.

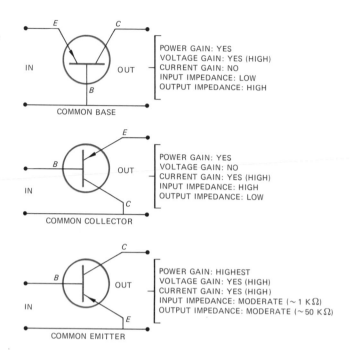

POWER GAIN: YES
VOLTAGE GAIN: YES (HIGH)
CURRENT GAIN: NO
INPUT IMPEDANCE: LOW
OUTPUT IMPEDANCE: HIGH

COMMON BASE

POWER GAIN: YES
VOLTAGE GAIN: NO
CURRENT GAIN: YES (HIGH)
INPUT IMPEDANCE: HIGH
OUTPUT IMPEDANCE: LOW

COMMON COLLECTOR

POWER GAIN: HIGHEST
VOLTAGE GAIN: YES (HIGH)
CURRENT GAIN: YES (HIGH)
INPUT IMPEDANCE: MODERATE ($\sim 1\ K\Omega$)
OUTPUT IMPEDANCE: MODERATE ($\sim 50\ K\Omega$)

COMMON EMITTER

The three different transistor configurations are labeled according to the terminal that is common to both the input and output circuits.

The three different transistor amplifier configurations with their characteristics are shown in Figure 23.6. In each case, a small current injected to the base region results in the output current or voltage amplification. The common emitter configuration is a good compromise between power gain and output impedance and thus is the most widely used transistor circuit.

Let us consider a specific use for each of the transistor configurations shown in Figure 26.6. A *thermocouple* is a temperature transducer that produces a small voltage output ($\sim$ millivolt) at low output impedance. For the measurement of thermocouple output with a typical voltmeter, the common-base configuration provides an ideal amplifier to use between the thermocouple and the meter.

A *photomultiplier* is a light transducer that provides a small current signal at a high output impedance. It is frequently desirable to transmit this signal over a coaxial cable of low impedance (50 Ω) to an output device or recorder. The common-collector configuration is the ideal choice as a preamplifier at the photomultiplier tube to match the impedance of the tube with the coaxial cable for optimum signal transfer.

The *photo-relay system* is a typical application for the common-emitter configuration. In this case a solar cell is the light detector that is used to control a relay that in turn controls lights, counters, or some other power requiring device. The common-emitter circuit provides maximum power gain to activate the relay which controls the output device.

FIGURE 26.7
Schematic diagram of an operational amplifier set up as an inverting amplifier.

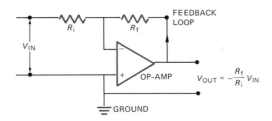

These are only three simple applications of the transistor circuits shown in Figure 26.6.

26.5 Operational Amplifiers

Let us consider a particular integrated circuit that is called an *operational amplifier* (op-amp). The op-amp is a high-gain amplifier that exhibits a frequency response from DC to at least 30 MHz. These devices are relatively low in cost and very versatile in applications. The ideal op-amp has very high input impedance, very low output impedance, and very high amplification. (The DC voltage gain is between 10^4 and 10^9 for maximum amplification.) The very high gain means that the output of the op-amp is determined by the negative feedback connection in the system (Figure 26.7). The amplification of this configuration shown in Figure 26.7 is given by the voltage ratio,

$$\text{amplification} = \frac{V_{\text{out}}}{V_{\text{in}}} = -\frac{R_{\text{f}}}{R_{\text{i}}} \qquad (26.2)$$

where V_{out} is the output voltage and V_{in} is the input voltage, and R_{i} and R_{f} are the resistances shown in the figure.

The op-amp can be used as a *differential amplifier*. A differential amplifier is designed to amplify the difference between two input signals. This form of amplifier is used in life science applications where it can greatly reduce the interference noise that is common to both input signals. For example, to measure an ECG you may connect an electrode to one wrist and one ankle. The two input signals are the inputs to a differential amplifier as shown in Figure 26.8, the interference noise that is common to these signals will not be amplified.

The operational amplifier is a very versatile circuit. With proper

FIGURE 26.8
Schematic diagram for an operation amplifier used to amplify the difference between two input signals.

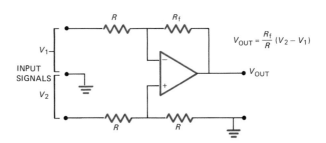

FIGURE 26.9
Typical circuits for using an operational amplifier as an (a) adder,
(b) integrator, and (c) differentiator.

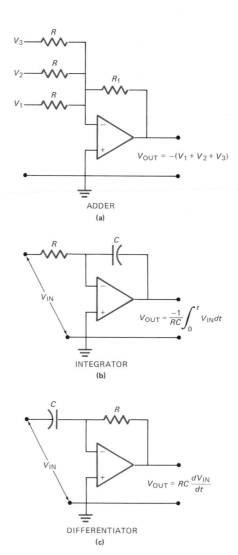

$V_{OUT} = -(V_1 + V_2 + V_3)$

ADDER
(a)

$V_{OUT} = \dfrac{-1}{RC} \displaystyle\int_0^t V_{IN}\,dt$

INTEGRATOR
(b)

$V_{OUT} = RC\,\dfrac{dV_{IN}}{dt}$

DIFFERENTIATOR
(c)

modifications in the external circuit elements, an op-amp can be used to
add, integrate, or differentiate input voltages as shown in Figure 26.9.*

26.6 The Oscilloscope

The oscilloscope is designed to display voltages as a function of time.
Because of its variable voltage sensitivity and wide range of response
speeds, the oscilloscope is a very useful instrument in the laboratory

---

* An excellent reference for anyone interested in learning more about using operational
amplifiers and other solid state devices is S. A. Hoenig and F. L. Payne, *How to Build and
Use Electronic Devices Without Frustration, Panic, and Mountains of Money, or an Engineering Degree,* Boston: Little, Brown and Company, 1973.

FIGURE 26.10
A schematic diagram of an oscilloscope.

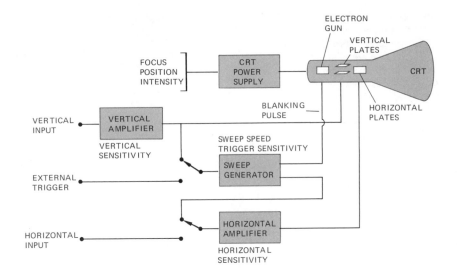

(Figure 26.10). The central element of the instrument is the cathode-ray
tube upon which the waveform of the signal is displayed. This waveform
is traced by a beam of electrons writing on the phosphorescent coating
of the tube face. The signal under study can be monitored visually by
the experimenter or photographed for a permanent record. The position
of the electron beam is controlled by electric fields applied across two
pairs of deflecting plates, called the horizontal and vertical deflection
plates as shown in Figure 26.10.

For most applications in the life sciences the internal time base of the
oscilloscope is used. The horizontal deflection is provided by using an
internal sweep generator connected internally to the horizontal amplifier. This sweep generator is a sawtooth wave that has a slope linear in
time. A control switch on the scope allows the operator to select the appropriate sweep rate, denoted by time per centimeter deflection on the
switch. For example, 1 msec/cm is a setting for which each centimeter of
horizontal deflection corresponds to 1 millisecond. The signal to be
studied as a function of time is the input to the vertical amplifier; a switch
allows selection of the proper vertical sensitivity designated in volts
per centimeter. A vertical sensitivity of 10 mV/cm means that each centimeter of vertical deflection corresponds to a signal amplitude of 10 millivolts. Let us consider two specific applications to illustrate the use of the
oscilloscope.

Suppose we wish to measure the conduction velocity of a nerve impulse along the sciatic nerve of a frog. The nerve is extracted and placed
in a wet nerve cell where it rests on silver electrodes equally spaced
along the chamber. A typical set up for this experiment is illustrated in
Figure 26.11. The nerve is stimulated by a voltage pulse from a stimulator
(an electronic pulse generator with variable pulse amplitude, frequency,
and duration). This stimulation (if above threshold) will produce a nerve
impulse that travels along the nerve. As this impulse passes the elec-

FIGURE 26.11
An experimental set up using an oscilloscope to measure the conduction velocity of a nerve impulse. The stimulator pulse triggers the horizontal sweep as it stimulates the nerve. When the nerve pulse reaches the pick up electrodes connected to vertical input, a pulse appears on the scope.

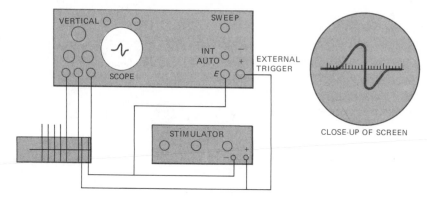

CLOSE-UP OF SCREEN

trodes, it produces a small voltage (a few millivolts) across adjacent electrodes. This is the input signal for the vertical input (set on 10 mV/cm).

In order to use the time base of the oscilloscope as an interval timer, it is necessary to start the sweep of the beam with the stimulator pulse. The time base should be set for 0.5 msec/cm, and the stimulator should be at a frequency of 50 pulses/sec. Each pulse is of 0.1 msec in duration. The amplitude of the stimulator pulse is increased until a signal is noted, that is, until a vertical deflection is noted on the oscilloscope tube when the impulse passes the electrodes connected to vertical input. The horizontal distance from the start of the sweep to the vertical impulse can be converted to the time it took the impulse to travel from the stimulator electrode to the vertical pickup electrode, that is, distance (cm) × 0.5 msec/cm. The distance between the stimulator electrode and vertical pickup electrode divided by this time gives the speed of the nerve impulse between these two points. In this example it was necessary to start the sweep with an external signal. This flexibility of the trigger source for the sweep of the oscilloscope adds much to its versatility.

As another example of oscilloscope use, consider the study of the heart potentials (ECG). In this case we are studying a waveform (ECG potential) that is periodic, and we wish to measure the period and observe the shape of the waveform. In this application it is necessary to trigger the scope on one of the voltage pulses that serves as the vertical input. The set-up that might be used is shown in Figure 26.12.

The synchronization of the sweep with the periodic input signal is

FIGURE 26.12
A schematic diagram illustrating a set up for measuring ECG on an oscilloscope.

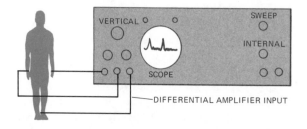

DIFFERENTIAL AMPLIFIER INPUT

achieved by switching the trigger selection switch to *internal,* which means trigger voltage will be provided by internal vertical amplifier of the oscilloscope shown in Figure 26.10. The trigger level can be adjusted to start the horizontal beam sweep at the initiation of a vertical signal. The time between heart beats can be determined by reading the distance between waveforms along the horizontal axis and multiplying this distance in centimeters by the time base setting. The vertical sensitivity should be 500 μV/cm.

Finally, consider the evaluation of a set of specifications provided by a manufacturer of oscilloscopes:

"Vertical bandwidth of DC to 15 MHz." This means that the vertical amplifier will faithfully amplify signals whose frequency may range from dc to 15 MHz.

"Vertical sensitivity of 10 mV/cm to 50 V/cm in twelve calibrated positions." This gives the range of input signals that can be studied on the cathode-ray tube screen. This screen is typically 6 cm vertically by 10 cm horizontally.

"Vertical input impedance of 1 meg ohm." This means the effective impedance across the input terminals is 1,000,000 Ω, and thus the oscilloscope is a good voltage measuring device.

"May have as many as 22 time bases from 2 sec/cm to 0.2 μsec/cm." This gives the range of horizontal time bases available.

"Trigger sensitivity of internal, 1 cm display, external, 0.5 V peak to peak." This information tells you that on internal trigger a vertical signal of 1 cm on any vertical sensitivity will trigger the time base sweep and that on the external trigger a voltage pulse of 0.5 V is needed to trigger the sweep.

These are the most important specifications to consider when selecting an oscilloscope, and they should be matched with the demands of your work.

Questions

1. For a heart rate of 72 beats/min, what would be a good time base setting?

SUMMARY

Use these questions to evaluate how well you have achieved the goals of this chapter. The answers to these questions are given at the end of this summary with the number of the section where you can find related content material.

Definitions

1. Suppose you are trying to measure a small-amplitude, short-time pulse in a biological system. Fill in the following blanks. The sample will be placed in a

copper cage, grounded to a water pipe, to reduce the a. _____ and thereby increase the b. _____ before sending the signal into an electronic device for c. _____. The wide d. _____ of the detector is desirable so that the pulse shape will not be changed. To make quantitative measurements of the amplitude of the pulse the e. _____ of the system must be calibrated. To make consistent repeated measurements, the f. _____ of the system will be improved by using a negative g. _____ control feature.

Electronic Devices

2. List two uses of a diode.
3. What kind of circuit element is a forward-biased diode?
4. What kind of circuit element is reversed-biased diode?
5. Match the following characteristics to these transistor configurations: common base, common collector, common emitter
   a. low input impedance
   b. high input impedance
   c. low output impedance
   d. high output impedance
   e. high power gain
   f. low power gain
   g. high voltage gain
   h. no voltage gain
   i. high current gain
   j. no current gain
   k. moderate output impedance
   l. moderate input impedance
6. List the ideal characteristics of an operational amplifier.

7. List four ways an op-amp can be used to process signals.
8. What does a differential amplifier amplify?
9. What is displayed on an oscilloscope tube?

Oscilloscopes

10. List at least four characteristics that are specified for an oscilloscope.

Answers

1. a. interference noise
   b. signal-to-noise ratio
   c. amplification
   d. frequency response
   e. linearity
   f. stability
   g. feedback (Section 26.2)
2. rectifier, temperature transducer (Section 26.3)
3. low impedance, high current flow (Section 26.3)
4. high impedance almost no current flow (Section 26.3)
5. common base: f, g, j, a, d; common collector: f, h, i, b, c; common emitter: e, g, i, k, l (Section 26.4)
6. infinite input impedance, zero output impedance, infinite amplication (Section 26.5)
7. to amplify signals, to amplify the difference between two signals, to add signals, to integrate signals, to differentiate signals (Section 26.5)
8. the difference between two signals (Section 26.5)
9. the voltage versus time curve for a signal (Section 26.6)
10. bandwidth, voltage sensitivity, time base range, input impedance, trigger sensitivity (Section 26.6)

ALGORITHMIC PROBLEMS

Listed below are the important equations from this chapter. The problems following the equations will help you learn to translate words into equations and to solve single-concept problems.

Equations

$$\text{output signal} = (\text{constant}_1)(\text{input signal}) + \text{constant}_2 \qquad (26.1)$$

$$\text{amplification} = \frac{V_{out}}{V_{in}} \qquad (26.2)$$

Problems

1. A linear electronic device gave output voltages of 25 V and 75 V for input volt-

ages of 5 V and 10 V respectively. What are the values of the linearity constants for this device?

2. A common-base transistor amplifier is reported to have an amplication factor of 50. What input voltage is needed to obtain a 1-V output signal?

Answers

1. $\text{constant}_1 = 10$, $\text{constant}_2 = -25$ V   2. 0.02 V

EXERCISES

These exercises are designed to help you apply the ideas of a section to physical situations. Where appropriate, the quantitative answer is given in brackets at the end of the exercise.

Section 26.3

1. What can you conclude about the ohmic nature of a diode? (Compare a I vs V plot of an ohmic resistor with Figure 26.2).
2. A thermocouple is a linear temperature transducer that generates an emf of about 1 mV when the reference junction is at 0°C, and the probing junction is at room temperature (27°C). Compare the sensitivity of a diode thermometer with a thermocouple.   [a diode is about 60 times more sensitive]

Section 26.4

3. Describe which of the transistor configurations you would use for each of the following situations (show a sketch of each):
   a. crystal microphone (high-impedance output) to audio-amplifier through a coaxial cable

   b. ECG potentials for display on scope with 10 mV/cm sensitivity
   c. photocell with small current output (microamps) to be monitored with milliamp meter.

Section 26.5

4. Given a voltmeter (1 V full scale), an op-amp, resistors of 10, 100, 1,000, 10,000, and 100,000 Ω, and a thermocouple, you wish to monitor the thermocouple with the meter. The thermocouple output is about 1 mV, sketch the system you could use showing the resistors used in the op-amp set up   [Op amp: $R_f = 10^5$ Ω; $R_i = 10^2$ Ω]

Section 26.6

5. Given four unknowns (one each of a capacitor, inductor, resistor, and diode) and a 100-V AC and 100-Volt DC source, explain how you could determine which unknown is which by using an oscilloscope with both AC and DC amplifiers.

PROBLEMS

The following problems may involve more than one physical concept.

6. Design an experiment to measure human reaction time to either a light or sound stimulus. Show a sketch of your set up. (Be sure to show clearly trigger mode selection).

7. Design a biofeedback system that might be used to condition a subject to reduce the temperature of a fingertip.
8. Using diodes and op-amps, design a differential thermometer system that measures small differences between two temperatures.

GOALS

When you have mastered the contents of this chapter you should be able to achieve the following goals:

Definitions
Define each of the following terms, and use it in an operational definition:

Planck's constant	length contraction
Planck's radiation law	time dilation
Wien's law	mass-energy equivalence
photon	Compton scattering
photoelectric effect	complementarity principle
work function	deBroglie wave
relativistic mass	uncertainty principle

Quantum and Relativistic Problems
Solve problems involving Wien's law, photoelectric effect, Compton scattering, deBroglie waves, the uncertainty principle, and relativistic physics formulations.

Tunnel Effect
Define and explain the physical significance of the tunnel effect.

PREREQUISITES

You should have mastered the goals of Chapter 4, Forces and Newton's Laws, Chapter 5, Energy, and Chapter 21, Electrical Properties of Matter, before starting this chapter.

27

Quantum and Relativistic Physics

27.1 Introduction

The topics we have presented in the proceeding chapters have been in the realm of classical physics. The concepts of mass, velocity, momentum, and energy were applied to macroscopic systems, and we expected the variables of these systems to have continuous ranges of values. It was assumed that classical physics based on Newton's mechanics, Maxwell's electromagnetic theory, and thermodynamics held the answers to all the problems of the physical universe. Some physicists at the end of the nineteenth century believed that the development of physics was complete. On the contrary, the new models developed in physics during the first half of the twentieth century may well signify one of the most exciting periods of human intellectual history. Quantum physics was born and showed that classical physics failed to describe physical phenomena of particles of 10^{-10} m dimensions. When velocities approach the speed of light, Einstein's special theory of relativity replaced classical physics as the working theory. In this chapter you will study the foundations of quantum and relativistic physics and some consequences of the new physics.

27.2 Planck's Radiation Law and Its Derivations

Perhaps you have noticed the change of color that occurs when you dim the incandescent lights of a room using a dimmer switch. When the switch is completely on, the lights are bright and give off white light. As you dim the lights, you can see that the light given off is not only less bright but is also of a different color. The dim light is red-orange in color. We can conclude that the color, or frequency, of light emitted from an incandescent bulb changes as we change the temperature of the filament in the bulb. This is one phenomenon that classical physics failed to explain, that is, the frequency distribution of the electromagnetic radiation emitted from heated objects. This radiation is emitted in a continuous range of frequencies in a characteristic way that depends on the temperature of the radiation source (Figure 27.1). Classical physics as applied by Lord Rayleigh failed to describe this distribution. In fact, Rayleigh's work predicted the "ultraviolet catastrophe," unlimited energy output at the wavelengths approaching zero.

Prior to Rayleigh's results Max Planck reported his solution to the problem. He combined classical equations that described each end of the distribution curve with the result being an empirical equation that fit the entire distribution curve. In order to explain his empirical formula, Planck had to postulate that vibrating objects, or oscillators, were responsible for the emission and absorption of thermal radiation. Furthermore, these oscillators could only have discrete energies. They could exist only in energy states that were integral multiples of a fundamental oscillator energy or quantum, when the quantum energy is given by hf, where h is *Planck's constant*, and f is the frequency of the oscillator.

The value of h necessary to make Planck's equation fit the radiation

FIGURE 27.1
The electromagnetic radiation of a perfect radiator as a function of temperature and wavelength. The electromagnetic radiation is measured as total energy radiated per unit area per unit time in the wavelength region λ to $\lambda + \Delta\lambda$. The ultraviolet catastrophe curve was Lord Rayleigh's theoretical attempt to fit the experimental data. Note the direction of the shift of the maximum as the temperature increases.

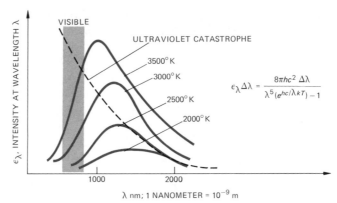

curves shown in Figure 27.1 is $h = 6.63 \times 10^{-34}$ joule-seconds. According to Planck's postulate, the various oscillators in a radiation source could only have energies given by

$$E_n = nhf \quad \text{or} \quad E_n = nhc/\lambda \tag{27.1}$$

since $f\lambda = c$ (see Chapter 16) where $n = 1,2,3, \ldots$, c is the speed of light in meters per second, f is the frequency in cycles per second, and λ is the wavelength in meters.

The equation that Planck found to describe the intensity or radiation from a perfect radiator as a function of the temperature of the radiator and the wavelength of the radiation is

$$E_\lambda = \frac{8\pi hc}{\lambda^5 \left(e^{hc/\lambda kT} - 1\right)} \text{watt/m}^3 \tag{27.2}$$

where c is the speed of light, λ is the wavelength of the radiation, k is Boltzmann's constant, T is the absolute temperature of the radiator, and E_λ is the energy intensity (watts/m²) emitted per unit wavelength (m).

Planck's theory was not an immediate success as its postulates were quite a break from some of the ideas of classical physics. This was much to the dismay of Planck who had hoped to explain his quantum postulate in classical terms. After all, he had used classical electromagnetic waves coupled to some special oscillators within the radiator. Two important results of Planck's theory were the derivation of the Stefan-Boltzmann law (Equation 11.8) and Wien's law (Equation 27.5). The Stefan-Boltzmann law, which was discussed in Chapter 11, describes the rate at which energy is radiated from a body as a function of the absolute temperature of a body. The mathematical equation is

$$I = \frac{P}{A} = \sigma T^4 \text{ watts/m}^2 \tag{11.8}$$

where σ is Stefan's constant $= 5.67 \times 10^{-8}$ watts/m²–°K⁴, T is the absolute temperature in degrees Kelvin, and I is the total energy per unit time per unit area emitted by a perfect thermal radiator, or the ratio of the emitted power P divided by area A. For radiators other than these perfect blackbody radiators we can write

$$I = \epsilon \sigma T^4 \tag{27.3}$$

where ϵ is the *emissivity* of the object $(0 < \epsilon < 1)$.

If an object of emissivity ϵ is maintained at a temperature T_1 in a chamber at a temperature of T_2, the net radiation from the object is equal to the difference between the radiation emitted from the object and absorbed from the chamber,

$$I = \epsilon \sigma (T_1{}^4 - T_2{}^4) \tag{27.4}$$

Wien's law is an empirical equation for the relationship between the temperature of a perfect radiator and the wavelength of the radiation of maximum intensity. It can be derived from Planck's law by finding the wavelength for maximum radiation intensity from Equation 27.2. The temperature of a radiating object and the wavelength of the maximum intensity of the emitted radiation have a constant product,

$$\lambda_{max}T = 2.88 \times 10^{-3}\text{m-}{}^\circ\text{K} \tag{27.5}$$

EXAMPLES

1. Given that the sun has its intensity peak output at 500 nm, find its effective blackbody temperature.

   Using Wien's law we can write:

   $$\lambda_{max}T = 2.88 \times 10^{-3}\text{m-}{}^\circ\text{K}$$

   $T = 2.88 \times 10^{-3}(\text{m}{}^\circ\text{K})/5.00 \times 10^{-7}$ m, $T = 5760{}^\circ\text{K}$ This is the effective *surface* temperature of the sun.

2. The emissivity of the human skin is close to one. Compare the rate of energy loss of a person (skin area $= 1.5$ m^2) inside when the temperature is $27{}^\circ$C, and outside when the temperature is $-23{}^\circ$C. (Assume skin temperature is $32{}^\circ$C.)

   From Equation 27.4 above you can write,

   $$\frac{\Delta E}{\Delta t} = IA = \epsilon \sigma A(T_1{}^4 - T_2{}^4)$$

   where T_1 is body temperature $= 350{}^\circ$K and T_2 is either the inside temperature $(300{}^\circ\text{K})$ or the outside temperature $(250{}^\circ\text{K})$,

   $$\frac{\Delta E_1/\Delta t}{\Delta E_2/\Delta t} = \frac{\sigma A \ (305^4 - 300^4)}{\sigma A \ (305^4 - 250^4)} = 0.116$$

   The rate of energy lost from the body by radiation is greatly reduced in a warm environment.

27.3 Photoelectric Effect

While Heinrich Hertz was investigating the properties of electromagnetic radiation in 1887, he observed that the metallic electrodes of his system were more easily discharged when the electrodes were illumi-

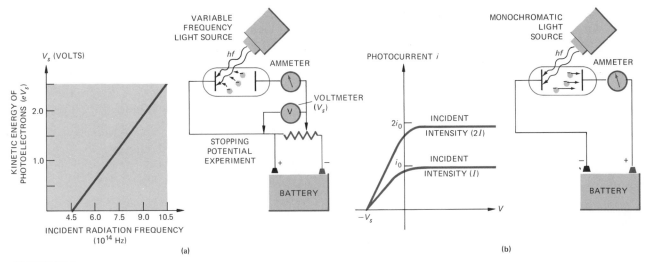

FIGURE 27.2
(a) The stopping potential as a function of incident radiation frequency in the photoelectric effect.
(b) The photocurrent as a function of applied voltage and incident radiation intensity in the photoelectric effect.

nated by ultraviolet light. This interaction between light and metal surfaces was subsequently studied in more detail and became called the *photoelectric effect*. The photoelectric effect is explained by a model of light energy interacting with a metal surface to cause electrons to be emitted from the surface. The light energy is partially absorbed by the metal and partially transformed into the kinetic energy of the emitted electrons.

The photoelectric effect is another example of quantitative experimental results as shown in Figure 27.2 that could not be explained by the models of classical physics. On the basis of the wave model of light, we found in Chapter 19 that the energy in a light beam is proportional to its intensity, and thus an intense red light should be able to provide enough energy to free an electron from the metal surface. The facts did not support this; the observations showed that below some threshold frequency, no matter how intense the light, electrons were not emitted from the surface (Fig. 27.2a). If the intensity of the light is increased, the photoelectric current is increased if the frequency of the light is above the threshold value (Figure 27.2b). The kinetic energy of the emitted photoelectrons (as measured by the voltage it takes to stop them from reaching the anode) is proportional to the frequency of the incident electromagnetic radiation (Figure 27.2a).

In 1905 Einstein offered his theory to explain the photoelectric effect. He applied Planck's quantum postulate to the incident light. He argued that the incident light was made up of quanta of light, which are called *photons*, each of energy hf. The intensity of the light beam is a measure of the number of photons per unit area per second. Einstein reasoned that the quantum interaction is an all or none interaction. If the photons do not have an energy equal to the energy with which electrons are bound to the metal (this binding energy is called the *work function*), there is no electron emission no matter how intense the light may be. There is no way for the electrons in the metal to store subthreshold photons until the threshold energy is reached. To support this, experiments show no observable time lag between time when light is turned on and when electrons appear.

If the photon has energy greater than the *work function,* the excess energy goes into the kinetic energy of the photoelectron. This is expressed by the following equation:

energy of the incident photon of light = kinetic energy
of the photoelectron + the work function of the metal

$$E = hf = \text{KE} + W \qquad (27.6)$$

where E is the photon energy, $hf = hc/\lambda$, W is the work function of the metal, and KE is the kinetic energy of the photoelectron. The electron volt (eV = 1.60×10^{-19} J) is a convenient energy unit for microscopic physics in this and the remaining chapters and is the energy increase of an electron when it is accelerated through a potential of one volt.

EXAMPLE

Light of wavelength $\lambda = 400$ nm strikes a metal and produces photoelectrons that are brought to rest by 1.30 V. Find the work function of the metal.

The energy of the photons is $E = hf = hc/\lambda = (6.63 \times 10^{-34}$ J–sec$) \times (3.00 \times 10^8$ m/sec$)/4.00 \times 10^{-7}$ m

$$E = 4.97 \times 10^{-19} \text{ J} = 3.11 \text{ eV}$$

where 1 eV = 1.60×10^{-19} J. The retarding potential of 1.30 V represents the stopping potential for the photoelectrons. Thus the kinetic energy of the photoelectrons is $e \times V_s = 1.30$ eV. From Equation 27.6, we find the work function,

$$W = hf - \text{KE} = 3.11 \text{ eV} - 1.30 \text{ eV}$$

$$= 1.81 \text{ eV} = 2.90 \times 10^{-19} \text{ J}$$

27.4 Relativistic Physics

Before we continue our discussion of the quantum properties of light, let us digress to consider some other work that Albert Einstein did during 1905 when he was a 26-year-old patent examiner in the patent office in Berne, Switzerland. Einstein was studying the electrodynamics of moving objects. He wrote a paper in which he proposed the two postulates that form the basis for the special theory of relativity. Einstein's special theory of relativity has some important consequences for the realm of quantum physics. These consequences are derived from Einstein's two fundamental postulates.

Postulate 1: The laws of physics are the same in all reference systems at rest or moving at a constant velocity.

Postulate 2: The speed of light is always measured to be the same value in a vacuum, and it is independent of the state of motion of the source or observer.

We will now give some of the results of Einstein's special theory of relativity along with the physical significance of these results.

Length Contraction: *All moving lengths appear to shrink in the direction of motion.*

This length contraction is frequently called *Lorentz contraction*. There is no contraction of lengths perpendicular to the velocity **v**. This phenomenon seems to contradict everyday experience, but it has been verified many times in modern physics experiments. The equation for the apparent length L in the direction of motion for an object moving with speed v is given by:

$$L = L_0 \sqrt{1 - \frac{v^2}{c^2}} \tag{27.7}$$

where L_0 is the length of the object in a system in which it is at rest, v is the speed of the moving object and c is the speed of light.

Time Dilation: *Moving clocks run slow.*

Again this result seems to contradict common experience, but it too has been verified many times in modern experimental physics. The equation for a time interval from the point of view of a stationary observer Δt_0 is:

$$\Delta t_0 = \Delta t \sqrt{1 - (v^2/c^2)} \tag{27.8}$$

where Δt is the time period shown by the clock moving with the apparatus, v is the speed of the motion of the apparatus, and c is the speed of light.

Consider the following experiment which illustrates Lorentz contraction and time dilation. This experiment has been repeated in several ways and in each case there is complete agreement with the consequences of Einstein's postulates.

EXAMPLE

Muons are a common component of cosmic rays that strike the top of the earth's atmosphere. They are unstable particles that decay with an average lifetime of 2.00×10^{-6} sec when at rest. An experiment was carried out in which the number of muons incident at the top of a mountain 3000 m above sea level was counted. These particles were moving with a speed of $0.998c$. In an average lifetime a muon would travel a distance of only about 600 m ($\sim c \times 2 \times 16^{-6}$ sec), and thus no muons would be expected to reach sea level. When the experiment was carried out it was found that most of the muons do reach sea level. The explanation for these observations comes from Einstein's equation as follows: From the point of view of a person in a lab on the mountain, the actual decay time of the muon is slowed down because the muon is a moving clock: its average lifetime moving at speed 0.998 c was

$$\Delta t = \frac{\Delta t_0}{\sqrt{1-(v^2/c^2)}} = \frac{2.00 \times 10^{-6} \text{ sec}}{\sqrt{1-0.998^2}} = 31.6 \times 10^{-6} \text{ sec}$$

and the distance the muons traveled in this time was $0.998c \times 31.6 \times 10^{-6} = 9500$ m, which is greater than the mountain height, thus explaining why so many muons reached sea level. From the point of view of the moving muon the physical result must be the same, but here we see that the mountain (which is moving at speed $0.998c$ with respect to the muon) contracts to a height of: $H = 3000 \sqrt{1-0.998^2}$ or $H = 190$ m and thus the 600 m traveled is enough to reach sea level. Here we see an example how equations of special relativity give the correct answer in each reference system.

Einstein's special theory of relativity has several important modifications for the dynamics of particles. In order to preserve the form of Newton's second law, namely that impulse is equal to the change in momentum (Equation 6.2), the relativistic momentum is given the following definition

$$\mathbf{p} = \frac{m_0}{\sqrt{1-(v^2/c^2)}} \mathbf{v} \tag{27.9}$$

where m_0 is the mass of the particle when it is at rest, $\mathbf{v}$ is the velocity of the particle, v is the speed, or magnitude of the velocity, of the particle, and c is the speed of light. You may notice that we can write Equation 27.9 in exactly the same form as the usual momentum equation if we invent a quantity m that is equal to the rest mass m_0 divided by $\sqrt{1-(v^2/c^2)}$. We call this quantity the relativistic mass m,

$$m = \frac{m_0}{\sqrt{1-(v^2/c^2)}} \tag{27.10}$$

Mass is a measure of inertia. This equation shows us that the inertia of a particle becomes indefinitely large as its speed approaches the speed of light. This means that it becomes more and more difficult to accelerate a relativistic object and that it is physically impossible to accelerate an object to the speed of light.

Einstein showed that as a result of his postulates the total energy of a relativistic particle could be expressed in the now famous equation,

$$E = mc^2$$

$$= \frac{m_0 c^2}{\sqrt{1-(v^2/c^2)}} \tag{27.11}$$

where E is the total energy, m is the relativistic mass, and c is the speed of light. If the rest energy of the particle is defined as $m_0 c^2$, then the kinetic energy of a relativistic particle can be written as

$$KE = E - m_0 c^2 = m_0 c^2 \left[\frac{1}{\sqrt{1-(v^2/c^2)}} - 1 \right] \tag{27.12}$$

The mass-energy equivalence formula as expressed in Equation 27.11 shows that in a fundamental way mass must be considered in a new formulation of the conservation of energy. The difference in masses of objects due to changes in energy in classical physics are too small to be measured, but this mass-energy equivalence is the basis of the nuclear binding energy processes which we will discuss in Chapter 31.

EXAMPLE

Consider the mass equivalent to a potential energy increase of 40,000 J. (This is about the energy required to lift a 10 kg mass from sea level to the top of the Empire State Building.) The equivalent mass would be given by

$$m = \frac{4.00 \times 10^4 \text{ J}}{9.00 \times 10^{16} \text{ (m/sec)}^2} = 0.444 \times 10^{-12} \text{ kg}$$

$$= 4.44 \times 10^{-10} \text{g}$$

This is a mass much too small to be detected with our best analytical balances.

In many atomic and nuclear processes the relativistic form of the kinetic energy of a particle should be used. In problem 23 at the end of the chapter you are asked to show that in the classical region of physics (where v is much smaller than c), the kinetic energy reduces to $\frac{1}{2}m_0v^2$.

27.5 Compton Effect

The photon model of light gives light particlelike characteristics instead of the wave characteristics we discussed in Chapter 19. In 1922 A. H. Compton reported that he had observed elastic collisions between x-ray photons and electrons. He was able to explain his results by assigning a momentum to the photons given by the equation:

$$p = h/\lambda = hf/c \tag{27.13}$$

where p is the momentum of a photon, h is Planck's constant, λ is the wavelength of the x rays, c is the speed of light, and f is the x-ray frequency.

An experimental set up to illustrate the Compton effect is shown in Figure 27.3a. The intensity of x rays scattered from a carbon block is measured as a function of the scattering angle and scattered x-ray wavelength. Loosely bound electrons may be freed from the carbon by the incident x-ray photons. The freed electrons are called photoelectrons.

The scattered photons have less energy than the incident photons. Thus, in addition to the unmodified x rays, scattered x rays with a slightly greater wavelength are detected at all angles except the zero scattering angle.

Compton observed collisions like the ones shown in Figure 27.3, and he could explain the results by using the conservation of momen-

FIGURE 27.3
(a) An experimental set up designed to study the Compton effect. (b) The scattering diagram for the inter-action of the incident photon with the scattered photon and recoiling electron in the Compton effect.

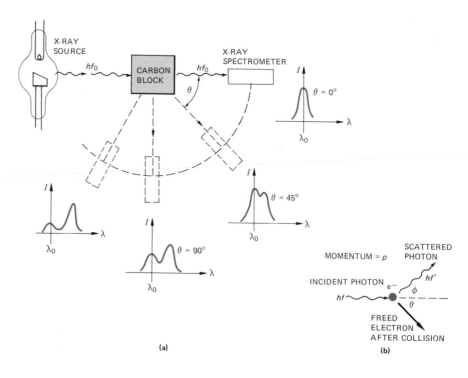

(a)

(b)

tum applied to the photon-electron collision. The following equations are derived from the application of the conservation of energy and momentum to this collision (Figure 27.3b).

MOMENTUM CONSERVATION

Resolving the momentum vector into its components,

x-components:
$$p_{\lambda x} = p_{\lambda' x} + p_{ex}$$
$$hf/c = (hf'/c) \cos \phi + p_e \cos \theta$$

y-components:
$$p_{\lambda y} = p_{\lambda' y} - p_{ey}$$
$$0 = (hf'/c) \sin \phi - p_e \sin \theta$$

ENERGY CONSERVATION

$$hf = hf' + KE$$

where KE is the kinetic energy of the electron.

By the time of Compton, it was known that the classical equations for the interchange momentum and kinetic energy of a moving object had to be modified for objects traveling at high velocities. The modified equations for the kinetic energy and the momentum for relativistic electrons are given by Equations 27.9 and 27.12. These equations can be solved to obtain the following equation for the wavelength of the scattered x-ray photon in terms of scattering angle and incident photon wavelength:

$$\lambda' - \lambda = \frac{h}{m_0 c}\left(1 - \cos\phi\right) \qquad (27.14)$$

where λ' is the scattered photon wavelength, λ, the incident photon wavelength, and $h/m_0 c$, is a constant called the Compton wavelength, which equals 2.43×10^{-12} m.

EXAMPLE

Given that the incident photon had a wavelength of 12.4×10^{-12} m, and the scattered photon recoiled at $180°$, find the energy transferred to the electron. (*Note:* when $\phi = \pi$, then the energy transferred is maximum.

First calculate the energy lost by the photon because of the scattering:

$$\lambda' - \lambda = \frac{h}{m_0 c}\left(1 - \cos 180°\right) = \frac{2h}{m_0 c}$$

$$\lambda' = \lambda + \frac{2h}{m_0 c}$$

$$= (12.4 + 4.86) \times 10^{-12}\text{ m} = 17.3 \times 10^{-12}\text{ m}$$

$$\text{kinetic energy of the electron} = \frac{hc}{\lambda} - \frac{hc}{\lambda'} = \text{energy lost by the photon}$$

$$= \left(\frac{10^{12}}{12.4} - \frac{10^{12}}{17.3}\right) hc$$

$$= (6.63 \times 10^{-34} \times 3.00 \times 10^{8})\left(\frac{4.9}{12.4 \times 17.3} \times 10^{12}\right)$$

$$= 45 \times 10^{-16}\text{ J}$$

$$= 28.1\text{ keV}$$

where $1\text{ keV} = 1.60 \times 10^{-16}$ J.

27.6 *Complementarity Principle*

The Compton effect and the photoelectric effect evoke a particle model for light. The particle model of light is used to describe the interaction of electromagnetic radiation with matter; the classical wave model is used to explain propagation phenomena. We use two models for light, one model is particlelike and the other is wavelike. The aspect of light that we observe in a particular experiment depends on the nature of the experiment. Niels Bohr would cite this as an example of the *complementarity principle*. The wave and particle models of light complement each other and give us a theory that encompasses the many potentialities of both particle and wave behavior. Bohr applied the complementarity principle to areas outside of physics. Bohr's goal was to discover the interrelationships among all areas of knowledge. He viewed the complementarity principle as an important step toward his goal. In biology, Bohr presented the molecular model of living systems and the teleological approach to the causality model of living systems as complementing components of the organization of the living system. In psychology, the complements may be behaviorism and self-conscious freedom of choice. (What complementary models would you suggest for a political system or organization?)

27.7 The deBroglie Wave

In 1924 Louis deBroglie proposed that material particles such as electrons might have a dual nature. He suggested that the wavelength of a particle should be defined by the same equation that applies to the photon:

$$\lambda = h/p \qquad (27.15)$$

where h is Planck's constant and p is the magnitude of the momentum of the particle. The waves associated with matter are not electromagnetic waves. These waves are probability waves. The intensity of the probability wave at a point is a measure of the probability of finding the particle at the point. The test of this idea would be to do a wave experiment with particles. One experiment that depends entirely on the unique features of waves is a diffraction experiment. Such an experiment was done by Clinton J. Davisson and L. H. Germer in 1924. They found that electrons were diffracted by crystals. The results were explained by using Equation 27.15 for the electron wave diffracted by a crystal lattice with spacing the same order of magnitude as the electron wavelength. The wave model of electrons can be used to explain the operation of the electron microscope. Small electron wavelengths give high resolving power for the electron microscopes.

EXAMPLES

1. Davisson and Germer used a beam of 54.0-eV electrons in their electron diffraction experiment. Find the wavelength of these electrons.
    The wavelength λ and the momentum p are given by

    $$\lambda = h/p \qquad p^2/2m_0 = KE = eV$$

    where e is the charge of the electron, V is the voltage, h is Planck's constant, and m_0 is the electron rest mass of $9.11 \times 10 = 10^{-31}$ kg.

    $$p = (2\ m_0 eV)^{1/2}$$

    $$\lambda = \frac{h}{(2\ m_0 eV)^{1/2}}$$

    $$\lambda = \frac{6.63 \times 10^{-34} \text{ J-sec}}{[2 \times (9.11 \times 10^{-31} \text{ kg}) \times 1.60 \times 10^{-19} \text{ C} \times (54.0 \text{ V})]^{1/2}}$$

    $$= \frac{6.63 \times 10^{-34} \text{ J-sec}}{3.97 \times 10^{-24} \text{ kg-m/sec}}$$

    $$= 1.67 \times 10^{-10} \text{ m} = 0.167 \text{ nm}$$

2. Find the speed v of a neutron that has a deBroglie wavelength of 0.100 nm.

$$mv = \frac{h}{\lambda} = \frac{6.63 \times 10^{-34} \text{ J-sec}}{1.00 \times 10^{-10} \text{ m}}$$

$$v = \frac{6.63 \times 10^{-24}}{1.67 \times 10^{-27}} = 3.97 \times 10^3 \text{ m/sec}$$

This is a slow neutron with a kinetic energy of 0.08 eV. This is an energy comparable to thermal energies at room temperature and thus such neutrons are called *thermal neutrons*. Thermal neutrons are used in neutron diffraction studies to measure the distances between lattice planes in solids. Neutron diffraction is especially useful in studying organic crystals containing hydrogen. (Can you explain why neutrons are useful for such studies while x rays and electrons are not very useful?)

27.8 Uncertainty Principle

We have now arrived at an understanding of a property of a system called a particle (e.g., an electron) that can be thought of as a particle or as a wave. At one instant we can think of the electron as a small chunk of matter with an electrical charge e located at some point in space. The next instant we can think of the electron as a cloud of probability waves that extends throughout space. What sense then can we make of the idea of the position of a particle?

Werner Karl Heisenberg formulated his now famous *uncertainty principle* in 1927 in order to clarify the use of the terms *position* and *velocity* when referring to quantum phenomena. The various forms of the uncertainty principle express limits to the preciseness of our knowledge of the magnitudes of pairs of physical variables. Algebraically stated the Heisenberg uncertainty principle can be written as

The product of the uncertainty in momentum and position is greater than Planck's constant

$$\Delta p_x \, \Delta x \geq h \tag{27.16}$$

where Δp_x is the uncertainty in the x-component of the momentum, Δx is the uncertainty in the position of the object in the x-direction and h is Planck's constant. Another equivalent form is:

$$\Delta E \Delta t \geq h \tag{27.17}$$

Here ΔE is the uncertainty in the energy, and Δt is the uncertainty in the time. These uncertainty relationships must be understood as a part of our models to explain nature; they are not based on limitations of measuring instruments. As a result of these relationships, the certainties of classical physics are impossible in quantum physics. In fact, quantum theory is a theory based on a probabilistic understanding of events and physical variables. The uncertainty principle is a consequence of this fundamental model of physical reality as it is now formulated by physicists.

EXAMPLES

1. Find the uncertainty in velocity of a 1.0-gm particle if its position is measured to within 0.1 mm.

$$m\Delta v = \Delta p \simeq \frac{h}{\Delta x}$$

$$\Delta v \simeq \frac{h}{m\Delta x} = \frac{6.63 \times 10^{-34} \text{ J–sec}}{1.0 \times 10^{-3} \text{ kg} \times 10^{-4} \text{ m}} = 6.63 \times 10^{-27} \text{ m/sec}$$

This is an uncertainty far below any measurement capability.

2. Find the uncertainty in the energy of photons emitted in the time of 10^{-8} sec. (This is a characteristic time for excitation of many atomic systems.)

$$\Delta E \ \Delta t \simeq h$$

$$\Delta E \simeq \frac{6.63 \times 10^{-34} \text{ J–sec}}{10^{-8} \text{ sec}} = 6.6 \times 10^{-26} \text{ J} = 4.1 \times 10^{-7} \text{ eV}$$

27.9 Tunnel Effect

One of the results of quantum theory is the prediction of events that are forbidden by classical physics. A potential-energy barrier exists wherever a repelling force acts to restrict a particle from a region of space. Figure 27.4 shows an energy well formed by a finite barrier on one side and an infinite barrier on the other side. (An infinite barrier designates a forbidden region of space.)

In this case let us assume we can confine electrons to a region of 1.00 nm using a 2.00-eV energy barrier. This uncertainty in x gives a corresponding uncertainty in momentum of

$$h/\Delta x = 6.63 \times 10^{-34} \text{ J–sec}/1.00 \times 10^{-9} \text{ m} = 6.63 \times 10^{-25} \text{ kg–m/sec}$$

The corresponding uncertainty in energy is given by the equation,

FIGURE 27.4
A potential barrier system where an electron is initially inside the potential well. Experiments show that some electrons may be found outside the barrier. This is known as the quantum physics tunnel effect in which the electron "tunnels" through an energy barrier.

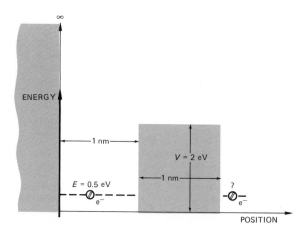

$$\Delta E = \frac{(\Delta p)^2}{2m} = \frac{[6.63 \times 10^{-25} \text{ kg-(m/sec)}^2]^2}{2 \times 9.11 \times 10^{-31} \text{ kg}} = 2.41 \times 10^{-19} \text{ J}$$

$$\Delta E = 1.51 \text{ eV}$$

This uncertainty means that an electron with energy of 0.5 eV or more has an uncertainty in energy sufficient to allow it to get over the top of the 2.00-eV barrier. Since the uncertainty in x is equal to the width of the barrier we cannot say with certainty on which side of the barrier the electron is. In fact, a rigorous quantum mechanical treatment of this problem shows that the electron has a probability of tunneling through the barrier. As the barrier height and/or barrier width is increased, the probability for tunneling through it decreases. This quantum tunneling phenomena is helpful in understanding alpha decay in radioactive nuclei and electron behavior in semiconductors.

SUMMARY

Use these questions to evaluate how well you have achieved the goals of this chapter. The answers to these questions are given at the end of the summary with the number of the section where you can find related content material.

Definitions

1. Planck's radiation law provides a correct equation for the graph of radiation _____ as a function of _____.

2. Wien's law could be used with the spectral distribution of a radiation source to find
   a. Planck's constant
   b. velocity of light
   c. temperature of source
   d. deBroglie wavelength
   e. source size

3. The photon is defined as the smallest energy packet for a given frequency and is equal to
   a. h/λ
   b. hf
   c. hc/λ
   d. c/λ
   e. hc

4. Compton scattering gives experimental evidence that the photon has particlelike
   a. wavelength
   b. frequency
   c. velocity
   d. momentum
   e. mass

5. The photoelectric effect shows that the electron emitted from a metal has energy that depends on
   a. metal material
   b. photoelectric current
   c. photon energy
   d. light intensity
   e. none of these

6. The uncertainty principle states that it is impossible to know exactly both a particle's momentum and
   a. energy
   b. position
   c. velocity
   d. acceleration
   e. mass

7. If a proton, which has an electric charge of $+e$ and a rest mass equal to about 1900 times the rest mass of the electron, and an electron are accelerated from rest through the same potential difference, the ratio of proton de Broglie wavelength to electron de Broglie wavelength will be
   a. one
   b. zero
   c. greater than 1
   d. less than 1
   e. none of these

8. An example of the complementarity principle is

a. particle-wave nature of matter
b. particle-wave nature of radiation
c. momentum-energy of particles
d. time-position of particles
e. none of these

9. Einstein's mass-energy equivalence formulation predicts a mass change associated with energy change ΔE to be

a. $\dfrac{\Delta E}{hf}$

b. $\dfrac{\Delta E}{\Delta p}$

c. $\dfrac{hf}{\Delta E}$

d. $\dfrac{\Delta E}{c^2}$

e. $\dfrac{\Delta p}{\Delta e}$

10. A long stick is observed to have a length L as it moves past at a speed v; its length in its own rest frame is

a. $\dfrac{L}{\sqrt{1-(v^2/c^2)}}$

b. $L\sqrt{1-(v^2/c^2)}$

c. L

d. $L(V/c)$

e. $L(c/v)$

11. A clock moving past an observer at speed v appears to her to have a period t. The period of a clock in a rest frame is

a. $\dfrac{t}{\sqrt{1-(v^2/c^2)}}$

b. $t\sqrt{1-(v^2/c^2)}$

c. t

d. $t(v/c)$

e. $t(c/v)$

Quantum and Relativistic Problems

12. The peak wavelength for a 10,000°K source will be
    a. 2.9×10^{-7} m
    b. 2.88×10^{-1} m
    c. 1.2×10^{-12} m
    d. 2.4×10^{-12} m
    e. 4.8×10^{-12} m

13. The maximum increase of a photon wavelength undergoing Compton scattering is (in meters)
    a. 0.24×10^{-12}

b. 0
c. 1.2×10^{-12}
d. 2.4×10^{-12}
e. 4.8×10^{-12}

14. If a photon has an energy equal to the rest energy of an electron, then its wavelength is

a. $\dfrac{2m_0c^2}{hf}$

b. $\dfrac{h}{m_0c}$

c. $\dfrac{hf}{2m_0c^2}$

d. $\dfrac{h}{2m_0c}$

e. 0

15. If the uncertainty in position of a microbe (10^{-9} kg) is 10^{-6} m, the uncertainty in its speed will be (in m/sec)
    a. 6.63×10^{-34}
    b. 6.63×10^{-28}
    c. 6.63×10^{-19}
    d. 6.63×10^{-25}
    e. zero

16. The length of a meter stick moving with speed $0.8c$ in the direction of its length will appear how long to an observer at rest.
    a. 80 cm
    b. 64 cm
    c. 36 cm
    d. 125 cm
    e. 60 cm

17. The period of a clock in its rest frame is one second. When moving at a speed of $0.6c$ the period will appear to a laboratory observer to be
    a. 1.67 sec
    b. 0.6 sec
    c. 0.64 sec
    d. 1.25 sec
    e. 0.8 sec

Tunnel Effect

18. The physical significance of the tunnel effect of quantum physics is that particles are found to tunnel through regions where
    a. PE < 0
    b. KE < 0
    c. KE + PE < 0
    d. KE > 0
    e. none of these

Answers

ALGORITHMIC PROBLEMS

Listed below are the important equations from this chapter. The problems following the equations will help you learn to translate words into equations and to solve single-concept problems.

Equations

$$E_n = nhf \qquad n = 1, 2, 3, \ldots \tag{27.1}$$

$$E_\lambda = \frac{8\pi hc^2}{\lambda^5(e^{hc/\lambda kt} - 1)} \tag{27.2}$$

$$I = \sigma \epsilon T^4 \tag{27.3}$$

$$\lambda_{max} T = 2.88 \times 10^{-3} \text{ m-}^\circ\text{K} \tag{27.5}$$

$$hf = \text{KE} + W \tag{27.6}$$

$$L = L_0 \sqrt{1 - (v^2/c^2)} \tag{27.7}$$

$$\Delta t_0 = \Delta t \left(\sqrt{1 - (v^2/c^2)} \right) \tag{27.8}$$

$$\mathbf{p} = \frac{m_0 \mathbf{v}}{\sqrt{1 - (v^2/c^2)}} \tag{27.9}$$

$$m = m_0 / \sqrt{1 - (v^2/c^2)} \tag{27.10}$$

$$\Delta E = mc^2 = \left(\frac{m_0}{\sqrt{1 - (v^2/c^2)}} \right) c^2 \tag{27.11}$$

$$p = \frac{hf}{c} \tag{27.13}$$

$$\lambda' - \lambda = \frac{h}{m_0 c} (1 - \cos \phi) \tag{27.14}$$

$$\lambda = \frac{h}{mv} = \frac{h}{p} \tag{27.15}$$

$$\Delta p_x \, \Delta x \geq h \tag{27.16}$$

$$\Delta E \cdot \Delta t \geq h \tag{27.17}$$

Problems

1. The frequency of a given source of green light is 5.5×10^{14} Hz. What is the energy of this radiation, i.e., the energy of each photon?
2. If the absolute temperature of a filament in a light bulb is 2500°K, what is the wavelength for the maximum intensity of the radiation from the hot filament? What is its corresponding energy per photon of radiation? Is this bulb useable as a source of light for reading?
3. Two bodies are exactly alike except in temperature, one is at absolute temperature of T and the other is at absolute temperature of $1.5\ T$. What is the ratio of the rates of energy emission from the two bodies?
4. The value of the work function of potassium is 2.25 eV. What is the longest wavelength that will produce a photoelectron from a potassium surface?
5. What is the momentum of a photon that has a frequency of 5.50×10^{14} Hz?
6. What is the wavelength that is associated with you when you are walking at 5 km/hour? What do you think of the prospects of detecting this wavelength?
7. Suppose you know your momentum (as for example in problem 6) to within 0.1 percent. What is the limit upon knowing your exact location at that time? Comment upon this result.

Answers

1. 3.6×10^{-19} J = 2.3 eV
2. 1.15×10^{-6} m, 1.73×10^{-19} J = 1.08 eV
3. 1:5.1
4. 553 nm

5. 1.22×10^{-27} kg–m/sec
6. $\sim 10^{-36}$ m
7. $\sim 10^{-32}$ m

EXERCISES

These exercises are designed to help you apply the ideas of a section to physical situations. When appropriate the numerical answer is given in brackets at the end of each exercise.

Section 27.2

1. A prominent line emitted by mercury street lights is green and has a wavelength of 5460 Å or 546 nm. What is its frequency and its energy in joules and in electron volts? [5.49×10^{14} sec^{-1}, 3.64×10^{-19} J, 2.27 eV]
2. One can express the energy of any photon radiation in terms of electron volts. A very useful rule of thumb relationship is that the product of energy in terms of electron volts and wavelength in nanometers is equal to 1240 eV–nm. Can you verify this?
3. The wavelength of maximum energy of radiation from the sun is 480 nm. What is the frequency of radiation of this wavelength and what is the energy of this photon in electron volts? [6.25×10^{14} sec^{-1}, 2.58 eV]

4. Find the temperature of a perfect thermal radiator that has its peak output at 400 nm. How much would this temperature have to be raised to increase the radiated power per unit area by 10 percent? [7200°K, 173°K]

Section 27.3

5. In studying wave motion, you found that the intensity of a wave depends upon the square of its amplitude. What does the intensity of light mean in terms of quantum concepts?
6. If 5 watts of light of the yellow line of a sodium emission spectrum (589 nm) fall upon a surface, how many photons strike the surface per second? [1.48×10^{19} photons/sec]
7. In quantum physics we say energy = $h \times$ frequency. What are the dimensions of h?
8. If sodium has a work function of 2.28 eV, what is the longest wavelength that will produce a photoelectron from sodium? If one doubles the intensity of light of

560 nm, will a photoelectron be emitted from sodium? [545 nm]

9. Potassium has a photoelectric work function of 2.24 eV. If it is illuminated by the 436-nm mercury line, what will be the maximum kinetic energy of the photoelectrons? [0.60 eV]

10. It is noted that the threshold wavelength for a certain metal is 500 nm. Find the energy of the photoelectrons emitted when this metal is radiated with 450 nm light. [0.28 eV]

11. The work function of cesium is 2.0 eV. The stopping potential in an experiment is found to be 1.1 V.
    a. Find the threshold wavelength for cesium.
    b. Find the wavelength of light used in the experiment. [a. 6.2×10^{-7} m; b. 4.0×10^{-7} m]

Section 27.4

12. Show that when $v/c \ll 1$, the relativistic kinetic energy reduces to the classical form, $\frac{1}{2} m_0 v^2$.

13. A particle with an average lifetime of 10^{-6} sec at rest moves with a speed of $0.9c$. Show that an observer in the lab would measure a lifetime of 2.3×10^{-6} sec.

14. Find the distance the particle in problem 13 will move in the lab. [6.4×10^2 m]

15. By solving Equation 27.9 for the magnitude of the velocity, v, eliminate v from Equation 27.11 to obtain a relationship between total energy, E, momentum, p, and rest energy, $m_0 c^2$. Show that the following statement is correct: $E^2 = p^2 c^2 + m_0^2 c^4$.

Section 27.5

16. Calculate the increase in wavelength of the scattered photon for each of the following scattering angles:
    a. 0°
    b. 37°
    c. 53°
    d. 90°
    e. 150°
    f. 180° [a. 0; b. 4.86×10^{-4} nm; c. 9.72×10^{-4} nm; d. 2.43×10^{-3} nm; e. 4.53×10^{-3} nm; f. 4.86×10^{-3} nm]

17. If a photon behaves as a particle and obeys particle mechanics, what happens if a photon and a particle collide head on? Set up equations obeying conservation of momentum and conservation of kinetic energy for a collision between a photon and an electron. Explain in words what happens.

PROBLEMS

Each of the following problems may involve more than one physical concept. When appropriate, the answer is given in brackets at the end of the problem.

18. A photon of wavelength 7.2×10^{-12} m is scattered at $\pi/2$ radians.
    a. Find the energy of the scattered photons.
    b. Find the energy of the electrons after collision. [a. 129 keV; b. 43.5 keV]

19. Derive the equation for an electron deBroglie wavelength after the electron has been accelerated from the rest through V volts. (Assume nonrelativistic speeds.)

20. Find the energy of an electron with a wavelength equal to the circumference of the Bohr orbit in the hydrogen atom. (*Hint:* circumference = $2\pi \times 5.3 \times 10^{-11}$ m) [13.8 eV]

21. Find the ratio of the de Broglie wavelength of an electron to that of a proton with the same energy. [$\sqrt{m_p/m_e}$]

22. The resolving power of a microscope is proportional to the wavelength divided by the numerical aperture. Calculate the ratio of the resolving power of an electron microscope using 10,000-V electrons to that of a microscope using 500-nm light with the same numerical aperture for both. [4.07×10^4]

23. Find the uncertainty in the momentum and energy of an electron within the nucleus ($\Delta x = 10^{-14}$ m). What does your answer suggest regarding electrons in the nucleus [6.63×10^{-20} kg–m/sec, 1.51×10^{10} eV]

24. Assume the uncertainty in the lifetime of a particle is the time it takes light to travel across the nucleus. Find the uncertainty in the energy of this particle. (This was a calculation used to estimate the rest mass energy of the π meson, the nuclear force field particle.) [1.99×10^{-11} J, 1.24×10^8 eV]

25. Suppose that a 40-g bullet has its position known to within the diameter of the rifle (1 cm). Find its uncertainty in momentum. If it is aimed at a target at a distance that it takes 3 sec to reach, by how much will it miss its target due to the uncertainty principle? [6.63×10^{-32} kg m/sec, 4.98×10^{-30} m]

26. If the energy of a particle of mass m is fV where V is the barrier height energy (Figure 24.4), and f is a fraction less than one, find the relation between V

and the barrier width (Δx) that will just allow tunneling using the uncertainty principle. [$\Delta x \leq h/\sqrt{2mfV}$]

27. Show that if an electron's deBroglie wavelength equals the width of a barrier 10^{-10} m wide, the electron can just pass over a 145-eV barrier.

28. Use the uncertainty principle to find the energy uncertainty for an electron confined to a space corresponding to a virus D nm in diameter. What physical significance can you give this energy value?

$$\left[\Delta E = \frac{(h/D)^2}{2m} = \frac{(hc)^2}{2mc^2D^2} = \frac{(1240 \text{ eV nm})^2}{2 \times 0.51 \times 10^6 \text{ eV}}\left(\frac{1}{D^2}\right)\right]$$

29. Calculate the ratio of photon energy to electron energy if both the photon and electron have the same wavelength. [$E_p/E_e = v/c = \sqrt{p^2/(p^2 + m_0^2c^2)}$]

30. The mass of a body is a function of velocity in accordance with Equation 27.10. Plot the mass of an electron for velocities of $0.1c$ to $0.99c$. (*Rule of thumb:* We neglect change in mass for velocities below $0.1c$.)

31. What is the mass of an electron accelerated through a potential of
    a. 100 kV?
    b. 2×10^6 V? [a. $1.19m_0$; b. $4.8m_0$]

32. What is the velocity of an electron accelerated through
    a. 100 kV
    b. 2×10^6 V [a. $v \simeq 0.54\ c$; b. $v \simeq 0.98c$]

GOALS

After you have mastered the contents of this chapter, you will be able to achieve the following goals:

Definitions
Define each of the following terms and use it in an operational definition:

atomic number   ionization energy
Bohr radius   Beer's law
quantum number   Lambert's law
energy level   optical density
spectra   absorbance
laser

Bohr Model
State the assumptions and predictions of the Bohr model of the hydrogen atom. Deduce the concept of ionization potential.

Pauli Exclusion Principle
Use the Pauli exclusion principle and the four quantum numbers; n, l, m_l, and m_s to account for the periodic chart of the elements.

Bohr Model Problems
Solve problems using the Bohr equations for hydrogen.

Energy Level Diagrams
Use energy level diagrams to explain the emission and absorption of radiation by atomic systems.

Spectrometery and Lambert's Law
Sketch a simple spectrometer, and discuss how it can be used in measuring the concentration of an element in a solution.

Laser
Explain the principle of laser action.

PREREQUISITES

Before you begin this chapter you should have achieved the goals in Chapter 4, Forces and Newton's Laws, Chapter 5, Energy, Chapter 21, Electrical Properties of Matter, and Chapter 27, Quantum and Relativistic Physics.

28

Atomic Physics

28.1 Introduction

The realm of quantum physics is not an obvious part of your daily experience. Where do you expect to find observable evidence of the quantum model? Although the quantum model was originated to solve the problem of thermal radiation, it was the atomic model of matter that gave rise to the impressive early successes of the quantum model. Do you know the atomic basis for the operation of the common fluorescent light?

Emission and absorption spectroscopy play a central role in the chemical analysis of specimens in medical laboratories. The atomic and quantum models have been combined to provide physical explanations for these and other atomic phenomena. In this chapter you will be introduced to a quantum model of the atom and some of its important applications.

28.2 Rutherford's Nuclear Atom Model

The early model of the atom proposed by J. J. Thomson in 1904 consisted of a positively charged spherical volume with point particle electrons distributed throughout the volume. (This was referred to as the "plum pudding" model of the atom, with the electrons playing the role of raisins.) This model, with its various symmetrical distributions of electrons, attempted to explain the periodicity of chemical properties of elements as well as the electromagnetic spectra of atoms. The empirical equations for the various atomic spectral series could not be derived from the then existing models.

The model met with very little success, and in 1911 Lord Rutherford published a paper which showed that the atom was much different than the Thomson model. Rutherford's model was based on the scattering of alpha particles (discovered to be helium nuclei of atomic mass 4) by gold foil (Figure 28.1). Since the alpha particle is much more massive than the electron, $(m_\alpha/m_e \cong 7360)$, you should expect very little deflection of the alpha particle by the electrons in the gold. Rutherford found, however, that some of the alpha particles were scattered at

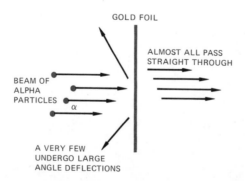

FIGURE 28.1
Rutherford's experiment in which helium nuclei were scattered by a thin gold foil indicated to him that the atom had a massive nucleus.

GOLD FOIL

ALMOST ALL PASS
STRAIGHT THROUGH

BEAM OF
ALPHA
PARTICLES
α

A VERY FEW
UNDERGO LARGE
ANGLE DEFLECTIONS

very large angles, as though they had collided head on with a small, positively charged particle. Rutherford concluded that the atom must be made up of a small, positively charged nucleus surrounded by the electrons. Rutherford estimated the positive nucleus radius to be about 1/3000 times the atomic radius. (He was about a factor of 3 too large in his estimate.) Rutherford imagined the electrons circling the nucleus, held in circular orbits by the Coulomb force. The atom as a whole has a neutral charge; that is, the positive charge on the nucleus is equal in magnitude to the sum of the electron charges. In his early work Rutherford represented the nuclear charge as Ze, where Z is the *atomic number* of the element and e is the electron charge. (Z is also equal to the number of electrons in the atom.) Elements are arranged in the periodic table in accordance with the individual atomic number.

28.3 Bohr's Model for the Hydrogen Atom

Niels Bohr, a young Danish theoretical physicist, had the insight and courage to propose a model that applied quantum physics concepts to the hydrogen atom in 1913. Planck's quantum hypothesis had accounted for blackbody radiation. Einstein had used the quantum concept and the photon to explain the photoelectric effect. Bohr's atomic model was even more remarkable. His model was designed to explain the stability of the atom and to account for the characteristic line spectra of hydrogen. Bohr's model incorporated Ernest Rutherford's nuclear model of the atom. Rutherford's alpha-particle scattering experiments had shown that most of the mass was associated with the positive charge of the atom in a small volume at the center of the atom. In Rutherford's nuclear model, electrons revolved around the positive nucleus. However, there was a serious flaw in this classical physics model. In classical electromagnetism, any accelerating charge becomes a source of electromagnetic radiation. Rutherford's model described the electron as being under constant centripetal acceleration, and classical physics predicted the ultimate spiral of the electron into the nucleus. To solve this dilemma, Bohr presented the following postulates:

Postulate 1: *Classical dynamical equilibrium of the atom involves stationary states given by the quantum condition on the angular momentum for stable orbits of the electron.*

$$L = \frac{nh}{2\pi}$$

(28.1)

where n is the quantum number, $n = 1, 2, 3, \ldots$, and L is angular momentum, mvr. These stationary-state orbits are not required to satisfy the radiation laws of classical physics.

Postulate 2: *Radiation is emitted or absorbed when the system makes a transition from one stationary state to another.*

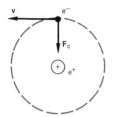

FIGURE 28.2
A schematic diagram of the Bohr model of the hydrogen atom. The negative electron is pictured moving in a circular orbit around the positive proton.

$$hf = E_f - E_i \qquad (28.2)$$

where E_f is the final energy and E_i is the initial energy of the states involved when a photon of energy hf is emitted or absorbed. Postulate 1 coupled with classical physics can be used to derive equations for the energy value and the circular orbit radius associated with Bohr's allowed stationary states for the hydrogen atom.

The energy of the nth state of the hydrogen atom is derived as follows (Figure 28.2). From Newton's and Coulomb's laws

force of attraction = centripetal force

$$F_c = \frac{1}{4\pi\epsilon_0}\frac{e^2}{r^2} = \frac{mv^2}{r}$$

From conservation of energy

kinetic energy + potential energy = E_{total}

$$\frac{1}{2}mv^2 - \left(\frac{1}{4\pi\epsilon_0}\right)\left(\frac{e^2}{r}\right) = E_{total}$$

From the force equation we have

$$mv^2 = \left(\frac{1}{4\pi\epsilon_0}\right)\left(\frac{e^2}{r}\right)$$

Thus

$$E_{total} = \frac{1}{2}\left(\frac{1}{4\pi\epsilon_0}\right)\left(\frac{e^2}{r}\right) - \left(\frac{1}{4\pi\epsilon_0}\right)\left(\frac{e^2}{r}\right)$$

$$= -\frac{1}{2}\left(\frac{1}{4\pi\epsilon_0}\right)\left(\frac{e^2}{r}\right)$$

The negative sign for E_{total} indicates that electron is bound to the nucleus. The quantum condition of angular momentum is

$$L = mvr = n\frac{h}{2\pi} \quad \text{where } n = 1,2, \ldots$$

and from the force equation we have

$$mv = \left(\frac{1}{4\pi\epsilon_0}\right)\left(\frac{me^2}{r}\right)^{1/2} \quad \text{and} \quad \left(\frac{me^2r}{4\pi\epsilon_0}\right)^{1/2} = n\frac{h}{2\pi}$$

Thus, the radius of the nth electron orbit of hydrogen is given by

$$r_n = n^2\left[\left(\frac{h}{2\pi}\right)^2\left(\frac{4\pi\epsilon_0}{me^2}\right)\right] = n^2a_0 \qquad (28.3)$$

where $a_0 = h^2\epsilon_0/\pi me^2 = 0.529 \times 10^{-10}$ m and is called the *Bohr radius* of the hydrogen atom. The quantized energies for the hydrogen atom are given by

$$E_n = \frac{1}{n^2}\left(\frac{me^4}{8\epsilon_0^2 h^2}\right) = -\frac{13.6 \text{ eV}}{n^2} \qquad (28.4)$$

where $n = 1, 2, \ldots$ (Bohr's quantum number), m is the electron mass, e, the electron charge, h, Planck's constant, and ϵ_0, the permittivity of free space. (What is the significance of the negative sign for this energy?)

EXAMPLE

The energy level of the ground state ($n = 1$) for the hydrogen atom is given by $E_1 = -13.6$ eV/1^2. If the hydrogen atom receives this much energy from the external source, the electron will be freed from the atom, and we say the atom is ionized. The *ionization energy* of hydrogen in its lowest energy or ground state is 13.6 eV.

The second postulate of Bohr's model for the hydrogen atom states that if the orbit of the electron remains the same, the atom will neither gain nor lose energy. In order for the atom to either absorb or emit energy, the principle quantum number (n) must change to some other positive integer, say m. The energy change is given by

$$E_m - E_n = -\frac{13.6}{m^2} - \frac{-13.6}{n^2} = -13.6\left(\frac{1}{m^2} - \frac{1}{n^2}\right) \text{eV} \qquad (28.5)$$

On an energy level diagram an arrow pointing vertically upward indicates an increase in the quantum number for the atom. This corresponds to the absorption of energy, say a photon of energy hf. What energy must be absorbed to raise the hydrogen atom from the $n = 1$ state to the $n = 3$ state?

On an energy level diagram an arrow pointing vertically downward indicates a decrease in the quantum number of the atom. This corresponds to the emission of energy, a photon of light.

How much energy is emitted when an $n = 5$ hydrogen atom goes into the $n = 2$ state?

Transitions are shown for the various energy levels from $n = 1$ to $n = 6$ in Figure 28.3. We calculate the absorption and emission energies given by the arrows 1 through 6.

FIGURE 28.3

An energy level diagram for the hydrogen atom.

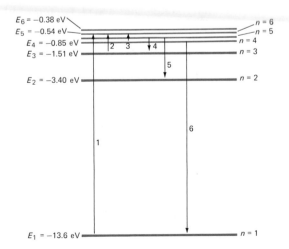

Assume that the energy changes above correspond to the absorption or emission of a photon. Remember that λ (nm) $\simeq 1240(eV\text{-}nm)/(eV)$.

The measured wavelengths of the six photons emitted are:

$\lambda_1 = 93.80$ nm $\lambda_4 = 1278.35$ nm
$\lambda_2 = 1097.35$ nm $\lambda_5 = 433.57$ nm
$\lambda_3 = 2638.30$ nm $\lambda_6 = 97.25$ nm

Which of these photons correspond to visible light?

Bohr's model was successful because it predicted experimentally verifiable atomic spectra. Bohr's model could be used to derive the previously-known empirical equations for spectral series. Unlike Planck's model, Bohr had *no* adjustable parameters to use to fit the experimental data!

28.4 Atoms Other than Hydrogen

The Bohr model served well as a first approximation for a single electron atom. Bohr himself supported his model as only the first step toward a comprehensive quantum theory. In spite of refinements (such as elliptical orbits), the Bohr model could not account for some of the fine structure, and line intensities of atomic spectra of systems with more than one electron such as helium remained unexplained. On the philosophical level, the Bohr model was a mixture of classical (orbits) and quantum (photons) concepts. The quantum jumps as electrons go between orbits were very mysterious indeed. In 1925 Schrödinger and Heisenberg developed their equivalent versions of modern quantum theory of atoms. Schrödinger's formulation has become known as *wave mechanics*, reflecting the mathematical form of its equations. In modern quantum theory the electron in the atom is characterized by a probability amplitude (known as a *wave function*). These wave functions, which are solutions to Schrödinger's quantum wave equation, give stationary states that are characterized by four quantum numbers (n, l, m_l, and m_s) instead of the single Bohr quantum number. Like the Bohr model the stationary states have definite energies and are the only allowed states for the electron in the atom. We can assign the following significance to the quantum numbers: n, the *principle quantum number*, corresponds most closely to Bohr's quantum number. n largely determines the energy of the stationary states; l, the *orbital angular momentum quantum number*, determines the spatial symmetry of the electron probability distribution in the atom; m_l, the *magnetic quantum number*, determines the orientation of the electron distribution in an external magnetic field; m_s, the *spin quantum number*, determines the intrinsic magnetic property of the electron. The word "spin" is derived from a model of the electron as an electric charge spinning on its axis. These quantum numbers are found to be governed by the following rules:

1. The principle quantum number n must be an integer, $n = 1, 2, \ldots$
2. The orbital angular momentum quantum number l must be an integer; it may take all values up to $n - 1$, $l = 0, 1, 2, \ldots, n - 1$
3. The magnetic quantum number m_l must be an integer, taking all values from $-l$ to $+l$, $m_l = -l, -l + 1, \ldots 0, 1, \ldots +l$
4. The spin quantum number m_s is either $+\frac{1}{2}$ or $-\frac{1}{2}$, corresponding to spin up and spin down.

Wolfgang Pauli formulated the *exclusion principle* in 1922. The exclusion principle states that no two electrons in a given atom can be in the same quantum state. This means that no two electrons can have identical sets of quantum numbers. We can think of the set of four quantum numbers as a set of unique coordinates for each electron in an atom (n, l, m_l, m_s).

Consider the following example. The lowest state in an atom corresponds to $n = 1$, $l = 0$, $m_l = 0$ with m_s being either $+\frac{1}{2}$ or $-\frac{1}{2}$. We see that the $n = 1$ states are $(1, 0, 0, +\frac{1}{2})$ and $(1, 0, 0, -\frac{1}{2})$. Thus two electrons exhaust the available $n = 1$ states consistent with exclusion principle. The two electron atom is helium, an inert gas.

Pauli used the exclusion principle to explain the periodic table of elements. The periodicity in the appearance of chemical properties of an atom is primarily associated with electrons in the highest quantum states. These electrons are said to be in outer orbits and more loosely bound than "inner" electrons. When the electrons of an atom fill the allowed states for a given atom for a given n value, the atom is relatively inert (as is helium). Atoms with one electron outside an inert core of electrons are said to be hydrogenlike. Sodium is an example of such an atom. List three other hydrogenlike atoms.

The order in which atomic systems fill the allowed quantum states is determined by the rule that lowest energy states are filled first. The situation is complicated by the fact that lower n values have lower energies, but smaller l values have lower energy values than larger l values. For example, the $n = 4$, $l = 0$ states have lower energy than the $n = 3$, $l = 2$ states.

EXAMPLE

Find the element in which atoms have all $n = 2$ states filled. From our previous example, we know that there are two $n = 1$ states. The $n = 2$ states are: $(2, 1, 0, +\frac{1}{2})$, $(2, 1, 0, -\frac{1}{2})$, $(2, 1, 1, +\frac{1}{2})$, $(2, 1, 1, -\frac{1}{2})$, $(2, 1, -1, +\frac{1}{2})$, $(2, 1, -1, -\frac{1}{2})$, $(2, 0, 0, +\frac{1}{2})$, $(2, 0, 0, -\frac{1}{2})$. The total number of states for an $n = 2$ system is thus 10. The atom with 10 electrons is neon, another inert gas.

The charge distribution for electron clouds associated with a given l value are called orbitals. Each value of m_l for a given l value provides a different orientation for the orbital. The directional and shape properties of atomic orbitals provides a basis for a chemical bonding theory as developed by Heitler, London, Slater, and Pauling. Figure 28.4 shows some examples of electron orbitals.

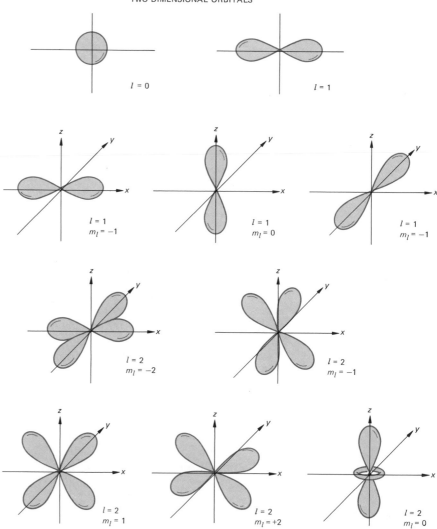

FIGURE 28.4
Electron orbitals for several quantum states.

28.5 Spectra: The Fingerprints of Nature

The basic model is the same for all the elements. Every atom has a series of energy levels that may be occupied by electrons. In its lowest energy state the atom will have all of its electrons in the lowest possible electron states. If an atom is to emit or absorb energy, the electrons must change from one electron state to another, and so the atom is only capable of absorbing or emitting energy of certain values. It is not possible for hydrogen to emit or absorb energy of 12 eV. The other elements can emit or absorb energy only at certain discrete energies. The energy level diagram for sodium is shown in Figure 28.5.

FIGURE 28.5
An energy level diagram for the sodium atom. The outer electron has higher energy in its ground state than hydrogen ground state electron. The outer sodium electron is shielded from the nucleus and its orbit has a larger radius than that of the hydrogen atom.

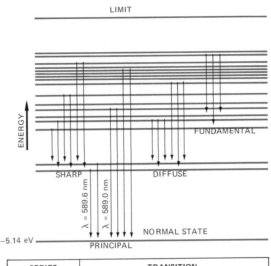

SERIES	TRANSITION
PRINCIPAL	P-LEVEL TO 1S LEVEL
SHARP	S-LEVEL TO 2P LEVEL (I.E., 3S OR HIGHER)
DIFFUSE	D-LEVEL TO 2P LEVEL (I.E., 3D OR HIGHER)
FUNDAMENTAL	F-LEVEL TO 3D LEVEL (I.E., 4F OR HIGHER)

The energy level diagram for atoms of each element is absolutely unique. It is not duplicated by the atoms of any other element. So the emission and absorption characteristics of a material are indications of the kinds of atoms that are present in that material. The chemical composition of a material can be deduced from an examination of the wavelengths of optical photons (spectra), which are emitted or absorbed by a material. In general, to reduce the interaction effects that can occur between neighboring atoms, these spectral properties of materials are best examined with the material in the gaseous state, so that the atoms of the material are as far apart as feasible.

28.6 Emission Photometry

To observe the emission spectra of an atom, energy must be absorbed by the atom so that at least one of its electrons is in an excited energy state. Then as these electrons make a transition to a lower energy level, characteristic photons will be emitted, and these photons can be observed (Figure 28.6).

In the usual laboratory setting the excitation energy is supplied by a heat source such as a flame. A schematic diagram of a typical flame photometry system is shown in Figure 28.7.

The physics of such a system is now within your grasp. The atomizer is used to spray a fine mist of a solution containing the material to be observed into the flame. The thermal energy of the flame excites some

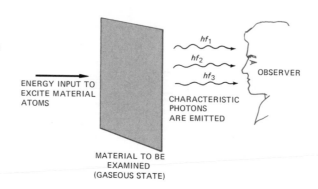

of the electrons in the atoms in the solution. Some of the excited atoms will make transitions to lower energy states with the emission of characteristic photons. The observation of the numbers and energies of the emitted photons gives an indication of the kinds of atoms (elements) that are present in the material as well as the concentration of each element.

In order to perform quantitative analyses using a flame photometer, an internal standard element is introduced into the solution in known amount C_s. The unknown concentration of the element C_x is calculated from the ratio of the intensity of the emission of the unknown concentration to the intensity of the light emitted by the standard:

$$\frac{C_x}{C_s} = \frac{I_x}{I_s} \tag{28.6}$$

where I_x is the intensity of emission from the unknown concentration at wavelength λ_s and I_s is the intensity of emission from the known concentration of the internal standard at wavelength λ_s.

FIGURE 28.7
Schematic diagram of a flame photometry system.

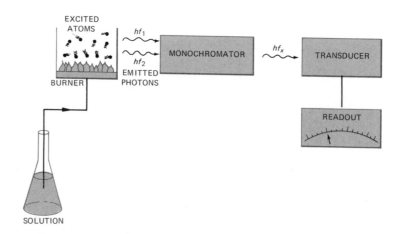

28.7 *Absorption Photometry*

Most of the atoms in a flame remain in their lowest energy states; so the sample as a whole is more likely to absorb radiation than to emit it. Therefore, the measurement of the amount of radiation *absorbed* by a substance in the flame will be an order of magnitude more sensitive than the measurement of flame emission. Once again on the basis of your study of the hydrogen atom, the physics of *absorption photometry* can be understood. Just as all elements emit characteristic wavelengths of light when returning from an excited state to a lower energy state, elements will only absorb characteristic wavelengths of light as they are excited from their lowest energy states. The quantitative use of absorption photometry relies upon the use of Beer's law:

The change in light intensity is proportional to the product of the concentration of the absorbing atoms and the sample thickness.

Atomic absorption photometry may be used for the chemical analysis of blood and urine samples. This technique not only makes use of the inherent sensitivity of the absorption technique, but uses the spectral characteristics of a cold-cathode excitation lamp to predetermine the element whose concentration will be measured. For example, when it is desired to know the amount of sodium in a sample, a sodium lamp is used to irradiate the vapor, and only the absorption of a characteristic sodium line is measured. The presence of other substances in the host materials will not influence the measurement of the sodium concentration. Atomic absorption measurements usually proceed in three steps. First, the sample to be examined is vaporized, usually by a flame, so that most of its constituents are in atomic form. Second, the atomic vapor is irradiated by light characteristic of the element being sought. Finally, the absorption of the characteristic light is related to the concentration of the element. An example of an atomic absorption system used to measure sodium concentration is sketched in Figure 28.8.

FIGURE 28.8
Schematic diagram of an atomic absorption system.

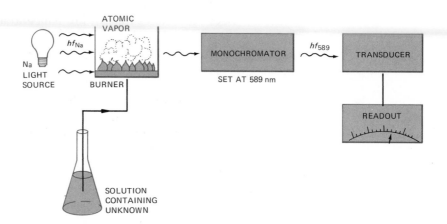

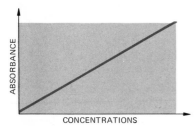

ABSORBANCE

CONCENTRATIONS

FIGURE 28.9
Absorbance as a function of concentration of absorbing element in the sample.

According to Beer's law the absorbance or optical density is proportional to the concentration of material in the solution (Figure 28.9). The absorbance is given by,

$$A = \log_{10} \frac{I_0}{I} \qquad (28.7)$$

where I_0 is the intensity of the light with none of the unknown element present, and I is the intensity with some of the unknown material present. Hence, I is a smaller number than I_0, and the ratio of I_0 to I is a positive number larger than one. Therefore, A is always a number greater than zero. In practice, an atomic absorption system is calibrated with a series of known solutions (C_s). The concentrations of an unknown (C_x) is determined directly from a graph or ratio equation:

$$C_x = \frac{A_x C_s}{A_s} \qquad (28.8)$$

where A_s is the absorbance of the known solution and A_x is the absorbance of the unknown.

28.8 The Colorimeter and the Spectrophotometer

As we have noted, the absorption of light takes place when the atoms in the sample absorb photons. In this process the photon energy is used to excite atomic electrons to higher energy levels. Each atomic system has its own unique absorption spectrum which can be used in the chemical analysis of complex systems.

For a given wavelength of incident light, the intensity of light that is transmitted through a sample of thickness x is given by *Lambert's law* which is written as follows:

$$I = I_0 e^{-\mu x} \qquad (28.9)$$

where I_0 is the incident intensity, μ, the absorption coefficient, x, the thickness of sample.

Beer's law states that for low concentrations the absorption coefficient is proportional to the concentration of the absorbing atomic system in the sample. This can be expressed in equation form as:

$$\mu = \alpha C$$

where α is a constant depending on the material and the wavelength of light used and C is the concentration of the absorbing system. From Lambert's law we find the following equation for x:

$$x = \frac{1}{\alpha C} \ln \frac{I_0}{I}$$

or

$$\alpha C = \frac{1}{0.434x} \log_{10} \frac{I_0}{I} \qquad (28.10)$$

where $\log_{10}(I_0/I)$ is called the *optical density* (OD) of the sample, or the *absorbance*.

The colorimeter is an instrument designed to measure the absorption of light at certain selected wavelengths. The wavelength used is determined by the particular atomic system for which you are looking in the sample. In many colorimeters the output is the optical density of the sample for the wavelength selected. In a simple colorimeter the output may simply be a linear intensity scale. In this case the solvent is first placed in the sample cell and the output reading is the value to be used as I_0. The sample is then placed in the instrument and the value of I is obtained. Consider the following example.

EXAMPLE

The sample cell and pure solvent give a reading of 95 (linear scale), while the sample cell with absorbing atomic system NaCl added gives a reading of 38.
a. Find the optical density of the absorbing sample.

$$OD = \log_{10}\frac{I_0}{I} = \log_{10}\frac{95}{38} = .39$$

b. If solvent is added to the sample until the reading is 45, find the ratio of the original concentration of NaCl to the final concentration.

$$\alpha C_1(0.434) = \log_{10}\frac{I_0}{I_1} = \log_{10}\frac{95}{38} = 0.39$$

$$\alpha C_2(0.434) = \log_{10}\frac{I_0}{I_2} = \log_{10}\frac{95}{45} = 0.32$$

$$\frac{C_1}{C_2} = \frac{.39}{.32} = 1.23$$

The spectrophotometer is an instrument that is designed to measure and display the absorption spectra, that is, the transmission intensity versus wavelength, for different samples. Figure 28.10 shows both a single beam and a double beam spectrophotometer. In the single beam instrument a blank sample cell is used to give an I_0 reading for each wavelength, and the sample gives a corresponding value for I (transmitted intensity). The more sophisticated (and also much more expensive) double beam instrument uses a chopped beam and electronics that determine the ratio of I/I_0 for each wavelength as the instrument scans through a wavelength range.

28.9 *Quantum Efficiency in Photobiology*

Interactions between light and matter seem to involve single photon absorption. This model predicts a direct relationship between the number of incident photons and the photoproducts resulting from the inter-

FIGURE 28.10
Schematic diagrams (a) single-beam
and (b) double-beam spectrophotometer.

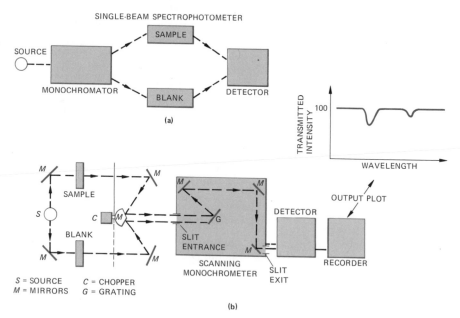

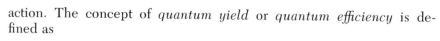

(b)

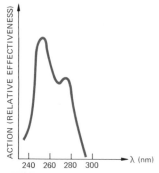

FIGURE 28.11
The action spectrum of a photosynthesis reaction. The relative
effectiveness for different wavelengths of ultraviolet light is shown.

action. The concept of *quantum yield* or *quantum efficiency* is defined as

$$\text{quantum efficiency} \equiv \frac{\text{photo-reaction rate}}{\text{photon incidence rate}} \tag{28.11}$$

The photon incidence rate can be determined from the intensity of the
incident radiation as follows:

$$\text{intensity} = \frac{\text{energy}}{\text{photon}} \times \frac{\text{photons}}{\text{unit area} \times \text{sec}}$$

Then we find

$$\frac{\text{photons}}{\text{sec}} = \frac{\text{intensity}}{(\text{energy/photon})} \times \text{area of sample}$$

or

$$\frac{\Delta N}{\Delta t} \left(\frac{\text{photons}}{\text{sec}} \right) = \frac{I(\text{watts/m}^2)A(\text{m}^2)}{hf(\text{joules/photon})} \tag{28.12}$$

A determination of quantum efficiency is very helpful in studying the
relative importance of light and living matter reactions. An example is
given in Figure 28.11 which shows the relative effectiveness of different photons in producing photosynthesis reactions in a sample of
Euglena cells.

28.10 A Model for Laser Operation

As we pointed out in the chapter on the wave properties of light, the
laser is a most significant light source. We can now describe an energy
level model that serves as the basis for many laser systems.

FIGURE 28.12

A typical energy level diagram for a laser system. An input source of energy (either a radiofrequency field or a flash source) excites atoms to an excited state E_2. Some atoms in this state decay to state E_1 which is metastable (i.e., it has a long lifetime before decaying to ground state). Atoms in the metastable state are stimulated to decay coherently to the ground state by a photon of proper energy.

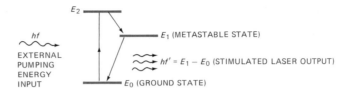

Atoms of lasing material are excited by an external energy source (usually by a flash tube or radio-frequency excitation). This input energy raises electrons to an excited state. Some of these electrons end up in an excited state with an unusually long lifetime (called a *metastable* state). This produces a population inversion with more electrons in the metastable state than normal. When an electron does drop back to the ground state, the emitted photon stimulates other excited atoms to decay. The ends of the laser cavity have reflecting coatings so that these photons make several passes down the tube producing an avalanche of photons. This represents a great amplification of the initial photon by emitted radiation. The unique feature of this radiation is that the atoms emit radiation in phase with each other. The high intensity is due to the collective behavior of the atoms, and the monochromatic nature of the radiation is due to the sharpness of the energy level transitions involved in the emission process. An energy level diagram of a typical laser system is shown in Figure 28.12.

ENRICHMENT
28.11 Mathematical Derivation of Lambert's Law

We consider a beam of light passing through a light absorbing solution. Imagine a layer of this solution dx thick (perpendicular to the incident beam). If the intensity of the incident beam is I, and I' is the intensity of the transmitted beam, then the fraction of light absorbed is given as

$$\frac{I - I'}{I}$$

This should be proportional to the number of absorbing systems in the beam in the solution. If C is the concentration, that is the amount of solute per unit volume, then the quantity of absorbers per unit area of solution is

$$\frac{\text{amount of solute (absorbers)}}{\text{unit area}} = C \, dx$$

Thus

$$\frac{I - I'}{I} \propto C \, dx$$

or, since $I - I' = dI$ and I' is less than I, we have

$$\frac{-dI}{I} = \alpha C \, dx$$

where α is a proportionality constant. If we let $\alpha C = \mu$, the absorption coefficient,

$$\frac{-dI}{I} = \mu\, dx \quad \text{or} \quad \ln I = -\mu x + \text{constant}$$

If I_0 is the intensity of incident beam $(x = 0)$, then the constant equals $\ln I_0$, and we have

$$\ln I = \ln I_0 - \mu x \quad \text{or} \quad I = I_0 e^{-\mu x}$$

If the solution contains several different absorbing solutes with different concentrations we get the following equation:

$$\frac{-dI}{I} = (\alpha_1 C_1 + \alpha_2 C_2 + \cdots)dx$$

$$I = I_0 \exp\left[-(\alpha_1 C_1 + \alpha_2 C_2 + \cdots)x\right]$$

The optical density OD or absorbance is defined as follows:

$$\text{OD} = \log_{10}\frac{I_0}{I_x} = \frac{\alpha}{2.3} Cx$$

since $\ln y = 2.3 \log_{10} y$. For multiple absorbing solutes this becomes

$$\text{OD} = \sum_{i=1}^{N} \frac{x}{2.3} \alpha_i C_i$$

where x is the thickness of the absorbing solution.

SUMMARY

Use these questions to evaluate how well you have achieved the goals of this chapter. The answers to these questions are given at the end of this summary with the number of the section where you can find related content material.

Bohr Model

1. Bohr's quantum postulate stated that one of the following quantities came in discrete units of $h/2\pi$:
   a. linear momentum
   b. energy
   c. position
   d. angular momentum
   e. mass

2. Bohr's model for the hydrogen atom predicts that the absorption spectra involves
   a. accelerating electrons
   b. same wavelengths as emission spectra
   c. electrons going to higher energy levels
   d. electrons dropping to lower energy levels
   e. deaccelerating electrons

3. Which of the following characteristics of hydrogen atoms was most difficult for classical physics to interpret:
   a. mass
   b. charge
   c. line spectra
   d. ionization
   e. size

Pauli Exclusion Principle

4. Give the possible sets of quantum numbers for $n = 2$.

Bohr Model Problems

5. The series of hydrogen spectral lines corresponding to energy transitions ending on the $n = 2$ level is called the Balmer series. The shortest wavelength of the Balmer series is
   a. 122 nm
   b. 91.2 nm
   c. 656 nm
   d. 3.40 nm
   e. 365 nm
6. The longest wavelength of the Balmer series will be
   a. 122 nm
   b. 91.2 nm
   c. 656 nm
   d. 13.4 nm
   e. 365 nm
7. Since room temperature thermal energy corresponds to about 0.03 eV, you would expect hydrogen atoms at room temperature to be
   a. ionized
   b. in $n = 2$ state
   c. in $n = 20$ state
   d. in $n = 1$ state
   e. in $n = 10$ state

Energy Level Diagrams

8. Which of the following transitions would give the shortest wavelength *emission* line:
   a. $n = 1$ to $n = 2$
   b. $n = 1$ to $n = 3$
   c. $n = 3$ to $n = 1$
   d. $n = 4$ to $n = 2$
   e. $n = 5$ to $n = 4$
9. Which of the following transitions would give the longest wavelength absorption line:

   a. $n = 1$ to $n = 2$
   b. $n = 2$ to $n = 3$
   c. $n = 1$ to $n = 3$
   d. $n = 4$ to $n = 2$
   e. $n = 5$ to $n = 4$

Spectrometry and Lambert's Law

10. For low concentrations Beer's law predicts that the absorption coefficient of a solution will be linear with
    a. wavelength
    b. frequency
    c. concentration
    d. energy
    e. intensity
11. If the optical density of a sample is 1.0, then the ratio of transmitted to incident radiation is
    a. 0.1
    b. 10
    c. 100
    d. 0.01
    e. 100
12. The spectrophotometer is an instrument designed to measure and display
    a. frequency *vs.* wavelength
    b. transmission intensity *vs.* wavelength
    c. intensity *vs.* thickness
    d. intensity *vs.* concentration
    e. absorption coefficient *vs.* concentration

Answers

1. d (Section 28.3)
2. b, c (Section 28.3)
3. c (Section 28.3)
4. $l = 1$, $m_l = -1, 0, +1$, $m_s = \pm\frac{1}{2}$; $l = 0$, $m_l = 0$, $m_s = \pm\frac{1}{2}$; total of 8 (Section 28.4)
5. e (Section 28.3)
6. c (Section 28.3)
7. d (Section 28.3)
8. c (Section 28.3)
9. e (Section 28.3)
10. c (Section 28.8)
11. a (Section 28.8)
12. b (Section 28.8)

ALGORITHMIC PROBLEMS

Listed below are the important equations from this chapter. The problems following the equations will help you learn to translate words into equations and to solve single-concept problems.

Equations

$$L = \frac{nh}{2\pi} = mvr \qquad n = 1,2,3, \ldots \tag{28.1}$$

$$hf = E_\text{f} - E_\text{i} \tag{28.2}$$

$$r_n = n^2 \left(\frac{h^2 \epsilon_0}{\pi m e^2} \right) = n^2 a_0 \qquad \text{where } a_0 = 0.529 \times 10^{-10} \text{ m} \tag{28.3}$$

$$E_n = \frac{-13.6 \text{ eV}}{n^2} \tag{28.4}$$

$$E_m - E_n = -13.6 \left(\frac{1}{m^2} - \frac{1}{n^2} \right) \text{ eV} \tag{28.5}$$

$$A = \text{OD} = \log_{10} \frac{I_0}{I} \tag{28.7}$$

$$C_x = A_x \frac{C_\text{s}}{A_\text{s}} \tag{28.8}$$

$$I = I_0 e^{-\mu x} \tag{28.9}$$

Problems

1. What would be the velocity of an electron in an orbit for $n = 1$ according to the Bohr model? The radius $r_1 = .053$ nm. How does it compare with the velocity in an orbit for $n = 2$?
2. If radiation is emitted by an atom in going from an energy level of -1.90 eV (3.04×10^{-19} J) to an energy level of -3.40 eV (5.44×10^{-19} J), what is
   a. the frequency of the radiation
   b. its wavelength
3. What is the absorbance or optical density of a sample in which the ratio of I_0/I is 3?
4. What is the ratio of I/I_0 for a specimen 0.5 cm thick of a material which has an absorption coefficient of 2 per centimeter?

Answers

1. 2.19×10^6 m/sec; $v_1/v_2 = 2$ 3. 0.477
2. a. 3.62×10^{14} Hz; b. 829 nm   4. 0.37

EXERCISES

These exercises are designed to help you apply the ideas of a section to physical situations. When appropriate the numerical answer is given in brackets at the end of each exercise.

Section 28.3

1. Suppose you could observe the hydrogen atom through a magnifying system such that the hydrogen

nucleus appeared to be the size of a baseball ($r \simeq$ 4 cm). What would the apparent radius of the $n = 1$ electron orbit? Assume the radius of a proton $= 1.5 \times 10^{-15}$ m. [1.42×10^5 cm]

2. Assume you want to find an electron orbit that is as far away from the $n = 1$ electron orbit as is the nucleus. What is the n value for such a fictitious orbit? [$\sqrt{2}$]

3. It has been suggested that the volume of an atom is mostly empty space. What percentage of the volume of an $n = 1$ hydrogen atom is occupied by matter? Radius of proton $= 1.5 \times 10^{-15}$ m [Volume of atom about 4×10^{13} times as large as volume of proton]

4. What is the velocity of the electron in a hydrogen atom for $n = 2$? The average time in the excited state is usually about 10^{-8} sec. How many revolutions does this electron make during this time? [$v = 0.6 \times 10^6$ m/sec; 4.73×10^6 revolutions]

5. Compute the energy associated with a hydrogen atom for $n = 1, 2, 3,$ and 4. Draw an energy level diagram. [$n = 1, -13.6$ eV; $n = 2, -3.4$ eV; $n = 3, -1.51$ eV; $n = 4, -0.86$ eV]

6. A large number of hydrogen atoms are excited to state $n = 4$. What transitions and spectra lines are then possible? [$4 \to 3, 4 \to 2, 4 \to 1, 3 \to 2, 3 \to 1, 2 \to 1$]

Section 28.4

7. Find the n and l numbers for the final electron in ground state of krypton ($Z = 36$). [$n = 4, l = 1$.]

Section 28.7

8. An atomic absorption instrument is being used to measure the concentration of sodium in an aqueous solution. The following calibration measurements are taken:

Solution	Readout Current (mA)
Distilled water	272.
10 ppm Na in H_2O	100.
20 ppm Na in water	36.8
40 ppm Na in water	4.97
60 ppm Na in water	0.67
80 ppm Na in water	0.09

What are the concentrations of sodium in unknown solutions whose readout currents are given by: unknown solution 1, 200 mA; unknown solution 2, 40.0 mA; unknown solution 3, 8.00 mA.

Na in 1 = _____ ppm
Na in 2 = _____ ppm
Na in 3 = _____ ppm
[4, 19, 35]

9. The absorbance of a 10-ml sample is found to be 3. The sample is accidently diluted. The lab technician (who had completed this physics course) ran another absorbance spectrum and found the new absorbance to be 2.7.
   a. Find the ratio of sample concentrations and volume of the diluted sample.
   b. Find the ratio of transmitted intensities for these two measurements. [a. $C_2/C_1 = 0.9, V_2 = 11.1$ ml; b. $I_2/I_1 = 2$]

10. A blood sugar study shows the absorbance of a morning sample to be 2 and that of an evening sample to be 1.8. Find the ratio of sugar concentration in the two samples. [$C_E = 0.9 C_M$]

PROBLEMS

Each of the following problems may involve more than one physical concept. The answer is given in parentheses at the end of the problem.

11. Find the atom that completes the third period in the periodic table (all states up to $n = 3, l = 1$ states filled). [Argon]

12. A helium He^+ ion is similar to a hydrogen atom. Compare the radii of orbits for $n = 1$ for He^+ ion and the hydrogen atom. (Use reduced the mass correction from problem 14.) [$r_{He} = \frac{1}{2} r_H$]

13. Assume the sodium atom is hydrogen type (a nucleus with $+11e$ charge surrounded by the 10 inner elec-

trons) and that it has an ionization potential of 5.12 eV. What is the radius of the eleventh electron orbit? Compare this with the size of the Na atom. [$r_{11} = 0.14$ nm]

14. One of the first corrections that was made in the simple Bohr model was the correction for the relative motion of the proton and the electron. This correction involves the relative motion of the proton and electron about the center of mass of the system. The results show that this correction calls for the use of the reduced mass in place of the electron mass. The reduced mass μ is given by $\mu = m_e M_p / (m_e + M_p)$ where $m_e =$ electron mass, $M_p =$ proton mass $= 1840\, m_e$. Use

this correction to compare the Balmer series limit wavelength ($n = \infty \to n = 2$) of deuterium with the corresponding hydrogen line. [Differs by about 30 parts in 109,707]

15. Imagine a new kind of atom in which the electron is attached to the nucleus of the hydrogen atom by a quantum mechanical spring. Hence the force of attraction is given by $F = -kr$ where k has a value of $32.4\pi^2$ N/m. (This atom is *not* the usual Bohr atom where the force of attraction is given by Coulomb's law!)

   a. Compute the value of r in terms of m, k, and v for an electron traveling in a circular orbit of radius r with a velocity v in this new kind of atom.

   b. Use Bohr's postulate to quantize the angular momentum of the electron, and express the radius of the electron orbit in this new atom as a function of n, the principle quantum number where $n = 1, 2, 3, \ldots$.

   c. Since you know that for a force of the form kr, the potential energy is given by $\frac{1}{2}kr^2$, calculate the total energy of the electron in terms of n.

   d. Sketch the energy level diagram for the new atom. How does it differ from the energy level diagram for the Bohr model of the hydrogen atom?

   [a. $r = (\sqrt{m/k})v$; b. $r = (n\hbar\sqrt{1/mk})^{1/2} = [(nh/2\pi)\sqrt{1/mk}]^{1/2}$; c. $E = -E_0 + n\hbar\sqrt{k/m} = -E_0 + (nh/2\pi)\sqrt{k/m}$;

New Atom		n		Bohr's Model
[sketch of energy levels to be inserted by student]	$E_4 = -E_0 + 4\hbar\sqrt{k/m}$	4	$E_4 = -0.85$ eV	[sketch of energy levels to be inserted by student]
	$E_3 = -E_0 + 3\hbar\sqrt{k/m}$	3	$E_3 = -1.5$ eV	
	$E_2 = -E_0 + 2\hbar\sqrt{k/m}$	2	$E_2 = -3.4$ eV	
	$E_1 = -E_0 + \hbar\sqrt{k/m}$	1	$E_1 = -13.6$ eV	

16. Light from an ultraviolet light source consists of all wavelengths between 100 nm and 200 nm. If this light is incident upon a cell containing hydrogen gas, discuss quantitatively the physical phenomena you expect to result from this interaction.

17. Beer's law may be stated in the following form:

$$I_{transmitted} = I_{incident}e^{-kx}$$

where x is length of path in absorbing medium and k is the absorption coefficient. What are the units of k? If $I_{transmitted}$ is 75 percent of $I_{incident}$ for a given thickness x, what thickness is required to reduce $I_{transmitted}$ to 50 percent? $\left[x_1 = x + \frac{1}{k} \ln \frac{3}{2}\right]$

GOALS

When you have mastered the content of this chapter, you will be able to achieve the following goals:

Definitions
Define each of the following terms, and use each term in an operational definition:

ionic bonding extrinsic semiconductor
covalent bonding semiconductor
fluorescence superconductivity
phosphorescence nuclear magnetic resonance
bioluminescence

Band Theory of Solids
State the band theory of solids, and use it to explain the optical and electrical properties of different solids.

Molecular Absorption Spectra
Explain the basis of vibrational and rotational absorption spectra of molecules.

Solid-State Problems
Solve problems involving vibrational and rotational energy states of molecules and the band theory of solids.

PREREQUISITES

Before beginning this chapter you should have achieved the goals of Chapter 21, Electrical Properties of Matter, Chapter 27, Quantum and Relativistic Physics, and Chapter 28, Atomic Physics.

29
Molecular and Solid-State Physics

29.1 Introduction

What factors determine the structure of molecules? Molecular construction is important in understanding the chemical properties and reactions of matter. The field of organic chemistry involves the study of specific atoms and the way they are joined together to form a molecule. The structure of the molecules is closely related to the chemical behavior of materials. In this chapter we will study the physical basis for molecular bonding and emission and absorption spectra of molecules.

Have you ever wondered why some solids are opaque and mirrorlike while others are translucent or transparent? What is the physical basis for the optical properties of solids? Have you noticed the difference in the thermal conductivity between stainless steel tableware and silver tableware? Have you noticed that good electrical conductors are usually good thermal conductors? In this chapter we will develop a model that will enable us to understand the behavior of solids in a qualitative way.

29.2 Molecules and Bonding

Molecules are stable configurations of atoms. As in other naturally stable systems, the stability of molecules indicates that the energy of the molecule is lower than the energy of the system of separate atoms. We can formulate the following guide: if interacting atoms can combine and thereby reduce the total energy of the system, then the atoms will form a molecule. What can you conclude about monatomic gases (such as helium, He) as compared with diatomic gases (such as hydrogen, H_2)? A useful model for describing diatomic molecules is the "spring model," in which two atoms are bound by a Hooke's law force.

The electronic states of atoms determine the nature of the bonding that occurs in the formation of molecules. As atoms are brought together there are three limiting situations that can occur. These define the pure cases of chemical bonding (Figure 29.1).

1. NO BOND IS FORMED BETWEEN INTERACTING ATOMS   If all of the lowest electron states of the interacting atoms are filled, the atoms repel each other when these electronic shells overlap. This is a quantum force resulting from the Pauli exclusion principle, which does not allow two electrons to occupy the same quantum state. Helium atoms behave in this way.

2. IONIC BOND IS FORMED   In this case one or more electrons are stripped from one atom and captured by another atom. The first atom is left with a positive charge and the second atom becomes negatively charged. They are held together by the electrostatic attraction of these two ions. Sodium chloride, Na^+Cl^-, is an example of an ionic compound. The sodium atom gives up an electron to the chlorine atom, and the resulting Na^+ and Cl^- attract each other to form the NaCl system. Ionic compounds are usually soluble in water. Can you give a plausible physi-

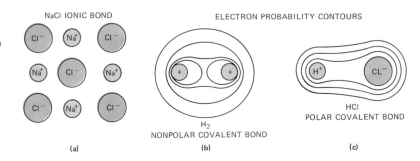

FIGURE 29.1
(a) Ionic bonding of NaCl, (b) Non-polar covalent bonding for hydrogen molecules, (c) Polar covalent bond for hydrogen chloride.

cal explanation for this solubility, remembering that the dielectric constant of water is about 80?

3. COVALENT BOND IS FORMED   In this case the atoms share electrons. This sharing produces a high probability electron distribution between the atoms. These shared electrons produce an attracting force between the atoms. Hydrogen, H_2, is an example of *nonpolar* covalent bonding, and HCl is a *polar* covalent bond. A *nonpolar* bond (such as that in H_2, Cl_2) is electrically symmetrical producing no external electric field. In general, nonpolar molecules are usually gases or liquids at room temperature. If nonpolar, covalent molecules form solids, they are soft and easily vaporized. What physical explanation can you give for this observation?

On the other hand, *polar molecules* exhibit relatively strong external electric fields due to their nonsymmetrical charge separation. For example, in HCl the shared electrons are closer to the chlorine nucleus than to the hydrogen nucleus, thus making a polar molecule as shown in Figure 29.1. Polar substances usually have higher melting and boiling points than nonpolar substances. Hydrogen gas, H_2, represents a typical molecule with a covalent bond, and sodium chloride, NaCl, represents a typical molecule with an ionic bond, there are many molecules that represent intermediate bonding, with some characteristics of these two pure cases.

29.3 *Absorption Spectra for Molecules*

The way in which a substance absorbs light provides information about the molecular structure of the substance. The spectrophotometer is an instrument designed to measure and record the absorption spectra of molecular and atomic samples. The details of the spectrophotometer were discussed in Chapter 28. As with atoms, the absorption spectra of molecules are the result of photon absorption by the system (in this case the molecules of the sample). The absorption spectra of molecules are more complex than those of atoms because in addition to the electronic energy transitions available, there are additional vibrational and rotational energy transitions available in molecules. The quantum equation that applies to the photon absorption process is given by Planck's equation,

$$E = hf \tag{29.1}$$

where h is Planck's constant $= 4.13 \times 1 0^{-15}$ eV-sec and f is the frequency of the photons absorbed.

The output of the spectrophotometer is a plot of the percent of light transmitted as a function of wavelength. The absorption of visible and ultraviolet wavelengths is the result of electronic transitions from a lower to a higher energy level in the molecules of the sample. These energy transitions are the order of electron volts in magnitude. The vibrational and rotational energy states of the molecules are quantified just as the electronic states are. These energy levels are much closer together, and the energy differences are the order of 10^{-2} eV for the vibrational states and 10^{-4} eV for the rotational states. The wavelengths absorbed by the transitions between vibrational states lie in the infrared region 1.5 to 30 microns ($1\mu = 10^{-6}$ m), while the absorption that arises from transitions between rotational states occurs in the infrared and microwave regions (wavelengths from 30 μ to 1 cm).

29.4 Energy Levels for Vibrational States

The energy level differences for vibrational states depends upon the stiffness of the bond (analogous to the spring constant of a spring) and upon the effective mass of the vibrating atoms making up the molecule. For a diatomic molecule we can use a dumbbell model, Figure 29.2. The equation for the energy levels for the vibrating diatomic dumbbell is:

$$E_{\text{vib}} = (n + \tfrac{1}{2})hf = (n + \tfrac{1}{2})\left(\frac{h}{2\pi}\right)\left(\frac{k}{m}\right)^{1/2} \tag{29.2}$$

where k is the effective spring constant of the bond, m is the effective mass of the molecule given by $m = M_1 M_2 / (M_1 + M_2)$ and n is the vibrational quantum number ($n = 0, 1, 2, \ldots$). The expression for f in an SHM system was developed in Chapter 15.

EXAMPLE

Find E_0 (the ground state) for the vibrational levels of carbon monoxide, CO. For CO molecules, $k = 1870$ N/m and

$$m = \frac{12 \times 16}{12 + 16} \times 1.67 \times 10^{-27} \text{kg}$$

$$m = 1.14 \times 10^{-26} \text{kg}$$

Thus,

$$f = \frac{1}{2\pi}\left(\frac{k}{m}\right)^{1/2} = \frac{1}{2\pi}\left(\frac{1870 \text{ N/m}}{1.14 \times 10^{-26}\text{kg}}\right)^{1/2}$$

FIGURE 29.2
The vibrational and rotational motion of a diatomic molecule using a dumbbell model.

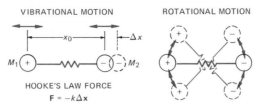

VIBRATIONAL MOTION   ROTATIONAL MOTION

HOOKE'S LAW FORCE
$F = -k\Delta x$

$$f = 6.45 \times 10^{13} \text{Hz}$$

$$E_0 = \tfrac{1}{2} hf = 3.3 \times 10^{-34} \text{ J-sec} \times 6.45 \times 10^{13} \text{ Hz}$$

$$= 21.29 \times 10^{-21} \text{J}$$

$$= 0.13 \text{ eV}$$

29.5 Energy Levels for Rotational States

The energy levels for rotational states depend upon the effective mass of the molecules and the separation of the atoms making up the molecule. The equation for the rotational energy level is:

$$E_{\text{rot}} = J(J+1)\frac{\hbar^2}{2mR^2} = J(J+1)\frac{\hbar^2}{2I} \tag{29.3}$$

where $\hbar = h/2\pi$ and where m is the effective mass, R, the separation of atoms, and J, the rotational quantum number ($J = 0, 1, 2, \ldots$). Note that mR^2 is the moment of inertia, I, of the molecule about an axis through the center of mass.

EXAMPLE

Find the energy difference ($J = 0$ to $J = 1$) for rotational transitions in CO molecules.

From the previous problem we have $m = 1.14 \times 10^{-26}$ kg. For CO, $R = 1.13 \times 10^{-10}$ m. For $J = 1$,

$$E_1 = \frac{1(1+1)\hbar^2}{2mR^2} = \frac{(1.05 \times 10^{-34} \text{ J-sec})^2}{1.14 \times 10^{-26} \text{ kg} \times (1.13 \times 10^{-10}\text{m})^2}$$

$$E_1 = 7.57 \times 10^{-23} \text{J} = 4.75 \times 10^{-4} \text{ eV}$$

$$E_0 = 0$$

$$\Delta E = 4.73 \times 10^{-4} \text{eV}$$

29.6 Identifying Molecular Structure

In practice absorption spectra give the wavelengths of the absorbed photons. This information is then used to calculate the energy differences for the vibrational and/or rotational states. The energy differences can then be used to calculate the effective spring constants for the bonds and the separation of the atoms making up the molecules. (The masses of the constituent atoms are known quantities in this case.) In complex molecules there are different bonds and different separations in different directions. These considerations lead to a complicated set of energy levels that can be used with the spectrophotometer data to analyze the molecule's structure. No two molecules have the same absorption spectra. The vibrational and rotational spectra are unique characteristics of the molecules. Certain functional groups are found to have characteristic absorption band spectra that can be used to identify new molecular structures and to help distinguish between isomers (compounds involving the same kinds of atoms, but having different structure.)

29.7 Fluorescence and Phosphorescence

The phenomena of fluorescence and phosphorescence arise from the complexity of molecular energy level systems and interactions. In each case the system is excited by external radiation—usually visible or ultraviolet light. In the case of *fluorescence,* the system gives up some of its energy (through vibrational excitation) in radiationless collisions with other molecules. The molecule then emits a photon from a lower vibrational state, and thus a longer wavelength photon is emitted. In *phosphorescence* there is a radiationless relaxation from one excited state to another excited state with a long lifetime (usually called a metastable state). The subsequent decay to the ground state is relatively slow and gives a characteristic afterglow associated with phosphorescence.

29.8 Solids

The solid state of matter is characterized by its relative rigidity and fixed volume. We can use as a model for a solid a system of atoms, molecules, or ions rigidly held close together. The binding forces in solids may be ionic, covalent, van der Waals, or metallic. Ionic binding results in solids where positive and negative ions are involved as in ionic molecules, and the binding energy is electrostatic energy. An example of an ionic solid is NaCl (table salt). The solid state ionic bond is strong, and ionic solids are usually hard crystals with high melting points. Would you expect ionic solids to be brittle or ductile? Ionic crystals are usually soluble in polar liquids.

Covalent solids result from the sharing of electrons just as in the molecular covalent bond. Examples of covalent crystals are the semiconductors silicon and germanium. The solid-state covalent binding force is very strong, and these solids are very hard (diamond, for example, the hardest substance known, is a covalent solid) and have high melting points. Covalent solids are insoluble in most liquids.

FIGURE 29.3
Schematic diagram of dipole-dipole interactions.

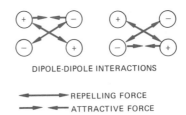

DIPOLE-DIPOLE INTERACTIONS

REPELLING FORCE
ATTRACTIVE FORCE

The van der Waals force in solids results when the atoms involved have dipole moments (nonsymmetric charge distribution) and display primarily dipole-dipole interactions (Figure 29.3). This dipole-dipole interaction results in a weak binding force. Molecular crystals having van der Waals bonding have low melting points and boiling points. Solid methane (melting point $= -259°C$) is an example of such a solid.

The metallic bond results from the attraction of the free electrons (made up of the valence electrons of the atoms) that are shared by all of the positive ion cores. This gives rise to a very strong metallic bond. This metallic bond is similar to the covalent bond but it is unsaturated — that is, it could accommodate more electrons with little increase in the energy of the bond. The unsaturated nature of the metallic bond makes it possible to form mixtures of metals called alloys that vary in properties as the composition is changed. Modern technology has made good use of alloys with special electrical, magnetic, and thermal properties. Can you think of alloys that are used for specific applications?

29.9 *Electron Behavior in Solids*

Classical physics explained the electrical conductivity of solids through the use of the electron gas model. According to this classical model good conductors have loosely bound electrons that move through the solid. The resistance of the conductor is accounted for by the collisions between the electrons and the vibrating positive ion cores. Using this model, we can explain why the resistance of a conductor should increase as the temperature is increased. In this classical model the insulators are represented as systems with tightly bound electrons. This model provides a qualitative picture of electrical phenomena in solids, but it fails to give good quantitative predictions.

Quantum physics provides us with a much more comprehensive understanding of the electron behavior in solids. The question that needs answering is what happens to the quantized energy levels of the atoms as they condense into the solid state? The exact solution of this N-atom (where N is in the order of 10^{23}) problem is not known exactly, but quantum physics can be used to give us a model for solids. As the N atoms condense into the solid state, the individual electronic states form bands of closely spaced energy levels. The energy levels are so close within a band that for practical purposes they can be treated as a con-

FIGURE 29.4
Schematic diagram of energy band structure for (a) conductors, (b) semiconductors, and (c) insulators.

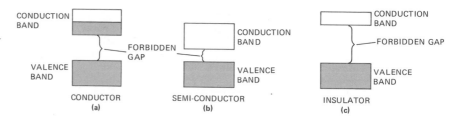

tinuum. But just as with the atomic energy levels, there are certain forbidden energy gaps between allowed energy bands. According to the Pauli exclusion principle there is a set number of electrons in any band. The band theory model applied to a conductor, semiconductor, and an insulator is illustrated in Figure 29.4. The highest energy, completely filled band is called the valence band. The next higher energy band, which is at most partially filled, is called the conduction band.

29.10 Conductors

For a conductor, such as silver, the band of energy levels that contain the conduction electrons, called the conduction band, is only partially filled. This means that there are empty energy states very near ($< 10^{-4}$ eV) the filled electron states in the conduction band. When energy is supplied to a conductor, an electron at the top of the distribution in the conduction band can absorb this energy and move into an empty state moving through the solid like a "free" electron. This external energy may be electrical in nature; thus a battery will produce an electrical current in the conductor. The energy may be electromagnetic (photons, in the language of quantum physics); the electron absorbs the photons and accelerates through the solid, thus producing reflected light and the opaque appearance common to conductors. If the energy is thermal in nature, the electrons contribute to the thermal conductivity of the conductor.

Question

1. Use the band model for a conductor to explain the relationship between thermal and electrical conductivity.

29.11 Insulators

An insulator has a large (~ 10 eV), forbidden, energy gap between the filled valence band and an empty conduction band. Many insulators are solids made up of atoms with an even number of electrons. In this case external energy cannot move an electron to an empty state unless the energy supplied to an electron is sufficient to cause the electron to jump the forbidden energy gap. Thus we see that a battery connected to an insulator does not produce a current. Many insulators are transparent to visible light as the photon energy passes through the conductor without being absorbed by the valence band electrons. Insulators are typically poor thermal conductors.

29.12 Intrinsic Semiconductors

Intrinsic semiconductors, silicon and germanium, for example, are characterized by a filled valence band with a small (~ 1 eV) energy gap between the valence band and empty conduction band. At room temperature there is some tunneling through the forbidden gap by a few electrons. This gives rise to a small electrical conductivity by the few thermally excited electrons in the conduction band. Unlike conductors, the resistance of semiconductors decreases with increases in temperature. The reason for this behavior is that the number of carriers in the conduction band increases exponentially with temperature and dominates the increase in electron–atom collisions that occur in conductors. Since the forbidden gap is small, semiconductors are opaque to visible light, but transparent to infrared light.

Question

2. Use the band model of a semiconductor to explain photoconductivity for semiconductors.

29.13 Extrinsic Semiconductors

Thus far we have considered pure materials. The basic materials in the development of solid-state electronic devices are the impure semiconductors. By adding appropriate impurities to silicon and germanium, it is possible to produce additional energy states within the forbidden energy gap.

By doping the host semiconductor with an impurity that has one more valence electron than the host has, it is possible to create donor states in the forbidden gap near the bottom of the conduction band. Electrons from these donor states are easily excited into the conduction band and thereby provide negative carriers, creating an n–type semiconductor (Figure 29.5a). By doping the semiconductor with an impurity that has one fewer electron than the host semiconductor, it is possible to create acceptor states in the forbidden gap near the top of the valence band. Electrons are easily excited from the valence band into these acceptor states where they are trapped. The resulting hole in the valence band serves as a positive carrier moving through the solid. Acceptor doping creates a p-type semiconductor (Figure 29.5b).

FIGURE 29.5
Energy band diagrams for n-type and p-type semiconductors.

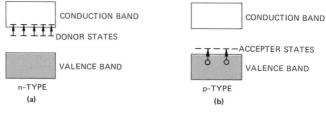

FIGURE 29.6
Schematic diagram illustrates the charge separation and potential barrier at a pn junction. Forward bias moves electrons and holes away from junction potential barrier.

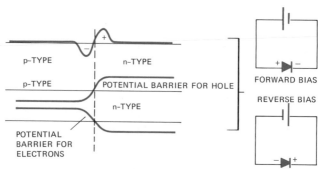

In solid-state devices the junction of the p- and n-type materials is the basic element. In the establishment of equilibrium at the junction, the mobile charges (holes for p-type material and electrons for n-type material) diffuse across the junction leaving behind the fixed charge of the impurity state, that is, negative for acceptors with their trapped electrons and positive for donors with their lost electrons. Thus the pn junction is a rectifying junction, the basis of the semiconductor diode operation. The forward bias applied voltage lowers the barrier for charge flow, and the reverse bias voltage raises the barrier and prohibits charge flow (Figure 29.6).

29.14 Transistors

Transistors consists of two pn junctions in series, that is, either a pnp sandwich or a npn sandwich. An npn transistor, set up as a simple amplifier, is illustrated in Figure 29.7.

Note that the first junction is forward biased, and the second junction

FIGURE 29.7
A circuit diagram for a common base npn transistor amplifier.

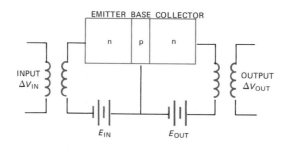

is reversed biased. A small input signal (voltage) produces a change in the current flowing through the very thin middle layer (called the *base* and composed of p material in this example). This current flows through the large resistance second junction (reversed bias) thus producing a large output voltage signal. This voltage amplification is achieved through the use of bias voltage supplies.

29.15 Solar Batteries

The solar battery is also based on the pn junction. Electron-hole pairs are created by the incident photons absorbed near the junction. These electrons and holes move under the influence of the potential difference at the barrier. The minority carriers (electrons in the p material and holes in the n material) create an appreciable change in the minority current through the cell. For the cell to be efficient the absorbing layer must be of large area and very thin (microns thick). For the operation of a solar battery see Figure 29.8.

29.16 Luminescence and Bioluminescence

Luminescence is the emission of light due to the excitation of electrons. Luminescent devices using the pn junction include the light emitting diode (LED) in which recombination of electrons and holes at the junction releases photons in the visible region of the spectrum, characteristic of the energy difference involved in the process. The solid-state laser is also a luminescent device.

Bioluminescence is the term used when referring to light production by living systems. Bioluminescence is common in a wide range of animals, insects, bacteria, and fungi. Probably the most common example of bioluminescence is that exhibited by fireflies. In this particular case there is strong evidence that bioluminescence plays a very important part in the mating process of the species. Other systems exhibiting bioluminescence are fish, marine worms, and some luminous sea bacteria.

FIGURE 29.8
A schematic diagram of a solar battery. Incident photons provide energy to set up a current across the pn junction of the cell.

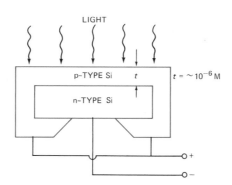

Bioluminescent production of light involves transformation of chemical energy into the electromagnetic energy of photon emission. Most of these processes involve enzyme-catalyzed reactions. Recent research has made the chemistry and physics of bioluminescence much clearer, but the biological significance of the phenomena for many of the systems remains quite mysterious.

29.17 Superconductivity

Superconductivity was discovered by Heike K. Onnes in 1911. Now it is receiving significant technological attention. Most metals lose some of their electrical resistance as their temperatures are lowered toward absolute zero, but finally their resistances become constant. A superconductor abruptly loses all of its resistance at some critical temperature that is characteristic of the superconducting material. The critical temperatures of superconductors range from tenths of a degree above absolute zero to temperatures in the neighborhood of 20°K. Tin, lead, and mercury are three well-known superconductors. Modern technology has developed many superconducting alloys.

The physical explanation of superconductivity is provided by the Bardeen-Cooper-Schreiffer (or BCS) theory. According to this theory quantum principles make it possible for two electrons to couple in pairs of zero momentum. These pairs are coupled through vibrations of the lattice (the ion cores of the metal) in such a way that there is an actual attraction between the electrons that overcomes their electrostatic repulsion. The electron-lattice-electron interaction produces an energy gap between the superconducting state and normal state for the metal. At the critical temperature for a given superconductor the system makes an abrupt transition into the superconducting state. Since the lattice vibrations are the source of resistance at ordinary temperatures, it is not surprising that superconductors are not among the best conductors at room temperature.

In addition to their zero resistance property (which has obvious energy saving potential for electrical energy transfer), superconductors have very low thermal conductivity. A new class of alloy superconductors are now used to make solenoids that are capable of maintaining magnetic fields over 5 tesla continuously. These superconducting solenoids offer the potential for many applications of strong magnetic fields, such as magnetic levitation transportation systems, whole-body nuclear magnetic resonance, and efficient ferromagnetic ore separation.

29.18 Nuclear Magnetic Resonance

Nuclear magnetic resonance, or NMR as it is commonly called, has become a very important tool in chemistry. The physical basis for the phenomenon is the resonant absorption of electromagnetic energy by

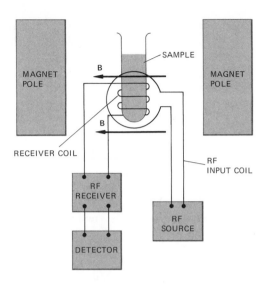

protons in a magnetic field. The sample (liquid or solid) is placed in a
homogeneous magnetic field and subjected to radio-frequency (in the
megahertz region) radiation. The absorption takes place when the proton absorbs energy as the radio frequency equals the natural precession
frequency of the proton in the field. The proton magnetic moment has a
natural precession frequency about the external magnetic field given by,

$$f = \frac{\boldsymbol{\mu} \cdot \mathbf{B}}{2\pi s} \tag{29.4}$$

where s is the spin angular momentum of the proton and is equal to
one half of $\hbar$, or 5.28×10^{-35} J/sec, $\boldsymbol{\mu}$ is the magnetic moment of proton
$(1.41 \times 10^{-26}$ J/T); and $\mathbf{B}$ is the external magnetic field.

A schematic diagram of a NMR apparatus is shown in Figure 29.9.
The detection system measures the radio-frequency energy absorbed by
the sample. A typical NMR spectrum for ethanol (CH_3CH_2OH) is shown
in Figure 29.10. The fine structure is shown in the high resolution spectrum (Figure 29.10b). Each of the different proton groupings in the
sample is in a different magnetic environment, and thus there is a
slightly different resonant frequency for each grouping. The area under
each absorption band is proportional to the number of protons in the
absorbing group for that band. The chemist and biochemist are able to

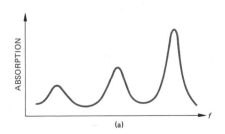

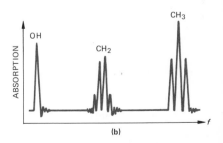

distinguish each different chemical grouping and the number of protons in each group for a particular sample. The structural information revealed in NMR spectra is highly detailed, and it has provided great insight into the nature of chemical bonding.

SUMMARY

Use these questions to evaluate how well you have achieved the goals of this chapter. The answers to these questions are given at the end of this summary with the number of the section where you can find the related content material.

Definitions

1. Covalent bonding is characterized by an attractive force involving
   a. dipole-dipole interactions
   b. shared electrons
   c. magnetic interaction
   d. gravitational interaction
   e. none of these
2. Fluorescence and phosphorescence are distinguished from each other by the difference in the time of emission. The _____ radiation is present as long as sample is the irradiated. _____ radiation continues after stimulating radiation has been removed.
3. Bioluminescence involves the emission of light from biological systems. The energy source for this light is primarily
   a. heat
   b. chemical
   c. magnetic
   d. gravitational
   e. none of these
4. A p-type semiconductor is one that has been doped with an impurity that has valence electrons that number _____ (*greater than/less than/equal to*) host semiconductor.
5. In order to operate a superconducting solenoid magnet it is necessary to maintain
   a. high current
   b. temperature below 20°K
   c. zero magnetic field
   d. low current
   e. none of these

6. NMR is a valuable tool for chemists because it yields molecular data that show
   a. electron distribution
   b. proton groups
   c. numbers of protons
   d. fluorescence
   e. none of these

Band Theory of Solids

7. The band theory of solids can be used to explain which of the following properties of solids?
   a. mechanical
   b. optical
   c. thermal
   d. electrical
   e. none of these

Molecular Absorption Spectra

8. The vibrational spectrum of a molecular system can be used to determine the molecular
   a. moment of inertia
   b. rotational frequency
   c. force constant
   d. shape
   e. vibration frequency
9. The rotational spectrum of a molecular system can be used to determine the molecular
   a. moment of inertia
   b. vibration frequency
   c. force constant
   d. shape
   e. rotational frequency

Solid-State Problems

10. The effect of the mass change that occurs when tritium, $_{1}^{3}H$, is substituted for hydrogen in $_{1}^{3}H_2$ is to multiply the vibrational frequency of $_{1}^{3}H_2$ by

a. 3
b. $\sqrt{3}$
c. $\frac{1}{3}$
d. $1/\sqrt{3}$
e. 6
11. The effect of this substitution (problem 10) on the rotational frequency will be to multiply the rotational frequency of $_1^3H_2$ by
a. 3
b. $\sqrt{3}$
c. $\frac{1}{3}$
d. $1/\sqrt{3}$
e. 6

Answers

1. b (Section 29.2)
2. fluorescent, phosphorescent (Section 29.7)
3. b (Section 29.16)
4. less than (Section 29.13)
5. b (Section 29.17)
6. b, c (Section 29.18)
7. b, d (Sections 29.9, 10)
8. c, e (Section 29.4)
9. a, e (Section 29.5)
10. d (Section 29.4)
11. c (Section 29.5)

ALGORITHMIC PROBLEMS

Listed below are the important equations from this chapter. The problems following the equations will help you learn to translate words into equations and to solve single-concept problems.

Equations

$$E = hf \tag{29.1}$$

$$E_{vib} = (n + \tfrac{1}{2})hf = (n + \tfrac{1}{2})\left(\frac{h}{2\pi}\right)\left(\frac{k}{m}\right)^{1/2} \tag{29.2}$$

$$E_{rot} = J(J + 1)\frac{\hbar^2}{2mR^2} \tag{29.3}$$

$$f = \frac{\boldsymbol{\mu} \cdot \mathbf{B}}{2\pi s} \tag{29.4}$$

Problems

1. Find the energy in the vibrational state for $n = 1$ for carbon monoxide (CO) molecules. (See the example in Section 29.4 for constants.)
2. Draw a vibrational energy level diagram for $n = 0, 1, 2$ for CO molecules.
3. Calculate the energy of rotation for CO molecules for $J = 0, 2, 3$. Compare the energy of rotation with the energy of vibration.
4. For CO molecules with $n = 1$ place the levels for $J = 1, 2, 3$ on the energy level diagram drawn in problem 2. (The $J = 1$ rotational energy is calculated in the example of Section 29.4.)

Answers

1. 40.0×10^{-2} eV
3. 0, 14.3×10^{-4} eV, 28.5×10^{-4} eV

EXERCISES

These exercises are designed to help you apply the ideas of a section to physical situations. When appropriate the numerical answer is given in brackets at the end of the exercise.

Section 29.2

1. Discuss the solubility of a polar and a nonpolar substance in water (a polar solvent). Give a possible physical explanation for this behavior.
2. The ionization energy of a system is the energy needed to remove the most loosely bound electron in the ground state of the system in the gaseous phase. The ionization energy of the hydrogen atom is 13.6 eV, and that of the H_2 molecule is 15.7 eV. Give the physical explanation for this observation.
3. Rank the four potassium halides (KBr, KCl, KI, KF) in order from most to least ionic, and give your reasons for ranking.

Section 29.4

4. The H_2 molecule absorbs infrared light of 2.3 μ wavelength when the vibrational state changes by $\Delta n = 1$. Find the effective spring constant for the H_2 molecule. [562 N/m]
5. The frequency of vibration of the $^1H^{19}F$ molecule is 8.72×10^{13} Hz.
   a. Find the photon wavelength for a $\Delta n = 1$ vibrational transition.
   b. Find the effective spring constant for the HF molecule. [a. 3.4×10^{-6} m; b. 476 N/m]
6. Find the ratio of the $\Delta n = 1$ transitions for H_2 and D_2 vibrational energy levels. (Deuterium, D, is heavy hydrogen, ^2_1H.) [$(\Delta E)_D/(\Delta E)_H = 1/\sqrt{2}$]

Section 29.5

7. The separation of the H atoms in the H_2 molecule is 0.074 nm. Find the smallest energy level separation for the rotational states of the H_2 molecule. Find the photon wavelength for this energy. [1.06×10^{-2} eV, 1.22×10^{-6} m]
8. Find the ratio of the $\Delta J = 1$ transitions for H_2 and D_2 rotational energy levels. [$\Delta E_H/\Delta E_D = 2$]

Section 29.8

9. Construct a model based on the deBroglie wave property of electrons in solids that will explain the existence of allowed energy bands and forbidden energy gaps.

Section 29.13

10. Indium is used to dope germanium. Is the result a n- or p-type semiconductor? Explain. Sketch a band structure for germanium doped with indium.
11. Antimony is used to dope germanium. Is the result a n- or p-type semiconducting material? Explain. Sketch a band structure for germanium doped with antimony.

Section 29.14

12. Sketch a pnp transistor amplifier circuit that would function like the npn amplifier shown in Figure 26.8.
13. Find the minimum energy gap in a semiconductor LED which emits visible light if the photons emitted correspond to recombination of electrons and holes. [1.77 eV]

Section 29.17

14. Make a list of elements that are well-known superconductors and locate them on the periodic table. What conclusions can be drawn about superconductors?

PROBLEMS

The following problems may involve more than one physical concept. The numerical answer is given in brackets at the end of each problem.

15. The energy gap in germanium is 0.75 eV. Find the wavelength at which Ge begins to absorb light. [1650 nm]
16. The energy gap of an insulator is 20 eV. What wavelength radiation would begin to excite electrons into the conduction band? [61.7 nm]
17. The energy required to break the bond holding atoms together in a molecule is called the dissociation energy. The dissociation energy for hydrogen gas at 25°C at constant pressure is 104,220 cal/mole. Find the

dissociation energy in electron volts for a single molecule of H_2. [4.56 eV/molecule]

18. Consider a proton that has a magnetic moment of magnitude equal to 1.41×10^{-26} J/T.

    a. What is the natural precession frequency of the proton in the earth's magnetic field ($B = 6.00 \times 10^{-4}$ T)? What kind of electromagnetic radiation has this frequency?

    b. What is the precession frequency of the proton in the magnetic field of a superconducting magnet ($B = 5.00$ T)? What kind of electromagnetic radiation has this frequency? [a. 2.55×10^4 Hz, long radio waves; b. 2.13×10^8 Hz, short radio waves]

19. Tantalum becomes a superconductor at a temperature of $4.39°$K. What is the approximate binding energy of the coupled pairs of electrons in tantalum? [3.78×10^{-4} eV]

GOALS

When you have mastered the material in this chapter, you will be able to:

Definitions
Define each of the following terms, and use it in an operational definition:

hard and soft x rays	Roentgen
Bremsstrahlung	rad
characteristic x rays	rem
absorption coefficient	reb
	RBE

X-ray Problems
Solve problems involving the generation, absorption, and detection of x rays.

X-ray Interactions
List and discuss the interactions of x rays with matter — particularly those with humans.

PREREQUISITES

Before beginning this chapter you should have achieved the goals of Chapter 21, Electrical Properties of Matter, Chapter 27, Quantum and Relativistic Physics, and Chapter 28, Atomic Physics.

30

X Rays

30.1 Introduction

What are the characteristics of x rays that make them useful in medical diagnosis and therapy? How is it possible for some color television sets to generate x rays? If you cannot see x rays, how can they be detected? X rays are electromagnetic radiation with wavelengths of the order of 10^{-10} m. This corresponds to photon energies in the kiloelectron volt region. In this chapter you will be introduced to the physical basis of x-ray generation and detection. You will study the characteristics of x rays and their interaction with matter. Your understanding of the physical basis of x-ray phenomena will enable you to understand the value of x rays in the health sciences.

30.2 X Rays

Wilhelm K. Roentgen, who in 1901 was the first man to receive the Nobel Prize in physics, first observed x rays in 1895. He was studying the light produced when electricity was passed through a gas in a tube at low pressure. He noted that a paper screen coated with a fluorescent material glowed when it was in the vicinity of the tube under operation. We now know that x rays are produced if electrons are accelerated through a potential of the order of 10^4 or more volts and then allowed to strike a metal target. Classical electromagnetic theory indicates that the deceleration of an electric charge causes it to radiate energy. In this case the form of electromagnetic radiation is called x rays. Early observations showed that this newly detected radiation had greater penetration power than any other electromagnetic radiation known at that time. It was also observed that these x rays affected a photographic film and would ionize atoms. These effects are utilized in detecting devices for x rays. Since x rays are electromagnetic waves, they are not deflected by either an electric or a magnetic field.

The usual way of producing x rays involves the use of the thermionic x-ray tube as shown in Figure 30.1. The heated filament supplies electrons that are accelerated through a large potential difference (generally of the order 10^4 and 10^5 volts) between the cathode and the anode. The electron beam is focused by the use of electric fields onto the anode. The wavelength of the x rays is controlled by the applied voltage between the cathode and anode. For the higher potential differences (short wavelengths) the term *hard x rays* is used and for the lower

FIGURE 30.1

Diagram of an x-ray tube. The heated filament emits electrons, which are accelerated through a potential of more than 10,000 V. The electrons strike the anode and cause it to emit energy as x-ray photons. The anode serves as a transducer changing the kinetic energy of the electrons into heat energy and electromagnetic radiation energy in the x-ray region of the spectrum.

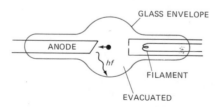

FIGURE 30.2

A typical x-ray spectrum with the intensity I of the x rays in arbitrary units plotted against the wavelength λ of the x rays. From the Planck relationship ($E = hc/\lambda$), note that the highest-energy x-rays are those with low wavelengths. The Bremsstrahlung x-ray radiation forms the smooth continuous spectra. The sharp peaks are the characteristic x-ray spectral lines. These lines are characteristic of the kind of metal used as a target. This figure represents an emission spectrum of tungsten. The value of λ_{min} is given by $hc/(\text{voltage})e$, where e is the charge on an electron.

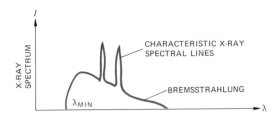

potential differences (long wavelengths) the term *soft x rays* is used to describe the quality of the radiation. Since the tube is highly evacuated, the electron beam current (usually in the 10 milliampere range) is determined by the filament current. When these electrons strike the metal anode target some of them generate x rays as they are abruptly brought to rest. This deceleration radiation is called *Bremsstrahlung (braking radiation) which appears as the continuous x-ray background shown in Figure 30.2.*

Much of the electron energy goes into heating up the anode. Some of the electrons in the beam interact with the innermost, most tightly, bound electrons in the target and "knock" them into excited states. When these excited atoms return to their ground state, photons are emitted. Since the target material is a metal with many electrons, the innermost electron energy levels are of the order of thousands of electron volts (keV). The photons emitted due to this excitation are characteristic of the target material and are called the *characteristic spectra.*

The maximum energy x rays correspond to the conversion of the maximum electron beam energy into a photon,

electron beam energy = photon energy maximum

$$eV = hf_{max} = \frac{hc}{\lambda_{min}} \tag{30.1}$$

where V is the accelerating voltage for the tube, f_{max} is maximum x-ray frequency, h is Planck's constant, and e is the charge of the electron.

EXAMPLE

An x-ray tube is operated at a potential of 62,000 V. Find the short wavelength limit of the x rays emitted.

$$hf_{max} = \frac{hc}{\lambda_{min}} = eV = 62 \text{ keV}$$

$$\lambda(\text{nm}) = \frac{1240 \text{ eV–nm}}{62,000 \text{ eV}} = 0.02 \text{ nm}$$

where hc/e = constant = 1240 eV–nm (to three significant figures). This is a good number to remember for quick conversion of wavelengths and voltages. That is,

$$\lambda V = 1240 \text{ eV–nm} \tag{30.2}$$

30.3 Interaction with Matter

Early experiments showed that x rays would penetrate matter. In fact, within one month after the discovery of x rays, two French physicists had produced a photograph showing the bones in a human hand. From experiments it has been found that the number of x-ray photons is diminished by passage through matter, particularly if they are passing through a material of high atomic number. It has also been found that some of the photons that emerge have the same energy as the incident energy, even though the number that emerge decreases as the thickness increases. For photons in x-ray energy range there are two physical processes that are of importance in reducing the number of x rays that pass through a material. These processes are the photoelectric effect and the Compton effect (scattering), which were discussed in Chapter 27.

First, the photoelectric effect as you recall is an interaction between the photon and an electron that is bound in an atom. The larger the number of electrons in an atom, the greater the probability that the photoelectric effect will occur. In the photoelectric effect the incident photon gives up its energy to a bound electron in an atom. The basic

Medical x-ray unit. (Courtesy of General Electric Company.)

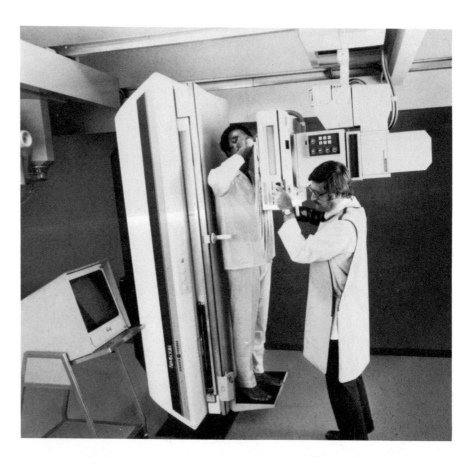

X-ray tooth diagnosis—Enlargement reproduction of tooth x ray. This is a bite wing of the molar section of the left side of the patient's mouth. The dark areas on the teeth are fillings. There are also light areas on the edges of four teeth indicating decay. A dentist can learn additional information on these teeth from the x rays. (Courtesy of Dr. Carl James, Rolla, Missouri)

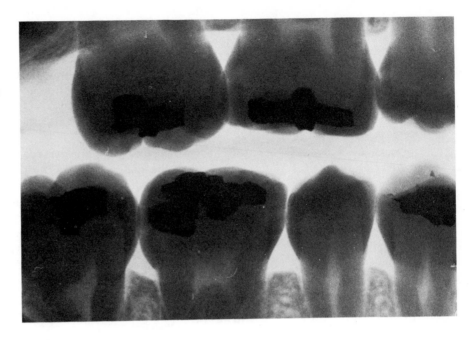

energy equation for this interaction is that the kinetic energy of the ejected electron is equal to the energy of the incident photon energy minus the energy required to remove the electron from its atom. In many cases, the work function is much smaller than x-ray photon energy, and the photoelectron has nearly the same energy as the incident photon. The photoelectron may also interact with matter. Secondly, the Compton effect, which we described in Chapter 27, is the interaction between a photon and a free electron. In this interaction the photon looses only a fraction of its energy. This loss of energy is imparted to the free electron. In general, the photon and the free electron are scattered at angles relative to the incident direction. The energy of the free electron depends upon the energy of the incident photon and the angles of scattering. As the atomic number of the material increases, the photoelectric effect becomes predominant over the Compton effect in reducing the x-ray intensity.

Both of the above effects are consistent with the fact that electromagnetic waves behave as particles, the photon concept. In each case the electron, first interacts with the incident x-ray photon, then interacts with matter as a charged particle.

30.4 Absorption Coefficient

The x-ray photons interact with matter through the photoelectric effect and Compton scattering. The photoelectric effect increases rapidly with Z, the atomic number of the target material. The photoelectric interaction probability is proportional to Z^5 and is the greatest for low-energy photons. The Compton effect is relatively independent of energy and the atomic number of the absorber. These two processes determine the *absorption coefficient* of the material for x rays. The intensity of x rays passing through a material of thickness x can be expressed as an exponential function,

$$I = I_0 e^{-\mu \Delta x} \tag{30.3}$$

I is the x-ray intensity at a distance x in the material, μ is absorption coefficient of the material, Δx is the thickness of the material, and I_0 is the incident intensity of the x rays in joules per unit area per second.

For soft x rays (those with energy less than 50 keV) the photoelectric effect is the important factor in x-ray absorption, while for hard x rays (those with energy greater than 100 keV) the Compton effect contributes significantly to the absorption. The absorption coefficient as a function of energy for water (approximately the same as that of biological material) and for lead are shown in Figure 30.3.

EXAMPLE

Find the thickness of water required to reduce the intensity of 100 keV x rays by a factor of 1000.

From Figure 30.3, $\mu = 0.035$ cm. We want the value of Δx that gives $I = 10^{-3} I_0$; thus we have

$$10^{-3} = e^{-0.035 \Delta x} \quad \text{or} \quad 0.035 \, \Delta x = \ln 1000$$

$$\Delta x = (6.9078/0.035) \text{ cm} = 197.4 \text{ cm} \sim 2 \text{ m}$$

It is worth noting that the relative absorption of bone to water is approximately 150. That is, bone is 150 times more opaque to x rays than is water. Lead is about 75 times more opaque than bone.

30.5 Detection of X Rays

The detection of x rays results from the effects of the energy absorbed in the detector. The fluoroscope uses the emission of visible light from a material when it is subjected to x rays. The photons emitted by the fluoroscope screen are in the visible region, and the observer sees a picture of the material through which the x rays pass before reaching the fluoroscope.

Photographic film is also used to record x rays. Small film badges are worn by persons in potential radiation areas as reliable detectors of radiation exposure.

An ionization chamber can be used to monitor the presence of x rays. A simple ionization chamber is represented in Figure 30.4. The chamber between a pair of parallel plates is filled with a gas at low pressure. The plates are connected in series with a source of DC potential of the

FIGURE 30.3
X-ray absorption curves for (a) water and (b) lead as a function of energy. To get actual transmission one uses the total absorption curve. To get actual absorption one uses true absorption values. The difference between the two is due to scattering. Notice that the value of μ_{Pb} is about an order of magnitude larger than μ_{H_2O}. So the absorption of lead is about e^{10} or 22,000 times that of water. If these two absorption curves are normalized for the density of the material they are similar since $\rho_{Pb} \simeq 10 \ \rho_{H_2O}$.

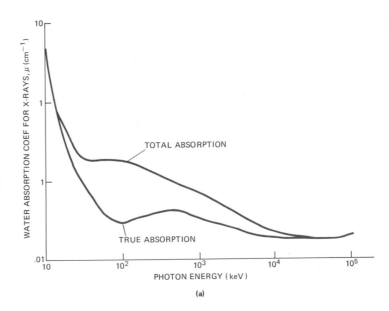

(a)

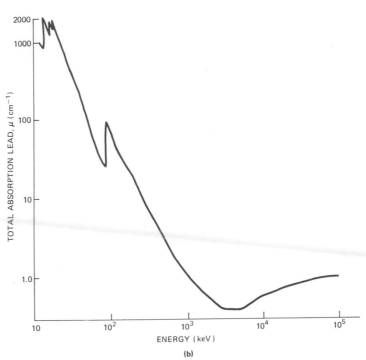

(b)

order of 1000 V and with a current measuring device. A beam of x rays is directed along the long axis of the tube. The x rays are an ionizing agent, and as the atoms of the gas are ionized, the ions and the free electrons are subjected to an electric field. The positive ions are accel-

FIGURE 30.4
Schematic diagram of a simple
ionization chamber. Incident x-rays
create an ionization current pulse
that is detected by an external meter.

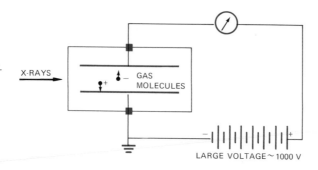

erated toward the cathode and the electrons and negative ions are
accelerated toward the anode. This flow of ions produces an electric
current between the plates and hence through the circuit. The current
is proportional to the ions produced and thus to the energy of x rays
absorbed.

Different types of ionization chambers are used in the detection of
x rays. These differ in the geometry of the chamber and the magnitude
of the applied voltage between the electrodes. One which is very useful in the detection of x rays is the Geiger-Müller (G-M) counter. These
counters are relatively economical, sensitive to x rays, rugged, and
portable. They are generally made in a glass tube. The inside of the
tube is coated to serve as the cathode and a coaxial wire serves as the
anode (Figure 30.5). The thickness of the entrance window determines
the minimum energy of entering radiation for ionization. The tube is
generally filled with an ionizable gas (argon and neon are often used)
at a reduced pressure of about 0.1 atmosphere. Added to the major gas
is a quenching gas (about 0.1 percent of which may be halogen). The
ionizing radiation enters the counter, and ions are produced. The ions
move in the radial field which is much stronger near the anode than
the cathode. The electrons move more rapidly to the anode than the
positive ions do to the cathode. The avalanche of ions occurs along the
electrode. The quenching gas and the circuitry are designed to impede
a continuous discharge. The G-M counter is a pulse counting device,
it has a high gas amplification (10^6–10^9), and the operating voltage is in
the neighborhood of 1000 volts.

Another x-ray detecting system is a combination of a scintillation

FIGURE 30.5
A cross-section diagram of a Geiger-
Müller tube. Radiation entering the
tube through the entrance window
creates an ionization pulse that is
amplified by the high voltage applied
to the tube. The resulting current
pulse is detected by an external
circuit and meter.

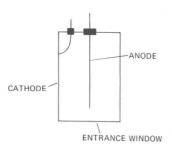

FIGURE 30.6
A schematic diagram of a photo-multiplier tube.

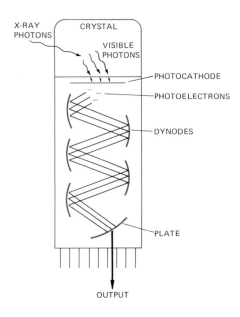

phosphor used in series with a photomultiplier tube. A photomultiplier tube has a photocathode which gives up electrons when struck by a photon. These electrons are focused on an electrode which will emit a multiple number n of electrons when struck with an electron having an energy of about 100 eV. If the photomultiplier has ten electrodes, it has a gain of n^{10}, where n is usually 2 to 4. The final output from photo-multiplier is conducted to the plate of the tube which is part of the electrical counting circuit. A diagram of photomultiplier is shown in Figure 30.6. Most of the phosphors have a very short decay time so this system is very useful for counting incident photons.

30.6 Radiation Units

Several units of radiation have been defined for use in working with the effects of radiation on living systems. The old unit of dosage of x rays is the *roentgen* (R). The roentgen is defined as that quantity of x rays that produces 1.61×10^{15} ion pairs per kilogram in air. This is equivalent to 89 ergs ($= 89 \times 10^{-7}$ joule) per gram of air. The roentgen is an *exposure* dose. The *rad* is the accepted *absorbed* dose that applies to any absorber and any type of radiation. The rad is defined as the radiation that re-leases 10^{-5} joules per gram of absorbing material. Because different types of radiation produce different biological effects for the same amount of energy absorbed, other dosage units are defined as follows:

rem = rad equivalent man; the absorbed dosage that produces in man the effect equivalent to one rad of x rays

reb = rad equivalent biological; the absorbed dosage that produces in

TABLE 30.1
Typical Radiation Doses

Source	Dose
Background	0.075 rem/year (varies widely) (up to 1 rem/year in India and Brazil)
Ra^{226} in body	0.005 rem/yr
K^{40} in body	0.020 rem/yr
Diagnostic x rays	0.2 rem/yr
Fall out	0.1 rem/yr (decreasing now)
Local dose of irradiation of tumors	(3000-7000) rem/treatment period
Lethal dose (whole body)	$\sim$ 400 rem at one time

some biological material the same effect as one rad of x rays in that material

The ratio of the rem to the rad is called the relative biological effectiveness (RBE) of the radiation. For example, fast neutrons have a RBE of 10, slow neutrons have a RBE of 2 to 5, alpha particles of 5 MeV have a RBE of 20, and electrons of 1 MeV have a RBE of 1. Some typical doses in rem are shown in Table 30.1.

30.7 Biological Effects of X Rays

The responses of living systems to ionizing radiation are complex and subtle. X rays are only one component of the radiations to which people are exposed.

The interaction of radiation with matter depends upon the absorption of energy, and the extent of biological damage depends upon the energy absorbed per gram in the organism.

Absorbed radiation produces physical changes in the cell. Some specific cellular changes that may occur are break-up of molecules, production of free radicals, inactivation of enzymes, change of deoxyribonucleic acid synthesis, and break-up of chromosomes. In addition gross physical changes may occur. Among these are increase of the viscosity of cellular fluids, increase of permeability of cellular membranes, swelling, and death. In many cases the effects are known, but the exact steps that occur between the absorption of energy and final results are not known. Also the absorption of a constant amount of energy does not produce the same effect in all types of cells. As a sort of rule of thumb, cells that are growing rapidly are most susceptible to radiation. The most sensitive cells in the body are the bone marrow, lymphoid, and epithelial tissues. Cells of a fetus are very sensitive to radiation. The cells of bone, muscle, and blood vessels are less sensitive, and the nerve cells are the least sensitive, most resistant, to radiation.

It has not been established whether there is or is not a minimum absorption of radiation energy that is necessary to produce biological damage. There are different hypotheses relative to threshold. However, in general, once a level is reached in which the effect has been established, the total effect seems to be directly proportional to the absorbed energy.

The set of characteristic symptoms shown by individuals who have a large dosage in a short period of time is called acute radiation syndrome. You should realize that the response varies a great deal among individuals so the dosage size will suggest an overall average effect. For a whole body dosage of 25 rad, there are essentially no detectable effects. For dosages of 25 to 100 rad, the person shows little or no effects other than a detectable change in the blood count; lymphocyte count drops. Bone marrow, lymph nodes, and spleen are damaged. At 100–300 rad, one has blood changes, experiences vomiting, malaise, fatigue, and loss of appetite. With this dosage you would probably have to take antibiotics, but recovery would be expected. For dosage of 300–600 rad, the effects of lower dosage plus hemorrhaging, infection, diarrhea, loss of hair, and temporary sterility can be anticipated. A person receiving a dosage in this range will have to take antibiotics, have blood transfusions, and perhaps bone marrow transplant. About 50 percent of individuals exposed to this dosage range recover. Dosages above 600 rad produce the symptoms indicated above plus damage to the nervous system. Dosages of about 1000 rad and above produce almost certain complete incapacitation and death. In any case, death may be caused by a break down of either the circulatory or the respiratory system, or both.

In addition to the acute effects discussed in the previous paragraphs, there are delayed effects. These delayed effects may be in the exposed individual (*somatic effects*) or in his progeny (*genetic effects*). We will first consider the somatic effects.

The delayed somatic effects due to radiation are generally evident in the statistical study of a population.

The first known delayed effects occurred in the case of the bombings or Hiroshima and Nagasaki in the summer of 1945. Many of the survivors of these bombings developed leukemia, with the peak incidence of this disease developing in the period 1950–1952. There was also evidence of an increased incidence of other types of abnormal tissue growth among the survivors. Another example of diseases produced in humans as a result of exposure over a long period of time includes bone cancers, which occurred among workers who painted radium watch dials. These effects have been demonstrated in animal subjects in laboratory experiments.

The authors had a friend who considered his poor health in later years and eventually his death to be brought about by exposures which he received while working with the Manhattan Project for the development of an atomic bomb in the mid-forties. This was a delay of about 20 years.

Some effects which have been produced by irradiation in animals are:

shortened life time, lens opacities and cataracts in the eyes, and effects upon pregnant mothers, such as still-births, infant mortalities, mental retardation, abnormal physical development, and deformities.

In 1927 H. J. Müller discovered that ionizing radiation produced gene mutation. Only the mutations which are produced in a gene cell can be transmitted to its progeny. These are known as genetic effects of radiation. The genetic changes in an offspring may be from minor changes to severe handicaps. The frequency of mutations induced by radiation depends upon both the dosage and the dose rate. It apparently becomes noticeable with dosage of the order of a 20–50 rads, and then increases linearly with total dosage. Also the frequency and type of mutation depends upon time of mating relative to the time of exposure. If a gene cell is already mature at time of mating, the irradiated cell carries a relatively high percentage of lethal mutations. If the mating occurs when the cell is immature the mutations will be predominantly recessive.

30.8 Radiation Protection Standards

The above discussion has probably raised the question in your mind, do we have any protective standards? The answer to the question is yes. For a comprehensive set of recommendations, see the report of the International Commission on Radiological Protection, ICRP Publication No. 9, Pergamon Press, London (1966). A brief statement of permissible dosage equivalents recommended by ICRP follows in Table 30.2.

Studies indicate that medical and dental x rays constitute the greatest single source for radiation for inhabitants of the United States. These are not under the control of the federal government. This means that risks are essentially the judgment of the licensed physician. However, some states have laws to control the acceptable standards of operation of irradiating units. In some states these regulations also apply to machines

TABLE 30.2
Permissible X-ray Dosage
Equivalents

Exposed Group	Blood Forming Organs, Gonads, Lens of Eye	Other Organs
Radiation workers	5 rem/yr after age 18; no more than 2.5 rem in any 3-month period	Skin, bone, thyroid, 30 rem/yr; hands, feet, forearms, 75 rem/yr; other organs, 15 rem/yr
Members of the public in vicinity of a controlled area	0.5 rem/yr	Skin, bone, thyroid, 30 rem/yr for adults, 1.5 rem/yr for children to age 16; hands, feet, forearms, 7.5 rem/yr; other organs, 1.5 rem/yr

for shoe fitting, etc. However, these regulations do not cover radiation that you may get from other sources.

30.9 *Medical Applications of X Rays*

Diagnostic uses of x rays involve the differential absorption of different body parts for the x rays used. Almost all tissue will stop some x rays and cast a shadow on the fluoroscope. Diagnostic x-ray machines operate at energies less than 150 keV. For greater contrast it is sometimes necessary to insert a material with greater absorption than the organ. Barium salts and iodine compounds are either fed or injected into patients for this purpose.

The therapeutic value of x rays rests in their potential for killing living tissue. If a parallel beam of x rays is directed at a tumor with dosages of 2000 to 7000 rem, much if not all of the tumor can be killed. Therapeutic x rays may have energies of 5 MeV for deep-seated tumors. The energy deposited by the x-ray beam in the target tissue is designed to be lethal for that volume. The direction of the beam is usually altered during treatment so that tissue other than target tissue receives only small dosages of radiation.

EXAMPLE

A dentist uses an x-ray machine with 100 kV peak voltage. The machine operates at an exposure rate of 3 R/hr at a distance of 1 m from the machine.

a. Find the new exposure rate if the target current is increased by a factor of 10.

The exposure rate is directly proportional to the production rate and the target current. Thus the new exposure rate will be 30 R/hr.

b. Find the thickness of lead required to bring the exposure rate back to 3 R/hr at the higher current.

Using data from Figure 30.3, we find the absorption coefficient for lead (for 100 keV photons) to be $56.8/cm^{-1}$. Thus for $I = 0.10\,I_0 = I_0\,e^{-56.8x}$, we can solve for the thickness of lead, x.

$$x = \frac{\ln 10}{56.8/cm} = \frac{2.303}{56.8/cm} = 0.04\ cm = 0.4\ mm$$

c. Find the distance from the machine that would also reduce the exposure to 3 R/hr assuming the inverse square law applies for the intensity in air. (What conditions are necessary for this to be a valid assumption?)

Assuming the intensity falls off as I/x^2, we have the following relation:

$$\frac{3\ R/hr}{1\ m^2} = \frac{30\ R/hr}{x^2}$$

Thus

$$x^2 = 10.0\ m^2$$

$$x = 3.16\ m$$

SUMMARY

Use these questions to evaluate how well you have achieved the goals of this chapter. The answers to these questions are given at the end of this summary with the number of the section where you can find related content material.

Definitions

Indicate whether the following statements are true or false. For the false statements, write a true statement using the *italicized* words.

1. The German railroad engineer who changes trains at *Bremsstrahlung* is called an *RBE*, railroad bremsstrahlung engineer.
2. The *characteristic x rays* from tungsten form a continuous background of *hard x rays* for the sharp *roentgen* lines of the source.
3. The number of *soft x rays* that stick to a ball of cotton is determined by the *absorption coefficient* of cotton.
4. The *absorption coefficient* of lead is computed for a thickness of 10 cm of lead as $0.1 \ln I_0/I$.
5. The radiation equivalent basic (*reb*) is equal to the radiation equivalent man (*rem*) divided by the relative basic effectiveness (*RBE*) of the radiation.
6. A civil war army, *reb*, had two kinds of soldiers, men (*rem*) and women (*rew*) (radiation equivalent woman).

X-Ray Problems

7. An x-ray tube is operated at 50 kV.

a. Find the highest frequency for emitted radiation.
b. How much heat is produced by each accelerated electron that does not produce an x ray?
8. If a beam of x rays is incident upon material that has an absorption coefficient of 200 cm^{-1}, what thickness is necessary to reduce the number of emerging x-ray photons to one-half of the number of incident photons?

X-Ray Interactions

9. a. What are the basic sources of energy losses of x rays in matter?
   b. List at least three late somatic effects produced by x rays.

Answers

1. F (Sections 30.2, 30.6)
2. F (Sections 30.2, 30.6)
3. F (Sections 30.2, 30.4)
4. T (Section 30.4)
5. F (Section 30.6)
6. F (Section 30.6)
7. a. 1.21×10^{19} Hz;
   b. 8×10^{-15} J (Section 30.2)
8. 0.347×10^{-3} cm (Section 30.4)
9. a. photoelectric effect, Compton effect (Section 30.3)
   b. leukemia, bone cancer, cataracts (Section 30.7)

ALGORITHMIC PROBLEMS

Listed below are the important equations from this chapter. The problems following the equations will help you learn to translate words into equations and to solve single concept problems.

Equations

$$eV = hf_{max} = \frac{hc}{\lambda_{min}} \qquad (30.1)$$

$$\lambda V = 1240 \text{ eV–nm} \qquad (30.2)$$

$$I = I_0 e^{-\mu \Delta x} \qquad (30.3)$$

Problems

1. If f is the frequency, 5.12×10^{18} Hz, for an x ray incident upon your body, what is its energy in electron volts and its wavelength in meters?
2. The potential across an x-ray tube used for diagnosis may be 25,000 V. What is maximum frequency of emitted x rays?
3. For a given wavelength (0.154 nm) of x rays, the absorption coefficient of aluminum is 132 per cm and that of lead is 2610 per cm. How thick should a sheet of aluminum be to give the same shielding effect as 1 mm of lead?

Answers

1. 2.12×10^4 eV, 5.86×10^{-11} m   3. 1.98 cm
2. 6.03×10^{18} Hz

EXERCISES

These exercises are designed to help you apply the ideas of a section to physical situations. When appropriate, the numerical answer is given in brackets at the end of each exercise.

Section 30.2

1. The most energetic characteristic x rays from copper have a wavelength of 1.54 Å. Find the energy of these x rays. $[E \simeq 8050 \text{ eV}]$
2. An x-ray tube is operated at 40 kV. Find the wavelength of the most energetic x rays emitted from this tube. $[\lambda = 0.0309 \text{ nm}]$.

Section 30.4

3. The absorption coefficient for 0.154-nm x rays in nickel is 439 cm^{-1}. Find the thickness of nickel required to reduce the intensity of the incident x rays to 1/1000 of the incident value. $[x = 1.57 \times 10^{-2} \text{ cm}]$
4. The linear absorption coefficients of 0.154-nm x rays in aluminum, nickel and lead are 132 cm^{-1}, 427 cm^{-1}, and 2600 cm^{-1} respectively. What thickness of each absorber is necessary to reduce the intensity of the beam to 10.0 percent of its original value? $[174 \times 10^{-2} \text{ cm Al}, 5.4 \times 10^{-3} \text{ cm Ni}, 8.85 \times 10^{-4} \text{ cm Pb}]$

5. If 1 cm of aluminum reduces a certain x-ray beam by a factor of 10 $(I = I_0/10)$, what thickness is needed, if the incident intensity is doubled, to maintain the same final intensity? [1.3 cm]
6. The intensity of an x-ray beam may be controlled by absorbers. Suppose that you have 10 identical sheets of absorbing material and that the intensity of the beam from the tube is I_0. There is a loss of 10 percent in intensity when the beam passes through one sheet. Plot the intensity of the x-ray beams as a function of the number of absorbing sheets used.

Section 30.6

7. Find the amount of charge produced (either + or − charge) and 1 R of x-ray radiation in 1 g of air. $[2.6 \times 10^{-7} \text{ C/g}]$
8. Find the energy needed to create an ion-pair in air. Use data from the definition of the roentgen. [34 eV]

PROBLEMS

The following problems may involve more than one physical concept. When appropriate, the numerical answer is given in brackets at the end of each problem.

9. If a spherical virus cell (assume same density as water) is destroyed by 500 keV x-ray photons (in single photon interactions), find the dose needed to kill 50 percent of a virus of 10-nm radius. $[.95 \times 10^9$ rad.]
10. Find the energy per second deposited in a cube of water 10 cm on a side for 60-keV x rays in a 300-watt/

m^2 beam. The absorption coefficient for these x rays is 0.1 cm^{-1}. [1.9 joules/sec.]

11. Discuss each of the following factors as it relates to diagnostic x-ray use: voltage to x-ray tube, body type of patient, distance of patient from x-ray machine, part of body under study.

12. An x-ray examination is often used in diagnosing broken bones. What is the basic physical principal involved when using x rays in this capacity?

13. Many TV sets have been criticized because they emitted x rays. Why? If you have a set that does produce x rays, what safety factors would you recommend? Why?

14. What is the shortest wavelength that will be emitted by an x-ray target if the voltage across the tube is 10 kV? [$\lambda = 0.124$ nm]

15. Calculate the power supplied to an x-ray tube if it is operated at 50 kV and the current is 100 mA. [5 kw]

16. In an x-ray tube, the current to the target is 1.5 mA, and the voltage across the tube is 10,000 V. What is the wattage? How many electrons strike the target per second? [15 watts, 9.375×10^{15} electrons/sec

17. What is the charge in coulombs passing through an x-ray tube for one-half second and a current of 80 mA? How many electrons strike the target per second? [4×10^{-2} C, 5×10^{17} electrons/sec]

18. What is the velocity of an electron that is accelerated through a potential of 10,000 V as it reaches the target of an x-ray tube? [nonrelativistic, $v = 5.93 \times 10^9$ cm/sec]

GOALS

When you have mastered the contents of this chapter,
you should be able to achieve the following goals:

Definitions
Define each of the following terms, and use each term
in an operational definition:

atomic mass number   alpha, beta, and gamma
isotope   radiation
isobar   half-life
isotone   nuclear fission
nuclear binding energy   nuclear fusion
radioactivity

Carbon-14 Dating
Describe and explain the physical basis of carbon-14
dating.

Radioactivity Problems
Solve problems involving radioactivity.

Binding-Energy Problems
Solve nuclear binding-energy problems.

PREREQUISITES

Before beginning this chapter you should have achieved
the goals of Chapter 5, Energy, and Chapter 28, Atomic
Physics.

31

Nuclear Physics

31.1 Introduction

What is the process that makes nuclear reactors a useful source of energy? What is the basis of the concern over the radioactive waste from reactors and fallout from nuclear weapons testing? What is the difference between nuclear fission and fusion? What is the physical basis for the carbon-14 dating that was used to date the Dead Sea Scrolls? The answers to these questions involve the physics of the nucleus. In this chapter you will be introduced to the concepts and principles of nuclear physics that will enable you to answer these questions and will provide you with an understanding of the applications of nuclear physics.

31.2 The Nucleus

What do we know about the nucleus?

As you may recall from Chapter 28, the need for a nuclear model of the atom was indicated by the alpha particle-scattering experiments performed around 1910. Since that time numerous experiments have probed the properties of nuclei. Nuclei seem to be more complex than a simple, massive, uniform sphere of positive electric charge.

We currently think of the nucleus as being composed of *nucleons* of two kinds called protons and neutrons. The proton is a particle that has a positive charge which is equal in magnitude to the charge on the electron. We designate the number of protons in the nucleus by Z so the charge on the nucleus is equal to Ze. (Z is also equal to the atomic number of the element.) A neutron, the other constituent of the nucleus, is a particle that has no charge. We shall designate the number of neutrons in the nucleus by N. The total number of nucleons in the nucleus is $N + Z = A$. The total A is called the *atomic mass number*.

Isotopes are nuclei with the same number of protons, same atomic number, but with a different number of neutrons, and different values of A. Isotopes are similar in their chemical properties because chemical changes involve only the electrons of the atom and the number of electrons is the same as the number of protons. All of the isotopes of one kind of atom have the same chemical properties. Isotopes are symbolized by writing the atomic mass number as a superscript before the atomic symbol. For example, ^{1}H and ^{2}H represent the hydrogen isotopes of mass 1 and 2 respectively. The atomic number Z is written as a subscript directly under the value of A. So the complete designation for these hydrogen nuclei would be:

$^1_1$H   and $^2_1$H

The nuclei,

$^{10}_{6}$C $^{11}_{6}$C $^{12}_{6}$C $^{13}_{6}$C $^{14}_{6}$C $^{15}_{6}$C

TABLE 31.1
The Fundamental Forces

Kind of Interaction	Source	Range	Relative Strength
Nuclear	Nucleons	$< 10^{-15}$ m	1
Electromagnetic	Charged particles	Infinite ($\sim 1/r^2$)	10^{-2}
Weak	Beta decay	Short	10^{-15}
Gravitational	Mass	Infinite ($\sim 1/r^2$)	10^{-38}

are all isotopes of carbon. How many neutrons are in the nucleus of each carbon isotope? The nuclei of different elements which have the same number for A are called *isobars*. This means they must have different values of Z and N. Some isobars are,

$$^{23}_{10}\text{Ne} \qquad ^{23}_{11}\text{Na} \qquad ^{23}_{12}\text{Mg}$$

The nuclei of different elements with equal values of N are called *isotones*. Some isotones are:

$$^{13}_{5}\text{B} \qquad ^{14}_{6}\text{C} \qquad ^{15}_{7}\text{N} \qquad ^{16}_{8}\text{O}$$

The two basic nucleons have approximately the same mass, the neutron mass being 1.008665 atomic mass units (amu) and the proton mass being 1.007276 amu. An atomic mass unit (amu) is defined to be one-twelfth of the carbon-12 atomic mass or 1.660566×10^{-27} kg. Research is still being done to find out about the composition of nucleons and to formulate a model for the nuclear force that holds the nucleus together. We know that this is the strongest of all of the fundamental forces. The fundamental forces and their relative strengths and ranges are listed in Table 31.1. The nuclear force is very short range, decreasing rapidly with separation of particles and becoming insignificant for distances greater than a fermi (1 fermi (F) $= 10^{-15}$m). But for nucleons within the nuclear volume, the nuclear force is strong enough to hold protons together in spite of their Coulomb repulsive force.

The nuclear force appears to be independent of the electric charge. Thus the neutron-neutron and the neutron-proton forces are equal. Nuclear forces are the only forces in nature that exhibit a *saturation effect*. That is, the ability of nuclear forces to act upon other nucleons reaches a saturation value when a given nucleon is completely surrounded by other nucleons. This means that nucleons located outside the surrounding nucleons do not experience an interaction with the surrounded nucleon.

31.3 Properties of the Nucleus

From the previous definitions and your own experience it would seem appropriate to think the mass of the nucleus should be given by the sum of mass of protons and the mass of the neutrons.

$$\text{mass} = Zm_p + Nm_n \tag{31.1}$$

where m_p is the mass of the proton and m_n is the mass of the neutron. However, experiments show that the mass of the nucleus is less than the above. That is,

$$\text{mass} < (Zm_p + Nm_n) \tag{31.2}$$

We shall designate the difference between the computed mass in Equation 31.1 and the real mass as the mass defect.

$$\Delta m = Zm_p + Nm_n - M \tag{31.3}$$

where M is the measured mass of the nucleus. We shall now make use of Albert Einstein's famous mass-energy equation, mass is equivalent to energy,

$$\Delta E = (\Delta m)c^2 = (Zm_p + Nm_n - M)c^2 = E_{\text{binding}} \tag{31.4}$$

This energy is the *binding energy* that holds the neutrons and protons together in the nucleus. It represents the depth of the energy well that the nucleons experience within a nucleus.

This indicates that when Z protons and N neutrons combine there is a release of energy (usually in the form of photons). Also it follows that energy must be supplied to break a stable nucleus into its constitutent nucleons.

The results of scattering experiments using charged particles as projectiles and the nucleus as a scatterer indicate that the radius of the nucleus is given by:

$$R = r_0 A^{1/3} \tag{31.5}$$

where $r_0 = 1.2 \times 10^{-15}$m $= 1.2$ fermi (F).

When neutrons are scattered by the nucleus, a somewhat larger value of R is obtained. For this case,

$$r_0 = 1.5 \times 10^{-15} \text{ m} = 1.5 \text{ F}$$

In the solution of problems, use a value of 1.3 F for r_0.

We can determine the density of the nucleus by dividing its mass by the volume. We need to make some assumption about the shape of the nucleus. For strong forces such as nuclear forces one would expect a configuration which would give a minimum volume for a given number of particles, that is, a spherical shape. The nuclear volume V would be given by the usual expression for the volume of a sphere,

$$V = \tfrac{4}{3}\pi R^3 = \tfrac{4}{3}\pi [r_0 A^{1/3}]^3 = \tfrac{4}{3}\pi r_0^3 A \tag{31.6}$$

We can approximate the nuclear mass by taking the product of A and an approximate mass of a nucleon, let's say $\bar{m} = 1.67 \times 10^{-27}$ kg. Then, the density of the nucleus ρ

$$\rho = \frac{A\bar{m}}{\tfrac{4}{3}\pi r_0^3 A} = \frac{\bar{m}}{\tfrac{4}{3}\pi r_0^3}$$

Substituting values gives

$$\rho = \frac{1.67 \times 10^{-27}}{\frac{4}{3}\pi(1.3 \times 10^{-15})^3} \approx 2 \times 10^{17} \text{ kg/m}^3$$

(A dime's volume of nuclear matter would weigh over 60 million tons!) Let's compare this with other densities: first, atomic densities. The radius of an atom is about 10^4 times that of a nucleus. So the volume of an atom is 10^{12} times that of the nucleus, and you have learned that most of the mass of an atom is concentrated in the nucleus so

$$\rho_{\text{atom}} = \frac{\rho_{\text{nucl}}}{10^{12}} \approx 2 \times 10^5 \text{ kg/m}^3$$

Now consider densities of bulk materials. For instance the density of water is,

$$\rho_{\text{H}_2\text{O}} = 10^3 \text{ kg/m}^3$$

Are the above values consistent with what you expect from the forces of interaction? This value of atomic density is too high. How could you obtain a more realistic answer?

31.4 Nuclear Binding Energy

We have learned how to determine the mass defect (Equation 31.3), and we have learned how to convert this mass defect into energy (Equation 31.4). This energy, called the binding energy (BE) of the nucleus, is given by

$$\text{BE} = (Zm_{\text{p}} + Nm_{\text{n}})c^2 - Mc^2 \tag{31.7}$$

where M is the mass of the nucleus as determined by mass spectrometry.

In a typical periodic table of the elements, the masses which are given are usually the atomic masses instead of the nuclear masses. We can convert the atomic masses into nuclear masses that can be used in binding energy calculations by subtracting the masses of the electrons.

$$m_{\text{p}} = m_{\text{H}} - m_{\text{e}} \tag{31.8}$$

where m_{H} is the atomic mass of hydrogen atom and m_{e} is mass of the electron. Hence, $Zm_{\text{p}} = Z(m_{\text{H}} - m_{\text{e}})$, and the mass of the nucleus is given by the mass of the atom minus the total mass of all the electrons in the atom,

$$M = m_{\text{a}} - Zm_{\text{e}} \tag{31.9}$$

where m_{a} is the mass of the atom.

Substituting these expressions in Equation 31.7 we have a complete expression for the binding energy:

$$\text{BE} = (Zm_{\text{H}} + Nm_{\text{n}} - m_{\text{a}})c^2 \tag{31.10}$$

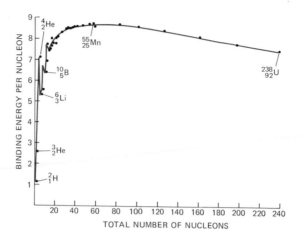

FIGURE 31.1
The binding energy per nucleon in MeV as a function of the total number of nucleons in the nucleus of the element.

The binding energy may be expressed in atomic mass units if we divide by c^2, and express the masses in amu, thus

$$BE_{amu} = Zm_H + Nm_n - m_a \qquad (31.11)$$

We also find that often the binding energy is expressed in terms of million electron volts (MeV). The conversion factor is

$$1 \text{ amu} = 931.5 \, \frac{MeV}{c^2}$$

If the binding energy is greater than zero, the nucleus is stable, and it is necessary to supply energy in order to break the nucleus into its constituent parts. If the binding energy is less than zero, the nucleus is unstable, and it will disintegrate spontaneously.

We also find the binding energy per nucleon to be a useful value. This value is found by dividing the total binding energy by the number of nucleons. A graph of the binding energy per nucleon as a function of the number of nucleons is shown in Figure 31.1.

EXAMPLES

1. What is the binding energy of deuterium? The deuterium nucleus ^2_1H has one proton and one neutron. Thus, $BE = m_H + m_n - m_D$.

$$m_H = 1 \text{ mass } ^1_1H = 1.007825 \text{ amu}$$

$$m_\lambda = 1 \text{ mass neutron} = \underline{1.008665} \text{ amu}$$
$$2.016490 \text{ amu}$$

$$m_D = \text{mass of } ^2_1H = \underline{2.014102} \text{ amu}$$

$$BE = 0.002388 \text{ amu} = 2.224 \text{ MeV}$$

$$BE/\text{nucleon} = 1.112 \text{ MeV/nucleon}$$

2. What is the binding energy of the most common helium nucleus? Helium nucleus has two protons and two neutrons.

TABLE 31.2
Atomic Masses of Some Isotopes

Name	Symbol	Atomic Mass in amu
Proton	p	1.007277
Hydrogen	$_1^1\text{H}$	1.007825
Neutron	n	1.008665
Deuterium	$_1^2\text{H}$	2.014102
Helium three	$_2^3\text{He}$	3.016029
Tritium	$_1^3\text{H}$	3.016049
Helium four	$_2^4\text{He}$	4.002603
Carbon twelve	$_6^{12}\text{C}$	12.000000
Potassium 39	$_{19}^{39}\text{K}$	38.96371
Radon 222	$_{86}^{222}\text{Rn}$	222.01761
Radium 226	$_{88}^{226}\text{Ra}$	226.02536
Uranium 235	$_{92}^{235}\text{U}$	235.0439

1 amu $= 1.660566 \times 10^{-27}$ kg $= 931.5016$ MeV

$$\text{BE} = 2m_\text{H} + 2m_\text{n} - m_\text{He}$$

$$2\ m_\text{H} = 2.015650 \text{ amu}$$

$$2\ m_\text{n} = \underline{2.017330} \text{ amu}$$
$$4.032980 \text{ amu}$$

$$m_\text{He} = \underline{4.002603}$$

$$\text{BE} = 0.030377 \text{ amu} = 28.30 \text{ MeV}$$

$$\text{BE/nucleon} = 7.07 \text{ MeV/nucleon}$$

The binding energy per nucleon is many orders of magnitude greater than the binding energy of the electron in the atom. Recall that binding energy of the outermost electron (the ionization energy) is of the order of electron volts and for inner electrons of heavy atoms it is the order of thousands of electron volts. You may notice that the binding energy per nucleon is greater for helium than for deuterium. What would you conclude about the relative stability of $_1^2\text{H}$ and $_2^4\text{He}$ from these calculations?

31.5 Natural Radioactivity

In 1896 Henri Becquerel found that certain uranium salts emitted penetrating radiations. He observed that photographic plates which were stored in a drawer with uranium salts were blackened. The original discovery indicated that the results were similar to exposure to x rays which had been discovered the preceding year. The importance of this discovery was not initially apparent. A few years later Pierre and Marie Curie separated from pitchblende two materials which were many times more radiation active than uranium. These new substances were shown to be two new elements—polonium and radium.

The NaI (Tl) crystal detector used for measurement of γ radiation from radium in the human body. The detector is placed equidistant from various points of the body. The output of the NaI detector is fed into a standard NaI (Tl) γ-ray spectrometer. (Courtesy of Argonne National Laboratory.)

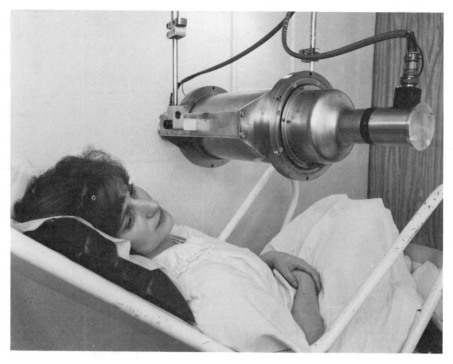

These active materials were called natural radioactive elements. These elements emit radiation spontaneously without the addition of energy to them. Many studies were made to determine the characteristics of the radiation. The results of these experiments showed that there were three types of radiation which were called *alpha* (α), *beta* (β), and *gamma* (γ) *radiation*. Characteristics of each type of radiation were determined. The alpha radiation was found to be positively charged particles each with a mass of the helium nucleus. They have low penetrating ability. In fact, they are stopped by a sheet of paper. They are heavy ionizers. The beta radiation carries a negative charge and has the mass of an electron; they are high-energy electrons. They were more penetrating than the α particle but could be stopped by thin aluminum sheets. The β particle was less effective as an ionizer than an α particle. The γ radiation was not deflected by a magnetic field; hence it carried no charge. The γ radiation was very penetrating and behaved very much like x rays. Gamma radiation would pass through a lead plate. In fact it was found that γ radiation exhibited the following properties: produced fluorescence in certain crystals; blackened a photographic paper; ionized gases; produced a small effect when absorbed; was diffracted by crystals or ruled grating and refracted in crystals; reflected from crystal planes; propagated rectilinearly; and produced secondary γ rays.

In addition to the natural radioactive nuclei there are *induced radioactive nuclei*. These are produced by bombarding a stable nucleus with some particle that is captured by and combines with the original nucleus, which then breaks up into new constituents.

31.6 Stable and Unstable Nuclei

A plot of the number of neutrons and number of protons for the stable nuclei as shown in Figure 31.2. For light nuclei, the most stable nuclei are those in which the number of neutrons and the number of protons are the same. For the heavier stable elements the number of neutrons exceeds the number of protons.

An unstable nucleus spontaneously decays into another nucleus. An isotope made up of such nuclei is called a *radioactive nuclide*. The decay of the individual nuclei in a radioactive sample is a random process. The equation for radioactive decay is given as follows:

$$N = N_0 \, e^{-\lambda t} \tag{31.12}$$

where N_0 is the number of nuclei at $t = 0$, N is the number of nuclei left after time t, and λ is the decay constant for the nuclide.

It is usual to use the *half-life* as a measure of radioactivity. The half-life $t_{1/2}$ is the time required for one-half of the nuclei to decay. We can express this in equation form as follows:

$$N = \tfrac{1}{2} N_0 = N_0 e^{-\lambda t_{1/2}}$$

FIGURE 31.2

A graphical presentation of the stable nuclei on a plot of neutron number N versus proton number Z.

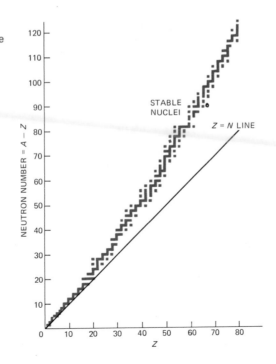

$$\tfrac{1}{2} = 1(e^{-\lambda t_{1/2}})$$

$$t_{1/2} = \frac{\ln 2}{\lambda}$$

$$t_{1/2} = \frac{0.693}{\lambda} \tag{31.13}$$

The half-lives of radioactive nuclides vary from small fractions of a second to billions of years. Likewise, the decay modes of different radioactive nuclides vary. There are two principle modes of radioactive decay for natural elements. These modes are called alpha and beta decay. Alpha decay involves the emission of a ^{4}He nucleus (α particle) from the radioactive nuclide. For example,

$$^{239}_{94}\text{Pu} \rightarrow \,^{235}_{92}\text{U} + \,^4_2\text{He} \qquad (t_{1/2} = 24{,}360 \text{ yr})$$

Notice that the total atomic mass numbers (superscripts) and atomic numbers or positive charges (subscripts) are equal across the reaction. All nuclear reaction equations are balanced in this way. A general form for alpha decay can be written as follows:

$$^A_Z\text{X} \rightarrow \,^{A-4}_{Z-2}\text{Y} + \,^4_2\text{He} + \gamma + Q$$

where X represents the parent nucleus, Y, the daughter nucleus, γ, a gamma ray, and $Q =$ energy released as kinetic energy of the daughter nucleus and the alpha particle.

The *gamma radiation* that is present with many natural radioactive nuclides is due to the nuclei changing energy levels. If the alpha decay leaves the daughter nucleus in an excited state, it will decay to its ground state by emitting a high energy photon (nuclear energy levels are mega-electron volts apart). These high energy photons are the *gamma rays*. The Q value for this general alpha decay can be calculated by finding the mass-energy difference for the reaction.

If the daughter nucleus is in the ground state, the Q value for the reaction is given by the following equation

$$Q = [m_X - (m_Y + m_\alpha)]c^2 \tag{31.14}$$

where m_X is the mass of the parent nucleus, m_Y is the mass of the daughter nucleus, and m_α is the mass of the α particle.

The distribution of the energy Q between the daughter nucleus and the alpha particle is determined by the conservation of momentum in the process.

Beta decay is a term that applies to the emission from an unstable nucleus of either an electron or a *positron*, a particle with the same rest mass of an electron but positive charge. At first beta-decay processes baffled the physicists because the observed particles did not seem to conserve momentum and energy. In 1934 Enrico Fermi proposed a theory that involved a new massless, chargeless particle that he called the *neutrino* (ν). The neutrino represented a fudge factor that would satisfy the conservation of energy and of linear and angular momentum for the process. (It was not until 1953 that the existence of the neutrino was experimentally verified. What does this say about the faith of physicists in their conservation laws?)

The neutron (when not in a nucleus) decays via beta decay into a proton, an electron and an antineutrino $\bar{\nu}$ as follows:

$$^1_0n \rightarrow {}^1_1p + e^- + \bar{\nu}$$

Twelve minutes is the half-life for free neutrons. Beta decay may alternately involve the emission of a positron instead of an electron. In this case, the third particle is a neutrino.

An example of positron emission is given in the following:

$$^{20}_{11}Na \rightarrow {}^{20}_{10}Ne + e^+ + \nu$$

The e^+ is the symbol for a positron which has the same mass as electron but positive charge. Another important beta decay process is that of carbon-14.

$$^{14}_6C \rightarrow {}^{14}_7N + e^- + \bar{\nu}$$

Carbon-14 dating is a technique that can be used to determine the age of relics containing carbon. The ratio of $^{14}_6C/^{12}_6C$ in the atmosphere is assumed to have remained constant as the cosmic rays incident on the earth maintain the equilibrium amount of $^{14}_6C$. Living plants take in natural carbon, maintain a constant ratio of $^{14}_6C/^{12}_6C$, and exhibit a radioactivity of about 16 counts per minute for each gram of carbon. This activity is the number of nuclear decays per minute. The activity of a radioactive sample is governed by the following equation:

$$A = A_0 e^{-\lambda t} = A_0 e^{-0.693t/t_{1/2}} \tag{31.15}$$

where A_0 is the activity at $t = 0$, A is the activity at time t, and λ is the decay constant. By Equation 31.13 $\lambda = 0.693/t_{1/2}$ where $t_{1/2}$ is the half-life of the decay.

Thus, if the activity of a gram of carbon from the relic is measured to be A, we know that this is a measure of the amount of carbon-14 left after the relic died and stopped taking in carbon-14 from its environment. We can obtain the following equation from Equation 31.15 for the time elapsed since the sample stopped taking in carbon: Take the natural logarithm of both sides of Equation 31.15, and then solve for t to obtain:

$$t = \frac{\ln (A_0/A)}{\lambda} = \frac{t_{1/2} \ln (A_0/A)}{0.693}$$

$t_{1/2}$ for $^{14}_6C$ is ≈ 5700 years

Since A_0 is the activity of the same amount of living carbon (about 16 counts per minute per gram) t can be calculated. Carbon-14 dating is limited to about 50,000 years because of its low activity per gram of carbon. Other radioactive dating processes make use of $^{238}_{92}U$ and $^{40}_{19}K$.

EXAMPLE

A relic is found to give an activity count of 12 cpm (counts per minute) for each gram of carbon. If living trees give a count of 16 cpm, find the approximate age of the relic.

$$t = \ln \frac{16/12}{0.693/5700} \, \text{yr} = \ln \left[\frac{(1.33)(5700)}{0.693} \right] \text{yr} = 2345 \text{ yr}$$

A general equation for beta decay can be written as follows:

$$_Z^A X \rightarrow \, _{Z+1}^A Y + e^- + \bar{\nu} \quad \text{electron emission}$$

or

$$_Z^A X \rightarrow \, _{Z-1}^A Y + e^+ + \nu \quad \text{positron emission}$$

In these equations you may notice that atomic mass number A and electric charge are conserved. In addition, total energy, mass, and momentum are conserved. The same principles of disintegration are applicable to all nuclei.

The activity of a radioactive sample is expressed in terms of the unit known as the *curie* (Ci). The curie is defined as 3.70×10^{10} disintegrations per second. The curie is a very high activity, and it is much more common to find samples whose activity is measured in millicuries (mCi) or microcuries (μCi).

31.7 Interactions of Charged Particles with Matter

There are three charged particles with which we are concerned when we consider general interactions with matter. These are protons, alpha particles, and beta particles. Protons are constituents of the cosmic rays that come from outer space, in which case they may have energies above 1 GeV (10^9 eV). Protons may also be produced in machines that ionize hydrogen. Energies in these cases depend upon the characteristics of the machine. There are a few accelerators that may produce protons of energies of the order of magnitude of 10^9 eV. However, most of the humanly accelerated protons only reach energies of the order of 10^6 eV.

Alpha particles occur in nature as the result of the decay of naturally radioactive elements. The energies of the alpha particles from radioactive sources are generally in the range of 4 to 10 MeV. Alpha particles in these energy ranges are not very penetrating; they have a range in air of a few centimeters and are stopped by metal foil.

Beta particles have a wide range of energies as emitted from a radioactive source. The energy of any one electron from a given source may be from zero to the maximum possible value for the source. High energy electrons may also come from cosmic rays or accelerators, and they have exactly the same properties as beta particles. For a given energy range, beta particles or electrons are more penetrating than either protons or

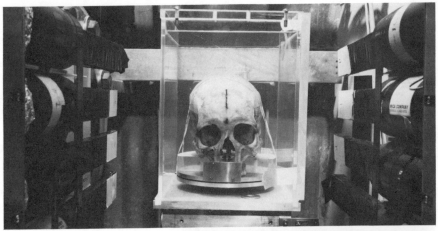

(a)

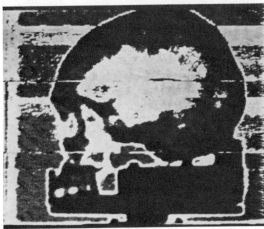

(b)

High energy medicine. (a) A brain called "Charley" was placed in the skull, and the unit shown above was placed in a beam of 200 MeV protons. By firing a 1-mm square beam of these 200 MeV protons through Charley, measuring the energy loss at each point and feeding the data to a computer, the researchers hoped to get a black and white picture showing the clot. (Courtesy of Argonne National Laboratory.)

(b) Charley's radiograph. The brain had a known blood clot in the right parietal-temporal lobe. The blood clot would absorb more energy of protons than the surrounding tissue and hence a dark spot should appear. The radiograph shows a dark area where the clot should have been but the results were unclear.

In order to understand why proton radiographs are preferred over x ray ones, you need to know something about how x rays and protons act when passing through the human body. Radiation is the greatest danger in the human tissues where the greatest amount of energy is absorbed. In x rays the greatest absorption occurs at first skin contact. Thus, to cut the patients radiation absorption one must reduce the x-ray intensity, which gives poorer x-ray radiographs. Protons lose energy gradually as they pass through the body. They tend to be absorbed when they have lost all their energy. If their energy is sufficiently large, they will pass through the body with little absorption of radiation. The protons are given enough energy to pass through the body, assuming all the tissues are normal. The protons that pass through the denser tissues, such as tumors, will lose enough energy to be absorbed. The tumor will then show as a hole in the beam of emerging protons. (Courtesy of Argonne National Laboratory.)

alpha particles. For comparison, a 20 MeV electron will penetrate about 10 cm of water and a 20 MeV proton will penetrate about 0.3 cm of water. The ranges of 10 MeV particles in water are: electrons, 3.0 cm; protons, 0.1 cm; and alpha particles, 0.01 cm. In general the greater the specific ionization (ions produced per cm) the lower the penetration. On the average, each pair of ions produced requires an energy loss of 30 to 35 eV.

31.8 Interactions of Neutrons with Matter

The interaction of neutrons with matter is much more complicated than either that of charged particles or x rays (gamma rays). For charged particles we speak in terms of range. For x rays we have an exponential law of penetration.

Neutrons have no charge, and hence they do not experience repulsive coulomb forces. They encounter large forces only when they are within range of nuclear forces. Also neutron interactions depend a great deal upon the energy of the neutron. We can classify the neutron energy in different ranges. There are thermal, intermediate, fast and relativistic neutron energies. The thermal energy range is that in which the energy is characterized by the ambient temperature. This is an energy range of a few hundredths of an electron volt. For order of magnitude one can use 0.025 eV as the room temperature energy of a neutron. Fast neutrons are those energies within the 10 keV to 10 MeV range. Between thermal and 10 keV we have intermediate energy range. Above 10 MeV are the relativisitic neutrons. The type of nuclear interaction that occurs with the incident neutron also depends upon the type of target nucleus, the particular isotope of a given element. In general, three types of neutron capture reactions have been found:

1. Gamma radiation is emitted. For example

$$_0^1 n + {}_{13}^{27}Al \rightarrow {}_{13}^{28}Al + \gamma$$

2. A charged particle is emitted. For example

$$_0^1 n + {}_5^{10}B \rightarrow {}_3^7Li + {}_2^4He \text{ (neutron detection)}$$

3. Fission of the nucleus follows:

$$_0^1 n + {}_{92}^{235}U \rightarrow {}_{92}^{236}U \rightarrow {}_{54}^{140}Xe + {}_{38}^{94}Sr + 2{}_0^1 n + Q$$

where Q is the energy released.

Using the method shown in Section 31.4, you should be able to calculate Q for the neutron capture by uranium 235 using the following two sequences of beta decay:

$$_{54}^{140}Xe \xrightarrow{\beta^-} {}_{55}^{140}Cs \xrightarrow{\beta^-} {}_{56}^{140}Ba \xrightarrow{\beta^-} {}_{57}^{140}La \xrightarrow{\beta^-} {}_{58}^{140}Ce$$

and

$$\beta^- \qquad \beta^-$$
$$^{94}_{38}\text{Sr} \rightarrow ^{94}_{39}\text{Y} \rightarrow ^{94}_{40}\text{Zr}$$

The atomic mass of $^{140}_{58}\text{Ce}$ is 139.908013 amu; the atomic mass of $^{94}_{40}\text{Zr}$ is 93.906140 amu; you get a value for Q slightly greater than 200 MeV.

31.9 Fission

O. Hahn and F. Strassman in 1938 showed by chemical analysis that an isotope of barium was present when uranium was bombarded with neutrons. In early 1939, O. Frisch and L. Meitner suggested that the uranium was undergoing nuclear fission. They said a single nucleus was being fragmented into two particles of comparable size. On the basis of Hahn's analysis of $^{141}_{56}\text{Ba}$ and $^{92}_{36}\text{Kr}$, a quick calculation showed that relatively speaking a large amount of energy was available, about 200 MeV. The total mass of the components is less than that of the heavy nucleus, and the mass difference appears as kinetic energy of the produced nuclei. One fission process that is triggered by a thermal neutron hitting a uranium-235 nucleus was shown as the third example of neutron processes in Section 31.8. Such fission reactions were the basis of the nuclear bombs dropped on Japan in 1945 and of the first nuclear reactors.

31.10 Fusion

In 1938 Hans Bethe suggested that the source of energy in the sun and other stars came from a nuclear reaction in which two light nuclei combined to form a heavier nucleus. These reactions are called *fusion*. Again, the product heavy nucleus has less mass than the sum of the two fusing nuclei, and the mass difference appears as the kinetic energy of the product. In order for a fusion reaction to take place, it is necessary that the fusing nuclei get close enough together that the nuclear force overcomes their repulsion. The temperature needed to get the nuclei close enough together for fusion is in the neighborhood of 10^7 degrees Kelvin. In the sun gravitational collapse provides the necessary energy. In nuclear fusion bombs the energy can be provided by fission reactions. It is the problem of generating these high temperatures and the associated containment problems that have made fusion energy reactors so elusive.

Bethe suggested the following cycle as one for release of energy by fusion.

a. $^1_1\text{H} + ^{12}_6\text{C} \rightarrow ^{13}_7\text{N} + \gamma$
b. $^{13}_7\text{N} \rightarrow ^{13}_6\text{C} + e^+ + \nu$
c. $^1_1\text{H} + ^{13}_6\text{C} \rightarrow ^{14}_7\text{N} + \gamma$

d. $^1_1H + {}^{14}_7N \rightarrow {}^{15}_8O + \gamma$

e. $^{15}_8O \rightarrow {}^{15}_7N + e^+ + \nu$

f. $^1_1H + {}^{15}_7N \rightarrow {}^{12}_6C + {}^4_2He$

In checking we see that $^{12}_6C$ acts as a catalyst and the net result is that four ^1_1H atoms are consumed to form one ^4_2He atom, γ rays, two neutrinos, and two positrons. The cycle can be represented by

$$4\,^1_1H \rightarrow {}^4_2He + 2e^+ + 2\nu + Q$$

Again, we can use the method of Section 31.4 and calculate the value of Q, which turns out to be 26.7 MeV. That is combining four protons to form one alpha particle in this way yields 26.7 MeV of energy.

Another fusion cycle that combines hydrogen to produce helium, called the proton-proton cycle, is possible:

a. $^1_1H + {}^1_1H \rightarrow {}^2_1H + e^+ + \nu$

b. $^2_1H + {}^1_1H \rightarrow {}^3_2He + \gamma$

These two steps repeated yield $2\,^3_2He$.

c. $^3_2He + {}^3_2He \rightarrow {}^4_2He + 2^1_1H$

In this cycle six protons produce ^4_2He, two positrons, two protons, two neutrinos, and γ rays. The energy released is 26.7 MeV. Only certain nuclei are suitable for fission or fusion reactions. What test can you suggest that would be useful in determining candidate nuclei for each of these reactions? (*Hint:* see Figure 31.1.)

31.11 Dangers of Radiation

The beginning of nuclear weapons testing in the atmosphere significantly increased the background radiation level. The fallout debates among scientists during the 1950s and 1960s were heated and confusing to many laymen. How can scientists with the same data disagree on the dangers of nuclear fallout? The answer to this question is worth our consideration. According to one school of thought any increase in ionizing radiation produces undesirable genetic damage to the population. Another school of thought argues that radiation effects in human systems are primarily threshold phenomena, and that as long as the radiation level is below threshold levels, there need not be any concern over increased background radiation. The problem arises because there is no clear evidence to refute the threshold position. But if the genetic damage is long-term in its effects, to ignore increased background radiation is to ask for future trouble. To heighten the concern, one of the radioactive components of the fallout is strontium-90 (half-life, 28 yr) which can substitute chemically for calcium in many of the body processes. When this happens, it amounts to putting a radioactive source into the human body. The issue is not dead; there are scientists today

Micrographs of cross-sections of human bone. At the left is a normal-bone sample, and at the right is a bone sample from a person with an extremely high body burden of radioactive (γ radiation) radium. (Courtesy of Argonne National Laboratory.)

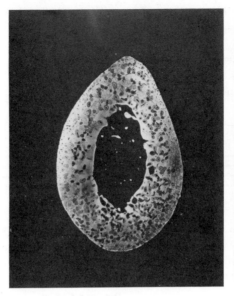
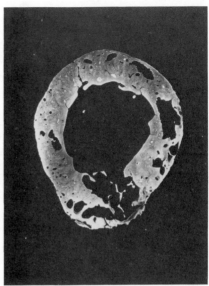

still arguing about the safe levels that should be set for radiation. Nuclear power plants have become the new focal point of concern. What are the risks? We need more scientists working on the problem and better communication between people everywhere concerning the risks versus the benefits of nuclear power plants.

ENRICHMENT
31.12 Derivation of the Radioactivity Decay Law

Given a large number N of radioactive atoms, the average number decaying in time interval dt will be dN. This number will be proportional to the number present at time t. Thus we can write

$$-dN = \lambda N\, dt \quad \text{or} \quad \frac{dN}{N} = -\lambda\, dt \tag{31.16}$$

where λ is the decay constant. Integrating this equation we get

$\ln N = -\lambda t + \text{constant}$

At $t = 0$, let $N = N_0$. Thus,

$\ln N - \ln N_0 = -\lambda t$

$\ln \dfrac{N}{N_0} = -\lambda t$

$\dfrac{N}{N_0} = e^{-\lambda t}$

$N = N_0 e^{-\lambda t} \tag{31.17}$

The activity of a sample is defined as its rate of decay. Thus, activity $A = dN/dt$, and we can write

$$A = \frac{dN}{dt} = -\lambda N_0\, e^{-\lambda t} = A_0 e^{-\lambda t} \tag{31.18}$$

where

$$A_0 = \lambda N_0 = \lambda \times \frac{\text{mass of sample}}{\text{atomic weight}} \times \text{Avogadro's number}$$

SUMMARY

Use these questions to evaluate how well you have achieved the goals of this chapter. The answers to these questions are given at the end of this summary with the number of the section where you can find related content material.

Definitions

1. a. Nuclear binding energy = (_____)c^2
   b. Radioactivity is a process characterized by the _____.
   c. Half-life is equal to the time required for _____.
   d. Fission occurs when a _____ nucleus _____ into _____.
   e. Fusion occurs when a $_1^2$H nucleus is _____ by _____.
   f. α particle = _____; β^- particle = _____; γ ray = _____.

Carbon-14 Dating

2. a. What is the count per minute for a gram of living tree?
   b. What type of radiation is responsible for these counts?
   c. What does the statement, "The half-life of $^{14}_6$C is 5700 years," mean?

Radioactivity

3. The activity of a certain radioisotope is observed to decrease by a factor of eight in 30 hours.
   a. What is the half-life of the isotope?
   b. What is the disintegration constant?
   c. If the present activity of the sample is 300 millicuries, what will be its activity after ten hours?

4. The half-life of $^{234}_{90}$Th is 24.1 days. How long after a sample of this isotope has been isolated will it take for 90 percent of it to change to $^{234}_{91}$Pa?

Binding Energy

5. Given that $m_H = 1.007825$ amu, $m_n = 1.00865$ amu, and $m_{16_O} = 15.994915$ amu, what is the binding energy of $^{16}_8$O? Find the binding energy per nucleon.

6. Given the masses $^{20}_{10}$Ne $= 19.992442$ amu, $^{21}_{10}$Ne $= 20.993849$ amu, $^{22}_{10}$Ne $= 21.991385$ amu, and $m_n = 1.00865$ amu, what is the binding energy of the last neutron in ^{22}Ne?

Answers

1. a. $(Zm_p + Nm_n - M)c^2$
      (Section 31.3)
   b. random decay of nuclei (Section 31.5)
   c. one-half of the nuclei to decay (Section 31.6)
   d. heavy, splits, lighter nuclei (Section 31.9)
   e. formed, $_1^1$H + $_1^1$H getting close enough for nuclear force to produce $_1^2$H (Section 31.10)
   f. $\alpha = _2^4$He nucleus, $\beta^- =$ electron, $\gamma =$ electromagnetic radiation (Section 31.6)

2. a. 16
   b. β^- ray; c, half the carbon-14 atoms have disintegrated in 5700 years (Section 31.6)

3. a. 10 hr
   b. 0.0693 hr^{-1}
   c. 150 millicuries (Section 31.6)

4. 80 days (Section 31.6)

5. 127.51 MeV, 7.97 MeV/nucleon (Section 31.4)

6. 10.35 MeV (Section 31.4)

ALGORITHMIC PROBLEMS

Listed below are the important equations from this chapter. The problems following the equations will help you learn to translate words into equations and to solve single-concept problems.

Equations

$$E = (\Delta m)c^2 \tag{31.4}$$

$$BE = (Zm_p + Nm_n - M)c^2 \tag{31.4}$$

$$R = r_0 A^{1/3} \tag{31.5}$$

$$BE = (Zm_H + Nm_n - m_a)c^2 \tag{31.10}$$

$$N = N_0 e^{-\lambda t} = N_0 e^{-0.693t/t_{1/2}} \tag{31.12}$$

$$t_{1/2} = 0.693/\lambda \tag{31.13}$$

$$Q = [m_X - (m_Y + m_\alpha)]c^2 \tag{31.14}$$

$$A = A_0 e^{-\lambda t} = A_0 e^{-0.693t/t_{1/2}} \tag{31.15}$$

Problems

1. If you had a system whereby you could completely convert 2 g of gasoline into energy, how much energy would be produced?
2. In certain experiments one finds that if one has sufficient energy, an electron and positron are produced. What is the minimum energy that would be needed? Assume the mass of the electron and positron to be the same.
3. What is the value for λ if one-half of the nuclei in a sample disintegrate in 1 day?
4. If the activity of a given radioactive source is reduced to one-half in one day, when will the activity be reduced to one-fourth?
5. What is the Q value for the radioactive disintegration? $^{226}_{88}\text{Ra} \rightarrow {}^{222}_{86}\text{Rn} + {}^{4}_{2}\text{He}$

Answers

1. 1.8×10^{14} J.
2. 1.02 MeV $= 1.64 \times 10^{-13}$ J
3. 8×10^{-6} sec^{-1}
4. 2 days
5. 0.00515 amu $= 4.79$ MeV

EXERCISES

These exercises are designed to help you apply the ideas of a section to physical situations. When appropriate the numerical answer is given in brackets at the end of the exercise.

Section 31.2

1. About 93 percent of potassium is the isotope with an atomic number 19 and mass number 39. How many protons and neutrons are there in the nucleus? [19, 20]

Section 31.4

2. What is the binding energy of the $^{39}_{19}\text{K}$ nucleus? What is the binding energy per nucleon? [333.7 MeV, 8.55 MeV]

3. Find the binding energy per nucleon for tritium ($^3_1$H). [2.83 MeV]

4. Is it energetically possible for a proton to decay into a neutron and a positron?

Section 31.6

5. A possible neutron source is called the beryllium α particle sponge in which the $^9_4$Be nucleus absorbes an α particle and a neutron is emitted. Write this nuclear equation? What are the end products?

6. In the uranium-radium family there are 7 α-particle emissions and 4 β-particle emissions from $^{238}_{92}$U to the stable end product. What is the end product? Give atomic number and atomic mass.

PROBLEMS

The following problems may involve more than one physical concept. Problems requiring the material from the enrichment section are marked with a dagger †. When appropriate the answer is given in parentheses at the end of each problem.

7. A neutron may have an energy from thermal energy 0.02 eV up to an energy of several MeV (say, 10 MeV). What are the relative velocities of the neutron for these energies? [6.19×10^3 m/sec, 4.4×10^7 m/sec]

8. An alpha particle has a kinetic energy of 6.08 MeV. What is the velocity of the particle? [1.71×10^7 m/sec]

9. Assume that the parent nucleus ^A_ZX in alpha decay is at rest. Show that the kinetic energy of the emitted alpha particle is given by $(A-4)Q/A$. (*Hint:* Use conservation of momentum, and assume the daughter nucleus is in the ground state.)

10. Find the kinetic energy of the alpha particle in the following reaction:

$$^{226}_{88}\text{Ra} \rightarrow {}^{222}_{86}\text{Rn} + {}^4_2\text{He}$$

Use the results of problem 9. [4.79 MeV]

11. Find the minimum energy needed to break a carbon-12 nucleus into three alpha particles. [7.27 MeV]

12. Fill in the missing component for each of the following reactions:
   a. $^{10}_5\text{B} + {}^4_2\text{He} \rightarrow ? + {}^1_0\text{n}$
   b. $^{14}_6\text{C} \rightarrow {}^{14}_7\text{N} + ?$
   c. $^{16}_8\text{O} + {}^{16}_8\text{O} \rightarrow {}^{28}_{14}\text{Si} + ?$
   d. $^1_0\text{n} + {}^{235}_{92}\text{U} \rightarrow {}^{107}_{43}\text{Tc} + ? + 5\text{n}$ [$^{13}_7$N, β^-, $^4_2$He, $^{124}_{49}$In]

13. Find the energy released in the following fusion reaction: $^2_1\text{H} + {}^2_1\text{H} \rightarrow {}^3_1\text{H} + {}^1_1\text{H}$ [3.84 MeV]

14. Find the activity of 1 g of $^{226}_{88}$Ra (half-life = 1620 yr). Compare this number with the definition of the curie. [3.60×10^{10} sec^{-1}]

15. Find the activity of a milligram of $^{227}_{90}$Th (half-life 1.9 yr) in curies.   [0.85 curies]

†16. Given a 10-millicurie sample of $^{24}_{11}$Na (half-life = 15 hr), find the mass of the sample.   [7.5×10^{-10} g]

17. Iodine-131 has a half-life of 8 days. If its activity on July 1 is 6 millicuries, find its activity 16 days later. [1.5 millicuries]

†18. The $^{134}_{55}$Cs nucleus is a beta emitter which has a half-life of 2.30 year. What size source would one need for 10 disintegrations per second?   [2.34×10^{-13} g]

†19. The specific activity of an isotope whose atomic mass is 212 is 2.98×10^{-17} curie/g. What is its half-life?   [3050 sec]

20. A lab radiologist finds a radioactive sample with no date of acquisition on it. She places the sample in front of a counter and records an activity of 1600 cpm. She then places a similar sample that has just arrived in front of the counter and records an activity of 12,800 cpm. How long has the unlabeled sample been in the lab if the half-life of the material is six days?   [18 days]

†21. Carbon-14 is a beta emitter and has a half-life of 5.7×10^3 year. It is used in radioactive dating. If one finds 45 disintegrations per minute from a sample of 12 g of carbon from a bone sample, how many atoms of $^{14}_6$C are in the sample? Other measurements have indicated that the equilibrium ratio of $^{14}_6$C to $^{12}_6$C in the atmosphere is 1.3×10^{-12}. If we assume this was the ratio at the time the bone was formed, what would you conclude was the age of the bone sample?   [2 half-lives or 11,400 yrs.]

22. In doing an experiment to determine the half-life of an unknown sample one might get the following data: 30 days ago the activity was recorded as 920 disintegrations per minute. Today, the activity is 230 disintegrations per minute. What is the half-life of the sample?   [15 days]

23. The radioactivity of carbon from trees felled in Wisconsin glaciation is about $12\frac{1}{2}$ percent as intense as

activity from trees cut today. When did this last ice age occur according to carbon dating? (Use 5700 years as half-life for carbon-14.) [3 half-lives]

24. Given the half-life of $^{14}_{6}C$ as 5700 years. One gram of carbon from a living tree has an activity of 720 counts/hour. A gram sample of a wooden relic has an activity of 45 counts/hour. Find the approximate age of the relic. [4 half-lives]

†25. Using calculus, show that the average lifetime of a radioactive isotope is equal to $(1/\lambda)$. *Hint:* the average life is the total of all lifetimes divided by the number of particles:

$$t_{ave} = \frac{1}{N_0} \int_0^{N_0} t \, dN$$

GOALS

When you have mastered the content of this chapter, you will be able to achieve the following goals:

Definitions
Define each of the following terms, and use each term in an operational definition:

fission	thermonuclear reaction
chain reaction	plasma
critical mass	"magnetic bottle"
moderator	tracer
breeder reactor	biological half-life
neutron activation	effective half-life
fusion	

Fission Reactors
Explain the operation of a nuclear fission reactor.

Radiation Detectors
Compare different nuclear radiation detectors and their uses.

Radiation Problems
Solve problems involving radiation applications using biological half-life and dosage data.

Tracer Applications
Outline the use of radioactive tracers in medical applications.

PREREQUISITES

Before beginning this chapter you should have achieved the goals of Chapter 21, Electrical Properties of Matter, Chapter 27, Quantum and Relativistic Physics, Chapter 30, X Rays, and Chapter 31, Nuclear Physics.

32

Applied
Nucleonics

32.1 Introduction

What are the conditions necessary for a self-sustaining chain reaction? What are the basic requirements for a nuclear reactor? Why is nuclear fusion a process that is considered promising for meeting future energy needs? The answers to these questions involve the applications of the nuclear fission and fusion processes. In this chapter you will study such applications.

What are the basic ways of producing artificial radioactive materials? What are some of the uses of radiation in medical diagnosis and treatment? Some techniques and the physical basis for the use of radiation in nuclear medicine will be presented in this chapter. Your understanding of these nuclear applications will provide a basis for your evaluating the potential of new developments in nuclear medicine and technology.

32.2 Nuclear Fission

The discovery of nuclear fission in 1938 by Hahn and Strassman suggested the possibility of tapping the energy of the nucleus. Recall that it is a conversion of some of the nuclear binding energy into kinetic energy that characterizes both fission and fusion. The basis of this conversion can be seen from the binding energy per nucleon curve which was shown in Figure 31.1. It was noted that if a heavy nucleus is split, the resultant nuclei are in the middle of the curve. In this middle region the average binding energy per nucleon is about 8.5 MeV as compared with 7.5 MeV per nucleon for the heavy nuclei like uranium. This corresponds to the mass difference that accounts for an energy release of about 1 MeV for each nucleon involved in the process.

The fission reactions that were studied first involved the use of the neutron bombardment of the fissionable material. The theoretical explanation for the process was first suggested by Niels Bohr and John A. Wheeler in 1939. They suggested that the fission of uranium was associated with the $^{235}_{92}U$ isotope rather than the much more abundant $^{238}_{92}U$. They suggested that the $^{235}_{92}U$ nucleus has low stability, and that after capturing a slow neutron the $^{235}_{92}U$ divides into two nuclei of roughly the same size. The key to the usefulness of the fission reaction is its potential for a *self-sustaining* chain reaction. The chain reaction is possible because each fission releases some neutrons. If these neutrons are captured by other fissionable nuclei, the chain reaction can become self-sustaining with a tremendous release of energy.

The criterion that must be met for a nuclear fission explosion is that the surface area to volume ratio must be such that enough neutrons produce fissions to sustain the chain reaction. This condition leads to the concept of *critical mass* for the fissionable material. This is the mass of material that will capture just enough neutrons to maintain the chain reaction. It should be noted that the fission process is analogous to the break up of a liquid drop. The product nuclei (droplets) are not unique, and the number of neutrons released during fission depends upon the particular fission reaction that occurs. In each fission reaction two or more neutrons are emitted. The average for the $^{235}_{92}U$ reaction is about 2.5 neutrons per fission. If the fission reaction is to be used in slow energy release applications, it is necessary to be able to control the rate of energy release. There are three controllable physical parameters: the purity of the fissionable material (impurities tend to absorb neutrons and thereby reduce chances for a chain reaction); the production of thermal neutrons; and the mass of the fissionable material. For a bomb it is a matter of combining two subcritical masses (each of high purity) to make a critical mass for a nuclear explosive chain reaction (this takes approximately 10^{-6} sec).

For a nuclear-fission reactor the key factor lies in the ability to insert control rods of neutron absorbing material (such as boron or cadmium) which have very high interaction probabilities for thermal neutrons. With these control rods the neutron flux can be maintained at the desired level. A *moderator* material is used to slow down the fast neutrons from the fission reaction. The properties of the moderator that are important are its low nuclear mass—so that neutron collisions rapidly reduce the neutron energy—and low neutron absorption probability. Two good moderator materials are heavy water (D_2O), and carbon (graphite blocks). A diagram of a nuclear fission reactor is shown in Figure 32.1.

The nuclear fission reactor has been used as a source for heat energy. The most common use of this heat energy is in the generation of electrical energy using conventional steam turbine technology (Figure 32.2). The nuclear reactor has also been developed into the energy source for submarines and large ships. Unfortunately, development of small, portable nuclear-energy sources for use in homes and automobiles seems unlikely. In fact, many people are concerned about the hazards of nuclear electrical generators and the benefit *vs* risk debate on the use of nuclear energy to generate electrical power still rages.

FIGURE 32.1
Schematic diagram of a nuclear fission reactor.

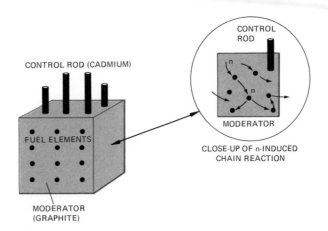

CONTROL ROD (CADMIUM)

CONTROL ROD

n

n

MODERATOR

CLOSE-UP OF n-INDUCED
CHAIN REACTION

FUEL ELEMENTS

MODERATOR
(GRAPHITE)

FIGURE 32.2
Schematic diagram of a nuclear power plant.

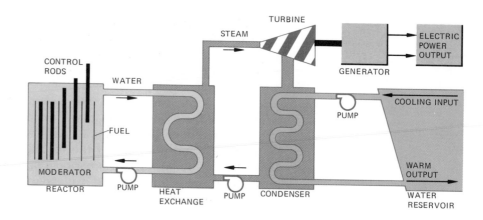

In addition to the energy output of the fission reactor, it is possible for a reactor to produce more fuel (fissionable material) than it consumes. These *breeder reactors* are possible because of neutron capture by $^{238}_{92}\text{U}$ (the most abundant isotope of uranium) leads to the creation of $^{239}_{94}\text{Pu}$. Plutonium-239 is fissionable by slow neutrons and becomes new nuclear fuel. The half-life of $^{239}_{94}\text{Pu}$ is 24,000 years. What can you conclude about the presence of $^{239}_{94}\text{Pu}$ in natural ores?

32.3 Nuclear Activation Analysis

A nuclear fission reactor can be used as a neutron source. The neutron flux from a reactor is a useful research tool. The neutrons can be used to bombard nonradioactive materials and thereby create artificial radioactive isotopes by neutron capture. Such neutron capture is the basis of neutron activation analysis. Neutron activation analysis can be used to identify constituents of a sample. The sample is bombarded with neutrons, and the components capture neutrons and become radioactive beta emitters. The radiation is monitored and is characteristic of a particular radioisotope. For example, a sample of $^{23}_{11}\text{Na}$ will capture a neutron to produce $^{24}_{11}\text{Na}$, which has a half-life of 15 hours. The presence of $^{23}_{11}\text{Na}$ is detected by the measurement of the $^{24}_{11}\text{Na}$ present in the sample. Quantities as small as 10^{-9} g of some elements can be detected by neutron activation analysis.

32.4 Nuclear Fusion

Although the energy from both the fission and fusion reactions results from conversion of nuclear binding energy to kinetic energy, the two processes have some important differences. The Coulomb repulsive force assists the fission reaction while it represents an energy barrier for the fusion reaction. The fusion reaction requires energy input equivalent to thermal energies corresponding to 100 million degrees celsius. For this reason the fusion reaction is also called a *thermonuclear reaction*. At these high temperatures all of the electrons are stripped off of the atoms of the reactants, and the result is a *plasma* of bare nuclei (positive) and electrons (negative). For a thermonuclear bomb it is possible to generate the necessary temperatures for fusion with a small fission bomb. The result is a bomb with greater energy release than possible from just a fission bomb.

This solution is obviously not applicable to a controlled fusion reactor. The sun uses gravitational collapse to generate the temperatures necessary for fusion, but this is not possible in the laboratory. One approach has been to create a plasma by stripping off electrons and to squeeze the plasma in a strong magnetic field (called a *magnetic bottle*) in an attempt to get densities great enough to sustain a fusion reaction. Thus far, reaction times have been limited to the order of microseconds. There is currently much interest in the possibility of using a high-energy laser beam focused on a fuel pellet, such as solid hydrogen.

The fusion reactor, if perfected, offers several advantages over fission. These advantages include:

1. an abundance of fuel (^2_1H from water)
2. very low levels of radioactivity
3. the possibility of direct conversion into electrical energy

The problems associated with fusion reactors are difficult to solve, but the potential of fusion is beginning to receive more attention of research funding government bodies everywhere. (During the 1960s more money was spent for white side wall automobile tires in one year in the United States than was spent for fusion research.)

32.5 *Applications of Radioactive Isotopes*

There are many applications of artificially produced radioactive isotopes (see Table 32.1). Such isotopes are made by neutron bombardment of stable nuclides as previously described. One of the most widely used isotopes is $^{60}_{27}Co$ which is produced by the capture of a neutron by stable $^{59}_{27}Co$. Cobalt-60 is a beta emitter which emits gamma radiation of 1.17 MeV and 1.33 MeV in the course of beta decay. These gamma rays can be used for radiation therapy in the treatment of tumors. In some cases the $^{60}_{27}Co$ is implanted in the tumor in the form of a needle. Such implants usually have an activity of 100 millicuries or less. Cobalt-60 is used to some extent for radiographs. It is also used in the diagnosis of certain blood disorders. Vitamin B_{12} which has been tagged with $^{60}_{27}Co$ is administered orally and the urine is checked for radioactivity. In patients with chronic myelogenous leukemia the clearance of B_{12} from the digestive tract is slowed. This can be checked with the $^{60}_{27}Co$ tagged vitamin.

TABLE 32.1
Characteristics of Some Radioisotopes

Isotope	Radiation	Radiation Energy (MeV)	Half-life	Some Uses and Applications
$^{3}_{1}H$	β^{-}	0.0180 particle	12.26 yr	Tagging drugs, hormones, vitamins
$^{14}_{6}C$	β^{-}	0.156 particle	5770 yr	Tagging drugs, hormones, vitamins
$^{18}_{9}F$	β^{+} (97%)	0.65 particle 0.51 particle	1.87 hr	Detection of bone disorders
$^{24}_{11}Na$	β^{-} γ	1.39 particle 2.753 γ rays 1.37 γ rays	15 hr	Radiocardiography, blood flow
$^{32}_{15}P$	β^{-}	1.71	14.3 d	Treatment of diseases of the blood, bone marrow; wide application
$^{35}_{16}S$	β^{-}	0.167 particle	86.7 d	Tagging amino acids
$^{42}_{19}K$	β^{-} (82%) γ	3.53 particle (82%) 2.03 particle (18%) 1.516 γ-ray (18%)	12.4 hr	Breast neoplasm; potassium composition in body parts
$^{47}_{20}Ca$	β^{-} γ	0.66 particle (83%) 1.94 particle (17%) 0.50, 0.234, 0.48, 0.83, 1.31 $\}$ γ rays	4.7 d	Detection of bone disorders
$^{59}_{26}Fe$	β^{-}	0.46 particle (53%) 0.27 particle (46%) 1.56 particle (0.3%) 0.145, 0.191, 0.337, 1.102, 1.290 $\}$ γ rays	45 d	Physiology of the blood; plasma increases in persons with anemia, polycythemia, and leukemia
$^{62}_{30}Zn$	Electron capture β^{+}	0.67 particle (19%) 0.04 to 0.51 γ rays 0.04 γ rays 0.60 γ rays	9.3 hr	Scanning
$^{63}_{30}Zn$	β^{+} Electron capture	2.36 particle (93%) 0.67 to 2.9 γ rays (7%)	38.3 m	Scanning
$^{68}_{31}Ga$	β^{+} Electron capture	1.89 particle (86%) 0.81 γ rays 1.06 γ rays 1.24 γ rays 1.88 γ rays	68 min	Diagnosing and locating brain tumors
$^{74}_{33}As$	γ	0.280	8 sec	Diagnosing and locating brain tumors
$^{75}_{34}Se$	Electron capture	0.024 to .280 γ rays	120 d	Cancer of pancreas
$^{85}_{38}Sr$	Isomer transfer Electron capture	0.150 γ rays (86%) 0.225 γ rays (13%) 0.233 γ rays (1%) 0.008 γ rays (13%)	70 min	Detection of bone disorders
$^{89}_{38}Sr$	β^{-}	1.46 particle (100%)	50.4 d	Treatment of bone tumors
$^{90}_{38}Sr$	β^{-}	0.54 particle (100%)	28 yr	Treatment of benign tumors on white of eye

TABLE 32.1
(Continued)

Isotope	Radiation	Radiation Energy (MeV)		Half-life	Some Uses and Applications
$^{90}_{39}$Y	β^-	2.27	particle (100%)	64.2 h	Scanning
$^{125}_{51}$Sb	β^-	0.30	particle + γ (43%)	2 yr	Diagnosis
		0.12	particle + γ (60%)		
		0.61	particle + γ (3.5–64%)		
	γ	0.035 to .64 (10 γ's range)			
$^{131}_{53}$I	β^-	0.61, .25, .81 particles		8.05 d	Very widely used; see Section 32.5
	γ	0.08 to .73 (7 γ's range)			
$^{137}_{55}$Cs	β^-	0.52	particle (92%)	30 yr	Radiation; tagged amino acids
		1.18	particle (8%)		
	γ	0.66	γ ray		
$^{197}_{80}$Hg	Electron capture	0.0775	γ ray	65 hr	Study kidney, liver, spleen disorders; granulocytic leukemia
		0.1916	γ ray		
		0.269	γ ray		

Another approach used in radioactive isotope therapy is based on the biochemistry of the body and the selective localization of specific isotopes. For example, $^{131}_{53}$I is localized in the thyroid, $^{32}_{15}$P is concentrated in bone, $^{42}_{19}$K is localized in muscle, $^{35}_{16}$S is localized in the skin, and $^{59}_{26}$Fe is concentrated in the blood. In addition to radiation therapy, these isotopes can be used as *tracers*. Tracer studies allow the scientist to track biochemical processes by following the radioactive tracer through the body. For example, Melvin Calvin was awarded a Nobel Prize for his work using $^{14}_{6}$C to unravel the photosynthesis reactions. The uptake of iodine-131 by the thyroid can be monitored, and the functioning of this organ can be evaluated.

Ingested iodine-131 is also used therapeutically to destroy cancerous tissue in the thyroid and to treat hyperthyroidism and certain cases of heart disease.

Other radioisotopes are used to diagnose and treat abnormal situations. It is possible, by selecting the proper isotope that emits the desired radiation at the correct energies, to selectively destroy certain tissues or organs and leave adjacent areas with little or no damage.

An important consideration in using injected radioactive isotopes is the biological half-life of the isotope. The biological half-life is the time required for one-half of the biochemically active isotope to be eliminated by natural processes. Assuming the biological process is exponential, we find the following equation giving *effective half-life T* in terms of the biological half-life T_b and the radioactive half-life T_r:

$$\frac{1}{T} = \frac{1}{T_b} + \frac{1}{T_r} \tag{32.1}$$

An aerial view of the experimental breeder reactor-II (EBR II) at Argonne-West located at the Idaho National Engineering Laboratory near Idaho Falls. (Courtesy of Argonne National Laboratory.)

The activity of a sample is equal to the number of nuclei present N times the number of disintegrations the isotope makes in one second,

$$A = N\lambda \tag{32.2}$$

where λ is the disintegration constant from Chapter 31

$$N = N_0 e^{-\lambda t} \tag{31.12}$$

EXAMPLES

1. Find the amount of cobalt-60 necessary for a 100-millicurie needle. The half-life of $^{60}_{27}\text{Co}$ is 5.24 yr.

$$A = N\lambda$$

$$\text{where } \lambda = \frac{0.693}{t_{1/2}}$$

$$= \frac{0.693}{5.24 \times 3.17 \times 10^7 \text{ sec}} = 4.2 \times 10^{-9} \text{ sec}^{-1}$$

Fission track autoradiograph of human bone. The short, heavy black marks are fission tracks, indicating the presence of plutonium, and the dark gray areas are marrow spaces. (Courtesy of Argonne National Laboratory.)

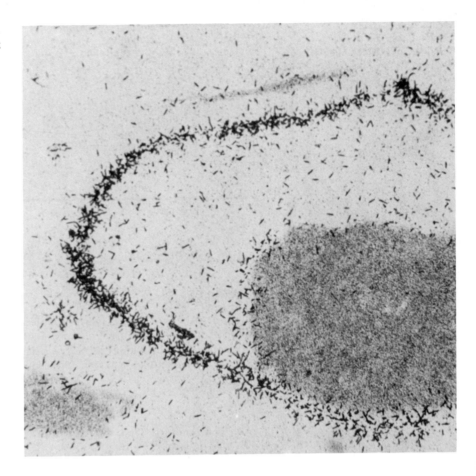

Thus

$$A = 100 \times 3.7 \times 10^7 \text{ dis/sec} = N \times 4.2 \times 10^{-9} \text{ sec}^{-1}$$

$$N = \frac{3.7}{4.2} \times 10^{18} \text{ nuclei} = (M/60) \text{ g} \times 6.02 \times 10^{23}$$

$$M = 8.78 \times 10^{-5} \text{ g for 100 millicurie of } {}^{60}_{27}\text{Co}$$

2. The effective half-life of iron-59 is 27 days, and its radioactive half-life is 46 days. Find the biological half-life of ${}^{59}_{26}\text{Fe}$.

$$\frac{1}{T} = \frac{1}{T_b} + \frac{1}{T_r} = \left(\frac{1}{T_b} + \frac{1}{46}\right) \text{ days}^{-1}$$

$$\frac{1}{T_b} = \left(\frac{1}{27} - \frac{1}{46}\right) \text{ days}^{-1}$$

$$T_b = 65.4 \text{ days}$$

Some of the radioactive isotopes with their effective half-lives are

TABLE 32.2
Effective Half-Life of Some Radioisotopes Which May Be Found or Used in the Human Body

Isotope	Organ of Concentration	Effective Half-life (days)	Radioactive Half-life
${}_{1}^{3}\text{H}$	Total body	19	12.3 yr
${}_{6}^{14}\text{C}$	Fat	35	5700 yr
${}_{6}^{14}\text{C}$	Bone	180	5700 hr
${}_{15}^{32}\text{P}$	Bone	14	14.2 days
${}_{19}^{42}\text{K}$	Muscle	0.5	0.5 days
${}_{16}^{35}\text{S}$	Proteins (skin)	18	87 days
${}_{20}^{45}\text{Ca}$	Bone	151	164 days
${}_{26}^{59}\text{Fe}$	Blood	27	46 days
${}_{53}^{131}\text{I}$	Thyroid	7.5	8 days

listed in Table 32.2. Note that certain isotopes tend to concentrate in particular parts of the body.

EXAMPLES

1. Radioactive iodine, ${}_{53}^{131}\text{I}$, is used in medical diagnosis. Iodine-131 has a half-life of approximately eight days. A sample of this isotope has an activity of 20 microcuries when administered to a patient.

   a. Find the activity of this material four days later.

   $$A_4 = A_0\, e^{-6.93\ t/t_{1/2}} = 20\, e^{-0.693\, \times\, 4/8} = 20 \times .707$$

   $$A_4 = 14.1 \text{ microcuries}$$

   b. How long will it take for the activity to drop to 1 percent of its initial value?

   $$A_t = 0.01\, A_0 = A_0\, e^{-0.693\, \times\, t/8}$$

   $$e^{-0.693t/8} = 0.01$$

   $$e^{0.693t/8} = 100$$

   $$0.693t/8 = \ln 100$$

   $$t = 53.2 \text{ days}$$

2. By injecting a known volume of a specific activity serum into the blood of a subject, it is possible to determine the volume of the subject's blood. After allowing a short time for the serum to mix completely (10 to 30 minutes), a blood sample is taken. The specific activity of the sample is measured. This specific activity is equal to the total activity divided by the total blood volume of the subject. From this relation the blood volume can be determined. Four cubic centimeters of serum with 100-microcurie/cm³ specific activity is injected into a subject. Twenty minutes later a blood sample is taken and its specific activity is found to be 0.038 microcuries/cm³. Find the blood volume of the subject.

   total activity of injected serum $= 4 \text{ cm}^3 \times 100\ \mu\text{C/cm}^3 = 400\ \mu\text{C}$

   final specific activity $= 0.038\ \mu\text{C/cm}^3 = 400\ \mu\text{C}/V$

where V is the total volume of blood. Thus the volume of blood is

$$V = 400/0.038 \text{ cm}^3 = 1.05 \times 10^4 \text{ cm}^3$$

Note: We have assumed that the activity at the time of sampling is the same as when the serum was injected. What conditions are necessary for this approximation to be acceptable? Iodine-131 is frequently used for these measurements. Its half-life is 8 days. Is the approximation a good one for this case?

The applications of radioisotopes are not limited only to medical uses. They are broadly used in other fields of endeavor. Some examples are:

AGRICULTURE
a. animal husbandry: nutrition studies by tracers, diet additives, biochemistry of milk production
b. plant physiology: mechanism of photosynthesis, soil fertility, uptake of fertilizers

ARCHAEOLOGY
Carbon dating of deposits

ASTRONOMY
solar energy from nuclear processes

BOTANY
transport of fluids
photosynthesis research

CHEMISTRY
synthesis of new elements
alchemy
reaction mechanisms

ENGINEERING
radiology
thickness tests
friction studies

FOOD PROCESSING
preservation of foods

METALLURGY
atomic diffusion studies

GEOLOGY
well logging
mineralogy: color change of crystals by radiation

ZOOLOGY
mutations
destruction of life by radiation

FIGURE 32.3

A Geiger-Müller tube detection system. Each ionization pulse created in the G-M tube by radiation is recorded on the digital counter.

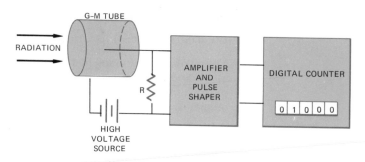

32.6 Geiger-Müller Tube

All of the uses of radiation require the ability to measure radiation levels. One of the most widely used detectors is the Geiger-Müller tube described briefly in Chapter 30. The G-M tube is gas filled and has a high voltage applied between its cylindrical conducting wall and a coaxial central electrode, as shown in Figure 30.5 on page 676.

When radiation enters the tube, it produces ionization in the gas inside the tube. The ions created are accelerated by the high voltage applied across the tube. These accelerating ions will create other ions upon collisions with gas molecules, and thus an amplification of 10^6 to 10^7 is achieved. This ionization pulse creates a current pulse across the resistor for each incoming particle that produces ionization. The G-M tube is designed to measure radiation intensity. It is capable of detecting all forms of ionizing radiation. The G-M tube is relatively slow with a resolving time (time between just resolvable pulses) of about 10^{-4} seconds. (This means a 5 percent loss at count rates of only 100 counts per second.) The G-M tube is not as efficient for gamma ray photons as it is for charged particles. A block diagram for a G-M tube detecting system is shown in Figure 32.3.

32.7 Scintillation Counter

A much more efficient photon detector is the scintillation counter which we introduced in Chapter 30. A typical scintillator and photomultiplier tube apparatus is shown in Figure 30.6 on page 677. The scintillator crystal of sodium iodide doped with thalium absorbs the high-energy photons and radiates light in the visible or ultraviolet region. These secondary photons enter the photomultiplier (PM) tube where they strike a photocathode, and upon the absorption of the photons, the photocathode emits electrons.

These photoelectrons are accelerated by the high voltage between the electrodes (called *dynodes*) in the photomultiplier tube. The overall amplification of the PM tube may be the order of 10^7. This scintillator-PM tube combination not only detects gamma rays and x rays, but it can also be calibrated so that the current pulses are proportional to the energy of the original photon. There are some applications for which this energy discrimination ability is quite useful.

32.8 Thermoluminescence Dosimetry

Some insulators have traps for electrons created by radiation. One such insulating crystal is lithium fluoride LiF doped with manganese or europium. In thermoluminescence dosimetry (TLD) the crystal is exposed to the radiation to be measured. Then at a selected time the crystal is heated. The thermal energy releases the electrons from the traps. As the electrons make transitions to lower energy states the crystal emits photons. The total number of photons emitted is proportional to the intensity of the radiation incident on TLD crystals. The LiF crystal has radiation absorption characteristics similar to those of living tissue.

32.9 Charged-Particle Detectors

Specially doped germanium and silicon crystals can be made into very sensitive charged-particle detectors. A schematic diagram of a typical surface barrier silicon detector is shown in Figure 32.4. The incoming charged particle creates electron-hole pairs that in turn create a pulse in the external circuit current as shown in the figure. These solid-state detectors are very efficient and can be calibrated to give excellent energy discrimination for the incoming charged particles. In addition, they have very fast response times ($\sim 10^{-9}$ sec).

32.10 Film Exposure

One of the oldest detectors of radiation is the photographic emulsion. The incoming radiation exposes the film and produces an optical density that is proportional to the intensity of radiation exposing the film

FIGURE 32.4
Schematic diagram of a surface barrier silicon detector.

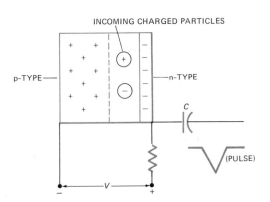

FIGURE 32.5
Block diagram of an experimental system used to measure optical density of film used to detect ionizing radiation. A plot of the optical density versus the $\log_{10}$ (intensity × time) reveals a linear region to use for best results.

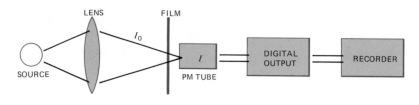

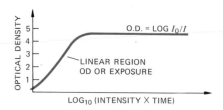

The optical density is defined in the following equation:

$$OD = \log_{10} \frac{I_0}{I} \tag{32.3}$$

where I_0 is the reference light intensity and I is the intensity of the light that passes through the exposed film.

A diagram of a densitometer used to measure the optical density of the exposed film is shown in Figure 32.5. Note there is a region where OD is proportional to the exposure (intensity × time).

EXAMPLE

A radiogram (photographic record of radiation) shows an optical density of 2 for a known exposure E, an unknown exposure produces an OD of 2.5. Find the unknown exposure E' in terms of the given value E.

$$2 = \log_{10} \frac{I_0}{I} \quad \text{and} \quad 2 = k \times E$$

Thus,

$$2.5 = k \times E'$$

and

$$E' = \frac{2.5}{2.0} E = 1.25 \, E$$

32.11 Shielding

The term *shielding* has come to mean a system to control radiation hazards around nuclear facilities and other radiation producing devices. The purpose of the shielding is to reduce the intensity of the radiation to an acceptable (nonhazardous) level. The biological dose uses the concept of flux. The flux is expressed of number per square centimeter per second. For example, for x rays it is the number of photons per square centimeter per second, and for neutrons it is the number of neutrons per square centimeter per second. The shielding should be designed to attenuate the radiation or to convert the energy to an innocuous form, generally heat.

Different processes are used in shielding, and the specific process for each case depends upon the type of radiation and its energy. The nature of radiation suggests the type of shielding to be used. In general the radiation can be divided into the following four areas:

1. HEAVY CHARGED PARTICLES: All heavy charged particles (α-particles, protons, etc.) have similar interaction with matter, principally with electrons. High density materials can be used for shielding. The major concern of biological shielding for these types would be high-energy particles from accelerators.

2. BETA PARTICLES: The beta particles (both $+$ and $-$) of the energy range of most radioisotopes are fairly easily to control by even low-Z materials. That is, a thickness of 7 mm ($\frac{1}{4}$ inch) of such material as aluminum, Lucite, or Plexiglas is sufficient. However, high-energy electrons from accelerators present a much greater problem. For these the control of the Bremsstrahlung produced in the shield becomes the major problem. The methods of photon shielding are used for these rays.

3. PHOTONS: To shield for x and γ rays the most effective materials are those with high electron density, that is, high Z. The economics of the situation then becomes important: You want to select the material which is cheapest per unit mass which will do the job. You then calculate the needed thickness. Some of the materials that are used for photon shields are concrete, earth, iron, and lead.

4. NEUTRONS: The shielding desired for neutrons depends upon the neutron energy. Fast neutrons are best attenuated by interaction with heavy nuclei. For low-energy neutrons you need material made of materials with low Z, hence something that has a high concentration of hydrogen nuclei. Paraffin is used for low-energy shielding. For neutrons of mixed energies you need mixed materials. Concrete is a definite possibility as it has both the heavy nuclei and the hydrogen nuclei. Water is also used for neutron shielding.

In cases where shielding is impractical, sources of radiation can be placed at greater distances from human populations. The plans to bury nuclear reactor wastes in salt mines or under the ocean is based on the shielding properties of earth and water and on the inverse square law that relates the energy density of radiation at some location to the distance from that location to the source of radiation. The energy density at a point in joules per unit area is proportional to the reciprocal of the square of distance from that point to the source of radiation. For example, if you double the distance between your body and a radiation source, the amount of radiation energy you receive will be reduced to a level of one-fourth of that at the original distance.

$$\frac{\text{energy}}{\text{area}} \propto \frac{1}{(\text{distance to source})^2} \qquad (32.4)$$

SUMMARY

Use these questions to evaluate how well you have achieved the goals of this chapter. The answers to these questions are given at the end of this summary with the number of the section where you can find related content material.

Definitions

Write the correct word in each blank.

1. In order for a nuclear fission reaction to be self-sustaining there must be a _____ of fissionable material present, and a _____ must occur in which each fission of a nucleus releases some neutrons to continue the process. If the nuclear fission process used in a reactor process is one that produces more nuclear fuel, it is called a _____.
2. A fusion reaction which can be confined by large magnetic fields in a _____ is also called a _____ and involves the _____ of _____ nuclei to produce _____ nuclei.
3. The rate of fission energy release can be _____ by the use of _____.
4. Nuclear fission reactors can be used to produce _____, _____ and _____.
5. The _____ from a reactor can be used to produce _____ and to cause substances to become beta emitters so that their presence can be detected down to _____ g by means of _____ _____.
6. The biological half-life of a system is _____.

Fission Reactors

7. What are the basic principles of a nuclear fission reactor? Sketch a block diagram one as you envision it.
8. List two detectors for each of the following: α-particles, β-particles, γ-rays, protons, neutrons.

Radiation Problems

9. What energy absorption rate results from an implant of 1.2 millicuries of radioactive palladium ($^{103}_{46}$Pd) per gram of tissue? The half-life of palladium-103 is 17 days. This isotope decays by electron capture with an emission of 0.57 MeV in γ radiation per disintegrating atom. Assume all energy is absorbed in adjoining tissue.

Tracer Applications

10. List three specific applications of radioisotopes as tracers.

Answers

1. critical mass, chain reaction, breeder reactor (Section 32.2)
2. magnetic bottle, thermonuclear reaction, combining, light, heavy (Section 32.4)
3. controlled, moderators (Section 32.2)
4. heat energy, additional fuel, neutrons (Sections 32.2, 32.3)
5. neutrons, artificial isotope tracers, 10^{-9}, neutron activation analysis (Sections 32.3, 32.5)
6. time required for one-half of the isotope to be eliminated from a biological process (Section 32.5)
7. see Section 32.2
8. see Sections 32.6, 32.7, 32.8, 32.9, 32.10)
9. 4.05×10^{-6} J/g-sec. (Section 32.5)
10. see Section 32.5 and Tables 32.1 and 32.2)

ALGORITHMIC PROBLEMS

Listed below are the important equations from this chapter along with two from Chapter 31. The problems following the equations will help you learn to translate words into equations and to solve single-concept problems.

Equations

$$\frac{1}{T} = \frac{1}{T_b} + \frac{1}{T_r} \tag{32.1}$$

$$A = N\lambda \tag{32.2}$$

$$OD = \log_{10} \frac{I_0}{I} \tag{32.3}$$

$$\frac{E}{A} \propto \frac{1}{d^2} \tag{32.4}$$

$$0.693 = \lambda t_{1/2} \tag{31.13}$$

$$A = A_0 e^{-\lambda t} \tag{31.15}$$

Problems

1. What is the biological half-life for $^{131}_{53}I$? (See Table 32.2)
2. What is the disintegration constant for a nuclide ($^{32}_{15}P$) which has a half-life of 14.3 days?
3. If one has originally the same amount of two radioactive materials, one finds that sample A originally has much greater activity than sample B. A few days later he finds that sample B has a greater activity than sample A. Explain.
4. A radioactive capsule of Iodine-131 is rated 2.00 microcuries. When should the rating become 0.500 microcurie?
5. Two patients are placed at distances of 1 m and 3 m from a radioactive point source. How will their dosages compare?

Answers

1. 120 days
2. 5.61×10^{-7} sec^{-1}
4. 16 days
5. 9:1

EXERCISES

These exercises are designed to help you apply the ideas of a section to physical situations. When appropriate the numerical answer is given in brackets at the end of the exercise.

Section 32.2

1. Outline the operation of a nuclear fission reactor, and suggest where you would put safety controls and indicate the nature of controls you would employ.

Section 32.5

2. Using Table 32.2, find the biological half-life for each of the following isotopes: $^3_1$H, $^{35}_{16}$S, and $^{59}_{26}$Fe. [19.1 days, 22.7 days, 65.4 days]

3. Outline a procedure that might be used to measure the red blood cell count in a subject. (*Hint:* red blood cells concentrate iron biochemically.)

4. Suppose that a woman has 10^{-9} curie of strontium-90 in her bones. Strontium-90 emits beta particles with a total energy of 2.75 MeV. Assume all of this energy is absorbed in the 1.00-kg bone. Calculate the dosage this woman will get in one year. (Assume radiation is constant since half-life is 28.0 yr.) [4.32×10^{-4} J, 43 mrem]

5. Derive Equation 32.1 using the assumption that the biological half-life of an isotope is governed by the exponential decay law analogous to radioactivity.

6. Find the mass of $^{131}_{53}$I giving an activity of 100 microcuries. The half-life of $^{131}_{53}$I is 8 days. [0.8×10^{-9} g]

Section 32.11

7. What are the ways in which neutrons may interact as they pass through a tissue? What method of protection would you recommend against a powerful source of neutrons?

8. What would you recommend as shielding for
   a. γ rays
   b. β particles
   c. neutrons
   d. α particles

PROBLEMS

The following problems may involve more than one physical concept. When appropriate the numerical answer is given in brackets at the end of the problem.

9. Given the following information, how many curies are there in 1 gram of substance?

Radioactive Isotope	Type of Emission	Half-life
$^{32}_{15}$P	β-particle	14.3 days
$^{24}_{11}$Na	β-particle	15 hours
$^{90}_{38}$Sr	β-particle	28 years
$^{131}_{53}$I	β-particle	8.05 days

[P, 2.9×10^5 curies; Na, 8.7×10^6 curies; Sr, 140 curies; I, 1.27×10^5 curies]

10. A beam of β-rays or γ-rays can be used to locate a foreign body in a tissue. As an example, the source (small) to film distance is 80.0 cm and two exposures are taken on the film. A marker is placed on the upper skin of the patient between the foreign body and the source. If the source is moved a horizontal distance of 10.0 cm, the image of the foreign body and marker are displaced 10.7 cm and 2.5 cm respectively. What is the vertical distance between the marker and foreign body? [16 cm]

11. Zinc-69 decays from an excited nuclear state by gamma emission with a half-life of 1.38 hours and a gamma ray energy of 0.400 MeV.

   a. Find the mass difference between the ground state and the excited state of this nucleus.
   b. Find the maximum energy that such a gamma ray might transfer to an electron through Compton scattering. [a. 7.1×10^{-31} kg; b. 0.243 MeV]

12. What is the activity of potassium-40 in an 80-kg person under the following conditions: 0.35 percent of the person is potassium, and 1.2 percent of the postassium is $^{40}_{19}$K. Potassium-40 is a β emitter and has a half-life of 1.27×10^9 years. [8.6×10^5 disintegrations/sec]

13. If only two percent of a given radiation is transmitted through 20 cm of soft tissue, what is the effective absorption coefficient of the tissue? Assume homogeneous radiation. What thickness would reduce the beam to half its original intensity? [1.96×10^{-1} cm^{-1}, 3.5 cm]

14. What is the photon flux at a distance of 1 m from a 500-millicurie point source gamma ray emitter? Assume one photon per disintegration. [1.45×10^9 photons/m^{2}sec]

15. Radioactive cobalt-60 is used in the treatment of some diseases. It decays by emitting two gamma rays of energy 1.17 MeV and 1.33 MeV. The half-life is 5.26 years. Assume there is 1.00-g source and that one percent of the emitted gamma rays strike the patient, the patient's mass is 70.0 kg, and exposure lasts for 3.00 minutes. What is the energy absorbed per kilogram per second? What is this in rads? [2.38×10^6 J/kg–sec, 42.9 rad]

16. One of the newest isotopes to receive considerable use is technetium-99 ($^{99}_{43}$Tc). This isotope has a 6-hour

half-life as it decays by 0.14 MeV gamma rays from an excited state to its ground state. Assume that 50 millicuries of $^{99}_{43}Tc$ is injected into the body, and that one-half of the radiation is absorbed in the 50-kg body of the patient. Find the dosage received by the patient. [4.1×10^{-7} J/kg-sec]

17. The average energy released in a fission of $^{235}_{92}U$ nucleus is about 180 MeV. Suppose we had 1.00 g of $^{235}_{92}U$, and assume fission of all of the nuclei. How much energy would be released? Now reduce this amount to any percentage efficiency that you think might be practical. Estimate how much coal or gasoline it would take to produce the same amount of energy? (*Hint:* What is the order of magnitude for energy involvement in chemical reactions?) [7.35×10^{10} J, 10^3 kg]

18. How much energy would be released by complete conversion of one mole of deuterium to helium? The reaction is $^2_1H + ^2_1H = ^3_2He + ^1n$. [$5.14 \times 10^{11}$ J]

19. The heat from the sun at the earth's surface is 1.35 kilowatts per square meter, and the distance between the sun and the earth is 1.5×10^8 km. The mass of a hydrogen atom is 1.08144 amu and that of a helium atom is 4.00387 amu. Assume that the source of the solar energy is a process in which four hydrogen atoms are converted into one helium atom. How much hydrogen is used up per second? [3.18×10^{37} hydrogen atoms per second]

20. If a uranium-235 power plant has a conversion efficiency of 16 percent, and the average release per fission of the uranium-235 is 180 MeV, how much uranium would it take to produce 1 gigawatt hour (one million kWh) of electrical energy? [7.85×10^{23} atoms of uranium or 0.30 kg of uranium-235]

21. Compare fission and fusion as potential energy sources for the future.

22. Nuclear radiation has effectively been used to measure the thickness of metallic sheets. Outline a procedure that might be used for this purpose.

23. Using the value of 200 MeV as the energy released in one $^{235}_{92}U$ fission, find the TNT equivalent for a 1 kg of $^{235}_{92}U$ if one ton of TNT releases 10^9 cal of energy. [1.96×10^4 tons of TNT]

Appendix

Mathematical Background for Introductory Physics

A.1 Introduction

Physics is a quantitative science. The models we use in physics make use of the quantitative tools of mathematics. You will be able to master physics more easily if you have facility with the following mathematical skills and concepts.

A.2 Powers of Ten Notation

In physics we have occasion to use such numbers as the mass of proton

$m_p = 0.000,000,000,000,000,000,000,000,00167265$ kilograms

and the number of molecules in a kilogram-mole of an element

$n_0 = 602,204,500,000,000,000,000,000,000$

as well as more usual numbers such as the number of inches in a meter (39.370) and the number of kilometers in a mile (1.60935). Therefore, it is important for us to be able to use the convenient *powers of ten notation* for writing numbers. In this notation any number is written as a number between one and ten multiplied by ten raised to some power— for example, 23 million is written as 2.3×10^7. For working with large and small numbers the power of ten notation offers the following distinct advantages:

1. It is less cumbersome to write.
2. It is easier to read.
3. It is less prone to errors during transcription.
4. It provides simplified comparisons of measurement.
5. It simplifies computations and estimations that involve computation.
6. It avoids ambiguity in precision of measurements.

The powers of ten notation is simply the counting of the number of tens multiplied together and the writing of that number as an exponent of ten. For example:

$$1,000 = 10 \times 10 \times 10 = 10^3$$

$$10,000 = 10 \times 10 \times 10 \times 10 = 10^4$$

$$1,000,000 = 10 \times 10 \times 10 \times 10 \times 10 \times 10 = 10^6$$

In these examples the numbers 3, 4, and 6 are called exponents. The number 10^6 is read as ten to the sixth power. How is 10^4 read? 10^{10}? (Note that there are two exceptions: 10^3 is read as 10 cubed and 10^2 is read as 10 squared.) It follows that 2000 is written as 2×10^3, 560 as 5.6×10^2 and 204,956,271 as 2.04956271×10^8.

Practice Problem

Write the following populations in powers of ten notation.

Country	Population	Powers of Ten Notation
Burma	29 million	_____
China	850 million	_____
Greece	9 million	_____
Liechtenstein	23 thousand	_____
Oman	750 thousand	_____
USA	210 million	_____
USSR	250 million	_____

Using the principles of multiplication and division that you already know, you should be able to deduce the rules for multiplication and division using the powers of ten notation. Consider:

$$30 \times 20 = 600$$

or

$$3.0 \times 10^1 \times 2.0 \times 10^1 = 6.0 \times 10^2$$

What happened to the exponents? Again,

$$410 \times 5000 = 2,050,000$$

or

$$4.10 \times 10^2 \times 5.00 \times 10^3 = 20.5 \times 10^5 = 2.05 \times 10^6$$

Can you state a multiplication rule? Test your rule on these:

$$32 \times 150 =$$

$$352,000 \times 12,000,000 =$$

Now consider division:

$400 \div 50 = 8$

or

$$\frac{4.0 \times 10^2}{5.0 \times 10^1} = 0.8 \times 10^1 = 8$$

Again,

$$56,000 \div 280 = \frac{5.6 \times 10^4}{2.8 \times 10^2} = 2.0 \times 10^2$$

Can you state the rule for division? Try it out for these:

$380,000 \div 190 =$

$720,000,000 \div 360,000 =$

The division rule can be used to derive the expression for ten to the zeroth power as well as negative exponents. For example,

$$1 = \frac{1000}{1000} = \frac{10^3}{10^3} = 10^{3-3} = 10^0$$

What is 10^{-1} equal to in decimal form? $5 \times 10^{-3} =$ _____ in decimals. Now consider simple addition and subtraction: $10^3 - 10^2 = 9 \times 10^2$. Why? $10^3 + 10^2 = 1.1 \times 10^3$. Why? Try these on your own:

$7.8 \times 10^3 - 18.78 \times 10^4 =$ _____ $4.17 \times 10^{-5} - 3.18 \times 10^{-4} =$ _____

$4.92 \times 10^5 + 9.400 \times 10^6 =$ _____ $7.810 \times 10^{-3} + 8.41 \times 10^{-4} =$ _____

Practice Problems

1. $\dfrac{10^{-7} \times 10^8}{10^{-18}} =$ _____

2. $\dfrac{10^{-18} \times 7.6 \times 10^9}{10^{-36}} =$ _____

3. Solve for A:

$$A = \frac{1.0 \times 10^{-8} \times 7.8 \times 10^9}{1.0 \times 10^9 \times 1.0 \times 10^{-9}}$$

4. Solve for A:

$$10^{-18} = \frac{7.8 \times 10^7 \, A}{3.9 \times 10^{-6}}$$

5. Using the powers of ten notation calculate the number of seconds in one year.

The answers to these practice problems are given at the end of this appendix.

A.3 Significant Figures

The use of significant figures is a method of designating the reliability of measured quantities. For example, a length given as 1.2 m means that the length lies between 1.15 m and 1.25 m. This distance 1.2 m is a two significant figure quantity. The use of powers of ten notation to present significant figures is convenient and unambiguous. The previous length could be expressed as 1.2×10^2 cm as a two significant figure quantity or 1.200×10^2 cm as a four significant figure quantity.

In calculations, the number of significant figures in a result usually cannot exceed the least number of significant figures used in the calculation. Consider the area of a square whose side is given as 1.2×10^2 cm. This calculation gives

$$A = l^2 = (1.2 \times 10^2)^2 \text{ cm}^2$$

$$A = 1.44 \times 10^4 = 1.4 \times 10^4 \text{ cm}^2 \text{ (2 significant figures)}$$

For the four significant figure case you get:

$$A = l^2 = (1.200 \times 10^2)^2 = 1.440 \times 10^4 \text{ cm}^2$$

Practice Problems

6. Give the number of significant figures in each of the following numbers:
   a. 3.14 sec
   b. 86,400 sec/day
   c. 3.1558×10^7 sec/year
   d. $5 \times 10^7 y$
   e. 0.314×10^1
7. Make the following calculations for a cube of edge length equal to 0.40×10^2 cm to correct significant figures:
   a. area of a side
   b. volume

The answers to these practice problems are given at the end of this appendix.

A.4 Cartesian Graphs

Quantitative experiments are performed to understand the relationship that exists between the measured variables in the experiment. Throughout the performance of an experiment, one variable called the *independent variable* is carefully manipulated. The value of another variable which responds to each manipulation is recorded and this quantity is referred to as the *dependent variable*. Both values generally make up the data. A graph can now be plotted from the data. In plotting graphs, it is customary to use the *horizontal axis for the independent variable and the vertical axis for the dependent variable*.

The type of graph that we shall use is called the cartesian graph,

FIGURE A.1
A set of cartesian coordinate axes labeled *x* and *y*. The point (4, 2) is located as shown.

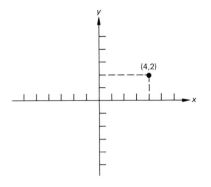

FIGURE A.2
These curves in graphs A and B show different relationships between the *y* and *x* variables.

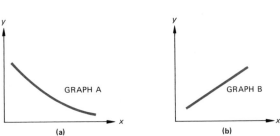

GRAPH A

GRAPH B

(a)

(b)

since it is based upon the model of space suggested by Rene Descartes, a seventeenth century French mathematician. Each axis is marked off on a linear scale. The scales on the two axes need not be the same. An ordered pair of numbers (x, y) names a particular point on the coordinate system. The first number gives the location along the horizontal axis. The second number represents the vertical location of the point. Point (4, 2) is plotted for you in Figure A.1. Plot the following points on the axes in Figure A.1: (0, 0), (0, 1), (0, −3), (3, 0), (2, 0), (4, 3). Graphs are used to show the relationships between the variables. Graphing is a way of picturing how one variable changes in response to changes in another variable. Answer the following questions from the graphs in Figure A.2.

a. From graph A determine what happens to *y* as *x* is increased. _____
b. From graph A, determine what happens to *x* as *y* is decreased. _____
c. What type of relationship exists between *x* and *y* in graph A? _____
d. From graph B determine what happens to *y* as *x* is increased. _____
e. From graph B determine what happens to *x* as *y* is decreased. _____
f. What type of relationship exists between *x* and *y* in graph B? _____

All straight lines belong to the family of linear relationships. Let us see what we can discover about their slopes.

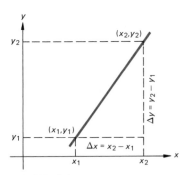

FIGURE A.3
A linear relationship between *x* and *y* appears as a straight line.

$$\text{slope} = \frac{\text{change in vertical}}{\text{change in horizontal}} = \frac{\text{change in } y}{\text{change in } x} = \frac{\Delta y}{\Delta x} = \frac{y_2 - y_1}{x_2 - x_1} \qquad (\text{A.1})$$

where (x_2, y_2) and (x_1, y_1) are points on the linear curve (Figure A.3). *Note:* When drawing a graph, your choice of scales for the axes should be such that your graph fills the space designated for the graph.

FIGURE A.4
Four linear relationships are shown.

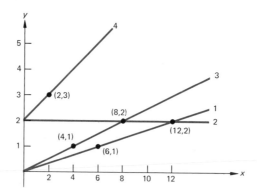

Examine the graph in Figure A.4, and answer the following questions.

a. Compare the lines. For which one does y increase the most with increasing x? _____; increase the least? _____

b. For which one does x increase the most with increasing y? _____; increase the least? _____

c. Which one has zero slope? _____

d. Calculate the slope of curve 1 between points $(0, 0)$ and $(6, 1)$. _____; between points $(6, 1)$ and $(12, 2)$. _____. What did you notice about the slope?

e. Calculate the slope of curve 3 between points $(4, 1)$ and $(8, 2)$. _____; between points $(0, 0)$ and $(8, 2)$ _____. What did you notice about the slope?

f. Make a generalization about the slope of a linear graph.

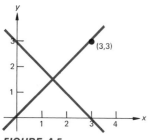

FIGURE A.5
Two lines which intersect at $(1\frac{1}{2}, 1\frac{1}{2})$.

Consider the two lines shown in Figure A.5, and answer the following questions.

a. Find the slope of these two lines.

b. Make a general statement about the slope of a straight line.

Let us examine a curve that has an equation of the form $y = kx^2$. For this curve the dependent variable varies as the square of the independent variable.

In Figure A.6, graph the following relationships:

FIGURE A.6
Two different choices for the variable to be plotted along the x-axis.

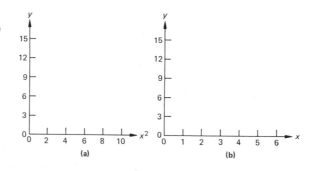

FIGURE A.7
The distance in meters plotted against
the travel time for a moving car.

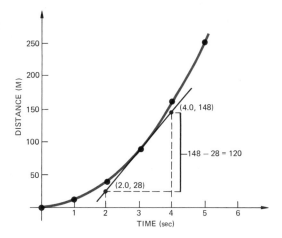

a. Make a graph of y vs. x^2 (graph A).
b. Make a graph of y vs. x, where $y = x^2$ (graph B).
c. Describe the relationship that exists in the graph A.

You will notice that changing the independent variable or the quantity plotted in the horizontal direction can change the shape of the curve. Since the linear relationship between variables is especially simple, you may find it convenient to manipulate your experimental data in such a way as to have the graph of the data appear as a linear relationship between the variables plotted along the two axes.

Another aspect of the curve that you plotted in graph B is its slope at a specific point. The definition of slope used for a straight line is valid here, only the application is more involved.

The slope of the curve at a point is found by constructing a tangent to the curve at the specified point. A tangent is a straight line that touches the curve at only one point. Use a ruler and move it around until it touches the curve at only one point, draw a straight line, then find the slope of this line. This slope is the slope of the curve at the point where the tangent line touches the curve.

EXAMPLE

As a car races to town, data on time and distance are recorded. The data are plotted on a graph (Figure A.7). We want to find the speed at $t = 3$ sec. The slope of the tangent line will give the speed at three seconds.

$$\text{slope} = \frac{(148 - 28) \text{ m}}{(4.0 - 2.0) \text{ sec}} = \frac{120 \text{ m}}{2.0 \text{ sec}}$$

$\text{slope} = 60$ m/sec

In the following graph (Figure A.8) the position of a moving object is plotted vs. the time. Draw the tangent lines at A, B, and C, and estimate the slopes of

FIGURE A.8
The positive of a moving object as a function of time.

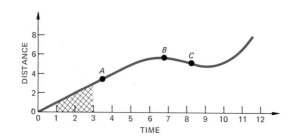

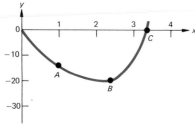

FIGURE A.9
A plot of *y* as a function of *x*.

these lines (signs are important). Can you describe the motion of this object in words?

In some physical situations the area under the curve is important. The area under the curve between $t = 1.0$ and $t = 3.0$ (Figure A.8) is shown by the cross-hatched area. It can be measured with a planimeter, estimated by counting small graph squares, or by the use of geometry. For the area shown in this figure, which seems to be a trapezoid, it may be calculated by a simple equation,

$$A = \tfrac{1}{2}h(b + b') = \tfrac{1}{2}(2.0)(1.0 + 3.0) = 4.0$$

Practice Problems

8. Find the slopes at points *A*, *B*, and *C* on the graph in Figure A.9.
9. Graph the following data for time (*x*) and speed (*y*): (0, 0), (1.0, 4.0), (2.0, 8.0), (3.0, 12), (4.0, 16), (5.0, 20), (6.0, 20), (7.0, 20), (8.0, 20).
    a. What is the slope at three seconds?
    b. What is the slope at seven seconds?
10. Graph the following data for time (*x*) and distance (*y*): (0, 0), (1.0, 2.0), (2.0, 8.0), (3, 18), (4, 32), 5, 50).
    a. What is the slope at two seconds?
    b. What is the slope at four seconds?
    c. What is the area under the curve between three and five seconds?

The answers to these practice problems are given at the end of this appendix.

A.5 Dimensional Analysis

Nearly all quantities that we measure in physics have two properties — a size (or magnitude) and a dimension. We can define four fundamental dimensions. They are length, mass (or quantity of matter), time, and electrical charge (or quantity of electricity). All other dimensions can be written in terms of these four fundamental dimensions.

In practice, two other physical quantities are defined in the SI — temperature and luminous intensity. The reason for this is that it is more convenient to define them separately rather than in terms of the basic four.

The algebraic manipulation performed on a physical quantity *must* be performed on *both* its magnitude and its dimension. For example, John drove his car a distance L in a time t. What was his speed? Equation $d = rt$, distance = speed × time, applies; so the dimension of $r =$ distance/time. What is his speed? Often the dimension of a calculated quantity can give you information about your calculation procedures. For example, suppose you wanted to calculate the area of a square, and obtained an answer in units of length³. You made an error. Why?

In order to compare the measurements we make with those made by other people we must agree with them upon some standard units to be used for comparison. Some years ago, representatives from many nations gathered together and accepted an international system of units, called the Système International (SI).

The SI is in almost universal use now. The United States of America is almost alone in the world in its use of the English system of units, which even the British have abandoned. We have avoided using the English units in this course. You should begin to develop a "physical feeling" for the sizes of common SI units.

Information to Know and Use

The standard SI unit of length is the *meter*.

1 meter (m) = 100 centimeters (cm)
1 cm = 10 millimeters (mm)
1 kilometer (km) = 1000 meters (m)
1 meter = 1.094 yards (yd)

The standard SI unit of mass is the *kilogram*.

1 kilogram (kg) = 1000 grams (g)
the force of gravitational attraction on 1 kg at sea level is
9.8 newtons (N), or 2.2 lb

The standard unit of time in the SI is the second.
The standard SI unit of electric charge is the coulomb.

You can analyze the dimensions of the quantity of a physical system by using a symbol for the fundamental dimensions — M for mass or quantity of matter, L for length or quantity of space, T for time, Q for quantity of electric charge. In terms of these symbols, the time it takes to get to physics class has the dimensions of T, how much food you eat for dinner has the dimensions of M, how far it is from here to where you are going to sleep tonight has the dimension of L, the useful life of the new automobile battery that you buy has the dimensions of Q. The rate at which you eat your dinner may have the dimensions of M/T. What are the dimensions for the speed with which you walk across the street? What are the dimensions for the area of a table top? For the volume of your favorite mug? For the speed of your automobile? For the density of water? You may find it convenient to begin a table in your physics notebook that enables you to keep a record of the dimensions and units for the various quantities you are studying in this course.

You may find that the manipulation of units can help you solve many physics problems. Let us take a simple example. Suppose you wish to compute the number of centimeters that are equivalent to the length of one foot. If you know that 2.54 cm = 1 inch and that 12 inches = 1 foot, then you may proceed as follows:

$$1 \text{ ft} = 1 \text{ ft} \times \frac{12 \text{ inches}}{1 \text{ ft}} \times \frac{2.54 \text{ cm}}{1 \text{ inch}} = \left(\frac{1 \times 12 \times 2.54}{1 \times 1}\right)\left(\frac{\text{ft} \times \text{inch} \times \text{cm}}{\text{ft} \times \text{inch}}\right)$$

$$= 30.5 \text{ cm}$$

Notice that each factor is simply a fraction whose numerical value is 1, and which is the ratio of two equivalent ways of measuring the same fundamental quantity.

Now try this technique on the following problem: The density of an object is defined as the ratio of the mass of the object to its volume. Suppose you have a quantity of water that has a mass of 88 g and a volume of 88 cm^3. What is the density of the water in g/cm^3? What is the density of the water in kg/m^3? What is your density? What is your volume? (*Hint:* In water a human body floats primarily because of air in the lungs. Therefore, the human body must have approximately the same density as water. You should be able to calculate your volume.)

Practice Problems

11. Find the dimensions for weight. $W = mg$, m = mass, g = acceleration. Acceleration is velocity/time, velocity = distance/time.
12. From the definition of kinetic energy; that is, kinetic energy is equal to one-half the product of the mass times the square of the velocity

    $$KE = \tfrac{1}{2}mv^2$$

    deduce the connection between the SI unit of energy, the joule, and the SI units of mass, length, and time (kg, m, and sec).
13. The viscous force acting on a sphere of radius r falling through a fluid with a speed v (m/sec) is given by the equation $F = 6\pi\eta rv$, where η is called the coefficient of viscosity. What are the units of η in the SI system? (*Hint:* the units of force in the SI system are called newtons. 1 N = kg-m/sec^2)
14. The Stefan-Boltzmann law of radiation is given by $Q/At = kT^4$, where Q has the dimensions of joules, A is area, t is time, and T is the temperature in degrees Kelvin. What are the dimensions of the constant, k?
15. The gravitational force is given by $F = G(Mm/r^2)$ where F is in newtons, M and m are in kilograms, and r is in meters. What are the SI units for G?
16. The properties of a fluid in motion are related to the Reynolds number, a dimensionless quantity given by $R = rv\rho/\eta$ where r is the linear dimension, v is the speed of flow, ρ is the density of the fluid, and η is viscosity. Derive the SI units for the viscosity and compare with the results of problem 13.
17. A swimming coach tells her team of swimmers that "muscle times distance

equals energy." What dimensions does she envisage for muscle? (See problem 12 for energy dimensions.)

18. As weights are added to a spring it stretches in a linear way. The "force constant" for a spring is Δ force/Δ stretch, the slope of the force vs. stretch curve. What are the dimensions of the spring force constant? (See problem 13 for force units.

19. The equation for the conduction of heat through a cubical solid is

$$\frac{Q}{At} = K\frac{(T_2 - T_1)}{d}$$

where Q is the quantity of heat in joules, A is the area of a cube face, t is the time of heat flow, T_2 and T_1 are the temperatures of the cube sides in °C, d is the thickness of the cube, and K is coefficient of thermal conductivity. Find the units of K in the SI.

20. Many physical phenomena are described by exponential functions, such as e^{-kt} or e^{-cx}, where t = time and x = distance. Since exponents must be unitless, it is easy to discern the units of the constants k and c. Find the units for each constant in the following exponential equations:
    a. $N = N_0e^{-kt}$
    b. $I = I_0e^{-cx}$
    c. $P = Ae^{-\beta v^2}$ where v = speed

21. Find the conversion factors for each of the following units:
    a. Force = 1 newton = 1 N = 1 kg-m/sec^2 = _____ dynes, where 1 dyne = 1 g-cm/sec^2
    b. Length = 1 micron = 10^{-6} m = _____ cm
    c. Volume = 1 m^3 = _____ cm^3
    d. Area = 1 m^2 = _____ cm^2
    e. Pressure = force/area = N/m^2 = _____ dyne/cm^2
    f. 1 year = _____ sec

22. There are several dimensionless quantities used in physics. One of these is the radian: radian = arc length/radius. Find the ratio that defines each of the following:

specific gravity	reflection coefficient
index of refraction	transmission coefficient
coefficient of friction	engine efficiency

The answers to these practice problems are given at the end of this appendix.

A.6 Right Triangles

Many of the basic applications of physics to the mechnical properties of the human body require a thorough knowledge of the basic properties of right triangles, triangles that have one angle that is equal to 90°. Furthermore, many of the problems of rotational motion depend upon the radian measurement of angles. First examine the properties of the right triangles in Figure A.10.

Use a protractor and metric ruler to complete the following data table for the triangles I, II, and III.

FIGURE A.10
Three similar right triangles.

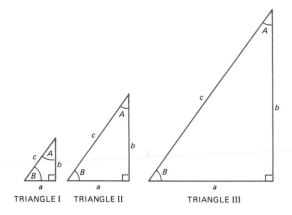

TRIANGLE I   TRIANGLE II   TRIANGLE III

	Triangle I	Triangle II	Triangle III
A	_____ °	_____ °	_____ °
B	_____ °	_____ °	_____ °
C	_____ °	_____ °	_____ °
a	_____ cm	_____ cm	_____ cm
b	_____ cm	_____ cm	_____ cm
c	_____ cm	_____ cm	_____ cm

Now perform the simple computations required to fill in the following table based upon the measurements you have made on the three triangles, I, II, and III.

What can you say about the angles A, B, and C of each triangle? Does the difference in size of the sides make a difference on the value of angle B? What determines the difference between angle B and angle A? Take a look at the ratio of b (side opposite angle B) to c (hypotenuse) of each triangle. What do you discover about these ratios? In the trigonometry table on page 753, find the value of the sine B. Compare this with the above ratio value. What have you discovered? State your discovery as an operational definition of the sine of an angle of a right tri-

	Triangle I	Triangle II	Triangle III
$A + B$	_____ °	_____ °	_____ °
$A + B + C$	_____ °	_____ °	_____ °
a/c	_____	_____	_____
b/c	_____	_____	_____
a/b	_____	_____	_____

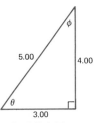

FIGURE A.11
A right triangle with sides of lengths 3, 4, and 5.

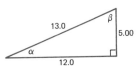

FIGURE A.12
A right triangle with sides of lengths 5, 12, and 13.

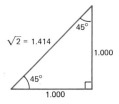

FIGURE A.13
An isosceles right triangle.

FIGURE A.14
A set of three isosceles right triangles.

FIGURE A.15

angle. Similarly, compare ratios of sides of triangles with table values of cosine and tangent values and state an operational definition of the cosine and tangent for angles of a right triangle. Add angles A and B together. What is your result? Make a statement about the two acute angles of a right triangle. These angles are called *complementary angles*. All three angles of a triangle add up to _____.

Look at the triangle in Figure A.11. $\sin \theta =$ _____; $\cos \theta =$ _____; $\cos \phi =$ _____; $\sin \phi =$ _____.

Look at the triangle in Figure A.12. $\sin \alpha =$ _____; $\cos \alpha =$ _____; $\cos \beta =$ _____; $\sin \beta =$ _____.

From these two examples can you express a general statement about complementary angles and their trigonometric functions? Look at the triangles in Figures A.11 and A.12. Now square each leg of the right triangle and add them together. Then square the hypotenuse of each triangle. Express in words that you notice. Write a quantitive mathematical equation about your statement. Your general statement should represent the Pythagorean theorem. Find a definition of the Pythagorean theorem, and see if your equation is correct.

Three special right triangles will come up again and again, so you need to give them special consideration.

The isoceles right triangle is a right triangle with two equal sides. It is also known as the 45°-45°-90° triangle, Figure A.13. Fill in the blanks: $\sin 45° =$ _____; $\cos 45°$ _____; $\tan 45° =$ _____.

From the triangles in Figure A.14 furnish the missing side and angle in Figure A.15. From the above example state some general characteristics of the isosceles right triangle.

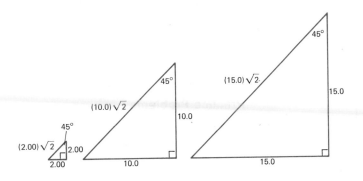

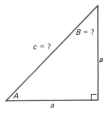

FIGURE A.16
Two 30°-60°-90° triangles.

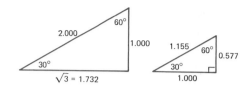

The 1:2:√3 triangle has one side whose length is half the length of the hypotenuse. It is known as the 30°-60°-90° triangle (Figure A.16). Fill in the blanks: sin 30° = _____ ; cos 30° = _____ ; tan 30° = _____ ; sin 60° = _____ ; cos 60° = _____ ; tan 60° = _____ .

The 3:4:5 triangle has sides whose lengths are in the ratio of 3:4:5. It is known also as the 37°-53°-90° triangle (Figure A.17). Fill in the blanks: sin 37° = _____ ; cos 37° = _____ ; tan 37° = _____ ; sin 53° = _____ ; cos 53° = _____ ; tan 53° = _____ .

Do you recognize triangle Z in Figure A.18? If not, divide each side of the triangle by 0.6. Do you recognize the triangle now? What about triangle W in Figure A.19? Fill in the missing side by looking only at the other two sides. Can you stretch your knowledge of angles to some extreme situations? Deduce the values of the following trigonometry functions:

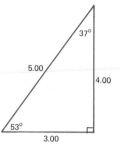

FIGURE A.17
A 3:4:5 triangle.

sin 0° = _____ ;   sin 90° = _____ ;   cos 0° _____ ;

cos 90° _____ ;   tan 0° _____ ;   tan 90° _____ .

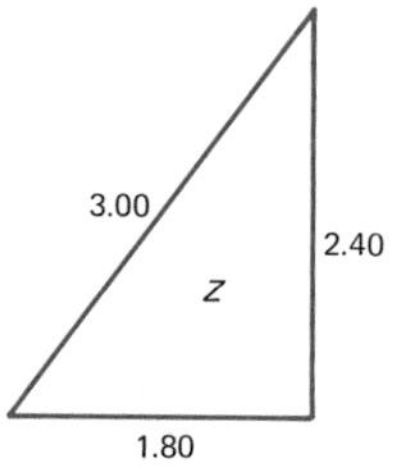

Let us take a look at radian measure. The number of radians in an angle at the center of a circle is the number of times the radius will divide into the arc subtended by the angle. See Figure A.20.

To travel completely around a circle with a radius of 1 m, you will go in an arc of length 6.28 m (2π radians). But once around the circle is equal to 360°. Derive a formula to convert back and forth between angular measurements in degrees and in radians. Check your formula by converting 45° to radian measure.

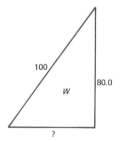

FIGURE A.18

Practice Problems

23. Convert the following angles to radian measures, and find their sine, cosine, and tangent values from a trigonometry table or an electronic calculator.
   a. 60°
   b. 53°
   c. 37°
24. Find the unknown angles and sides of triangles A and B in Figure A.21.
   a. triangle A: $B =$ _____ ; $a =$ _____ ; $b =$ _____
   b. triangle B: $B =$ _____ ; $a =$ _____ ; $b =$ _____
25. One acute angle of a right triangle is 20°. The length of the hypotenuse is 6 cm. Use trigonometry to calculate the lengths of the two sides.
26. What do you know about a 45°-45°-90° right triangle?
   a. length of the hypotenuse;
   b. lengths of the sides

FIGURE A.19

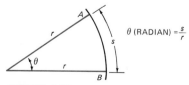

FIGURE A.20
A segment s of an arc of radius r.

FIGURE A.21

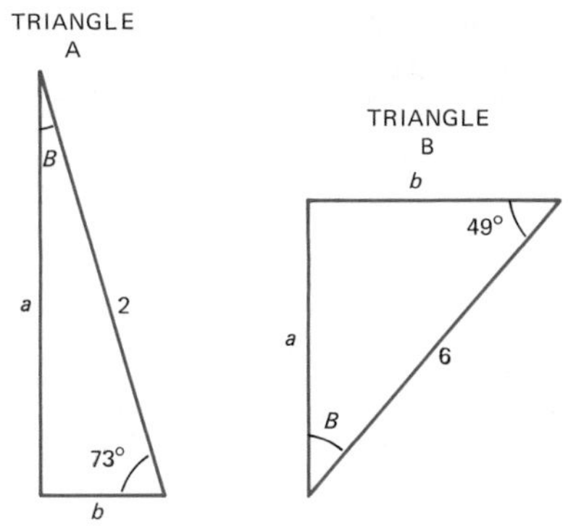

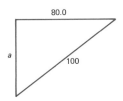

TRIANGLE A

TRIANGLE B

FIGURE A.22

27. State what a in Figure A.22 is equal to without using a trigonometry table or the Pythagorean theorem.
28. A car was traveling northeast. If it went a total distance of 42.42 km, how far north did it actually go?
29. Jim was traveling at an angle of 60° east of north for a distance of 60 m. How far was his actual northerly progress?

The answers to these practice problems are given at the end of this appendix.

A.7 Exponents

Since the notation of $2 \times 2 \times 2 \times 2 \times 2 = 32$ is inconvenient, we have developed the exponent notation to write $2 \times 2 \times 2 \times 2 \times 2$ as 2^5; hence $2^5 = 32$. The number 5 is called the power to which the base number 2 is raised to give the number 32. The number 5 is called the *exponent of the base*, 2. The number 10 multiplied by itself three times, $10 \times 10 \times 10$, is equal to 1000, so $10^3 = 1000$, where the base number is 10 and the exponent is the number 3 and the resulting number is 1000.

Using what you already know about multiplication and the definitions of exponents, deduce the multiplication and division rules for the use of the exponent notation.

$16 \times 4 = 64$ or $2^4 \times 2^2 = 2^6$

$64 \div 32 = 2$ or $2^6 \div 2^5 = 2^1$

Try other examples. Write a rule for multiplication. Write a rule for division. *Note:*

$64 \div 64 = 2^6 \div 2^6 = 2^0 = 1$

Define the number 1 in exponent notation. Show that the following relationships are valid: Since $64 = 2^6$, then $2 = 64^{1/6} = 64^{0.666}$. Since $1/64 = 2^0/2^6$, then $1/64 = 2^{-6}$.

What are numbers with positive exponents less than 1? What are numbers with negative exponents?

A.8 Logarithms

The *logarithm* of a number is the power to which some positive *base* (except 1) must be raised in order to equal the number. Consider the quantity $10,000 = 10^4$. In terms of logarithms, we write the equality as $\log_{10} 10,000 = 4$, and we read this as "the logarithm of 10,000 to the base 10 equals 4." Similarly, $100 = 10^2$ may be written as $\log_{10} 100 = 2$. How would you write $1000 = 10^3$ in terms of logarithms?

A logarithm is another way of writing exponents. The logarithm or exponent may be an integer, a decimal fraction or a combination of both.

Let us examine in further detail the relationship between the logarithm and the exponential forms. Fill in the blanks.

a. $8 = 2^3$, $\log_2 8 = 3$; $81 = 3^4$, $\log_3 81 =$ _____
b. $10,000 =$ _____, $\log_{10} 10,000 = 4$
c. $0.01 = 10^{-2}$, $\log_{10} 0.01 = -2$
d. $0.001 = 10^{-3}$, $\log_{10} 0.001 =$ _____
e. $1 = 10^0$, $\log_{10} 1 =$ _____

In general, if $N = b^x$ then $\log_b N = x$.

Using the special case of the zero exponent, write the following in terms of logarithms $1 = 2^0$, $1 = 11^0$, $1 = 8^0$, $1 = 6^0$. What do you discover? Write the following in terms of logarithms. $2 = 2^1$, $7 = 7^1$, $b = b^1$. What do you discover? *Note:* In our definition of logarithms, the number 1 was excluded for a logarithmic system. Why can 1 not serve as the base for a meaningful logarithmic system?

Logarithms are exponents and naturally follow the law of exponents:

1. A logarithm of a product equals the sum of the logarithms of the factors.

$$\log (20) \times (15) = \log 20 + \log 15$$

$$= 1.3010 + 1.1761 = 2.4771 = \log 300$$

Taking antilog we see that

$$20 \times 15 = 300$$

2. A logarithm of a quotient equals the logarithm of the numerator minus logarithm of the denominator.

$$\log \frac{20}{5} = \log 20 - \log 5$$

$$= 1.3010 - 0.6990$$

$$= 0.6021 = \log 4$$

Taking the antilog we see that

$$\frac{20}{5} = 4$$

3. A logarithm of a number to a power equals the power times logarithm of the number.

$$\log 2^2 = 2 \log 2 = 2(0.3010) = 0.6021 = \log 4$$

$$\log \sqrt{16} = \tfrac{1}{2} \log 16 = \tfrac{1}{2} (1.2042) = 0.6021 = \log 4$$

There are two widely used systems of logarithms called *natural logarithms* and common logarithms. The natural logarithm uses the so-called natural number e as the base. The natural number e can be expressed by the sum of numbers as follows:

$$e = 1 + \frac{1}{1} + \frac{1}{2} + \frac{1}{6} + \frac{1}{24} + \frac{1}{120} + \frac{1}{720} + \frac{1}{5040} + \cdots$$

and has the numerical value of $e = 2.71828\ 18284\ 59045\ 23536\ \ldots$. In the system of natural logarithms we will replace the symbol $\log_e x$ by $\ln x$. For example, suppose some number N can be expressed by the equation $N = e^x$ where x is some real number. Then

$$N = 1 + \frac{x}{1} + \frac{x^2}{2} + \frac{x^3}{6} + \frac{x^4}{24} + \frac{x^5}{120} + \frac{x^6}{720} + \frac{x^7}{5040} + \cdots$$

and

$$\ln N = x$$

$$\ln 10 = 2.303; \text{ so } 10 = e^{2.303} = 1 + 2.303 + (2.303)^2/2 + \cdots.$$

The *common*, or base 10, *logarithms* use the abbreviation $\log N$ for $\log_{10} N$. The common logarithm of a number is the exponent of the power to which 10 must be raised in order to obtain the given number.

The common logarithms are composed of two parts: the *mantissa*, or decimal part, and the *characteristic*, or integer part. Mantissas are always positive and may be obtained from logarithm tables.

The characteristic of the common logarithm of any positive number greater than one is numerically one less than the number of digits to the left of the decimal point of the number. The characteristic of the common logarithm of a positive number less than one is always negative and is numerically one greater than the number of zeros immediately to the right of the decimal point.

Make a generalization about the characteristic of a common logarithm and the exponents in powers of ten notation.

Sometimes it is convenient to change from one base to another.

$$2.303 \log_{10} y = \ln y. \text{ Let } y = e^x$$

Since e is the base of the natural logarithms what do you think ln y is equal to? _____ log y = _____ . We take the natural logarithm of y and set it equal to x. Why?

EXAMPLE

$y = e^{-2x}$

$\ln y = \ln (e^{-2x}) = -2x$

So

$$x = -\frac{\ln y}{2}$$

Practice Problems

30. Solve for x: $10 = e^{3x}$. (*Hint:* ln $10 = 2.303$)
31. Solve for x by using logarithms
    a. $x = \frac{1}{2} at^2$
    b. $a = 9.80$ m/sec²
    c. $t = 38.0$ sec
    (*Hint:* log $cd^2 = \log c + 2 \log d$)
32. Solve for x: $1000 = e^{2x}$. (*Hint:* log $1000 = 3$; ln $y = 2.303 \log y$)

A.9 The Exponential Function

The importance of the exponential function in medical applications can hardly be overstressed. Many of the properties of life systems and their interactions can be described by the exponential function. For example, consider the growth of population of a country. If we define ΔN as the change in the number of people in the country for a given time interval Δt, say a year, then ΔN can be related directly to the total population N of the country at the beginning of the time interval Δt by the use of the birth rate constant b and the death rate constant d where b is the number of births per 1000 of the population in time Δt and d is the number of deaths per 1000 of the population in time Δt.

Then, if N is written as so many thousands of people, ΔN is given by the following expression: $\Delta N = (b - d)\ N\ \Delta t$. Can you verify this expression using dimensional analysis?

Now let us define a population growth constant $c = b - d$ and solve for the ratio of $\Delta N/\Delta t$:

$\Delta N/\Delta t = cN$

In words this equation says that the time rate of change of the population $\Delta N/\Delta t$ is directly proportional to the total population N and the proportionality constant is c. This equation has been extensively studied by mathematicians and it can be shown that one solution for this equation is given by the following equation:

$$N = N_0 e^{ct}$$

where N_0 is the population of the country at the beginning of the time interval and t is the time at which the country has a population of N (see Figure A.23).

EXAMPLE

1. Between 1800 and 1860 the population of the United States increased about 35 percent every ten years. During that period of time, how long did it take for the population of the United States to double?

   Let us begin by finding c as the population growth per year.

   $$1.35\,N_0 = N_0 e^{10c}$$

   So $\ln(1.35) = 10c$ and $c = 0.03/\mathrm{yr}$. Then $2N_0 = N_0 e^{0.03\,t_d}$, where t_d is the doubling time in years. So, $t_d =$ _____.

Practice Problems

33. The population of the United States in 1840 was almost exactly 17 million people. If the population growth had continued at a constant rate ($c = 0.03/\mathrm{yr}$) until 1970, what would the population of the United States have been in 1970?

34. Between 1960 and 1970 the population of the United States increased by 13.3 percent. What was the annual population growth constant? What was the daily population growth constant? How many more people would have lived each day in a city of 100,000?

FIGURE A.23
An exponential function of population N at a time t when the population was N_0 at the beginning of time, $N = N_0 e^{ct}$. Note that population doubles every $0.69/c$ units of time.

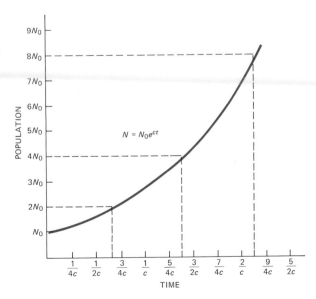

Answers

1. 10^{19}
2. 7.6×10^{27}
3. $A = 7.8 \times 10^1$
4. $A = 5.0 \times 10^{-32}$
5. for a year of 365.25 days = 3.1558 $\times 10^7$ seconds
6. a. three
   b. 3 or 4 or 5, ambiguous
   c. 5; d. one
   e. three
7. a. 2.2×10^3 cm², b. 10×10^5 cm³
8. a. -9
   b. zero; c. $+20$
9. a. 4
   b. zero
10. a. 8
    b. 16
    c. approximate 131 squares
11. ML/T^2
12. KE is ML^2/T^2, 1 joule = 1 kg m²/sec²
13. M/LT or N sec/m² or kg/m-sec
14. mass/(time)³ (temperature)⁴
15. newton (meter)²/(kilogram)²
16. M/LT, kg/m-sec, same as in 13
17. ML/T².
18. M/T^2
19. joules/(meter/sec °C)
20. a. T^{-1}
    b. L^{-1}
    c. T^2/L^2
21. a. 1 N = 10^5 dynes
    b. 1 micron = 10^{-4} cm
    c. 1 m³ = 10^6 cm³
    d. 1 m² = 10^4 cm²
    e. 1 N/m² = 10 dyne/cm²
    f. 1 year = 3.16×10^7 sec
23. a. 1.05 radians, 0.866; 0.500; 1.732
    b. 0.925 radians; 0.799, 0.602, 1.33
    c. 0.646 radians; 0.602, 0.799, 0.754
24. $< B = 27°$; $a = 1.91$; $b = 0.585$ for Triangle A; Triangle B, $< B = 41°$, $a = 4.53$; $b = 3.94$
25. 2.05 and 5.64
26. a. equal to $\sqrt{2}$ times length of a side
    b. sides of equal length
27. 60
28. 30 km
29. 30 m;
30. 7.68×10^{-1}
31. 7080; $\log x = \log \frac{1}{2} + \log a + 2 \log t$
32. 3.454;
33. 840 million
34. 1.25×10^{-2}, 3.419×10^{-5}, 3.4 people

Tables

Basic Quantity	Unit	Derived Quantity	Unit
Length	meter (m)	Force	newton (N)
Time	second (sec)	Work (energy)	joule (J)
Mass	kilogram (kg)	Power	watt (W)
Temperature	Kelvin degree (K)	Frequency	hertz (Hz)
Current	ampere (A)	Pressure	Pascal (Pa)
		Viscosity	Poiseuille (Pl)
		Charge	coulomb (C)
		Potential	volt (V)
		Capacitance	farad (F)
		Resistance	ohm (Ω)
		Magnetic induction	Tesla (T) $= (1 \text{ Wb/m}^2)$
		Magnetic flux	weber (Wb)
		Inductance	henry (H)
		Wavelength	nanometer (nm)

The international unit of length, meter, is equal to 1,650,763.73 wavelengths in vacuum of a particular orange radiation, identified as $2p_{10} \to 5d_5$ transition, of an atom isotope of krypton known as ^{86}Kr.

The second is equal to the sum of 9,192,623,770 periods of the radiation corresponding to the transition between two hyperfine levels of the ground state of ^{133}Cs atom.

The international standard of mass is kilogram and the standard mass is a platinum-iridium cylinder which is kept in the International Bureau of Weights and Measures at Sevres, France (near Paris).

The Kelvin degree (°K) is the international standard for temperature and is equal to 1/273.16 of temperature of the triple point of water.

The ampere (A) is that current in two very long parallel conductors one meter apart which given a rise to a magnetic force of 2×10^{-7} N/m. (See Section 23.8)

SI Prefixes

Prefix	Symbol	Value
tera	T	10^{12}
giga	G	10^{9}
mega	M	10^{6}
kilo	k	10^{3}
hecto	h	10^{2}
deca	da	10^{1}
deci	d	10^{-1}
centi	c	10^{-2}
milli	m	10^{-3}
micro	μ	10^{-6}
nano	n	10^{-9}
pico	p	10^{-12}
femto	f	10^{-15}
atto	a	10^{-18}

Physical Constants

Constant	Symbol	Value
Gravitational constant	G	6.67×10^{-11} N-m^2/kg^2
Avogadro's number	N_0	6.02×10^{23} molecules/gm-mole
Boltzmann's constant	k	1.38×10^{-23} J/molecule K$^\circ$
Universal gas Constant	R	8.31 J/$^\circ$K mol = 1.99 cal/mol K$^\circ$
Charge of electron (basic unit of charge)	e	1.60×10^{-19} C
Velocity of light	c	3.00×10^{8} m/sec
Mass of electron (rest)	m_e	9.11×10^{-31} kg
Mass of proton (rest)	m_p	1.672×10^{-27} kg
Mass of neutron (rest)	m_n	1.675×10^{-27} kg
Atomic mass unit	amu	1.661×10^{-27} kg = 931.5 MeV/c^2
Permittivity of free space	ϵ_0	8.85×10^{-12} C^2/N-m^2
Coulomb constant	$1/4\pi\epsilon_0$	8.99×10^{9} N-m^2/C^2
Permeability of free space	μ_0	1.26×10^{-6} Wb/m-A
Planck's constant	h	6.63×10^{-34} J-sec
1 atmosphere	atm	76 cm Hg = 1.01×10^{5} N/m^2
Mass of earth		5.98×10^{24} kg
Mean radius of Earth		6.37×10^{6} m
Mass of sun		1.99×10^{30} kg
Radius of sun		6.96×10^{8} m
Mass of moon		7.34×10^{22} kg
Radius of moon		1.74×10^{6} m
Mean earth-sun distance		1.50×10^{11} m
Mean earth-moon distance		3.84×10^{8} m

Trigonometric Functions

Degrees	Sine	Cosine	Tangent	Cotangent	Degrees
0	.0000	1.0000	.0000	∞	90
1	.0175	.9998	.0175	57.29	89
2	.0349	.9994	.0349	28.64	88
3	.0523	.9986	.0524	19.08	87
4	.0698	.9976	.0699	14.30	86
5	.0872	.9962	.0875	11.430	85
6	.1045	.9945	.1051	9.514	84
7	.1219	.9925	.1228	8.144	83
8	.1392	.9903	.1405	7.115	82
9	.1564	.9877	.1584	6.314	81
10	.1736	.9848	.1763	5.671	80
11	.1908	.9816	.1944	5.145	79
12	.2079	.9781	.2126	4.705	78
13	.2250	.9744	.2309	4.332	77
14	.2419	.9703	.2493	4.011	76
15	.2588	.9659	.2679	3.732	75
16	.2756	.9613	.2867	3.487	74
17	.2924	.9563	.3057	3.271	73
18	.3090	.9511	.3249	3.078	72
19	.3256	.9455	.3443	2.904	71
20	.3420	.9397	.3640	2.748	70
21	.3584	.9336	.3839	2.605	69
22	.3746	.9272	.4040	2.475	68
23	.3907	.9205	.4245	2.356	67
24	.4067	.9135	.4452	2.246	66
25	.4226	.9063	.4663	2.144	65
26	.4384	.8988	.4877	2.050	64
27	.4540	.8910	.5095	1.963	63
28	.4695	.8829	.5317	1.881	62
29	.4848	.8746	.5543	1.804	61
30	.5000	.8660	.5774	1.732	60
31	.5150	.8572	.6009	1.664	59
32	.5299	.8480	.6249	1.600	58
33	.5446	.8387	.6494	1.540	57
34	.5592	.8290	.6745	1.483	56
35	.5736	.8192	.7002	1.428	55
36	.5878	.8090	.7265	1.376	54
37	.6018	.7986	.7536	1.327	53
38	.6157	.7880	.7813	1.280	52
39	.6293	.7771	.8098	1.235	51
40	.6428	.7660	.8391	1.192	50
41	.6561	.7547	.8693	1.150	49
42	.6691	.7431	.9004	1.111	48
43	.6820	.7314	.9325	1.072	47
44	.6947	.7193	.9657	1.036	46
45	.7071	.7071	1.0000	1.000	45
Degrees	**Cosine**	**Sine**	**Cotangent**	**Tangent**	**Degrees**

Periodic Table of the Elements

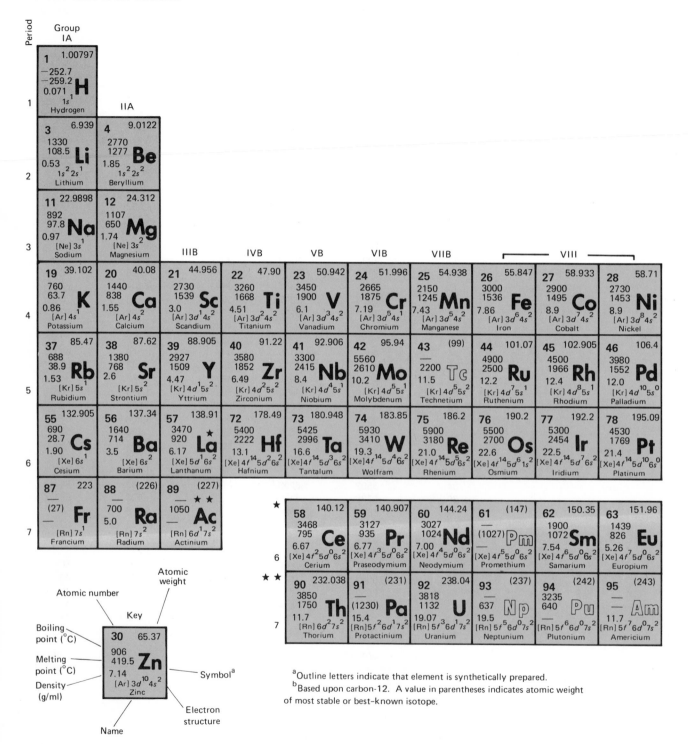

Inert
Gases

					IIIA	IVA	VA	VIA	VIIA	

			2	4.0026
			−268.9	
			−269.7	**He**
			0.126	
			$1s^2$	
			Helium	

5	10.811	6	12.0111	7	14.0067	8	15.9994	9	18.9984	10	20.183
—		4830		−195.8		−183		−188.2		−246	
(2030)	**B**	3727g	**C**	−210	**N**	−218.8	**O**	−219.6	**F**	−248.6	**Ne**
2.34		2.26		0.81		1.14		1.11		1.20	
$1s^2 2s^2 2p^1$		$1s^2 2s^2 2p^2$		$1s^2 2s^2 2p^3$		$1s^2 2s^2 2p^4$		$1s^2 2s^2 2p^5$		$1s^2 2s^2 2p^6$	
Boron		Carbon		Nitrogen		Oxygen		Fluorine		Neon	

					13	26.9815	14	28.086	15	30.9738	16	32.064	17	35.453	18	39.943
					2450		2680		280w		444.6		−34.7		−185.8	
IB		IIB			660	**Al**	1410	**Si**	44.2w	**P**	119.0	**S**	−101.0	**Cl**	−189.4	**Ar**
					2.70		2.33		1.82w		2.07		1.56		1.40	
					[Ne]$3s^2 3p^1$		[Ne]$3s^2 3p^2$		[Ne]$3s^2 3p^3$		[Ne]$3s^2 3p^4$		[Ne]$3s^2 3p^5$		[Ne]$3s^2 3p^6$	
					Aluminum		Silicon		Phosphorus		Sulfur		Chlorine		Argon	

29	63.54	30	65.37	31	69.72	32	72.59	33	74.922	34	78.96	35	79.909	36	83.80
2595		906		2237		2830		613*		685		58		−152	
1083	**Cu**	419.5	**Zn**	29.8	**Ga**	937.4	**Ge**	817	**As**	217	**Se**	−7.2	**Br**	−157.3	**Kr**
8.96		7.14		5.91		5.32		5.72		4.79		3.12		2.6	
[Ar]$3d^{10}4s^1$		[Ar]$3d^{10}4s^2$		[Ar]$3d^{10}4s^24p$		[Ar]$3d^{10}4s^24p$		[Ar]$3d^{10}4s^24p$		[Ar]$3d^{10}4s^24p^4$		[Ar]$3d^{10}4s^24p$		[Ar]$3d^{10}4s^24p$	
Copper		Zinc		Gallium		Germanium		Arsenic		Selenium		Bromine		Krypton	

47	107.870	48	112.40	49	114.82	50	118.69	51	121.75	52	127.60	53	126.904	54	131.30
2210		765		2000		2270		1380		989.8		183		−108.0	
960.8	**Ag**	320.9	**Cd**	156.2	**In**	231.9	**Sn**	630.5	**Sb**	449.5	**Te**	113.7	**I**	−111.9	**Xe**
10.5		8.65		7.31		7.30		6.62		6.24		4.94		3.06	
[Kr]$4d^{10}5s^1$		[Kr]$4d^{10}5s^2$		[Kr]$4d^{10}5s^25p$		[Kr]$4d^{10}5s^25p^2$		[Kr]$4d^{10}5s^25p^3$		[Kr]$4d^{10}5s^25p$		[Kr]$4d^{10}5s^25p$		[Kr]$4d^{10}5s^25p^6$	
Silver		Cadmium		Indium		Tin		Antimony		Tellurium		Iodine		Xenon	

79	196.967	80	200.59	81	204.37	82	207.19	83	208.980	84	(210)	85	(210)	86	(222)
2970		357		1457		1725		1560		254		—		(−61.8)	
1063	**Au**	−38.4	**Hg**	303	**Tl**	327.6	**Pb**	271.3	**Bi**	254	**Po**	(302)	**At**	(−71)	**Rn**
19.3		13.6		11.85		11.4		9.8		(9.2)				
[Xe]$4f^{14}5d^{10}6s^1$		[Xe]$4f^{14}5d^{10}6s$		[Xe]$4f^{14}5d^{10}6s^26p$		[Xe]$4f^{14}5d^{10}6s^26p^2$		[Xe]$4f^{14}5d^{10}6s^26p^3$		[Xe]$4f^{14}5d^{10}6s^26p$		[Xe]$4f^{14}5d^{10}6s^26p$		[Xe]$4f^{14}5d^{10}6s^26p$	
Gold		Mercury		Thallium		Lead		Bismuth		Polonium		Astatine		Radon	

64	157.25	65	158.924	66	162.50	67	164.930	68	167.26	69	168.934	70	173.04	71	174.97
3000		2800		2600		2600		2900		1727		1427		3327	
1312	**Gd**	1356	**Tb**	1407	**Dy**	1461	**Ho**	1497	**Er**	1545	**Tm**	824	**Yb**	1652	**Lu**
7.89		8.27		8.54		8.80		9.05		9.33		6.98		9.84	
[Xe]$4f^75d^16s^2$		[Xe]$4f^95d^06s$		[Xe]$4f^{10}5d^06s^2$		[Xe]$4f^{11}5d^06s$		[Xe]$4f^{12}5d^06s$		[Xe]$4f^{13}5d^06s$		[Xe]$4f^{14}5d^06s$		[Xe]$4f^{14}5d^16^2$	
Gadolinium		Terbium		Dysprosium		Holmium		Erbium		Thulium		Ytterbium		Lutetium	

96	(247)	97	(247)	98	(251)	99	(254)	100	(253)	101	(256)	102	(254)	103	(257)	104		105	(260)
—		—		—		—		—		—		—		—		—		—	
—	**Cm**	—	**Bk**	—	**Cf**	—	**Es**	—	**Fm**	—	**Md**	—	**No**	—	**(Lw)**	—		—	**Ha**
[Rn]$5f^76d^17s^2$		[Rn]$5f^76d^27s^2$		[Rn]$5f^96d^17s^2$															
Curium		Berkelium		Californium		Einsteinium		Fermium		Mendelevium		Nobelium		(Lawrencium)		Rutherfordium		Hahnium	

From Acosta, V., Cowan, C. L., and Graham, B. J. In *Essentials of Modern Physics.* Copyright © 1973 by Harper & Row, Publishers, Inc.

Glossary

Aberration, lens
Departure of actual image formation from simple thin lens model predictions. (Section 18.12)

Absorption coefficient   (of light, of x rays)
The physical parameter that determines the absorption of electromagnetic radiation (light or x rays) per unit length of absorber. (Section 30.4)

Acceleration
The time rate of the change in velocity. (Section 3.6)

Accommodation
The ability of the eye to focus on objects between the near point and far point of the eye. (Section 20.2)

Acoustic impedance
The acoustical impedance of a medium is given by the product of the density of the medium and the velocity of sound in the medium. (Section 17.3)

Actual mechanical advantage
Ratio of the load to the applied force. (Section 5.10)

Adhesion
Attraction of unlike molecules. (Section 14.6)

Adiabatic process
A process that takes place with no change in heat energy in the system. (Section 12.8)

Alpha radiation
The radiation that consists of helium nuclei emitted by heavy radioactive nuclei. (Section 31.6)

Amplification
The ratio of output signal to input signal of a device. (Section 26.2)

Amplitude
Maximum displacement from equilibrium position. (Sections 15.2, 16.4, and 17.2)

Angular acceleration
Time rate of change of angular velocity. Usually given in radians per (second)^2. (Section 7.2)

Angular displacement
Angle between two positions of a rigid body rotating about a fixed axis. Usually given in radians. (Section 7.2)

Angular momentum
The product of moment of inertia and angular velocity gives the magnitude of the angular momentum. A right-hand rule can be used to determine its direction. (Section 7.8)

Angular velocity
Time rate of change of angular displacement. Usually given in radians per second. (Section 7.2)

Archimedes' principle
Any object wholly or partially immersed in a fluid is buoyed up by a force equal to the weight of the displaced fluid. (Section 8.9)

Astigmatism
The condition produced by a nonspherical cornea surface. Point objects do not form point images on the retina. (Section 20.4)

Atomic mass unit
The number of nucleons (protons and neutrons) in an atomic nucleus. (Section 31.2)

Atomic number
The number of protons (or electrons) in an atom. The symbol Z represents the atomic number. (Section 28.2)

Bernoulli's theorem
A statement of conservation of energy. In streamline flow Bernoulli's theorem may be represented in the equation form $P + \rho gh + \frac{1}{2}\rho v^2 =$ constant. (Section 8.11)

Beta radiation
The radiation that results in electron or positron emission from a radioactive nucleus. (Section 31.6)

Bioluminescence
Certain biological systems produce light as a consequence of their biochemical processes. (Section 29.16)

Birefringence
Property of crystals that have different velocity of light for different directions. (Section 19.10)

Bohr radius
The radius of the first Bohr orbit for the hydrogen atom 0.53×10^{-10}m. (Section 28.3)

Breeder reactor
A nuclear fission reactor that produces additional fissionable fuel material in addition to usable nuclear energy. (Section 32.2)

Bremsstrahlung
"Braking radiation" is the name given to the continuous x-ray spectrum from an x-ray source. (Section 30.2)

Bulk modulus
Elastic constant of proportionality for a deformation of volume. (Section 13.5)

Buoyant force
Resultant of all of the pressure forces acting on the body. (Section 8.9)

Calorie
The mean amount of heat required to increase the temperature of one gram of water one degree celsius. (Section 10.6)

Capacitance
Defined as the ratio of charge to voltage for an electrical system unit of capacitance 1 farad = coulomb/volt. (Section 21.8)

Capillarity
The rise of water in a capillary above the water level in the surrounding area. (Section 14.10)

Carnot cycle
Cycle of an ideal engine which includes an isothermal expansion at higher temperature, an adiabatic expansion, an isothermal compression at lower temperature, and an adiabatic compression. (Section 12.11)

Center of mass
The balance point of an object; the force of gravity (weight) produces zero torque about the center of mass. (Section 7.4)

Centripetal force
The unbalanced force acting on the body towards the center of rotation in circular motion. (Section 4.6)

Chain reaction
The fission reaction sustained by neutrons from preceding nuclear fission reactions. The basis of nuclear fission reactions and bombs. (Section 32.2)

Characteristic x rays
The x rays produced by inner electron energy state transitions that result in x-ray photons. (Section 30.2)

Chromatic aberration
The effect produced by the dependence of focal length on the frequency (color) of light. (Section 18.12)

Coefficient of friction
Ratio of the maximum frictional force to the net force pressing the surface together. (Section 4.4)

Coefficient of linear expansion
Change in length per unit length per degree change in temperature. (Section 10.3)

Coefficient of performance of a refrigerator
Ratio of heat absorbed to the amount of work supplied to the refrigerator. (Section 12.12)

Coefficient of volumetric expansion
Change in volume per unit volume per degree change in temperature. (Section 10.4)

Coherent waves
Waves that always maintain a constant phase relationship are coherent. (Section 19.3)

Cohesion
Attraction of like molecules. (Section 14.4)

Complementarity principle
This principle defined by N. Bohr uses both the wave and particle nature of matter and radiation. (Section 27.6)

Compton effect
The phenomena that results in scattered electrons when photons of incident radiation interact with weakly bound electrons in metals. (Section 27.5)

Conservation laws
There is no change in a physical property with a change in time. (Section 2.6)

Conservation of momentum
The total momentum of an isolated system is constant in magnitude and direction. (Section 6.3)

Conservative and nonconservative systems
If a body makes a round trip— *i.e.*, *a* to *b* and back to *a*— and net work is expended, a system is nonconservative and the forces acting are nonconservative. Hence if no net work is done the system is conservative. (Section 5.7)

Continuity of matter
Time rate of change of the amount of material in unit volume of material is equal to the change of the current density with respect to distance. (Section 9.6)

Convection
Transfer of heat energy by motion of matter (particles). (Section 11.3)

Converging lens
A lens that bends parallel light toward the axis. Focal length is positive. (Section 18.7)

Covalent bonding

The bond produced by shared electrons between atoms making up covalent systems, *e.g.*, H_2. (Section 29.2)

Critical mass

The necessary mass of fissionable material to sustain a chain reaction. (Section 32.2)

Current

A steady and onward moment of a physical quantity. The ratio of the change in a quantity to the change in time. (Section 2.5)

Current density

Quantity of an item that passes through a unit area in unit time. (Section 9.6)

Current sensitivity

The proportionality constant between the current in a coil and its angle of deflection in a magnetic field. (Section 23.9)

Damping

Loss of energy of vibration due to friction (Section 15.5)

de Broglie waves

The matter waves that were postulated by de Broglie to have a wave length equal to Planck's constant divided by the momentum of the particle. (Section 27.7)

Decibel

The unit of sound level that is given by ten times the $\log_{10}$ of the ratio of I (sound intensity) to I_0 (standard intensity of 10^{-12} watt/m²). (Section 17.4)

Density

Mass per unit volume. (Section 8.3)

Dichromat

Color vision resulting from only two of the three pigments required for normal color vision. (Section 20.5)

Dielectric constant

The ratio of the permittivity of a material to permittivity of a vacuum. (Section 21.2)

Diffraction

Occurs if a wave front encounters an object and there results a superposition of wave fronts to produce a resultant wave bending around the object. (Sections 16.10 and 19.7)

Discrimination

Ability to distinguish between two different stimuli. (Section 1.3)

Dispersion

A medium will produce a dispersion of waves if the wave velocity is a function of the frequency. (Sections 16.13 and 19.2)

Displacement

Change in location of an object specified by both magnitude and direction from a given origin. (Section 3.2)

Distortion

The property of a system in which the input and output are not proportional to each other. (Section 17.7)

Diverging lens

A lens that bends parallel light away from the axis. Focal length is negative. (Section 18.7)

Doppler effect

Change in frequency of a wave resulting from the relative motion of the source and observer. (Section 17.10)

Effective values   (for current and voltage)

The AC current and voltage values that produce the same heating effects as the given DC values. For sinusoidal AC the effective value = rms value = $amplitude/\sqrt{2}$. (Section 25.2)

Efficiency

Ratio of work accomplished to work supplied. (Section 5.10)

Efficiency of heat engine

Ratio of useful work output divided by energy input. (Section 12.10)

Elastic body

Any material or body which is deformed by an applied force and returns to its original shape after the distorting force is removed. (Section 13.2)

Elastic collision

Both momentum and kinetic energy of bodies involved in an elastic collision are the same before and after impact. (Section 6.4)

Elastic limit

Limit of distortion for which deformed body returns to original shape after deforming force is removed. (Section 13.3)

Electrical conductivity

The proportionality constant between current density and potential gradient. (Section 22.2)

Electric dipole

Equal (+) and (−) charges. q, separated by a distance s have a dipole moment equal to qs. (Section 21.6)

Electric field

A property of space defined as the electric force per unit charge at each point (newtons/coulomb). (Section 21.2)

Electromotive force   (emf)

The potential difference across a source of electrical energy with no current flow, $\mathscr{E}$ (volts). (Section 22.3)

Energy

Property of a system which causes changes in its own state or state of its surroundings; measure of ability to do work (physical). (Sections 1.3 and 5.3)

Energy level

The allowed energy values permissible for a system in accord with quantum theory. (Section 28.4)

Equilibrium

All the influences acting on the system are cancelled by others, resulting in a balanced unchanging system. (Section 2.2)

Equipotential surface

The surface defined by constant potential coordinates. (Section 21.4)

Extrinsic semiconductor

A semiconductor that has been doped with a donor material (*n* type) or an acceptor material (*p* type). (Section 29.13)

Farad

The unit of capacitance equal to a coulomb per volt. (Section 21.8)

Faraday's law of induction   (law of electromagnetic induction)

Predicts that a change in magnetic flux with time produces an induced emf. (Section 24.2)

Far point

The greatest object distance for sharp image on the retina (lowest power of accommodation). (Section 20.2)

Feedback

A portion of the output from the system is returned as input into the same system. (Sections 2.7 and 26.2)

Ferromagnetism

The result of cooperative interactions of magnetic dipoles in such materials as iron, nickel, and cobalt. The result is very strong internal magnetic fields. (Section 23.13)

Field

A region of space characterized by a physical property with a determinable value at every point in the region, *e.g.*, gravitational field, magnetic field, etc. (Section 1.5)

Fluids

Substance which takes the shape of the container and flows from one location to another. (Section 8.2)

Fluorescence

The light emitted by a substance while it is irradiated. Fluorescent light has longer wavelength than incident light and ceases when incident light is removed. (Section 29.7)

Focal length

Distance from lens or mirror to point of convergence for parallel light. (Sections 18.6 and 18.7)

Focal point

The location on the axis of a lens or mirror where parallel light converges to a point. (Sections 18.6 and 18.7)

Force

A measure of the strength of an interaction; a push or pull; the effect of a force is to alter the state of motion of a body. A vector quantity with a magnitude measured in newtons (N). (Sections 1.4 and 4.1)

Fourier's theorem

Any wave form may be produced by the superposition of sine waves of specific wavelengths and amplitudes. (Section 16.12)

Frequency

Number of complete oscillations per second. (Sections 2.4, 15.2, 16.4, and 17.2)

Frequency response

Refers to the range of input frequencies that yield linear outputs for an instrument or device. (Section 26.2)

Frictional force

Non-conservative force that opposes the motion of an object. (Section 4.4)

Gamma ray   (electromagnetic radiation)

γ-ray photon radiation results when a nucleus drops from a higher to lower energy state. Energies are typically measured in million electron volt units. (Section 31.6)

Gas

Fluid which takes on the shape and volume of the container.

Gradient

Rate of change of a physical quantity relative to position in direction of maximum rate of change.

Half-life

Biological

The time required for one-half the biochemically active sample to be eliminated by natural processes. (Section 32.5)

Effective

The resultant of radioactive and biological half-life. (Section 32.5)

Radioactive

The time required for one-half the radioactive nuclei in a sample to decay. (Section 31.6)

Heat

Thermal energy of transfer in which the internal energy of a system is changed. (Section 10.6)

Heat capacity

Product of the mass of a body in grams and specific heat, *i.e.*, amount of heat required to raise the temperature of the body one degree Celsius. (Section 10.6)

Heat of combustion

Energy produced per unit measure of quantity of material for complete oxidation. (Section 10.8)

Heat conduction

Process by which internal energy travels from one part of a solid to another. (Section 11.2)

Heat engine

Absorbs a quantity of energy from higher temperature reservoir, does work, and rejects a quantity of heat energy to lower temperature reservoir, and returns to its original state. (Section 12.10)

Henry
The unit for inductance (self and mutual) is the henry. (Section 24.5)

Hooke's law
Within elastic limits strain is proportional to the stress. (Section 13.3)

Hyperopia
The hyperopic eyeball has parallel light focused behind the retina (farsighted condition). (Section 20.4)

Image distance
The axial distance from the center of a lens or mirror to the image position. (Sections 18.6 and 18.7)

Image, real
Light rays converge to form real image that can be focused on a screen. (Sections 18.6 and 18.7)

Image, virtual
Light rays appear to diverge from virtual image, and image cannot be focused on a screen. (Sections 18.6 and 18.7)

Impedance
In an AC circuit, the impedance, Z, determines current, I, for a given voltage, V. (Section 25.4)

Impulse
Product of the net force and the time interval over which it acts is called the impulse of the force. (Section 6.2)

Index of refraction
Ratio of the velocity of light in a vacuum to the velocity of light in the refractive material. (Section 18.3)

Inelastic collision (perfectly)
Bodies stick together during impact; momentum is conserved, but kinetic energy is not conserved. (Section 6.4)

Inertia
Property of system that is a measure of the system's resistance to change. (Section 2.3)

Infrasonic
Sound with frequencies below 20 hertz/ sec. (Section 17.5)

Intensity
Energy transported through a unit area in one second. (Sections 1.5, 16.6, and 17.3)

Interaction
Action or influence exerted between systems. (Section 1.2)

Interference
The superposition of two or more waves with constant phase differences produces interference. (Sections 16.9 and 19.3)

Internal reflection (total)
When angle of incidence is equal to or greater than a critical angle θ_c in going from a dense to a less dense medium, light is totally reflected back into incident medium where $\sin \theta_c = \frac{n_2}{n_1}$. (Section 18.5)

Intrinsic semiconductor
A material that in a pure state has a small energy gap between an empty conduction band and the valence band, *e.g.*, Si or Ge. (Section 29.12)

Inverse Square Law
A property of a system whose magnitude is inversely proportioned to the square of the independent variable, *e.g.*, gravitational force, electrostatic force, energy intensity of a spherical wave. (Sections 4.5, 21.2, and 16.15)

Ionic bonding
Attractive force between positive and negative ions makes a strong bond for ionic materials like $Na^+ Cl^-$. (Section 29.2)

Isobar
Nuclei that have the same atomic mass number, A, are called isobars. (Section 31.2)

Isobaric process
Takes place at constant pressure. (Section 12.6)

Isochoric process
Takes place at constant volume. (Section 12.5)

Isothermal process
Takes place at constant temperature. (Section 12.7)

Isotone
Nuclei with equal numbers of neutrons are called isotones. (Section 31.2)

Isotope
Refers to atoms that have the same atomic number but different numbers of neutrons. (Section 31.2)

Joule's law
The power supplied to an electrical circuit is given by Joule's law. $P = VI$ (watts). (Section 22.4)

Kinetic energy
Energy of motion of a body. (Sections 1.3 and 5.4)

Laser
An acronym for light amplifier by stimulated electromagnetic radiation. A highly monochromatic coherent light source. (Section 19.13)

Latent heat of fusion
Heat required to change unit mass of material from the solid state to the liquid at the melting point. (Section 10.7)

Latent heat of vaporization
Heat required to convert unit mass of material from liquid state to vapor state at the boiling point temperature. (Section 10.7)

Laws of thermodynamics
Zeroth law
If system A is in thermal equilibrium with system B, and system B is in thermal equilibrium with system C, then systems A, B, and C are in equilibrium with each other, and they are at the same temperature. (Section 12.2)
First law
The change in the internal energy of a system is equal

to the heat added to the system minus the work done by the system. (Section 12.4)

Second law

It is not possible to transfer heat from a lower temperature body to a higher temperature body without the expenditure of work. (Section 12.9)

Third law

As the temperature approaches absolute zero, the entropy approaches zero. (Section 12.15)

Length: relativistic contraction

Lorentz contraction of special relativity gives an apparent contraction of a moving length in its direction of motion, $L = L_0 \sqrt{1 - v^2/c^2}$. (Section 27.4)

Lenz's law

Predicts that the induced emf of Faraday's law gives a current that opposes the flux change producing the emf. (Section 24.2)

Lever

A simple machine. Essentially a rigid body with a fixed pivot. (Section 5.12)

Light rays

Radii of spherical waves. Imaginary lines drawn in the direction of the light wave propagation. (Section 18.2)

Linear expansion

Change in length accompanied by a change in temperature. (Section 10.3)

Linearity

Refers to the condition when the output (O) has a linear relationship with the input (I). (Section 26.2)

Linear system

A system in which the principle of superposition is valid. (Sections 2.4, 15.6, 16.8, 17.6)

Liquid

A fixed volume fluild takes on shape of the container to the limit of its volume. (Section 8.2)

Lissajou's figures

Figures produced in a plane by the superposition of two simple harmonic motions at right angles. (Section 15.6)

Longitudinal waves

The individual particle vibrates in a direction parallel to the direction of propagation of the wave. (Section 16.3)

Magnetic bottle

The magnetic field confinement system used to contain plasma in fusion reactors. (Section 32.4)

Magnetic field

The magnetic (induction) field, **B**, is a quantity introduced to explain interaction between moving charges. (Units are weber/m^2 or tesla). (Section 23.3)

Magnetic force

The force on a moving charge ($+q$) with velocity (v) in a magnetic induction field (**B**) has the value of $q\mathbf{v} \times \mathbf{B}$. (Section 23.3)

Magnification

The ratio of image to object size. (Sections 18.6 and 18.7)

The ratio of the angle subtended by the image with lens to an angle subtended by the object at the near point. (Section 18.8)

Mass-energy relation

Einstein's special theory of relativity predicts the famous mass-energy equivalence given by $E = mc^2$. (Section 27.4)

Mechanical equivalent of heat

The amount of mechanical energy which is equivalent to one unit of heat or thermal work. (Section 10.6)

Moderator

The material used in fission reactors to slow down fast neutrons so that the chain reaction can be maintained. (Section 32.2)

Modulus of rigidity

Elastic constant of proportionality for a shear deformation. (Section 13.6)

Moment of inertia

Measure of the ability of a body to resist a change in rotation. It depends upon the distribution of its mass relative to axis of rotation. (Section 7.5)

Momentum

Product of the mass of a body and its velocity, a vector property whose direction is the same as the direction of the velocity. (Section 6.2)

Myopia

The myopic eyeball has parallel light focused in front of the retina (near-sighted condition). (Section 20.4)

Natural frequency

The frequency of a freely vibrating system. (Section 2.4)

Near point   (of the eye)

The closest object distance for sharp image on the retina (maximum power of accommodation). (Section 20.2)

Neutron activation

The process resulting in induced radioactivity of samples when subjected to neutron bombardment. (Section 32.3)

Newton's law of gravitation

Any two particles attract each other by a force which is proportional to the product of their masses and inversely proportional to the square of the distance between the particles. (Section 4.5)

Newton's law of motion

First law

Every body persists in its state of rest or of uniform linear motion unless it is acted upon by an unbalanced force. (Section 4.2)

Second law

The time rate of change in velocity of an object is proportional to the net unbalanced force acting upon it. (Section 4.3)

Third law
For every action there is an equal and an opposite reaction. (Section 4.2)

Noise
Interference noise is the unwanted environmental signals always present in a physical measurement, *e.g.*, 60 cycle noise or thermal noise. (Section 26.2)

Noncoherent radiation
Waves that have random phase relationships. (Section 19.3)

Nuclear binding energy
The energy that binds neutrons and protons in a nucleus is equal to the mass-energy difference between separate nucleons and the particular nucleus. (Sections 31.3 and 31.4)

Nuclear fission
The process occurring when a heavy unstable nucleus splits to form two lighter nuclei with the release of considerable energy and a few neutrons. The process is exploited in nuclear reactors. (Sections 31.9 and 32.2)

Nuclear fusion
The process that occurs when two light nuclei get close enough together that the nuclear force causes them to fuse to form a heavier nucleus with a release of energy. The basic source of energy in stars. (Sections 31.10 and 32.4)

Nuclear magnetic resonance
The phenomena characterized by energy level resonance associated with the proton magnetic dipole in materials. (Section 29.18)

Object distance *(p)*
The axial distance from the center of a lens or mirror to the object position. (Sections 18.6 and 18.7)

Ohm (unit of resistance)
The ohm is the resistance that has a potential difference of one volt across it when it carries a current of one ampere. (Section 22.2)

Ohm's law
The law that describes the linear relationship between voltage and current for many materials $V = IR$. (Section 22.2)

Operational definition
A definition given in terms of a specific experience or operation. This definition is not an operational definition. (Section 2.1)

Optical activity
Phenomenon produced by certain materials that rotate the plane of polarization as light passes through. (Section 19.12)

Optical axis
The line of symmetry through the center of a lens or mirror. (Sections 18.6 and 18.7)

Optical path length
Equals thickness of sample multiplied by index of refraction for wavelength in the sample (Section 19.3)

Oscillatory motion
Characteristic periodic motion of systems having a linear restoring force. (Section 2.4)

Osmosis
Motion of fluid through a semipermeable membrane until the potential energy on the two sides of the membrane is the same. (Section 14.11)

Pascal's principle
A pressure applied to a confined liquid is transmitted undiminished to all parts of the liquid and the walls of the containing vessel. (Section 8.9)

Peltier effect
The process characterized by the generation or absorption of heat at a junction of dissimilar metals when a current passes through the junction. (The reverse of the Seebeck effect.) (Section 22.9)

Period
Time required for one cycle of motion. (Section 15.2)

Periodic motion
Motion which repeats itself in fixed time intervals. (Section 15.2)

Phase angle
Angular displacement at the starting time or angle relative to fixed reference. (Section 15.2)

Phosphorescence
Characterized by light emitted after incident radiation is removed from the source material. (Section 29.7)

Photoelectric effect
The phenomena that results in electron emission from materials when radiated with electromagnetic radiation. (Section 27.3)

Photon
The quantum of electromagnetic energy given by the product of Planck's constant and the frequency of the radiation. (Section 27.3)

Photopic vision
Vision of bright-light adapted eye. (Section 20.5)

Piezoelectricity
The ability of certain crystals (*e.g.*, quartz) to generate an emf when subjected to mechanical strain. (Section 22.10)

Planck's constant
The proportionality constant for energy quanta equal to 6.64×10^{-34} joules/sec. (Section 26.2)

Planck's radiation law
Predicts the thermal radiation intensity as a function of wavelength for a black body at temperature T (°K).

Plasma
The state of matter consisting of completely ionized atoms and electrons as characterized by fusion reactions. (Section 32.4)

Poiseuille's law
The rate of flow of a viscous liquid through a tube is pro-

portional to the pressure gradient, inversely proportional to the viscosity of the liquid, and proportional to the fourth power of the radius of the tube. (Section 8.12)

Polarization plane
Plane of oscillation of electric field in a light wave. (Section 19.10)

Potential difference
The work per unit electrical charge required to move the charged object from one point to another (volts). (Section 21.4)

Potential energy
Ability to do work as a result of position or configuration. (Section 5.5)

Potential gradient
The rate of change of potential in space; a vector quantity oriented in the direction of maximum change of potential. (Section 21.4)

Power
Time rate of doing or using energy. (Section 5.9)

Power factor
Equals the cosine of the phase angle, ϕ, between voltage and current in an AC circuit. (Section 25.4)

Presbyopia
Near point recession with age. This is due to gradually diminishing range of accommodation as the crystalline lens loses its flexibility. (Section 20.4)

Pressure
Normal force per unit area. (Section 8.5)

Projectile motion
An object has constant velocity in one direction and uniform acceleration in a direction at right angles to the constant velocity. This motion represents the idealized motion of objects near the surface of the earth when spin and wind are neglected. (Section 3.8)

P-V diagram
Graphical representation of a process using pressure-volume axes. (Section 12.3)

Q-factor
A measure of the sharpness of resonance. (Section 25.7)

Quantum number
The discrete nature of quantum theory is characterized by four quantum numbers (n, l, m_l, m_s) that determine the energy structure and quantum state of atoms. (Section 28.4)

Rad
Radiation that releases 10^{-5} joules per gram of absorbing material. (Section 30.6)

Radial acceleration
Acceleration directed toward center of curvature of motion, *i.e.*, perpendicular to the curve of the movement of the object in space. (Section 3.9)

Radiation
Transfer of heat energy by electromagnetic waves. (Section 11.4)

Radioactivity
The name given to the process occurring when unstable nuclei decay with characteristic radiation. (Section 31.5)

RBE
The ratio of the radiation doses of 1 rem to 1 rad. (relative biological equivalent). (Section 30.6)

Reactance
The proportionality factor (measured in ohms) between the voltage and current for capacitors and inductors in AC circuits. (Section 25.3)

Reb
Radiation dosage that produces the same effect in biological material as one rad of x rays. (Section 30.6)

Reflection
A phenomenon which occurs at the interface between two media and the wave from the interface is in the same media as the incident wave. (Section 16.7)

Reflection coefficient of light
For normal incidence, the fraction of light intensity reflected. (Section 18.3)

Refraction
Bending or change in direction which occurs at the interface between two media and the direction of the wave in the second medium is different than in the incident medium resulting from different velocities of propagation in the two media. (Section 16.7)

Refrigerator
A heat engine that operates by the input of work when heat energy is extracted from a lower temperature resorvior and transferred to a higher temperature reservoir (Section 12.12)

Relativistic mass
The special theory of relativity predicts that the mechanical inertia, m, of an object increases as its speed, v, approaches the speed of light, c, in accord with
$$m = \frac{m_o}{\sqrt{1 - v^2/c^2}}$$ (Section 27.4)

Rem
Radiation dose unit that produces in man the equivalent effect as one rad of x rays. (Section 30.6)

Resistivity
Resistance of material of unit length with unit cross-sectional area, with units of ohms-meter, ρ. (Section 22.2)

Resonance
Occurs when the frequency of the external interaction equals a natural frequency of the system. (Sections 2.9, 15.5, and 25.5)

Restoring force
Force acting to return a displaced body to equilibrium position. (Sections 2.4 and 15.3)

Rocket propulsion
Practical example of conservation of momentum. (Section 6.6)

Roentgen
The quantity of x rays that produces 1.61×10^{15} ion pairs per kg in air. This is equivalent to 89×10^{-7} joule/g in air. (Section 30.6)

Root mean square velocity
The square root of the sum of the squares of the velocities divided by the number of the particles. (Section 14.3)

Rotational kinetic energy
Energy of rotation equal to one-half the product of moment of inertia and angular velocity squared. (Section 7.7)

Scalar quantity
Specified by a magnitude and dimensional unit. (Section 3.2)

Scotopic vision
Vision of the dark-adapted eye. (Section 20.5)

Seebeck effect
The physical process that produces the emf generated by junctions of dissimilar metals when they are heated or cooled, (e.g., thermocouples). (Section 22.8)

Self-inductance
Self inductance, L, determines induced emf in a coil of wire due to change of current through the wire. (Section 24.5)

Sensitivity
Ability to respond to changes in a variable stimulus. (Section 1.3)

Signal-to-noise ratio
Refers to the ratio of the signal amplitude to the undesirable noise signal in a measuring system. (Section 26.2)

Simple harmonic motion
If the restoring force is proportional to the displacement and oppositely directed, the system will execute SHM when displaced from equilibrium. (Section 15.2)

Sound wave
The longitudinal wave propagated through matter, characterized by a periodic variation of pressure (or density) in a medium. (Section 17.2)

Spectra
Each atom has a unique set of emission lines, these spectra are the "fingerprints" of atoms. (Section 28.5)

Specific gravity
Ratio of density of a substance to the density of water. (Section 8.3)

Specific heat
Heat energy required to raise the temperature of a unit mass of material one Celsius degree. (Section 10.6)

Spherical aberration
The effect resulting in different focal lengths for off-axis rays when incident on spherical lenses or mirror surfaces. (Section 18.6 and 18.12)

Stability
Refers to the condition of a system when it returns to a stable base line when the input is zero. (Sections 2.2, 26.2)

Standing waves
These result from the superposition of two traveling waves of equal frequency moving in opposite direction in the medium and $180°$ out of phase. For a given physical system standing waves result at its resonance frequency. (Sections 16.11 and 17.8)

State
Particular form or condition of a system. (Sections 1.2, 12.2, 12.3)

Strain
Ratio of change in a given physical dimension to the original dimension, i.e., change in length to original length, or change in volume to original volume. (Section 13.2)

Streamline flow
Every particle in a flow which passes a given point will flow through all the points in the line of flow. (Section 8.10)

Stress
Ratio of the applied force to the area. (Section 13.2)

Superconductivity
The phenomena characterized by materials that show resistivity going to zero at a critical temperature between absolute zero ($0°K$) and $30°K$. (Section 29.17)

Superposition principle
The resultant effect is equal to the sum of the individual independent effects. A principle that holds true for linear systems. (Sections 2.8 and 16.8)

Surface tension
Force necessary to break a unit length of surface. (Section 14.7)

System
A whole entity. (Section 1.2)

Tangential acceleration
Acceleration directed along the direction of motion, tangent to the curve of the path of movement of the object in space. (Section 3.9)

Temperature
Measure of the relative hotness or coldness of an object. which depends upon the average kinetic energy of the particles. (Section 10.2)

Temperature gradient
Change of temperature per unit distance in direction of maximum change. (Sections 2.5 and 9.2)

Theoretical mechanical advantage

Ratio of distance through which the applied force acts to the distance through which the load moves. (Section 5.10)

Thermodynamics

Study of phenomena that result from energy changes produced by a transfer of heat and performance of work. (Section 12.1)

Thermometer

Device (transducer) for determining the temperature of an object. (Section 10.2)

Thermonuclear reaction

The name given to all fusion reactions and the energy source of stars. (Section 32.4)

Threshold

Minimum value of a stimulus or physical variable that produces a detectable response. (Section 1.3)

Time dilation

The special theory of relativity predicts that moving clocks run slow $\Delta t_o = \Delta t \sqrt{1 - v^2/c^2}$. Section 27.4)

Torque   (moment of force)

A vector product of a force and its perpendicular distance from line of action of force to the axis of rotation. A right-hand rule can be used to determine its direction $\tau = \mathbf{r} \times \mathbf{F}$. (Section 7.3)

Tracers

Radioactive tracers are samples that contain radioactive isotopes of chemically active materials. Their activity can be used to study or trace a process in a biological system. (Section 32.5)

Transducer

A sensor or detector that transforms a change in one physical variable to a change in another. (Section 1.3)

Transverse wave

The plane of vibration is perpendicular to the direction of propagation of the wave. (Section 16.3)

Ultrasonic

Sound of frequencies above 20,000 hertz/sec. (Section 17.5)

Uncertainty principle

Heisenberg's formulation of the limits of simultaneous knowledge of position and momentum or of energy and time in modern quantum theory. (Section 27.8)

Uniform circular motion

Objects moving in a circle with constant speed. (Section 3.9)

Uniformly accelerated angular motion

Rotational motion with a constant angular acceleration. (Section 7.2)

Uniformly accelerated motion

Motion of an object whose acceleration is constant, *i.e.*, both the *magnitude* and *direction* of the acceleration remain the same for all times. (Section 3.7)

Variable

Quantity that is subject to change. (Section 1.2)

Vector quantity

Specified by a magnitude, a direction, and a dimensional unit. (Section 3.2)

Velocity

Time rate of change of displacement. (Section 3.5)

Viscosity

Liquid pressure coefficient of viscosity is a relative measure of liquid friction and is equivalent to the ratio of the force per unit area to the change in velocity per unit length perpendicular to the direction of flow. (Section 8.12)

Visual acuity of the eye

The minimum separation between two equidistant point objects that can be resolved as separate objects. (Section 20.6)

Wavelength

The distance a wave moves during one time period or the distance between two particles in a wave of the same displacement and phase. (Sections 16.4 and 17.2)

Weight

The product of the mass and the acceleration due to gravity. (Section 4.3)

Weightlessness

Gravitational attraction of the earth is precisely equal to the centripetal force of the body in its orbit. (Section 4.6)

Wien's displacement law

Predicts the relationship between the absolute temperature and wavelength for maximum thermal radiation for a perfect radiator. (Section 27.2)

Work

Applied force produces a displacement; magnitude of work done is given by product of force component parallel to the displacement and the magnitude of the displacement. (Section 5.2)

Work function

The energy required to free an electron from a metal. (Section 27.3)

X rays

Electromagnetic radiation in the wavelength range 0.001 to 0.1 nm or energy range of about 10 keV to 1 meV. Section 30.2)

Young's modulus

Elastic constant of proportionality for a linear deformation. (Section 13.4)

References for Further Study

I. GENERAL REFERENCES

A. General Physics and Basic Concepts

1. Bauman, R.P., "The Meaning of Conservation Laws," *Phys. Teacher*, vol. 9, p. 186, 1971.
2. Huntoon, R.D., "The Basis of Our Measurement System," *Phys. Teacher*, vol. 4, p. 113, 1966.
3. "International System of Units," *Phys. Teacher*, vol. 9, p. 379, 1971.
4. Karplus, R., *Introductory Physics—A Model Approach*. New York: Benjamin, 1969.

   This text is designed for nonscience majors and clearly introduces fundamental concepts along with basic physical and mathematical models.

B. Physical Data and Formulas

1. *Bioastronautics Data Book*, 2nd ed., NASA–US Government Printing Office.

   This book contains a wealth of data from research on the human body including chapters on temperature sensing, vision, the auditory system, motion effects and radiation effects.
2. Tuma, J.J., *Handbook of Physical Calculations*. New York: McGraw-Hill, 1976.

   This is an excellent presentation of physical formulas, tables and example calculations.

C. Physiology

1. Tortora, G.J., and Anagnostakos, N.P., *Principles of Anatomy and Physiology*. New York: Harper & Row, 1975.

   A physiology book that points out connections between physics and physiology. Chapter 14 contains a good treatment of physiological basis of human senses. Chapter 11 points out the lever systems found in the human body.

D. Biophysics and Physics for Biologists

1. Campbell, G.S., *An Introduction to Environmental Biophysics*. New York: Springer-Verlag, Heidelberg Science Library, 1977.

   This paperback has many applications (some using calculus) to environmental problems. Topics introduced are temperature, moisture, wind, radiation, transfer problems for animals, humans, and plants.
2. Davidovits, P., *Physics in Biology and Medicine*. New York: Prentice-Hall, 1975.

   This is a noncalculus text that includes biological examples in every chapter. These examples include quantitative data. A good bibliography is included at the end of the book. The basic physics concepts are presented in the Appendix.
3. Duncan, G., *Physics for Biologists*. New York: Halsted Press (Division of Wiley), 1975.

   This paperback text contains biological examples in every chapter as well as problems with quantitative biological data. Some unique examples are found in this book.
4. Edington, J.A., and Sherman, H.J., *Physical Science for Biologists*, Hutchinson University Library, 1971.

   This is a compact presentation of some applications of physics to biological systems. Some interesting biological examples are included. Electrochemistry and osmosis sections are particularly worthwhile.

5. Holwill, M.E., and Silvester, N.R., *Introduction to Biological Physics*. New York: Wiley, 1973.

This is a calculus based text that includes biological examples throughout. A good general reference book—very few examples and problems.

6. Richardson, I.W., and Neergaard, E.B., *Physics for Biology and Medicine*. New York: Wiley-Interscience, 1972.

This is a calculus based textbook. Chapter 2 deals with fluids in rigid and elastic vessels; Chapter 3 deals with thermodynamics; Chapter 5 deals with diffusion and osmosis; Chapter 6 treats acoustics; and Chapter 7 deals with optics.

E. Environmental Physics

1. Lowry, W.P., *Weather and Life—An Introduction to Biometeorology*. New York: Academic Press, 1969.

This is a good introduction to biometeorology. The book includes many examples of heat transfer and energy budget calculations.

2. Shonle, J.I., *Environmental Applications of General Physics*. New York: Addison-Wesley, 1975.

An interesting and practical collection of general physics applications to environmental problems. Calculus is used when needed.

3. Thorndike, E.M., *Energy and Environment, A Primer for Scientists and Engineers*. New York: Addison-Wesley, 1976.

This book is designed to help make science and engineering students broadly knowledgeable in the areas of energy and environmental problems. Calculus is used when needed. Excellent references are given at the end of each chapter.

F. History and Philosophy of Physics

1. Gamow, G., *Biography of Physics*. New York: Harper & Row, 1961.

A classic written by one of the authors of 20th Century physics.

2. Heisenberg, W., *Physics and Beyond*. New York: Harper & Row, 1971.

This is an account of a brilliant Nobel lauerate's life, his physics, his friends, and his values. An excellent book for anyone interested in the philosophical implications of modern physics.

3. Hoffman, B., *Albert Einstein: Creator and Rebel*. New York: Viking Press, 1972.

This is an excellent biography about one of the world's greatest minds. The author is a physicist and he communicates clearly the great physical insights of Einstein.

II. SPECIFIC REFERENCES

A. Human Senses

1. Dethier, V.G., "The Taste of Salt," *Amer. Sci.*, vol. 65, no. 6, p. 744, Nov.–Dec. 1977.
2. Land, E., "The Retinex Theory of Color Vision." *Sci. Amer.*, p. 108, Dec. 1977.
3. Wrightman, F.L., and Green, D.M., "The Perception of Pitch," *Amer. Sci.*, vol. 62, no. 2, p. 208, Mar.–Apr. 1974.

B. Mechanics

1. Bartlett, A.A., "Which Way is UP," *Phys. Teacher*, vol. 10, p. 425, 1972.
2. Benedek, G.B., and Villars, F.M., *Physics—With Illustrative Examples From Biology*, vol. 1—Mechanics. New York: Addison-Wesley, 1973.

This is an excellent calculus based textbook with many mechanics, fluid dynamics and thermodynamics examples taken from biology.

3. Hay, J.G., *The Biomechanics of Sports Techniques*. New York: Prentice-Hall, 1973.

This book contains some interesting applications of physics in sports. This book should provide many ideas for experiments involving mechanics and sports.

4. Keller, J.B., "Theory of Competitive Running," *Phys. Today*, vol. 26, vol. 9, p. 42, 1973.
5. Northrip, J.W., Logan, G.A., and McKinney, W.C., *Introduction to Biomechanic Analysis of Sport*, Dubuque, Iowa: Brown, 1974.

Mechanics applied to sports in a straight-forward manner.

6. Page, P.I., "Mechanics of Swimming and Diving," *Phys. Teacher*, vol. 14, p. 72, 1976.
7. Shonle, J.I., and Nordick, D.L., "The Physics of Ski Turns," *Phys. Teacher*, vol. 10, p. 491, 1972.
8. Sutton, R., "Physics of Walking," *Amer. J. Phys.*, vol. 23, p. 490, 1955.
9. Swinson, D.B., "Skiing and Angular Momentum: A Proposed Experiment," *Phys. Teacher*, vol. 11, p. 415, 1973.
10. Tricker, R.A.R., and Tricker, B.J.K., *The Science of Mechancis*. New York: Elsevier, 1967.

Many examples of mechanics as applied to athletics and the human body.

11. Tucker, V.A., "The Energetic Cost of Moving About," *Amer. Sci.*, vol. 63, no. 4, p. 431, 1975.

C. Thermodynamics, Thermal Physics, and Fluids

1. Hobbie, R.K., "Osmotic Pressure in Physics Course for Students of Life Sciences," *Amer. J. Phys.*, vol. 42, p. 188, 1974.
2. Horowitz, H.J. *Entropy for Biologists—An Introduction to Thermodynamics.* New York: Academic Press, 1970.

   This is an excellent book pointing out the significance of thermal physics in many areas of biology. Calculus is used, but the author does a good job of explaining the meaning of the mathematics used.

D. Sound and the Physics of Music

1. Benade, A.H., *Fundamentals of Musical Acoustics.* New York: Oxford University Press.

   This is an excellent introduction into musical acoustics by a long time researcher in the field. It is written with the same clarity of his *Scientific American* articles on the same subjects.
2. Rigden, J.S., *Physics and the Sound of Music.* New York: Wiley, 1977.

   An excellent introduction into the physics of music. It includes an in depth discussion of sound systems and suggestions for buying systems.
3. Rossing, T.D., "Acoustics of Percussion Instruments," Part I, *Phys. Teacher*, vol. 14, p. 546, 1976; Part II, vol. 15, p. 288, 1977.
4. Stickney, S.E., and Englert, T.J., "The Ghost Flute," *Phys. Teacher*, vol. 13, p. 518, 1975.
5. Sundberg, J., "The Acoustics of the Singing Voice," *Sci. Amer.*, vol. 82, Mar. 1977.

E. Light

1. Clayton, R.K., *Light and Living Matter*, vol. 1—The Physical Part, vol. 2—The Biological Part. New York: McGraw-Hill, 1970.

   These two small volumes are excellent references for the interaction of light with biological systems. Both theoretical and experimental treatments are clearly presented.
2. Kock, W.E., "Sound Visualization and Holography," *Phys. Teacher*, vol. 13, p. 14, 1975.
3. Tinker, R.F., "The Safe Use of Laser Beams," *Phys. Teacher*, vol. 11, p. 455, 1973.
4. Thumm, W., "Some Thoughts on Teaching About Ultraviolet," *Phys. Teacher*, vol. 13, p. 135, 1975.
5. Whitaker, R.J., "Physics of the Rainbow," *Phys. Teacher*, vol. 12, p. 283, 1974.

F. Electromagnetism, and Electrical and Electronic Measurement

1. Bartlett, A.A., "Paths of Charged Particles in Magnetic Field," *Phys. Teacher*, vol. 13, p. 499, 1975.
2. Becker, R.O., "Electromagnetic Forces and Life Processes," *Tech. Rev.*, vol. 32, Dec. 1972.
3. Blatt, F.J., "Nerve Impulses in Plants," *Phys. Teacher*, vol. 12, p. 455, 1974.
4. Chedd, G., *How to Grow a New Leg*, Time-Life Year Book (Nature-Science), 1976.
5. Cohen, D. "Magnetic Fields of the Human Body," *Phys. Today*, vol. 28, no. 8, p. 34, 1975.
6. Cronwell, L., Weibell, F.J., Pfeiffer, E.A., and Usselman, L.B., *Biomedical Instrumentation and Measurements.* New York: Prentice-Hall, 1973.

   This is an excellent book on biomedical instrumentation.
7. Hobbie, R.K., "The Electrocardiogram as an Example of Electrostatics," *Amer. J. Phys.*, vol. 41, p. 824, 1973.
8. Lion, K.S., *Elements of Electrical and Electronic Instrumentation.* New York: McGraw-Hill, 1975.

   This is a good introductory book for those interested in learning more about electronic instrumentation. Excellent treatment of transducers and lab exercises are outstanding features of this book.
9. Schwartz, B.B., and Foner, S., "Large Scale Applications of Superconductivity," *Phys. Today*, vol. 30, p. 34, 1977.
10. Swez, J.A., "Research in Biophysics: The High Resolution Scanning Electron Microscope," *Phys. Teacher*, vol. 13, p. 240, 1975.
11. Weber, L.J., and McLean, D.L., *Electrical Measurement Systems for Biological and Physical Scientists.* New York: Addison-Wesley, 1975.

    This is a good book for learning the basic principles behind electrical measurements and modern solid state electronic instrumentation. Experiments are not given but examples include component values.
12. Wadsworth, R.M., Ed., *The Measurement of Environmental Factors in Terrestrial Ecology*, Blackwell Scientific Publications, 1968.

    A fine reference for anyone desiring to learn techniques and instrumentation for measuring environmental physical variables.

G. Quantum Physics and Atomic and Molecular Physics

1. Barrow, G.M., *The Structure of Molecules*. New York: Benjamin, 1964.

   This is a lucid introduction to molecular spectroscopy. It should be especially appreciated by chemistry and biophysics students.
2. Mehra, J., "Quantum Mechanics and the Explanation of Life," *Amer. Sci.*, vol. 61, no. 6, p. 722, Nov.–Dec. 1973.

H. Nuclear Physics and Applications of Radioactivity

1. Bushong, S.C., "Radiation Exposure in our Daily Lives," *Phys. Teacher*, vol. 15, p. 135, 1977.
2. Hurst, G.S., and Turner, J.E., *Elementary Radiation Physics*. New York: Wiley, 1970.

   This is a well-written text on radiation physics. Excellent problems provided at the end of each chapter.
3. Kowalski, L., "Radioactivity and Nuclear Clocks," *Phys. Teacher*, vol. 14, p. 409, 1976.
4. Lindenfeld, P., *Radioactive Radiations and Their Biological Effects*, (Issue Oriented Module) Amer. Assoc. Phys. Teachers, 1977.

   This is an excellent introduction to the biological effects of ionizing radiation. It includes a good bibliography, a section on experiments, a summary of properties, and a good set of problems.
5. McGinley, P.H., "Fast Neutron Radiotheraphy," *Phys. Teacher*, vol. 11, p. 73, 1973.
6. McNeil, K.G., "Photonuclear Reactions in Medicine," *Phys. Today*, vol. 27, no. 4, p. 75, 1974.
7. Ralph, E.K., and Michael, H.M., "Twenty-five Years of Radiocarbon Dating," *Amer. Sci.*, vol. 62, no. 5, p. 553.
8. Swindell, W., and Barrett, H.H., "Computerized Tomography: Taking Sectional X-Rays," *Phys. Today*, vol. 30, no. 12, p. 32, 1977.
9. Thumm, W., "Röntgen's Discovery of X-Rays," *Phys. Teacher*, vol. 13, p. 207, 1975.
10. Widman, J.C., Brnetich, J., Pousner, E.R., "Radioactive Decay Flow Diagram," *Phys. Teacher*, vol. 13, p. 554, 1975.

III. MISCELLANEOUS REFERENCES

1. Bason, F.C., *Energy and Solar Heating* (Issue Oriented Module), Amer. Assoc. Phys. Teachers, 1977.
2. Brown, Jr., F.A., Hastings, J.W., and Palmer, J.D., *The Biological Clock — Two Views*. New York: Academic Press, 1970.

   In chapter 2, F.A. Brown, Jr. presents his case for environmental timing of the biological clock (e.g., synchronization with geophysical fields or stimuli). In chapter 3, Hastings presents the cellular-biochemical clock hypothesis.
3. Hendee, W.R., and Siegel, E., "Career Opportunities in Medical Physics," *Phys. Teacher*, vol. 15, p. 215, 1977.
4. Hobbie, R.K., "Physics Useful to a Medical Student," *Amer. J. Phys.*, vol. 43, p. 121, 1975.
5. Riezemann, M.J., "Doctors Call for Emphasis Shift in Designing Electronics for Hospital," *Electronics*, p. 99, Sept. 1973.
6. Schwartz, G.E., "Biofeedback, Self-Regulation and the Patterning of Physiological Processes," *Amer. Sci.*, vol. 63, no. 3, p. 314, May–June 1975.
7. Whorton, J., "Some Historical Interactions of Physics and the Biomedical Sciences," *Phys. Teacher*, vol. 12, p. 159, 1974.

Index

79 80 81 82 9 8 7 6 5 4 3 2